Environmental Science

ENVIRONMENTAL SCIENCE

EARTH AS A LIVING PLANET

Third Edition

Daniel B. Botkin

Director, Program on Global Change
George Mason University, Fairfax, Virginia

President,
The Center for the Study of the Environment
Santa Barbara, California

Edward A. Keller

University of California, Santa Barbara

with assistance from:
Dorothy B. Rosenthal
Nicholas Pinter
Mel S. Manalis
Harold Ward
Marc J. McGinnes
Harold Morowitz

JOHN WILEY & SONS, INC.
New York • Chichester • Weinheim
Brisbane • Singapore • Toronto

ACQUISITIONS EDITOR   David Harris

FREELANCE DEVELOPMENTAL EDITOR   Susan Day

MARKETING MANAGER   Katherine Beckham

SENIOR PRODUCTION EDITOR   Patricia McFadden

SENIOR DESIGNER   Karin Kincheloe

SENIOR PHOTO EDITOR   Hilary Newman

ILLUSTRATION EDITOR   Anna Melhorn

COVER PHOTO   © Carr Clifton

COVER DESIGNER   Karin Gerdes Kincheloe

TITLE PAGE PHOTO   Frans Lanting/Mindon Pictures, Inc.

This book was set in 10/12 ITC Garamond Light by LCI Design
and printed and bound by Von Hoffmann Press.
The cover was printed by Lehigh Press, Inc.

This book is printed on acid-free paper.

The paper in this book was manufactured by a mill whose forest management programs
include sustained yield harvesting of its timberlands. Sustained yield harvesting principles
ensure that the numbers of trees cut each year does not exceed the amount of new growth.

ISBN 0-471-32173-7

Printed in the United States of America

10 9 8 7 6 5 4 3 2

DEDICATIONS

FOR VALERY
who contributed so much to this book

AND

FOR ERENE,
who gave and cared so much

Daniel B. Botkin is President of The Center for the Study of Environment and Professor of Biology at George Mason University. From 1978 to 1993, he was Professor of Biology and Environmental Studies at the University of California, Santa Barbara, serving as Chairman of the Environmental Studies Program from 1978 to 1985. For more than three decades, Professor Botkin has been active in the application of ecological science to environmental management. He is the winner of the Mitchell International Prize for Sustainable Development and the Fernow Prize for International Forestry, and he has been elected to the California Environmental Hall of Fame.

Trained in physics and biology, Professor Botkin is a leader in the application of advanced technology to the study of the environment. The originator of widely used forest gap-models, his research has involved endangered species, characteristics of natural wilderness areas, the study of the biosphere, and attempts to deal with global environmental problems. During his career, Professor Botkin has advised the World Bank about tropical forests, biological diversity, and sustainability; the Rockefeller Foundation about global environmental issues; the government of Taiwan about approaches to solving environmental problems; and the state of California on the environmental effects of water diversion on Mono Lake. He served as the primary advisor to the National Geographic Society for their centennial edition map on "The Endangered Earth." He recently directed a study for the states of Oregon and California concerning salmon and their forested habitats.

He has published many articles and books about environmental issues. His latest books are *Passage of Discovery: The American Rivers Guide to the Missouri River of Lewis and Clark* (Penguin-Putnam), *The Blue Planet* (Wiley), *Our Natural History: The Lessons of Lewis and Clark* (Putnam), *Discordant Harmonies: A New Ecology for the 21st Century* (Oxford University Press), and *Forest Dynamics: An Ecological Model* (Oxford University Press).

Professor Botkin was on the faculty of the Yale School of Forestry and Environmental Studies (1968–1974) and was a member of the staff of the Ecosystems Center at the Marine Biological Laboratory, Woods Hole, MA (1975–1977). He received a B.A. from the University of Rochester, an M.A. from the University of Wisconsin, and a Ph.D. from Rutgers University.

Edward A. Keller was chair of the Environmental Studies and Hydrologic Sciences Programs from 1993 to 1997 and is Professor of Geological Sciences at the University of California, Santa Barbara, where he teaches geomorphology, environmental geology, environmental science, river processes, and engineering geology. Prior to joining the faculty at Santa Barbara, he taught geomorphology, environmental studies, and earth science at the University of North Carolina, Charlotte. He was the 1982–1983 Hartley Visiting Professor at the University of Southampton, England.

Professor Keller has focused his research efforts into three areas: studies of Quaternary stratigraphy and tectonics as they relate to earthquakes, active folding, and mountain building processes; hydrologic process and wildfire in the chaparral environment of southern California; and the role of fluid pressure in the earthquake cycle of active fold-and-thrust belts. He is the recipient of various Water Resources Research Center grants to study fluvial processes and U.S. Geological Survey and Southern California Earthquake Center grants to study earthquake hazards.

Professor Keller has published numerous papers, and is the author of the textbooks *Environmental Geology, Introduction to Environmental Geology* and (with Nicholas Pinter), *Active Tectonics* (Prentice-Hall). He holds bachelors degrees in both geology and mathematics from California State University, Fresno; an M.S. in geology from the University of California; and a Ph.D. in geology from Purdue University.

ᴘREFACE

A study of the environment on the cusp of the twenty-first century is an exciting endeavor as we attempt to move from confrontation to cooperative problem-solving and place the study of the environment on a sound scientific basis. The enthusiasts of the 1960s have matured into today's environmental professionals: executives in alternative energy corporations, applied scientists who work on international projects to spread appropriate technologies to our inner cities and to developing nations, economists who calculate cost and benefits of pollution controls, environmental lawyers who mediate problems and help write laws to promote sustainable use of our resources, and other people in many related fields.

In recent years our understanding of many aspects of the environment has greatly increased. This has ranged from our understanding of the transport and fate of pollutants and toxins in the air and water, to appreciation of what is really necessary to save endangered species, to understanding how the Earth as a system operates.

Study of our environment has undergone tremendous change during the past four decades. In the 1960s and 1970s a grass roots movement to protect the environment began with the recognition of potential and real adverse effects of modern civilization on our environment. Views became polarized. Some environmentalists argued that everything about the environment was good and should be protected without change—that is, all development of natural resources was bad. Some environmentalists believed that the world would be destroyed if we did not change our approach to the environment. Opponents argued that these extreme "environmentalists" were opposed to progress and possibly to everything good stemming from civilization and technology. Nevertheless, during this period progress was made in dealing with environmental issues and problems. Enthusiasm for the environment remained high.

Environmentalism in the 1980s was characterized by a conscious shift from enthusiastic rhetoric to development of alternative ways to solve environmental problems associated with local, regional, and global issues such as human population, hazardous waste, acid precipitation, global warming, and stratospheric ozone depletion. People and institu-

tions began putting more energy and resources into solving environmental problems and learning more about how the earth works as a system.

During the 1990s there was a return to considerable confrontation and emotionalism surrounding the environment. Some activists place the Earth's life support system, the biosphere, at the top of the moral pyramid, to be protected at all cost from adverse human interference. At the other extreme some people see environmentalists and environmentalism as a threat to private property rights and their "way of life." Finding solutions to environmental problems in the twenty-first century will require that a spectrum of potential solutions be carefully evaluated in light of our values and scientific knowledge, while at the same time insuring social justice.

GOALS OF BOOK

The purpose of *Environmental Science* is to provide an up-to-date introduction to the most important and useful concepts in the study of the environment. Information is presented from an analytical and interdisciplinary perspective from which we must view environmental issues in order to deal successfully with them. The goal is to teach the student *how* to think through environmental issues.

Critical Thinking

We must do more than simply identify and discuss environmental problems and solutions. To be effective, we must think critically about them. Critical thinking is so important that we have made it the focus of its own chapter, Chapter 2. With this in mind we have also developed *Environmental Science* to present the material in a factual, unbiased format. Our goal is to help you think through the issues, not tell you what to think. To this purpose, at the end of each chapter, critical thinking exercises, called "Environmental Issues," are provided.

Interdisciplinary Approach

The approach of *Environmental Science* is interdisciplinary in nature. Environmental science integrates many disciplines and includes some of the most important topics of modern civilization as well as some of the oldest philosophical concerns of human beings—that of the nature of our relationship with our environment. Applied and basic aspects of environmental sciences require a solid foundation in the natural sciences, in addition to fields such as anthropology, economics, history, sociology, and philosophy of the environment. Not only do we need the best ideas and information to deal successfully with our environmental problems, but we also must be aware of the cultural and historical context in which we make decisions about the environment and understand ways in which choices are made and implemented. Thus, the field of environmental science integrates the natural sciences with environmental ethics, environmental economics, environmental law, environmental impact, and environmental planning. As a result, *Environmental Science* provides an introduction to the entire spectrum of relationships between people and the environment.

THEMES

Environmental Science is based on the philosophy that several threads of inquiry are of particular importance to environmental science. These key themes are woven throughout the book.

Human Population

Underlying nearly all environmental problems is the rapidly increasing human population. Ultimately, we cannot expect to solve these other problems unless we can limit the total number of people on Earth to an amount the environment can sustain. We believe that education is important to solving the population problem. As people become more educated, and as the rate of literacy increases, population growth tends to decrease.

Sustainability

Sustainability is a term that has gained much popularity recently. Speaking generally, it means that a resource is used in such a way that it continues to be available. However, the term is used vaguely and it is something we are struggling to clarify. Some would define it as insuring that future generations have equal opportunities to the resources that our planet offers. Others would argue that sustainability refers to types of developments that are economically viable, do not harm the environment, and are socially just. We all agree that we must learn how to sustain our environmental resources so they con-

tinue to provide benefits for people and other living things on our planet.

A Global Perspective

Until recently we generally believed that human activity caused only local, or at most regional, environmental change. We now know that effects of human activity on Earth are of such an extent that we are involved in a series of unplanned planetary experiments. The main goal of the emerging science known as *Earth System Science* is to obtain basic understanding of how our planet works as a system. This understanding can then be applied to help solve global environmental problems. The emergence of Earth System Science has opened up a new area of inquiry for faculty and students. Understanding the relationships between biological and physical sciences requires interdisciplinary cooperation and education.

The Urban World

An ever-growing number of people are living in urban areas. Unfortunately our urban centers have long been neglected and the quality of the urban environment has suffered. It is here we experience air pollution, waste disposal problems, social unrest, and other stresses of the environment. In the past we have centered our studies of the environment more on wilderness than the urban environment. In the future we must place greater focus on towns and cities as livable environments.

Values, Knowledge, and Social Justice

Finding solutions to environmental problems involves more than simply gathering facts and understanding the scientific issues of a particular problem. It also has much to do with our systems of values and issues of social justice. To solve our environmental problems, we must understand what our values are, and which potential solutions are socially just. Then, we can apply scientific knowledge about specific problems and find acceptable solutions.

These five key themes or threads of inquiry are discussed in more detail in Chapter 1. They are also revisited at the end of each chapter where we discuss some of the pertinent material relative to these themes.

ORGANIZATION

We believe a real strength of *Environmental Science* is the systematic and in-depth coverage of the multitude of subjects that comprise the field of Environmental Science. An important objective is to integrate physical and biological processes within a social framework. We recognize that environmental education is a life-long process and any one course may not be able to cover in depth all of the subjects presented in *Environmental Science*. The goal is to provide an instrument of learning useful to educators and students today so that future generations of students will be more informed and able to make judgments concerning the environment based upon sound scientific knowledge. We believe that the understanding of, and critical thinking about, environmental problems is much more important than mere presentation of facts and information.

To support this goal, our text is divided into eight parts. *Part I* provides a broad overview of the key themes in *Environmental Science,* the scientific method, and thinking critically about the environment. *Part II* presents the study of the Earth as a system, emphasizing how systems work and the basic biochemical cycles of our planet. *Part III* focuses on life and the environment and includes subjects such as human population, ecosystems, biological diversity, biological productivity and energy flow, and restoration and recovery of ecosystem response to disturbance. *Part IV* presents living resources from a sustainability viewpoint, and topics covered include world food supply, agriculture and environment, plentiful and endangered species, forest ecology, conserving and managing life in the oceans, and environmental health and toxicology. *Part V* introduces and discusses a wide variety of topics related to energy including basics necessary for understanding energy, fossil fuels and environment, alternative energy, and nuclear energy. *Part VI* presents the water environment on Earth in terms of water supply use and management, and water pollution treatment. *Part VII* concerns the air environment, from global issues such as climate, global warming, and stratospheric ozone depletion to regional issues such as acid rain, to local issues including urban air pollution and indoor air pollution. *Part VIII* is concerned with relationships between environment and society. Topics include environmental economics, the urban environment, integrated waste management, minerals and environment, environmental impact and planning, and integrating values and knowledge.

SPECIAL FEATURES

In writing *Environmental Science* we have designed a text that incorporates a number of special features that we believe will help teachers to teach and students to learn. These include:

- A **Case Study** introduces each chapter. The purpose is to interest the reader in the subject being discussed and to raise important questions on the subject matter. For example, in Chapter 14, which deals with Environmental Health and Toxicology, the Case Study introduces the problem of lead toxicity and asks the question, "Is lead in the urban environment contributing to antisocial (criminal) behavior?"

- **Learning Objectives** are introduced early in the chapter to help students focus on what is important in the chapter and what they should achieve after reading and studying the chapter.

- **A Closer Look** is the name of special learning modules presented in most of the chapters. The purpose here is to present more detailed information concerning a particular concept or issue. For example, A Closer Look 3.1, in the chapter on Systems and Change, discusses the important concept of electromagnetic radiation. Many of these special features contain figures and other data to enrich the reader's understanding.

- Near the end of each chapter, an **Environmental Issue** is presented as a method of encouraging critical thinking about the environment and to help students understand how these issues may be studied and evaluated. For example, Chapter 19 presents the environmental issue of how wet is a wetland? The issue in Chapter 16 examines the important environmental question of whether or not we should raise the gasoline tax.

- Following the Summary, a special section, **Re-examining Themes and Issues** reinforces the five major themes of the textbook. Here we also make value judgments concerning important environmental issues and questions.

- **Study Questions** for each chapter are provided. The purpose is to provide a study aid for students and utilize critical thinking skills.

- **Further Reading** and **Internet Resources** are provided with each chapter so that students may expand their knowledge and reading through major sources of information (both print and electronic) on the environment.

- **References** cited in the text are provided at the end of the book as notes for each chapter. These are numbered according to their citation in the text. We believe it's very important that introductory textbooks carefully cite sources of information used in the writing. These are provided to recognize those scholars whose work we depend upon, and so that the reader may draw upon these references as needed for additional reading and research.

IMPROVEMENTS IN THE THIRD EDITION

Environmental Science is a rapidly developing set of fields. Populations grow; species become threatened or released from near-extinction; our actions change. To remain comtemporary, a textbook in Environmental Sciences requires frequent updating. For example, human populations continue to grow and the distribution changes, as does the rate of urbanization. Data have been brought up to date throughout. As information and ideas change, references also require updating, and this has been done throughout.

Augmentation of Web Site References

Valid information is becoming increasingly available over the Web, and easy access to these data are of great value. Government data that used to take weeks of library search are available almost instantly over the Web. For this reason, we have greatly augmented the number of Web site references at the end of each chapter.

Integration of Themes

The third edition continues to integrate five major themes, adding more references in the text to the connection between a specific discussion and one of these themes.

New Case Studies

Each chapter begins with a case study that helps the student learn about the chapter's topic through a specific example. A major improvement in the third edition is the replacement of some older case studies with new ones that discuss current issues and are more closely integrated into the chapter. New case studies occur in the following chapters: (1) "How Can We Preserve the World's Coral Reefs?"; (11) "Clean Water Farms"; (16) "Fuel Efficiency of U.S. Vehicles"; (18) "Nuclear Energy and Public Policy"; (21) "El Niño 1997–98"; (22) "London Smog and Indonesian Fires"; (24) "Epidemic of Skin Cancer"; (27) "Fresh Kills Landfill, New York City"; (26) "The Ecological Capital of Brazil."

New Environmental Issues

Each chapter ends with a discussion of an environmental issue, with critical thinking questions for the students. This is one of the ways that the text is designed to help the student learn to think for oneself about the analysis of environmental issues. In the third edition, some older environmental issues

have been replaced with new ones and these have been more closely integrated into the text. These occur in the following chapters: (2) "How Do We Decide What to Believe about Environmental Issues?"; (4) "How Are Human Activities Affecting the Nitrogen Cycle?"; (6) "How are the Borders of an Ecosystem Defined?" (9) "How Can We Evaluate Constructed Ecosystems?"; (10) "Will There Be Enough Water to Produce Food for a Growing Population?"; (12) "Should Wolves Be Re-Established in the Adirondack Park?"

Other Inprovements

In Chapter 4 there is an expanded discussion of the hydrological cycle as part of the discussion of biogeochemical cycles; in Chapter 5 there are updated projections and new graphs for the human population; Chapter 7 has a new graphic comparing estimates of the total number of species on the Earth; Chapter 8 has an expanded discussion about ocean upwelling; Chapter 9 has new information about air pollution in Eastern Europe; Chapter 10 includes a new "Closer Look" concerning new genetic strains and hybrids; Chapter 11 has a much revised section on rangeland; Chapter 12 has an updated section on categories of endangered species; Chapter 13 has updated sections on both forestry and fishery; Chapter 14 has a new discussion of the concept of a toxic dose and a new table about toxic substances; Chapter 21 adds a discussion of the recently discovered "ocean conveyor belt" and causes of global climate change; Chapter 26 has an updated section on total consumption of products and wastes produced by people; Chapter 29 has updated information about the capping of ponds at Kesterson Reservoir; the final Chapter, 30, has a revised conclusion and a discussion of how a student can become involved in environmental issues.

SUPPLEMENTS

A variety of supplements for both students and instructors is available for *Environmental Science,* Second Edition:

- **Student Review Guide with Internet Companion,** prepared by Joseph Luczkovich and David Knowles, of East Carolina University. This exciting new supplement integrates the power of the World Wide Web for the first time! The Review Guide contains concise chapter summaries, chapter concepts, and chapter review questions. In addition, it contains an Internet component, which provides students with "hot links" to related environmental Web sites and other relevant points of interest on the World Wide Web.

- **Regional Case Books** supplement the case studies that are featured within the text. Cases have been carefully selected to allow instructors to tailor their courses specifically for their own locale. Case Books, compiled by Dorothy Rosenthal for the Eastern, Western, Southern, and Northern regions of the United States, are available. Monica Mulrennan, of Concordia University, designed a Case Study specifically for Canada, which provides a wide range of case studies for a diverse selection of regions.

- **Student Review CD** will provide students with an interactive environment in which they can access all of the information contained in the printed version of the Review Guide. Bonus features include on-line testing and on-line assessment, immediate access to related web sites through embedded hot links, and full color graphs and illustrations that will reinforce important topics and concepts.

- **Instructor's Manual,** prepared by Ann S. Causey, of Auburn University. The Manual contains teaching suggestions for all instructors, both experienced as well as the first-time lecturer; lecture and outline notes; a list of suggested supplemental readings, and lead-ins and critical thinking activities to give instructors ideas on how to begin and finish lectures. In addition, most chapters contain a list of five

to ten ways individuals can minimize their impact on Earth's available resources.

- **Test Bank,** prepared by Nicholas Pinter of University of California, Santa Barbara. This Test Bank contains approximately 1,700 multiple-choice, short-answer, and essay questions.

- **Transparency Acetates** of figures from the text with large bold-face labels aid in classroom presentations.

- **CD-ROM** contains an electronic library of the illustrations found in the text.

- **Environet** *(http://www.wiley.com/college/ environet)* is an exciting website designed to enhance any course using Botkin and Keller's *Environmental Science: Earth as a Living Planet, 2e.* For each chapter of the text, the site expands on either the *Case Study,* the *Environmental Issue,* or *A Closer Look,* providing current news items, reports, historical context, and a multimedia component. The site also contains frequently asked questions relating to the key concepts of environmental science along with links to other sites for help in answering those questions. The correct answers are provided in the *Student Review Guide with Internet Companion.* Both students and instructors will find this a rich site for interactive learning and exploration.

Daniel B. Botkin
Edward A. Keller

ACKNOWLEDGMENTS

Completion of this book was only possible due to the cooperation and work of many people. To all those who so freely offered their advice and encouragement in this endeavor, we offer our most sincere appreciation. We are indebted to our colleagues who made contributions: Dorothy A. Rosenthal for writing Chapter 2 and the environmental issues at the end of each chapter; Mel S. Manalis for assistance in developing the energy chapters, global warming, and stratospheric ozone depletion; Marc J. McGinnis for assistance in helping develop discussions concerning environmental law; Harold Ward for writing parts of Chapter 29, particularly relating to environmental legislation and review; Bill Kuhn for assistance in research; Joan Melcher for her excellent editing of the manuscript; and Nicholas Pinter who helped in many ways including review of the manuscript, suggested text revisions, and development of suggested readings for students. Finally, we would like to thank Harold Morowitz for helpful suggestions on how to introduce the basic energy concepts.

We are indebted to our editor, David Harris, with John Wiley & Sons for his support and encouragement as well as his very professional work. We also extend special thanks to our production editor, Elizabeth Swain, as well as to the following Wiley staff for their help: designer Laura Boucher, illustration coordinator Anna Melhorn, photo editor Mary Ann Price, and program assistant Cathy Donovan.

Of particular importance to the development of the book were the individuals who read the book chapter by chapter and provided valuable comments and constructive criticism. This was a particularly difficult job given the wide variety of topics covered in the text and we believe that the book could not have been successfully completed without their assistance. These reviewers are offered our special gratitude:

First Edition Reviewers

Marc Abrams, Pennsylvania State University

Michele Barker-Bridges, Pembroke State University (NC)

Susan Beatty, University of Colorado

David Beckett, University of Southern Mississippi

Mark Belk, Brigham Young University

Kristen Bender, California State University, Long Beach

Gary Booth, Brigham Young University

Grace Brush, Johns Hopkins University

Kelley Cain, University of Wisconsin, River Falls

John Campbell, Northwest Community College (WY)

Ann Causey, Prescott College (AZ)

Simon Chung, Northeastern Illinois State

Thomas B. Cobb, Bowling Green State University

Jim Dunn, University of Northern Iowa

Robert Feller, University of South Carolina

Andrew Friedland, Dartmouth College

Douglas Green, Arizona State University

James H. Grosklags, Northern Illinois University

Bruce Hayden, University of Virginia

David Hilbert, San Diego State University

Peter Kolb, University of Idaho

Henry Levin, Kansas City Community College

Hugo Lociago, University of California, Santa Barbara

Tom Lowe, Ball State University

Timothy Lyon, Ball State University

Mel Manalis, University of California, Santa Barbara

Earnie Montgomery, Tulsa Junior College, Metro Campus

Walter Oechel, San Diego State University

C. W. O'Rear, East Carolina University

Stephen Overmann, Southeast Missouri State University

Martin Pasqualetti, Arizona State University

David Pimental, Cornell University

Brian Reeder, Morehead State University

C. Lee Rockett, Bowling Green State University

Joseph Simon, University of South Florida

Lloyd Stark, Pennsylvania State University

Laura Tamber, Nassau Community College (NY)

Bruce Wyman, McNeese State University

Ann Zimmerman, University of Toronto

Second Edition Reviewers

Marvin Baker, University of Oklahoma

Mary Benbow, University of Manitoba

Grady Blount, Texas A&M University-Corpus Christi

John Bounds, Sam Houston State University

Vincent Breslin, SUNY-Stony Brook

Bonnie Brown, Virginia Commonwealth University

Annina Carter, Adirondack Community College

Peter Colverson, Mohawk Valley Community College

Harry Corwin, University of Pittsburgh

Craig Davis, Ohio State University

Paul Grogger, University of Colorado

David Johnson, Michigan State University

S. B. Joshi, York University

Stephen Malcolm, Western Michigan University

James Melville, Mercy College

Chris Migliaccio, Miami-Dade Community College-Wolfson

Clayton Penniman, Central Connecticut State University

Jeffery Schneider, SUNY-Oswego

Student Reviewers

One of our goals in writing this book was to help students to think critically about the environment. In developing the Second Edition, we solicited feedback from students who studied from the First Edition. Their response has been gratifying and also constructive. We wish to thank the following students for their input.

Student Evaluators

Bonnie Carvelli, Mohawk Valley Community College

C. Ehitz, Miami-Dade Community College-Wolfson

Dari Loreno, Adirondack Community College

Joan Petrie, Mohawk Valley Community College

Balkis Sierra, Miami-Dade Community College-Wolfson

Kelly Toloza, Miami-Dade Community College-Wolfson

Student Focus Groups

Professor Craig Fusaro at Santa Barbara City College coordinated a group of his students:

John Barthel

Tino Gutierrez

Jon Humfrey

Thomas Haug

Phillip Hellyer

Darren Hudson

Linda Martin

Patrick Park

Mark Wilde

Elizabeth Williams

Professor Vincent Breslin at SUNY-Stony Brook coordinated a group of his students

Heather Bittner

Lesia Clarke

Deborah Luby

Jessie Mayer

Orly Pinchas

Jason Terhune

Valerie Tierce

Andy Winslow

Richard Worthington

Dongqiang Zheng

Completion of this project required an incredible amount of word processing and other administrative assistance. We are very grateful to Ellie Dzuro for transcribing and typing many chapters of this text. We acknowledge Amy Weiss and Claudia Tyler for their contribution in researching and handling permissions for tables and illustrations. We are also indebted to Valery A. Rivera who reviewed a number of chapters of this book with the careful eye and mind that only a dedicated teacher could bring to the subject.

The University of California, Santa Barbara, and George Mason University provided the stimulating atmosphere necessary for writing. Parts of the text were also written while E. A. Keller was on leave at the Departimento de Geodinamica, Universidad de Granada, España, working with Professor Jose Chacon as well as at the Mediterranean Agronomic Institute of Chania on the island of Crete in Greece. Fellowships to Daniel B. Botkin from the East–West Center Honolulu and Rockefeller Bellagio Study and Conference Center, Bellagio, Italy, and from the Woodrow Wilson International Center for Scholars, Washington, D.C., provided time and stimulating environments in which ideas in the text underwent growth and enrichment.

Daniel B. Botkin
Edward A. Keller

$\mathcal{B}$RIEF CONTENTS

CONTENTS

PART IV
SUSTAINING LIVING RESOURCES, 183

PART VIII
ENVIRONMENT AND SOCIETY, *529*

Case Studies, Closer Looks, and Environmental Issues

CASE STUDIES

A CLOSER LOOK

ENVIRONMENTAL ISSUES

A SPECIAL NOTE TO STUDENTS

We have written this book with the belief that students really want to be challenged to think about the important environmental problems that face the world today. We are attempting to bring the environment to the forefront of education. This is in response to tremendous public concern about the environment. Public opinion polls confirm that the environment is an important social and political issue. We also believe that scientific literacy and critical thinking are very important. As a result, we have chosen to discuss important environmental issues, some of which are multifaceted, difficult, and challenging. We all need to recognize that solutions to the many environmental problems facing us today are not easy to find. There are differing opinions on which problems we should address and what solutions should be attempted. This is something we cannot teach you. We can present information and arguments, but ultimately you must think for yourself and develop solutions based on your value system and understanding. If this appears as a vague concept—it is! You are responsible for future generations just as are your instructors. Making informed decisions requires a lot of study as well as a commitment to be responsible in helping to insure that our planet maintains a healthy productive environment. We believe that if our book, *Environmental Science,* helps you in that endeavor, then it has been successful.

If our book has helped you to think more critically about environmental problems, we would like to hear from you. Please share your ideas or questions with us at DBotkin@Wiley.com.

Daniel B. Botkin
Edward A. Keller

STUDENT-TO-STUDENT PREFACE

When I came to college, my goal was to learn all I could about the environment and the problems facing it. *Environmental Science* is a text that I found to be essential for students looking to learn about the environment and current problems facing that environment. This was the first text dealing with the environment that I had ever used, and it was in a class entitled "Environmental Problems and Solutions." Both the class and the text helped to inform me of the problems facing the environment and gave me the opportunity to explore solutions to these problems. The text offers a broad description of environmental problems, while giving you a detailed explanation of specific problems. It is important for you, as a student, to obtain a sound basis in your study of the environment. A thorough understanding will better prepare you to make intelligent decisions in the future. This text gives you the opportunity to do just that by successfully introducing you to a problem, fully explaining that problem, and encouraging you to formulate your own opinion of the subject at hand.

When using this text, I suggest that you take advantage of everything that *Environmental Science* has to offer. Within each chapter, you will come across examples of specific issues that have occurred or are occurring in the environment. These sections are entitled "Case Study" at the beginning of each chapter, "A Closer Look" within the chapter, and "Environmental Issue" at the end of each chapter. You will find it useful to read through these sections as they will help give you a more focused definition of the issue or issues dealt with in the chapter. I would also recommend going over the Study Questions at the end of each chapter. They are designed to help you in your understanding of the more important points discussed in each chapter.

When studying for any environmental class, I suggest that you first look to your notes from the class itself. This will provide an outline for possible test topics that you may need to know. After you have obtained an outline of topics, go to the text to fine tune your understanding of the topics being dealt with in the notes. The text will provide you with the information that your instructor did not, as it is impossible for them to give you all of the facts in the time allotted to them. This is why the text will be essential to your studies, just as it was to mine.

In closing let me say that it has been over a year since I took this class and *Environmental Science* still sits on my desk and is used for many of my reference needs. You will find this text to be a good investment, not only now, but in the future as well.

Richard J. Worthington

A senior at SUNY Stony Brook studying Marine and Environmental Sciences, Rich is an avid surfer and scuba diver, which have both had a strong influence on his opinions about the environment. He hopes to achieve a Masters Degree in Coastal Marine Management and Policy and eventually seek employment in an environmental consulting firm.

Button blanket, Village Island.

Environment as an Idea

The rich diversity of life on and around coral reefs is illustrated in this photo from the Red Sea, Egypt.

CASE STUDY

Shrimp, Mangroves, and Pickup Trucks: Local and Global Connections Reveal Major Environmental Concerns

Maitri Visetak, who owns a small plot of land along the coast of southern Thailand, wanted to improve life for his family, and he succeeded. A growing demand for shrimp as a luxury food and overfishing of wild shrimp fueled growth of the world market for farm shrimp from a $1.5 billion industry 30 years ago to an $8 billion business today. In the early 1990s, Mr. Visetak began farming shrimp in two small ponds (0.2 hectare or 0.5 acre; Figure 1.1*a*). In 2 years, Mr. Visetak accumulated enough capital to purchase two pickup trucks, a clear indication of financial success in Thailand. But, by then, his ponds were contaminated with shrimp waste, antibiotics, fertilizers, and pesticides. Shrimp could no longer live in the ponds. But there was an even more widespread effect: Pollutants escaping from the ponds threatened survival of the area's mangrove trees (Figure 1.1*b*). Like thousands of other shrimp farmers in southeast Asia, India, Africa, and Latin America, Mr. Visetak considered abandoning these ponds and moving on to others.

As a result of urbanization and development, half of the world's mangrove forests have been destroyed and with them a major source of food for local human population and the breeding grounds for much of the tropical world's sea life. The United Nations Environment Program has estimated that one-fourth of the destruction of mangroves can be traced to shrimp farming. Environmentalists have become alarmed, and in many areas local people have staged protests against shrimp farming. With the world's population expected to increase from 5.9 billion to 9 to 10 billion by the middle of the next century, concern over the world's mangrove forests is growing. Maitri Visetak is trying to feed his family in the best way

(a)

(b)

FIGURE 1.1 (*a*) Shrimp farms such as this one threaten survival of mangrove forests. (*b*) Mangroves on the banks of Indian River, Isle of Dominica, West Indies. Mangrove trees grow in coastal wetlands. Their specialized roots can survive immersion in ocean water at high tide and exposure to the drying sun at low tide. Swamps formed by mangroves provide habitat for many kinds of ocean life and are important for commercial fisheries in many parts of the world.

he knows how, but multiplied by the thousands of shrimp farmers in the world, he is unwittingly contributing to destruction of one of the world's valuable ecosystems.

REFERENCES

1. Gordon, B. B. 1993. "Pampering Our Coastlines," *Sea Frontiers* 39(2):5.

2. Mydans, S. 1996 (April 28). "Thai Shrimp Farmers Facing Ecologists Fury," *The New York Times*.

3. "Pollution Wiping Out Shrimp Farms on Main Indonesian Island of Java." 1996. *Quick Frozen Foods International* 37(3):50.

4. Quarto, A. 1994. "Rainforests of the Sea: Mangrove Forests Threatened by Prawn Aquaculture," *E*,5(1):16–19.

5. Wickramayanake, S. D. 1995. "East Coast Shrimp Farms Face Trouble from Both Nature and Protest Groups." *Quick Frozen Foods International* 37(1):108–109.

The story of Maitri Visetak illustrates the major themes of environmental science. First, human population increase is a major contributor to environmental problems. Second, industrial development and urbanization have serious environmental consequences. Third, unsustainable use of resources must be replaced with sustainable practices. Fourth, local changes can have global effects. Finally, environmental issues involve values and attitudes as well as scientific understanding. The story of Maitri Visetak illustrates an important question that we all must face: Which individual actions contribute to environmental degradation, and what actions can people, both as individuals and as groups, take to limit environmental damage?

LEARNING OBJECTIVES

Certain issues are basic to the study of environmental science. After reading this chapter, you should understand:

- Why rapid human population growth is *the* fundamental environmental issue.

- Why we must learn to sustain our environmental resources so that they will be available in the future.

- How human beings affect the environment of the entire planet and why we must take a global perspective on environmental problems.

- Why urban environmental issues and the effects of urban areas on environments elsewhere need to be given primary focus.

- Why developing solutions to environmental problems requires making value judgments based on knowledge of scientific facts.

1.1 KEY THEMES

Today we stand at the threshold of a major change in our approach to environmental issues. Two paths lie before us. One is business as usual, the approach to environmental issues we have taken for the past 30 years, an approach that has produced many advances but also many failures. This path has emphasized confrontation, emotionalism, a lack of understanding of basic facts about the environment and of how natural ecological systems function and a willingness to base solutions on political ideologies and on ancient myths about nature.

The second path offers the potential for long-lasting, successful solutions to environmental problems. This approach seeks to move from confrontation to cooperative problem solving, from explaining the environment in terms of ancient myths about nature to providing a sound scientific basis from which to view environmental issues. The purpose of this book is to take the student down the second pathway. A little historical explanation will help clarify what we seek to accomplish.

Before 1960, few people had ever heard the word *ecology* and the term *environment* meant little as a political or social issue. Then, with the publication of Rachel Carson's landmark book, *Silent Spring* (Houghton Mifflin, Boston, 1960, 1962) as well as the occurrence of major environmental events, such as oil spills along the coasts of southern California and Massachusetts, and the highly publicized threats of extinction to many species, including whales, elephants, and songbirds, the environment became a popular issue.

As with any new social or political issue, at first only a minority recognized its importance, and there was a need to stress the problems—to emphasize the negative—in order to bring public attention to environmental concerns. The early days of modern environmentalism were dominated by confrontations between those labeled environmentalists and those labeled developers. Each group viewed itself as the savior of the world and stereotyped the point of view of the other.

These two approaches can be contrasted in very simplistic terms. Environmentalists believed that the world would be destroyed if people did not change their approach to the environment. Thus they believed that they held the key to the salvation of the world in a new worldview, which depended only secondarily on facts, understanding, and science. To these environmentalists, economic and social development meant the destruction of the environment and therefore, ultimately, the end of civilization, the extinction of many species, and, potentially, the extinction of human beings.

The anti-environmentalists, on the other hand, believed that social and economic health and progress were necessary if people and civilization were to prosper. From their perspective, environmentalists represented a dangerous and extreme view, with a focus on the environment to the detriment of people, a focus that they thought would destroy the very basis of civilization and lead to the ruin of our modern way of life.

Adding to the limitations of the early approach to environmental issues was a lack of scientific knowledge and practical know-how. Environmental sciences were in their infancies. Some people even saw science as part of the problem.

Today, the situation has changed considerably. The environment is widely accepted as a major social and political issue. Public opinion polls repeatedly show that people around the world rank the environment among the most important issues. There is no longer a need to prove that environmental problems are serious. The time for confrontation for its own sake has passed. There is now a general recognition that real solutions to environmental problems include and depend on human beings, that we must seek sustainability not only of the environment but also of our economic activities so that humanity and the environment may persist together into the future. Among many, the dichotomy of the last generation is giving way to a new unity: the idea that a sustainable environment and sustainable economy may be compatible.

Although our scientific understanding of the environment lags greatly behind our need to know, significant progress has been made in many areas of environmental science. Advances have also been made in the creation of legal frameworks for the management of the environment, thus providing a new basis from which to make progress in solving environmental issues. Now that environmental issues are accepted as important, the time is ripe to seek truly lasting solutions. It is time to get away from rhetoric, which is often emotionally moving but empty of content, and to proceed with the development of rational solutions.

The environment is complex and multifaceted. This book covers a wide range of topics, from environmental ethics to the chemistry of the ozone layer in the atmosphere. This wide range of topics is connected by five unifying themes. The purpose of this chapter is to acquaint you with these themes so that the material you study later will have a basic unity. With this basis of understanding, you will be ready to learn about a great variety of subjects that can provide the foundation for better environments in the future and for a sustainable world and its resources.

Human Population Problem

The rapidly increasing human population underlies all environmental problems. Ultimately, we cannot solve our environmental problems unless we can learn to limit the total number of people on the Earth to the number that the environment can sustain.

Sustainability

In the past, we have used living environmental resources such as forests, fish, and wildlife faster than they could replenish. We have extracted minerals, oil, and groundwater without concern for their limits or the need for recycling. As a result, many of these resources are no longer in abundance. We must learn how to sustain our environmental resources so that they continue to provide benefits to us and to the larger environment of which we are a part.

A Global Perspective

Life on our planet is characterized by complex relationships among living things—the atmosphere, the water, and the land. Human activity has begun to change the environment on a global scale in ways harmful to people and other living things. As a result, the next generation of environmental scientists as well as all citizens must have a global perspective.

An Urban World

Most of us live in cities and towns, and it is in these urban settings that many solutions to environmental problems must be found. Cities and towns are increasing in size and spreading over the landscape, sometimes eliminating valuable farmland and natural areas. In the past, environmental activism usually centered on wilderness. In the future, however, it will be important to focus on towns and cities as livable environments and to develop harmony between industrial and commercial activities and the quality of urban environments.

Values and Knowledge

Although environmental issues are often portrayed as simply a question of facts, central conflicts about the environment have to do with values and knowledge. Knowledge, which includes knowing the scientific data as well as understanding the issues, tells us what solutions are possible. What we ultimately decide to do—which of the possible solutions we choose—depends on our values. A central theme of this book is how we value the environment. Given the size of our population and our great technological power to change the environment, we can no longer avoid the connection between values and knowledge. (See this chapter's Environmental Issue.)

These five issues—the human population problem, sustainability, a global perspective, an urban world, and values and knowledge—are central to our understanding of environmental science and recur frequently throughout the text. They are the threads that tie together our ideas about the environment and provide the focus for our study of environmental science.

1.2 HUMAN POPULATION GROWTH

The John Eli Miller Family

John Eli Miller was a farmer like many other farmers in the United States, except for one thing—when he died on his farm in Middlefield, Ohio, in the mid-twentieth century, he was the head of the largest family in the United States (Figure 1.2). He was survived by 5 children, 61 grandchildren, 338 great-grandchildren, and 6 great-great-grandchildren. Within his lifetime, John Miller witnessed a family population explosion. What was perhaps even more remarkable was that the explosion started with a family of just 7 children—not all that unusual for the nineteenth-century United States.

During most of John Miller's life, his family was not unusually large. It is just that he lived long enough to find out what simple multiplication can do, and he lived in a time when the death rate among infants, children, and young adults was very small compared with typical death rates during the history of most human populations.[1] Of 7 children born to John Miller, 5 survived him; of 63 grandchildren, 61 survived him; and of 341 great-grandchildren (born to 55 married grandchildren—an average of slightly more than 6 children per parent), 338 survived him.

John Miller's family emphasizes a major factor in our modern population explosion. Modern technology, modern medicine, and the supply of food, clothing, and shelter have decreased death rates and accelerated the net rate of growth. As a result, human population has increased greatly, threatening the environment.

The Population Bomb

In the last part of the twentieth century we have seen the most dramatic increase in the history of the human population. In merely 35 years the human

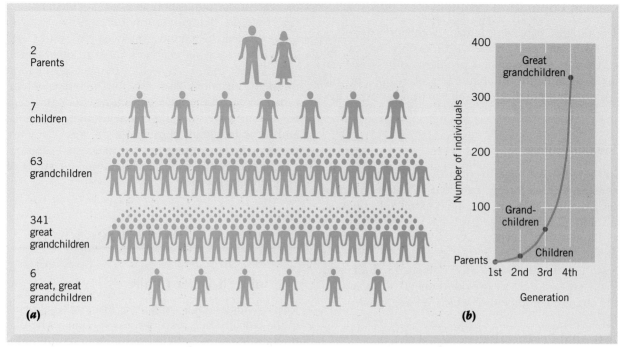

FIGURE 1.2 (*a*) A simplified family tree of the John Eli Miller family. (*b*) The population explosion of the John Eli Miller family shown in graphic form.

population of the world has more than doubled, increasing from 2.5 billion to over 5.9 billion.[2] Figure 1.3 illustrates the rapid explosion of the human population, sometimes referred to as the population bomb.[3]

FIGURE 1.3 Population change since 1950 projected to the year 2150 for major areas of the world, medium fertility scenario. The population of Africa will nearly quadruple. The only major area's population projected to drop over time is Europe—from 728 million to 595 million—a decline of 18% over 155 years. (*Source:* Population Division of the Department of Economic and Social Affairs at the United Nations Secretariat, *World Population Projections to 2150*, United Nations, New York, 1998.)

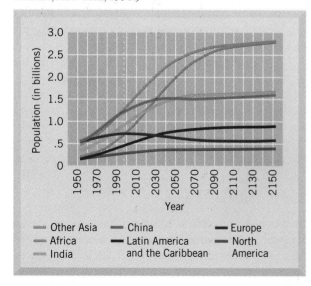

The human population issue is the underlying issue of the environment, because most current environmental damage results from the very large number of people on the earth. Human population is discussed in detail in Chapter 5; an outline of the problem is given here because it is so fundamental to an analysis of environmental issues.

Our species has not always grown as rapidly as it is growing today. For most of human history the total population was small, with an extremely low long-term rate of increase compared with that of today's population. (See A Closer Look 5.2: Growth of the Human Population.)[4,5] In addition, the growth of the human population has not been a steady march. Although it is customary to think of the human population as increasing continuously without declines or fluctuations, this has not been the case; for example, there were great declines during the time of the Black Death and the recent famines in Africa. (See A Closer Look 1.1.)

The African Famine

The world already has examples of what happens when a human population exceeds its environmental resources. Starvation in African nations that continues today first gained worldwide attention in the mid–1980s.[7,8] Drought and political unrest in Africa have led to major episodes of famine. In one year, as many as 22 African nations suffered catastrophic food shortages and 150 million Africans faced starvation.

CLOSER LOOK 1.1

THE BLACK DEATH

The epidemic disease bubonic plague, commonly known as the Black Death, spread throughout Europe during the fourteenth century. The greatest episodes occurred between 1347 and 1351, but there were many recurrences throughout the century (Figure 1.4).[6] The disease, caused by the bacteria *Yersinia pestis* and spread by fleas that live on rodents, was first recorded in Western history as a major human problem when it occurred in the seventh century in the Roman Empire and in North Africa. The plague probably first appeared in India in the seventh century and spread rapidly north and west. It did not cause another major epidemic until the fourteenth century, when it again spread rapidly, reaching Italy in 1348 and Spain, France, Scandinavia, and central Europe within 2 years. In England one-fourth to one-third of the population died within a single decade, but mortality varied widely in different regions, generally ranging between one-eighth and two-thirds of the population. Entire towns were abandoned, and the production of food for the remaining population was jeopardized.

The Black Death had many environmental and economic consequences. For example, the great reduction in the labor force led to an increase in wages and is believed to have been a contributing factor to a subsequent increase in the standard of living. Much agricultural land was abandoned because no one was available to work it.

This example illustrates that human populations have not always increased continuously but have suffered setbacks and declines. The bubonic plague is one of the best recorded and most well-known setbacks in human history, but it is likely that other such episodes, resulting from changes in climate, a decline in the food supply, or environmental catastrophes, have happened many times in human history.

FIGURE 1.4 (*a*) The change in the population of Europe during the time of the Black Death. (*b*) During the fourteenth century, the bubonic plague, known as the Black Death, killed many people in Europe. The reduced human population resulted in economic and environmental effects. This fourteenth-century illustration of two plague victims is from a miniature from the Toggenburg Bible. Note the swellings over the bodies, characteristic of the plague.

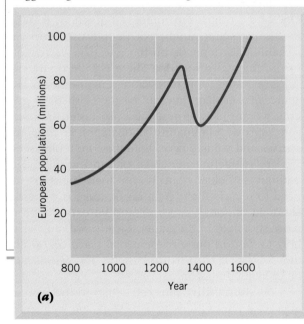

(a)

(b)

Four million tons of emergency food were needed. Five million refugees sought food and shelter.

Ten years earlier, a similar catastrophic situation developed following the 1973–1974 drought in the Sahel in Africa. At that time 500,000 Africans starved to death and several million more were permanently affected by malnutrition, with Chad, Ethiopia, Ghana, São Tomé, and Mozambique the most severely affected countries.[9]

Drought is not new to Africa, but the size of the population affected by drought is new. Human population growth in Africa has exceeded the increase in food production. In political disruptions the control and destruction of food has been used as a weapon. Massive starvation in Somalia was due in part to political disruption as well as to food shortages. Deserts in Africa appear to be spreading, in

Social conditions affect the environment and the environment affects social conditions, as illustrated by a Somalian boy with a gun (left photo). Political disruption in Somalia interrupted farming and food distribution, leading to starvation. Overpopulation, climate change, and poor farming methods also lead to starvation, which in turn promotes social disruption. Famine has been common in parts of Africa since the 1980s, as illustrated by gifts of food from aid agencies in southern Sudan, as shown in photograph at right.

part the result of changing climate but also as a result of human activities. Poor farming practices have increased erosion, and deforestation may be helping to make the environment drier.

Famines in Africa illustrate that population crises feed on themselves. People affect the environment; the environment affects people. The environment affects agriculture, but agriculture also affects the environment. In parts of Africa in the 1980s, human population growth severely stretched the capacity of the land to provide food and threatened its future productivity. This situation leads to a key question about values: *Which is more important, the survival of people alive today or the conservation of the environment, on which future food production and human life depend?*[10]

Information and knowledge are necessary to make value-based decisions. For example, we must determine whether we can continue to increase agricultural production without destroying the very environment on which that agriculture and, indeed, the persistence of life on Earth are based. This is a technical, scientific investigation. The answer provides a basis for a judgment that in turn depends on our values.

The human population is doubling every few decades, but our effects on the environment are growing at an even faster rate. Human beings cannot escape the laws of population growth. The question is: What will we do about our own species' increase and its impact on our planet and our future?

1.3 SUSTAINABILITY AND CARRYING CAPACITY

One of the central environmental questions is: What is the maximum number of people the Earth can sustain? This depends on the rate that we use our environmental resources. There is little doubt that we are using our renewable environmental resources faster than they can be replenished. In general, we are using forests and fish faster than they can regrow, eliminating habitats of endangered species and wildlife faster than they are replenished. We are extracting minerals, oil, and groundwater without sufficient concern for their limits or the need to recycle them. As a result, there is a present shortage of some resources and an expectation of more shortages in the future. Clearly, we must learn how to *sustain* our environmental resources so that they continue to provide benefits for people and other living things on our planet.

Sustainability must be achieved, but we are unclear at present how to achieve it, and we use the word to mean different things, often confusing ourselves by talking at cross purposes. There are several kinds of sustainability. It is important to distinguish these, so that we know which we are discussing. Two kinds of sustainability refer to the resources and their environment: (1) A *sustainable resource harvest*, such as a sustainable supply of timber, means that the same quantity of that resource can be taken each year (or other harvest in-

CLOSER LOOK 1.2

DETERMINING CARRYING CAPACITY OF THE CHINOOK SALMON

The problem of determining carrying capacity is illustrated by the history of commercial fishing of chinook salmon in the Columbia River of the Pacific Northwest. Figure 1.5 graphs the annual catch of salmon from 1866 (just after the Civil War, when commercial fishing began in earnest on the river) to 1966.[15] The catch increased rapidly from 1860 to 1880, then declined and fluctuated between 16 and 36 million pounds until 1920, when it declined again in a highly varying pattern. What, if any, levels of chinook salmon are sustainable? The high catches between 1890 and 1920 are about the same, so a person in charge of managing the salmon fishery in 1920 might reasonably have concluded that the salmon had a sustainable catch of between 16 and 36 million pounds. However, a person who had only the information after 1950, when the catch represents the tail of a declining curve, would conclude that the approximate level of catch in the early 1960s was sustainable. This graph illustrates the difficulty in determining a truly sustainable harvest.

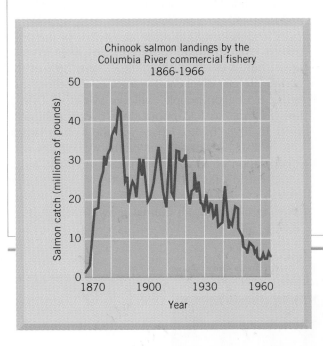

FIGURE 1.5 The commercial catch of chinook salmon on the Columbia River between 1866 and 1966. [*Source*: J. M. Van Hyning, 1973, "Factors Affecting the Abundance of Fall Chinook Salmon in the Columbia River," *Research Reports of the Fish Commission of Oregon*, 4(1), Figure 14, p. 38.)]

terval) for an unlimited or specified amount of time. (2) A *sustainable ecosystem* is an ecosystem from which we are harvesting a resource that is still able to maintain its essential functions and properties. Other kinds of sustainability refer to human societies. (3) A *sustainable economy* is an economy that maintains its level of activity over time in spite of uses of environmental resources. (4) *Sustainable development* typically is used to mean that a society is able to continue to develop its economy and social institutions and maintain its environment for an indefinite time.

Carrying Capacity of the Earth

Carrying capacity is a concept related to sustainability. It is usually defined as *the maximum number of individuals of a species that can be sustained by an environment without decreasing the capacity of the environment to sustain that same amount in the future.* When we ask "What is the maximum number of people that the Earth can sustain?" we are asking what is the carrying capacity, and we are also asking about sustainability.

For the human population, the carrying capacity depends in part on our values. Do we want those who follow us to live short lives in crowded surroundings without a chance to enjoy the Earth's scenery and diversity of life? Or do we hope that our descendants will have a life of high quality and good health? Once we choose a goal for the quality of life, we can use scientific information to understand what the carrying capacity might be and how we might achieve it. (See A Closer Look 1.2.)

1.4 A GLOBAL PERSPECTIVE

The recognition that, worldwide, civilization can change the environment at a global level is relatively recent. As discussed in detail in later chapters, scientists now believe that emissions of modern chemicals are changing the ozone layer high in the atmosphere and that the burning of fossil fuels increases the concentration of greenhouse gases in the atmosphere, which may change the Earth's climate. These atmospheric changes suggest that the actions of many groups of people at many locations affect the environment on a global level.[16,17]

Another new idea explored in later chapters is that nonhuman life affects the environment of our planet at a global level and has changed it over the course of several billion years. These two new ideas have profoundly affected our approach to environmental issues.

Awareness of the global interactions between life and the environment has led to the development of the **Gaia hypothesis,** originated by British chemist James Lovelock and American biologist Lynn Margulis.[16] The Gaia hypothesis (discussed in greater detail in Chapter 3) proposes that the environment at a global level has been profoundly changed by life over the history of life on the Earth and that these changes have tended to improve the environment for life—to increase the chances for the continuation of life. Because life affects the environment at a global level, the environment of our planet is different from that of a lifeless one. Because human actions have begun to change the environment at a global level, the next human generation must embrace a global perspective concerning environmental issues.

1.5 AN URBAN WORLD

In part as a result of the rapid growth of the human population and in part as a result of changes in technology, we are becoming an urban species, and our effects on the environment are more and more the effects of urban life (Figure 1.6).[11] With economic development comes urbanization; people move from farms to cities and then perhaps to suburbs. Cities and towns increase in size. Because cities are commonly located near rivers and along coastlines, urban sprawl often overtakes the good agricultural land of river floodplains and the coastal wetlands, which are an important habitat for many rare and endangered species. As urban areas expand, wetlands are filled in, forests cut, and soils covered over and removed from productive use.

In developed countries about 75% of the population lives in urban areas and 25% lives in rural areas, but in developing countries only 36% of the people are city dwellers. By 2025, 62% of the population, 6.5 billion people, will live in cities.[12] Only a few urban areas had populations over 4 million in 1950. In 1998, Toyko, Japan, was the world's largest city. In 2015, Toyko will still be the world's city with a population of 28.7 million. Today, Bombay, India, is the world's fastest growing city, with an annual growth rate of 4.22%. New York City is the slowest growing city, with an annual growth rate of .34%.

FIGURE 1.6 The United States viewed at night from space. The urban areas show up as bright lights, and the number of urban areas reflect the urbanization of our nation.

There are currently 12 megacities (urban areas with more than 10 million people)—but by the year 2010, there will be 22. No U.S. city will be among the top ten large cities in 2015. Asia will have seven megacities, more than any other continent, by 2015.[13]

Much of the activity and concern of environmental organizations has been directed toward nonurban areas—wilderness, endangered species, and natural resources, including forests, fisheries, and wildlife. Although these areas will remain important, in the future we must place more emphasis on urban environments and on the effects of urban environments on the rest of the planet.

1.6 VALUES AND KNOWLEDGE

As previously mentioned, one theme of this book is that *decisions concerning solutions to environmental problems require both values and knowledge.* The acceptable condition of the environment is not apparent to us simply from an inspection of facts and cannot be determined totally on the basis of feelings without knowing the facts. We must choose what we want the environment to be. To make this choice, we must first know what is possible, which requires knowing and understanding the implications of the scientific data. Once we know our options, we can select from among them. What we choose of the possible options is determined by our values. An example of a value judgment regarding the world's human environmental problem is the choice between the desire of an individual to have many children and the need to find a way to limit the human population worldwide.

Once we have chosen a goal based on knowledge and values, we have to find a way to attain that goal. This step also requires knowledge, and the more technologically advanced and powerful our civilization, the more knowledge is required. For example, to determine today whether chinook salmon are sustainable requires that we know how many there are now and how many there have been in the past. Our knowledge not only includes facts such as these, but an understanding of the processes of birth and growth of this fish, its food requirements, its habitat, its life cycle, and so forth, all the factors that ultimately determine the abundance of salmon in the Columbia River. Contrast the situation almost two centuries ago when Lewis and Clark first made an expedition to the Columbia and found many small villages of Native Americans who depended in large part on the fish in the river for food. As long as the human population was small and the methods of fishing simple, the maximum number of fish the people could catch probably posed no

threat to the salmon. In this case, the people could fish without scientific understanding of numbers and processes. (This example does not suggest that pre-scientific societies lacked an appreciation for the idea of carrying capacity or sustainability. On the contrary, many so-called primitive societies held strong beliefs about the limits of harvests.)

Placing a Value on the Environment

How do we place a value on some aspect of our environment? How do we choose between two different concerns? Environmental values can be based on four categories of justification: utilitarian, ecological, aesthetic, and moral. Two of these, utilitarian and ecological, concern practical reasons that have to do with our own survival or economic benefit. A **utilitarian justification** sees some aspect of the environment as valuable because it provides individuals with economic benefit or is directly necessary to their survival. For example, fishermen obtain their livelihood from the ocean and need a supply of fish so that they can continue to earn a living. An **ecological justification** is based on the value of some factor that is essential to larger life support functions, even though it may not benefit an individual directly. For example, as discussed in the opening case study, mangrove trees provide habitat for marine fish. Although we do not eat mangroves, we may eat the fish that depend on mangroves. As another example, burning coal and oil adds greenhouse gases to the atmosphere, which may lead to a change in climate that could affect the entire Earth. These ecological reasons form a basis for the conservation of nature that is essentially enlightened self-interest.

Aesthetic arguments have to do with our appreciation of the beauty of nature. For example, many people find wilderness scenery beautiful and would rather live in a world with wilderness than without it. The aesthetic justification is gaining a legal basis. The state of Alaska acknowledges that sea otters have an important role related to recreation: People observe and photograph the otters and enjoy viewing them in a wilderness setting. Many examples illustrate the importance of the aesthetic values of the environment. When people grieve following the death of a loved one, they typically seek out places with grass, trees, and flowers, and thus we decorate our graveyards. Although popular discussions of environmental issues might make aesthetic justifications seem superficial, in fact, beauty in their surroundings is of profound importance to people. Frederick Law Olmsted, the great American landscape planner, argued that plantings of vegetation provide medical, psychological, and social benefits and are essential to city life.[19]

How Can We Preserve the World's Coral Reefs?

Coral reefs, which are among the largest, oldest, most diverse, and most beautiful communities of plants and animals, have been seriously damaged or at risk. Scientists estimate that approximately 10% have already been destroyed while another 30% are threatened. The major threat to reefs are the direct or indirect results of human activities. If human impacts on these delicate ecosystems are not controlled, some scientists estimate that 60% of the world's reefs could collapse in the next 20 years.

The pieces of coral that most people know from souvenir and jewelry shops are the limestone skeletons secreted by colonies of living animals related to sea anemones and jellyfish. Like their relatives, each small individual coral animal, or polyp, uses tentacles equipped with stinging cells to capture food. In addition, polyps obtain nourishment from photosynthetic algae that live in their cells. When polyps die, their skeletons remain while the next generation of individuals secrete new material. Thus, reefs grow slowly by accretion; coral reefs that exist today are 5,000 to 10,000 years old. By taking the brunt of the force of waves, coral reefs protect coastlines from erosion, a function that has been estimated to have a value of $50,000 a year per square foot.

Coral reefs provide homes for a vast variety of plants and animals. Approximately 25% of all marine life, about 1 million species, is associated with coral reefs. Reef organisms are the source of many useful chemicals and medicines, and scientists are currently searching for others. The plant and animal species found around coral reefs are linked in intricate ways so that removing only one or two key elements may cause a catastrophic collapse. For example, overfishing in the waters off the Cook Islands in the South Pacific in the 1980s removed most of the reef's parrot fish and sea urchins, both of which feed on algae. Soon algae overgrew the reef and the entire community of reef life collapsed.

Coral reefs have long been the main source of protein for tropical people, which today number approximately 1 billion. With modern transportation and preservation methods, fish and other food organisms from coral reefs are now eaten by many other people. In fact, reef fish constitute about 15% of the entire worldwide catch. Unfortunately, many of the world's reefs are being overfished so that some species are now rare and endangered. Recently, some consumers, especially in Asian countries, have placed a high premium on eating fish that are alive when they reach restaurants. The demand for fish for aquariums also fuels the demand for live fish. Consequently, the price of some reef fish, such as grouper and wrasse, has increased from $5 a pound to $60.

To obtain live fish, many fishermen use dynamite to stun fish or cyanide to poison them temporarily. Both methods can kill or damage other organisms while dynamite can destroy the reef material itself. When fish are lodged in crevices in the reef, fishermen may use crowbars to pry apart the coral so they can reach the fish. And the limestone material that forms the bulk of a reef is sometimes mined to use as construction material. Millions of tourists from around the world who flock to reef areas to fish, swim, dive, and enjoy their beauty pose an additional threat.

But, perhaps the greatest threat to coral reefs comes from increasing population in the tropics. Population densities greater than that of the New Jersey coast (about 500 people per square kilometer) are found in parts of tropical Asia and the Caribbean. In many areas, raw or inadequately treated sewage is released nearshore while runoff from land development and deforestation adds to the burden of sediment and pollution. Degradation of the coral reef in Kaneohe Bay in Hawaii due to sewage and other runoff was dramatically reversed in the 1980s by diverting the sewage to the open ocean. But with increasing urbanization and population growth around the Bay in the 1990s, recovery slowed, or perhaps even reversed.

Critical Thinking Questions

1. How does the current state of the world's coral reefs illustrate each of the five key themes of this textbook?

2. What are the utilitarian, ecological, aesthetic, and moral justifications for preserving coral reefs?

3. If Maitri Visetak were making his living from fishing rather than farming shrimp, how might he view the preservation of coral reefs? What arrangements could be made to meet his needs while at the same time preserving coral reefs in his area?

4. What things can you do in your everyday life to contribute to the preservation of coral reefs?

References

Coles, S.L., and L. Ruddy. 1995. Comparison of water quality and reef coral mortality and growth in southeastern Kaneohe Bay, Oahu, Hawaii, 1990 to 1992, with conditions before sewage diversion. *Pacific Science*, vol. 49, no. 3, pp. 247–265.

Hinrichsen, D. 1997 (Oct.). "Coral Reefs in Crisis." *Bioscience*, vol. 47, no. 9, pp. 554–558.

Jameson, S.C., McManus, J.W., and M.D. Spalding. 1995 (May). "State of the Reefs: Regional and Global Perspectives." International Coral Reef Initiative Executive Secretariat (Background Paper).

Moral justification has to do with the belief that aspects of the environment have a right to exist and that it is our moral obligation to allow them to continue or help them to persist. An example of a moral justification is the assertion that Nine Mile Prairie, located near Lincoln, Nebraska, one of the few remaining prairie preserves, has a right to exist. Moral arguments have been extended to many non-human organisms, to entire ecosystems, and even to inanimate objects.[20] For example, the historian Roderick Nash has written an article entitled "Do Rocks Have Rights?" which discusses such moral justification.[21] And the United Nations General Assembly World Charter for Nature, signed in 1982, states that species have a moral right to exist.

A new discipline known as *environmental ethics* analyzes these issues. Another concern of environmental ethics is our obligation to future generations: Do we have a moral obligation to leave the environment in good condition for our descendants or are we at liberty to use environmental resources to the point of depletion within our own lifetimes?

SUMMARY

- Five threads or themes run through this text: the urgency of the population issue, the importance of urban environments, the need for sustainability of resources, the importance of a global perspective, and the role of values in the decisions we face.

- The human population is growing ever more rapidly; one way or another, this increase cannot continue indefinitely. The human population has grown at a rate unprecedented in history in the twentieth century; it is *the* environmental problem.

- When the impact of technology is combined with the impact of population, the result is a multiplied impact on the environment.

- In an increasingly urban world, we must focus much of our attention on the environments of cities and on the effects of cities on the rest of the environment.

- Determining the Earth's carrying capacity for people and levels of sustainable harvests of resources is difficult but important if we are to plan effectively to meet our needs in the future.

- Awareness of how the impact of people at a local level affects the environment globally gives credence to the Gaia hypothesis; future generations will require a global perspective concerning environmental issues.

- Placing a value on aspects of the environment requires knowledge and understanding of the science involved but also depends on how we judge the uses and aesthetics of the environment and on our moral commitments to other living things and to future generations.

REEXAMINING THEMES AND ISSUES

This chapter lays out the five themes that run through this book. These are:

- Human population growth is the fundamental environmental issue.
- Sustainability is the key to future availability of resources and environmental qualities.
- Our modern civilization is having global environmental effects.
- Our world is becoming increasingly urban, and urban areas have their own kinds of environmental effects, both inside the city and on the surrounding countryside.
- Any solution to environmental issues requires consideration of both knowledge and values.

These lead to key questions, including:

Human Population: What is more important: the quality of life of people alive today or the future life of a human fetus?

Sustainability: What is more important: abundant resources today—as much as we want and can obtain—or the persistence of these resources for future generations?

Global Perspective: What is more important: the quality of your local environment or the global environment—the environment of the entire planet?

Urban World: What is more important: human creativity and innovation, including arts, humanities, and science, or the persistence of certain endangered species? Must this always be a trade-off or are there ways to have both?

Values and Knowledge: Does Nature know best, so that we never have to ask what is the environmental goal we should seek, or do we need to have knowledge about our environment, so that we can make the best judgments given available information?

KEY TERMS

aesthetic arguments *11*
carrying capacity *9*
ecological justification *11*

Gaia hypothesis *10*
moral justification *13*
sustainability *8*

utilitarian justification *11*

STUDY QUESTIONS

1. Why is the density of people who live by hunting and gathering lower than the density of people who practice agriculture?
2. Refer to Figure 1.5. How would you respond to the statement, "The catch of salmon in the Columbia River is sustainable at 1960s levels?"
3. In what ways do the effects on the environment of a resident of a large city differ from the effects of someone living on a farm? In what ways are the effects similar?
4. Programs were established to supply food from Western nations to feed the starving people in Somalia. Some people argue that such food programs, which may have short-term benefits, actually increase the threat of starvation in the future. What are the pros and cons of international food relief programs?
5. What are the values involved in deciding whether to create an international food relief program? What are five kinds of information required to determine the long-term effects of such programs?
6. Which of the following are global environmental problems? Why?

a. The growth of the human population.
b. The furbish lousewort, a small flowering plant found in the state of Maine that is so rare it has been seen by few people, is considered endangered.
c. The blue whale is listed as an endangered species under the U.S. Marine Mammal Protection Act.
d. A car that has air-conditioning.
e. Almost all major ocean ports have seriously polluted harbors and coastlines.
7. How could you determine **(a)** the carrying capacity and **(b)** the sustainable yield of salmon in the Columbia River? (Refer to Fig. 1.5)
8. Is it possible that all the land on the Earth will sometime in the future become one big city? If not, why not? To what extent is the answer dependent on:
a. Global environmental considerations?
b. Scientific information?
c. Values?

FURTHER READING

Botkin, D. B. 1995. *Our Natural History The Lessons of Lewis and Clark.* New York: Putnam. What we can learn about the character of nature and people's relationship to it from the journals of America's great explorers.

Botkin, D. B. 1990. *Discordant Harmonies: A New Ecology for the 21st Century.* New York: Oxford University Press. An analysis of the myths that lie behind attempts to solve environmental issues.

Kent, M. M. 1990. *World Population: Fundamentals of Growth.* Washington, D.C.: Population Reference Bureau. Facts and data about the growing human population.

Kessler, E., ed. 1992. "Population, Natural Resources and Development," *AMBIO* 21(1). A special issue of the journal *AMBIO*, addressing many problems concerning human population growth and its economic and environmental implications.

Leopold, A. 1949. *A Sand County Almanac.* New York: Oxford University Press. Perhaps, along with Rachel Carson's *Silent Spring,* one of the most influential books of the post–World War II and pre–Vietnam War era about the value of the environment. Leopold defines and explains the land ethic and writes poetically about the aesthetics of nature.

Lutz, W. 1994. *The Future of World Population.* Washington, DC.:

Population Reference Bureau, Summary of current information on population trends and future scenarios of fertility, mortality, and migration.

Nash, R. F. 1988. *The Rights of Nature: A History of Environmental Ethics*. Madison: University of Wisconsin Press. An introduction to environmental ethics.

INTERNET RESOURCES

International Network for Sustainable Energy (INFORSE): *http://www.inforse.dk*—INFORCE is a worldwide NGO (nongovernmental organization) network formed at Global Forum in Rio de Janeiro, Brazil, June 1992. The mission is to promote sustainable energy and social development. The Web site provides links to other sites and on-line access to the publication *Sustainable Energy News*.

Population Reference Bureau: *http://www.prb.org/prb/*—Information on U.S. and international population statistics and trends.

United Nations Development Programme: *http://www.undp.org/*—Information on sustainable human development and links to the UN Environment Programme and other UN sites.

United Nations Population Information Network: *http://www.undp.org/popin/popin.htm*—Contains information on world demographic trends.

The World Bank: *http://www.worldbank.org*—The World Bank is an international development organization that is owned by its more than 180 member countries—both rich and poor. Its role is to reduce poverty by lending money to the governments of its poorer members—often called "developing" countries—and to those countries whose economies are in transition. The Web site provides access to World Bank publications and data for educators, students, business people, journalists, and the general public about developing countries and countries in transition.

The World Resources Institute's (WRI): *http://www.wri.org*—WRI's mission is to move human society to live in ways that protect Earth's environment and its capacity to provide for the needs and aspirations of current and future generations. WRI provides information and proposals for policy and institutional change for environmentally sound, socially equitable development. The Web site contains publications, news articles, links to other Web sites, and support services for educators.

Zero Population Growth: *http://www.zpg.org/zpg/index.html*—Home page of ZPG, the organization concerned with impacts of rapid population growth and wasteful consumption. Contains links to lists of population-related Web sites and gophers.

CHAPTER 2

THINKING CRITICALLY ABOUT THE ENVIRONMENT

Forester recording measurements as part of the scientific method.

CASE STUDY

The Case of the Mysterious Crop Circles

For 13 years, circular patterns appeared "mysteriously" in grain fields in southern England (Figure 2.1). Proposed explanations included aliens, electromagnetic forces, whirlwinds, and pranksters. The mystery generated a journal and a research organization headed by a scientist as well as a number of books, magazines, and clubs devoted solely to crop circles. Scientists from Great Britain and Japan brought in scientific equipment to study the strange patterns. Then, in September 1991, two men confessed to having created the circles by entering the fields along paths made by tractors (to disguise their footprints) and dragging planks through the fields. When they made their confession they demonstrated their technique to reporters and some crop circle "experts."[1]

FIGURE 2.1 (*a*) A series of crop circles and other shapes at the Vale Pwesey in southern England in July 1990. (*b*) Crop circles seen from the air make distinctive patterns.

(a)

(b)

How is it that so many people, including some scientists, took the crop circles seriously? The answer is that they misunderstood science and engaged in fallacious reasoning. The failure of some people to think critically about crop circles resulted in no harm, but the same type of thinking applied to other, more serious environmental issues can have serious consequences.

LEARNING OBJECTIVES

Science is a process of refining our understanding of nature by continual questioning and active investigation of questions. Students should approach science in this manner, rather than as a collection of facts to be memorized. After reading this chapter, you should understand:

- That thinking about environmental issues involves thinking scientifically.
- That scientific knowledge is acquired through observations of the natural world that can be tested through additional observations and experiments.
- The difference between deductive and inductive reasoning and how both are used in scientific thinking.

- That every measurement involves some degree of approximation, that is, uncertainty, and that a measurement without a statement about its degree of uncertainty is meaningless.
- That scientific discovery involves a number of processes, including the scientific method, but that science and scientists are too diverse to be described by just one method.
- That technology is not science but science and technology interact, stimulating growth in each other.
- That decision making about environmental issues involves society, politics, culture, economics, and values as well as scientific information.

2.1 THINKING ABOUT ENVIRONMENTAL SCIENCE

Thinking about the environment is as old as our first human ancestors. Their survival depended on knowledge of it. Today the environment plays a crucial role in the development of each of us; normal human development does not occur in the absence of environmental stimuli. In modern technological society, the environment for most of us is what we encounter in our everyday life.

Thinking *scientifically* about the environment, on the other hand, is only as old as science itself. Science had its beginnings in the ancient civilizations of Babylonia and Egypt, where observations of the environment were carried out primarily for practical reasons, such as planting crops, or for religious reasons, such as using the positions of the planets and stars to predict human events. These ancient practices differed from modern science in that they did not distinguish between science and technology or between science and religion.

These distinctions first appeared in classical Greek science. Because of their general interest in ideas, the Greeks developed a more theoretical approach to science, in which knowledge for its own sake became the primary goal. At the same time, their philosophical approach began to move science away from religion and toward philosophy.

Modern science is usually considered to have its roots in the end of the sixteenth and the beginning of the seventeenth centuries, with the development of the **scientific method** by Gilbert (magnets), Galileo (physics of motion), and Harvey (circulation of blood). Unlike earlier classical scientists who asked *why?* in the sense of "for what purpose?" they made important discoveries by asking *how?* in the sense of "how does it work?" Galileo also pioneered the use of numerical observations and mathematical models. The scientific method, which quickly proved to be very successful in advancing knowledge, was first described explicitly by Francis Bacon in 1620. Although not a practical scientist himself, Bacon recognized the importance of the scientific method, and his writings did much to promote scientific research.[2]

Our cultural heritage, therefore, gives us two ways of thinking about the environment: the kind of thinking we do in everyday life and the kind of thinking scientists do (see Table 2.1). There are

TABLE 2.1 A Comparison of Knowledge in Everyday Life with Science

Parameter	Everyday Life	Science
Goal	To lead a satisfying life (implicit)	To know, predict, and explain (explicit)
Requirements	Context-specific knowledge; no complex series of inferences; can tolerate ambiguities and lack of precision	General knowledge; complex, logical sequences of inferences; must be precise and unambiguous
Resolution of questions	Through discussion, compromise, consensus	Through observation, experimentation, logic
Understanding	Acquired spontaneously through interacting with world and people; criteria not well defined	Pursued deliberately; criteria clearly specified
Validity	Assumed, no strong need to check; based on observations, common sense, tradition, authorities, experts, social mores, faith	Must be checked; based on replications, converging evidence, formal proofs, statistics, logic
Organization of knowledge	Network of concepts acquired through experience; local, not integrated	Organized, coherent, hierarchical, logical; global, integrated
Acquisition of knowledge	Perception, patterns, qualitative; subjective	Plus formal rules, procedures, symbols, statistics, mental models; objective
Quality control	Informal correction of errors	Strict requirements for eliminating errors and making sources of error explicit

Source: Based on F. Reif and J. H. Larkin, "Cognition in Scientific and Everyday Domains: Comparison and Learning Implications," *Journal of Research in Science Teaching*, 28(9); pp. 733–760. Copyright 1991 National Association for Research in Science Teaching. Reprinted by permission of John Wiley & Sons.

many similarities between the two, which helps us to realize that thinking scientifically is something all of us can do. On the other hand, there are crucial differences, and ignoring these can lead to invalid conclusions in science, in general, and to serious errors in making critical decisions about the environment, in particular.[3]

2.2 UNDERSTANDING WHAT SCIENCE IS (AND WHAT IT ISN'T)

Science is one way of looking at the world. It begins with **observations** about the natural world from which scientists formulate hypotheses that can be tested. Modern science does not deal with things that cannot be tested by observation, such as the ultimate purpose of life or the existence of a supernatural being. Science also does not deal with questions that involve values, such as standards of beauty or issues of good and evil. The criterion by which we decide whether a statement is in the realm of science is whether it is possible, at least in principle, to test the statement by observation.

Although there are many other ways of looking at the world, such as the religious, aesthetic, and moral views, they are not science. The assertions made in those worldviews are not testable; they are based ultimately on faith, beliefs, and cultural and personal values. To say that these other ways of looking at the world are not science is not to denigrate them. Each way of viewing the world gives us

a different way of perceiving and of making sense of our world, and each is valuable to us.

Assumptions of Science

Science makes certain assumptions about the natural world; it is important to be aware of these assumptions in order to understand what science is:

1. Events in the natural world follow patterns that can be understood by people through careful observation and analysis.

2. The basic patterns, or rules, that describe the behavior of events in the natural world are the same throughout the universe.

3. Science is based on a type of reasoning known as induction; it begins with specific observations about the natural world and extends to generalizations.

4. Generalizations can be subjected to tests that disprove them. If such a test cannot be devised, then a generalization cannot be treated as a scientific statement.

5. Although new evidence can disprove existing scientific theories, science can never provide absolute proof of the truth of its theories.

The Nature of Scientific Proof

One source of serious misunderstanding about science is the use of the word *proof,* which most students encounter in mathematics, particularly in geometry. Proof in mathematics and logic involves reasoning from initial definitions and assumptions. If a conclusion follows logically from these **premises,** the conclusion is said to be proved. This process is known as **deductive reasoning.**

One example of deductive reasoning is the following:

> *A straight line is the shortest distance between two points.*
> *The line A–B is the shortest distance between points A and B.*
> *Therefore, A–B is a straight line.*

Note that the conclusion from this syllogism (series of logically connected statements) follows directly from the premises (the first two statements).

Deductive proof does not require that the premises be true, only that the reasoning be foolproof. Logically valid but untrue statements can result from false premises, as in the following example:

> *Humans are the only toolmaking organisms.*
> *The woodpecker finch uses tools.*
> *Therefore, the woodpecker finch is a human being.*

In this case, the concluding statement *must* be true if both the preceding statements are true; however, we know that the conclusion is not only false but ridiculous (Figure 2.2). If the second statement is true (which it is), then the first cannot be true and the conclusion must be false. The conclusion that a bird, the woodpecker finch, which uses cactus spines to remove insects from tree limbs, is human follows logically but defies common sense.

Generally, false conclusions are not very useful, especially in science. Because science is based on observations, its conclusions are only as true as the premises from which they are deduced. The rules of deductive logic govern only the process of moving from premises to conclusion. Science, on the other hand, requires not only logical reasoning but also correct premises. To be scientific, the three statements should be expressed conditionally (with reservation):

> If humans are the only toolmaking organisms

and

> The woodpecker finch is a toolmaker,

then

> The woodpecker finch is a human being.

When we formulate generalizations based on a number of observations, we are engaging in **inductive reasoning.** To illustrate, let us define birds with particular characteristics as swans. If we always observe that such birds are white, we may make the *inductive statement* "All swans are white." What we really mean is, "All of the swans we have seen are white." We never know when our very next observation will turn up a bird that is like a swan in all ways except that it is black. [Occasionally, mutations of black sheep occur, so it is not impossible that some swan, somewhere, may be black (Figure 2.3).]

Proof in inductive reasoning is, therefore, very different from proof in deductive reasoning. In induction, we can only state that it is *usually* true that all swans are white; we cannot say so with absolute certainty. When we say something is proved in induction, what we really mean is that it has a very high degree of probability. *Probability* is a way of expressing our certainty (or uncertainty)—our estimation of how good our observations are, how confident we are of our predictions. When we have a fairly high degree of confidence in our conclusions in science, we often forget to state the degree of certainty or uncertainty. Instead of saying, "There is a 99.9% probability that . . .," we say, "It has been proved that. . . ." Unfortunately, many people interpret this as a deductive statement, meaning the conclusion is absolutely true, which has led to much misunderstanding about science.

Although science begins with observations and, therefore, inductive reasoning, deductive reasoning is useful in helping scientists analyze whether conclusions based on inductions are logi-

Figure 2.2 A woodpecker finch on the Galapagos uses a twig to remove insects from a hole in a tree, demonstrating tool use by nonhuman animals.

(a)

(b)

(c)

FIGURE 2.3 (*a*) A white swan and offspring, (*b*) a black lamb with its mother, and, (*c*) an albino member of the species *Sciurus carolinensis*. The normal color of these squirrels is gray.

cally valid. Scientific reasoning combines induction and deduction—different, but complementary ways of thinking.[4]

2.3 MEASUREMENTS AND UNCERTAINTY

People put more faith in accuracy of measurements than do scientists. Scientists realize that *every measurement is only an approximation.* Measurements are limited; they depend on the instruments used and the people who use the instruments. Measurement uncertainties are inevitable; they can be reduced, but they can never be completely eliminated (see Environmental Issue: Should Minke Whales Be Hunted?). Because all measurements have some degree of uncertainty, a measurement is meaningless unless it is accompanied by an estimate of its uncertainty.

Imagine a simplified scenario in which an engineer is given a rubber O-ring used to seal fuel gases in a space shuttle. She is asked to determine the flexibility of the O-rings under different temperature conditions to help answer the question, "At what temperature(s) is it unsafe to fire the shuttle?" In other words, "At what temperature do the O-rings become brittle and subject to failure?"

After doing some tests, the engineer says that the rubber becomes brittle at −1°C (30°F). Is it safe to fire the shuttle at 0°C (32°F)? At this point, you do not have enough information to answer the question. You assume that the temperature data may have some degree of uncertainty, but you have no idea how large it is. What if the uncertainty is ±1°C, or ±5°C, or ±½°C? In order to make a reasonably safe and economically sound decision about whether to fire the shuttle, you must know the uncertainty of the measurement.

Dealing with Measurement Uncertainties

Measurement uncertainties and other errors that occur while doing experiments are called *experimental errors*. Errors that occur consistently, such as those resulting from incorrectly calibrated instruments, are systematic errors, in contrast to random errors. Scientists traditionally include a discussion of experimental errors when they report results. Often error analysis leads to greater understanding of a subject and sometimes even to important discoveries. For example, the existence of the eighth planet in our solar system, Neptune, was discovered when scientists investigated apparent inconsistencies in the orbit of the seventh planet, Uranus.

Measurement uncertainties can be reduced by improving the instruments used and by requiring a standard procedure in making measurements. They can also be reduced by using carefully designed experiments and appropriate statistical procedures. However, uncertainties are inherent in every measurement and can never be completely eliminated.

As difficult as it is for us to live with uncertainty, that is the nature of measurement and of science. We need to use our understanding of the uncertainties in measurements to read reports of scientific studies critically, whether they are in science journals or in magazines and newspapers. (see A Closer Look 2.1: Measurement of Carbon Stored in Vegetation).

Accuracy and Precision

Certain measurements have been made very carefully by many people over a long period of time and accepted values have been determined. Accuracy is the extent to which a measurement agrees with the accepted value. Precision is the degree of exactness with which a quantity is measured. A very precise measurement need not and may not be accurate. If the accepted value for the temperature of boiling water at sea level is 100.000°C (212.000°F) and you measure it as 99.875°C (211.775°F), your measurement is just as precise as the accepted value (i.e., it is measured to the nearest 1/1000 degree), but it is still inaccurate (although not very!).

Although it is important to make measurements as precisely as possible, it is equally important not to report measurements with more precision than they warrant. Doing so conveys a misleading sense of precision. A measurement based on a ruler divided into 10ths of a centimeter should be precise to the nearest 100th of a centimeter, but no more. A small bone that appears to end exactly at the 2.1-cm mark should be reported as 2.10 cm long, not as 2.100 cm, which would imply that the first zero was read directly, when in fact it was estimated.

2.4 THE METHODS OF SCIENCE

Observations, the basis of science, are made through any of the five senses or by instruments that measure beyond what we can. The accuracy of observations is checked by comparing observations made by many scientists. When an observation is agreed upon by all, or almost all, it is often called a **fact.**

An important distinction needs to be made between observations and **inferences,** or ideas based on observations. An observation about a substance might be that it is a white, crystalline material with a sweet taste; one might infer that the substance is sugar. Before this inference can be accepted as fact, it must be subjected to further tests. Confusing observations with inferences and accepting untested inferences is the kind of sloppy thinking often described by the phrase "thinking makes it so."

When scientists wish to test an inference, they convert it into a statement that can be disproved. This type of statement is known as a **hypothesis.** A hypothesis continues to be accepted until it is disproved. A hypothesis that has not been disproved has not been proved to be true in the deductive sense; it has only been found to be probably true until, and unless, evidence to the contrary is found.

The most typical hypotheses are those in the form of if–then statements. For example, "If I apply more fertilizer, then the tomato plants will produce larger tomatoes." The statement relates two conditions: amount of fertilizer applied and size of tomatoes. Each condition is called a variable, that is, something that can vary. The size of tomatoes is called the **dependent variable** because it is assumed to depend on the amount of fertilizer, the **independent variable.** The independent variable is also sometimes called a **manipulated variable,** because the scientist deliberately changes or manipulates it. The dependent variable would then be referred to as a **responding variable,** one that responds to changes in the manipulated variable.

In growing tomato plants, many variables may exist. Some, like the position of the North Star, can be assumed to be irrelevant; others, like the duration of daylight, are potentially relevant. In testing a hypothesis, a scientist tries to keep all relevant variables constant, except for the independent and dependent variables, an approach known as *controlling variables.* In a **controlled experiment,** the experiment is compared to a standard, or control, an exact duplicate of the experiment except for the condition of the one variable being tested (the independent variable). Any difference in outcome (dependent variable) between the experiment and the control can be attributed to the effect of the independent variable.

CLOSER LOOK 2.1

MEASUREMENT OF CARBON STORED IN VEGETATION

Recently, a number of people have suggested that a partial solution to global warming might be a massive worldwide program of tree planting. Trees take carbon dioxide (an important greenhouse gas) out of the air in the process of photosynthesis. Because they live a long time, trees can store carbon for decades, even centuries. But how much carbon can be stored? Many books and reports published during the past 20 years have contained numbers representing the total stored carbon in the vegetation of the Earth, but all were presented without any estimate of error (Table 2.2). Without an estimate of that uncertainty the figures are meaningless, yet important environmental decisions have been based on them.

Recent studies have reduced the error by replacing guesses and extrapolations with scientific sampling techniques, similar to the procedures used to predict the outcome of elections. Even these improved data would be meaningless, however, without an estimate of error. The new figures show that the earlier guesstimates were three to four times too large, grossly overestimating the storage of carbon in vegetation and, therefore, the contribution tree planting could make in offsetting global warming.

TABLE 2.2 **Estimates of Above-Ground Biomass in the North American Boreal Forest**

Source	Biomass[a] (kg/m^2)	Carbon[b] (kg/m^2)	Total Biomass[c] $(10^9$ metric tons$)$	Total Carbon[c] $(10^9$ metric tons$)$
This study[d]	4.2 ± 1.0	1.9 ± 0.4	22 ± 5	9.7 ± 2
Previous estimates[e]				
1	17.5	7.9	90	40
2	15.4	6.9	79	35
3	14.8	6.7	76	34
4	12.4	5.6	64	29
5	5.9	2.7	30	13.8

Source: D. B. Botkin and L. Simpson, 1990, "The First Statistically Valid Estimate of Biomass for a Large Region," *Biogeochemistry,* 9; pp. 161–274. Reprinted by permission of Kluwer Academic, Dordrecht, The Netherlands.

[a]Values in this column are for total above-ground biomass. Data from previous studies giving total biomass have been adjusted using assumption that 23% of the total biomass is in below-ground roots. Most references use this percentage; Leith and Whittaker use 17%; we have chosen to use the larger value to give a more conservative comparison.

[b]Carbon is assumed to be 45% of total biomass following R. H. Whittaker, 1974, *Communities and Ecosystems,* Macmillan, New York.

[c]Assuming our estimate of the areal extent of the North American boreal forest: 5,126,427 km² (324,166 mi²).

[d]Based on a statistically valid survey; aboveground woodplants only.

[e]Lacking estimates of error: (1) G. J. Ajtay, P. Ketner, and P. Duvigneaud. 1979. "Terrestrial Primary Production and Phytomass," in B. Bolin, E. T., Degens, S. Kempe, and P. Ketner, eds., *The Global Carbon Cycle.* New York: Wiley, pp. 129–182. (2) R. H. Whittaker and G. E. Likens. 1973. "Carbon in the Biota," in G. M. Woodwell and E. V. Pecan, eds., *Carbon and the Biosphere,* Springfield, Va.: Nat. Tech. Infor. Center, pp. 281–300. (3) J. S. Olson, H. A. Pfuderer, and Y. H. Chan. 1978. *Changes in the Global Carbon Cycle and the Biosphere.* ORNL/EIS-109, Oak Ridge, Tenn.: Oak Ridge National Laboratory. (4) J. S. Olson, I. A. Watts, and L. I. Allison, 1983. *Carbon in Live Vegetation of Major World Ecosystems.* ORNL-5862, Oak Ridge, Tenn.: Oak Ridge National Laboratory. (5) G. M. Bonnor. 1985. Inventory of Forest Biomass in Canada, Petawawa, Ontario, Can. For. Serv., Petawawa Nat. For. Institute.

An important aspect of science, but one often overlooked in descriptions of the scientific method, is the need to define or describe variables in very exact terms that can be understood by all scientists. The least ambiguous way to define or describe variables is in terms of what one would have to do to duplicate their measurement. Such definitions are called **operational definitions.** Cooking recipes are examples of operational definitions.

Before carrying out an experiment, both the independent and dependent variables must be defined operationally. Operational definitions allow other scientists to repeat experiments exactly and to check on the results reported.

While performing an experiment, a scientist must record the values of the input (independent variable) and the output (dependent variable). These values are referred to as *data* (singular *datum*), which may be numerical, **quantitative data,** or nonnumerical, **qualitative data.** In our example, qualitative data would be the size of tomatoes reported as small, medium, or large; quantitative data would be the mass in grams or diameter in centimeters.

Knowledge grows as more hypotheses are supported in an area of science. Scientists use accumulated knowledge to develop explanations, or models, that are consistent with currently accepted hypotheses. A *model* is "a deliberately simplified construct of nature."[5] It may be an actual working model, a pictorial model, a mental model, a computer model, a laboratory model, or a mathematical model. As new knowledge accumulates, the models may have to be revised or replaced, with the goal of finding models more consistent with nature. Because hypotheses in science are continually being tested for consistency with nature and evaluated by other scientists, science has a built-in self-correcting feedback system.[5]

Models may be *conceptual, mathematical,* or *numerical.* A conceptual model is a set of concepts expressed qualitatively. The phrase "The world is like a watch" is an example of a simple conceptual model. Often in science, models are explicitly mathematical or numerical. A mathematical model is a set of equations describing how things work. A computer simulation is a numerical model. A computer simulation allows models to be developed of very complex systems. Because environmental systems are complex, computer simulations are becoming more and more useful in environmental science. Computer simulation of the atmosphere has become important in scientific analysis of the possibility of global warming. Computer simulation is becoming important for biological systems as well, such as simulations of forest growth (Figure 2.4). Models that offer broad, fundamental explanations of many observations are called **theories.**

The ideas discussed in this section are usually referred to as the scientific method and can be presented as a series of steps:

1. Make observations and develop a question about the observations.
2. Develop a tentative answer to the question—a hypothesis.
3. Design a controlled experiment to test the hypothesis (implies identifying and defining independent and dependent variables).
4. Collect data in an organized manner, such as a table.
5. Interpret the data through graphic or other means.

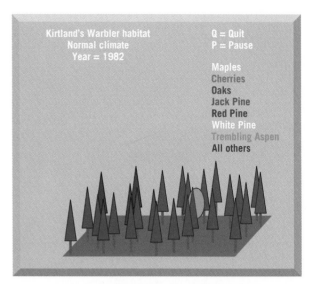

FIGURE 2.4 A computer simulation of forest growth. The color of the name of the tree at the right is the color the tree appears in the drawing below.) This kind of model is becoming increasingly important in environmental sciences. Shown here is the screen display of individual trees, whose growth is forecasted year by year, depending on environmental conditions. (*Source:* From *JABOWA-II* by D. B. Botkin. Copyright © 1993 by D. B. Botkin. Used by permission of Oxford University Press, Inc.)

6. Draw a conclusion from the data.
7. Compare the conclusion with the hypothesis and determine whether the results support or disprove the hypothesis (Figure 2.5).
8. If the hypothesis is accepted, conduct further tests to support it. If the hypothesis is rejected, make additional observations and construct a new hypothesis (Figure 2.5).

2.5 MISUNDERSTANDINGS ABOUT SCIENCE

Attempts to teach students the scientific method, as though all types of research carried out by scientists fit into a clearly defined stepwise process, lead to misunderstandings. Although the scientific method is aspired to by scientists and applies to much research, it is not universally applicable. It gives too little attention to the roles played by accidental discoveries (serendipity), creativity, and flashes of insight. The myth of a single scientific method does not take into account the differences among the disciplines of science; the logic of research in physics is not the same as in biology. In fact, within biology itself, research on evolution differs in significant ways from research on ecology. It is much more realistic to speak of the methods of science than of the scientific method.

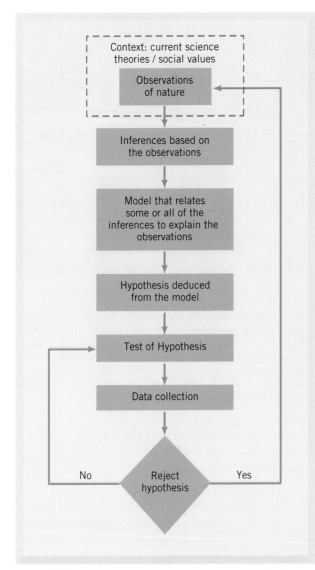

FIGURE 2.5 Flowchart. Scientific investigation as a feedback process. (*Source:* Modified from C. M. Pease and J. J. Bull, April 1992, *Bioscience,* 42, pp. 293–298.)

Theory in Science and Language

A common misunderstanding about science arises from confusion between the use of the word *theory* in science and its use in everyday language. A scientific theory is a grand scheme that relates and explains many observations and is supported by a great deal of evidence. In everyday usage, a theory can be a guess, a hypothesis, a prediction, a notion, a belief. You often hear the phrase "It's *just* a theory," which may make sense in everyday language but not in the language of science. In fact, theories have tremendous prestige and are considered the greatest achievements of science.[6]

Further misunderstanding arises when scientists use the word *theory* in several different senses[4]:

for example, a currently accepted, widely supported theory, such as the *theory of evolution by natural selection*; a now discarded theory, such as the *theory of inheritance of acquired characteristics*; a new theory, such as the *theory of evolution of multicellular organisms by symbiosis*; and a model dealing with a specific or narrow area of science, such as the *theory of enzyme action*.[6]

One of the most important misunderstandings about the scientific method is the relationship between research and theory. Although theory is usually presented as growing out of research, in fact, theories also guide research. The very observations a scientist makes occur in the context of existing theories. At times, discrepancies between observations and accepted theories become so great that a scientific revolution occurs; old theories are discarded and replaced with new or significantly revised theories.[7]

Science and Technology

Another misunderstanding about science occurs when science is confused with technology. Science is a search for understanding of the natural world, whereas technology is the control of the natural world for the benefit of humans. Science often leads to technological developments, just as new technologies lead to scientific discoveries. The telescope began as a technological device, but when Galileo used it to study the heavens, it became a source of new scientific knowledge. That knowledge stimulated the technology of telescope making so that better telescopes were produced, which in turn led to further advances in astronomy.

Science is limited by the technology available. Before the invention of the electron microscope, scientists were limited to magnifications of 1000 times and to studying objects about the size of 10ths of a micrometer. (A micrometer is 1/1,000,000 of a meter, or 1/1000 of a millimeter.) The electron microscope enabled scientists to view objects far smaller by magnifying more than 100,000 times. The electron microscope, a basis for new science, was also the product of science. Without prior scientific knowledge about electron beams and how to focus them, the electron microscope could not have been developed.

Most of us do not come in direct contact with science in our daily lives; instead, we come in contact with the products of science—technological devices such as cars, toasters, and microwave ovens. Thus there is a tendency for people to confuse the products of science with science itself. As you study science, it will help if you keep the distinction between science and technology in mind.

Science and Objectivity

One myth about science is the myth of objectivity, or value-free science: that scientists are capable of complete objectivity independent of their personal values and the culture in which they live and that science deals only with objective facts. Objectivity is certainly a goal of scientists, but it is unrealistic to think that they can be totally free of influence from their social environment. A more realistic view is to admit that scientists do have biases and to try to identify these explicitly rather than ignore them. In some ways, this is a similar situation to that of measurement error; it is inescapable, and we can best deal with it by recognizing it and estimating its effects.

We have only to look at recent controversies about environmental issues, such as whether or not to adopt more stringent automobile emission standards, to find examples of how personal and social values affect science. Genetic engineering, nuclear power, and the preservation of threatened or endangered species such as the Minke Whale discussed in this Chapter's Environmental Issue all involve conflicts among science, technology, and society. (The place of values in science is discussed in Chapter 30.)

The idea that scientists cannot always reach the goal of complete objectivity and that science is not entirely value free should not be taken to mean that fuzzy thinking is acceptable in science. It is still important to think critically and logically about science and related social issues. Without the high standards of evidence held up as the norm for science, we run the danger of accepting unfounded ideas about the world, of believing **pseudoscientific** (*pseudo* = "false") notions. When we confuse what we would like to believe with what we have the evidence to believe, we have a weak basis for making critical environmental decisions, decisions that could have far-reaching and serious consequences.

Science and Pseudoscience

Evidence for models at the frontiers of science is not as strong as for those ideas accepted by the scientific community. With more research, some of these frontier models may move into the realm of accepted science and new ideas will take their place at the advancing frontier.[8] However, research may not support other hypotheses at the frontier, and these will be discarded by scientists. Some people continue to believe in discarded scientific ideas, or pseudoscience. For example, although scientists long ago discarded the notion that the movements of the heavenly bodies could affect the personalities

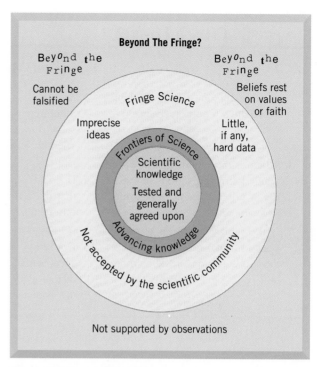

FIGURE 2.6 Beyond the fringe?

and fates of humans, a substantial number of Americans believe in or are not sure about astrology (25% believed and 22% were not sure).[9] Many of the things that astrologers learned about the movements of the stars and planets were so accurate that they became the basis for astronomy, the scientific study of the heavens. Parts of astrology moved into accepted science, parts into pseudoscience.

Accepted science may merge into frontier science, which in turn may merge into more far-out ideas, or fringe science (Figure 2.6). Really wild ideas may be considered beyond the fringe, or pseudo science. Although scientists have no trouble distinguishing between accepted science and pseudoscience, they do have trouble identifying ideas at the frontier that will become accepted and those that will be relegated to pseudoscience. That trouble arises because science is a process of continual investigation and an open system. This ambiguity at the frontiers of science leads many people to accept some frontier science before it has been completely verified and to confuse pseudoscience with frontier science (see the discussion of the Gaia hypothesis in Chapter 3).

Science and Media Coverage

Most media reports on scientific issues deal with new discoveries, fringe science, and pseudoscience. (See A Closer Look 2.2: Evaluating Media Coverage.)

CLOSER LOOK 2.2

EVALUATING MEDIA COVERAGE

The following questions will help you evaluate a report on science[10]:

1. Where is the report? Is it in a scientific journal, a serious newspaper, a popular science magazine, or a tabloid?

2. Is the report based on observations of actual occurrences? Were those observations made by more than one person? More than a few?

3. Are the sources of the report identified specifically? Are the sources specific, named scientists, scientific journals, or scientific organizations?

4. Does the report provide evidence that the claims are supported by other members of the scientific community?

5. Does the evidence for the claims seem sufficient? Is there contradictory evidence that might offset the evidence given for them?

6. Do the claims follow logically from the evidence? Do the claims violate reason (for example, "99-year-old woman gives birth to baby believed to be dead for 50 years")? Is there a simpler explanation for the observations?

7. Is there a clear basis for suspecting bias on the part of the source or the writer of the report?

8. Can you describe the line of reasoning that leads from the evidence to the claims?

It is important to analyze the claims in these reports to decide in which category they should be placed. Critical reading and listening to reports about science require thinking carefully about whether a statement is based on observations and data, an objective interpretation of data, an interpretation based on experience of an expert in the area, or a subjective opinion.

The need for critical examination of statements applies to those of scientists as well as to other sources. On the one hand, scientists' training in analyzing data may make them more qualified than the general public to decide complex issues; on the other, as discussed before, scientists' own values, interests, and cultural backgrounds may influence their interpretations of data and bias their pronouncements. That some scientists become media scientists, whose opinions are sought on a wide variety of topics, not all of which they have studied as scientists, can complicate the situation.

Many people accept any statement by a scientist as fact, although it may be anything from a gross misstatement to a casual opinion. The prestige invested in scientists and physicians by the public is capitalized on by advertisers. The practice of using a person in a white coat holding a test tube (a scientist) or wearing a stethoscope (a doctor) to promote products is so widespread as to have become a cliché ("9 out of 10 doctors …"). The opinions of experts do need to be heard, but appealing to scientists as authorities outside their area of expertise is contrary to the antiauthoritarian nature of science. The difficulties for the readers and viewers become even greater when scientists themselves cannot agree on the facts of an issue.

2.6 SCIENCE AND DECISION MAKING

The process of making decisions is sometimes presented, like the scientific method, as a series of steps:

1. Formulate a clear statement of the issue to be decided.

2. Gather the scientific information related to the issue.

3. List all alternative courses of action.

4. Predict the positive and negative consequences of each course of action and the probability that each consequence will occur. (This is known as risk–benefit analysis, a topic covered in Chapter 30.)

5. Weigh the alternatives and choose the best solution.

Such a procedure is a good guide to rational decision making, but it assumes a simplicity not often found in real-world issues. It is difficult to anticipate all the consequences of a course of action, and unintended consequences are at the root of many environmental problems. Often the scientific information is incomplete and even controversial. For example, the fact that the insecticide DDT causes eggshells of birds that feed on insects to be so thin as to result in the death of unhatched birds was a consequence that was not predicted. Only when populations of species such as the brown pelican became seriously endangered did people become aware of this consequence. The best solution often depends on one's point of view. The desirability of cutting down more trees in the Pacific Northwest, for example, is different for loggers than for environmentalists.

How Do We Decide What to Believe about Environmental Issues?

When you read about an environmental issue in a newspaper or magazine, how do you decide whether to accept the claims made in the article? Are they based on scientific evidence and are they logical?

Scientific evidence is based on observations, but media accounts often rely mainly on inferences (interpretations) rather than evidence. Distinguishing inferences from evidence is an important first step in evaluating articles critically.

Secondly, it is important to consider the source of a statement. Is the source a reputable scientific organization or publication? Does the source have a vested interest that might bias its claims? When sources are not named, it is impossible to judge the reliability of claims.

If a claim is based on scientific evidence presented logically from a reliable, unbiased source, it is appropriate to accept the claim tentatively, pending further information. Practice your critical evaluation skills by reading the article in the box and answering the questions.

Critical Thinking Questions

1. What is the major claim made in the article?
2. What evidence does the author present to support the claim?
3. Is the evidence based on observations, and is the source of the evidence reputable and unbiased?
4. Is the argument for the claim, whether based on evidence or not, logical?
5. Would you accept or reject the claim?
6. Even if the claim were well supported by evidence based on good authority, why would your acceptance be only tentative?

References

Ford, R. 1998 (March). "Critically Evaluating Scientific Claims in the Popular Press." *The American Biology Teacher*, vol. 60, no. 3, pp. 174–180.

Marshall, E. 1998 (May 15). "The Power of the Front Page of *The New York Times*." *Science*, vol. 280, pp. 996–997.

CLUE FOUND IN DEFORMED FROG MYSTERY

BY MICHAEL CONLON
Reuters News Agency (as printed in the *Toronto Star*)
November 6, 1996

A chemical used for mosquito control could be linked to deformities showing up in frogs across parts of North America, though the source of the phenomenon remains a mystery. "We're still at the point where we've got a lot of leads that we're trying to follow but no smoking gun," says Michael Lannoo of ball State University in Muncie, Ind. "There are an enormous number of chemicals that are being applied to the environment and we don't understand what the breakdown products of these chemicals are," says Lannoo, who heads the U.S. section of the world-wide Declining Amphibian Population Task Force.

He says one suspect chemical was Methoprene, which produces a breakdown product resembling retonoic acid, a substance important in development. "Retonoic acid can produce in the laboratory all or a majority of the limb deformities that we're seeing in nature," he says. "That's not to say that's what's going on. But it is the best guess as to what's happening." Methoprene is used for mosquito control, among other things, Lannoo says.

Both the decline in amphibian populations and the deformities are of concern because frogs and related creatures are considered "sentinel" species that can provide early warnings of human risk. The skin of amphibians is permeable and puts them at particular risk to agents in the water.

Lannoo says limb deformities in frogs had been reported as far back as 1750, but the rate of deformities showing up today was unprecedented in some species. Some were showing abnormalities that affected more than half of the population of a species living in certain areas, he adds. He says he doubted that a parasite believed to have been the cause of some deformities in frogs in California was to blame for similar problems in Minnesota and nearby states. Deformed frogs have been reported in Minnesota, Wisconsin, Iowa, South Dakota, Missouri, California, Texas, Vermont, and Quebec. The deformities reported have included misshapen legs, extra limbs, and missing or misplaced eyes.

In the face of incomplete information, scientific controversies, conflicting interests of different groups, and emotionalism, how can environmental decisions be made? We need to begin with the scientific evidence from all relevant sources, with estimates of the uncertainties in each. Where scientists disagree about the interpretation of the data, it may be possible to develop a consensus or a series of predictions based on the different interpretations. The impacts of the scenarios need to be identified and the risks associated with each analyzed in comparison with the benefits. Avoiding emotionalism and resisting slogans and propaganda are essential to developing sound approaches to environmental issues. Ultimately, however, environmental decisions are policy decisions negotiated through the political process. The makers of policies are rarely professional scientists, but political leaders and ordinary citizens. The scientific education of those in government and business as well as of all citizens is a crucial element in this process.

2.7 LEARNING ABOUT SCIENCE

Science is an open-ended process of finding out about the natural world. In contrast, science lectures and texts are usually summaries of the answers to this process, and science homework and tests are exercises in finding the right answer.

It is not difficult to understand, therefore, why students often perceive science as a body of facts to be memorized and lectures and texts as authoritative sources of absolute truths about the world. In contrast, scientists view scientific knowledge as *currently* accepted truth, always subject to change as new observations and interpretations are made. Students tend not to question the material in textbooks and lectures, whereas the essence of science is questioning and critical examination of accepted truths.

Viewing science as a collection of facts leads students to see science as difficult to understand and remember, whereas scientists emphasize the collection of facts into coherent pictures of the world that have the power to explain many phenomena in nature. These pictures (models, theories) are so powerful that scientists find them easy to remember and expect that students will also. Learning science also frequently involves solving problems, but here, too, the attitudes of students and scientists differ. Where students look to formulas, type problems, and algorithms to help solve problems, scientists look to general principles, critical thinking, and creativity.[3]

We encourage you to learn environmental science in an active mode, to be critical of what you hear, what you see, and what you read, including this textbook. We encourage you to try to understand what science is, its assumptions, limitations, and methods, and to apply this understanding to your study of environmental science. Most of all, we hope that you will see, not isolated facts to be learned by rote, but the connections among the facts that reflect the interdependence and unity in the environment.

SUMMARY

- Science has a tradition of critical thinking about the natural world, with the goal of understanding how nature works. Decisions on environmental issues must begin with an examination of the relevant scientific evidence. However, environmental decisions also require careful analysis of the economic, social, and political consequences. Solutions will reflect religious, aesthetic, and ethical values as well.

- Science begins with careful observations of the natural world, from which scientists formulate hypotheses. Whenever possible, scientists test hypotheses with controlled experiments.

- The process of formulating hypotheses based on observations or prior knowledge and testing the hypotheses through controlled experiments is referred to as the scientific method. Although this is often taught as a prescribed series of steps, it is better to think of it as a general guide to scientific thinking, with many variations.

- Scientific knowledge is acquired through inductive reasoning, in which general conclusions are based on specific statements. If the specifics are true, then the generalizations are probably true, but conclusions arrived at through induction can never be proved with certainty. Because of the inductive nature of science, it is possible to disprove hypotheses, but it is not possible to prove them with 100% certainty.

- Science also makes use of deductive reasoning, in which conclusions follow necessarily from the premises, provided that the rules of logic are followed. If the premises are true, then the conclusion is certainly true.

- Measurements are approximations that may be more or less exact depending on the measuring instruments and the people who use the instruments. A measurement is meaningful when it is accompanied by an estimate of the degree of uncertainty, or error.

- Accuracy in measurement is the extent to which the measurement agrees with an accepted value. Precision is the degree of exactness with which a measurement is made. A precise measurement need not be accurate. The estimate of uncertainty provides information on the precision of a measurement.

- A general statement that relates and explains a great many hypotheses is called a theory. Theories are the greatest achievements of science, but, like all scientific knowledge, they may be modified or discarded as new information is acquired. Existing theories provide a framework for the observations and hypotheses of science; new observations and hypotheses inconsistent with existing theories stimulate the development of new theories.

- Critical thinking can help us distinguish science from pseudoscience. It can also help us recognize possible bias on the part of scientists and in the media. Critical thinking involves questioning and synthesizing what is learned in order to achieve knowledge, rather than merely acquire information.

REEXAMINING THEMES AND ISSUES

This chapter summarizes the scientific method, which is essential to the analysis and solution to environmental problems. The scientific method is critical to understanding the **human population** problem and to developing sound approaches to **sustainability**. The **global perspective** on environment arises out of new findings in environmental science. The importance of our increasingly **urbanized world** is best understood with the assistance of scientific investigaton. Although often great faith is given to anything said by a scientist, and often we turn to scientists for the complete solution to an environmental problem, in reality solutions to environ-

mental problems require both **values** and **knowledge.** Ultimately, environmental decisions are policy decisions, negotiated through the political process, and involve both values and scientific knowledge. Understanding the scientific method is especially important if we are going to understand the connection between values and knowledge. Often, sufficient understanding of the scientific method is lacking among policymakers, leading to false conclusions. Uncertainty is part of the nature of measurement and science. We must learn to accept uncertainty as part of our attempts to conserve and use our natural resources.

KEY TERMS

controlled experiment *21*	inductive reasoning *19*	pseudoscientific *25*
deductive reasoning *19*	inferences *21*	qualitative data *23*
dependent variable *21*	manipulated variable *21*	quantitative data *23*
fact *21*	observations *18*	responding variable *21*
hypothesis *21*	operational definitions *22*	scientific method *17*
independent variable *21*	premises *19*	theories *23*

STUDY QUESTIONS

1. Which of the following are scientific statements and which are not? What is the basis for your decision in each case?
   a. The amount of carbon dioxide in the atmosphere is increasing.
   b. Picasso was the greatest painter of the twentieth century.
   c. Helping terminally ill people to die is wrong.
   d. The Earth is flat.
   e. The good will reap their reward in the afterlife.

2. What is the logical conclusion of each of the following syllogisms? Which conclusions correspond to reality, as we observe?
   a. All men are mortal.

   Socrates is a man.
   Therefore,
   b. All sheep are black.
   Mary's lamb is white.
   Therefore,
   c. All Amazons are women.
   No men are Amazons.
   Therefore,
   d. All elephants are animals.
   All animals are living beings.
   Therefore,

3. Which of the following statements are supported by deductive reasoning and which by inductive reasoning?
   a. The sun will rise tomorrow.

b. The sum of the squares of the sides of a right triangle is equal to the square of the hypotenuse.

c. Only male deer have antlers.

d. If $A = B$ and $B = C$, then $A = C$.

e. The net force acting on a body equals its mass times its acceleration.

4. The accepted value for the number of centimeters in an inch is 0.3937. Two students mark off an inch on a piece of paper and then measure the distance using a metric ruler. Student A finds the distance equal to 0.3827 cm and student B finds it equal to 0.39 cm. Which measurement is more accurate? Which is more precise? If student B measured the distance as 0.3900 cm, what would your answer be?

5. a. A teacher gives five students each a metal bar and asks them to measure the length. The measurements obtained are 5.03, 4.99, 5.02, 4.96, and 5.00 cm. How can you explain the variability in the measurements? Are these systematic or random errors?

b. The next day, the teacher gives the students the same bars but tells them that the bars have contracted because they have been in the refrigerator. In fact, the temperature difference would be too small to have any measurable effect on the length of the bars. The students' measurements, in the same order as in part (a), are 5.01, 4.95, 5.00, 4.90, and 4.95 cm. Why are the students' measurements different from those of the day before? What does this illustrate about science?

6. What are the independent and dependent variables in each of the following?

a. Change in the rate of breathing in response to exercise.

b. The effect of study time on grades.

c. The likelihood that people exposed to smoke from other people's cigarettes will contract lung cancer.

7. a. What technological advance resulted from a scientific discovery?

b. What scientific discovery resulted from a technological advance?

c. What technological device have you used today? What scientific discoveries were necessary before the device could be developed?

8. What is fallacious about each of the following statements?

a. A fortune cookie contains the statement, "A happy event will occur in your life." Four months later you find a hundred dollar bill. You conclude that the fortune was correct.

b. A person claims that aliens visited the Earth in prehistoric times and influenced the cultural development of humans. As evidence, the person points to the ideas among many groups of people about beings who came from the sky and performed amazing feats.

c. A person observes that light-colored forms of an animal almost always occur on light surfaces, whereas dark forms of the same species occur on dark surfaces. The person concludes that the light surface causes the light color of the animals.

d. A person knows three people who have had fewer colds since they began taking vitamin C on a regular basis. The person concludes that vitamin C prevents colds.

9. Using a newspaper article on a controversial topic, what are some loaded words used in articles, that is, words that convey an emotional reaction or a value judgment?

10. What are some of the social, economic, aesthetic, and ethical issues involved in a current environmental controversy?

FURTHER READING

American Association for the Advancement of Science (AAAS). 1989. *Science for All Americans.* Washington, D.C.: AAAS. This report focuses on the knowledge, skills, and attitudes a student needs to be scientifically literate.

Furnham, A. 1991 (January). "Hooked on Horoscopes," *New Scientist*, pp. 33–36.

Gould, S. J. 1996. *The Mismeasure of Man.* New York: W.W. Norton. The author takes on the controversial concept of biological determinism.

Grinnell, F. 1992. *The Scientific Attitude.* New York: Guilford Press. Examples from biomedical research are used to illustrate the processes of science (observing, hypothesizing, experimenting) and how scientists interact with each other and with society.

McCain, G., and Segal, E. M. 1982. *The Game of Science.* Monterey, Calif.: Brooks/Cole Publishing. The authors present a lively look into the subculture of science.

Sagan, C. 1995. *The Demon-Haunted World.* New York: Random House. The author argues that irrational thinking and superstition threaten democratic institutions and discusses the importance of scientific thinking to our global civilization.

PART II

Giraffes by Montas Antoine, Haitian

Earth As a System

The land and people in Africa, as in the rest of the world, interact in a world of change. Shown here are Masai people with some of their cattle, Kenya, Africa.

CHAPTER 3

SYSTEMS AND CHANGE

CASE STUDY

Amboseli National Park

Environmental change is often a result of a complex web of interactions. The most obvious answer to a problem may not be the right answer. Amboseli National Park is a case in point. The park, located in southern Kenya at the foot of the northern slope of Mount Kilimanjaro, is a game reserve that has supported an active tourist industry. Long-term management of the park and its resources depends on an understanding of the physical, biological, and human-use linkages that characterize this interesting landscape. Figure 3.1 shows the boundary of the park as well as the major geologic units. Particularly important is the fact that the park is

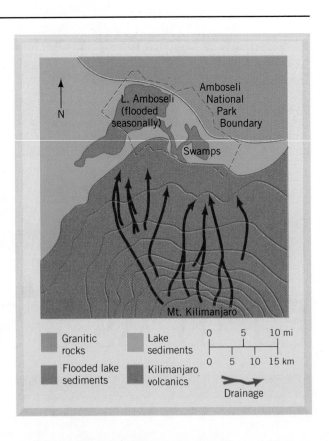

FIGURE 3.1 Amboseli National Park. Generalized geologic and landform map of Amboseli National Park, southern Kenya, Africa, and Mount Kilimanjaro. (*Source:* After T. Dunn and L. B. Leopold, 1978, *Water in Environmental Planning.* W. H. Freeman, San Francisco.)

Granitic rocks
Lake sediments
Flooded lake sediments
Kilimanjaro volcanics

0   5   10 mi
0   5   10   15 km

Drainage

centered on an ancient lake bed, remnants of which include the seasonally flooded Lake Amboseli and some swampland. Mount Kilimanjaro is a well-known volcano, composed of alternating layers of volcanic rock and ash deposits. To the north of the lake the bedrock consists of granite. Precipitation that falls on the slopes of Mount Kilimanjaro infiltrates the volcanic material (becomes groundwater) and moves slowly down the slopes to saturate some of the ancient lake beds, eventually emerging at springs located in the swampy, seasonally flooded land. The groundwater becomes very saline (salty) as it percolates through the lake beds, the sediments of which contain salt that easily dissolves.

The climate of the park area is arid, with only 350 to 400 mm (13.8 to 15.7 in.) of rainfall per year, sufficient to support dry savanna grasslands and brush. Prior to the mid–1950s, the park area was characterized by fever tree woodlands that supported mammal species such as kudu, baboons, vervet monkeys, leopards, and impalas (Figure 3.2). Starting in the 1950s, and particularly during the 1960s, the fever tree woodlands rapidly disappeared and were replaced by short grass and brush with an accompanying change to typical plains animals, such as zebras and wildebeest. Loss of the woodland habitat was prematurely blamed on overgrazing of cattle by the Masai people and damage to the trees from elephants (Figures 3.3 and 3.4). Of course, overgrazing can cause serious erosion problems and change of vegetation, and it is well known that herds of elephants can and do damage woodlands in their feeding behavior. In the case of Amboseli National Park, however, careful work showed that changes in climatic variables and soils were the primary culprits, rather than people.[1,2] The role of elephants is being investigated further, and they may be a more significant factor than previously believed.

Let us look at the evidence more closely. Those studying the problem were surprised to note that most dead trees were located in an area that had been free of stock since 1961, which predated the major decline in the woodland environment. Furthermore, some of the woodlands that suffered the least decline were those that had the highest density of people and cattle. These observations suggested that livestock and overgrazing were not responsible for loss of the trees.[1] Initially it appeared that elephant damage was a major factor; more than 83% of the trees in some areas showed some debarking. Also, some trees were pushed over, but these tended to be the younger, smaller trees. However, careful research concluded that elephants apparently played a secondary, or supporting, role in causing habitat change: as the den-

FIGURE 3.2 Amboseli National Park, Kenya, Africa. The mountain in the background is Mount Kilimanjaro.

sity of woodland species decreased, the incidence of damage caused by elephants increased.

Research that examined the rainfall and groundwater history and the soils suggested that the Amboseli National Park area is very sensitive to changing climatic conditions. During dry periods the depth of groundwater (which is very salty) increases and soils with relatively low salt content

FIGURE 3.3 Masai people grazing cattle in Amboseli National Park, Kenya. Grazing activities were prematurely blamed for loss of fever tree woodlands.

FIGURE 3.4 Elephant feeding on a yellow-bark acacia tree. Elephant damage to trees is considered to be a factor in loss of woodland habitat in Amboseli National Park. However, elephants probably have a relatively minor role in loss of woodlands compared to effects of oscillations in climate and groundwater conditions.

are present near the surface. These conditions favor the development of woodland environments. During wet periods the groundwater rises close to the surface, bringing with it salt that invades the root zones of trees. Loss of the trees is due then to *physiological drought* (intolerance to salt). As the woodlands dwindle, they are replaced by salt-tolerant grasses and low brush. During the most recent loss of woodland environment, the groundwater level rose as much as 3.5 m (11.4 ft) in response to unusually wet years in the 1960s. Analysis of the soils confirmed that the tree stands that were the most damaged were those associated with highly saline soils.[1,2]

Evaluation of the historic record from early European explorers, accounts from Masai herdsmen, and fluctuating lake levels in other East African lakes, such as Victoria and Rudolf, suggested that prior to 1890 there was another period of above-normal rainfall and sparse woodland environment. Thus we conclude that climatic cycles have a pronounced effect on the East African soils, vegetation patterns, and assemblage of animals present.[1]

This case history has important implications for long-term resource planning in East Africa. Cycles of wet and dry periods can be expected to continue, and associated with these will be changes in the soils, distribution of plants, and abundance and types of animals present. Furthermore, if groundwater tables rise close to the surface during wet periods, this factor should be taken into account in developing tourist facilities, such as roads and sewage systems, that may be adversely affected.

Amboseli National Park exemplifies how complex interactions among the geology, climate, and life in the park over decades converge to produce a recurring pattern of change that has profound effects on the regional environment. It also shows us that careful, detailed work is necessary to understand environmental systems and change.

LEARNING OBJECTIVES

Changes in systems can be natural or induced by humans, and many complex and far-reaching interactions can result. After reading this chapter, you should understand:

- Why solutions to many environmental problems involve the study of systems and rates of changes, how positive and negative feedback may destabilize or stabilize a system, and the role of nonlinearity and delays in systems response.
- What are the implications of exponential growth and the concept of doubling time.
- Why disturbances and changes in systems such as forests, rivers or coral reefs are important to their maintenance and how this affects the concept of balance of nature.

- What an ecosystem is and how sustained life on Earth is a characteristic of ecosystems, not of individual organisms, populations, or single species.
- Why the principle of environmental unity is important in studying environmental problems and why we must consider secondary and tertiary effects of particular changes.
- What the principle of uniformitarianism is, why it is an important concept in the physical and biological sciences, and how it may be used to anticipate future changes.
- What the Gaia hypothesis is and how life on Earth has affected Earth.

3.1 SYSTEMS AND FEEDBACK

A *system*, in a very general way, may be defined as any part of the universe that can be isolated for the purposes of observation and study. Some systems may be physically isolated—for example, chemicals in a test tube—or they may be isolated in our minds or in a computer database. In another sense, a system may be thought of as a set of components or parts that function together to act as a whole. A single organism may be thought of as a system, as may a sewage treatment plant, a city (Figure 3.5), a river, or your bedroom. On a much different scale, Earth is a system. Of particular interest and importance are the global systems related to Earth's energy balance and the global geologic cycle (see A Closer Look 3.1). We introduce Earth's energy balance and electromagnetic radiation with the discussion of systems and change because a modest understanding of the subject facilitates understanding of many Earth systems. Global geologic cycles are discussed in Chapter 4.

Understanding Systems

Solutions to environmental problems often involve an understanding of systems and rates of change occurring in systems. In environmental science at every level we must deal with a variety of systems that range from simple to complex. Regardless of how we approach environmental problems, we must be able to understand systems and how various parts of systems interact with one another.

Systems may be open or closed. A system that is open in regard to some factor exchanges that fac-

tor with other systems. The ocean is an **open system** in regard to water, which it exchanges with the atmosphere. A system that is closed in regard to some factor does not exchange that factor with other systems. Earth is an open system in regard to energy and a **closed system** (for all practical purposes) in regard to material.

FIGURE 3.5 Lake Michigan, Lincoln Park, and the city of Chicago, Illinois, comprise an example of a large complex urban system that includes air, water, and land resources. Urban systems are particularly important in environmental sciences because more and more people are living in urban areas.

CLOSER LOOK 3.1

ELECTROMAGNETIC RADIATION

The universe as we know it consists of two entities: matter and energy. **Matter** is physical material that forms the building blocks of the physical and biological environments (you are composed of matter, as is the chair or bed you are on while reading this); **energy** is the ability to do work. The *first law of thermodynamics*, or, as it is sometimes called, the first energy law, states that energy may not be created or destroyed but may change from one form to another. This is known as the *law of conservation of energy* and stipulates that the total amount of energy in the universe does not change. Our sun produces energy through nuclear reactions at high temperatures and pressures that change mass into energy. At first glance this may seem to violate the law of conservation of energy. How-

ever, this is not the case because energy and matter are interchangeable. Albert Einstein first described the equivalence of energy and mass (a measure of the amount of matter) in his now famous equation $E = mc^2$, where E is energy, m is mass, and c is the velocity of light (approximately 300,000 km/s or 186,000 mi/s). Because the velocity of light squared is a very large number, even a small amount of mass produces very large amounts of energy.[3]

In summary, energy may be thought of as a mathematical abstract quantity that is always conserved. This means that it is impossible to get something for nothing when dealing with energy; it is impossible to extract more energy from any system than the amount of energy that originally entered the system. In fact, the *second law of ther-*

modynamics, or the second energy law, states that you cannot break even. When energy is transformed from one form to another, it always moves from a more useful form to a less useful one. Thus, as energy moves through a real system and is transformed from one form to another, energy is conserved but it becomes less useful. We return to the concepts of the two energy laws when we discuss energy in ecosystems and in modern society (see Chapter 15).

Our Earth is part of a planetary energy system, receiving energy from the sun. This energy undergoes changes, affects life, oceans, atmosphere, climate, and sediments, and is eventually emitted as heat back into the depths of space. In this energy system, Earth is an intermediate part between the *source* (the sun) and

FIGURE 3.6 Annual energy flow to Earth from the sun and the relatively small component of heat from Earth's interior to the near-surface environment, measured in exajoules. An exajoule (10^{18} J) is a convenient measure to express large amounts of energy. One exajoule is approximately equivalent to 1 quad (quadrillion Btu). (*Source:* Modified from W. M. Marsh and J. Dozier, 1981, *Landscape*, Wiley, New York. Copyright 1981 by John Wiley & Sons. Reprinted by permission.)

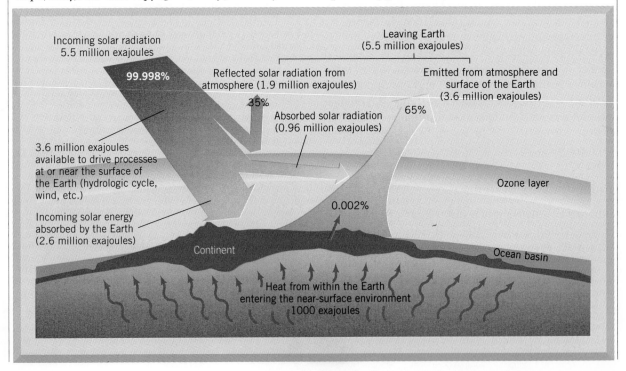

the *sink* (space). Energy flows continuously from the source to Earth to the sink.

Nearly all the energy available at Earth's surface comes from the sun (Figure 3.6), with small additional amounts coming from human activities, geothermal energy (from the interior of Earth), and tides. Because energy is conserved, the amount emitted to space matches that received from the sun plus the other small contributions. Although Earth intercepts only a very tiny fraction of the total energy emitted by the sun, solar energy sustains life on Earth and greatly influences climate and weather.

Electromagnetic Radiation

Energy is emitted from the sun in the form of electromagnetic radiation (EMR). Different forms of electromagnetic energy can be distinguished by their wavelengths. (See the special feature in Appendix A on EMR for a discussion of wavelength and two important laws concerning EMR.) The collection of all possible wavelengths of electromagnetic energy, considered a continuous range, is known as the **electromagnetic spectrum.** The electromagnetic spectrum is one of the most important phenomena in the physical sciences, and a modest knowledge of it is fundamental to understanding many environmental topics. Gamma rays, X rays, ultraviolet light, visible light, infrared radiation, television waves, radio waves, and radar are all different types of electromagnetic radiation. The relatively long wavelengths (greater than 1 m in the electromagnetic spectrum) include radio waves, and the shortest wavelengths are those of gamma rays and X rays. The electromagnetic radiation to which our eyes are sensitive, *visible electromagnetic radiation*, is only a very small fraction of the total spectrum, as you can see in Figure 3.7. Other types of electromagnetic radiation with environmental significance include radar, microwaves, and infrared radiation. These are useful for a variety of industrial pur-

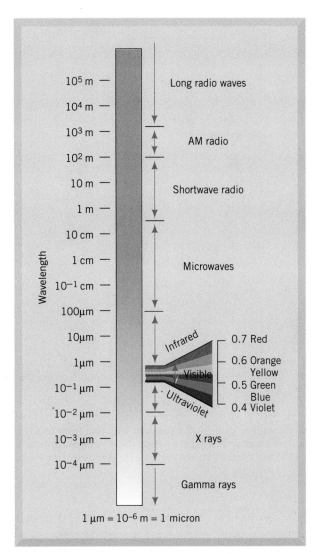

FIGURE 3.7 Types of electromagnetic radiation (EMR). Notice that the spectrum of wavelengths is over nine orders of magnitude, from radio waves to gamma rays (1μm = 10^{-6} m = 1 micron).

poses and for remote sensing (use of aerial photographs or images obtained from sensors on aircraft, satellites, or the space shuttle).

The amount of energy per unit time radiated from a body such as the sun or Earth varies with the fourth power of the absolute temperature of the body. Thus, if a body's temperature doubles, the energy radiated increases by 2^4, or 16 times. This phenomenon explains why the sun, with a temperature of 5800°C (10,500°F), radiates a tremendously greater amount of energy than does the surface of Earth, which radiates at an average temperature of 15°C (59°F). Figure 3.8 illustrates this and another important point: The sun

emits strongly in the visible region (relatively short-wave radiation of about 0.4 to 0.7 μm) whereas Earth emits in the infrared region (relatively long-wave radiation of about 10 μm). The hotter an object is, the more rapidly it radiates heat and the shorter the wavelength of its predominant radiation. (Thus a blue flame is hotter than a red flame.) Earth's surface and the surfaces of animals, plants, clouds, water, and rocks are cool enough that heat is radiated predominantly in the infrared wavelength, which is invisible to us.

Energy from the sun travels to Earth at the speed of light through the vacuum of space. Less than half the solar energy that reaches our at-

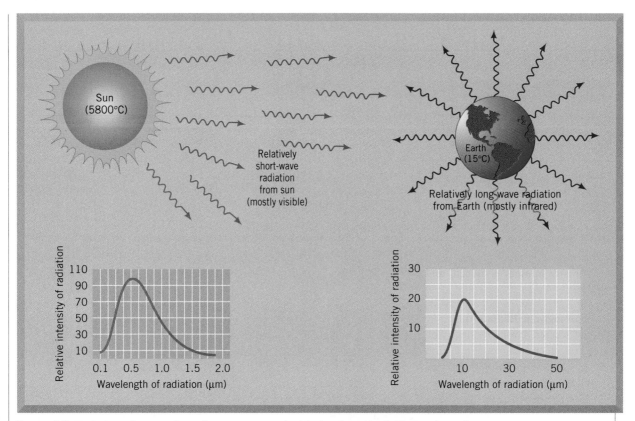

FIGURE 3.8 Emission of energy from the sun compared with that from Earth. Notice that solar emissions have a relatively short wavelength, whereas those from Earth have a relatively long wavelength. (*Source:* Modified from W. M. Marsh and J. Dozier, 1981, *Landscape*, Addison-Wesley, Reading, Mass.)

mosphere is absorbed near Earth's surface. The rest is either absorbed by the upper atmosphere or reflected back into space. Much of the harmful radiation, such as X rays and ultraviolet radiation, is filtered out by the upper atmosphere. The ozone layer, extending from approximately 15 to 45 km (9 to 28 miles) above Earth's surface, is particularly important in absorbing ultraviolet radiation from the sun and protecting living organisms at Earth's surface. The depletion of the ozone layer is discussed in Chapter 24.

Systems respond to inputs and have outputs. Your body, for example, is a complex system. If you are hiking and see a grizzly bear, the sight of the bear is an input. Your body reacts to that input: The adrenaline level in your blood goes up, your heart rate increases, and the hair on your arm may rise. Your response—perhaps moving away from the bear—is an output.

Feedback

A special kind of system response called feedback occurs when the output of the system also serves as an input and leads to changes in the state of the system. A classic example of feedback is human temperature regulation. If you go out in the sun and get hot, the increase in temperature affects your sensory perceptions (input). If you stay in the sun, your body responds physiologically: Your skin pores open, and you are cooled by evaporating water (you sweat). You may also respond behaviorally: Because you feel hot (input), you walk into the shade, and your temperature returns to normal. This is an example of **negative feedback:** The system's response is in the opposite direction from the output (an increase in temperature leads to a later decrease in temperature). With **positive feedback,** an increase in output leads to a further increase in the output. A fire starting in a forest provides an example of positive feedback. The wood may be slightly damp at the beginning and not burn well, but once a fire starts, wood near the flame dries out and begins to burn, which in turn dries out a greater quantity of wood and leads to a larger fire. The larger the fire, the more wood is dried, and the more rapidly the fire increases.

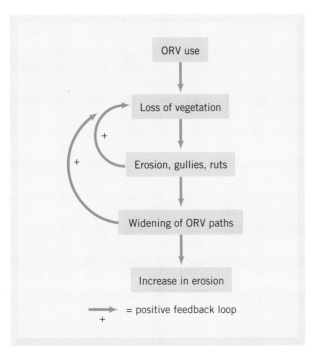

FIGURE 3.9 Diagram illustrating how use of off-road vehicles produces positive feedback to increase erosion.

Negative feedback is generally desirable because it is stabilizing: It usually leads to a system that remains in a constant condition. Positive feedback, sometimes called the vicious circle, is destabilizing. A serious situation can occur when our use of the environment leads to positive feedback. Off-road vehicle use (Figure 3.9), for example, may cause positive feedback with respect to erosion. As vehicle use occurs, churning tires erode the soil and uproot plants, increasing the rate of erosion. As more soil is exposed to erosive processes, such as running water, ruts and gullies are carved. People riding vehicles avoid the ruts and gullies, driving on adjacent sections that are not as eroded, thus widening paths and further increasing erosion. The gullies themselves cause an increase in erosion because they concentrate runoff and have steep side slopes. Once formed, gullies tend to grow in length and size, causing additional erosion (Figure 3.10). Eventually, an area with intensive off-road vehicle use may become a wasteland of eroded paths and gullies. In this example, the effect of positive feedback is that the situation gets worse and worse.

Changes in human population in large cities may be examined in terms of both positive and negative feedback loops, as illustrated on Figure 3.11. Positive feedback that increases the human population in cities results from people's perception that there are greater opportunities in cities and hope for a higher standard of living. If this is true, then positive feedback will occur and urban areas will grow. Of course, populations in cities grow for many other reasons as well, including people who are displaced from rural environments as a result of changes in agriculture that require fewer people to work on farms, or displacement of people following deforestation and environmental damage or change in rural areas that people formally lived in. Negative feedback loops in large cities may result from air and water pollution, along with disease, crime, and discomfort that encourages some portion of the population to migrate from large cities to fringe on more rural areas. This "urban flight" generally involves the more affluent portion of the population.

FIGURE 3.10 Off-road vehicle damage of rare plants living on these coastal dunes near San Luis Obispo, California. Note tire tracks extending into the dune field.

FIGURE 3.11 Potential positive and negative feedback loops for changes of human population in large cities. Modified after Maruyama, M. 1963. The second cybernetics: deviation-amplifying mutual casual processes. American Scientist 51:164–670. Reprinted by permission of *American Scientist,* magazine of Sigma Xi, The Scientific Research Society.

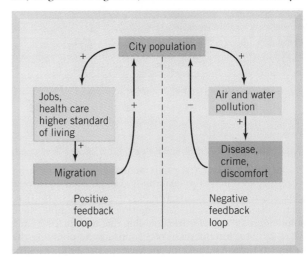

3.2 EXPONENTIAL GROWTH

A particularly important example of positive feedback occurs with **exponential growth.** Simply stated, growth is exponential when it occurs at a constant *rate* per time period (rather than a constant amount). For instance, suppose you have $1000 in the bank and it grows at 10% per year. The first year, $100 in interest is added to your account. The second year you earn more, because you earn 10% on the new total amount, $1100. The greater the amount, the greater the interest earned, so the money (or the population or whatever is increasing) increases by larger and larger amounts. This results in a J-shaped curve, or a skateboard ramp. The actual shape depends on the scale of the units of the curve. Figure 3.12 shows a typical exponential growth curve for a 7%-per-year growth rate and increase in human population for the past 2,000 years. Many systems, both human induced and natural, may exhibit exponential growth for some lengths of time.

Calculating exponential growth involves two related factors: the *rate of growth* measured as a percentage and the doubling time in years. The **doubling time** is the time necessary for the quantity being measured to double. A rule of thumb is that the doubling time is approximately equal to 70 divided by the annual percentage growth rate. To return to the bank account example, suppose that, at your birth, your parents put $1000 in a bank account that earns 10% interest per year and that you cannot touch the money until you are 50 years old. About how much will you have when you cash it in? It doubles every 7 years (70 ÷ 10%), so there will be just over seven doublings of the original investment in 50 years. At age 50 you will have over $128,000! Try graphing the increase in your funds with dollars ($1000 to $128,000) on the vertical axis and your age (1 to 50 years) on the horizontal, and examine the shape of the curve. Is it a J, or a skateboard ramp?

Exponential growth has some interesting (and sometimes alarming) consequences, as illustrated in a fictional story by Albert Bartlett.[4] Imagine a hypothetical strain of bacteria that has a division time of 1 minute (the doubling time is 1 minute). Assume that our hypothetical bacterium is put in a bottle at 11:00 A.M. and the bottle (its world) is full at 12:00 noon. When was the bottle half full? The answer is 11:59 A.M. If you were an average bacterium in the bottle, at what time would you realize that you were running out of space? There is no single answer to this question, but at 11:58 the bottle was 75% empty,

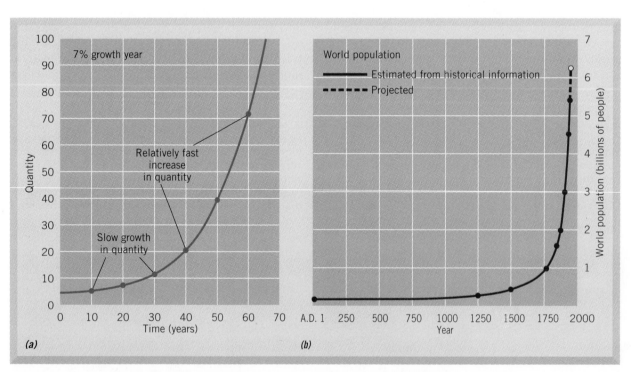

FIGURE 3.12 (*a*) Curve for exponential growth. Growth rate is constant at 7% and the time necessary to double the quantity is also constant at 10 years. Notice that for the first several doubling times the relative rate of growth of the quantity is slower than the relative increase that occurs after several doubling times. For example, the quantity changes from 2 to 4 (absolute increase of 2) for the doubling that took place from 10 to 20 years and increased from 32 to 64 (absolute increase of 32) during the doubling that took place between 50 and 60 years. (*b*) Human population increase for last 2000 years (data from U.S. Dept. of State.

and at 11:57 it was 88% empty. Now assume that at 11:58 some farsighted bacteria realized that the population was running out of space and started looking around for new bottles. Let's suppose that they were able to find three more bottles. How much time did they buy? Two additional minutes. They will run out of space at 12:02 P.M. What if they had found 15 additional bottles?

The preceding example about the growing population of bacteria, while obviously hypothetical, illustrates the power of exponential growth. We look at exponential growth and doubling time again in Chapter 5, when we consider the growth of the human population. Here, we simply note that many systems in nature display exponential growth for some periods of time, so it is important that we are able to recognize it. In particular, it is important to recognize exponential growth in a positive-feedback cycle, as accompanying changes may be very difficult to control or stop.

3.3 ENVIRONMENTAL UNITY

The discussion of positive and negative feedbacks introduces a fundamental concept in environmental science: **environmental unity.** Simply stated, environmental unity means that it is impossible to do only one thing; that is, everything affects everything else. Of course, this is not absolutely true; the extinction of a species of snails in North America, for instance, is hardly likely to change the flow characteristics of the Amazon River. On the other hand, many aspects of the natural environment are closely related. Disruptions or changes in one part of the system often have secondary and tertiary effects within the system or affect adjacent systems. Earth and its ecosystems are complex entities in which any action has several or many effects. The case history of Amboseli National Park at the beginning of this chapter emphasizes a series of effects resulting in part from periodic climate change and provides a good example of how the principle of environmental unity may operate.

An Urban Example

Consider, for example, the changes that accompany a major shift in land use from forests or agricultural land to a large urban complex, such as in the midwestern United States around Chicago or Indianapolis. Clearing the land for urban use increases the amount of sediment eroded from the land (soil erosion), which affects the form and shape of the river channel. The river carries more sediment, some of which is deposited on the bottom of the channel, reducing its depth. Eventually, as more land becomes paved or otherwise made impervious to water, the amount of sediment eroded from the land decreases, and the streams readjust to lower sediment load (the amount of sediment carried by the stream). The readjustment is a form of negative feedback interent in streams and rivers. The urbanization process is also likely to pollute the streams or otherwise change water quality (the increased fine sediment makes the water muddy), affecting the biological systems in the stream and adjacent banks. Thus, land-use conversion can set off a series of changes in the environment, and each change is likely to precipitate still others.

A Forest Example

The interaction among forests, streams, and fish in the Pacific Northwest provides another example of environmental unity. In the redwood forests of northern California and southern Oregon large pieces of organic debris, such as large stems and other pieces of redwood (tree trunks), are responsible for helping form and for maintaining nearly all the pool environments within small streams (Figure 3.13). The pools form because the large pieces of redwood that fall naturally into the stream reside there for hundreds of years and partially block the flow of the stream. The resulting pools provide much of the rearing habitat for young salmon, which spend part of their life cycle in the streams before migrating to the ocean. There would be far fewer pools in many of the steep streams in the redwood forest were it not for the large organic debris. It was formerly common practice to remove large organic debris from streams because it was thought that it blocked the migration of adult salmon attempting to return to spawning beds in the streams. We now know that this practice degrades the fish habitat. In fact, stream restoration projects now often include placing large organic debris back into channels to improve the fish habitat. The study of the role of large organic debris in stream processes and fish habitats in the redwood forest is a good example of environmental unity and the value of studying relations between physical and biological systems. Such studies are at the heart of environmental science (see "Science and Decision Making" in Chapter 2).

3.4 UNIFORMITARIANISM

Earth and its life-forms have changed many times, but certain processes necessary to sustain life and a livable environment have occurred throughout much of history.

The principle that physical and biological processes presently active in forming and modifying Earth can help explain the geologic and evolutionary history of Earth is known as **uniformitarian-**

FIGURE 3.13 Stream processes are significantly modified by fallen redwood trees. The large trunk or stem in the central portion of the photograph had produced through differential scour a small pool important for fish habitat. The site is Prairie Creek Redwood State Park, California.

ism. Simply stated as "the present is the key to the past," uniformitarianism was first suggested in 1785 by Scottish scientist James Hutton, known as the father of geology. Because Charles Darwin was impressed by the concept of uniformitarianism, it pervades his ideas on biological evolution. Today this doctrine is considered one of the fundamental principles of the biological and Earth sciences.

Uniformitarianism does not demand or even suggest that the magnitude and frequency of natural processes remain constant with time. Obviously, some processes do not extend back through all of geologic time. However, as long as the basic factors that rule biological evolution have not changed, we can infer that present processes have been operating for as long as the continents, oceans, and atmosphere have been essentially as they are today.

To be useful from an environmental standpoint, the principle of uniformitarianism has to be more than a key to the past. We must turn it around and say that a study of past and present processes may be the key to the future. That is, we can assume that in the future the same physical and biological processes will operate, but the rates will vary as the environment is influenced by natural change and human activity. Geologically ephemeral land-

forms, such as beaches and lakes, will continue to appear and disappear in response to storms, fires, volcanic eruptions, and earthquakes; extinctions of animals and plants will continue in spite of as well as because of human activity. What is important is to be able to better predict what the future may bring, and uniformitarianism can assist with this task.

3.5 CHANGES AND EQUILIBRIUM IN SYSTEMS

Changes in natural systems may be predictable and should be recognized by anyone looking for solutions to environmental problems. Where the input into the system is equal to the output (Figure 3.14*a*), there is no net change in the size of the pool, or reservoir of whatever is being measured and the system is said to be in a **steady state.** The steady state is really a dynamic equilibrium because material or energy is entering and leaving the system in equal amounts. That is, opposing processes (negative feedback) occurs at equal rates. In ecology, the term for this is **homeostasis** (where a cell or organism maintains a constant environment). Sustainable development (see chapter 1) is, in part, the result of

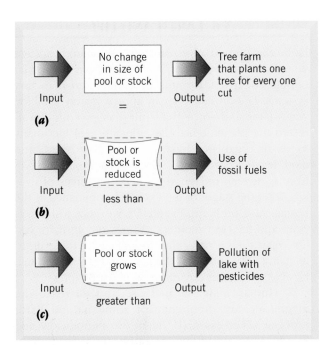

FIGURE 3.14 Systems and change. Major ways in which a pool or stock of some material may change. (*Source:* Modified after P. R. Ehrlich, A. H. Ehrlich, and J. P. Holvren, 1977, *Ecoscience: Population, Resources, Environment*, 3rd ed., W. H. Freeman, San Francisco.)

achieving an economic steady state. An approximate steady state may occur on a global scale, such as in the balance between incoming solar radiation and outgoing radiation from Earth, or on the smaller scale of a university, where new students are brought in and seniors graduate at a constant rate. When the input into the system is less than the output (Figure 3.14b), change in pool size occurs, as can happen with the use of resources such as groundwater or the harvesting of certain plants or animals. If the input is much less than the output, the groundwater may be completely used up or the plants and animals may die out. In a system where input exceeds output (Figure 3.14c), positive feedback may occur, and the *pool*, or reservoir of whatever is being measured, will increase. Examples are the buildup of heavy metals in lakes or the pollution of groundwater. By using rates of change or input–output analysis of systems, we can derive an average residence time for objects or material moving through a system.

The **average residence time** is a measure of the time it takes for any given part of the total pool or reservoir of a particular material to be cycled through the system. To compute the average residence time (when the size of the pool and the rate of throughput are constant), take the total size of the pool or reservoir and divide it by the average rate of transfer through that pool. For example, if our university (mentioned above) has 10,000 students and

each year 2,500 freshmen start and 2,500 seniors graduate, then the average residence time is 4 years (perhaps an impossible dream). If the average residence time for professors is much longer, perhaps about 10 years, this suggests (for a pool of 140 professors) that 14 professors per year enter and leave the university. A system such as a small lake with inlet and outlet and a high transfer rate of water has a short residence time for water, which makes the lake especially vulnerable to change if, for example, a pollutant is introduced. On the other hand, in this system the pollutant soon leaves the lake. Large systems with a slow rate of transfer of water, such as oceans, have a long residence time and are much less vulnerable to quick change. However, once polluted, large systems with slow transfer rates are difficult to correct.

An understanding of changes in systems is of primary importance in solving many problems in environmental science. In some cases, very small growth rates may yield incredibly large numbers in modest periods of time. With other systems, however, it may be possible to compute an average residence time for a particular resource and use this information to develop sound management principles. Recognizing positive- and negative-feedback systems and calculating growth rates and residence times, then, enables us to make predictions concerning environmental management. However, we also need to understand the ways in which physical and biological processes, with or without human interference, may modify Earth.

Inputs to systems may be thought of as causes and the output or response as an effect. For example, we may add a nitrogen fertilizer to an orange tree orchard. Adding the fertilizer is an input (or cause), and the output or response is the number of oranges the tree produces. It is important to understand that input and output coupled to feedback in a system may result in relationships between cause and effect that are *nonlinear* and there may be *delays* in the response.[5] If the relationship between a cause (input) and effect (output) is strictly proportional for all values, then we call this relationship linear. Some relationships in systems are linear over a particular range of input and then become nonlinear. For example, if you added 0.25 kg of fertilizer to an orange tree and the yield increased by 5% and you then added 0.50 kg and the yield increased by 10% and then 0.75 kg increased the yield by 15%, the relationship is linear over these values of input of fertilizer. What if you added 50 kg to increase the yield to 1000%? You would probably damage or kill the tree and the yield would be zero! Obviously, over the entire range from 0.25 to 50 kg the relationship between cause and effect is nonlinear. You might also note delays in response: When you add fertilizer, it takes time for it to enter the soil and be used by the tree. The important concept here is that many responses to systems with significant environmental importance (including human population change, pollution of the land, water, and air, and use of resources) experience nonlinearity and delays that need to be recognized if we are to understand and solve environmental problems.

Our discussion of changes in systems related to input and output provides an important framework for interpreting some of the changes that may affect systems. An idea that has been used and defended in the study of our natural environment is that natural systems that have not been affected by human activity tend toward some sort of steady state, or dynamic equilibrium. Sometimes this is called the *balance of nature*. Certainly, negative feedback operates in many natural systems and may tend to move the system toward an equilibrium. Nevertheless, it is worthwhile to question how dominant the equilibrium model really is.

If we examine natural systems in detail and perform our evaluation over a variety of time scales, it is evident that a steady state, or dynamic equilibrium, is seldom obtained or maintained for a very long period of time. Rather, systems are characterized by both natural and human-induced disturbances (sometimes called natural disasters, as for example floods and wildfires), and changes over time can be expected. For example, studies of such diverse systems as forests, rivers, and coral reefs suggest that disturbances owing to natural events are important in the maintenance of those systems. That is, processes such as storms that disturb reefs and fires and storms that disturb forests and floods that disturb rivers are important and necessary agents of change in the respective environments. We discuss this situation in more detail in Chapter 9. The environmental lesson here is that systems such as forests, rivers, and coral reefs change naturally. If we are going to manage systems for the betterment of the environment, we need to gain a better understanding of the following:

- types of disturbances and changes that are likely to occur,
- time periods over which changes occur, and
- the importance of the change to the long-term productivity of the system.

These concepts are at the heart of understanding the principles of environmental unity and sustainability. Management issues are significant: suppression of wildfire and building dams to control floods will likely cause unintended (and perhaps unwanted) environmental change.[6,7]

3.6 EARTH AND LIFE

Earth was formed approximately 4.6 billion years ago when a cloud of interstellar gas known as a solar nebula collapsed, forming protostars and planetary systems. Life on Earth began approximately 2 billion years later and since that time has profoundly affected the planet. Since the emergence of life, many kinds of organisms have evolved, flourished, and died, leaving only their fossils to record their place in history. Several million years ago, the forces of evolution set the stage for the eventual dominance of the human race on Earth. Eventually we, too, will disappear, and the brief moment of humanity in Earth history may not be particularly significant. However, to us living now and to all the generations still to come, how we affect our environment is important.

Human activities increase and decrease the magnitude and frequency of some natural Earth processes. For example, rivers rise periodically and flood the surrounding countryside regardless of human activities, but the magnitude and frequency of flooding may be greatly increased or decreased by human activity. Therefore, in order to predict the long-range effects of such processes as flooding, we must be able to determine how our future activities will change the rates of a physical process.

From a biological point of view, we know that the ultimate fate of every species is extinction. However, humans have accelerated this fate for many species. We see in Figure 3.15 that as the human population has increased, there has been a parallel increase in the extinction of species. These extinctions are closely related to land-use change—to agricultural and urban uses that change the ecological conditions of an area. Some species are domesticated or cultivated and their numbers grow; others are removed as pests.

Human activities now also affect Earth on a global scale, and these effects are increasing with technological advances. Our civilization has the potential to greatly alter our climate—the chemistry of our atmosphere, soil, and water—and even the chances that life will persist on Earth. We explore the interaction of life with its environment in the next section.

3.7 EARTH AS A SYSTEM

Earth as a planet has been profoundly altered by the life that inhabits it. Earth's air, oceans, soils, and sedimentary rocks are very different from what they would be on a lifeless planet. In many ways, life helps control the makeup of the air, oceans, and sediments.

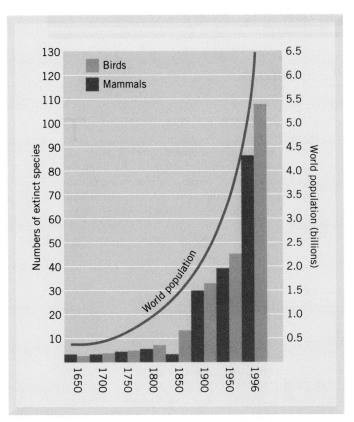

FIGURE 3.15 An increase in human population paralleled by an increase in the extinction of birds and mammals. (*Source:* Modified after U. Ziswiler, 1967, *Extinction and Vanishing Species*, Springer-Verlag, New York, with 1993 data from B. Groombridge, ed., and 1996, IUCN Red List of Threatened Animals, IUCN, The World Conservation Union.)

The Biosphere

Life interacts with its environment on many levels. A single bacterium in the soil interacts with the air, water, and particles of soil around it within the space of a fraction of a cubic centimeter. A forest extending hundreds of square kilometers interacts with large volumes of air, water, and soil. All of the oceans, all of the lower atmosphere, and all of the near-surface part of the solid Earth are affected by life as a whole.

A general term, **biota,** is used to refer to all living things (animals and plants, including microorganisms) within a given area—from an aquarium to a continent to Earth as a whole. The region of Earth where life exists is known as the **biosphere.** It extends from the depths of the oceans to the summits of mountains; most life exists within a few meters of Earth's surface. The biosphere includes all life as well as the lower atmosphere and the oceans, rivers, lakes, soils, and solid sediments that actively interchange materials with life. All living things require energy and materials. In the biosphere, energy is re-

ceived from the sun and the interior of Earth and is used and given off as materials are recycled.

To understand what is required to sustain life, consider the following question: How small a part of the biosphere could be isolated from the rest and still sustain life? Suppose you put parts of the biosphere into a glass container and sealed it. What minimum set of contents would sustain life? If you placed a single green plant in the container along with air, water, and some soil, the plant could make sugars from water and from carbon dioxide in the air and it could make many organic compounds, including proteins and woody tissue, from sugars and from inorganic compounds in the soil. But no green plant can decompose its own products and recycle the materials. Eventually, your green plant would die. We know of no single organism, population, or species that both produces all its own food and completely recycles all its own metabolic products. For life to persist, there must be several species within an environment that includes fluid media—air and water—to transport materials and energy.

3.8 ECOSYSTEMS

An **ecosystem** is a community of organisms and its local nonliving environment in which matter (chemical elements) cycles and energy flows. It is a fundamental principle that *sustained life on Earth is a characteristic of ecosystems*, not of individual organisms or populations or single species.

The term *ecosystem* is applied to areas of all sizes, from the smallest puddle of water to a large forest to the entire global biosphere. Ecosystems differ greatly in composition as well, that is, in the number and kinds of species, in the kinds and relative proportions of nonbiological constituents, and in the degree of variation in time and space. Sometimes the borders of an ecosystem are well defined, as in the transition from a rocky ocean coast to a forest or from a pond to the surrounding woods. Sometimes the borders are vague, as in the subtle gradation of forest to prairie in Minnesota and the Dakotas in the United States or from grasslands to savannas or forests in East Africa. What is common to all ecosystems is not physical structure—size, shape, variations of borders—but the existence of the processes we have mentioned—the flow of energy and the cycling of chemical elements.

Ecosystems can be natural or artificial. A pond constructed as part of a waste treatment plant is an artificial ecosystem. Ecosystems can be natural or managed, and the management can vary over a wide range of actions. Agriculture can be thought of as partial management of certain kinds of ecosystems.

Is the Gaia Hypothesis Science?

According to the Gaia hypothesis, Earth and all living things form a single system, with interdependent parts, communication among these parts, and the ability to self-regulate.

Is the Gaia hypothesis and/or its component hypotheses science, fringe science, or pseudoscience? Is it anything more than an attractive metaphor? Does it have religious overtones? Answering these questions is more difficult than answering the same questions about the examples of the crop circles or lifting the ban on hunting minke whales described in Chapter 2. Analyzing the Gaia hypothesis forces us to deal with our most fundamental ideas about science and life. If for no other reasons, the Gaia hypothesis is valuable.

Critical Thinking Questions

1. What are the hypotheses of the Gaia hypothesis? (Other hypotheses may be less obvious; examine the hypotheses carefully for those that are implied.)
2. What kind of evidence would support each hypothesis?
3. Which of the hypotheses can be tested?
4. Is each hypothesis a science, fringe science, or pseudoscience?
5. Some scientists have criticized James E. Lovelock, who formulated the Gaia hypothesis, for using the term Gaia. Lovelock responds that it is better than referring to a "biological cybernetic system with homeostatic tendencies." What does this phrase mean?
6. What is the Gaia hypothesis and what are its major strengths and weaknesses?

References

Barlow, C. 1993. *From Gaia to Selfish Genes*. Cambridge: MIT Press.

Kirchner, J. W. 1989. "The Gaia Hypothesis: Can It Be Tested?" *Reviews of Geophysics* 27:223–235.

Lovelock, J. E. 1995. *Gaia: A New Look at Life on Earth*. New York: Oxford University Press.

Lyman, F. 1989 (July). "What Hath Gaia Wrought?" *Technology Review* 92(5):55–61.

Resnik, D. B. 1992 (Summer). "Gaia: From Fanciful Notion to Research Program," *Perspectives in Biology and Medicine* 35(4):572–582.

Schneider, S. H. 1990 (May). "Debating Gaia," *Environment* 32(4):4–9, 29–32.

Natural ecosystems carry out many public service functions for us. Wastewater from houses and industries is often converted to drinkable water by passage through natural ecosystems such as soils. Pollutants, like those in the smoke from industrial plants or in the exhaust from automobiles, are often trapped on leaves or converted to harmless compounds by forests.

The Gaia Hypothesis

Our discussion of Earth as a system, life in its environment, the biosphere, and ecosystems leads us naturally to the question of how much life on Earth has affected our planet. In recent years the **Gaia hypothesis,** named for Gaia, the Greek goddess Mother Earth, has become a hotly debated subject.[8] (See the Environmental Issue).The hypothesis is that life manipulates the environment for the maintenance of life. For example, some scientists believe that gas- and scale-producing algae floating near the surface of the ocean influence rainfall at sea and the carbon dioxide content of the atmosphere, thereby significantly affecting the global climate. It follows then that the planet Earth, is capable of physiological self-regulation (a form of global homeostasis).

According to James Lovelock,[8] a British scientist who has been developing the Gaia hypothesis since the early 1970s, the idea of a living Earth is probably as old as humanity. James Hutton, whose theory of uniformitarianism was discussed earlier, stated in 1785 that he believed Earth to be a superorganism and compared the cycling of nutrients from soils and rocks in streams and rivers to the circulation of blood in an animal.[8] In this metaphor the rivers are the arteries and veins, the forests are the lungs, and the oceans are the heart of Earth.

The Gaia hypothesis is really a series of hypotheses, the first of which is that life, since its inception, has greatly affected the planetary environment. Few scientists would disagree. On the second level, the Gaia hypothesis asserts that life has altered Earth's environment in ways that have allowed life to persist. Certainly, there is some evidence that life has had, in a physiological fashion, an effect on Earth's climate. A popularized distortion of the Gaia hypothesis is that life deliberately (consciously) controls the global environment. Few scientists accept this. Rather, it appears that systems of positive and negative feedback that operate in growing populations in the oceans and atmosphere explain mechanisms by which life affects the environment.

The hypothesis has merit as it relates to the present and the future. Humans have evolved to a point of consciousness about our effects on the planet, and we influence changes in the global environment. Thus the concept that our form of life can consciously make a difference in the future of our planet is not as extreme a view as we once thought. The future status of the human environment may depend, in part, on actions we take now and those we will take in coming years. So, in fact, people living on Earth now make conscious decisions that affect the future of the planet. As discussed in Chapter 1, the decisions we make depend on how much we value the environment as well as the knowledge and understanding we have obtained on how Earth works.

SUMMARY

- A system may be defined as any part of the universe that may be isolated for the purpose of study. Environmental studies deal with complex systems at every level, and solutions to environmental problems often involve understanding systems and rates of change.

- Systems respond to inputs and have outputs. Feedback is a special kind of system response; positive feedback is often destabilizing, whereas negative feedback tends to stabilize or encourage more constant conditions in a system.

- Relationships between input (cause) and output (effect) of systems may be nonlinear and there may be delays.

- The principle of environmental unity, simply stated, holds that everything affects everything else. It emphasizes the inner relations between parts of systems. The principle of uniformitarianism can be used to help predict future environmental conditions on the basis of the past and the present.

- A particularly important aspect of positive feedback is exponential growth, in which the increase per time period is a constant fraction or percentage of the current amount. Exponential growth involves two important factors: the rate of growth and the doubling time.

- Changes in systems may be studied through input–output analysis. The average residence time is the average time it takes for the total stock or supply of a particular material to be cycled through the system.

- Life on Earth began more than 3 billion years ago and since that time has profoundly changed our planet. Sustained life on Earth is a characteristic of ecosystems—local communities of interacting populations and their local nonbiological environment—not of individual organisms or populations.

- The general term *biota* refers to all living things, and the region of Earth where life exists is known as the *biosphere*. An interesting hypothesis related to life on Earth is the Gaia hypothesis, which in its simplest form assumes that life on the Earth, through a complex system of positive and negative feedback, regulates the planetary environment to help sustain life.

REEXAMINING THEMES AND ISSUES

Human Population: Our discussion of positive and negative feedback in this chapter reminds us that the human population of Earth is undergoing a variety of positive-feedback mechanisms leading to an ever-increasing population. Of particular concern is local or regional increase of population density (number of people per unit area), which strains resources and leads to human suffering and economic damage. We are also very concerned that a population crash resulting from strong negative feedback may eventually result. A major environmental concern is to forestall or reduce the possibility of this.

Sustainability: Negative feedback is often desirable for systems because it leads to a stabilizing of the system, which is important in obtaining sustainability. That is, if we are to have a sustainable human population or use our resources consistent with sustainable development, then we must have a series of negative feedback within our agricultural, urban, and industrial systems.

Global Perspective: In this chapter we introduce Earth as a system and discuss the Gaia hypothesis. One of the most fruitful areas for research remains the investigation of relationships between physical and biological processes on Earth at the global scale. These relationships must be understood if we are to solve environmental problems related to such issues as potential global warming, ozone depletion, and disposal of toxic waste.

Urban World: We are reminded in this chapter that everything affects everything else and that this concept of environmental unity is particularly appropriate in our urban environments, where land-use changes result in a variety of secondary and tertiary changes that affect physical and biochemical processes.

Values and Knowledge: Our discussion and debate over the Gaia hypothesis and its various components reminds us that we still know very little about how our planet works and the interactions between physical, biological, and chemical systems. What we do know is that we need more knowledge and scientific understanding, and this is driven in part by the value that we place on our environment and the well-being of the other living things with which we share this planet.

KEY TERMS

average residence time *43*	electromagnetic spectrum *37*	matter *36*
biosphere *45*	energy *36*	negative feedback *38*
biota *45*	environmental unity *40*	open system *35*
closed system *38*	exponential growth *39*	positive feedback *38*
doubling time *39*	Gaia hypothesis *45*	steady state *42*
ecosystem *45*	homeostasis *42*	uniformitarianism *42*

STUDY QUESTIONS

1. How does the Amboseli National Park case history exemplify the principle of environmental unity?

2. What is the difference between positive and negative feedback in systems? Provide an example of each.

3. What is the environmental concern associated with exponential growth? Is it always good or bad?

4. Why is the study of equilibrium in systems somewhat misleading in looking at environmental questions? Is there really anything true about the establishment of a balance of nature?

5. Why is the concept of the ecosystem so important in the study of environmental science? In other words, should we be or not be worried about disturbing ecosystems? Under what circumstances?

6. Is the Gaia hypothesis a true statement of how nature works or is it simply a metaphor?

7. How might you use the principle of uniformitarianism to help evaluate environmental problems? Is it possible to use this principle to help evaluate the potential consequences of too many people on Earth?

FURTHER READING

Abelson, P. H. 1990. "Global Change," *Science* 249:1085. This editorial outlines what the United States is doing to address the challenges of global change and what it can do in the future.

Ehrlich, P. R., Ehrlich, A. H., and Holdren, J. P. 1970. *Ecoscience.* San Francisco: W. H. Freeman. Although this is an older book, Chapter 2 provides a good overview of the physical world and how systems and changes may affect our environment.

Kirchner, J. W. 1989. "The Gaia Hypothesis: Can It Be Tested?" *Reviews of Geophysics* 27:223–235. The Gaia theories are evaluated both as a colorful metaphor and as a scientific hypothesis.

Lovelock, J. 1988. *The Ages of Gaia.* New York: W. W. Norton. This small book states the Gaia hypothesis, presenting the case that life very much affects our planet and in fact may regulate it for the benefit of life.

Meadows, D. H., Meadows, D. L., and Randers, J. 1992. *Beyond the Limits.* Port Mills, Vt.: Chelsea Green.

CHAPTER 4

THE BIOGEOCHEMICAL CYCLES

Part of a biogeochemical cycle in action: Nevada Falls in snow melt flood, Yosemite National Park, California.

CASE STUDY

Lake Washington

The city of Seattle, Washington, lies between two major bodies of water—saltwater Puget Sound to the west and freshwater Lake Washington to the east (Figure 4.1a). Beginning in the 1930s, the freshwater lake was used for disposal of sewage. By 1954, 10 sewage treatment plants that removed disease-causing organisms and much of the organic matter had been built along the lake. With an additional treatment plant added in 1959, 76,000 m³ of effluent flowed per day from the treatment plants into the lake. Smaller streams that fed the lake brought in additional untreated sewage.

The lake's response to the effluents was a major bloom of undesirable algae and photosyn-

thetic bacteria, affecting fishing and the general aesthetics of the lake. Similar changes occurred in other lakes. For example, Medical Lake, also in the state of Washington, became clogged with algae and photosynthetic bacteria; the lake turned a dark, turbid green; fish and algae died and, blown by wind, formed masses of stinking dead algae and fish windrows on the lee shore. Once clear, the lake had become dark and dying.

Public concern increased immediately, and Seattle's mayor appointed an advisory committee to determine what might be done. Scientific research showed that it was phosphorus in sewage effluent that stimulated the growth of algae and bacteria. As is the case with many freshwater lakes, this chemical element had been the factor limiting

FIGURE 4.1(*a*) Thematic image (North to top) of Seattle, Washington region (about 24,000 km²): where dark, blue/black is water; (e.g. Puget Sound and Lake Washington); urban areas (e.g. Seattle) pink/light-blue/gray; forested land, dark green; clear cut (logged) lands (e.g. East, right on image, of Seattle urban area) pink/yellow; and snow and ice on Mt. Rainer and other areas, baby blue.

algae and bacteria. Although sewage had been cleaned of disease-causing organisms, it served as a potent fertilizer and posed a chemical problem to the lake ecosystem. The solution to unpleasant overgrowth of these organisms was to redirect sewage effluent. Fortunately for Seattle, Puget Sound is a large body of ocean water with a rapid rate of exchange with the Pacific Ocean. The committee advised the city to divert sewage from the lake to the Sound. This was done by 1968, and the lake improved rapidly. A year later the unpleasant algae decreased, and surface waters became two and one-half times clearer than they had been five years before. Oxygen concentrations in deep water increased immediately to levels greater than those observed in the 1930s, favoring an increase in fish.[1]

However, urbanization (Figures 4.1*b, c*) around the lake continues to be a concern because urban runoff is a potential source of water pollution that may degrade the water quality of streams or lakes it enters. Nevertheless, water quality of Lake Washington in the late 1990s remains good for a large urban lake.

FIGURE 4.1(*b*) Lake Washington with City of Seattle and mountains in background.

FIGURE 4.1(*c*) Lake Washington in summer of 1998 at a popular recreational beach on the eastern shoreline.

Studies at experimental lakes in Canada showed that phosphorus and no other chemical element caused the kind of changes that took place in Lake Washington. This research showed that specific chemical elements can have great effects on life within an ecosystem, and that the study of chemical cycles could provide the basis for the solution to some kinds of environmental problems. The story of Lake Washington illustrates the importance of knowledge about biogeochemical cycles, which is the subject of this chapter.

LEARNING OBJECTIVES

Life is composed of many chemical elements, and these are required in the right amounts, the right concentrations, and the right ratios to each other. If these conditions are not met, then life is limited. The study of chemical availability, called in the most general form biogeochemical cycles, is important to the solution of many environmental problems. After reading this chapter, you should understand:

- What major chemical cycles are.
- What are some of the factors that control these chemical cycles.
- Why some chemical elements cycle quickly and some slowly.
- How each major component of the Earth system—the atmosphere, waters, solid surfaces, and life—are involved in chemical cycles.
- How these major cycles interact.
- Which biogeochemical cycles are most important to life.
- How to analyze biogeochemical cycles.

4.1 INTRODUCTION

Earth is a peculiarly fit planet for life, especially from a chemical point of view. There is plenty of oxygen in the atmosphere, which we and other animals need to breathe. There is plenty of water. In many places, soils are fertile, containing all the chemical elements necessary for plants to grow. There are ores in the Earth's bedrock that contain valuable metals and fuels. Of course the whole surface of the Earth is not perfect for life: there are deserts without water; chemical deserts, like the middle regions of the oceans where nutrients necessary for life are not abundant; and certain soils, where some of the chemical elements required for life are lacking or others toxic to life are present.[2] In essence, the study of biogeochemical cycles is an attempt to answer the question: how did the chemical situation, which makes the persistence of life possible come about? In terms of environmental issues, the question becomes: what kinds of chemical alterations benefit or harm the environment, ourselves, and other life forms, as in the example of the opening case study about Lake Washington?

4.2 HOW CHEMICAL ELEMENTS CYCLE

In its most general form, a **biogeochemical cycle** is the complete pathway that a chemical element follows through the Earth system—from the atmosphere, waters, rock, or soils, to living organisms and back to the atmosphere, ocean, soils, or to other organisms. It is a *chemical* cycle because chemical elements are the form that we consider. It is *bio-* because these are the cycles that involve life. It is *geo-* because these cycles include atmosphere, water, rocks, and soils.

The simplest way to think of one of these cycles is as a "box and arrow" diagram, which shows where an element is stored (called reservoirs, compartments, or pools) and the pathways along which it is transferred from one reservoir to another (Figure 4.2) (see A Closer Look 4.1, "A Biogeochemical Cycle"). Sometimes it is useful to consider one of these cycles from a global perspective, for example, when we consider the problem of potential global warming (see Chapter 21) and need to understand the cycling of all the carbon that passes into and out of the atmosphere. Sometimes it is useful to consider a biogeochemical cycle at the ecosystem level, as in the case study of Lake Washington. We can consider a biogeochemical cycle at any spatial scale of interest to us, from a single ecosystem to the whole Earth. The key that unifies these cycles is the involvement of the four principal components of the Earth system: solid earth (rocks and soils), air, water, and life.

There are several kinds of environmental questions that require an understanding of biogeochemical cycles. These include:

Biological Questions
- What factor, including chemical elements (necessary for life), might be limiting abundance and growth?
- What toxic chemical element might be present that is limiting abundance and growth?
- What can people do to improve the production of a desired biological resource?
- What are sources of chemical elements required for life, and how might we make these more readily available?
- What problems occur when an element is too abundant, as in the case of Lake Washington?

Geological Questions
- What physical and chemical processes control the movement and storage of chemical elements in the environment?
- How are chemical elements transferred from the solid Earth to the water, atmosphere, or life forms?
- How does the long-term (1,000's of years or longer) storage of chemical elements in rocks and soils affect ecosystems at local to global scales?

Atmospheric Questions
- What determines the concentrations of elements and compounds in the atmosphere?
- Where the atmosphere is polluted as the result of human activities, how might we alter a biogeochemical cycle to lower the pollution?

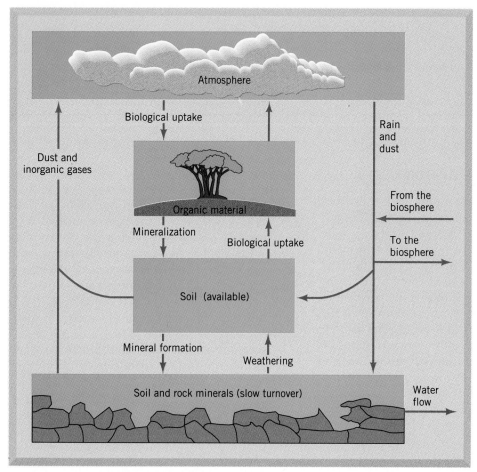

FIGURE 4.2 A generalized ecosystem mineral cycling. Chemical elements cycle within an ecosystem or exchange between an ecosystem and the biosphere. Organisms exchange elements with the nonliving environment; some elements are taken up from and released to the atmosphere, and others are exchanged with water and soil or sediments. The parts of an ecosystem can be thought of as storage pools for an element. The elements move among pools at different transfer rates and remain within different pools for different average lengths of time called residency times. For example, the soil in a forest has an active part, which rapidly exchanges elements with living organisms, and an inactive part, which exchanges elements slowly (as shown in the lower part of the diagram). Generally, life benefits if elements are kept within the ecosystem and are not lost by geologic processes, such as erosion, that remove them from the ecosystem.

Hydrological Questions

- What determines whether is body of water will be biologically productive?
- When a body of water becomes polluted, how can we alter the biogeochemical cycles that involve the pollutant, to reduce its level and its effects.

By their very basic characteristics, the four major components of the Earth system (the lithosphere (rock or soil), atmosphere, hydrosphere, and the biota) have different average rates of storage. In general, residence time of chemical elements is long in rocks, short in the atmosphere, and intermediate in the hydrosphere and the biota.

Biogeochemical Cycles and Life

All living things are made up of chemical elements, but of the 103 known chemical elements, only 24 are required by organisms (Table 4.1). These 24 are divided into the **macronutrients,** elements required in large amounts by all life; and **micronutrients,** elements required either in small amounts by all life, or in moderate amounts by some forms of life and not others. The macronutrients in turn include the "Big Six," the elements that form the fundamental building blocks of life. These are carbon, hydrogen, nitrogen, oxygen, phosphorus, and sulfur. Each element plays a special role in organisms. Carbon is the basic building block of organic com-

A Closer Look 4.1

A Biogeochemical Cycle

This simplest way to view a biogeochemical cycle is as a box and arrow diagram, where the boxes represent places where a chemical element is stored (called reservoirs, pools, or storage compartments) and the arrows represent pathways of transfer (Figure 4.3a). A biogeochemical cycle is generally drawn for a single chemical element, but sometimes it is drawn for a compound, as for example water (H_2O). Figure 4.3b shows these basic elements of a biogeochemical cycle, which represents three parts of the hydrological cycle. Water is stored temporarily in a lake (compartment B). It enters the lake from the atmosphere (compartment A) as precipitation, or from the land around the lake as runoff (compartment C). It leaves the lake through evaporation to the atmosphere, or by runoff to a surface stream or to subsurface flows. In each compartment, there is an average length of time that any atom of an element

is stored before it is transferred. This is called the **residence time.** The amount that leaves or enters the compartment per unit time is called the **flux** or **rate of transfer.** For example, suppose there is a salt lake with no surface runoff out of the

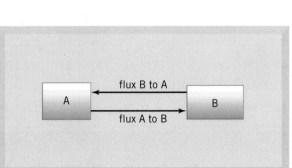

(a)

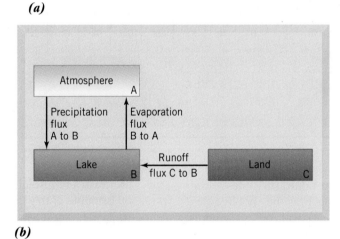

(b)

Figure 4.3 Basic parts of a biochemical cycle. *(a)* A and B are reservoirs or storage compartments. Materials flow from each compartment to the other. *(b)* Some components of the hydrologic cycle.

lake and the only transfer out is by evaporation. Suppose the lake contains 3,000,000 m³ (106 million ft³) of water and the evaporation is 3000 m³/day (106,000 ft³/day). Surface runoff into the lake is also 3000 m³/day so the volume of water in the lake remains constant. Then the average residence time of the water in the lake is 3,000,000 m³ divided by 3000 m³/day, which is 1000 days (or 2.7 years). The residence time—the average time a molecule of water spends in the lake—is therefore 1000 days.

Another crucial aspect of a biogeochemical cycle is the set of factors or processes that control the flow from one compartment to another. Often these are not diagramed, but described. To really understand a biogeochemical cycle, these factors and processes should be quantified and understood from basic principles. For example, how air temperature or wind velocity across a lake influence the evaporation rate of water in the lake.

pounds. Along with oxygen and hydrogen, carbon forms carbohydrates. Nitrogen, along with the other three, makes proteins. Phosphorus is the "energy element" occurring in the compounds called ATP and ADP, important in the transfer and use of energy within cells. As examples of the roles of other macronutrients, calcium is the structure element, occurring in bones of vertebrates, shells of shellfish, and the wood-forming cell walls of vegetation. Sodium and potassium are important to nerve signal transmission. Many of the metals required by living things are necessary for specific enzymes. An enzyme is a complex organic compound that acts as a

catalyst—it causes or speeds up chemical reactions as for example digestion.

For any form of life to persist, chemical elements must be available at the right times, in the right amounts, and in the right relative concentrations to each other. When this does not happen, then a chemical can become a **limiting factor,** preventing the growth of an individual, population, or species, or even causing its local extinction.

Other elements are toxic. Some, like mercury, are toxic even in low concentrations. Some, like copper, are required in low concentrations, but can be toxic if they are present in high concentrations.

TABLE 4.1 Periodic Table of the Elements

Atomic number → 20　　* ← Element relatively abundant in the Earth's crust

Ca ← Element symbol

Environmentally important trace elements → □ □

Calcium

Name

1 **H** Hydrogen	2 **He** Helium

3 **Li** Lithium	4 **Be**ₓ Beryllium

5 **B** Boron	6 **C** Carbon	7 **N** Nitrogen	8 **O** Oxygen	9 **F** Fluorine	10 **Ne** Neon

11 * **Na** Sodium	12 * **Mg** Magnes-ium		13 * **Al** Aluminum	14 * **Si** Silicon	15 **P** Phosphorus	16 **S** Sulfur	17 **Cl** Chlorine	18 **Ar** Argon

19 * **K** Potassium	20 * **Ca** Calcium	21 **Sc** Scandium	22 **Ti** Titanium	23 **V** Vanadium	24 **Cr** Chromium	25 **Mn** Manganese	26 * **Fe** Iron	27 **Co** Cobalt	28 **Ni** Nickel	29 **Cu** Copper	30 **Zn** Zinc	31 **Ga** Gallium	32 **Ge** Germanium	33 **As** Arsenic	34 **Se** Selenium	35 **Br** Bromine	36 **Kr** Krypton
37 **Rb** Rubidium	38 **Sr** Strontium	39 **Y** Yttrium	40 **Zr** Zirconium	41 **Nb** Niobium	42 **Mo** Molybde-num	43 **Tc** Technet-ium	44 **Ru** Ruthenium	45 **Rh** Rhodium	46 **Pd** Palladium	47 **Ag**ₓ Silver	48 **Cd**ₓ Cadmium	49 **In** Indium	50 **Sn** Tin	51 **Sb** Antimony	52 **Te** Tellurium	53 **I** Iodine	54 **Xe** Xenon
55 **Cs** Cesium	56 **Ba** Barium	57 **La** Lanthanum	72 **Hf** Hafnium	73 **Ta** Tantalum	74 **W** Tungsten	75 **Re** Rhenium	76 **Os** Osmium	77 **Ir** Iridium	78 **Pt** Platinum	79 **Au** Gold	80 **Hg**ₓ Mercury	81 **Tl**ₓ Thallium	82 **Pb**ₓ Lead	83 **Bi** Bismuth	84 **Po**ₓ Polonium	85 **At** Astatine	86 **Rn**ₓ Radon
87 **Fr** Francium	88 **Ra** Radium	89 **Ac**ₓ Actinium															

58 **Ce** Cerium	59 **Pr** Praseody-mium	60 **Nd** Neodym-ium	61 **Pm** Prometh-ium	62 **Sm** Samarium	63 **Eu** Europium	64 **Gd** Gadolin-ium	65 **Tb** Terbium	66 **Dy** Dyspros-ium	67 **Ho** Holmium	68 **Er** Erbium	69 **Tm** Thulium	70 **Yb** Ytterbium	71 **Lu** Lutetium
90 **Th** Thorium	91 **Pa** Protactin-ium	92 **U**ₓ Uranium	93 **Np**ₓ Neptun-ium	94 **Pu**ₓ Plutonium	95 **Am**ₓ Americium	96 **Cm**ₓ Curium	97 **Bk**ₓ Berkelium	98 **Cf**ₓ Californ-ium	99 **Es**ₓ Einstein-ium	100 **Fm**ₓ Fermium	101 **Md**ₓ Mendelev-ium	102 **No**ₓ Nobellium	103 **Lw**ₓ Lawren-cium

□ = Required for all life

▨ = Required for some life-forms

◣ = Moderately toxic: either slightly toxic to all life or highly toxic to a few forms

⊠ = Highly toxic to all organisms, even in low concentrations

Finally, some elements are neutral for life—either they are chemically inert, such as the "noble gases" like neon, which do not react with other elements, or they are present on the Earth in very low concentrations.

Although there are as many biogeochemical cycles as there are elements, certain general concepts hold true for these cycles.

- Some chemical elements cycle quickly and are readily regenerated for biological activity. Oxygen and nitrogen are among these. Typically, these elements have a gas phase and occur in the Earth's atmosphere and/or they are easily dissolved in water, and therefore are carried by the hydrologic cycle.

- Other chemical elements are easily tied up in relatively immobile forms and are returned only slowly, by geological processes, to where they can be reused by life. Typically, these elements lack a gas phase and are not found in significant concentration in the atmosphere, and also

are relatively insoluble in water. Phosphorus is an example of this kind of chemical cycle.

- Biogeochemical cycles that include a gas phase and are stored in the atmosphere tend to cycle rapidly. Those without an atmospheric phase are likely to end up as deep-ocean sediment and recycle slowly.

- Since life evolved, it has greatly altered biogeochemical cycles, and this alteration has changed our planet in many ways as, for example, in the development of fertile organic-rich soils that agriculture depends upon.

- The continuation of processes that control biogeochemical cycles is essential to the long-term maintenance of life on Earth.

- Through modern technology, we have begun to alter the transfer of chemical elements (such as sulfur, nitrogen, and carbon) among the air, waters, and soils, at rates comparable to natural biological ones. These activities can benefit society, as in improving crop production, but they can also pose environmental dangers, as illustrated by the opening case study. To live wisely with our environment, we must recognize the positive and negative consequences of activities. Then we must attempt to accentuate the first and minimize the second.

4.3 GENERAL ASPECTS OF CHEMICAL CYCLES

Throughout the nearly 4.6 billion years of Earth's history, chemical compounds that make up the surface and bedrock near the surface have been continuously created from chemical elements, maintained, and changed by physical, chemical, and biological processes. Collectively, the processes responsible for formation and change of Earth materials are referred to as the **geologic cycle** (Figure 4.4), which is actually a group of subcycles: tectonic, hydrologic, rock, and biogeochemical.

Tectonic Cycle

The **tectonic cycle** involves creation and destruction of the solid outer layer of Earth, known as the *lithosphere*. The lithosphere, about 100 km (60 miles) thick, is broken into several large segments called plates, which are moving relative to one another (Figure 4.5).[3] **Plate tectonics** is the slow movement of large segments of Earth's outermost rock shell. These plates "float" on denser material. They move at rates of 2 to 15 cm/year (0.8 to 5.9 inches/year). (Or about as fast as your fingernails grow.) This cycle is driven by forces originating deep within Earth. At the surface, Earth materials are deformed, producing ocean basins, continents, and mountains.

Plate tectonics has important environmental effects of two kinds: physical and chemical. Physical includes changes in location and sizes of continents that alter atmospheric and ocean circulation, thereby altering climate. Also, movement of continents has led to the breakup of continental areas, creating ecological islands. When this has happened, closely related life forms are isolated for millions of years, leading to evolution of new species. In this way, plate tectonics has greatly affected life. Chemical influences of plate tectonics result in the following way: as plates collide, materials they are made of are subjected to heat and pressure and are altered in chemical form. Materials are buried, including organic products such as seashells, which may be converted to limestone, and vegetation, which may be converted to coal or natural gas. It is also important that boundaries between plates are geologically active areas: most volcanic activity and earthquakes occur there.

Three types of plate boundaries occur: divergent, convergent, and transform faults. *Divergent plate boundary* occurs at a spreading ocean ridge where plates are moving away from one another and new lithosphere is produced, a process known as *seafloor spreading* which produces ocean basins. *Convergent plate boundary* occurs when plates collide. If one plate (composed of relatively heavy ocean basin rocks) dives or subducts beneath the leading edge of another (composed of lighter continental

FIGURE 4.4 The geologic cycle and its subcycles, including the tectonic, hydrologic, rock, and biochemical cycles.

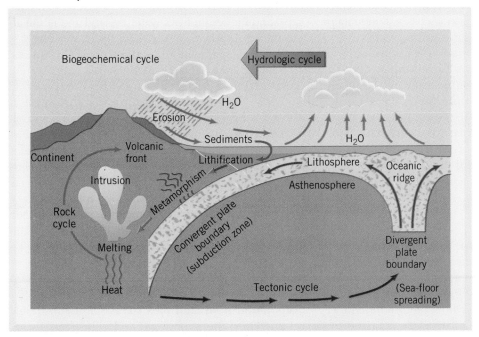

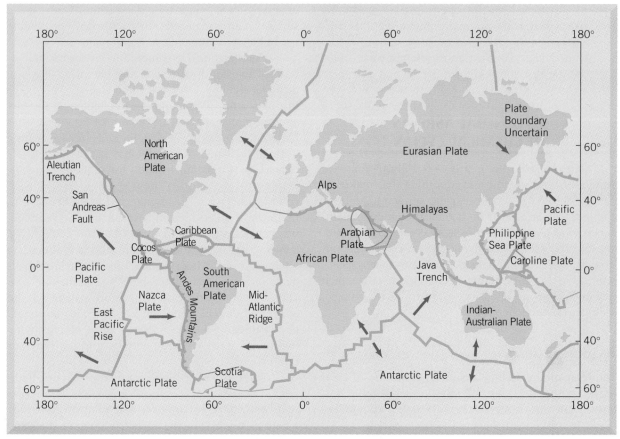

FIGURE 4.5 Generalized map of Earth's lithospheric plates. Divergent plate boundaries are shown as heavy lines, as, for example, the mid-Atlantic ridge. Convergent boundaries are shown as barbed lines, as, for example, the Aleutian trench. Transform fault boundaries are shown as thinner lines, as, for example, the San Andreas Fault. Arrows indicate directions of relative plate motions. (*Source:* Modified after B. C. Birchfiel, R. J. Foster, E. A. Keller, W. N. Melhorn, D. G. Brookins, L. W. Mintz, and H. V. Thurman, 1982, *Physical Geology: The Structures and Processes of the Earth,* Charles E. Merrill, Columbus, Ohio.)

rocks), a subduction zone is present and convergence may produce linear mountain ranges, such as the Andes in South America. If two plates, each composed of lighter continental rocks, collide, a mountain range such as the Himalayas in Asia may form.[3–5]

Transform fault boundary occurs where one plate slides past another as, for example, the San Andreas Fault in California, a boundary between North American and Pacific plates. The Pacific plate is moving north relative to the North American plate at about 5 cm/year (2 in/year). Los Angeles, about 500 km (300 miles) south of San Francisco, is moving slowly toward that city. If this motion continues, in about 10 million years San Francisco will be a suburb of Los Angeles.

The Hydrologic Cycle

The **hydrologic cycle** (Figure 4.6) is movement of water from oceans to atmosphere by evaporation, from atmosphere to oceans and land by precipitation, from land to oceans by runoff from streams and rivers and subsurface groundwater flow, and from land to atmosphere by evaporation. This cycle is driven by solar energy, which evaporates water from oceans, fresh waters, soils, and vegetation. Of the total water on Earth, 97% is in oceans, 2% in glaciers and ice caps, and the rest in fresh water on land and in the atmosphere. Although only a small fraction of water is on land; it is important in moving chemical elements, sculpturing landscape, weathering rocks, transporting sediments, and providing our water resources.

The total volume of water on Earth is huge at about 1.3 billion km[3]. We learned in A Closer Look 4.1 that biogeochemical cycles involve rates of transfer between storage compartments. For the water cycle, major compartments include the ocean, glaciers and ice caps, and ground and surface waters as shown in Figure 4.6. Also shown are annual rates of transfer from these storage compartments. We can see we are really talking about a water budget. If we sum arrows going up into the atmosphere with all going down, the two sums are the same at 577,000 km[3] per year. There is also a balance between precipitation on land (119,000 km[3] per year) which is balanced by evaporation from land plus surface and subsurface runoff. Thus, the hydrologic cycle illustrates fluxes

Storage Compartments of Water

Compartment	Vol. (thousands of km³)	Percentage of Total Water
Ocean	1,230,000,000	97.2
Glaciers and ice caps	28,600	2.2
Shallow groundwater	4,000	0.3
Lakes	123	0.009
Atmosphere	12.7	0.001
Rivers	1.2	0.001

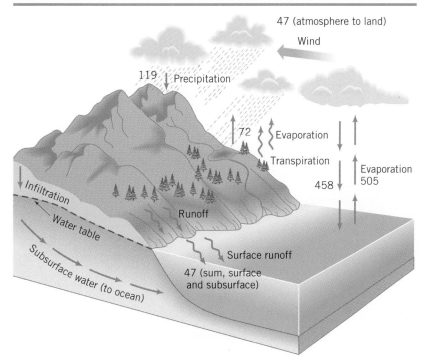

FIGURE 4.6 The hydrologic cycle, showing the transfer of water (thousands of km³/yr) from the oceans to the atmosphere to the continents and back to the oceans again. (From P. H. Gleick, 1993. Water in Crises. NY Oxford Univ Press.)

between storage compartments as ideally shown on Figure 4.3. More important from an environmental perspective is that rates of transfer on land are small relative to what's happening in the ocean. We see from Figure 4.6 approximately 60% of water that falls by precipitation on land each year evaporates to the atmosphere with a smaller component (about 40%) occurring as surface and subsurface runoff to the ocean. It's this small annual transfer of water that supplies resources for rivers and our urban lifestyle and agriculture. Unfortunately, distribution of water on land is far from uniform and as a result, shortages occur in some areas and some water scientists believe we are headed toward a water crisis in the future as human populations increase.

The Rock Cycle

The **rock cycle** consists of several processes that produce rocks and soils. It depends on the tectonic cycle for energy and on the hydrologic cycle for water. As shown in Figure 4.7, there are three classifications of rock: igneous, sedimentary, and meta-

morphic. These are involved in a worldwide recycling process. Internal heat from the tectonic cycle produces igneous rocks from molten material near the surface, such as lava from volcanoes. These new rocks weather when exposed. Freezing and thawing of water in cracks in the rocks breaks the material apart due to the unusual property of water becoming less dense and expanding as a solid (ice). Weak acids that form in water, including carbonic acid formed when carbon dioxide in the atmosphere becomes dissolved in water, dissolve some chemical elements and compounds from the rocks. This process of *weathering* produces sediments, including boulders, pebbles, sand, silt, and clay, as well as dissolved chemical elements. The sediments are transported by wind, water, or the movement of glaciers.

The weathered materials accumulate in depositional basins. These can be in the ocean, where the sediments are compacted by material deposited above them, and then converted to sedimentary rocks. After sedimentary rocks are buried to sufficient depth (usually tens to hundreds of kilometers), they may be altered by heat, pressure, or chemically

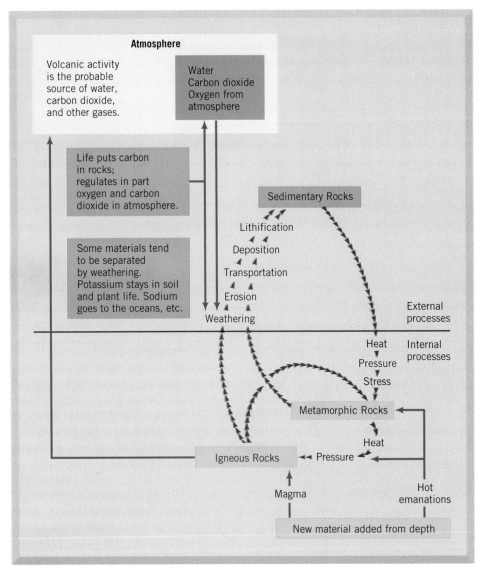

Atmosphere

Volcanic activity is the probable source of water, carbon dioxide, and other gases.

Water
Carbon dioxide
Oxygen from atmosphere

Life puts carbon in rocks; regulates in part oxygen and carbon dioxide in atmosphere.

Some materials tend to be separated by weathering. Potassium stays in soil and plant life. Sodium goes to the oceans, etc.

Sedimentary Rocks

Lithification

Deposition

Transportation

Erosion

Weathering

External processes

Heat

Pressure

Stress

Internal processes

Metamorphic Rocks

Heat

Igneous Rocks

Pressure

Magma

Hot emanations

New material added from depth

FIGURE 4.7 The rock cycle and major paths of material transfer from the solid Earth to the atmosphere as modulated by life. (*Source:* B. C. Birchfiel, R. J. Foster, E. A. Keller, W. N. Melhorn, D. G. Brookins, L. W. Mintz, and H. V. Thurman, 1982, *Physical Geology: The Structures and Processes of the Earth,* Charles E. Merrill, Columbus, Ohio.)

FIGURE 4.8 The Colorado River has eroded through the sedimentary rocks of the Colorado plateau to produce the spectacular scenery of the Grand Canyon. The erosion has been in response to the slow tectonic uplift of the region. The Colorado River in recent years has been greatly modified by dams and reservoirs above and below the Grand Canyon. Sediment once carried to the Gulf of California is now deposited in reservoirs. As a result of the Glen Canyon Dam, the water in the Colorado River through the Grand Canyon is clearer and colder, resulting in changes in the abundance of sand bars and species of fish that live in the river. The presence of the upstream dam has forever changed the hydrology and environment of the Colorado River in the Grand Canyon.

active fluids. They are transformed again, to metamorphic rocks. These may start the rock cycle again, by being transported to the surface by plate tectonics, and subjected to weathering. Variations in the idealized rock cycle are indicated by arrows in Figure 4.7. Notice that life processes play an important role in the rock cycle through the incorporation of carbon into rocks, through biosedimentary processes that produce rocks such as limestone, which is mostly a calcium carbonate (the material of seashells and bones), and our fossil fuel resources.

It is the weathering, erosion, and transport of sediment, with tectonic processes of uplift, that produce the tremendously varied topography on Earth. The spectacular Grand Canyon of the Colorado River in Arizona (Figure 4.8), sculptured from mostly sedimentary rocks, is one example. Another is the beautiful Tower Karst in China (Figure 4.9), resistant blocks of limestone that have survived chemical weathering and erosion that removed the surrounding rocks.

It is through the combined actions of plate tectonics, the hydrologic cycle, and life processes that rocks and minerals so important to modern civilization are formed and concentrated, and by which chemical elements required for life are returned to the surface, after they have been weathered, transported, and deposited in ocean basins. Because all life on the land, in rivers or lakes, near shore, or in the ocean is greatly affected by characteristics of rock substrate, an understanding of various aspects of the rock cycle (particularly surficial processes, such as erosion and transport of rocks, minerals, and soils) facilitates understanding our environment.

4.4 BASIC CONCEPTS OF BIOGEOCHEMICAL CYCLING

When we ask questions about what chemical elements might limit the abundance of a specific individual organism, population, or species, we first look for the answer at a local, or ecosystem, level. As explained in Chapter 3, an ecosystem is a community of different species and their nonliving environment. The boundaries that we choose for this investigation can be somewhat arbitrary and are selected for the convenience of measurement and analysis. On the land, it has become usual to consider ecosystem chemical cycling within a watershed, which is an area of land within which any drop of rain that falls, and does not evaporate, flows out through the same stream. Freshwater bodies—lakes, ponds, bogs—also are convenient units for analysis of ecosystem chemical cycling.

Chemical cycling in an ecosystem begins with inputs from outside. On land, chemical inputs to an ecosystem are from the atmosphere via rain and dust (called dry fallout) and from adjoining land via stream flow and subsurface water runoff. Ocean and fresh water ecosystems have the same atmospheric and land inputs, and in addition have inputs from ocean currents, and submarine vents (hot springs), especially at divergent plate boundaries.

Chemical elements cycle internally, from inorganic parts of the local environment—air, water, inorganic soil—through food chains to the different species. With the death of individual organisms, decomposition returns chemical elements to inorganic

FIGURE 4.9 This landscape in the People's Republic of China is termed "tower karst." These steep-sided hills or pinnacles are composed of the rock limestone, which has been slowly dissolving by chemical weathering. The pinnacles and hills are remnants of the weathering and erosion processes.

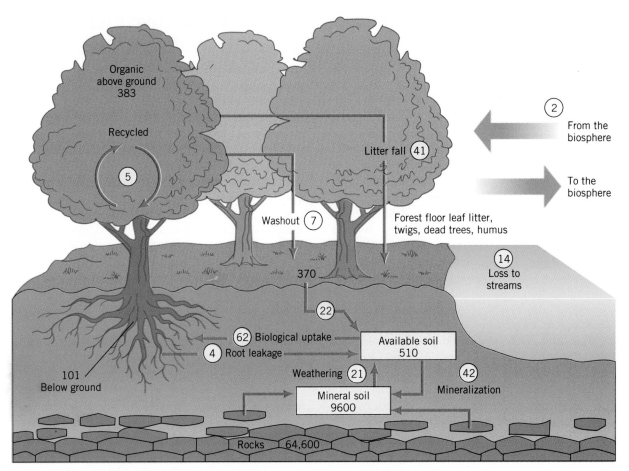

FIGURE 4.10 Annual calcium cycle in a forest ecosystem. In the circles are the amounts transferred per unit time (the flux rates) (kilograms per hectare per year). The other numbers are the amounts stored (kilograms per hectare). Unlike sulfur, calcium does not have a gaseous phase. The information in this diagram was obtained from Hubbard Brook Ecosystem. (*Source:* G. E. Likens, F. H. Bormann, R. S. Pierce, J. S. Eaton, and N. M. Johnson, 1995, *The Biogeochemistry of a Forested Ecosystem,* 2nd ed., Springer-Verlag, New York.)

parts of the ecosystem. Also, living organisms release some chemical elements directly.

An ecosystem can export chemical elements outside the system to other ecosystems. An ecosystem that has little loss of chemical elements can function in its current condition for longer periods than a "leaky" ecosystem that loses chemical elements rapidly. All ecosystems, however, leak and therefore require some external inputs of chemical elements.

Ecosystem Cycles of a Metal and a Nonmetal

Elements have different pathways, as illustrated in Figure 4.10 for calcium and in Figure 4.11 for sulfur. The calcium cycle is typical of a metallic element, and the sulfur cycle is typical of a nonmetallic element. An important difference between these cycles is that calcium, like most metals, does not form a gas on Earth's surface, and therefore calcium has no major phase in the atmosphere; it occurs only as a compound in dust particles. In contrast, sulfur forms several gases, including sulfur dioxide (a major air pollutant and component of acid rain; see Chapter 22) and hydrogen sulfide (swamp or rotten egg gas, usually produced biologically).

Because sulfur has gas forms, it can be returned to an ecosystem more rapidly than can calcium. Annual input of sulfur from the atmosphere to a forest ecosystem has been measured to be 10 times that of calcium. For this reason, elements without a gas phase are more likely to become limiting factors.

Calcium is highly soluble in water in its inorganic form and is readily lost from a land ecosystem in water transport.

Chemical Cycling and the Balance of Nature

To sustain life indefinitely within an ecosystem, energy must be continuously added and the storage of

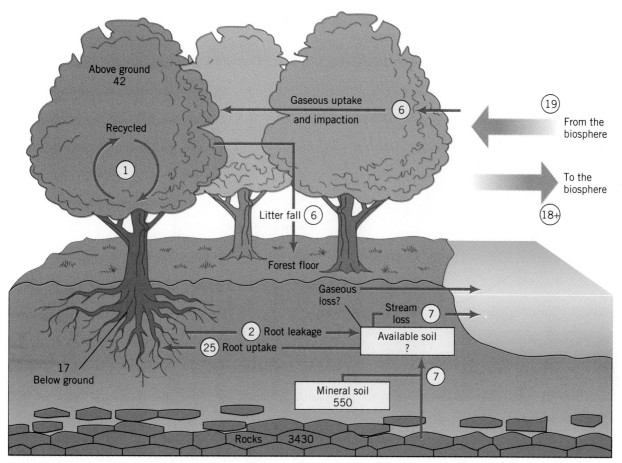

Figure 4.11 Annual sulfur cycle in a forest ecosystem. The circles show the amounts transferred per unit time (the flux rates) (kilograms per hectare per year). The numbers are the amounts stored (kilograms per hectare). Sulfur has a gaseous phase as H_2S and SO_2. The information in this diagram was obtained from Hubbard Brook Ecosystem. (*Source:* G. E. Likens, F. H. Bormann, R. S. Pierce, J. S. Eaton, and N. M. Johnson, 1995, *The Biogeochemistry of a Forested Ecosystem,* 2nd ed., Springer-Verlag, New York.)

essential chemical elements must not decline. As discussed in Chapters 1 and 3, it is a common belief that, without human influence, life would be sustained in a constant condition indefinitely. In Chapter 3, we also discussed the belief that life tends to function to preserve an environment beneficial to itself. Both beliefs presume that chemical elements necessary for life exist in a dynamic steady-state (see Chapter 3) within an ecosystem.[6] These beliefs can be rephrased as a scientific hypothesis: without human disturbance, the net storage of chemical elements within an ecosystem will remain constant over time. But as we discussed earlier, inevitably, some fraction of the chemical elements stored in an ecosystem is lost and must be replaced. This raises the question: are ecosystems ever in a dynamic steady-state in regard to chemical elements? The chemical steady-state hypothesis appears not to hold for cases where it has been studied, but the hypothesis has not been adequately tested.

4.5 SOME MAJOR GLOBAL CHEMICAL CYCLES

In recent years, it has become apparent that human activities can influence global biogeochemical cycles. We can also expand this idea and ask: what chemical elements limit the abundance of life? We begin with the ecosystem level, but, once having considered the question at this local level, we eventually begin to ask the question in a global context: what limits all life on the Earth? These questions bring us to a consideration of the global cycles of specific chemical elements, a few of which are presented in this chapter by way of illustrating general issues.

Earlier we said that the chemical elements required by life are divided into two major groups: macronutrients, which are those required by all forms of life in large amounts; and micronutrients, which are either required by all forms of life in small

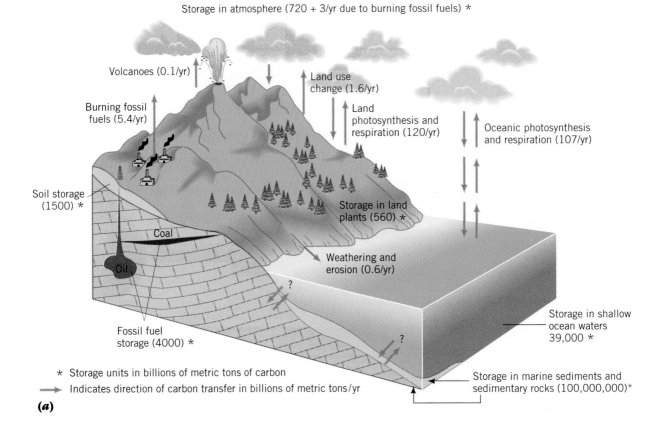

Storage in atmosphere (720 + 3/yr due to burning fossil fuels) *

Volcanoes (0.1/yr)

Land use change (1.6/yr)

Burning fossil fuels (5.4/yr)

Land photosynthesis and respiration (120/yr)

Oceanic photosynthesis and respiration (107/yr)

Soil storage (1500) *

Coal

Storage in land plants (560) *

Oil

Weathering and erosion (0.6/yr)

?

Fossil fuel storage (4000) *

?

Storage in shallow ocean waters 39,000 *

* Storage units in billions of metric tons of carbon

→ Indicates direction of carbon transfer in billions of metric tons/yr

Storage in marine sediments and sedimentary rocks (100,000,000)*

(a)

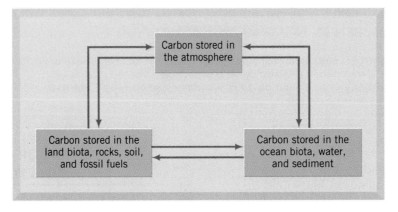

Carbon stored in the atmosphere

Carbon stored in the land biota, rocks, soil, and fossil fuels

Carbon stored in the ocean biota, water, and sediment

(b)

FIGURE 4.12 *(a)* Generalized global carbon cycle. *(b)* Parts of the carbon cycle simplified to illustrate the cyclic nature of the movement of carbon. [*Source:* Modified after G. Lambert, 1987, *La Recherche,* 18, pp. 782-783, with some data from R. Houghton, 1993, *Bulletin of the Ecological Society of America,* 74(4), pp. 355-356.]

amounts, or are required by only certain life forms. In the next sections of this chapter we consider the global cycles of carbon, nitrogen, and phosphorus. We focus on these in part because they are three of the "Big Six"—the elements that are the basic building blocks of life. But each also poses important environmental issues that have attracted attention over the past years and will continue to do so in the future.

The Carbon Cycle

Carbon is the building block of life. It is the element that anchors all organic substances, from coal and oil to DNA (deoxyribonucleic acid), the compound that carries genetic information. Although of central importance to life, carbon is not one of the most abundant elements in the Earth's crust. It is 14th by weight, contributing only 0.032% of the weight of the crust, ranking far behind oxygen (45.2%), silicon (29.5%), aluminum (8.0%), iron (5.8%), calcium (5.1%), and magnesium (2.8%.)[7,8]

The major pathways and storage reservoirs of the carbon cycle are shown in Figure 4.12. Carbon is one of the elements that has a gaseous phase as part of its cycle, occurring in the Earth's atmosphere as carbon dioxide (CO_2) and methane (CH_4), both greenhouse gases (See Chapter 21). Carbon enters the atmosphere through the respiration of living things, through fires that burn organic compounds, and by diffusion from the ocean. It is removed from the atmosphere by photosynthesis of green plants, algae, and photosynthetic bacteria (see A Closer Look 4.2, "Photosynthesis and Respiration").

Over the Earth's history, the rate of biological processes of removal of carbon dioxide from the atmosphere has exceeded the rate of addition, so that the Earth's atmosphere has much less carbon dioxide than would occur on a lifeless Earth, and much less than occurs in the atmospheres of Venus and Mars, where carbon dioxide is the primary constituent. In the atmosphere, carbon dioxide can dissolve in water droplets to form a mild acid called carbonic acid (H_2CO_3). As mentioned earlier, this mild acid is important in weathering rocks at and near the surface of the land.

Carbon occurs in the ocean in several inorganic forms, as dissolved carbon dioxide, as carbonate and bicarbonate, and it occurs, of course, in the organic compounds of marine organisms and their products, such as seashells. Carbon enters the ocean from the atmosphere by simple diffusion of carbon dioxide. The carbon dioxide then dissolves and then is converted to carbonate and bicarbonate. Marine algae and photosynthetic bacteria obtain the carbon dioxide they use from the water in one of these three forms. Carbon is also transferred from the land to the ocean in rivers and streams as dissolved carbon, including organic compounds, and as organic particulates (fine particles of organic matter). Winds also blow small organic particulates from the land to the ocean. The transfer via rivers and streams makes up a comparatively small fraction of the total carbon flux to the ocean, but it is a flux that is locally and regionally of great importance, influencing near-

 CLOSER LOOK 4.2

PHOTOSYNTHESIS AND RESPIRATION

Carbon dioxide enters biological cycles through **photosynthesis,** the process whereby the cells of living organisms (such as plants) convert energy from sunlight to chemical energy. In the process carbon dioxide and water are combined to form organic compounds such as simple sugars and starches, with oxygen as a byproduct (Figure 4.13). The rate of carbon dioxide uptake depends on the number and growth rate of photosynthetic organisms, on the concentration of carbon dioxide in the atmosphere, and other environmental conditions. Although most of the carbon in living tissue is stored in woody tissue of vegetation, particularly in trees, a larger amount of carbon is stored in soils and sediments.

Carbon leaves the living biota through **respiration,** the process by which organic compounds are broken down to release gaseous carbon dioxide. For example, animals (including people) take in air, which has a relatively high concentration of oxygen that is absorbed by blood in the lungs; through respiration, carbon dioxide is released into the at-

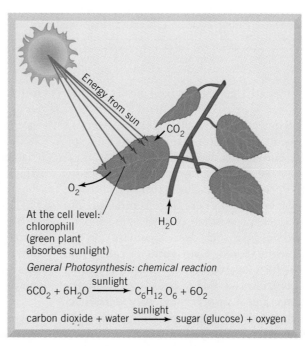

FIGURE 4.13 Idealized diagram illustrating photosynthesis for a green plant (tree) and generalized reaction.

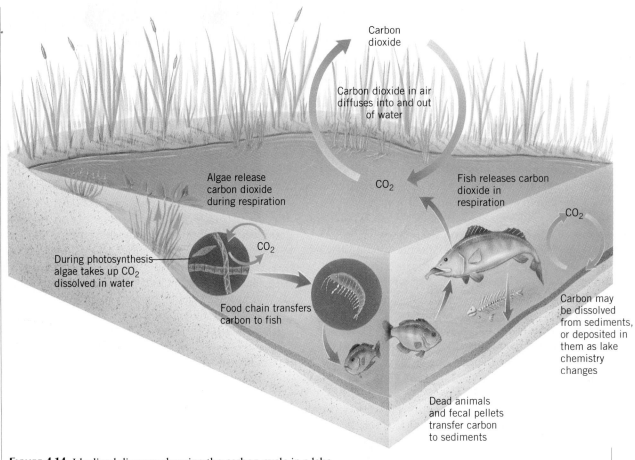

Carbon
dioxide

Carbon dioxide in air
diffuses into and out
of water

CO_2

Algae release
carbon dioxide
during respiration

Fish releases carbon
dioxide in
respiration

CO_2

CO_2

During photosynthesis
algae takes up CO_2
dissolved in water

Food chain transfers
carbon to fish

Carbon may
be dissolved
from sediments,
or deposited in
them as lake
chemistry
changes

Dead animals
and fecal pellets
transfer carbon
to sediments

FIGURE 4.14 Idealized diagram showing the carbon cycle in a lake.

mosphere, with a reduced concentration of oxygen. For example, Figure 4.14 shows the carbon cycle in a lake with transfers of carbon dioxide from the atmosphere to the water and living organisms in the lake through processes that include photosynthesis, respiration, and diffusion.

shore areas that are often highly biologically productive.

Carbon enters the biota through photosynthesis and is returned to the atmosphere or waters by respiration or by burning, as with wildfires. When an organism dies, most of its organic material decomposes to inorganic compounds including carbon dioxide, but some may be buried where there is no oxygen to make this conversion possible or where the temperatures are too cold for such decomposition, and the organic matter is stored. Over years, decades, and centuries, such storage occurs in wetlands including parts of floodplains, lake basins, bogs, swamps, deep-sea sediments, and near-polar regions. Over longer periods, some of this material may be buried under other sediments and form part of sedimentary rocks and become fossil fuels—coal, oil, and natural gas.

In addition to occurring in the solid Earth as fossil fuels, carbon also occurs in a few inorganic forms. Pure carbon is found in a few inorganic materials in rocks, including graphite, diamonds, and a recently discovered exotic molecule composed of 60 atoms of carbon. These are known as Buckyballs (named after Buckminster Fuller's geodesic domes, which have a similar shape, like a soccer ball). Researchers are considering these as possible new building blocks to carry other atoms important in bioengineering and new medicines. Nearly all of the carbon stored in the lithosphere exists as sedimentary rocks, and most of this is in the form of carbonates such as limestone, much of which has a direct biological origin.

The cycling of carbon dioxide between land organisms and the atmosphere is a large flux, with approximately 15% of the total carbon in the atmosphere taken up by photosynthesis and released by respiration annually. In this way, life has a large effect on the chemistry of the atmosphere.

Because carbon is the most important organic compound and because it forms two of the most important greenhouse gases, much research has been devoted to understanding the carbon cycle. However, at a global level some key issues remain unanswered. For example, notice in Figure 4.12 that our burning of fossil fuels releases roughly 5.4 units of carbon per year, and land use changes such as deforestation contribute roughly 1.6 units per year. (Deforestation leads to the decomposition and burning of trees and soils, thus converting organic carbon to carbon dioxide.) Monitoring of atmospheric carbon dioxide levels over the past 35 years suggests that of the approximate total amount of 7 units released per year by human activities into the atmosphere, approximately 3.2 units remain, increasing the carbon dioxide concentration in the atmosphere. It is estimated that approximately 2 units should diffuse into the ocean. This leaves 1.8 units unaccounted for.[9,10] Several hundred or more million tons of carbon are burned each year from fossil fuel and end up somewhere unknown to science. Inorganic processes do not account for the fate of this "missing carbon." Either marine or land photosynthesis, or both, must provide the additional flux. But at this time scientists do not agree which processes dominate, nor in what regions of the Earth this missing flux occurs. The "missing carbon" problem illustrates the complexity of biogeochemical cycles, especially ones where the biota play an important role.

The carbon cycle will continue to be an important area of research because of its significance to global climate investigations, especially to global warming.[11,12] To better understand this issue, we now discuss the interactions between the carbon and tectonic cycles through the carbonate-silicate cycle.

The Carbon-Silicate Cycle

Carbon cycles rapidly among the atmosphere, oceans, and life, and with these and some of the solid sediments. However, over geologically long periods, the cycling of carbon becomes intimately involved with the cycling of silicon, and the combined **carbon-silicate cycle** is therefore of geological importance, taking place over periods that exceed one-half billion years.[13] This cycle begins when carbon dioxide in the atmosphere dissolves in water to form carbonic acid (H_2CO_3), as explained earlier (Figure 4.15). Therefore, rain is typically slightly acidic, and as fresh rain water migrates through the ground and surface water systems, it weathers rocks and facilitates erosion of silicate-rich rocks. This is important because silicate minerals are the most common and abundant rock-forming minerals in Earth's crust.

Among other products, this weathering and erosion releases calcium ions (Ca^{++}) and bicarbonate (HCO_3^-) ions. These enter the ground and surface waters and eventually are transported to the ocean. Both calcium and bicarbonate ions make up

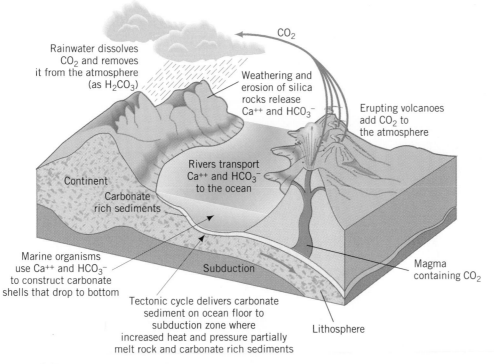

FIGURE 4.15 Idealized diagram showing the carbonate-silicate cycle. (*Source:* Modified after J. E. Kasting, O. B. Toon, and J. B. Pollack, 1988, "How Climate Evolved on the Terrestrial Planets," *Scientific American,* 258, p. 2.)

a major portion of the dissolved load that rivers deliver to the oceans.

Tiny floating marine organisms in the ocean take up calcium and bicarbonate and use it to construct shells. When these organisms die, the shells sink to the bottom of the ocean, where they accumulate as carbonate-rich sediments. When they eventually enter a subduction zone and are subjected to increased heat and pressure, there is a partial melting of the carbonate-rich sediments. The resulting magma gives off carbon dioxide, which rises in volcanoes and is released into the atmosphere, providing a lithosphere-to-atmosphere flux of carbon.

Nitrogen Cycle

Nitrogen is essential to life because nitrogen is necessary for proteins including DNA, the carrier of genetic information. Free nitrogen (N_2 uncombined with any other element) makes up approximately 80% of the Earth's atmosphere. However, organisms cannot use nitrogen directly. Some, like animals, require nitrogen in an organic compound. Others, including plants, algae, and bacteria, can take up nitrogen either as the nitrate ion (NO_3^-) or the ammonium ion (NH_4^+). In contrast to hydrogen, oxygen, and carbon, nitrogen is a relatively unreactive element and there are few processes that convert molecular nitrogen to one of these compounds. Lightning oxidizes nitrogen, producing nitrate. In nature, essentially all other conversions of molecular nitrogen to biological useful forms are conducted by bacteria.

Nitrogen is one of the most important and most complex global cycles (Figure 4.16). Briefly, inorganic nitrogen in the atmosphere is transformed by lightning or bacterial uptake to nitrate or ammonia. By far the greater amount (approximately 90%) is converted by biological activity. The process of converting inorganic, molecular nitrogen in the atmosphere to ammonia is called **nitrogen fixation.** Once in this form, it can be taken up on the land by plants and in the oceans by algae. Bacteria, plants, and algae then convert these inorganic nitrogen compounds into organic ones, and the nitrogen becomes available, through ecological food chains, as organic compounds. When organisms die, other bacteria are able to convert the organic compounds containing nitrogen back to nitrate, ammonia, or, by a series of chemical reactions, to molecular nitrogen, when it is then returned to the atmosphere. The process of releasing fixed nitrogen back to molecular nitrogen is called **denitrification.**

FIGURE 4.16 The global nitrogen cycle. Pools (⬚) and annual (→) flux in 10^{12} g N_2. Note that the industrial fixation of nitrogen is nearly equal to the global biological fixation. (*Source:* Data from R. Söderlund and T. Rosswall, 1982, in O. Hutzinger (ed.), *The Handbook of Environmental Chemistry,* Vol. 1, Pt. B, Springer-Verlag New York.

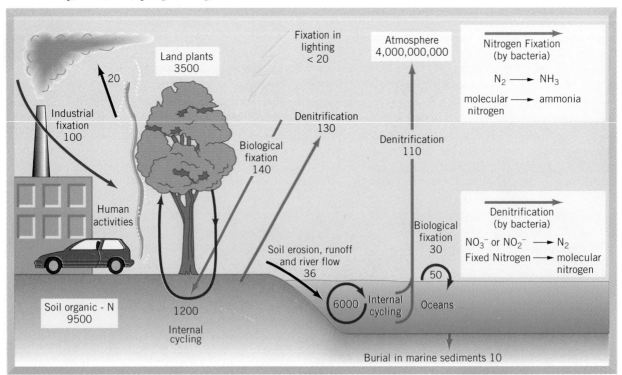

It follows that ultimately all other organisms depend on nitrogen-converting bacteria. Some organisms have evolved symbiotic relationships with nitrogen-converting bacteria. For example, plants of the pea family have nodules in their roots that provide a habitat for such bacteria. The bacteria obtain organic compounds for food from the plants, and the plants obtain useable nitrogen. Such plants can grow in otherwise nitrogen-poor environments. When these plants die, they contribute relatively nitrogen-rich organic matter to the soil, thereby improving the fertility of the soil. Alder trees also have nitrogen-fixing bacteria as symbionts in their roots. These trees grow along streams and their nitrogen-rich leaves fall into the streams and increase the supply of the element in a biologically useable form to fresh water organisms.

Nitrogen-fixing bacteria also are symbionts in the stomachs of some animals, particularly ruminants, the cud-chewing animals that include buffalo, cows, deer, moose, giraffes, and many of the gazelles of the African plains. These are called ungulates and have a specialized four-chambered stomach. The ruminant bacteria provide as much as half the total nitrogen needs of a ruminant, with the rest provided by protein in the green plants eaten by the animal. The bacteria feed on organic matter in the ruminant's stomach and also digest plant compounds, such as the woody tissue, that cannot be digested directly and otherwise could not be used by the ruminant.

In terms of availability for life, nitrogen is somewhere between carbon and phosphorus. Like carbon, nitrogen has a gaseous phase and is a major component of the Earth's atmosphere. However, unlike carbon, it is not very reactive and its conversion depends heavily on biological activity. Thus the nitrogen cycle is not only essential to life, it is primarily driven by life.

In the early part of the twentieth century, scientists discovered that electric sparks produced by industrial processes could convert molecular nitrogen into compounds useable by plants. This greatly increased the availability of nitrogen in fertilizers. Today industrial fixation is a major source of commercial nitrogen fertilizer and a large source of fixed nitrogen in the nitrogen cycle.[14]

Nitrogen combines with oxygen in a high temperature atmosphere. This occurs wherever the temperature and pressure conditions are appropriate. As a result, many modern industrial combustion processes produce oxides of nitrogen. This includes the burning of fossil fuels within gasoline and diesel engines. Thus oxides of nitrogen, one kind of air pollution, are an indirect result of modern industrial activity and modern technology, and these oxides play a significant role in urban smog (see Chapter 22).

In summary, nitrogen compounds are a bane and a boon for society and for the environment. Nitrogen is required for all life, and its compounds are used in many technological processes and in modern agriculture. It is also a source of pollution of air and waters.

The Phosphorus Cycle

Unlike carbon and nitrogen, phosphorus does not have a gaseous phase on the Earth (Figure 4.17). This leads to major differences in its cycle, and we have included its description for that reason and also because, as noted earlier, it is one of the "Big Six" elements required in large quantities by all forms of life. Phosphorus is often a limiting element for plant and algal growth, as we learned in the opening case history of Lake Washington. However, if phosphorus is too abundant, it can cause environmental problems, as we also learned from that case history.

Because phosphorus does not have a gaseous phase on the Earth, it exists in the atmosphere only as part of small particles of dust. In addition, phosphorus tends to form compounds that are relatively insoluble in water and therefore are not readily eroded as part of the hydrologic cycle. Phosphorus occurs commonly in its oxidized state as phosphate, which in turn usually combines with calcium, potassium, magnesium, and iron to form minerals found in soils and in waters. As a result, the rate of transfer of phosphorus tends to be slow in comparison to carbon and nitrogen.

Phosphorus enters the biota through uptake by plants, algae, and some bacteria. Plants can take up phosphorus in its oxidized form, as phosphate, a common ion. Because these phosphate minerals are relatively insoluble in water, phosphorus becomes available very slowly through the weathering of rocks or rock particles in the soil. In a relatively stable ecosystem, much of the phosphorus that is taken up by vegetation is returned to the soil. Some of the phosphorus, however, is inevitably lost to wind and water erosion. It is transported out of the soil in a water soluble form or as suspended particles and is transported by rivers and streams to the oceans.

As we have indicated, the return of phosphorus to the land from the ocean is slow. One important local way that phosphorus returns from the ocean to the land is by way of ocean-feeding birds, such as the Chilean pelican. These birds feed on small fish, especially anchovies that, in turn, feed on tiny ocean plankton. Plankton thrive where nutritionally essential chemical elements occur. These

Numbers in ☐ represent stored amounts in millions of metric tons (10^{12}g)

Numbers in ◯ represent flows in millions of metric tons (10^{12}g) per year

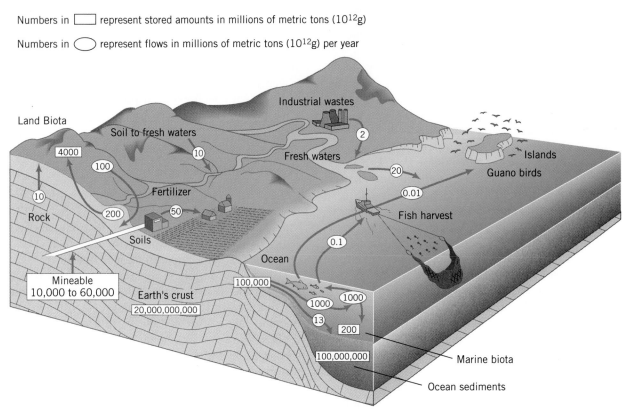

FIGURE 4.17 The global phosphorus cycle. Phosphorus is recycled to soil and land biota by geologic processes that uplift the land and erode rocks, by birds that produce guano, and by human beings. Although Earth's crust contains a very large amount of phosphorus, only a small fraction of it is minable by conventional techniques. Phosphorus is therefore one of our most precious resources. Values of the amount of phosphorus stored or in flux are compiled from various sources. Estimates are approximate to the order of magnitude. (*Source:* Based primarily on C. C. Delwiche and G. E. Likens, 1977, "Biological Response to Fossil Fuel Combustion Products," in W. Stumm, ed., *Global Chemical Cycles and Their Alterations by Man*, Abakon Verlagsgesellschaft, Berlin, pp. 73–88, and U. Pierrou, 1976, "The Global Phosphorus Cycle," in B. H. Svensson and R. Soderlund, eds, "Nitrogen, Phosphorus and Sulfur—Global Cycles," *Ecological Bulletin*, Stockholm, pp. 75–88.)

occur in areas of oceanic upwellings, which are rising currents. Upwellings carry nutrients, including phosphorus, from the depths to the surface. Thus upwellings are among the most fertile and biologically productive regions of oceans. Many major fisheries of the world occur where there are upwellings, as discussed in Chapter 12. Upwellings occur near to continents where the prevailing winds blow offshore. These push the surface waters away from the land, and deeper waters rise to replace those moved offshore.

The fish-eating birds nest on offshore islands where they are protected from predators. Over time their nesting sites become covered with their phosphorus-laden excrement, called guano. The birds nest by the thousands, and the deposits of guano accumulate over centuries.

In relatively dry climates, the guano hardens into a rocklike mass that may be up to 40 m (130 ft)

thick. These have provided some of the major sources of phosphorus for fertilizers. In the mid-1800s as much as 9 million metric tons per year of guano material was shipped to London from such islands off of Peru. This guano has been a valuable and plentiful fertilizer that is the result of a combination of biological and nonbiological processes. Without the plankton, fish, and birds, the phosphorus would have remained in the ocean. Without the upwellings, the result of the interactions between atmosphere and ocean circulation along the continent—ocean boundary, the phosphorus would not have been available.

Another, more recently used, source of phosphorus fertilizers is the mining of sedimentary rocks made up of fossils of marine animals. The richest such mine in the world is Bone Valley, 40 km east of Tampa, Florida. Between 10 and 15 million years ago, Bone Valley was the bottom of a shallow sea

How Are Human Activities Affecting the Nitrogen Cycle?

Scientists estimate that nitrogen deposition to Earth's surface will double in 25 years. What is causing this and how will it effect the nitrogen cycle?

The rate of natural nitrogen fixation on land is estimated to be 140 teragrams (Tg) of nitrogen a year. (One teragram = 1 million metric tons.) Human activities such as use of fertilizers; draining wetlands; clearing land for agriculture; and burning fossil fuels, forests, wood fuels, and grasslands are adding to the natural background rate.

Before the 20th century, fixed nitrogen was recycled by denitrifying bacteria with no net accumulation. Since 1900, however, the use of commercial fertilizers has increased exponentially (see graph). Nitrates and ammonia from burning fossil fuels have increased 20% in the last decade. These inputs have overwhelmed the denitrifying part of the nitrogen cycle and the ability of plants to use fixed nitrogen. Excess nitrates and ammonia wash into streams and are released to the atmosphere. Large portions are then redeposited on the surface. Currently, human activities are responsible for more than half of fixed nitrogen deposited on land.

The result is increases in amounts of ammonia, nitrous oxide, and nitric oxide. Nitric oxide contributes to acid rain and catalyzes formation of ground-level ozone. 80% of nitric oxide emissions results from human activities. Ozone is a known health hazard to plants and animals, including humans, and a greenhouse gas. Nitrous oxide is also a greenhouse gas as well as destructive to the ozone layer in the stratosphere that shields living organisms from dangerous levels of ultraviolet radiation.

Nitrate ions in soil leach out minerals important to plant growth, such as magnesium, and potassium. When these ions are depleted, toxic ones, such as aluminum, are released and cause damage to tree roots. Acidification of soil by nitrate ions is harmful to organisms, including those responsible for the nitrogen cycle. When toxic minerals wash into streams, they can kill fish. Excess nitrates in rivers and along coasts can cause algae to overgrow with effects similar to those described earlier for Lake Washington. High levels of nitrates in drinking water from streams or groundwater contaminated by fertilizers are a health hazard.

The nitrogen and carbon cycle are linked because nitrogen is a component of chlorophyll, the molecule plants use in photosynthesis. Because nitrogen is a limiting factor on land, some scientists predict that increasing levels of global nitrogen may increase plant growth. Increase in photosynthesis would result in greater absorption of carbon dioxide, the principal greenhouse gas. Some people

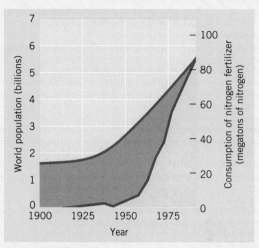

Source: From Smil, V., p. 77.

have, therefore, hoped that increased nitrogen levels might help offset global warming due to carbon dioxide.

Recent studies have suggested a beneficial effect will be short-lived because as plants use additional nitrogen, some other factor will become limiting. When that occurs, plant growth will slow and so will uptake of carbon dioxide. More research is needed to understand the interactions between carbon and nitrogen cycles and predict long-term effects of human activities.

Critical Thinking Questions

1. How does the total rate of human contributions to nitrogen fixation compare to the natural rate?
2. How does the change in fertilizer use compare to the change in world population? Why is this so?
3. Develop a diagram that illustrates the links between the nitrogen and carbon cycles.
4. Make a list of ways human activities could be modified to reduce human contributions to the nitrogen cycle.

References

Asner, G. P., Seastedt, T. R., and Townsend, A. R. 1997 (April). "The Decoupling of Terrestrial Carbon and Nitrogen Cycles." *Bioscience*, 47 (4): 226–234.

Hellemans, A. 1998 (February 13). "Global Nitrogen Overload Problem Grows Critical." *Science*, 279: 988–989.

Smil, V. 1997 (July). "Global Populations and the Nitrogen Cycle." *Scientific American*, 76–81.

Vitousek, P. M., Aber, J., Howarth, R. W., Likens, G. E., Matson, P. A., Schindler, D. W., Schlesinger, W. H., and Tilman, G. D. 1997. "Human Alteration of the Global Nitrogen Cycle: Causes and Consequences." *Issues in Ecology* (http://esa.sdsc.edu/).

where marine invertebrates lived and died.[15] Through tectonic processes, Bone Valley was uplifted. In the 1880s and 1890s, phosphate ore was discovered at Bone Valley and it provides more than one-third of the world's phosphate production and three-fourths of U. S. production.

Some experts believe that total U. S. reserves of phosphorus are about 2.2 billion metric tons, a quantity estimated to supply our needs for several decades. However, for a higher price, more phosphorus is available. Florida is thought to have 8.1 billion metric tons of phosphorus recoverable with existing methods.[15,16] However, mining processes have negative effects on the landscape. For example, at Bone Valley, huge mining pits and slurry ponds scar the landscape (Figure 4.18). Balancing the need for phosphorus with environmental impacts of mining is a major environmental issue. Bone Valley represents a slow return of phosphate to land, illustrating one part of the global phosphorus cycle.

FIGURE 4.18 Large, open-pit phosphate mine in Florida (similar to "Bone Valley" 40 km east of Tampa), with piles of waste material. The land in the upper part of the photograph has been reclaimed and is being used for pasture.

SUMMARY

- Biogeochemical cycles are the major way in which elements important to Earth processes and life are moved through the atmosphere, hydrosphere, lithosphere, and biosphere.

- The biogeochemical cycles can be described as a series of compartments, or storage reservoirs, and pathways between reservoirs.

- In general, some chemical elements may cycle quickly and are readily regenerated for biological activity. Biogeochemical cycles that include a gaseous phase in the atmosphere tend to have more rapid recycling than those that do not.

- Our modern technology has begun to alter and transfer chemical elements in the biogeochemical cycles at rates comparable to those of natural processes. Some of these activities are beneficial to society, but others pose grave dangers.

- In order to be better prepared to manage our environment, we must recognize both positive and negative consequences of activities that

transfer chemical elements and deal with them appropriately.

- All the biogeochemical cycles tend to be complex, and Earth's biota has greatly altered the cycling of chemicals between the air, water, and soil. Continuation of these processes is essential to the long-term maintenance of life on Earth.

- Chemical cycling is complex. Every individual requires a number of chemical elements, and these must be available in the appropriate forms and amounts, in amounts and ratios acceptable to the organism, and in a timely manner. Chemicals can be reused and recycled, but in any real ecosystem some elements are lost over time and must be replenished if life in the ecosystem is to persist.

- Change and disturbance of natural ecosystems are probably the norm. That is, a steady state where net storage of chemical elements in an ecosystem does not change over time can probably not be maintained.

REEXAMINING THEMES AND ISSUES

Human Population: We make the point in this chapter that by way of modern technology we are now transferring some chemical elements such as carbon, nitrogen, and sulfur through the air, water, soil, and biosphere at

rates that are comparable to natural processes. As our population increases, so does our utilization of resources, and these rates of transfer are increasing as well. This is a potential problem because eventually the rates of transfer may become so significantly large for a particular element that pollution of the environment results.

Sustainability: It is clear from our discussion of biochemical cycles that if we are to maintain a quality environment on Earth, the major biogeochemical cycles need to operate within ranges of transfer and storage necessary to sustain healthy ecosystems. This is one reason why understanding biogeochemical cycles is so important. For example, the release of sulfur into the atmosphere is thought to be seriously degrading the quality of our atmosphere. As a result, the United States is striving to control these emissions (see Chapter 22).

Global Perspective: The major biogeochemical cycles discussed in this chapter are presented from a global perspective. Through extensive ongoing research we are trying to better understand the carbon cycle and what

the transfer rates are, particularly as it realtes to burning of fossil fuels and storage of carbon in the biosphere or oceans of the world.

Urban World: Our urban society has greatly concentrated the use of resources and as a result the release of various compounds and chemicals into the biosphere, soil, water, and atmosphere. Thus, in urban areas increases in transfer rates of chemicals in many biogeochemical cycles tend to increase environmental problems.

Values and Knowledge: It is well recognized that our understanding of biogeochemical cycles is fairly primitive. There are very large uncertainties in the measurement of fluxes of chemical elements such as nitrogen, carbon, phosphorous, and others. Nevertheless, we are studying these intensively because we believe that understanding these cycles will allow us to better solve environmental problems.

KEY TERMS

biogeochemical cycle *51*
carbonate–silicate cycle *65*
carbon cycle *62*
denitrification *66*
flux *53*
geologic cycle *55*
hydrologic cycle *56*

limiting factor *53*
macronutrients *52*
micronutrients *52*
nitrogen cycle *66*
nitrogen fixation *66*
phosphorous cycle *67*
photosynthesis *63*

rate of transfer *53*
residence time *53*
plate tectonics *55*
respiration *63*
rock cycle *57*
tectonic cycle *55*

STUDY QUESTIONS

1. Why is an understanding of biogeochemical cycles important in environmental science? Justify your answer with two examples.
2. What are some of the general rules that govern biogeochemical cycles, especially the transfer of material from different parts of the cycle?
3. What are the major aspects of the carbon and

carbonate–silicate cycles and the environmental concerns associated with them?
4. What are differences in the geochemical cycles for phosphorus and nitrogen?
5. How can the carbonate–silicate cycle provide a negative-feedback mechanism to control the temperature of Earth's atmosphere?

FURTHER READING

Berner, R. A., and Berner, E. K. 1996. *Global Environment, Water, Air and Geochemical Cycles.* Upper Saddle River, NJ: Prentice-Hall. Good discussions of environmental geochemical cycles, focusing on Earth's air and water systems.

Kasting, J. F., Toon, O. B., and Pollack, J. B. 1988. "How Climate Evolved on the Terrestrial Planets," *Scientific American* 258(2):90–97. This paper provides a good discussion of the carbonate–silicate cycle and why it is important in environmental science.

Lerman, A. 1990. "Weathering and Erosional Controls of Geologic Cycles," *Chemical Geology* 84: 13–14. Natural transfer of elements from the continents to the oceans is largely accom-

plished by erosion of the land and transport of dissolved material in rivers.

Post, W. M., Peng, T., Emanual, W. R., King, A. W., Dale, V. H., and DeAngelis, D. L. 1990. "The Global Carbon Cycle," *American Scientist* 78:310–326. The authors describe the natural balance of carbon dioxide in the atmosphere and review why the global climate hangs in the balance.

Schlesinger, W. H. 1992. *Biogeochemistry: An Analysis of Global Change.* San Diego: Academic. This book provides a comprehensive and up-to-date overview of the chemical reactions on land, in the oceans, and in the atmosphere of Earth.

INTERNET RESOURCES

Biogeochemistry, Department of Geology and Geochemistry, Stockholm: *http://pcwww.natgeo.su.se*—Research and courses focus on element cycling and environmental processes. Fields of special interest are the biogeochemistry of the Baltic Sea, origin of life, and the rise of oxygen in the atmosphere. Special interest is also given to environmental issues associated with bacterial activity, e.g., acid mine drainage and bioremediation of soils. Web site has on-line publications on these topics.

Biogeochemistry Journal: *http://kapis.www.wkap.nl/kapis/CGI—BIN/WORLD/journalhome.htm?0168-2563*: *Biogeochemistry*, an international journal, publishes original papers and reviews on the chemistry of the environment and the geochemical control of the structure and function of ecosystems. An on-line edition of the journal is available on the Web site.

Environmental Sciences Division at Oak Ridge National Labora-tory (ESD ORNL): *http://www.esd.ornl.gov*—The Microbial Biogeochemistry Group focuses on basic microbial ecology and bioremediation. The Web site provides access to publications about these areas of research.

National Center for Atmospheric Research (NCAR): *http://www.ncar.ucar.edu*—NCAR conducts atmospheric and related research programs, including biogeochemistry. NCAR is sponsored by the National Science Foundation. The Web site provides access to NCAR sponsored research papers.

Oak Ridge National Laboratory (ORNL): *htto://www.ornl.gov*—ORNL is a multiprogram science and technology laboratory managed for the U.S. Department of Energy by Lockheed Martin Energy Research Corporation. Scientists and engineers at ORNL conduct basic and applied research and development. The Web site provides access to publications and journals.

PART III

Egyptian Pavement Fragment. Malqata. Dynasty 18, reign of
Amenhotep III, ca. 1390–1353 B.C.

Life and the
Environment

CHAPTER 5

THE HUMAN POPULATION AS AN ENVIRONMENTAL PROBLEM

Crowded temporary housing at a gold rush site on Mt. Dimata, Davao Del Norte, Mindanao, Philippines, illustrates the impact of human beings on the environment and the relationship between valuable natural resources and a growing human population.

CASE STUDY

Bangladesh

There are 123.4 million people in Bangladesh, where the annual population growth rate is 1.8%. This means that the annual increase in the population of that country is 2.22 million.[1] Recently, more than 100,000 people died in Bangladesh when storms pushed ocean waters onshore, flooding the low-lying coastal lands that make up most of this nation (Figure 5.1).[2] This great human tragedy is made even more tragic by the fact that, given the normal increase in population in Bangladesh, those 100,000 people were replaced in just 2 weeks: The growth curve of the population shows barely a ripple (Figure 5.2).[3]

Bangladesh is one of the poorest nations in the world, and this poverty affects human survival. The average number of calories of food available per person is only 85% of that required for good health. Less than half the population has access to

safe drinking water, and less than a fifth has access to adequate modern sanitation.[4] Average life expectancy is about 59 years. With inadequate resources for each individual and a rapid growth rate, Bangladesh struggles to maintain even its existing poor standard of living.

Some have argued that the low-lying coastal areas that make up most of this nation are fundamentally uninhabitable at high population densities for extended periods and are only inhabited now because of the huge numbers of people living on the Indian subcontinent.[5] For Bangladesh, it is difficult to talk about solving major environmental problems, conserving biological diversity, or optimizing production of fisheries and vegetation when people barely have sufficient resources to survive and the growth of the human population erases any advances.

FIGURE 5.1 A child floats in a large urn and thereby survives one of the catastrophic floods in Bangladesh that have claimed hundreds of thousands of lives. Overcrowding in Bangladesh increases the number of deaths from these floods.

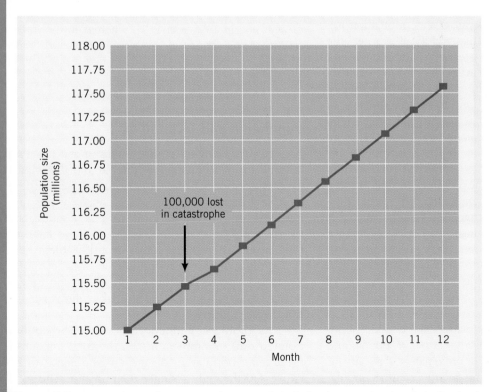

FIGURE 5.2 Projected population growth in Bangladesh for the months shortly before and after a loss of 100,000 people in a single catastrophe in the third month. The population is so large that growth continues with little slowdown and rapidly recovers the number lost.

Progress is being made in Bangladesh in family planning and reducing the growth rate. However, in the long run, there can be no solution to environmental issues in that nation, or elsewhere, as long as the population continues its momentous rise.

LEARNING OBJECTIVES

The current human population represents something unprecedented in the history of the world. Never before has one species had such a great impact on the environment in such a short time and continued to increase at as rapid a rate. These qualities make the human population issue *the* environmental issue. After reading this chapter, you should understand:

- Ultimately, there can be no long-term solutions to environmental problems unless there is a cessation in the growth of the human population.

- The rapid increase in the human population has occurred with little or no change in the maximum lifetime of an individual.

- Modern medical practices as well as improvements in sanitation, control of disease-spreading organisms, and supplies of human necessities have decreased death rates and accelerated the net rate of human population growth.

- Even under the best imaginable scenario—a smooth decline in population growth without catastrophes—the human population will double.

- Countries with a high standard of living have moved more quickly to a lower birth rate than have countries with a low standard of living.

- We cannot predict with absolute certainty what the future human carrying capacity of Earth will be.

5.1 BASIC CONCEPTS

The rapid regrowth of Bangladesh's population following devastating storms vividly illustrates the human population's great capacity for growth and the problems that this growth poses for the environment, perhaps even for the survival of our species on Earth. What can be done to reduce the rate of population growth in that country and in others? Knowledge of certain fundamental concepts about populations helps us to understand what approaches are open to us to solve this fundamental and crucial environmental problem. In this chapter, we explain these fundamental concepts. We also describe how the human population has changed over time and how human beings function as ecological factors. The basic concepts of population growth and change discussed here serve as background to discussions in later chapters about the conservation of endangered species, wildlife, fisheries, and forests.

Let us assume that in mid-1998 the world population numbers approximately 5.9 billion people and the annual growth rate is approximately 1.4%. At this rate, 82.6 million people are added to Earth's population in a single year, a number close to the 1998 population of Germany![3,6] About 90% of this growth takes place in developing countries, such as Bangladesh, and others in Africa, Asia, and Central and South America. In these regions, the average population growth rate in 1998 was 1.7%, whereas in the developed nations—those in Western Europe and North America, for example—the growth rate was less than 1% and in some cases much lower.[3]

Population and Technology

The current danger to the environment has two main factors: the number of people and the impact of each person on the environment. When there were few people on Earth and technology was limited, human impact was local. In that situation, the overuse of a local resource had few or no large or long-lasting effects. The fundamental problem now is that there are so many people and our technologies are so powerful that our effects are no longer local and unimportant. Our old habits, however, do not change quickly.

As biologist and educator Paul Ehrlich has pointed out, in simplest terms, the total impact of the human population on the environment can be expressed through a single relationship: The total environmental effect is the product of the impact per individual times the total number of individuals.[8] An increase in either the individual impact each of us

CLOSER LOOK 5.1

THE PROPHECY OF MALTHUS

Almost 200 years ago the English economist Thomas Malthus eloquently stated the human population problem. He based his argument on three simple premises[7]:

- Food is necessary for people to survive.

- "Passion between the sexes is necessary and will remain nearly in its present state" so that children will continue to be born.

- The power of population growth is "indefinitely greater than the power of the Earth to produce subsistence."

Malthus reasoned that it would be impossible to maintain a rapidly multiplying human population on a finite resource base. Malthus's projections of the ultimate fate of humankind were dire, as dismal a picture as that painted by the most extreme pessimists of today. The power of population growth is so great, he wrote, that "premature death must in some shape or other visit the human race. The vices of mankind are active and able ministers of depopulation, but should they fail, sickly seasons, epidemics, pestilence and plague, advance in terrific array, and sweep off their thousands and ten thousands." And worst of all, should these fail, "gigantic famine stalks in the rear, and with one mighty blow, levels the population with the food of the world."

Critics of Malthus continue to point out that his predictions have yet to come true; whenever things have looked bleak, technology has provided a way out, allowing us to live at greater densities. These critics have argued that our technologies will continue to save us from a Malthusian fate and that therefore we need not worry about the growth of the human population.

Who is correct? Ultimately, in a finite world, Malthus must be correct about the final outcome of unchecked growth. He may have been wrong about the timing; he did not anticipate the capability of technological changes to delay the inevitable. But, although some people believe that Earth can support many more people than it does now, *in the long run there must be an upper limit*. The basic issue that confronts us is this: How can we achieve a constant world population, or at least halt the increase in the population, in a way that is most beneficial to most people? This is undoubtedly one of the most important questions that faces, or has ever faced, humanity. The answer must be based on the application of sound environmental science coupled with arguments about rights, values, and ethics within a global perspective.

has on the environment or the total number of people results in an increase in the total human effect on the environment. Almost 200 years ago, Thomas Malthus foresaw the human population problem (see A Closer Look 5.1, "The Prophecy of Malthus"). One major difference between our situation today and that foreseen by Malthus is that the expansion of technological power, although perhaps delaying a population crisis, has ultimately increased our effect on the environment. The combination of rapid increases in both population and technology has exponentially increased our effect on the environment.

Technology not only increases the use of resources but also causes us to affect the environment in many new ways, compared with hunters and gatherers or with people who farmed with simple wooden and stone tools. For example, before the invention of chlorofluorocarbons (CFCs) to be used as propellants in spray cans and coolants in refrigerators and air conditioners, we were not causing depletion of the ozone layer in the upper atmosphere. Similarly, before we started driving automobiles there was much less demand for steel and little demand for oil—and much less air pollution.

Industrialized Nations

An important consequence of the population-times-technology equation is that the addition of each new individual to the population of an industrialized nation leads to a greater effect on the environment than does the addition of each new individual in the population of a poor, undeveloped nation. Today human population trends vary greatly among countries (Figure 5.3). Those in developed nations tend to criticize the undeveloped nations whose populations continue to grow at a rapid pace, failing to recognize that industrialized nations, even though their populations may not be increasing so rapidly, have greater per-capita effect owing to their higher standard of living and powerful technology. Nations that have large populations as well as high technology—such as the United States, Japan, and the combined European Union—have a huge effect on the environment.

Herein also lies a great irony between two standard goals of international aid: improving the standard of living and slowing the overall human population growth. Improving the standard of living increases the total environmental impact, countering the environmental benefits of a decline in population growth.

Human Demography

What can we do about this rapid increase in our numbers? Understanding the causes of the increases and learning how to calculate rates of increase are two key steps. People who study human population have defined a number of terms and concepts that help us to achieve this understanding. These include

1. age structure;
2. the demographic transition;
3. total fertility;
4. relationships between the human population and the environment;
5. factors that increase the death rate, including

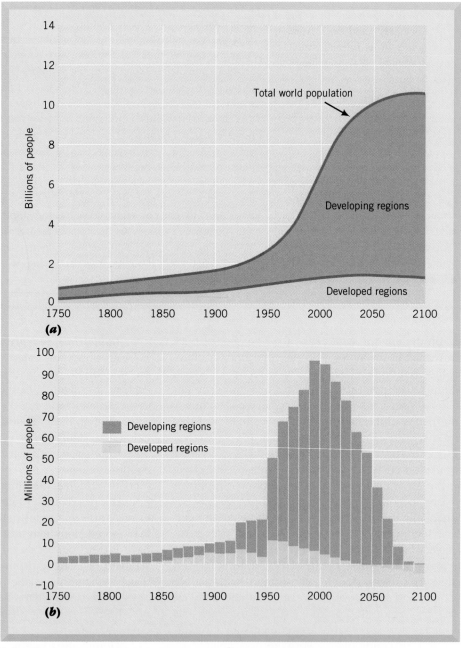

FIGURE 5.3 World population shown as total numbers (*a*) and growth per decade (*b*), by development status, 1800–2100. For example, during the decade from 1980 to 1990, 82 million people were added each year, for a total addition of 820 million people. (*c*) (see figure on facing page) World population distribution by region, 1800–2100. (*Source:* M. M. Kent and K. A. Crews, *World Populations: Fundamentals of Growth*, 1990, Population Reference Bureau, Washington, D.C.)

why death rates increase with crowding and why death rates will increase if we do not decrease birth rates; and

6. how a higher standard of living correlates with decreased birth rates, death rates, and growth rates.

The study of these and related concepts is called *human demography.* **Demography** means the study of populations. Demographers count a population (determine the actual size at some time) and attempt to project changes into the future.

The basic concepts of population growth and change, which together form the study of **population dynamics,** also apply to the study of all non-human populations, including endangered species; wildlife of commercial, recreational, and aesthetic value; timber resources; and commercial fisheries. The concepts discussed in this chapter provide important background for later discussions in Chapters 6 and 12.

5.2 HUMAN POPULATION GROWTH RATE

The history of the human population shown in A Closer Look 5.2, "Growth of the Human Population," can be viewed in four major periods (Figure 5.4): (1)

an early period of hunters and gatherers, when total human population was probably less than a few million; (2) a second period beginning with the rise of agriculture, which allowed a much greater density of people and the first major increase in the human population; (3) the industrial revolution, with improvements in health care and the supply of food, leading to a rapid increase in the human population; and (4) the present situation, where the rate of population growth has slowed in wealthy, industrialized nations but population continues to increase rapidly in poorer, less developed nations (Figure 5.4). It is also interesting to look at the total, cumulative population over the history of humans, which is explored in A Closer Look 5.3, "How Many People Have Lived on Earth?"

A surprising aspect of the second and third periods in the history of our population is that *human population growth has occurred with little or no change in the maximum lifetime.* What has changed are birth rates, death rates, population growth rates, age structure, and average life expectancy.

Population Dynamics

To understand the human population issue, we must understand certain basic principles of population dynamics. A **population** is a group of individuals of

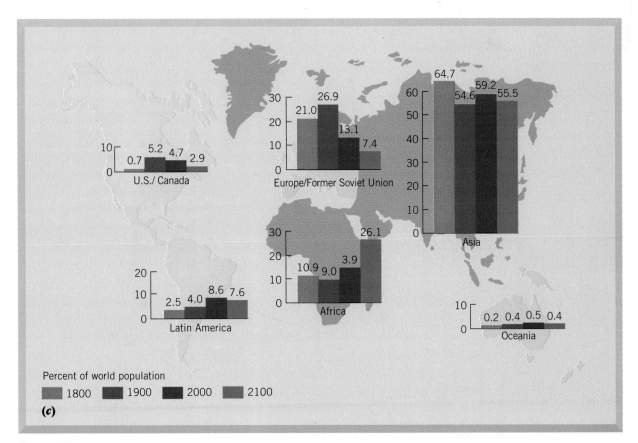

FIGURE 5.3 (c) See legend on facing page.

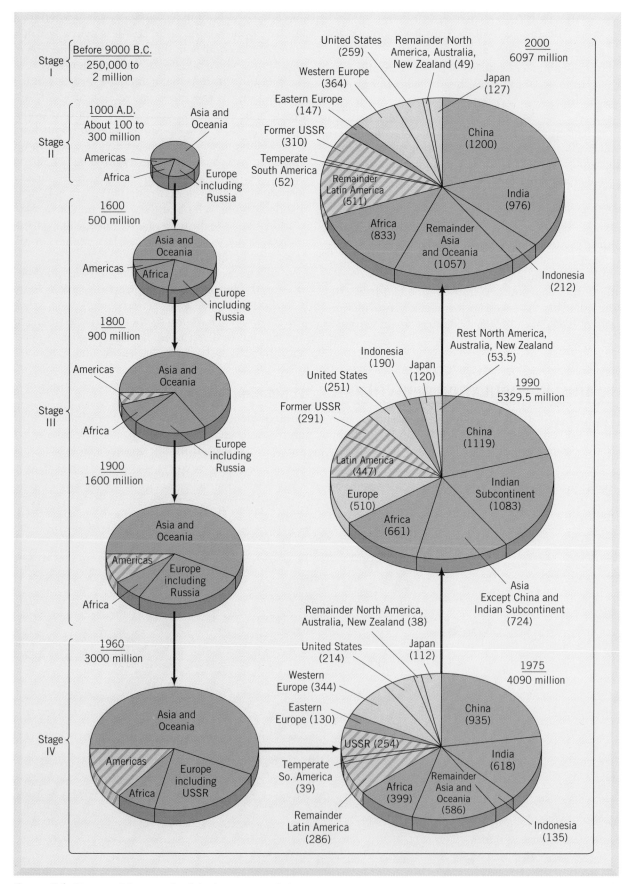

FIGURE 5.4 History of the growth of the human population.

CLOSER LOOK 5.2

GROWTH OF THE HUMAN POPULATION

Stage 1. Hunters and Gatherers: From the first evolution of humans to the beginning of agriculture.

Population density: About 1 person per 130 to 260 km² in the most habitable areas.[9]

Total human population: As low as one-quarter of a million, less than the population of modern small cities like Hartford, Connecticut, certainly less than a few million, which is fewer than now live in many of our largest cities.

Average rate of growth: *The average annual rate of increase over the entire history of human population is less than 0.00011% per year.*

Stage 2. Early, Preindustrial Agriculture: Beginning sometime between 9000 B.C. and 6000 B.C. and lasting approximately until the sixteenth century A.D.

Population density: With the domestication of plants and animals and the rise of settled villages, the human *population density* increased greatly, to about 1 or 2 people/km² or greater, beginning a second period in human population history. (Even today primitive people who practice agriculture have population densities greatly exceeding those of hunters and gatherers.)

Total human population: About 100 million by A.D. 1; 500 million by A.D. 1600. (See Figure 5.4.)

Average rate of growth: Perhaps about 0.03%, which was large enough to increase the human population from 5 million in 10,000 B.C. to about 100 million in A.D. 1; the Roman Empire accounted for about 54 million. From A.D. 1 to A.D. 1000, the population increased to between 200 and 300 million.

Stage 3. The Machine Age: Beginning about 1600 with the Renaissance in Europe. Some experts say that this time marked the transition from agricultural to literate societies, when better medical care and sanitation were factors in reducing the death rate.

Total human population: About 900 million in 1800, almost doubling in the next century and again (to 3 billion) by 1960.

Average rate of growth: By 1600, about 0.1% per year, with rate increases of about one-tenth of a percent every 50 years until 1950. This rapid increase occurred because of the discovery of causes of diseases, invention of vaccines, improvements in sanitation, other advances in medicine and health, and advances in agriculture that led to a great increase in the production of food, shelter, and clothing.

Stage 4. The Modern Era

Total human population: About 5.9 billion.

Average rate of growth: The growth rate of the human population reached 2% in the middle of the twentieth century and has declined slightly to 1.4%.[1]

 CLOSER LOOK 5.3

HOW MANY PEOPLE HAVE LIVED ON EARTH?

How many people have lived on Earth? Of course, before written history there were no censuses. The first estimates of population in Western civilization were attempted in the Roman era. During the Middle Ages and the Renaissance, scholars occasionally estimated the number of people. The first modern census was taken in 1655 in the Canadian colonies by the French and British.[11] The first series of regular censuses taken by a country began in Sweden in 1750, and the United States has taken a census every decade since 1790. Most countries began much later. The first Russian census, for example, was taken in 1870. Even today, many countries do not take censuses or do not do so regularly. The population of China has only recently begun to be known with any accuracy. By studying modern primitive peoples and applying principles of ecology, however, we can gain a rough idea of the total number of people who may have lived on Earth.[12,13]

Summing all the values, including those since the beginning of written history, the estimate is that about 50 billion people have lived on Earth.[12,13] *Surprisingly, the 5.9 billion people alive today represent more than 10% of all of the people who have ever lived.*

WORKING IT OUT 5.1
EQUATIONS OF POPULATION CHANGE

Populations change in size through births, deaths, immigration (arrivals from elsewhere), and emigration (individuals leaving to go elsewhere). We can write a formula to represent population change:

$$P_2 = P_1 + (B - D) + (I - E) \qquad (5.1)$$

where P_1 is the number of individuals in a population at time 1, P_2 is the number of individuals in that population at some later time 2, B is the number of births in the period from time 1 to time 2, D is the number of deaths from time 1 to time 2, I is the number entering as immigrants, and E is the number leaving as emigrants.

Ignoring for the moment immigration and emigration, we can state that how rapidly a population grows depends on the difference between the birth rate and the death rate. The birth rate (usually denoted by b) is the fraction or percentage of the population born in a unit of time. The death rate (d) is the fraction or percentage of the population that dies in a unit of time. The growth rate of a population (g), which is the net increase in a population in a unit of time, is simply the difference between the birth rate and the death rate, or

$$g = b - d \qquad (5.2)$$

Note that in all these cases the units are numbers per unit time.

Recall from Chapter 3 that doubling time, the time for a population to reach twice its present size, can be estimated by the formula

$$T = 70/\text{annual growth rate} \qquad (5.3)$$

where T is the doubling time and where the annual growth rate is expressed as a percentage. For example, a population growing at 2% per year would double in approximately 35 years.

---

the same species living in the same area or interbreeding and sharing genetic information. A **species** is all individuals that are capable of interbreeding. A species is made up of populations. Basic equations of population change are given in Working It Out 5.1, "Equations of Population Change"; some examples of calculations are given in Working It Out 5.2, "Examples of Population Calculations" for those who want to study these concepts in more detail.

How rapidly a population changes depends on the **growth rate,** which is the difference between the **birth rate** and the **death rate.** Rates for human populations are often stated as crude rates, which are rates per 1000 individuals, rather than as percentages, which are rates per 100 individuals (see Table 5.1). The base of 1000 is used for convenience, because small differences in rates can have large effects on future population sizes. For exam-

---

WORKING IT OUT 5.2
EXAMPLES OF POPULATION CALCULATIONS

Birth rates, death rates, and growth rates can be calculated from the number of events during some time period and the total population at some specific time during that period. Letting N equal the total number of individuals in the population, the birth rate (b) is the number of births per unit time (B) divided by the total population (N), or

$$b = B/N \qquad (5.4)$$

The death rate (d) is the number of deaths per unit time (D) divided by the total population (N), or

$$d = D/N \qquad (5.5)$$

The growth rate (g) is the result of the number of births minus the number of deaths per unit time di-

vided by the total number in the population (N), or

$$g = (B - D)/N \quad \text{or} \quad g = G/N \qquad (5.6)$$

In making these calculations, it is important to be consistent in using the population at the beginning, middle, or end of the period. Usually, the amount at the beginning or the middle is used.

Now we can consider an example. There were 18,700,000 people in Australia in mid-1998 and there were 261,800 births from 1998 to 1999. The birth rate b calculated against the mid-1998 population was 261,800/18,700,000, or 1.4%. During the same period, there were 130,900 deaths; the death rate d was 130,900/18,700,000, or 0.7%. The growth rate g was (261,800 − 130,900)/18,700,000, or 0.7%, just under 1% per year.[1]

TABLE 5.1 Common Terms

Crude birth rate: number of births per 1000 individuals per year, crude because population age structure not taken into account

Crude death rate: number of deaths per 1000 individuals per year

Crude growth rate: net number added per 1000 individuals per year; also equal to crude birth rate minus crude death rate

Fertility: refers to pregnancy or the capacity to become pregnant or to have children

General fertility rate: number of live births expected in a year per 1000 women aged 15 to 49 years, considered the childbearing years

Age-specific rate: number of births expected per year among a fertility specific age group of women in a population

Total fertility rate (TFR): average number of children expected to be born to a woman through her child-bearing years

Cause-specific death rate: number of deaths from one cause per 100,000 total deaths

Incidence rate: number of people contracting a disease during a time period, usually measured per 100 people

Prevalence rate: number of people afflicted by a disease at a particular time

Case fatality rate: percentage of people who die once they contract a disease

Morbidity: general term meaning the occurrence of disease and illness in a population

Rate of natural increase (RNI): birth rate minus death rate, implying annual rate of population growth not including migration

Doubling time: number of years for a population to double assuming a constant rate of natural increase

Infant mortality rate: annual number of deaths of infants under age 1 per 1000 live births

Life expectancy at birth: average number of years a newborn infant can expect to live under current mortality levels

GNP per capita: gross national product (GNP), which includes the value of all domestic and foreign output, per person

Source: C. Haub and D. Cornelius, 1998, World Population Data Sheet, Population Reference Bureau, Washington, D.C.

---

ple, in 1998 the *crude death rate* in the United States was 9, meaning that 9 of every 1000 people died. (The same information expressed as a percentage is a rate of 0.9%.) It follows that the *crude birth rate* is the number of births per 1000 individuals in a population. In 1998, the crude birth rate in the United States was 15. (The average for the entire population was 15 births per 1000 individuals.)[1] Then the *crude growth rate* is the net change: the birth rate minus the death rate. The crude growth rate in 1998 in the United States was 6: For every 1000 people at the beginning of 1998, there were 1006 at the end of the year. **Maximum lifetime** is the genetically determined maximum possible age to which an individual of a species can live. **Life expectancy** is the average number of years an individual can expect to live given his or her present age. (Note that often the term is applied, without qualification, to mean the life expectancy of a newborn.)

Exponential Growth

As discussed in Chapter 3, *exponential growth* simply means that a population increases by a *constant percentage* in each unit of time. The growth rate actually increased during the first part of the twentieth century, peaking during the years 1965 to 1970 at 2.1%, owing to improvements in health, medicine, and food production. Thus, the human population has actually increased at a rate faster than the rate of exponential growth. This trend is analogous to a bank savings account in which the interest rate in-

creases every month rather than remaining constant. This increase in the growth rate has stopped, however. Currently, the growth rate is declining globally and is now approximately 1.4%.[1]

5.3 PROJECTING FUTURE POPULATION GROWTH

With human population growth such a central issue, it is important that we develop methods to forecast what will happen to our population in the future: how quickly it will increase and whether and when it might stop increasing. One of the simplest approaches is to calculate the doubling time.

Doubling Time

Recall from our discussion in Chapter 3 that *doubling time*, a concept used frequently in discussing human population growth, is the time required for a population to double in size (see Working It Out 5.1, "Equations of Population Change"). The standard way to calculate doubling time is to assume that the population is growing exponentially (has a constant growth rate). We can estimate doubling time by dividing 70 by the annual growth rate stated as a percentage. The doubling time based on an exponential is very sensitive to growth rate; that is, it changes quickly as growth rate changes (Figure 5.5). A few examples demonstrate this sensitivity. The U.S. pop-

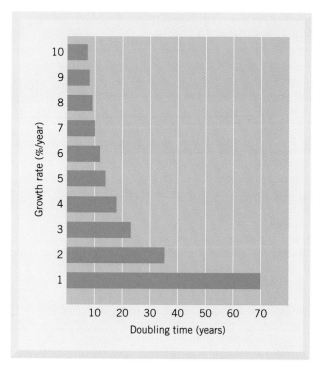

FIGURE 5.5 Doubling time changes rapidly with the population growth rate. Because the world's population is increasing at a rate between 1% and 2%, we expect it to double within the next 35 to 70 years.

ulation grew approximately 0.9% per year during the 1980s; we can calculate the doubling time as 70 divided by 0.9, or 78 years. During the same period, the Ivory Coast's population grew at 3.8% per year; 70 divided by 3.8 is approximately 18, so the doubling time for the Ivory Coast's population was a little more than 18 years.

Populations of Belgium, Austria, Great Britain, and Germany are growing at an annual rate of about 0.1%, or a doubling time of 700 years. The world's most populous country, China, has grown at an annual rate of 1.4%; its doubling time is 50 years. In the latter part of the twentieth century the world population growth rate was 1.7%, which would lead to a doubling of 41 years.[12]

Clearly, no population can sustain an exponential rate of growth indefinitely. Eventually the population will run out of food and space. Recall from Chapter 3 the example of bacteria cells in a test tube doubling every minute. Even if a cell divided only every 2 hours, if there were 2 cells at the beginning, there would be 4 cells after 2 hours, 8 cells after 4, 16 after 6, 4,096 after 1 day, 1 billion after 2 days, and 68 billion after 3 days. In a matter of weeks the number of cells would require all the matter in the universe.

In some countries the human population is growing at 5% per year, which means the popula-

tion doubles in 14 years. A population of 100 increasing at 5% per year would grow to 1 billion in less than 325 years. If the human population had increased at this rate since the beginning of recorded history, it would now exceed the matter available in the universe. These examples demonstrate that no population can grow exponentially forever.

The Logistic Growth Curve

If a population cannot increase forever, what can we expect its changes over time to be? One of the first suggestions made about the human population is that it would follow a smooth S-shaped curve known as the **logistic growth curve.** The population would increase exponentially only temporarily, after which the rate of growth would gradually decline (i.e., population would increase more slowly) until an upper population limit, called the **carrying capacity,** was reached (Figure 5.6). Once the carrying capacity was reached, the population would remain at the carrying capacity. This growth curve was first suggested in 1838 by a European scientist, P. F. Verhulst, as a theory for the growth of animal populations. It has been applied widely to the growth of many animal populations, including those important in wildlife management, endangered species, and fisheries (see Chapter 12).

Although the logistic growth curve has been used frequently to project eventual maximum human population size, there is little evidence that human populations—or any animal populations for that matter—actually follow this growth curve. The logistic curve involves assumptions that are unrealistic for humans and for other mammals. These assumptions include a constant environment, a constant carrying capacity, and a homogeneous population (all individuals identical in their effects on each other). The logistic curve is especially unlikely if death rates continue to decrease owing to improvements in health, medicine, and food supplies. Once a human population has benefited from these improvements, it must pass through what has become known as the **demographic transition** to achieve **zero population growth,** which means a stabilized population.

Forecasting Human Population Growth Using The Logistic Curve

In the past 50 years or so, most forecasts for the long-term size of human populations in specific nations have been based on the logistic curve. As explained earlier, the logistic curve allows a simple calculation of a future carrying capacity, but is unrealistic for many reasons. It ignores any changes in environment or, in the case of human popula-

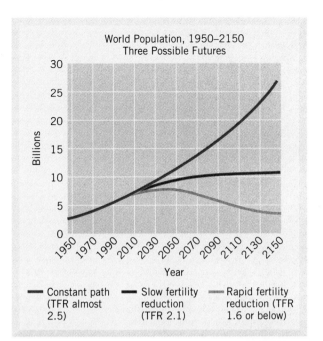

World Population, 1950–2150
Three Possible Futures

— Constant path (TFR almost 2.5) — Slow fertility reduction (TFR 2.1) — Rapid fertility reduction (TFR 1.6 or below)

FIGURE 5.6 Exponential and logistic growth curves. Three possible paths of future world population growth, as projected by the United Nations. The constant path assumes that the 1998 growth rate will continue unchanged, resulting in an exponential increase. The slow fertility reduction path assumes that the world's fertility will decline to reach replacement level by the year 2050 and the world's population would stabilize at about 11 billion by the twenty-second century. The rapid fertility reduction path assumes that the world's fertility will go into decline in the 21st century peaking at 7.7 billion in 2050 and dropping to 3.6 billion by 2150. These are theoretical curves. (*Source:* United Nations Population Division, 1998.)

tions, technology that might influence future growth. This S shaped curve first rises steeply upward then changes slope, curving toward the horizontal carrying capacity (Figure 5.6). The point at which the curve changes is the *inflection point*. Until a population has reached the inflection point, we cannot project the final logistic size.

Unfortunately for those who want to make this calculation, the human population has not made the bend around the inflection point. The typical way this has been dealt with is to assume that today's population is just reaching the inflection point. This standard practice, however, leads invariably to a great underestimate of the maximum number of people that will occur. For example, one of the first projections of the upper limit of the U.S. population, made in the 1930s, assumed that the inflection point occurred then. That assumption resulted in an estimate that the final population of the United States would be approximately 200 million, a number that has long since been exceeded; the U.S. population is now about 260 million.

5.4 THE DEMOGRAPHIC TRANSITION

The demographic transition is a three-stage pattern of change in birth rates and death rates that has occurred during the process of industrial and economic development of Western nations. It leads to a decline in population growth. The decline in the death rate is the first stage in the three stages of the demographic transition (Figure 5.7).[15] In a nonindustrial country, birth rates and death rates are high and the growth rate is low.[12] With industrialization, health and sanitation improve, and the death rate drops rapidly. The birth rate remains high, however, and the population enters stage II, a period with a high growth rate. Most European nations passed through this period in the eighteenth and nineteenth centuries.

As education and the standard of living increase and as family-planning methods become more widely used, stage III is reached. The birth rate drops toward the death rate and the growth rate therefore also decreases, eventually to a low or zero growth rate. However, it appears that this occurs only if individual families believe there is a direct connection between future economic well-being and funds spent on the education and care of the young of that family. When this situation exists, it is to the benefit of a family to have few children and put all their resources toward the education and well-being of those few.

As a result of a common desire among parents for large families, many nations face a problem in making the transition from stage II to stage III (Figure 5.7c), especially if, as in the case of Bangladesh, the growth rate is so high that it outpaces increases in economic development. The developed countries are now approaching stage III, but it is an open question whether some developing nations will make the transition before a serious population crash occurs. The key point here is that the demographic transition will only take place if parents come to believe that a small family size is to their benefit.

Although the demographic transition is traditionally defined as consisting of three stages, it is possible (with advances in treating chronic health problems such as heart disease) that a stage III country could experience a second decline in the death rate with improvements in treating chronic diseases and diseases of aging. This could bring about a second transitional phase of population growth (stage IV). A second stable phase of low or zero growth (stage V) would be achieved only when the birth rate declined to match the decline in the death rate. Thus there is a danger of a new spurt of growth, even in industrialized nations that have passed through the standard demographic transition.

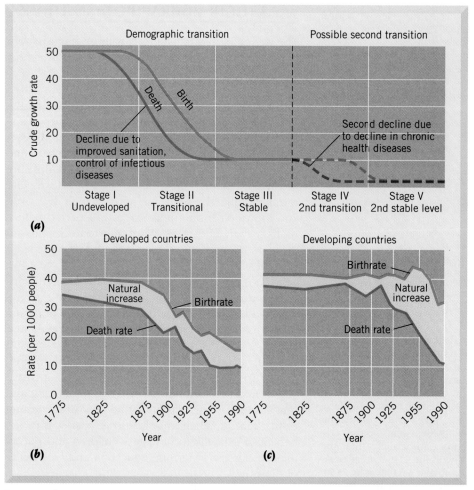

FIGURE 5.7 The demographic transition: (*a*) theoretical, including possible fourth and fifth stages that might take place in the future; (*b*) as has taken place for developed countries since 1775; and (*c*) as may be occurring in developing nations since 1775. (*Source:* M. M. Kent and K. A. Crews, 1990, *World Population: Fundamentals of Growth,* Population Reference Bureau, Washington, D.C., Copyright 1990 by the Population Reference Bureau, Inc. Reprinted by permission.

Human Death Rates and the Rise of Industrial Societies

We can get an idea of the first stage in the demographic transition by comparing a modern industrialized country, such as Switzerland, which has a crude death rate of 9 per 1000, with a developing nation, such as Sierra Leone, which has a crude death rate of 30. Modern medicine has greatly reduced death rates in Switzerland owing to diseases, with the first and greatest effect due to the reduction of death from acute or epidemic diseases.

An *acute* or *epidemic disease* appears in the population, affects a comparatively large percentage of it, and then declines or almost disappears for a while, only to reappear later. Epidemic diseases typically are rare but undergo occasional outbreaks during which a large fraction of the population is infected. Influenza, plague, measles, mumps, and cholera are examples of epidemic diseases. A

chronic disease, in contrast, is one that is always present in a population, typically occurring in a relatively small but relatively constant percentage of the population.

The great decrease in the percentage of deaths due to acute or epidemic diseases can be seen in a comparison of causes of deaths in Ecuador in 1987 and the United States in 1900 and 1987 (Figure 5.8).[16] In Ecuador, a developing nation, acute diseases and those listed as "all others" accounted for about 60% of the mortality, whereas in the United States in 1987 these accounted for only 20%. Chronic diseases, including heart disease, cancer, and stroke, accounted for about 70% of the mortality in the modern United States. In contrast, these accounted for less than 20% of the deaths in the United States in 1900 and about 33% in Ecuador in 1987.

One of the first to recognize the difference between acute (or epidemic) and chronic diseases was John Graunt, a British merchant who made the first

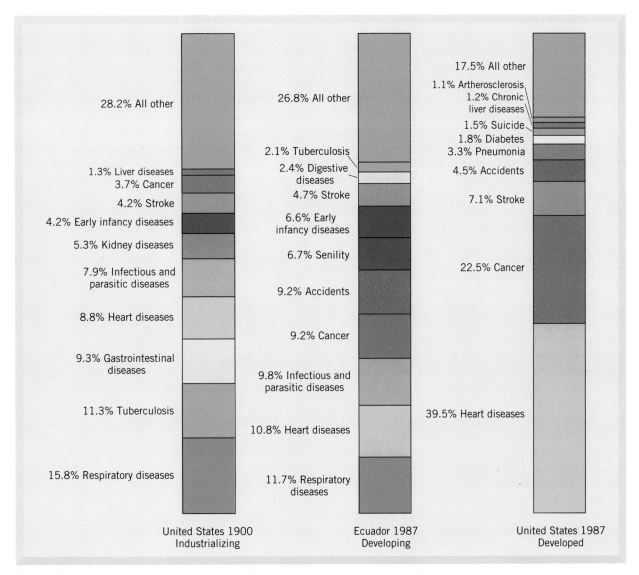

28.2% All other

1.3% Liver diseases
3.7% Cancer
4.2% Stroke
4.2% Early infancy diseases
5.3% Kidney diseases
7.9% Infectious and parasitic diseases
8.8% Heart diseases
9.3% Gastrointestinal diseases
11.3% Tuberculosis
15.8% Respiratory diseases

United States 1900
Industrializing

26.8% All other

2.1% Tuberculosis
2.4% Digestive diseases
4.7% Stroke
6.6% Early infancy diseases
6.7% Senility
9.2% Accidents
9.2% Cancer
9.8% Infectious and parasitic diseases
10.8% Heart diseases
11.7% Respiratory diseases

Ecuador 1987
Developing

17.5% All other
1.1% Artherosclerosis
1.2% Chronic liver diseases
1.5% Suicide
1.8% Diabetes
3.3% Pneumonia
4.5% Accidents
7.1% Stroke
22.5% Cancer
39.5% Heart diseases

United States 1987
Developed

FIGURE 5.8 Causes of mortality in industrializing, developing, and industrialized nation. (*Source:* M. M. Kent and K. A. Crews, 1990, *World Population: Fundamentals of Growth,* Population Reference Bureau, Washington, D.C., Copyright 1990 by the Population Reference Bureau, Inc. Reprinted by permission.)

modern study of the causes of death. In *Natural and Political Observations Made upon the Bill of Mortality* (1662), Graunt examined death records for more than 20 years in several parishes of London. He listed all the various causes of death and noted the number of deaths from each cause. He also classified the various diseases as "acute diseases (the plague excepted) . . . or chronical diseases."[11]

Although outbreaks of epidemic diseases have declined greatly during the past century in industrialized nations, there is now concern that there may be an increase in the incidence of these diseases. In part this is a result of the evolution of strains of disease organisms resistant to antibiotics and other modern methods of control. In part, spread of these diseases is a consequence of an increasingly dense population of poor people. The spread of AIDS is a

factor in this increase because individuals with AIDS lack resistance to these diseases. In 1997, AIDS killed 2.3 million people worldwide and HIV, the virus that causes AIDS, infected 30 million people. Ninety one percent of AIDS deaths occurred in 34 countries, 29 of them in sub-Sahara Africa. However, because fertility in African countries is high, populations will continue to grow.[18]

Carrying Capacity

When we talk about the limits to human populations, we are talking about the carrying capacity of the environment for human beings. Speaking generally, carrying capacity means the maximum population size that can be sustained indefinitely by the environment (see Chapter 1). A population that exceeds the carrying capacity will change the environment in a way that will decrease future population size. (Carrying capacity is sometimes used to mean the final equilibrium value of a logistic growth curve, but we are using its more general meaning here.)

How can we determine the actual carrying capacity for human beings either on a national basis or for the whole Earth? This is difficult. In order to calculate a carrying capacity, we have to agree on what the average standard of living and quality of life should be—to some extent a question of values.[18] In practice, several simple indexes are used to provide a general idea of a nation's carrying capacity, including the gross national product (GNP) per capita and the calories of food per capita.

Limiting Factors

Eventually human populations will be limited by some factor or combination of factors. We can group the limiting factors into those that could affect a population during the year in which they became limiting (short-term factors), those whose effects would be apparent after 1 year but before 10 years (intermediate-term factors), and those whose effects would not be apparent for 10 years after they became limiting (long-term factors). Some factors fit into more than one category, having, say, both short-term and intermediate-term effects.

The most important short-term factor is something that disrupts the distribution of food within a country. This disruption could be the result of political events, including wars, or of a local loss of current crops coupled with a shortage of energy needed to transport food. An abrupt change in weather, such as a drought, can cause such a disruption. Other short-term factors could include major world catastrophes, such as thermonuclear war, the worldwide spread of a major toxic chemical, and the outbreak of a new disease or a new strain of a previously controlled disease.

Intermediate-term factors include certain climatic changes; energy shortages that affect food production and distribution; desertification (see Chapter 11); wide dispersal of certain pollutants, such as the spread of toxic metals into waters and fisheries; disruption in the supply of nonrenewable resources, such as rare metals used in making steel alloys for transportation machinery; and a decrease in the supply of firewood or other fuels for heating and cooking.

Long-term factors include soil erosion; a decline in groundwater supplies; disruption in the supply of nonrenewable resources; climatic changes, such as global warming, resulting from changes in atmospheric chemistry; and wide dispersal of certain pollutants, such as acid rain.

One way to view limiting factors is to consider the resources available per person, referred to as the **per-capita availability.** It has been a common belief that the quality of life will get better and better in the future as a result of improvements in our technology, but an analysis of the change in per-capita resources suggests that we have already passed the peak in amount of biological resources available per person. Some people argue, therefore, that we have exceeded the long-term carrying capacity of Earth for people. For example, the per-capita production of wood peaked at 0.67 m^3/person in 1967 (0.88 yd^3/person), fish peaked at 5.5 kg/person (12.1 lb/person) in 1970, beef at 11.81 kg/person (26.0 lb/person) in 1977, mutton at 1.92 kg/person (4.21 lb/person) in 1972, wool at 0.86 kg/person (1.9 lb/person) in 1960, and cereal crops at 342 kg/person (754.1 lb/person) in 1977.[19] Before these peaks, the per-capita production of each had grown rapidly. Although part of the decline may be due to short-term economic forces, another part of the decline results from our nearing the limit of productive capacity with existing technology.

Age Structure

If doubling times, exponential growth, and logistic curves are of limited use to us in forecasting the future of the human population, what can we use that provides greater accuracy? One answer is the **population age structure,** which is the *proportion of the population in each age class.* The age structure of a population affects current and future birth rates, death rates, and growth rates; our impact on the environment; and current and future social and economic status.

Although age structure varies considerably by nation, we can discuss the general types (Figure 5.9). Kenya illustrates an age structure in a rapidly growing population heavily weighted toward youth and shaped like a pyramid. In developing countries today about 37% of the populations are under 15 years old.[1,12] Such an age structure indicates that the population will grow very rapidly in the future. The United States illustrates a population with slow growth. Austria is an example of a nation with a decline in growth. Elderly people are a small percentage of Kenya's population but a much larger percentage of the populations in the United States and Austria.

Age structure provides insight into a population's history, its current status, and its likely future. For example, the baby boom that occurred after World War II in the United States (a great increase in births from 1946 through 1964) forms a pulse in the population that can be seen as a bulge in the age structure, especially of those aged 35 to 45 in 1995. There is a secondary, smaller bulge as a result of a larger number of offspring from the first baby boom, which can be seen as a slight increase in the 5- to 15-year-olds. The baby boom pulse is moving

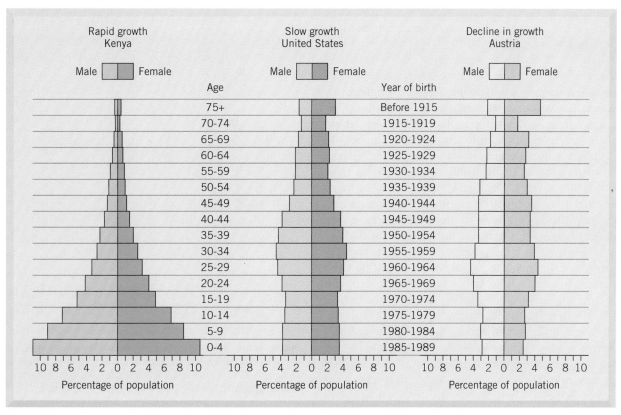

FIGURE 5.9 Age structure of Kenya, the United States, and Austria. (*Source:* M. M. Kent and K. A. Crews, 1990, *World Population: Fundamentals of Growth*, Population Reference Bureau, Washington, D.C.)

through the age structure. At each life stage, the baby boom has put pressure on social and economic resources; for example, schools were crowded when the baby boomers were of primary and secondary school age, during which the schools had an overcapacity.

Age Structure and Economic Factors

In preindustrial societies where average lifetimes are short, the young care for the old within the same family. With the development of modern technological societies, family size is reduced, and care for the elderly is distributed, through taxes, so that those who work provide funds to care for those who cannot. This provision for the elderly is a necessary factor in achieving zero population growth. Without it, parents will continue to rely on a large number of children to ensure their future well-being. However, a shift from a youthful age structure (like Kenya's) to that of an elderly age structure (like Austria's) means that a smaller percentage of the population works and therefore the taxes available for care for the elderly decrease (Figure 5.9).

This situation poses problems for a national government. The easiest way to increase tax income is to increase the percentage of the young and therefore to promote rapid population growth. Thus

short-term economic pressures at a national level can lead to political policies supporting rapid population growth not in the long-term interest of the nation.

Total Fertility Rate and the Lag in Response of Populations

Another important factor that can delay changes in population growth is the average number of children expected to be born to a woman during her lifetime; this is known as the **total fertility rate,** or TFR. The total fertility rate in the United States peaked at about 3.8 children per woman in the late 1950s and by 1998 was about 2.0.[1,17] This means that a woman who was age 15 in the late 1950s would have been expected to have about four children by the time she reached 44 years of age, whereas a 15-year-old woman in 1998 will be expected to have only about two children in her lifetime. A population that maintained a TFR of 3.8 over a long time would increase rapidly, whereas a population that had a TFR of 2.0 over a long time would decline.[1,17]

Replacement-level fertility is the TFR required for a population to remain constant. From the day that a population achieves replacement-level fertility, that population will continue to grow for several generations—approximately 50 to 200 years. This phenomenon is called *population momentum* or *population lag effect.*

How does this momentum occur? On the day that the women currently alive reach replacement-level fertility, the age structure is still weighted toward the young. Because previous fertility was greater than required for replacement levels, there are more women of childbearing age than would occur in a completely constant population. These women continue to bear children throughout their childbearing years. Because there are more of these women than necessary to replace the population, the population continues to grow. This is a time-lag effect of great importance to human populations.

Eventually, if the population maintains replacement-level fertility, the population size will become constant but the final size will be larger than the present size. Even though the U.S. TFR is below replacement level, the total population is still growing.

A comparison of nations suggests that the TFR declines as income increases. Bangladesh, which has an average income per person of a few hundred dollars per year, has a high total fertility rate—approximately 4.3 children born to each woman. Brazil, with an average income of more than $2,000, has a TFR of 2.9 (Figure 5.10).[14] Figure 5.11 shows that, worldwide, there has been a steady decrease in the TFR since 1950.

Future Population Trends

The World Bank, an international organization that makes loans and provides technical assistance to developing countries, has made a series of projections based on current birth rates and death rates and assumptions about how these will change. These projections form the basis for the logistic curves presented in Figure 5.6. Their critical assumptions are (1) mortality will fall everywhere and level off when female life expectancy reaches 82 years; (2) fertility will reach replacement levels everywhere between 2005 and 2060; and (3) there will be no major worldwide catastrophe. Even assuming a rapid achievement of replacement fertility, this approach projects an equilibrium world population of 10.1 to 12.5 billion.[12] Developed countries would only increase from 1.2 billion today to 1.9 billion, but developing countries would increase from 4.5 billion to 9.6 billion. Bangladesh would reach 257 million (in an area the size of Wisconsin); Nigeria, 453 million; and India, 1.86 billion. In these projections, the developing countries contribute 95% of the increase.[14]

Life Expectancy

Life expectancy is *the estimated average number of years a person of a specific age can expect to*

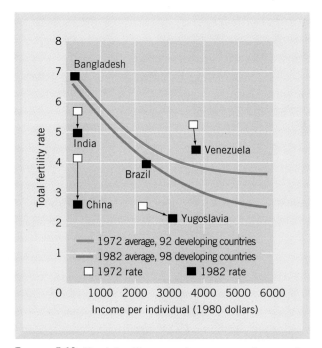

FIGURE 5.10 Total fertility rate decreases as income increases. Average curves are shown for developing nations in 1972 and 1982. Some specific values for selected countries are also shown. (*Source:* Modified from World Bank.)

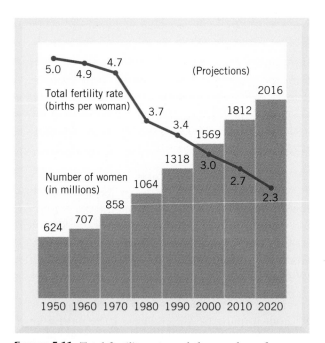

FIGURE 5.11 Total fertility rate and the number of women of childbearing age, 1950–2020. The total fertility rate dropped between 1950 and 1990 as the world population increased. Projecting present trends into the future suggests that by year 2020 a woman would have about two children during her lifetime, a major decline from an average of greater than three at the present time. (*Source:* M. M. Kent and K. A. Crews, 1990, *World Population: Fundamentals of Growth*, Population Reference Bureau, Washington, D.C.; data from the United Nations, 1991, *World Population Prospects 1990*, report no. 120, New York.)

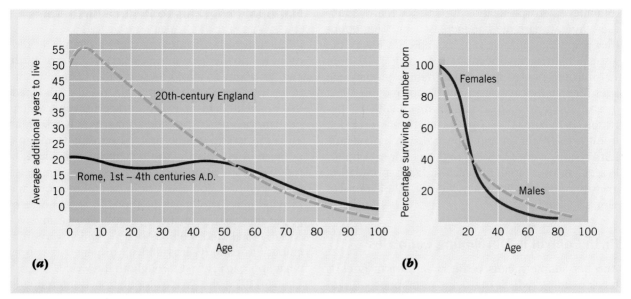

FIGURE 5.12 (*a*) Life expectancy in ancient Rome and modern England. This graph shows the average additional number of years one could expect to live having reached a given age. For example, a 10-year-old in England could expect to live about 55 more years; a 10-year-old in Rome could expect to live about 20 more years. Among the young, life expectancy is greater in modern England than it was in ancient Rome. However, the graphs cross at about age 60. An 80-year-old Roman could expect to live longer than an 80-year-old modern Briton. The graph for Romans is reconstructed from ages given on tombstones. (*b*) Approximate survivorship curve for Rome for the first four centuries A.D. The percent surviving decreases rapidly in the early years, reflecting the high mortality rates for children in ancient Rome. Females had a slightly higher survivorship rate until age 20, after which males had a slightly higher rate. (*Source:* Modified from G. E. Hutchinson, 1978, *An Introduction to Population Ecology*, Yale University Press, New Haven, Conn. Copyright 1978 by Yale University Press. Reprinted by permission.)

live. Technically, it is an age-specific number; each age class within a population has its own life expectancy. For general comparison, however, the life expectancy at birth is used. As mentioned earlier, the human population's great rate of increase has occurred despite the fact that life expectancy has not increased. In fact, the chances of a person 75 years old living to be age 90 were greater in ancient Rome than they are today in the United States.

Life expectancy differs by nation and by sex, age, and other factors (Figure 5.12). By studying the dates on tombstones, we can reconstruct the life expectancy of other periods in history. The age at death can also be used to infer the survivorship curve of the population (Figure 5.12*b*). These reconstructions suggest that death rates were much higher among young people in Rome and medieval Europe than they are now. In ancient Rome, the life expectancy of a 1-year-old was about 22 years, whereas in twentieth-century England it has been about 50 years. The life expectancy in a primitive hunter-gatherer society is short; among the !Kung bushmen of Botswana, life expectancy at birth is 33 years.[13] Life expectancy in modern England is greater than it was in ancient Rome for all ages until about age 55, when the life expectancy values ap-

pear higher for ancient Romans than for a modern Briton, suggesting that the hazards of modern life may be concentrated more on the aged. Pollution-induced diseases are one factor in this change.

5.5 HOW CAN WE STOP POPULATION GROWTH?

The simplest and one of the most effective means of slowing population growth is to delay the age of first childbearing by women.[20] As more women enter the work force and as education levels and standards of living increase, this delay tends to occur naturally. Social pressures that lead to deferred marriage and childbearing can be very effective.

Countries with high growth rates have early marriages. In south Asia and in Africa south of the Sahara, about 50% of women marry between the ages of 15 and 19. In Bangladesh women marry at age 16 on the average, whereas in Sri Lanka the average age for marriage is 25. The World Bank estimates that if Bangladesh adopted Sri Lanka's marriage pattern, families could average 2.2 fewer children.[14] Increases in the marriage age could account for 40% to 50% of the drop in fertility required to achieve zero population growth for many countries. Age at marriage has increased in some countries, especially in Asia. For example, in Korea the average marriage age went from 17 in 1925 to 24 in 1975. China passed laws fixing minimum marriage ages, first at 18 for women and 20 for men in 1950, then at 20 for women and 22 for men in 1980.[12] Between 1972 and 1985 China's birth rate dropped from 32 to 18 per thousand people, and the average fertility rate went from 5.7 to 2.1 children. China's leaders have a goal of reaching zero population growth in the year 2000, at a level of 1.2 billion people.

Birth Control in Developing Countries

Another simple means of decreasing birth rates is breast feeding, which can delay resumption of ovulation.[21] This is used consciously as a birth control method by women in a number of countries; in the mid-1970s the practice of breast feeding provided more protection against conception in developing countries than did family planning programs, according to the World Bank.[12]

The details of current birth control methods are beyond the scope of this book; we consider these only in terms of their consequences for the environment. Much emphasis is placed on the need for family planning.[22] Traditional methods range from abstinence to induction of sterility with natural agents. Modern methods include the pill, which prevents ovulation through control of hormone levels; surgical techniques for permanent sterility; and mechanical devices.

Contraceptive devices are used widely in many parts of the world, especially in eastern Asia, where surveys have suggested that about two-thirds of women use them. In Africa the percentage is below 10%, and in Central and South America it is about 40%. Abortion is also widespread. Although now safe medically in most cases, abortion is one of the most controversial methods from a moral perspective. Ironically, it is also one of the most important birth control methods in terms of its effects on birth rates. Approximately 45 to 60 million abortions are performed each year.[17]

Two terms are useful here: the *abortion ratio* and the *abortion rate*. The *abortion ratio* is the estimated number of abortions per 1000 live births in a given year.[22] The *abortion rate* is the estimated number of abortions per 1000 women aged 15 to 44 in a given year. For example, in 1992 in the United States, the abortion ratio was 335 and the abortion rate was 23. That is, for every 1000 live births, there were 335 abortions, and of every 1000 women of childbearing age, 23 had an abortion.[23]

National Programs to Reduce Birth Rates

Reduction in birth rates requires a change in attitude, a knowledge of the means to control birth, and the ability to afford these means. Change in attitude can arise simply with an increase in the standard of living. In many countries, however, it has been necessary to provide formal family planning programs to explain the problems arising from rapid population growth and the benefits to individuals of reduced population growth. These programs also provide information about birth control methods and provide access to these methods.[17] The choice of population control methods is an issue that involves social, moral, and religious beliefs, which vary from country to country. It is difficult to generalize about what approach is best throughout the world.

In 1974, the World Population Conference in Bucharest, Romania, was attended by representatives of 136 countries. They approved a world population plan that recognized that individuals have the right to decide freely the number and spacing of their children and to have access to information telling how to achieve their goals.[24]

The first country to adopt an official population policy was India in 1952. But few developing countries had official family planning programs before 1965. Since 1965 there has been widespread introduction of such programs. Over the last 27 years, the World Bank lent more than $3 billion to over 70 countries to support reproductive health projects. United States contributions to international family planning funds were $547 million in 1995 and $385 million in 1997.[25, 26] Although most countries now have some kind of program, the effectiveness varies greatly.

A wide variety of approaches have been used, from simply providing more information to promoting and providing some of the means for birth control to offering rewards and extending penalties. Penalties are usually in the form of taxes. For example, Ghana, Malaysia, Pakistan, Singapore, and the Philippines have used a combination of methods, including limits on several benefits such as tax allowances for children, maternity benefits, and tax deductions after one or two children. Tanzania has taken another approach, restricting paid maternity leave for women to a frequency of once in 3 years. Singapore does not take family size into account in allocating government-built housing, so larger families are more crowded. As an example of the use of rewards, Singapore increased the priority of school admission to children from smaller families.[12]

China has one of the oldest and most effective family planning programs. In 1978 China adopted an official policy to reduce the country's human population growth from 1.2% in that year to zero by the year 2000. An emphasis was placed on single-children families. The government uses education, a network of family planning that provides information and means for birth control, and a system of rewards and penalties. Women are given paid leave for abortions and for sterilization operations. There are benefits to families with a single child, including financial subsidies in some areas, and in some parts of China families that have a second child must return the bonuses received for the first. Other rewards and penalties vary from province to province.[12]

Other countries, including Bangladesh, India, and Sri Lanka, have paid people who have voluntarily been sterilized. In Sri Lanka, this practice has applied only to families with two children, and a voluntary statement of consent must be signed.

Migration

Another change in our current situation has to do with migration. In the past, if the population of one area or one country exceeded current resources, people would immigrate to a new area. Of course the story of America is the story of immigration, as are the stories of Australia and New Zealand, but people have always migrated. Migration is still an important movement of people in response to political situations, natural disasters, or other compelling factors, but the relief it provides to population problems is temporary and finite. The difference today is that we have run out of new territories on Earth. Some believe that our technologies will allow us to colonize previously unused areas, such as the Antarctic and even the deep sea; this may be possible to a limited extent. Others suggest that we can take care of Earth's population problems by creating space stations or migrating to other planets. But these solutions could take care of only a small fraction of the 87 million people added each year to the human population. We may be able to inhabit other planets, but this is not a solution for our world's population problems.

ENVIRONMENTAL ISSUE

How Many People Can Earth Support?

In mid–1992, the world population reached 5.5 billion. Assuming an annual growth rate of 1.7%, it is expected to reach 6.2 billion by the end of the century. Estimates of how many people the planet can support range from 2.5 billion (have we already exceeded the carrying capacity?) to 40 billion. Why do the estimates vary so widely?

An estimate of 2.5 billion assumes that we maintain current levels of food production and that everyone eats as well as Americans do now, that is 30% to 40% more calories than needed. The estimate of 40 billion assumes that all the remaining flat land of the world can be used to produce food, although most of it is too cold or too dry to farm. What is a realistic carrying capacity? What factors need to be considered to answer this question?

Food Supply

World grain production has apparently leveled off since reaching its highest levels in the mid-1980s. The production of grain from 1984 to 1994 remained at approximately 1.7 billion tons, after rising from 631 million tons in 1950. If the present harvest were distributed evenly and everyone ate a vegetarian diet, it could support 6 billion people. The remarkable increase in productivity since 1950 was the result of the development of high-yielding varieties, use of chemical fertilizers, application of pesticides, and doubling of cropland acreage. Although grain production continues to be high, as world population grows, the per-capita allotment of grain has been falling since 1984, when it stood at 346 kilograms per capita; in 1994 it had fallen to 311 kilograms per capita.

Land and Soil Resources

Almost all the usable agricultural land, approximately 1.5 billion hectares (ha; 3.7 billion acres) is already being cultivated. An increase of 13% in agricultural lands is possible but would be costly. Per-capita land area devoted to raising crops has dropped since 1950 to 1.7 ha (4.2 acres) and will continue to drop to approximately 1 ha (2.5 acres) by 2025 if present population projections hold. More soil is lost each year to erosion (about 26 billion metric tons) than is formed.

Water Resources

Water suitable for drinking and irrigation represents only a small proportion (less than 3%) of the water on Earth.

Underground reservoirs are being depleted on the order of feet per year but are being replaced in inches or even only fractions of inches per year. Per-capita water consumption varies from 350 to 1000 liters (371 to 1060 quarts) a day in the developed countries to 2 to 5 liters (2.1 to 5.3 quarts) a day in rural areas, where people obtain water directly from streams or primitive wells.

Net Primary Production

Human and domestic animals use about 4% of the net primary production of the world's land area and 2% of that of the oceans. Cultivated lands produce less than naturally vegetated areas. Net primary productivity on the planet has dropped about 13% since the 1950s.

Population Density

Population density varies greatly, from 1,163 people per square kilometer on the tiny island of Malta to 18 people/km² in Africa as a whole. Bangladesh has 915 people/km², the Netherlands, 456/km², and Japan, 332/km².

Technology

Carrying capacity is not merely a matter of numbers of people. It also involves the impact they have on the world's resources, most critically, energy resources. Multiplying population by per-capita energy consumption gives a relative measure of the impact people have on the environment. By that measure, each American has the impact of 35 people in India and 140 people in Bangladesh.

Critical Thinking Questions

1. Some people claim that agricultural technology can improve food production so that it continues to outpace, or at least to keep pace with, population growth, thereby increasing carrying capacity above 6 billion. What evidence supports this point of view? What evidence refutes it?

2. If humans and domestic animals use only 4% of the net primary productivity of the land and 2% of the oceans, there is obviously much primary productivity that is not used by humans. What would prohibit people from using a much greater proportion of the world's net primary productivity?

3. The land area of the world is approximately 150 million km² (57.9 million mi²). What would the world population be if the average density were 400 people/km²? Is this a sound basis for determining carrying capacity? Explain.

4. In recent decades, some scientists have debated whether population or technology is the basic limitation on carrying capacity. What is your position and how do you support it?

5. What factors, in addition to those six listed in this box, do you think should be considered in determining the planet's carrying capacity?

References

Ehrlich, P. R., and Ehrlich, H. A. 1990. *The Population Explosion.* New York: Simon & Schuster.

Environment. New York: Prentice-Hall.

Environmental Almanac. Boston: Houghton Mifflin.

SUMMARY

- The human population issue is *the* environmental issue, because most current environmental damages result from the very high number of people and our great power to change the environment.

- Throughout most of our history, the human population and its average growth rate have been small. The growth of the human population can be divided into four major phases. Although the population has increased in each phase, the current situation is unprecedented.

- Except during episodes such as the great plagues of the fourteenth century, the world's human population has grown rapidly and its rate of increase has become greater over time.

- Countries that have undergone a decrease in birth rates have experienced a demographic transition marked first by a decrease in death rates followed by a later decrease in birth rates.

- Many developing nations have experienced a great decrease in the death rate but still have a very high birth rate. It remains an open question whether some of these nations will be able to achieve a lower birth rate before reaching disastrously high population levels that will be subject to catastrophic causes of mortality.

- The maximum population Earth can sustain and what population will be attained by human beings are controversial questions. Standard estimates suggest that the human population will reach somewhere between 10 and 16 billion before stabilizing.

- Considerable lags in the responses of human populations to changes in birth rates and death rates occur because of the age structure of the population. A population that achieves replacement-level fertility will continue to grow for several generations.

REEXAMINING THEMES AND ISSUES

Human Population: Our discussion in this chapter shows the many ways that a rapidly increasing human population impacts the environment and reemphasizes the underlying point: there can be no long-term solution to our environmental problems unless the human population stops growing at its present worldwide rate. This makes the human population problem a top priority.

Sustainability: As long as the human population continues to grow, it is doubtful that our other environmental resources can be made sustainable.

Global Perspective: Although the growth rate of the human population varies geographically—from nation to nation—and although most people tend to live near where they were born, the overall environmental effects of the rapidly growing human population are global. For example, the increase in use of fossil fuels in Western nations since the beginning of the industrial revolution affects the entire globe. The growing demand for fossil fuels and growing increase in their use in developing nations, are having a global effect.

Urban World: One of the major patterns in the growth of the human population is the in-creasing urbanization of the world. Cities are not self-contained but are linked to the surrounding environment, depending on it for resources and affecting environments elsewhere. Urban development often leads to encroachment on highly valued natural areas, especially because cities are typically located where water, transportation, and material resources are readily available.

Values and Knowledge: The human population problem epitomizes the connection between values and knowledge. Increasing scientific and technological knowledge has helped us cure diseases, decrease death rates, and thereby increase the growth of the human population. Our present abilities to forecast human population growth provide much useful knowledge, but what we do with this is subject to intense debate around the world, because values are so important in relation to birth control and family size. We can only solve the human population problem by confronting the connection between the way we value human life and our scientific knowledge about the growth of the human population and its environmental effects.

KEY TERMS

age structure *88*	growth rate *82*	population age structure *88*
birth rate *82*	life expectancy *83*	population dynamics *79*
carrying capacity *84*	logistic growth curve *84*	replacement-level fertility *89*
death rate *82*	maximum lifetime *83*	species *82*
demographic transition *84*	per-capita availability *88*	total fertility rate *89*
demography *79*	population *79*	zero population growth *84*

STUDY QUESTIONS

1. What are the principal reasons that the human population has grown so rapidly in the twentieth century?

2. Given a world population in 1990 of 5.3 billion with a growth rate of 1.8%, what is the doubling time?

3. Why is it important to consider the age structure of a human population?

4. Three characteristics of a population are birth rate, growth rate, and death rate. How has each been affected by (a) modern medicine, (b) modern agriculture, and (c) modern industry?

5. What is meant by the statement "What is good for an individual is not always good for a population"?

6. Strictly from a biological point of view, why is it difficult for a human population to achieve a constant size?

7. What environmental factors are likely to in-

crease the chances of an outbreak of an epidemic disease?

8. Why is it so difficult to predict the growth of Earth's human population?

9. Before the beginning of the scientific and industrial revolutions, what factors tended to decrease the size of the human population?

10. To which of the following can we attribute the great increase in human population since the beginning of the industrial revolution: changes in human (a) birth rates, (b) death rates, (c) longevity, and (d) death rates among the very old? Explain.

11. What is meant by the demographic transition? When would one expect replacement-level fertility to be achieved: before, during, or after the demographic transition?

12. What are some arguments for or against the statement "With proper planning, we can achieve a constant carrying capacity for human beings on Earth"?

13. Based on the history of human populations in various countries, how would you expect the following to change as per-capita income increased: (a) birth rates, (b) death rates, (c) average family size, and (d) age structure of the population? Explain.

FURTHER READING

Brown, L. R., and Kane, H. 1994. *Full House: Reassessing the Earth's Population Carrying Capacity*. New York: Norton. Factors affecting Earth's human carrying capacity and how they change.

Cohen, J. E. 1995. *How Many People Can the Earth Support?* New York: Norton. A detailed discussion of world population growth, Earth's human carrying capacity, and factors affecting both.

Kent, M. M., and Crews, K. A. 1990. *World Population: Fundamentals of Growth*. Washington, D.C.: Population Reference Bureau. A clear and concise introduction to population concepts and historical and present trends in the human population, all well illustrated.

Kessler, E., ed. 1992. "Population, Natural Resources and Development," *AMBIO* 21(1). A special issue of the journal *AMBIO*, addressing many problems concerning human population growth and its economic and environmental implications. It includes a description of several case studies.

McFalls, J. A. 1991. "Population: A Lively Introduction," *Population Bulletin* 46(2):1–43. A good introduction to human population dynamics.

Postel, S. 1992. *Last Oasis: Facing Water Scarcity*. New York: W.W. Norton. Issues of water supply and demand facing the modern world.

World Bank. 1985. *Population Change and Economic Development*. Oxford: Oxford University Press. An excellent compilation of facts about human populations and methods for calculating population trends.

World Bank. 1992. *World Development Report 1992*. Oxford: Oxford University Press. Additional useful and updated material about the human population situation.

INTERNET RESOURCES

The Center for International Health Information: *http://www.cihi.com*—Provides information on the population, health, and nutrition sector in developing countries assisted by USAID, USAID programs, and health statistical reports.

Centers for Disease Control and Prevention: *http://cdc.gov/*—Contains health information and data and statistics about health-related issues.

International Planned Parenthood Federation: *http://www.ippf.org/*—Provides access to publications about family planning and sexual and reproductive health. IPPF links national autonomous Family Planning Associations in over 150 countries worldwide.

National Center for Health Statistics: *http://cdc.gov/nchswww/*—Provides access to publications and statistical tables, selected public-use data files, on-line database queries and searches. Also contains information about each NCHS survey and data system and links to other sites.

National Health Information Center: *http://nhic-nt.health.org/*—NHIC is a health information referral service. The Health Information Resource Database includes 1,100 organizations and government offices that provide health information upon request. Entries include contact information, short abstracts, and information about publications and services the organizations provide.

Planned Parenthood Federation of America: *http://www.plannedparenthood.org*—Provides educational resources, public affairs news, and indexed links about reproductive health and family planning services in the United States.

Population Reference Bureau: *http://www.prb.org/prb/*—Information on U.S. and international population statistics and trends.

United Nations Development Programme: *http://www.undp.org/*—Information on sustainable human development and links to the UN Environment Programme and other UN sites.

The World Bank: *http://www.worldbank.org*—Provides access to publications and data about developing countries and countries in transition.

World Health Organization: *http://www.who.int/*—Information on health topics, World Health Organization reports, and links to other United Nations Web sites.

Zero Population Growth: *http://www.zpg.org*—Home page of ZPG, the organization concerned with impacts of rapid population growth and wasteful consumption. Contains links to lists of population-related Web sites and gophers.

CHAPTER 6

ECOSYSTEMS AND ECOLOGICAL COMMUNITIES

The linkages between white-tailed deer, Lyme disease, mice, and gypsy moths in eastern forests illustrate the complexities of ecosystems and ecological communities. The deer share the ticks that carry Lyme disease with mice and people.

◢ CASE STUDY

The Acorn Connection

As young children, most people learn that acorns grow into oak trees without realizing that *most* acorns do not grow into trees but become food for mice, chipmunks, squirrels, and deer. In the woodlands of the northeastern United States, where oak trees abound, large crops of acorns (Figure 6.1*a*) are produced every 3 to 4 years. Acorns are rich in proteins and fats and an excellent source of nutrition. A steady supply of acorns would be an excellent food base for the woodland animals. For any oak tree, the production of acorns is affected by the amount of light and rain the tree receives, the

temperature patterns over the year, and the quality of the soil. Scientists reason that if oaks produced the same number of acorns each year, the populations of animals that feed on them would grow so large that very few acorns would survive.[1] In reality, the number of acorns produced varies from year to year, with "mast" years—years of high production—occurring occasionally. In the years between bumper crops of acorns, mice populations decline. With the next bumper crop, there are more acorns than can be eaten by the consumers of acorns, so many acorns survive to become oaks.

FIGURE 6.1 The tick that carries Lyme disease (*c*) feeds on both the white-footed mouse (*b*) and the white-tailed deer (*d*). Oak leaves (*f*) are an important food for the deer and for gypsy moths (*e*), while oak acorns (*a*) are important food for the mouse. But the mouse also eats the moths. The more mice, the fewer gypsy moths, but the greater the abundance of ticks.

Also, because of the abundance of food, the mice populations increase.[2]

White-footed mice (Figure 6.1*b*) also carry tick larvae (Figure 6.1*c*). As the ticks feed on the blood of the mice, they pick up the microorganisms responsible for Lyme disease. Mice populations are highest during the summer following a bumper crop of acorns and so are tick larvae. In later stages of their life cycle, ticks attach to other animals, including deer (Figure 6.1*d*). As deer brush against plants, ticks are deposited and can be picked up by people brushing against the plants as they walk through the same areas. If an infected tick bites a person, the person may contract Lyme disease.[1,2]

Since colonial times, the forests of the northeastern United States were cleared to make space for farming and settlements, to provide fuel, and to provide timber for commercial uses. As coal, oil, and gas replaced wood as a primary fuel, and as farming moved westward to the more fertile Great Plains, fields that had been cleared were abandoned. In many areas, the maximum clearing occurred around 1900, and since then forests have grown back.

As second-growth forest area has increased and deer populations have soared, Lyme disease has become the most common tick-borne disease in the United States. If left untreated, Lyme disease can have serious effects on the joints, nervous system, and heart. Fortunately, if caught early enough, Lyme disease can be treated successfully with antibiotics.

In addition to feeding on acorns and other grains, mice feed on insects, including larvae of the gypsy moth. Gypsy moth larvae (Figure 6.1*e*) feed on leaves of trees and are particularly fond of oak leaves. Studies suggest that in years when mice populations are low—the years between bumper crops of acorns—gypsy moth populations can increase dramatically. During these periodic outbreaks, gypsy moth larvae can virtually denude an area of forest. Oaks that have lost most or all of their leaves may not produce bumper crops of acorns.[1] Once the leaves are off the trees, more light reaches the ground, and seedlings of many plants that could not do well in deep forest shade begin to grow. As a result, other species of trees may gain a foothold in the forest and change its species profile. Of course, the next generation of gypsy moth larvae finds little to eat and the population of gypsy moths begins to decline again. Thus, when mice populations are high, gypsy moth damage is low, but the incidence of Lyme disease increases.[2]

Abundant acorns draw deer into the woods, where they browse on small plants and tree seedlings. Ticks drop off the deer and lay eggs in the leaf litter. When the eggs hatch, the larvae attach to mice and the cycle of Lyme disease continues. Deer do not eat ferns, however, and in areas where deer populations are dense, many ferns but few wildflowers and tree seedlings are found.

Predators of the mice, chipmunks, squirrels, and birds that feed on acorns are also affected by the periodic nature of acorn crops. For example, birds that feed on gypsy moth larvae lose a food source when moth populations are low. When moth populations are high, bird nests are more exposed to predation.[1]

The acorn connection illustrates many of the basic characteristics of ecosystems and ecological communities. First, all of the living parts of the oak forest community are dependent upon the nonliving parts of the ecosystem for their survival: water, soil, air, and the light that provides energy for photosynthesis. Second, the members of the ecological community affect the nonliving parts of the ecosystem. When gypsy moths denude an area, for example, more sunlight can reach the forest floor. Third, the living organisms in the ecosystem are connected in complex relationships that make it difficult to change one thing without having many effects. Fourth, the relationships among the members of the ecological community are dynamic and constantly changing and, as with the advantages provided oaks by a varying acorn production, many species are adapted to and benefit from a changing environment. Fifth, the implication for human management of ecosystems is that any management practice involves trade-offs. In this case, managing the forest to protect people against Lyme disease only results in more potential for gypsy moth damage.[2]

LEARNING OBJECTIVES

Life on Earth is sustained by ecosystems, which vary greatly but have certain attributes in common. After reading this chapter, you should understand:
- The basic characteristics of ecosystems that allow to them to sustain life: their structure and processes.
- The concepts of the ecosystem and the ecological community and why they are crucial to an understanding of many environmental issues.
- Trophic levels and how they work in food chains and food webs.
- Some of the direct and indirect interactions in ecological communities.
- Why the interactions and extent of particular ecosystems are difficult to understand and why artificial ecosystems are difficult to create.

6.1 WHAT SUSTAINS LIFE ON EARTH?

We tend to associate life with individual organisms, for the obvious reason that it is individuals that are alive. But *sustaining* life on Earth requires more than individuals or even single populations or species. Life is sustained by the interactions of many organisms functioning together, interacting through their physical and chemical environments. To understand important environmental issues, such as conserving endangered species, sustaining renewable resources, and minimizing the effects of toxic substances, we must know certain basic principles about a collection of organisms of different kinds that function together. That is, we must understand the concepts of the ecosystem (introduced in Chapter 3) and the ecological community. An ecosystem is an ecological community, along with its nonliving environment, functioning as a unit. An **ecological community** is a set of interacting species that occur

Closer Look 6.1

Yellowstone Hot Springs Food Chain

Perhaps the simplest natural ecosystem is a hot spring, such as those found in a geyser basin in Yellowstone National Park, Wyoming.[4] Few organisms can live in these hot springs, because the environment is so severe. Water in parts of the springs is close to the boiling point. Also, some springs are very acid and others are very alkaline; either extreme makes a harsh environment.

Some of the organisms that can live in hot springs are brightly colored and give these pools a striking appearance for which they are famous (Figure 6.2a). Typically, the springs have a wide range of water temperatures, from almost boiling near the source to much cooler near the edges, especially in the winter when there may be snow on the ground next to a spring. In a typical alkaline hot spring, the hottest waters, between 70 and 80°C (158 to 176°F), are colored bright yellow-green by photosynthetic blue-green bacteria, one of the few kinds of photosynthetic organisms that can survive in hot springs. In slightly cooler waters, 50 to 60°C (122 to 140°F), thick mats of bacteria and algae accumulate, some becoming 5 cm thick.

First trophic level. These mats are formed by long strings of photosynthetic bacteria and algae. As the flowing springwater passes over the mats, the long strings of cells trap and hold single-cell algae. Photosynthetic bacteria and algae make up the spring's first trophic level, which is composed of **autotrophs,** organisms that make their own food from inorganic chemicals and a source of energy. In the hot springs, as in most communities, the source of energy is sunlight (Figure 6.2b).

Second trophic level. Some flies, called *Ephydrid* flies, live in the cooler areas of the springs. One species, *Ephydra bruesi*, lays bright orange-pink egg masses on stones and twigs that project above the mat. Another species, *Aracoenia turbida*, lays white eggs in the mat. The fly larvae feed on the bacteria and algae. Since these flies eat only plants, they are herbivores, and they form the second trophic level.

Third trophic level. Another fly, called the *Dolichopodid* fly, is carnivorous and feeds on the eggs and larvae of the herbivorous flies. Dragonflies, wasps, spiders, tiger beetles, and one species of bird, the killdeer, also feed on the herbivorous flies. The herbivorous *Ephydrid* flies also have a parasite, a red mite, which feeds on fly eggs and travels by attaching itself to the bodies of the adult flies. Another parasite, a small wasp, lays its eggs within the fly larvae. The dolichopodid fly, dragonfly, wasp, spider mites, and killdeer are **predators** that feed on herbivores; these along with the parasitic mite form the third trophic level.

Fourth trophic level. Wastes and dead organisms of all trophic levels are fed on by **decomposers,** which in the hot springs are primarily bacteria. These form the fourth trophic level.

Trophic-level terminology seems clear and simple when it is applied to organisms that feed on only one trophic level, but it is more confus-

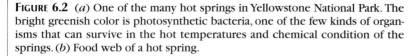

Figure 6.2 (*a*) One of the many hot springs in Yellowstone National Park. The bright greenish color is photosynthetic bacteria, one of the few kinds of organisms that can survive in the hot temperatures and chemical condition of the springs. (*b*) Food web of a hot spring.

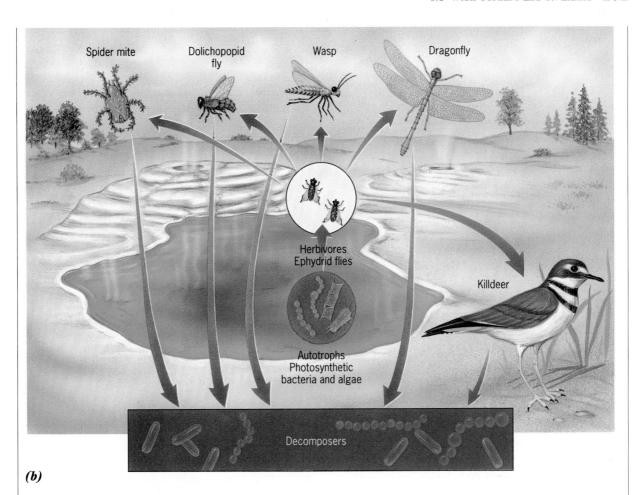

(b)

ing for the decomposers, because they feed on several trophic levels. If a species feeds on several trophic levels, the convention is to assign that species to the highest trophic level of which it is a member (the level above that of any of its food sources).

The entire hot springs community of organisms—photosynthetic bacteria and algae, herbivorous flies, carnivores, and decomposers—is maintained by two factors: (1) sunlight that provides an input of usable energy for the organisms and (2) a constant flow of hot water that provides a continual new supply of chemical elements required for life and a habitat in which the bacteria and algae can persist. Even though this is one of the simplest ecological communities in terms of the numbers of species, a fair number of species are found, including eight species of animals [two species of herbivorous flies, five species of car-

nivorous invertebrates (see Chapter 7), and one species of birds]. In high-temperature, alkaline springs there are two or three dominant species of photosynthetic bacteria and algae, especially *Synechococcus*, the blue-green bacteria that live at the highest temperatures, *Mastigocladus*, found in intermediate temperatures, and *Microcoleus*, found in the cooler (but still very hot) waters. The decomposing bacteria include a number of species. The total number of species important in this ecosystem is about 20. These form an ecological community, with a food web and trophic structure, which has been sustained for long periods in these unusual habitats.

Another interesting aspect of the hot springs ecosystem is species dominance. **Dominance** refers to the species that are most abundant or otherwise most important within the community. (We discuss this in connection with diversity in Chapter

7.) As noted earlier, in the hot springs community the species of photosynthetic bacteria or algae that is dominant changes with the temperature; one species dominates the hotter springs and hottest regions within a spring, and another species dominates cooler waters. Because the algae are so brightly colored, this spatial patterning in dominance is readily apparent to visitors. It was striking to one of the earliest explorers of Yellowstone, a trapper named Osborne Russell, who visited the springs in the 1830s and 1840s and wrote that one boiling spring about 100 m (328 ft) across had three distinct colors: "From the west side for one-third of the diameter it was white, in the middle it was pale red, and the remaining third on the east light sky blue."[5] This passage also illustrates spatial variations in the relative importance of species within springs and between springs.

in the same place. The purpose of this chapter is to develop these two important concepts as they relate to sustaining life on Earth.

6.2 THE ECOSYSTEM

The ecosystem concept is at the heart of the management of natural resources. When we try to conserve species or manage natural resources so that they are sustainable, we must focus on their ecosystem and make sure that it continues to function. If it doesn't, we must replace or supplement ecosystem functions with our own actions.

At its simplest, an ecosystem consists of several species—at least one species that produces its own food from inorganic compounds in its environment and one species that decomposes the wastes of the first species—plus a fluid medium (air, water, or both). Two basic kinds of processes must occur in the ecosystem: a cycling of chemical elements and a flow of energy.

Given these qualities needed for life, we recognize that *sustained life on Earth is a characteristic of ecosystems, not of individual organisms or populations.*[3] No species exists that both produces all its own food and decomposes all its wastes so that the materials can be reused to produce food. No individual cell, population, or community of populations forms a sufficient system to support life.

In the presence of light, green plants, algae, and photosynthetic bacteria produce sugar from carbon dioxide and water; from sugar and inorganic compounds they make many other organic compounds, including proteins and woody tissue. But no green plant can decompose woody tissue back to its original inorganic compounds. Living things, such as bacteria and fungi, that decompose organic matter do not produce their own food, but instead obtain their energy and chemical nutrition from the dead tissues on which they feed. For complete recycling of chemical elements to take place, several species must interact.

As discussed in Chapter 4, chemical cycling is complex. All chemical elements required for growth and reproduction must be made available to each organism at the right time, in the right amounts, and in the right ratios to each other. These chemical elements must be recycled, that is, converted to a reusable form by the system. So wastes are converted into food, which is converted into wastes, which must be converted into food once again.

Basic Characteristics of Ecosystems and Ecological Communities

Ecosystems have several fundamental characteristics. First, an ecosystem has structure: nonliving and living parts. Nonliving parts include rocks, water, and air. The living part is the community, which is a set of interacting species. Second, an ecosystem has processes, including the two already mentioned: Energy flows through it and chemical elements cycle within it. Third, an ecosystem changes over time and can undergo development through a process called succession, which is discussed in Chapter 9.

In practice, an ecological community—the set of interacting species that are the living part of an ecosystem—is defined by ecologists in two ways. The first, which is a simply pragmatic or operational definition, is that the community consists of all the species found in an area, whether or not they are known to interact and affect one another. Animals in different cages in a zoo could be called a community according to this approach.

The second method is to define the community as a set of *interacting* species found in the same place and functioning together to make possible the persistence of life. This seems more meaningful as an idea, but it is usually difficult in practice to know the entire set of interacting species.

Food Chains and Trophic Levels

One way individuals in a community interact is by feeding on one another. Energy, chemical elements, and some compounds are thus transferred from creature to creature along **food chains** (the linkage of who feeds on whom), which in more complex cases are called **food webs.**

Ecologists group the organisms in a food web into trophic levels. A **trophic level** consists of all those organisms in a food web that are the same number of feeding levels away from the original source of energy. The original source of energy in most ecosystems is the sun. Green plants produce sugars through the process of photosynthesis, using only the energy of the sun and CO_2 from the air, so they are grouped into the first trophic level. *Herbivores*, organisms that feed on plants, are members of the second trophic level; *carnivores* (meat eaters) that feed directly on herbivores are in the third trophic level; carnivores feeding on third-level carnivores are in the fourth trophic level; and so on.

Food chains and webs are often quite complicated and thus difficult to analyze. A detailed look at one of the simplest food chains is provided in A Closer Look 6.1, "Yellowstone Hot Springs Food Chain." We also look briefly at several more complicated food chains.

A Terrestrial Food Chain

An example of terrestrial food chains and trophic levels is shown in Figure 6.3a. This is a north temperate woodland food web that existed in

North America before European settlement and includes human beings. The first trophic level includes grasses, herbs, and trees. The second trophic level, herbivores, includes mice, the pine borer insect, and other animals such as deer not shown here; the third trophic level, carnivores, includes foxes and wolves, hawks and other predatory birds, spiders, and predatory insects. People are *omnivores* (eaters of both plants and animals) and feed on several trophic levels. In Figure 6.3a, people would be listed on the fourth trophic level, which is the highest trophic level, in which they would take part. Decomposers also feed on several trophic levels and are shown here on the fourth level since they feed on all the other trophic levels.

An Oceanic Food Chain

In the oceans, food webs involve more species and tend to have more trophic levels than they do in the hot springs or the terrestrial ecosystem just considered. In a typical oceanic ecosystem (Figure 6.3b), microscopic single-cell planktonic algae are in the first trophic level. Small invertebrates called *zooplankton* and some fish feed on the algae, forming the second trophic level. Other fish and invertebrates feed on these herbivores and form the third trophic level. The great baleen whales filter seawater for food, feeding primarily on small herbivorous zooplankton (mostly crustaceans), and thus the baleen whales are also in the third level. Some fish and marine mammals, such as killer whales, feed on the predatory fish and form higher trophic levels.

The Food Web of the Harp Seal

In the abstract, a diagram of trophic levels seems simple and neat, but in reality food webs are complex, because most creatures feed on several trophic levels. For example, consider the food web of the harp seal (Figure 6.3c). The harp seal is shown at the fifth level.[6] It feeds on flatfish (level 4), which feed on sand lances (level 3), which feed on euphausiids (level 2), which feed on phytoplankton (level 1). But the harp seal feeds at several trophic levels, from the second through the fourth, and it

FIGURE 6.3 Food webs: (*a*) a typical terrestrial food web. Roman numerals identify trophic levels.

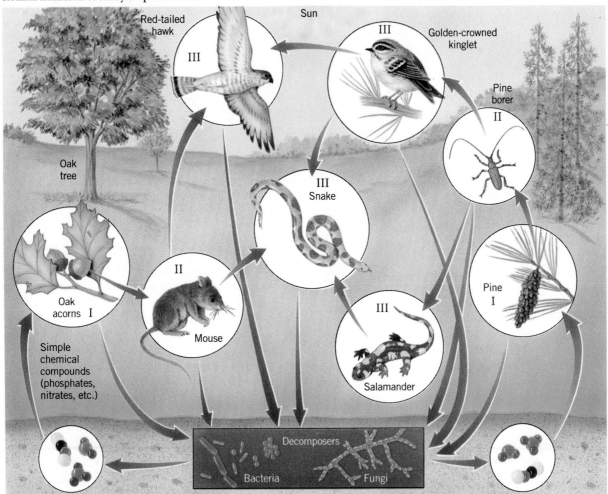

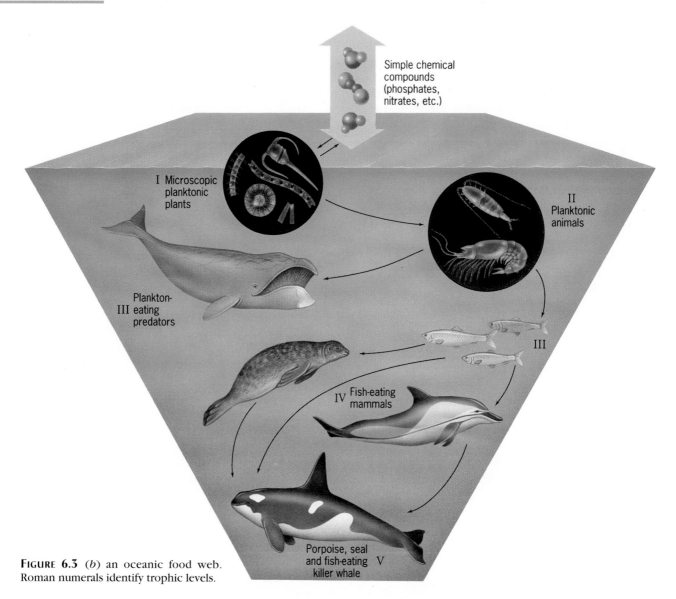

FIGURE 6.3 (*b*) an oceanic food web. Roman numerals identify trophic levels.

feeds on predators of some of its prey and thus is a competitor with some of its own food.[6] Since a species that feeds on several trophic levels typically is classified as belonging to the trophic level above the highest from which it feeds, we consider the harp seal on the fifth trophic level.

6.3 THE COMMUNITY EFFECT

Species can interact directly through food chains and in other ways, such as symbiosis and competition, which are discussed in the next chapter. One species can also affect other species indirectly, by influencing members of the community that, in turn, affect another set of species in the community. One species can also affect the environment, which then affects a group of species in the community. Changes in that second group affect a third group. Such indirect and more complicated interactions are referred to as **community-level interactions.**

Sea Otters and the Community Effect

Interactions at the community level are illustrated by the sea otters of the Pacific Ocean. In fact, the community-level interactions of the sea otter are at the heart of some arguments in favor of conservation of this species. Sea otters feed on shellfish, including sea urchins and abalone (Figure 6.4). Sea otters originally occurred throughout a large area of the Pacific Ocean coasts, from northern Japan, northeastward along the Russian and Alaskan coasts, and southward along the coast of North America to Morro Hermoso in Baja California, Mexico.[7] The otters were brought almost to extinction by commercial hunting for their fur during the eighteenth and nineteenth centuries: they have one of the finest furs in the world. By the end of the nineteenth century there were too few otters left to sustain commercial exploitation, and there was concern that the species would become extinct. A small population survived and has increased since then, so that today there are

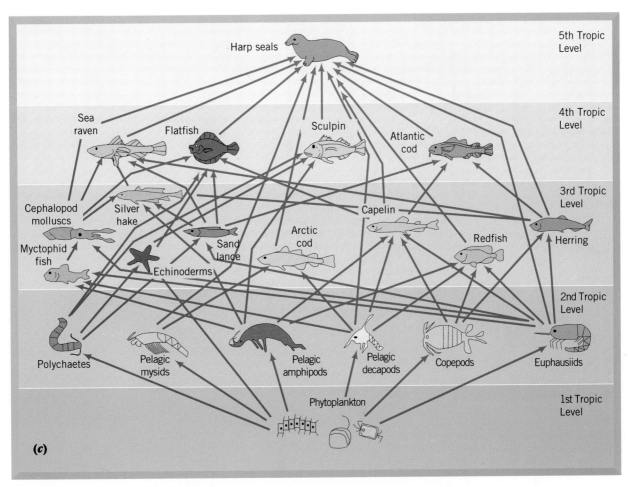

FIGURE 6.3 (*c*) food web of the harp seal.

approximately 161,500 sea otters. According to The Marine Mammal Center, in 1997 there were approximately 2,300 sea otters along the coast of California, a few hundred in Washington and British Columbia, and 150,000 along the Aleutian Islands of Alaska. The sea otter population in Russian waters was about 9,000.

Legal protection of the sea otter by the U.S. government began in 1911 and continues under the U.S. Marine Mammal Protection Act of 1972 and the Endangered Species Act of 1973. This animal has been a focus of controversy and research. On the one hand, fishermen argue that there are too many sea otters today and that they take large amounts of abalone and compete with commercial fishing.[8] On the other hand, conservationists argue that sea otters have an important community role, necessary for the persistence of many oceanic species. What is this important role?

One of the preferred foods of sea otters is sea urchin. Sea urchins feed on kelp, the large brown algae that form undersea forests. Kelp beds are important habitat for many species and are the location for reproduction of some of these species. Sea urchins do not eat the entire kelp. Instead, they graze along the bottoms of the beds, feeding on the base of kelp, called holdfasts. When holdfasts, which attach kelp to the bottom, are eaten through, the kelp float free and die.

Where sea otters are abundant, as on Amchitka Island in the Aleutian Islands, kelp beds are abundant and there are few sea urchins (Figure 6.4*b*). At nearby Shemya Island, which lacks sea otters, sea urchins are abundant and there is little kelp (Figure 6.4*c*).[7] Experimental removal of sea urchins led to an increase in kelp.[9]

Otters affect the abundance of kelp, but the influence is indirect. Sea otters neither feed on kelp nor protect individual kelp plants from attack by sea urchins. Sea otters reduce the number of sea urchins. With fewer sea urchins, there is less destruction of kelp. With more kelp, there is a larger habitat for many other species, so, indirectly, sea otters increase diversity of species.[9,10] Thus sea otters have a community-level effect.

This example shows that community-level effects can occur through food chains. These community-level effects can alter the distribution and abundance of individual species. When a species such as

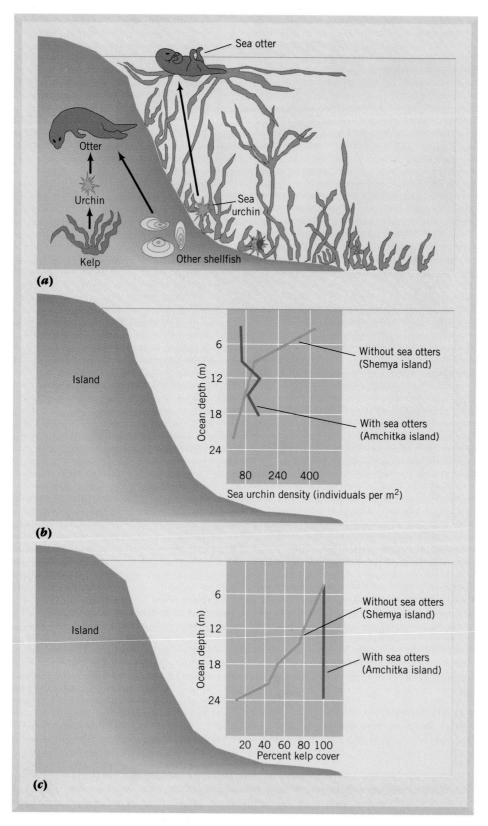

FIGURE 6.4 The effect of sea otters on kelp.

the sea otter has a large effect on its community or ecosystem, it is called a **keystone species,** or a key species.[11] The balance of the entire system is keyed to the activities of this species; its removal or the al-

teration of its role within the ecosystem would change the basic nature of the community.

The existence of community-level effects tells us that there is a reality to an ecological community;

there are processes that can take place only because of the existence of a set of species interacting along a food chain.

Holistic View of the Community

As we have indicated before, the ecological community is a difficult concept: How do we know if and how species interact? If every single species were completely essential to the organization of the community, the whole community could be viewed as a superorganism whose component species could not easily be replaced by others. However, this idea of the community, first developed by Frederick E. Clements at the beginning of the twentieth century, has been shown not to be realistic.[12,13]

Species associations are not constant; they vary from site to site and they also change over time. Nevertheless, every additional species in a community adds more species interactions: it may be prey or predator or contribute to chemical cycling; it may be a partner in a symbiosis or a competitor with other species. In this sense, even in a more realistic *individualistic view* of the community, the whole is indeed more than just the sum of its parts.[14] The view that the whole is more than the sum of its parts is often called the *holistic view* of the community. (Remember the discussion of environmental unity in Chapter 3.) The conflicting ideas about the nature of communities are of major importance in the discussion of ecological succession and biological conservation (see Chapter 9).[15] However, whatever view

one takes, it is clear that the idea of environmental unity—that everything is tied to everything else—is basic to the philosophy of ecosystems.

6.4 HOW DO YOU KNOW WHEN YOU HAVE FOUND AN ECOSYSTEM?

We have said that an ecosystem is the minimal entity that has the properties required to sustain life. This implies that an ecosystem is real and important and, therefore, that we should be able to find one easily. However, the key to ecosystems is in the processes that occur in them, and ecosystems vary greatly in structural complexity and in the clarity of their boundaries. Thus we have applied the term ecosystem to areas of Earth that differ greatly in size, from the smallest puddle of water to a large forest. We have also seen that ecosystems differ greatly in composition, from a few species in the small space of a hot spring to many species interacting over a large area of the ocean. In addition, ecosystems differ in the kinds and relative proportions of their nonbiological constituents and in their degree of variation in time and space.

Sometimes the borders of the ecosystem are well defined, such as the border between a lake and the surrounding countryside. Sometimes the transition is gradual, as the transition from desert to forest on the slopes of the San Francisco Mountains in Arizona (Figure 6.5). Sometimes the borders are vague, as in

FIGURE 6.5 (*a*) Sometimes the transition from one ecosystem to another is sharp and distinct, as in the transition from lake to forest at Lake Moraine in Banff National Park, Alberta, Canada. (*b*) Sometimes the transition is gradual and fuzzy, as in the transitions among vegetation types and their associated ecosystems on the slopes of Humphreys Peak, one of the San Francisco Peaks in Coconio National Forest near the Grand Canyon in Arizona.

(a)

(b)

FIGURE 6.6 The transition between boreal forest and tundra is often subtle and occurs over a large area. Here we see thinly scattered, small trees interspersed among tundra vegetation, within the transition in Kluane Park, Yukon, Canada.

the subtle gradations from grasslands to savannas in East Africa and from boreal forest to tundra in the far north, where the trees thin out gradually (Figure 6.6).

A commonly used practical delineation of the boundary of an ecosystem on land is the watershed. A **watershed** is defined most simply as follows: Within a watershed, any drop of rain that reaches the ground flows out in the same stream. Topography (the lay of the land) determines the watershed. When a watershed is used to define the boundaries of an ecosystem, the ecosystem is unified in terms of

chemical cycling. This, of course, is the idea introduced in the story about two drops of water in Chapter 4: They fall into different watersheds and thus enter different ecosystems with different chemical cyclings. Some classic experimental studies of ecosystems have been conducted on forested watersheds in U.S. Forest Service experimental areas, including the Hubbard Brook experimental forest in New Hampshire (Figure 6.7) and the Andrews experimental forest in Oregon.

What all ecosystems have in common is not a particular physical size or sharp boundaries, but the existence of the processes we have mentioned: the flow of energy and the cycling of chemical elements. The ecological community (the living part of an ecosystem) is characterized by food webs, food chains, and trophic levels.

Ecosystems can be natural or artificial or a combination of both. An artificial pond that is a part of a waste treatment plant is an example of an artificial ecosystem. Ecosystems can also be managed, and the management can vary over a large range of actions. Agriculture can be thought of as partial management of certain kinds of ecosystems (see Chapters 10 and 11), as can forests managed for timber production (see Chapter 13). Wildlife preserves are examples of partially managed ecosystems (see Chapter 12). Sometimes, when we manage or domesticate individuals or populations, we separate them from their ecosystems. When we do this, we must replace the ecosystem functions of energy flow and chemical cycling with our own actions. This is what happens in a zoo, where we must provide food and remove the wastes for individuals separated from their natural environments.

FIGURE 6.7 The V-shaped logged area in this picture is the famous Hubbard Brook ecosystem study. Here, a watershed defined the ecosystem, and the V shape is an entire watershed cut as part of the experiment.

How Are the Borders of an Ecosystem Defined?

The borders between ecosystems may be well-defined, gradual, or vague. Those considered well-defined are often, as in the case of freshwater streams, studied separately from surrounding ecosystems by researchers with different training and with different methods. Recent research on streams of southeast Alaska, in which salmon spawn, has raised questions about the usual practice of studying aquatic and terrestrial ecosystems separately.

Salmon are one group of anadromous fish—those that come from the ocean to spawn in freshwater streams. In southeast Alaska, enormous numbers of salmon spawn in over 5,000 streams. In 1985, 147 million salmon were harvested in that area of Alaska alone. Although salmon are born in freshwater, they migrate to the ocean where most of there growth occurs. After they return to their home streams, they spawn and die. In one sense, therefore, salmon are a means of transporting resources from the ocean to freshwater. Because of their large numbers, salmon have the potential of making significant contributions to organic and mineral content of streams.

Salmon have a high lipid content compared to many other fish and are thus a good energy source for animals that prey on them. In addition, their decay adds nitrogen, phosphorus, carbon, and other inorganic elements to freshwater. In one lake in western Alaska, its 24 million fish add 170 tons of phosphorus to the lake each year—an amount equal to or greater than recommended rates for applying fertilizers to trees. When the fish die, their carcasses decay and provide nourishment for algae, fungi, and bacteria. Invertebrates feed on these and decaying bits of fish. Other fish feed on the invertebrates. Finally, bears and other carnivores eat salmon, both live and dead, during their upstream migration. In that way, nutrients derived from salmon pass into the soil and vegetation surrounding the streams.

Because spawning fish have higher proportions of heavy isotopes of nitrogen and carbon (^{15}N and ^{13}C), they can be used to trace the relative contributions of anadromous fish to the nitrogen and carbon content of organisms in the food web. One such study showed that spawning salmon contributed 10.9% of the nitrogen found in invertebrate predators and 17.5% in the foliage of riparian plants.[3] While it is not surprising to find aquatic invertebrates, which feed on salmon eggs and juveniles, with large amounts of nitrogen derived from salmon, researchers were surprised at the high levels in stream side vegetation. When terrestrial mammals and birds feed on salmon, their feces and any uneaten salmon carcasses decay and add nutrients to the soil, where they can be taken up through the roots of plants. In southeast Alaska, over 40 species of mammals and birds feed on salmon. Salmon migrations attract large numbers of predators to streams and lakes. Salmon and other anadromous fish appear to link the ocean, freshwater, and land to an extent that is only beginning to be appreciated.

Critical Thinking Questions

1. Given the intricate connections between the aquatic and terrestrial ecosystems along salmon streams, how would you define the boundaries of the ecosystem?

2. Numbers of adult salmon that reach the spawning grounds and are above those needed to maintain the population have been considered excess. How might the research described here affect that view?

3. Some biologists have called salmon a keystone species. Given what you know about keystone species, how would you argue for or against this designation?

4. In recent years, the numbers of anadromous fish along the Pacific Coast of North America have declined precipitously due to overfishing and habitat destruction. What effects would you predict this might have on the ecology of freshwater streams and their adjoining land areas?

5. What types of management decisions about fish, wildlife, and forests would follow from recognizing the connection between aquatic and terrestrial ecosystems?

References

Bilby, R. E., Fransen, B. R., and P. A. Bisson. 1996. "Incorporation of nitrogen and carbon from spawning coho salmon into the trophic system of small streams: Evidence from stable isotopes," *Canadian Journal of Fisheries and Aquatic Sciences*, 53, pp. 164–173.

Spencer, C. N., McClelland, R. R., and J. A. Stanford. 1991 (January). "Shrimp stacking, salmon collapse, and eagle displacement," *BioScience* 41 (1) pp. 14–21.

Willson, M. F., Gende, S. M., and B. H. Marston. 1998 (June). "Fishes and the forest," *BioScience* 48 (6) 455–462.

Willson, M. F., and K. C. Halupka. June 1995. "Anadromous fish as keystone species in vertebrate communities," *Conservation Biology* 9 (3) 489–497.

SUMMARY

- An ecosystem is the simplest entity that can sustain life. At its most basic, an ecosystem consists of several species and a fluid medium (air, water, or both); it sustains two processes, the cycling of chemical elements and the flow of energy.

- The living part of an ecosystem is the ecological community, which is a set of species connected by food webs and trophic levels. A food web or chain is a diagram showing who feeds on whom. A trophic level consists of all the organisms that are the same number of feeding steps from the initial source of energy.

- Community-level effects are the result of indirect interaction between species, such as occurs when sea otters influence the abundance of urchins. These effects would not occur if there were only direct interactions between pairs of species, without involvement of the environment or of a set of species on different trophic levels.

- Ecosystems are real and important, but it is often difficult to define the limits of a system or to pinpoint all the interactions that take place. Creating an artificial ecosystem requires an understanding and balancing of many complex interactions.

REEXAMINING THEMES AND ISSUES

Human Population: The human population depends on many ecosystems that are widely dispersed around the globe. Although our modern technology gives us the appearance that we are independent of these natural systems, the greater our global connections through modern transportation and communication, the more the kinds of ecosystems on which we depend. Therefore the ecosystem concept is one of the most important we will learn about in this book.

Sustainability: The ecosystem concept is at the heart of managing for sustainability. When we try to conserve species or manage living resources so that they are sustainable, we must focus on their ecosystem and make sure that it continues to function (Figure 6.3).

Global Perspective: Our planet has sustained life for approximately 3.5 billion years. To understand how Earth as a whole has sustained life for such a long time, we must understand the ecosystem concept, because the environment at a global level must meet the same basic requirements as any local ecosystem.

Urban World: Cities are embedded in larger ecosystems. But like any life-supporting system, a city must meet basic ecosystem needs. This is accomplished through the connections between cities and surrounding environments. Together, these function as ecosystems or a set of ecosystems. To understand how we can create pleasant and sustainable cities, we must understand the ecosystem concept.

Values and Knowledge: The introductory case study about acorns, mice, deer, and Lyme disease illustrates the interactions between values and knowledge about ecosystems. It does not seem possible to have abundant deer and beautiful, fully leafed forests, without imposing actions on mice or on the deer. Science can tell us how these organisms interact. This knowledge confronts us with choice. Which choice we make depends on our values. The ecosystem concept tells us about opportunities and constraints, within which our values can determine much of the specific features (Figure 6.1).

KEY TERMS

autotrophs *100*	ecological community *99*	predators *100*
community-level interactions *104*	food chains *102*	trophic level *102*
decomposers *100*	food webs *102*	watershed *108*
dominance *101*	keystone species *106*	

STUDY QUESTIONS

1. What is the difference between an ecosystem and an ecological community?

2. In what ways would an increase in the number of sea otters and a change in their geographic distribution benefit fishermen? In what ways would these changes be a problem for fishermen?

3. Based on the discussion in this chapter, would you expect a highly polluted ecosystem to have many species or few species?

4. Is our species a keystone species? Explain.

5. What factors do you think are most important to monitor in Biosphere II?

6. Suppose you were asked to set up a monitoring system for a natural ecosystem, such as a forested watershed. Would you monitor the same factors that you listed in response to question 5?

7. Which of the following are ecosystems? Which are ecological communities? Which are neither?

a. Chicago

b. a 1000-ha farm in Illinois

c. a sewage treatment plant

d. the Illinois River

e. Lake Michigan

FURTHER READING

Bormann, F. H., and Likens, G. E. 1979. *Pattern and Process in a Forested Ecosystem.* New York: Springer-Verlag. A synthetic view of the northern hardwood ecosystem is presented, including its structure, function, development, and relationship to disturbance.

Brown, D. E., Reichenbacher, F., and Franson, S. E. 1998. *A Classification of North American Biotic Communities/Maps.* University of Utah Press. This volume describes a hierarchical classification system for biotic communities occurring from the Arctic Circle through Central America.

Diamond, J., and Case, T. J., eds. 1986. *Community Ecology.* New York: Harper & Row. This advanced text examines community ecology from several perspectives, including spatial and temporal scales, equilibrium theory, and community structuring mechanisms.

Duggins, D. O. 1980. "Kelp Beds and Sea Otters: An Experimental Approach," *Ecology* 61:447–453. The author discusses how experiments with sea urchin removal led to observation on kelp succession and diversity.

Gleason, H. A. 1926. "The Individualistic Concept of the Plant Association," *Bulletin of the Torrey Botanic Club* 53:7–26. This is a classic paper in which an alternative view to the then-popular concept of plant succession (the superorganism) was presented.

McIntosh, R. P. 1985. *The Background of Ecology: Concept and Theory.* New York: Cambridge University Press. The author, a leader in the field of ecology, gives an insider's history of the discipline.

Pahl-Wostl, C. 1995. *The Dynamic Nature of Ecosystems: Chaos and Order Entwined.* John Wiley & Sons. Presents the argument that it is the trade-off between the irregular, chaotic dynamics at the level of populations and the spatio-temporal organization at the level of the system as a whole that shapes ecological systems.

Strong, D. R., Simberloff, D., Abele, L. G., and Thistle, A. B., eds. 1984. *Ecological Communities: Conceptual Issues and the Evidence.* Princeton, N.J.: Princeton University Press. This collection of papers illuminates species interactions and the ecological effects of these interactions.

INTERNET RESOURCES

Channel Islands National Marine Sanctuary: *http://www.cinms. nos.noaa.gov*—Information and resources available on marine life in Channel Islands, California.

The Ecosystem Home Page: *http://www.reast.demon.co.uk//*—An environmental information service with full text updates organized by topic from the news service of The Environment Digest.

The Marine Mammal Center: *http://www.tmmc.org*—Organization concerned with protecting marine mammals. Educational and research information available on marine mammals.

World Wide Fund for Nature International: *http://www.panda.org*—Conservation organization with information available on climate change, oceans, forests, sustainability, various species. The site has an expert database and fact sheets on a variety of topics.

Researchers seeking new pharmaceutical drugs from tropical rain forest plants in Costa Rica illustrate one of the often cited values of biological diversity: the potential that yet undiscovered plants may provide chemicals to cure diseases.

CHAPTER 7

BIOLOGICAL DIVERSITY AND BIOGEOGRAPHY

CASE STUDY

Purple Loosestrife

An important component of biological diversity is the geographic distribution of living things, known as **biogeography.** Over Earth's history, species sometimes become isolated from one another by physical barriers. Introduction of long-separated species from one part of the world into another, a common human practice, has many consequences. Without the natural enemies of their native habitats, that is, predators or parasites, including disease-causing organisms, introduced species can spread unchecked and become a threat to biodiversity in their new habitats.

Purple loosestrife, a perennial plant of wetlands, was brought to North America from Europe in the early nineteenth century. This introduction illustrates some of the negative consequences of introducing alien species (Figure 7.1).[1] It has been responsible for eliminating many of the native plants that provide food and cover for wildlife, including waterfowl, in wetlands of Canada and the United States.[2]

Purple loosestrife was introduced accidentally when ships from Europe released ballast water contaminated with loosestrife seeds in North American waters. It was also introduced deliberately as an herbal remedy for diarrhea, dysentery, bleeding, ulcers, and sores. Without natural enemies and highly tolerant of variations in moisture and temperature, purple loosestrife spread rapidly along the New England coast. Construction of canals and waterways in the 1880s helped it spread into more inland areas of New York and the St. Lawrence River Valley. By the mid-1990s, it occurred throughout southeastern Canada and eastern and midwestern United States. It had started to appear in the western United States and southwestern Canada.

Purple loosestrife illustrates the great capacity of living things to grow and reproduce. A single plant can produce more than 2.5 million seeds each year.[3] The seeds, which can survive for long periods of time, are spread by water and in mud clinging to wildlife, livestock, and people. Disturbances such as fires or human disturbances of the environment, such as clearing of land, provide exposed sites in which the seeds can germinate. As many as 20,000 seedlings can be found in one square meter (1.2 square yards). Few native species can compete with this productivity. Once

FIGURE 7.1 (*a*) Purple loosestrife, a European plant introduced into North America, has caused many problems for the biological diversity of native plants (photographic closeup). (*b*) A wetland dominated by purple loosestrife, which has replaced native American vegetation. (*c*) An insect parasite of purple loosestrife, from its native habitat, may be useful in controlling this plant in North America (photograph of one of the insect parasites of purple loosestrife).

loosestrife matures, it shades other seedlings and its relatively tall height among wetland herbs and shrubs (approximately 2 meters or 6 feet) prevents other plants from growing. It has endangered such species as bulrushes in Massachusetts, dwarf spike rushes in New York, and bog turtles in the northeastern United States. In addition, livestock do not find it as palatable as native grasses and sedges, so it has reduced the carrying capacity of grazing areas and led to agricultural losses.

Roots of the purple loosestrife are exceptionally resistant and resilient when cut, burned, or sprayed with herbicides. The stems regrow readily from the roots. The only effective control is to remove every bit of the plant, both above and below ground, which is expensive and time consuming. Currently, a group of scientists from Canada and the United States are guiding research on the use of biological controls. The effort focuses on three species of European insects that are natural parasites of the loosestrife in its original habitat. The most promising of the insect species is being released with the goal of eliminating 75% to 80% of purple loosestrife in Canada and the United States (Figure 7.1*c*).

The story of the purple loosestrife illustrates many of the major topics of this chapter: biogeography and evolution, interactions among species, and inadvertent effects of people on biological diversity. One lesson of the purple loosestrife case history is that we must be very careful about inadvertent introductions of exotic species; in many cases such introductions threaten native species and cause other problems. In recent years, many problems have resulted from the introduction of exotic species by people, ranging from the introduction into South America of African killer bees, which have migrated north to the United States, to the introduction of cane toads into the wet tropics of Australia.

Learning Objectives

One of the oldest questions people have asked is how the amazing diversity of life on Earth came to be. This diversity has developed through biological evolution; it is affected by interactions among species and by the environment. After reading this chapter, you should understand:

- How the conservation of biological diversity involves an understanding of the intricate relationships among species and between species and their environments.

- How mutation, natural selection, migration, and genetic drift lead to evolution.

- Why so many species have been able to evolve and persist.

- How species interactions affect diversity.

- How large-scale global patterns and the environment affect diversity.

- Some of the ways in which people affect biological diversity.

7.1 UNDERSTANDING BIOLOGICAL DIVERSITY

Biological diversity attracts great public attention. Frequently, articles in newspapers and discussions on television address issues associated with disappearing species around the world and the need to conserve these species. Controversies over endangered species are sometimes seen as running counter to other uses of land and resources by people. Discussions are complicated by the fact that people can mean various things when they talk about the conservation of diversity. We can mean the conservation of a variety of habitats, a number of genetic varieties, a number of species, a relative abundance of species, or a specific endangered species. As we will see in this chapter, all these concepts are interrelated. Before we can intelligently discuss the issues involved in conserving Earth's diversity of organisms, we must understand how this diversity came to be. This chapter first addresses the principles of biological evolution that lead to the evolution of new species. We then turn to biological diversity itself: its various meanings, how interactions between species increase or decrease diversity, and how the environment affects diversity.

7.2 BIOLOGICAL EVOLUTION

Understanding biological evolution is the key to understanding biological diversity. It is also essential to understanding all the other issues in environmental sciences that involve life. Outside the realm of biology, the term *evolution* is used broadly to mean the history and development of something, but not the evolution of entirely new forms. For example, geologists sometimes talk about the evolution of Earth, which simply means Earth's history and geologic changes that have occurred over that history. Book

reviewers talk about the evolution of the plot of a novel, meaning how the story unfolds as one reads the book. These uses of the word evolution differ from the meaning of *biological evolution*. **Biological evolution** refers to the change in inherited characteristics of a population from generation to generation. It can result in new species—populations that can no longer reproduce with members of the original species.

Along with self-reproduction, biological evolution is one of the features that distinguish life from everything else in the universe. Biological evolution is a one-way process. Once a species is extinct, it is gone forever. You can run a machine, like a mechanical grandfather clock, forward and backward. But when a new species evolves, it cannot evolve backward into its parents. The one-way history inherent in biological evolution is a key way in which this kind of evolution differs from the more general meaning of the term.

Processes That Lead to Evolution

Four processes lead to evolution: mutation, natural selection, migration, and genetic drift.

Mutation

Genes, contained in chromosomes within cells, are inherited from one generation to the next. Genes are made up of a complex chemical compound called deoxyribonucleic acid (DNA), which is made up of chemical building blocks that form a code, or kind of alphabet of information. Biologists now understand the chemistry of these compounds and are beginning to unravel the code—the message each set of chemical compounds transmits—for some species.

When a cell divides, the DNA is reproduced so that each new cell gets a copy. Sometimes there is an error or a failure in the reproduction of the DNA, resulting in a change in the DNA and therefore in a change in inherited characteristics. At other times an external environmental agent, such as radiation or certain toxic organic chemicals, comes in contact with DNA and alters it. Radiation, such as X rays or gamma rays, can break the DNA apart or change its chemical structure. Toxic chemicals can react with a DNA molecule, causing a chemical change. When DNA changes in any of these ways, the DNA is said to have undergone **mutation.**

In extreme cases, cells or offspring with a mutation cannot survive. In less extreme cases, individuals with the mutation are so different from their parents that they cannot reproduce with the normal offspring. In this case, a new species has been created. In milder cases, the mutation can simply add variability to the inherited characteristics. Note that the change is not always for the better: *Mutation can result in a new species whether or not that species is better adapted than its parental species to the environment.*

Natural Selection

When there is variation in the characteristics of a species, some individuals may be better suited to the environment; as a result, variation may put the species at an overall advantage. **Natural selection** is a process by which organisms whose biological characteristics better fit them to the environment are better represented by descendants in future generations than are those whose characteristics are less fit for the environment.

The peppered moths in Great Britain described in A Closer Look 7.1, "Industrial Melanism and Natural Selection," illustrate the process of natural selection and its four primary characteristics (Figure 7.2): (1) inheritance of traits from one generation to the next and some variation in these traits, that is, genetic variability; (2) environmental variability; (3) differential reproduction that varies with the environment; and (4) an influence of the environment on survival and reproduction.

Natural Selection and Evolution of New Species

In the case of the peppered moth, only one trait, color, changed over a relatively short time period. If natural selection were to take place over a very long time, however, a number of characteristics could change. Sometimes, two populations of the same species become isolated completely from one another and cannot reproduce together for a long time. In such cases, some of these changes alter the populations so much that they can no longer reproduce together even when they were brought back into contact. In this case, two new species would have evolved from the original species. Thus natural selection and geographic separation of populations is a mechanism by which the evolution of new species can take place.

Migration

A third process that can lead to evolution is **migration.** The migration of one population of a species into a habitat previously occupied by another population is a process that can lead to changes in gene frequency. Even organisms that do not move themselves may have reproductive structures that migrate; seeds of flowering plants, for example, are blown by the wind and moved by animals.

Migration has been an important process over geologic time. For example, during intervals between recent ice ages and at the end of the last ice age, Alaska and Siberia were connected by a land bridge that permitted the migration of plants and animals. Great Britain was covered by ice during the glacial advances, after which seeds of plants were transported by wind (and later by people) to the islands to

Genetic Drift

Another process that can lead to evolution is genetic drift. **Genetic drift** refers to changes in the frequency of a gene in a population as a result of chance rather than mutation, selection, or migration. Chance may determine which individuals become isolated in a small group from a larger population and thus which genetic characteristics are most common in that isolated population. The individuals may not be better adapted to the environment; in fact, they may be more poorly adapted or neutrally adapted.

Suppose, for example, that you begin a study of the peppered moth described in A Closer Look 7.1, and go out to collect a sample with a butterfly net, taking the first dozen moths you find in a rural area. The moths occur in a white and a black form. You discover that by chance all of the 12 are the dark form, even though the dark form is the rarer one in the rural environment. Suppose that you then set up a preserve for the moths where they can no longer mix with any others. Even if you kept this population for many generations, the dark form would be the only one found (unless, of course, mutation led to a new color form). A change would have occurred in gene frequency because of chance isolation of individuals. This is an example of genetic drift.

Genetic drift can occur in any small population. It can be a problem for a rare or endangered species for two reasons: (1) characteristics that are less adapted to existing environmental conditions may dominate, making the chance of survival of the species less likely; (2) genetic variability of the species is greatly reduced, meaning that its ability to

CLOSER LOOK 7.1

INDUSTRIAL MELANISM AND NATURAL SELECTION

The principles of natural selection are illustrated by the history of the peppered moth, a medium-sized insect that lives in Great Britain, where it often alights on tree trunks.

Insect-eating birds try to catch the moths as they rest on the trees. Before the rise of industry, most of the moths were whitish, a color that melded with light tree bark and with lichens that covered much of the tree bark. The bark camouflaged the moths so that it was hard for predatory birds to see them. A few of the moths were black, but this color was easy to see against tree trunks and lichens, so black moths were readily caught by birds.[37] Black forms remained rare.

With the rise of industry in the nineteenth century, black soot from burning coal and other fuels became common near cities such as Birmingham, where a study of the moths was carried out. The soot fell on the trees, darkening the bark. Because lichens that grow on trees are sensitive to air pollution, especially to sulfur oxides, these may have declined in abundance as well, making the darker natural color of some tree trunks more apparent. Against the darker background, the whitish moths were readily visible, but the black ones were camouflaged.

In a study of the moths, the number of each color taken by birds was recorded over a certain time period in two different locations, rural and urban. In the rural woodlands where the bark surface appeared light, more than 80% of the moths caught by birds were dark. In the urban woodlands, polluted by soot so that the bark surface appeared dark, more than 70% of the moths caught were light (Figure 7.2). The relative advantage of color was reversed by pollution. In less than 100 years, most peppered moths living near cities were black, and those living in rural areas were mainly whitish.

In recent years, modern air pollution controls have improved air quality in and around British cities. Gas is burned instead of coal, and there is much less soot. Tree trunks are no longer covered with soot, and lichens again grow on the trunks. Thus whitish moths are again camouflaged against tree trunks and black moths are not; as we would expect, the whitish forms are becoming common again in the cities.

The increase in the relative abundance of the dark form of the pep-

FIGURE 7.2 Industrial melanism: moth color, pollution, and predation. The number of peppered moths of each color form, white and dark, that were caught by birds was counted in a study in both rural and urban woodlands in Great Britain. Results of the study of peppered moths demonstrate that the birds primarily caught the dark form in rural areas and the light form in polluted, urban areas. (*Source:* Redrawn from data in H. B. D. Kettlewell, 1959, *Scientific American,* 200, pp. 48–53.)

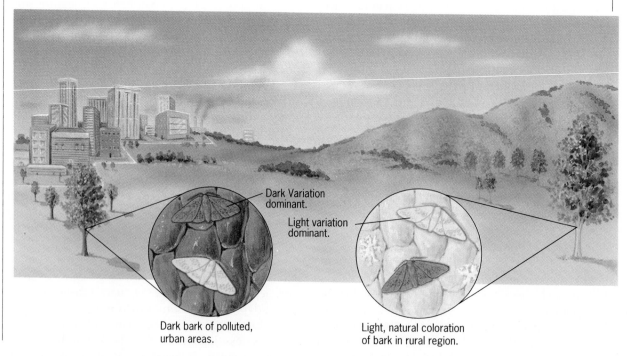

Dark Variation dominant.

Light variation dominant.

Dark bark of polluted, urban areas.

Light, natural coloration of bark in rural region.

pered moth when soot covered tree trunks is an example of **industrial melanism.** The change in the ratio of dark- to light-colored moths is a result of *differential selection* due to changes in the environment. This example illustrates two major features of biological evolution: variation and natural selection.

Here are the key attributes to keep in mind. First, in the case of the moth, there is *variation in coloration inherited from one generation to another:* whitish and black.

Second, there is *variation in the environment;* at any one time in any one location, tree trunks have been mainly light or mainly dark. Third, there is a *mechanism in the environment that affects survival (and therefore reproduction);* in the case of the peppered moth, this mechanism is the birds that prey on the moths.

Both forms of the moth have the potential for a high rate of reproduction, and the two forms compete for resources. Under varying conditions, there is a difference in survival and

therefore in success of reproduction of the two forms. When tree trunks are mainly light, predation keeps the black forms rare, and therefore there are few offspring with this color. When soot darkens the trees, the light forms are readily caught by birds, so fewer of this color reproduce. Meanwhile, the dark forms, well hidden, survive longer and have more offspring. When the environment changes, the relative reproductive success of the two forms changes.

---

adapt to future changes in the environment through natural selection has been reduced.

For example, bighorn sheep live in the mountains of the desert southwest of the United States. These sheep feed high up on the mountains in the summer, where it is cooler and wetter and there is more vegetation. Before human settlement of the region, the sheep could move freely and sometimes migrated from one mountain to another in the winter. With human settlement and the development of cattle ranches and other human activities, many populations of bighorn sheep have become isolated in very small numbers. Commonly, a dozen or so sheep are isolated on a mountaintop. In such cases, genetic drift may pose a problem for the survival of these populations.

7.3 THE EVOLUTION OF LIFE ON EARTH

With peppered moths described in A Closer Look 7.1, evolution occurred rapidly, but over most of the history of Earth, evolution has proceeded much more slowly. We know about the history of the evolution of species in part through the study of fossils in rocks and in other geologic deposits. The story of evolution begins, as far as we know it today, with the oldest fossils of microorganisms, which appear to be ancestral forms similar to bacteria and what some microbiologists now call Archea.[32] We believe they lived about 3.5 billion years ago. For 2 billion years there was only such microbial life on Earth, but these greatly changed the global environment.

Photosynthesis developed during that time, and early photosynthetic organisms removed carbon dioxide from Earth's atmosphere and added large amounts of oxygen to that atmosphere (illustrating our ongoing assertion that life has always changed the environment on a global scale). With the high concentration of oxygen in the atmosphere, the eco-

logical stage was set for the evolution of new forms of life. Free oxygen allowed the evolution of respiration, which paved the way for oxygen-breathing organisms, including, eventually, humans.

Among the oldest questions people have asked are these: How can we explain the origin of the wonderful complexity and adaptiveness of life on Earth? For example, how can we explain such things as the eye of a hawk that can see so far? Why are there so many species on Earth? The answer that biologists give to these ancient and fundamental questions is that biological evolution, over a very long time, has led to the forms of life we observe. Understanding how various factors have contributed to the evolution of this diversity is an important objective of this chapter; the answers provide insight into why our actions are leading to an increasing rate of extinction and what we can do to slow the rate of extinction.

The science of molecular biology and the practice of genetic engineering are creating a revolution in the way in which we think about and deal with species. There are important environmental implications of this revolution. Advances in genetic engineering reinforce the basis for confidence in the theory of evolution. Scientists can now manipulate DNA and can therefore manipulate inherited characteristics of crops, bacteria, and other organisms and give them new combinations of characteristics not found before (see Chapter 10). These new organisms can be to our benefit, but we must be careful not to release into the environment new strains that have great reproductive potential and could become unexpected pests.

Evolution as a Game

Biological evolution is so different from other processes that it is worthwhile to spend some extra time exploring this idea. The ecologist Lawrence

Slobodkin has called evolution a game in which the only rule is to stay in the game. (You're winning if you, as a species, are still there; you lose by becoming extinct.)

There are not any simple rules that species must follow to win or to stay in the game of life. Sometimes, when we try to manage species or to predict their future condition and probable persistence, we assume that evolution will follow simple rules. When we make these assumptions, we find that species play tricks on us; they adapt or fail to adapt over time in ways that we did not anticipate.

Such unexpected outcomes result from our failure to fully understand how species have evolved in relation to their ecological situation. Complexity is a feature of evolution; species have evolved many intricate and amazing adaptations that have allowed them to persist. It is essential to realize that these adaptations have not evolved in isolation, but in the context of their relationships to other organisms and to the environment. The environment sets up a situation or scene within which evolution, by natural selection, takes place. The ecologist G. E. Hutchinson referred to this situation in the title of one of his books, *The Ecological Theater and the Evolutionary Play*, meaning that the ecological situation—the condition of the environment and other species—is the scenery against which natural selection occurs, and natural selection results in a story of evolution during the history of life on Earth.[4]

7.4 BASIC CONCEPTS OF BIOLOGICAL DIVERSITY

To develop workable policies for the conservation of biological diversity, we must be clear about the meaning of this concept. It is often used loosely. As noted in the beginning of the chapter, sometimes conservation of biological diversity means simply the protection of a single endangered species, such as the African rhinoceros, the blue whale, or the spotted owl of the Pacific Northwest. At other times, the concern is with the conservation of the total number of species in a given area, regardless of what those species are. This is generally the case with the concern over the loss of tropical rain forests.

Biological diversity involves three concepts:

1. *Genetic diversity:* the total number of genetic characteristics, sometimes of a specific species, subspecies, or group of species.
2. *Habitat diversity:* the diversity of habitats in a given unit area.
3. *Species diversity,* which has three aspects:

   a. Species richness—the total number of species.
   b. Species evenness—the relative abundance of species.
   c. Species dominance—the most abundant species.

It is easy to confuse these related concepts, and sometimes people speak about conserving biological diversity without being sure which concept they have in mind. Whatever meaning is chosen, it can be said that people have greatly changed biological diversity, decreasing the number of species, increasing the rate of extinction of species, decreasing the genetic diversity of many species, decreasing the number of habitats, and greatly altering the geographic distribution of many species. In the balance of this chapter we will explore the factors that increase or decrease diversity. As we develop an understanding of how so many species came to be and why species become extinct, we will begin to understand what is involved in preserving this amazing diversity.

Species Diversity

As background to our study of species interaction, let us expand on the aspects of species diversity.

Species Richness, Species Evenness, and Dominance

Imagine two ecological communities, each with 10 species and 100 individuals (Figure 7.3). In the first community, 82 individuals belong to 1 species, and the remaining 9 species are represented by 2 individuals each. In the second community, all the species are equally abundant and each therefore has 10 individuals. Which community is more diverse? At first, one would think that they had the same species diversity. However, if you walked through both communities, the second would appear more diverse. If you walked through the first community, most of the time you would see individuals only of the predominant species, and there would be many species you probably would not see at all. In the second community, even a casual visitor would see many of the species in a short time. The first community would appear relatively low in diversity until it was subject to extremely careful study. You can test the differences in the probability of encountering a new species by laying a ruler down in any direction on Figures 7.3*a* and 7.3*b* and counting the number of species that it touches.

This example suggests that merely counting the number of species is not enough to describe biologi-

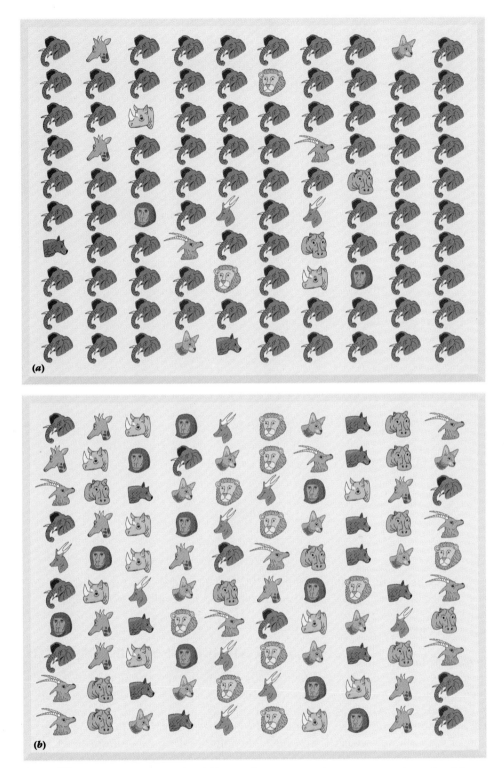

FIGURE 7.3 Diagram illustrating the difference between species evenness and species richness. Figures (*a*) and (*b*) have the same richness but different evenness. Lay a ruler across each diagram and count the number of species the edge crosses. Do this several times, then determine which figure has greater evenness. See text for explanation of results.

cal diversity. Diversity has to do with the relative chance of seeing species as much as it has to do with the actual number present. For a fixed number of species, a community in which each species has the same number of individuals as all others would appear to be the most diverse, whereas a community with one species making up most of the individuals would seem the least diverse. Ecologists divide these

TABLE 7.1 Comparison of Species Estimates

Kingdom/Phylum Subphylum/Class	Mader, 1997	Campbell, 1996	Raven, 1999	Margulis, 1988	E.O. Wilson, 1988
Virus					1000
Monera				100000	4760 (+Archea)
Archaebacteria			100		
Cyanobacteria			7500		
Fungi		100000	70000	100000	
Chytridiomycota			790		575
Zygomycota	600	600	1060	600	665
Ascomycota	30000	60000	32300	tens of thousands	28650
Basidiomycota	16000	25000	22300	25000	16000
Deuteromycota	25000			25000	
Protista		60000		thousands	
Ciliophora	8000		8000		
Myxomycota	560		700		500
Cryptophyta			200		
Dinophyta			2000–4000		
Oomycota	580		694		580
Euglenophyta	1000	800	900	800	800
Dinoflagellata	1000	1100		several thousand	1100
Sarcodina	40000				
Sporozoa	3600				
Chrysophyta	11000	850	1000		12500
Bacillariophyta		10000	100000	10000	
Phaeophyta	1500	1500	1500	1500	1500
Rhodophyta	4000	4000	4000–6000	4000	4000
Chlorophyta	7000	7000	17000	7000	7000
Plantae					
Bryophyta	12000	10000	9500	24000	16600
Hepatophyta	10000	6500	6000		
Anthocerophyta	100	100	100		
Psilophyta	several	10–13	several		9
Lycophyta	1000	1000	1000	1000	1275
Shenophyta		15	15	40	15
Filicinophyta				12000	10000
Peterophyta	12000	12000	11000		
Coniferophyta	550	550	550	550	
Cycadophyta	100	100	140	100	
Gneetophyta	70	70	70	70	
Anthophyta		235000	235000	230000	
Monocotyledones	65000				50000
Dicotyledones	170000				170000

aspects of species diversity, referring to the total number of species in an area as species richness, the relative abundance of species as species evenness, and the most abundant species as dominant.

Terms Related to Species Habitat

A species introduced into a new geographic area is called an **exotic species.** A species that is native to a particular area and not native elsewhere is called an **endemic species.** For example, Monterey pine is endemic to a portion of the California coast and exotic in New Zealand. A species with a broad distribution, occurring all over the world wherever the environment is appropriate, is called a **cosmopolitan species.** The moose is found in both North America and Europe and is therefore a cosmopolitan species of the northern boreal forests. The house mouse is cosmopolitan because it occurs many places where people provide a habitat for it. Species that are found almost anywhere are **ubiquitous species.** Humans are ubiquitous, as are some bacteria; for example, *Escherichia coli,* the common bacterium of human intestines, can probably be found almost anywhere on the land and in fresh waters.

7.5 THE NUMBER OF SPECIES ON EARTH

No one knows the exact number of species on Earth, because new species are being discovered all the time, especially in little-explored areas such as tropical rain forests. Estimations of total number of species inhabiting the earth range from the 1.4 million named to 100 million (see Table 7.1). It is common to think that there are two major kinds of life: animals and plants, but scientists categorize life into major groups based on evolutionary relationships—a biological genealogy. In the recent past, scientists classified life into five kingdoms: animals, plants, fungi, protists, and bacteria. New evidence from the fossil record and studies in molecular biology suggest that it may be more appropriate to describe life as existing in three major domains, one called Eukaryotes or Eucarya, which includes animals, plants,

TABLE 7.1 Comparison of Species Estimates (continued)

Kingdom/Phylum Subphylum/Class	Mader, 1997	Campbell, 1996	Raven, 1999	Margulis, 1988	E.O. Wilson, 1988
Animalia		>1,000,000			
Porifera	5000	9000		10000	5000
Cnidaria	9000	10000			9000
Platyhelminthes	13000	20000		15000	12200
Rotifera	2000	1800		2000	
Nematoda	12000	80000		80000	12000
Mollusca	110000	50000		110000	50000
Annelida	12000	15000		5400	12000
Arthropoda	>6 million	nearly 1,000,000		500000	989761
Cheliceriformes					
Uniramia					
Insecta					751000
Anoplura		2400			
Coleoptera		500000			
Dermaptera		1000			
Diptera		80000			
Hemiptera		55000			
Hymenoptera		90000			
Isoptera		2000			
Lepidoptera		140000			
Odonata		5000			
Orthoptera		30000			
Siphonaptera		1200			
Trichoptera		7000			
Crustacea		40000			
Brachiopoda		330		335	
Echinodermata	6550	7000		6000	6100
Chordata	45000			45000	
Vertebrata	44000				
Chondrichthyes	850	750			843
Osteichthyes	20000	30000			18150
Amphibia	3900	4000			4184
Reptilia	6000	7000			6300
Aves	9000	8600			9040
Mammalia	4500	4500			4000

Note: With ~15,000 new species described every year, a changing classification system, and no central data registry, it is not suprising that species estimates differ between publications and years.

Sources: E. O. Wilson, 1988; L. Margulis and K. V. Schwartz, 1988; Sylvia L. Mader, 1997; Neil A. Campbell, 1996; P. H. Raven, R. F. Evert, and S. E. Eichhorn, 1999.

fungi, and protists (mostly single-celled organisms); Bacteria; and Archaea.[32,33] Eucarya have a nucleus and other organelles; Bacteria and Archaea do not. Archaea used to be classified among Bacteria, but there are substantial molecular differences that suggest ancient divergence in heritage.

Insects and plants make up most of the species. Many of the insects are tropical beetles in rain forests. Mammals, the group to which people belong, are comprised of a comparatively small number of species, about 4000. As more explorations are carried out, especially in tropical areas, the number of identified invertebrates and plants will increase. Although there may be a few mammals yet undiscovered, it is not likely the number of known mammals will increase much in the future.

We have said that evolution is responsible for the development of species. But why are there so many? From our discussion of natural selection, we know that species compete with one another for resources. Competition would seem to result in a re-duction in the number of species. How then can so many different species survive? The next section looks at the interactions between species in an attempt to explain this apparent paradox.

7.6 INTERACTIONS BETWEEN SPECIES

There are three basic kinds of interactions between species: *competition,* in which the outcome is negative for both groups; *symbiosis,* which benefits both participants; and *predation-parasitism,* in which the outcome benefits one and is detrimental to the other. Scientists make more exact categories of interactions (see Table 7.2). Each type of interaction affects evolution, the persistence of species, and the overall diversity of life.

The histories of the purple loosestrife in America (discussed in the chapter opening case study) and the peppered moth in England (discussed in A Closer Look 7.1) show several of the ways species

TABLE 7.2 Types of Interactions Between Species

Interaction	Species A	Species B	Result
Neutralism	0	0	Neither affects the other.
Competition	–	–	A and B compete for the same resource; each has a negative effect on the other. Example: hummingbirds compete for the same flower nectar.
Inhibition	0	–	A inhibits B; A is not affected. Example: bacteria in a cow's rumen release a chemical that inhibits other bacteria.
Parasitism	+	–	A, the parasite or predator benefits. Example: heartworm in a reindeer.
Predation	+	–	A feeds on B, the host or prey, who suffers a direct negative effect. Example: wolves feeding on reindeer.
Symbiosis	+	+	A and B require each other to survive. Example: some bacteria and reindeer; hummingbirds and some flowers.
Commensalism	+	0	A requires B to survive; B is not affected significantly. Example: mistletoe on some trees.

Note: 0, no effect; +, positive effect on birth, growth, or survival; –, negative effect on birth, growth, or survival.

interact. The purple loosestrife competes for light with other plants. Some birds feed on its seeds and then spread the seeds in the mud stuck on their legs, resulting in a symbiotic relationship. The best way to control loosestrife appears to be with a natural insect parasite, a parasitic relationship.

In the case of the peppered moth, birds are predators of the moths, while the moths in turn compete among themselves to avoid predation by the birds. Birds, meanwhile, compete among themselves for their sources of food, including moths.

Dominance and Diversity: The Competitive Exclusion Principle

A consequence of natural selection is that when two species compete, the more fit species will win out and persist and the less fit species will lose or become extinct. This is known as the **competitive exclusion principle,** which states that *two species that have exactly the same requirements cannot coexist in exactly the same habitat.* Garrett Hardin expressed the idea most succinctly: "Complete competitors cannot coexist."[7]

Gray and Red Squirrels in Great Britain

These two species illustrate the competitive exclusion principle. Competition between species can result in one of the species becoming endangered or extinct. This is especially true when a new species is artificially introduced into a habitat. For example, the American gray squirrel was introduced into Great Britain because some people thought it was attractive. This was not an accidental introduction, but an intentional attempt. There were about a dozen attempts at this introduction, the first perhaps as early as 1830. By the 1920s, the American gray squirrel was well established in Great Britain, and its numbers expanded greatly in the 1940s and 1950s. Today the American gray squirrel is a problem; it competes with the red squirrel, which is native to Great Britain, and is winning the competition (Figure 7.4). As the American gray squirrel increased in numbers, the native red squirrel of Great Britain disappeared locally, losing in competition. At present the red squirrel population is approximately 160,000. Although red squirrels used to be found in deciduous woodlands throughout the lowlands of central and southern Britain, they now are common only in Cumbria, Northumberland, and Scotland. There are scattered populations in East Anglia, Wales, on the Isle of Wight, and in islands in Poole Harbor, Dorset.[8]

It is thought that one reason for the shift in the balance of these species is that the main source of food during winter for red squirrels is hazlenuts, whereas for gray squirrels it is acorns. Thus, red squirrels have a competitive advantage in areas with hazlenut, and gray squirrels have the advantage in oak forests. When gray squirrels were introduced, oaks were the dominant mature tree in Great Britain and about 40% of the timber planted was oaks.

If present trends continue, the red squirrel may disappear from the British mainland in the next 20 years. As in many cases, the introduced competitor threatens native species. Why this is often the case is useful for us to know; it will help us avoid causing the demise of native species in the future.

Why Hasn't the Competitive Exclusion Principle Limited Species Diversity?

According to the competitive exclusion principle, *complete competitors cannot coexist; one will always exclude the other.* Over a very long period of time, we might expect this to reduce the number of species over a large area. Taken to its logical ex-

(a)

(b)

FIGURE 7.4 The American gray squirrel (*Sciurus carolinensis*) (*a*) was introduced into Great Britain, where it is displacing the native European red squirrel (*Sciurus vulgaris*) (*b*), illustrating the competitive exclusion principle.

treme, we could imagine an Earth with very few species, perhaps one green plant on the land, one herbivore to eat it, one carnivore, and one decomposer. If we added 4 species for the ocean and 4 for fresh waters, we would have only 12 species on our planet.

Being a little more realistic, we could allow for the necessity of adaptations to major differences in climate and other environmental aspects. Perhaps we could specify 100 major climatic or environmental categories: cold and dry, cold and wet, warm and dry, warm and wet, and so forth. Even so, we would still expect that within each environmental category, competitive exclusion would result in a few species. Allowing 4 species per major environmental category would result in only 400 species. Yet about 1.5 million have been named and scientists speculate that many more millions may exist; the number is so great, we do not have even a good estimate of it. How can they all coexist?

Niches: How Species Coexist

Some classic studies of flour beetles (*Tribolium*) that live on wheat flour (Figure 7.5*a*) help explain how species with similar requirements can coexist. Flour beetles make good experimental subjects because they require only small containers of wheat flour to live. Populations of two species of beetles are placed in a small container with flour, and the number of individuals of each species is counted; the flour is passed through a sieve, separating the beetles, which are then identified by species

and put back into the container with the flour. Whenever two species are placed in a container that has uniform environmental conditions, one species of beetle always wins. Competitive exclusion takes place. However, it is not always the same species that survives. How do we explain this?

The answer is that the two beetle species have differential survival and reproduction under any set of environmental conditions. When it is warm and dry, one species wins: It has more offspring. When it is cool and moist, another species wins (Figure 7.5*b*). When conditions are in between there is some question about which species will survive, but one invariably persists whereas the second becomes extinct. In a mixed environment, however, both species of beetle may survive by using separate parts of the habitat: They do not coexist. Here, then, is the key to the answer about coexistence of many species. *Species that require the same resources can coexist by utilizing those resources under different environmental conditions.*[9] They are said to have different ecological niches.

Professions and Places: The Ecological Niche and the Habitat

Ecologist Charles Elton described the niche as a profession. He said that where a species lives is its **habitat,** but what it does for a living (its profession) is its **ecological niche.**[10] The general profession of a flour beetle is eating flour. A squirrel's profession is eating seeds of trees. Suppose you have a

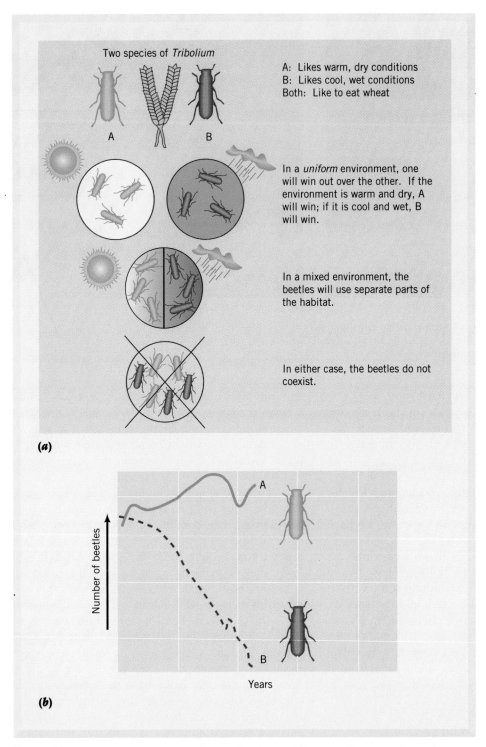

FIGURE 7.5 A classical experiment with flour beetles. (a) The general process illustrating competitive exclusion; (b) results of a specific, typical experiment under warm, dry conditions.

neighbor who is a bus driver. Where your neighbor lives and works is your town; what your neighbor does is drive a bus. Your neighbor's niche is bus driving; your neighbor's habitat is your hometown. Similarly, if someone says, "Here comes a wolf," you think not only of a creature that inhabits the northern forests (its habitat) but also of a predator that feeds on large mammals (its niche).

Understanding the niche of a species is useful in assessing the impact of development or of changes in land use. Will the change remove an essential requirement for some species' niche? A new highway that makes car travel easier might eliminate your neighbor's bus route and therefore eliminate his niche. In the same way, cutting a forest may drive away prey and eliminate the niche of the wolf.

Measuring Niches

When we think of the niche only in terms of a profession, it is difficult to quantify it: to determine exactly where it is and what its limits are for a given species. However, the niche can also be described as the set of all environmental conditions under which a species can persist. How can we measure the niche?

The idea of a measured niche is known as the *Hutchinsonian niche,* after G. E. Hutchinson, who first suggested it.[11] It is illustrated by the distribution of two species of flatworm, a tiny worm that lives on the bottom of freshwater streams. A study was made of two species of these small worms in Great Britain, where it was found that some streams contained one species, some the other, and still others both.[12] The stream waters are cold at their source in the mountains and become progressively warmer as they flow downstream. Each species of flatworm occurs within a specific range of water temperatures. In streams

where species A occurs alone, it is found from 6° to 17°C (42.8° to 62.6°F) (Figure 7.6*a*). Where species B occurs alone, it is found from 6° to 23°C (42.8° to 73.4°F) (Figure 7.6*b*). When they occur in the same stream, their temperature ranges are much reduced: Species A occurs in the upstream sections where the temperature ranges from 6° to 14°C (42.8° to 57.2°F), and B occurs in downstream areas where temperatures are warmer, from 14° to 23°C (57.2° to 73.4°F) (Figure 7.6*c*).

The range of temperature over which species A occurs when it has no competition from B is called its *fundamental temperature niche.* The set of conditions under which it persists in the presence of B is called its *realized temperature niche.* The flatworms show that species divide up resources along environmental gradients in nature. Of course, temperature is only one aspect of the environment. Flatworms have requirements in terms of the acidity of the water and other factors. We could create graphs for each of these factors, showing the range under which A and B occurred. The collection of all those graphs would constitute the complete description of the niche of each species.

The Importance of the Niche in Conservation

From the discussion of the competitive exclusion principle and the ecological niche, we learn something important about the conservation of species. If we are to conserve a species in its native habitat, we must make sure that all the requirements of its niche are present. Conservation of endangered species is more than a matter of placing many individuals of that species into an area; all the life requirements for that species must also be present.

Symbiosis

The discussion of biological evolution, which depends in part on competition between individuals, and the discussion of the competitive exclusion principle might leave the impression that species interact mainly by interfering with one another. Symbiosis is also important. The term derives from the Greek for "living together" and means a relationship between two organisms that is beneficial to both. Each partner in symbiosis is called a *symbiont.* Two species can benefit one another and enhance each other's chances of persisting.

Symbiosis is widespread and common; most animals and plants have symbiotic relationships with other species. Symbiosis therefore affects biological diversity. The widespread occurrence of symbiosis is illustrated by our own bodies. Microbiologists tell us that about 10% of a person's body weight is actually the weight of symbiotic microorganisms in the intestines. The resident bacteria help us in our diges-

FIGURE 7.6 The occurrence of freshwater flatworms in cold mountain streams in Great Britain. (*a*) The presence of species A in relation to temperature in streams where it occurs alone. (*b*) The presence of species B in relation to temperature in streams where it occurs alone. (*c*) The temperature range of both species in streams where they occur together. Inspect the three graphs; what is the effect of each species on the other?

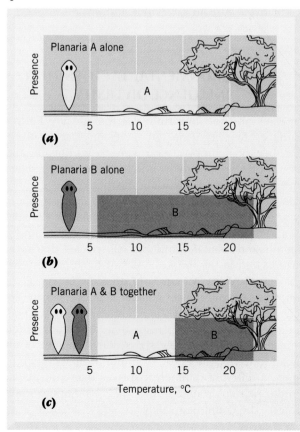

tion; we provide a habitat that supplies all their needs; both we and they benefit. We become aware of this intestinal community when it changes, for example, when we travel to a foreign country and ingest new strains of bacteria. Then we suffer the well-known traveler's malady: a gastrointestinal upset. As another example, our skin harbors a rich community of small organisms, some of which are beneficial to us.

Reindeer and Symbiotic Bacteria

Another important kind of symbiotic interaction occurs between certain mammals and bacteria. A reindeer on the northern tundra may appear to be alone but carries with it many companions. Like domestic cattle, the reindeer is a ruminant, with a four-chambered stomach (Figure 7.7) teeming with microbes (a billion per cubic centimeter). In this partially closed environment, the respiration of microorganisms uses up the oxygen ingested by the reindeer while eating. Other microorganisms digest

FIGURE 7.7 The stomach of a reindeer illustrates complex symbiotic relationships. For example, in the rumen, bacteria digest woody tissue the reindeer could not otherwise digest. The result is food for the reindeer and food and a home for the bacteria, which could not survive in the local environment outside.

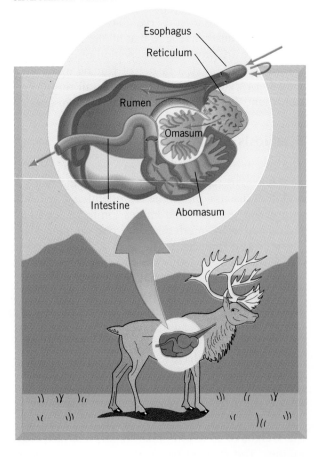

cellulose, take nitrogen from the air in the stomach, and make proteins. Still others, that is, those species that digest the parts of the vegetation that the reindeer cannot digest itself (in particular, the cellulose and lignins of cell walls in woody tissue), require a peculiar environment. They can survive only in an environment without oxygen. One of the few places on Earth's surface where such an environment exists is the inside of a ruminant's stomach.[14] The bacteria and the reindeer are symbionts, each providing what the other needs.

Thus we see that if we want to save a species from extinction, we must save not only its niche but also its symbionts. The attempt to save a single species almost invariably leads us to conserve a group of species, not just a particular habitat.

Predation

Predation-parasitism is the third way in which species interact. Predation can increase diversity of prey species by reducing the abundance of the dominant prey. In this way, predators can contribute to species evenness by keeping the dominant species from overwhelming others via competitive exclusion. For example, some studies have shown that a moderately grazed pasture has more species of plants than an ungrazed one. The same seems to be true about natural grasslands and savannas. Without grazers and browsers, there might be fewer species of plants in African grasslands and savannas.

7.7 ENVIRONMENTAL FACTORS THAT INFLUENCE DIVERSITY

Species are not uniformly distributed over Earth's surface; diversity varies greatly from place to place. For instance, suppose you were to go outside and count all the species in a field or any open space near where you are reading this book. The number of species would differ greatly depending on where you are. If you live in northern Alaska or Canada, Scandinavia, or Siberia, you would probably find a significantly smaller number of species than you would find in the tropical areas of Brazil, Indonesia, or central Africa. Variation in diversity is partially a question of latitude—in general, higher diversity occurs at lower latitudes. There are also many patterns of diversity within a local area. If you counted species in the relatively sparse environment of an abandoned city lot, for example, you would find quite a different number of species than you would in the rich environment of an old, long-undisturbed forest. The count in a valley would be different from

A **CLOSER LOOK 7.2**

BIOGEOGRAPHY, GLACIATION, AND PEOPLE

There are fewer native species of trees in Europe and Great Britain than in all other temperate regions of the world. Only 30 tree species are native to Great Britain (i.e., present on Great Britain prior to human settlement), but hundreds of species grow there today. Why are there so few native species in Europe and Great Britain? The answer is the combined effects of climatic change and the geography of mountain ranges. In Europe, major mountain ranges run east–west, whereas in North America and Asia the major ranges run north–south. During the past 2 million years Earth has experienced several episodes of continental glaciation, when glaciers several kilometers thick expanded from the arctic over the landscape. At the same time, glaciers formed in the mountains and expanded downward. Trees in Europe were caught between the ice from the north and the ice from the mountains and had few refuges; many species became extinct. In contrast, in North America and Asia, trees' seeds could spread southward as the ice advanced, where they became established and produced new seeds. Thus the tree species "migrated" southward and survived each episode or glaciation.[36]

Since the rise of modern civilization, there have been many practical effects of these ancient events. Starting in the sixteenth century, soon after the discovery of North America by Europeans, considerable effort was expended in bringing exotic species of trees and shrubs to Europe and Great Britain. These were used to decorate gardens, homesites, and parks and formed the basis of much of the commercial forestry in the region. For example, in the famous gardens of Alhambra in Granada, Spain, hedges that are grown and cut in elaborate shapes consist of Monterey cypress from North America. Douglas-fir and Monterey pine are important commercial timber trees in Great Britain and Europe. Knowledge of biogeography—being able to predict what will grow where based on climatic similarity—has been used for major economic benefits.

---

the count high up the side of a mountain. Let us look more closely at some of the environmental factors that influence diversity (see A Closer Look 7.2).

On a local scale on the land, the kinds of species and ecosystems that occur change with soil type and topographic characteristics: slope, aspect (the direction the slope faces), elevation, and relation to a drainage basin. These factors influence the number and types of plants; the plants, in turn, influence the number and types of animals. Some of the possible interrelationships are illustrated in Figure 7.8.[15]

Change in the relative abundance of a species over an area or a distance is referred to as an *ecological gradient*. In mountainous areas, changes in elevation lead to the same biogeographic changes that occur with changes in latitude. Figure 7.9 illustrates some of these changes for the Grand Canyon and the nearby San Francisco Mountains of Arizona. Although such patterns are most easily seen in vegetation, they occur for all organisms. For example, the pattern of distribution of representative African mammals on Mount Kilimanjaro is shown in Figure 7.10.

If environmental conditions were constant over time and space, the existing dominant species would become more abundant; they would tend to increase at the expense of rarer species, following the competitive exclusion principle. But a habitat that includes a variety of local environments is likely to offer a refuge to rarer species, thus leading to greater diversity. The more diverse habitat allows more niches, and more species persist.

Some habitats harbor little diversity because they are so severe or stressful to life. A comparison of vegetation in two areas of Africa illustrates the point. In eastern and southern Africa, the well-drained sandy soils have a diverse vegetation, including many species of *Acacia* and *Combretum* trees as well as many grasses. The well-drained soil is not stressful to plants. In contrast, on very heavy clay soils of wet areas near rivers, such as occur along the Sengwa River in Zimbabwe, the woodlands are composed almost exclusively of a single species called *mopane*. In this case, the very heavy clay soils impose a stress on plants; these soils hold water and allow little oxygen to reach plant roots. There, trees are restricted to those species with very shallow roots.

Moderate environmental disturbance can also increase diversity. For example, fire is a common disturbance in many forests and grasslands. Occasional light fires produce a mosaic of recently burned and unburned areas; these patches favor dif-

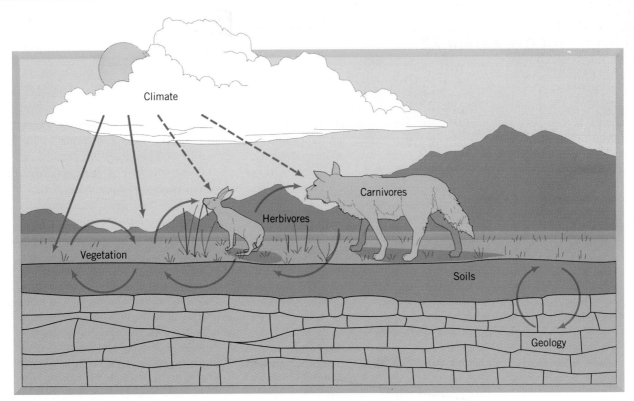

FIGURE 7.8 Interrelationships among climate, geology, soil, vegetation, and animals. What lives where depends on many factors. Climate, geologic features (bedrock type, topographic features), and soils influence vegetation. Vegetation in turn influences soils and the kinds of animals that will be present. Animals affect the vegetation. Arrows represent a causal relationship; the direction is from cause to effect. A dashed arrow indicates a relatively weak influence, and a solid arrow a relatively strong influence.

FIGURE 7.9 Changes in elevation lead to biogeographic changes similar to those that occur with latitude. As elevation above sea level increases, the climate becomes colder and wetter, creating a pattern in the distribution of life up a mountain that parallels changes that occur from equator to pole. The altitudinal zones of vegetation in the Grand Canyon of Arizona and in the nearby San Francisco Mountains are shown. (*Source:* From C. B. Hunt, 1974, *Natural Regions of the United States and Canada,* W. H. Freeman. Copyright 1974 by W. H. Freeman.)

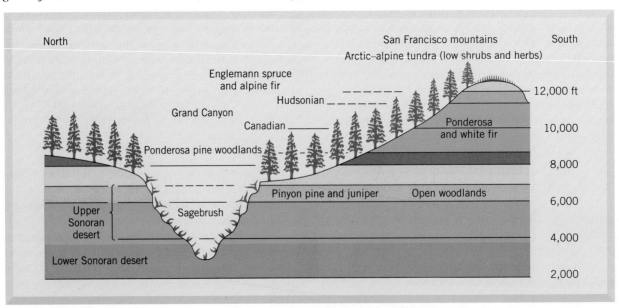

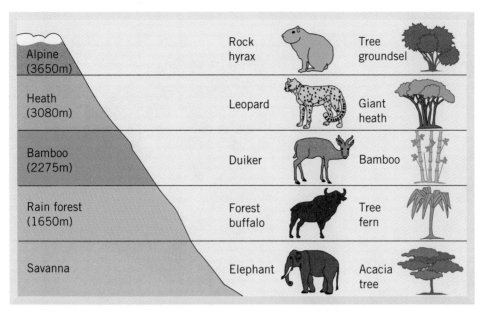

FIGURE 7.10 Changes in the distribution of animals with elevation on a typical mountain in Kenya. (*Source:* From C. B. Cox, I. N. Healey, and P. D. Moore, 1973, *Biogeography,* Halsted, New York.)

ferent kinds of species and increase the overall diversity. A variety of factors that increase and decrease diversity are shown in Table 7.3.

Of course, people also affect diversity. In general, urbanization, industrialization, and agriculture decrease diversity, reducing the number of habitats, and simplifying habitats. In addition, we intentionally favor specific species, manipulating populations for our own purposes, such as when a person plants a lawn or when a farmer plants a single crop over a large area.

In the past, people rarely thought about cities as having any beneficial effects on biological diversity. Indeed, the development of cities tends to reduce biological diversity, especially because cities are located at good sites for travel, such as along rivers or near oceans, where biological diversity is often high. However, in recent years there has been a growing realization that cities can contribute in important ways to the conservation of biological diversity. These topics are discussed in greater detail in Chapter 26.

TABLE 7.3 Some Major Factors that Increase and Decrease Biological Diversity

A. Factors that tend to increase diversity

1. A physically diverse habitat
2. Moderate amounts of disturbance (such as fire or storm in a forest or a sudden flow of water [from a storm] into a pond)
3. A small variation in environmental conditions (temperature, precipitation, nutrient supply, etc.)
4. High diversity at one trophic level, increasing the diversity at another trophic level (many kinds of trees provide habitats for many kinds of birds and insects)
5. An environment highly modified by life (for example, a rich organic soil)
6. Middle stages of succession
7. Evolution

B. Factors that tend to decrease diversity

1. Environmental stress
2. Extreme environments (conditions near to the limit of what living things can withstand)
3. A severe limitation in the supply of an essential resource
4. Extreme amounts of disturbance
5. Recent introduction of exotic species (species from other areas)
6. Geographic isolation (being on a real or ecological island)

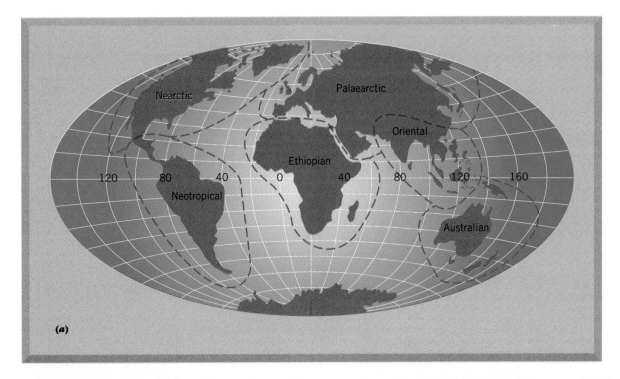

(b)

(c)

(d)

FIGURE 7.11 (*a*) The main biographic realms for animals are based on genetic factors. Within each realm the vertebrates are in general more closely related to each other than to those in other realms. Here herbivores from three realms are shown: (*b*) the eastern grey kangaroo from Australia; (*c*) the giraffe from Kenya, Africa; and (*d*) the agouti from South America.

of animals in the early Paleozoic era (beginning about 500 million years ago) was much lower than in the late Mesozoic (beginning about 130 million years ago) and the Cenozoic (beginning about 75 million years ago).[16]

The number of species per unit area varies over time as well as space. The changes occur over many time periods, from the very short (a year), to moderate (years, decades), to even longer (centuries or more in a forest), and finally to very long geologic periods. As an example of the latter, diversity

7.8 THE GEOGRAPHY OF LIFE

The kinds of species as well as the number vary greatly from place to place on Earth. People have long been fascinated by the great geographic variability in species. If we are interested in conserving biological diversity, it is important that we understand these large-scale, global patterns, which we refer to as patterns in **biogeography**.

Major global patterns in the distribution of species have long been recognized. Aristotle recorded some of the first principles of biogeography; he distinguished boreal, temperate, and tropical life zones. During the nineteenth century, global patterns became a subject of scientific study, partly because of explorations of the New World tropics, South America, the South Pacific, Australia, and parts of Asia.

Wallace's Realms: Biotic Provinces

In 1876, A. R. Wallace suggested that the world could be divided into six biogeographic regions on the basis of fundamental features of the animals found in those areas.[17] Wallace referred to these regions as realms and named them Nearctic (North America), Neotropical (Central and South America), Palaearctic (Europe, northern Asia, and northern Africa), Ethiopian (central and southern Africa), Oriental (the Indian subcontinent and Malaysia), and Australian. These have become known as *Wallace's realms* (Figure 7.11).

All living organisms are classified into groups called taxa, normally based on evolutionary relationships or similarity of characters. The hierarchy of these groups (from largest and most inclusive to smallest and least inclusive) is domain or kingdom, phylum (in animals) or division (in plants), class, order, family, genus, and species. An order may contain several families and a family may contain a few to many genera. Genera may contain one to many species.

In each geographic realm certain families or orders of animals are dominant. Animals filling the same ecological niches in each realm are of different genetic stock from those in the other realms. For example, bison and pronghorn antelope are among the large mammalian herbivores in North America; rodents such as the agouti fill those niches in South America, kangaroos fill them in Australia, and in central and southern Africa many species, including giraffes and antelopes, fill these niches.

The basic concept of Wallace's realms is still considered valid and has been extended to all life-forms,[18] including plants (Figure 7.12)[19,20] and invertebrates. These regions are now referred to as biotic provinces.[21] A **biotic province** is a *region inhabited by a characteristic set of taxa (species, families, orders), bound by barriers that prevent the spread of the distinctive kinds of life to other regions and the immigration of foreign species.*[22] In a biotic province, organisms share a common genetic heritage but may live in a variety of environments as long as they are genetically isolated from other regions.

Wallace did not have the benefit of our modern understanding of geologic processes to explain

FIGURE 7.12 The major vegetation realms are also based on genetic factors. Flowering plants within a realm are more closely related to each other than they are to flowering plants of other realms.

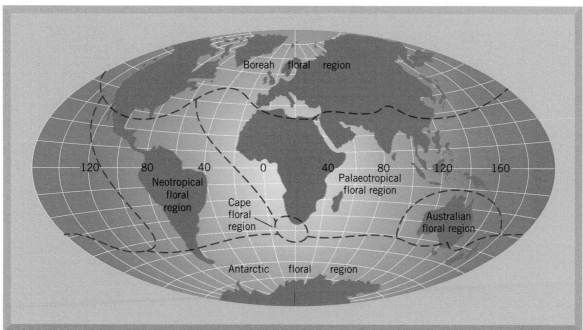

A CLOSER LOOK 7.3

A BIOGEOGRAPHICAL CROSS SECTION OF NORTH AMERICA

A generalized cross section of the United States shows the relationships among weather patterns, topography, and biota (Figure 7.13). Off the West Coast of the United States in the Pacific Basin occur the pelagic ecosystems: euphotic zones, which have sufficient light for photosynthesis and are occupied by small, mainly single-cell algae, and other zones, with too little light for photosynthesis, occupied by animals that depend for food on dead organisms that sink from above. Near the shore, particularly in areas of upwelling such as long the California coast, there is an abundance of algae, fish, birds, shellfish, and marine mammals. Where the tides and waves alternately cover and uncover the shore, a long, thin line of intertidal ecosystems is found, dominated by kelp and other large algae that are attached to the ocean bottom; by shellfish, such as mussels, barnacles, and abalone; by crabs and other invertebrates; and by shorebirds, such as sandpipers.

Weather systems move generally from west to east in the Northern Hemisphere. As air masses are forced over the coastal and Rocky Mountains, they are cooled, and the moisture in them condenses to form clouds and rain. The West Coast is an area of moderate temperature, because water has a high capacity to store heat and the Pacific Ocean modulates the land's temperature. Annual amounts of rainfall increase with elevation on the western slopes of the mountains. In the south, moving east from the Southern California coast, rainfall remains low until the mountains force the air high enough to condense much of its moisture. In general, the colder, wetter heights of the mountains support coniferous forests.

In the north, along the coasts of Washington and Oregon, cool temperatures year round lead to heavy rains near the shore, producing an unusual temperate-climate rain forest. The most well-known example occurs in the Olympia National Forest on the northwestern edge of the state of Washington.

The eastern slopes of the coastal ranges form a rain shadow. First, the air that passes over these eastern slopes gives up most of its moisture to the mountains; as a result, it is dry as it passes to the east. In addition, the air sinks to lower elevations, is warmed, and can hold more moisture. This dry air tends to take up moisture from the ground, producing the deserts of Utah, California, Arizona, and New Mexico. Whereas annual rainfall in the Olympic Peninsula of Washington reaches 375 cm/year (150 in./year), east of the Cascade Mountains it falls to 20 cm/year (8 in./year).

The same effect occurs in the Rockies. Less than 160 km (100 miles) west of Denver, in the Rocky Mountains, the annual rainfall is 100 cm (40 in.). In the Great Plains, 160 km east of Denver, the annual rainfall is only 30 to 40 cm (12 to 16 in.). Average annual rainfall increases steadily eastward: 50 cm (20 in.) at Dodge City, Kansas; 70 cm (27.5 in.) near Lincoln, Nebraska; and 90 cm (35.5 in.) near Kansas City, Missouri.[34]

The biomes reflect these changes in rainfall. Just east of Denver are

FIGURE 7.13 Generalized cross section of North America showing weather, landforms, and the geography of life. The weather patterns move from west to east.

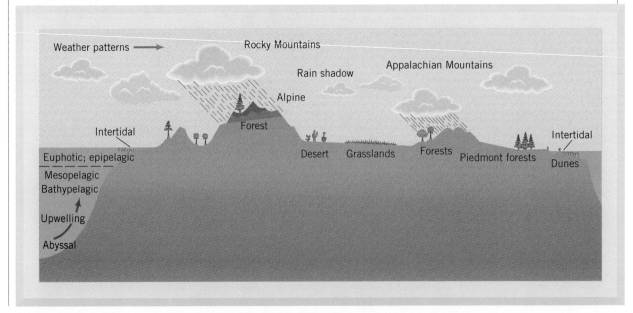

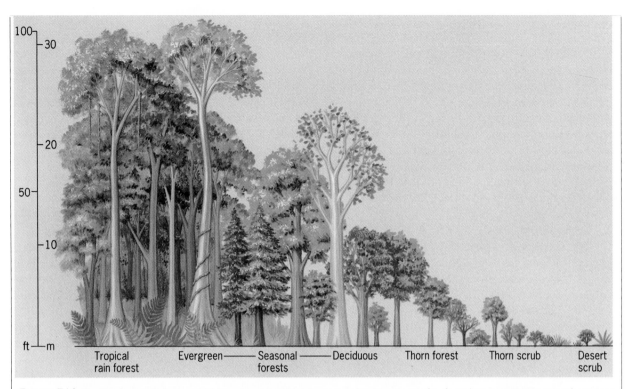

FIGURE 7.14 Environmental stress and biogeography. Certain general patterns can be found as an environment becomes more stressful. In this diagram, the effect of increasing water stress is shown. Where rainfall is plentiful, vegetation is abundant, and there are forests of tall trees of many species. As rainfall lessens, the size of the plants decreases to small trees, then shrubs and grasses, then scattered plants. The total biomass decreases, and, in general, the number of species decreases. Similar changes accompany increases in other kinds of stress, including the stress of certain pollutants. (*Source:* Adapted from Whittaker, R. H. 1975. *Communities and Ecosystems,* 2nd ed., New York: Macmillan.)

short-grass prairies, which become mixed-grass prairies (a mixture of short-grass and tall-grass prairies) and then tall-grass prairies as we continue eastward. Rainfall reaches levels sufficient to support forests farther east, near the South Dakota–Minnesota border in the north, where the annual rainfall reaches 50 to 64 cm (20 to 25 in.). From there to the East Coast the eastern deciduous forest (dominated by trees that lose their leaves during the winter) and the boreal forest of eastern North America predominate.

Sometimes these ecological borders, such as those between the short-grass and tall-grass prairies, are said to be subtle; but they were quite striking and visible to some of the early travelers in the West. One such traveler, Josiah Gregg, wrote in his journal in 1831 that to the west of Council Grove, Missouri, at the border of the tall-grass and short-grass

prairies, the "vegetation of every kind is more stinted—the gay flowers more scarce, and the scant timber of a very inferior quality," whereas to the east he found the prairies to have "a fine and productive appearance truly rich and beautiful."[35]

The patterns described for the United States occur worldwide. Changes with elevation from warm, dry-adapted woodlands to moist, cool-adapted woodlands are found in Spain, where beech and birch, characteristic of middle and northern Europe (Germany, Scandinavia), are found at high elevations and alpine tundra is found at the summits. Similar patterns are found in Venezuela, where a change in elevation from sea level to 5000 m (16,404 ft) at the summits of the Andes is equivalent to a latitudinal change from the Amazon basin to the southern tip of the South American continent. The seasonality of rainfall as well as the total amount

often determines which ecosystems occur in an area.

Two other general concepts of biogeography, illustrated by both the latitudinal patterns from the arctic to the tropics and the altitudinal patterns from mountaintops to valleys, are (1) *the number of species declines as the environment becomes more stressful* and (2) *on land the stature of vegetation decreases as the environment becomes more stressful* (Figure 7.14). These concepts apply to most stresses, including those that human beings impose by adding pollutants to the environment, decreasing the fertility of soils or otherwise impoverishing habitats, and increasing the rate of environmental disturbance. From these concepts we can predict that highly polluted and disturbed landscapes and seascapes will have few species and that, on the land, the dominant species of plants will have small stature.

how distinct biological groups could have evolved in major habitat regions. The modern explanation for the origin of Wallace's realms is that *continental drift*, caused by plate tectonics, has resulted in the periodic unification and separation of the continents (see the discussion in Chapter 4).[23,24] Unification of continents allowed genetic mixing, whereas separation imposed geographic isolation. Continental unification and land bridges spread genetic stock, which provided organisms with the potential to exploit new habitats. Continental separation led to genetic isolation and the evolution of new species.

Regional Patterns

The concept of continental drift leaves us with a picture of huge land masses moving ponderously over Earth's surface, periodically isolating and remixing groups of organisms and leading to an increase in the diversity of species. If each continent were a uniform plot of land in a uniform climate, then biological diversity within a continent would be low. Important patterns of distribution occur regionally and across continents. Proximity to oceans or other large bodies of water, ocean currents near shore, position relative to mountain ranges, latitude, and longitude all play an important part in determining climate. For an illustration of climates that occur from west to east across North America see A Closer Look 7.3, "A Biogeographical Cross Section of North America."

Biomes

Another way of looking at the biogeography of Earth is in terms of similarity of environments. Similar environments lead to the evolution of organisms similar in form and function (but not necessarily in genetic heritage or internal makeup) and to similar ecosystems. This is known as *the rule of climatic similarity* and leads to the concept of the biome. A **biome** is *a kind of ecosystem*, such as a desert, tropical rain forest, or grassland. It is common and con-

FIGURE 7.15 A pattern of types of vegetation in relation to humidity and temperature. Boundaries between types are approximate. For example, tropical rain forests occur in areas with an approximate annual rainfall of 250 to 450 cm and an approximate temperature range of 18° to 20°C. Note that deserts occur over a wide range of annual temperatures, from 30° to 25°C, as long as rainfall is less than 50 cm/year. The warmer the climate, the more rainfall is required to move from desert to another biome. (*Source:* Adapted from Whittaker, R.H. 1975. *Communities and Ecosystems,* 2nd ed., New York: Macmillan.)

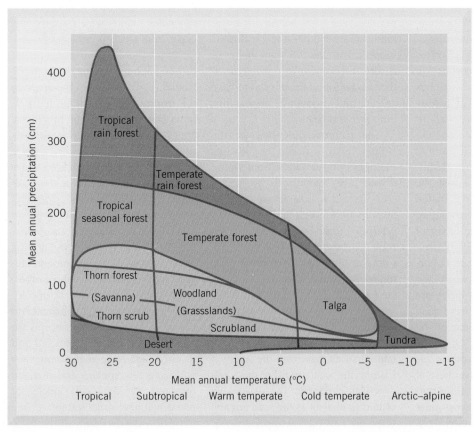

venient to divide the biosphere into biomes; parts of a biome can be completely isolated geographically from other parts.

The strong relationship between climate and life suggests that if we know the climate of an area, we can predict what biome will be found there, what its approximate biomass (amount of living matter) and production will be, and what the dominant kinds of organisms will be if the environment remains constant.[25,26] The general relationship between biome type and the two most important climatic factors—rainfall and temperature—is diagramed in Fig-

ure 7.15. (See also the special section on biomes that follows Chapter 9.)

Evolution and Diversity in Biomes

We can look at deserts or grasslands that are widely separated (in different biotic provinces) to examine the idea that similar environmental conditions eventually lead to the evolution of species with similar adaptations and to similar ecosystems. For example, certain plants of the American and African deserts may look similar but are not closely related, belonging to different biological families (Figure 7.16). Although geographically isolated for 180 million years, they have been subjected to similar climates, which imposed similar stresses and opened up similar ecological opportunities. The plants evolved to adapt to these stresses and potentials and have come to look alike and prevail in like habitats, a process known as **convergent evolution.** The

FIGURE 7.16 Given sufficient time and similar climates in two different areas, species with similar shape and form will tend to occur. The Joshua tree (*a*) and saguaro cactus (*b*) of North America look like the giant euphorbia (*c*) of East Africa. All three are tall, have green succulent stems that replace the leaves as the major sites of photosynthesis, and have spiny projections, but they are not closely related. The Joshua tree is a member of the agave family, the saguaro is a member of the cactus family, and the euphorbia is a member of the spurge family. Their similar shapes are a result of evolution under similar desert climates, a process known as a convergent evolution.

(a)

(b)

(c)

(a)

(b)

(c)

Figure 7.17 Divergent evolution. These three large flightless birds evolved from a common ancestor but are now in widely separated regions: (*a*) the ostrich in Africa, (*b*) the rhea in South America, and (*c*) the emu in Australia.

changing) organs and provide the best clues as to the genetic history of a species.

People make use of convergent evolution when they move decorative and useful plants around the world.

Cities around the world that lie in similar climates now share many of the same decorative plants. Bougainvillea, a spectacularly bright flowering shrub originally native to Southeast Asia, decorates cities as distant as Los Angeles and the capital of Zimbabwe. In New York City and its outlying suburbs, one can find the Norway maple from Europe, the tree of heaven and the gingko tree from China, and such native species as sweet gum, sugar maple, and pin oak.

Another important concept is divergent evolution. In **divergent evolution,** a population is separated, usually by geographic barriers; the separated subpopulations evolve separately but retain some common characteristics. It is now believed that the ostrich, the rhea, and the emu have a common ancestor but evolved separately (Figure 7.17). In open savannas and grasslands, a large bird that can run quickly but feed efficiently on small seeds and insects has certain advantages over other organisms seeking the same food. Thus, these species maintained the same characteristics in widely separated areas.

ancestral differences between these look-alike plants can be found in their flowers, fruits, and seeds, which are evolutionarily the most conservative (least

Island Biogeography

Islands have a special fascination for naturalists. Darwin's visit to the Galapagos Islands gave him his most powerful insight into biological evolution.[27] There he found many species of finches that were related to a single species found elsewhere. On the Galapagos, each species was adapted to a different niche.[28] Darwin suggested that finches, isolated from other species on the continents, eventually separated into a number of groups, each adapted to a more specialized role. This process is called **adaptive radiation.** On the Hawaiian Islands, a finchlike ancestor evolved into several species, including fruit and seed eaters, insect eaters, and nectar eaters, each with a beak adapted for its specific food (Figure 7.18).[29] We can make some generalizations about species diversity on islands. First, islands have fewer species than continents. Additionally, the smaller the island, the fewer the species, as can be seen in the number of reptiles and amphibians in various West Indian islands. And the farther the island is from the mainland, the fewer are the species (Figure 7.19).[30]

Why is this so? First, small islands tend to have fewer habitat types. Second, the two sources of new species on an island are migration from the mainland and evolution of new species in place (as with Darwin's finches on the Galapagos). It makes sense that there will be fewer migrants if an island is far from the mainland. A small island is also less likely than a large island to be found by migrating organisms. Also, every species is subject to risk of extinction; extinctions can be caused by predation, disease (parasitism), competition, climatic change, or habitat alteration. Generally, the smaller the population, the greater its risk of extinction. And the smaller the island, the smaller the population of a particular species that can be supported.

FIGURE 7.18 Evolutionary divergence among honeycreepers in Hawaii. Sixteen species of birds, eight of which are shown here, each with a beak specialized for its food, evolved from a single ancestor to fit ecological niches that, on the North Amercan continent, were previously filled by other species not closely related to the ancestor. (*Source:* From C. B. Cox, I. N. Healey, and P. D. Moore, 1973, *Biogeography,* Halsted, New York.)

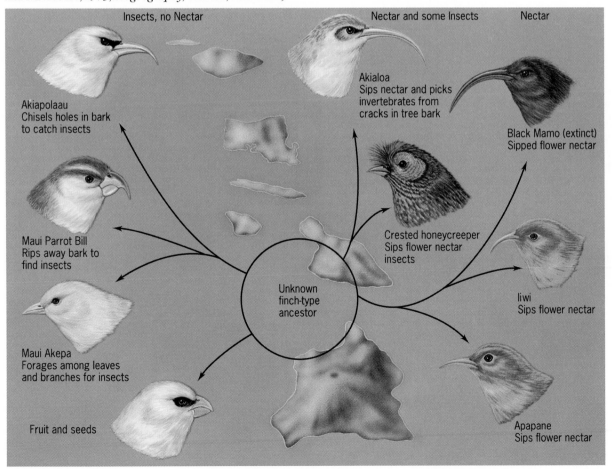

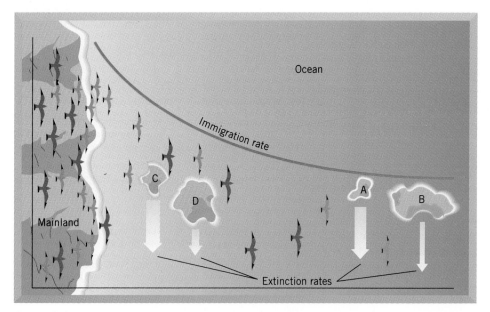

FIGURE 7.19 Idealized relation of an island's size, its distance from the mainland, and its number of species. The nearer an island is to the mainland, the more likely it is to be found by an individual, and thus the higher the rate of immigration. The larger the island, the larger the population it can support, and the higher the chance of persistence of a species; small islands have a higher rate of extinction. The average number of species, therefore, depends on the rate of immigration and the rate of extinction. Thus, a small island near the mainland may have the same number of species as a large island far from the mainland. The thickness of the arrow represents the magnitude of the rate. (*Source:* Modified from R. H. MacArthur and E. O. Wilson, 1967, *The Theory of Island Biogeography,* Princeton University Press, Princeton, N.J.)

Thus, over a long time, an island tends to maintain a rather constant number of species, which is the result of the rate at which species are added minus the rate at which they become extinct. These numbers follow the curves shown in Figure 7.20. For any island, the number of species of a particular life-form (birds, mammals, herbivorous insects, or trees) can be predicted from the island's size and distance from the mainland.

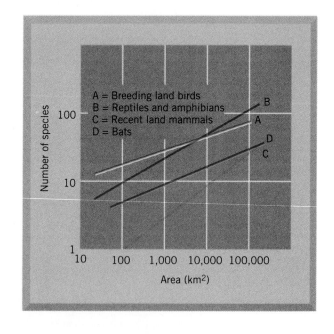

FIGURE 7.20 Islands have fewer species than do mainlands; the larger the island, the greater the number of species. This general rule is shown by a graph of the number of species of birds, reptiles and amphibians, land mammals, and bats for islands in the Caribbean. (*Source:* Modified from Wilcox.)

Why Preserve Biodiversity?

Preserving a diversity of life on Earth has come to be an accepted goal for many people. But when that goal comes into conflict with other goals, such as economic development, the question becomes, "How much diversity and at what cost?" To find the answer it is important to think carefully about the values of biological diversity and to separate those based on science from those based on other present or potential values, such as aesthetic, ethical, religious, or economic values.

In 1980, the International Union for Conservation of Nature and Natural Resources (IUCN) proposed a statement to form a basis for conserving biological diversity. That statement summarizes most of the arguments used as a rationale for conserving diversity. Read the statements reproduced below and answer the questions that follow. (Each sentence, or group of sentences, has been given a number for convenience in referring to it.)

An Ethical Basis for Preserving Biodiversity

1. The world is an interdependent whole made up of natural and human communities. The well-being and health of any one part depends on the well-being and health of the other parts.

2. Humanity is part of nature, and humans are subject to the same immutable ecological laws as are all other species on the planet.

3. All life depends on the uninterrupted functioning of natural systems that ensure the supply of energy and nutrients, so ecological responsibility among all people is necessary for the survival, security, equity, and dignity of the world's communities.

4. Human culture must be built on a profound respect for nature, a sense of being at one with nature, and a recognition that human affairs must proceed in harmony and in balance with nature.

5. The ecological limits within which we must work are not limits to human endeavor; instead, they give direction and guidance as to how human affairs can sustain environmental stability and diversity.

6. All species have an inherent right to exist. The ecological processes that support the integrity of the biosphere and its diverse species, landscapes, and habitats are to be maintained. Similarly, the full range of human cultural adaptations to local environments is to be enabled to prosper.

7. Sustainability is the basic principle of all social and economic development.

8. Personal and social values should be chosen to accentuate the richness of flora, fauna, and human experience. This moral foundation will enable the many utilitarian values of nature—for food, health, science, technology, industry, and recreation—to be equitably distributed and sustained for future generations.

9. The well-being of future generations is a social responsibility of the present generation. Therefore, the present generation should limit its consumption of nonrenewable resources to the level that is necessary to meet the basic needs of society and ensure that renewable resources are nurtured for their sustainable productivity.

10. All persons must be empowered to exercise responsibility for their own lives and for the life of Earth. They must therefore have full access to educational opportunities, political enfranchisement, and sustaining livelihoods.

11. Diversity in ethical and cultural outlooks toward nature and human life is to be encouraged by promoting relationships that respect and enhance the diversity of life, irrespective of the political, economic, or religious ideology in a society.

Critical Thinking Questions

1. Which of the statements are scientific, that is, are based on repeatable observations, can be tested, and are supported by evidence?

2. Using the table that follows, which statement from the argument for preserving biodiversity is implied by the type of value given? Write the number of the statement in the right column. A statement may fit into more than category.

Type of Value	Source of Value of Living Organisms	Statement
Ethical	The fact that they are alive	
Aesthetic	Their beauty and the rewards humans derive from their beauty	
Economic	The direct and indirect ways in which they benefit humans	
Ecological	Their contributions to the health of the ecosystem	
Intellectual	What they can contribute to knowledge	
Emotive	The sense of awe and wonder they inspire in humans	
Religious	Having been created by a supernatural being or force	
Recreational	Sport, tourism, and other recreations	

3. What is the rank of the types of values given?

4. Which type of value(s) could be used to support preservation of smallpox viruses? All 1200 (or more) species of beetles in the tropical rain forest? The northern spotted owl? Dolphins? Sharks?

Reference

McNeely, J. A., Miller, K. R., Reid, W. V., Mittermeier, R. A., and Werner, T. B. 1990. *Conserving the World's Biological Diversity*. Gland, Switzerland, and Washington, D.C.: International Union for Conservation of Nature and Natural Resources, World Resources Institute, Conservation International, World Wildlife Fund—U.S., World Bank.

SUMMARY

- Biological evolution—the change in inherited characteristics of a population from generation to generation—is responsible for the development of the many species of life on Earth. Four processes that lead to evolution are mutation, natural selection, migration, and genetic drift.

- Biological diversity involves three different concepts: genetic diversity (the total number of genetic characteristics), habitat diversity (the diversity of habitats in a given unit area), and species diversity. Species diversity, in turn, involves three ideas: species richness (the total number of species), species evenness (the relative abundance of species), and species dominance (the most abundant species).

- There are about 1.5 million species identified and named. Insects and plants make up most of the species. As more explorations are carried out, especially in tropical areas, the number of identified species, especially of invertebrates and plants, will increase.

- There are three basic kinds of interactions between species: competition, in which the outcome is negative for both groups; symbiosis, which benefits both groups; and predation–parasitism, in which the outcome benefits one and is a disadvantage for the other. Each type of interaction affects evolution, the persistence of species, and the overall diversity of life. It is important to understand that organisms have evolved together so that predator, parasite, prey, competitor, and symbiont have adjusted to one another. Human interventions frequently upset these adjustments.

- The competitive exclusion principle states that two species that have exactly the same requirements cannot coexist in exactly the same habitat; one must win. The reason that most species do not die out from competition is that they have developed a particular niche and thus avoid competition. It is also likely that the competitive exclusion principle does not work when there are several competing species rather than just two.

- The number of species in a given habitat is affected by many factors, including latitude, elevation, topography, the severity of the environment, and the diversity of the habitat. Predation and moderate disturbances, such as fire, can actually increase the diversity of species. The number of species also varies over time. Of course, people also affect diversity.

- Geographic isolation leads to the evolution of new species. Wallace's realms, or biotic provinces, are major geographic divisions (generally, continents) based on fundamental features of the animals found in them. Animals filling the same niches within each realm are of different stock from those in other realms.

- The rule of climatic similarity holds that similar environments lead to the evolution of biota and biological communities similar in external form and function but not in genetic heritage or internal makeup. These areas of climatic similarity with similar biota are known as biomes. A biome is a kind of ecosystem; examples are desert, grasslands, and rain forest.

- Convergent evolution occurs when two genetically dissimilar species that occur in geographically separate parts of a biome develop along similar lines and have similar external form and function. Divergent evolution occurs when several species evolve from a common ancestral species but develop separately because of geographic isolation.

REEXAMINING THEMES AND ISSUES

Human Population: The growth of human populations has led to decreases in biological diversity. As discussed in this chapter, people have had major effects on biological diversity, especially through habitat alteration, introduction of exotic species, and direct hunting and harvest. If the human population continues to increase, pressures will continue on endangered species, and it will be an increasingly greater challenge to maintain existing biological diversity.

Sustainability: This involves more than just having many individuals of a species. For a species to persist, its life requirements must be present and its habitat must be in good condition. A diversity of habitats allows sustainability of more species.

Global Perspective: For several billion years, life has affected the environment at a global scale. These global effects have in turn affected biological diversity. Life added oxygen to the atmosphere and removed carbon dioxide, thereby making animal life possible.

Urban World: People have rarely thought about cities as having any beneficial effects on biological diversity. However, in recent years there has been a growing realization that cities can contribute in important ways to the conservation of biological diversity; this will be discussed in Chapter 26.

Values and Knowledge: Perhaps no environmental issue causes more debate, is more central to arguments over values, or has greater emotional importance to people than biological diversity. Concern with specific endangered species has been at the heart of many political controversies. The path to resolving these conflicts and debates involves a clear understanding of the values at issue as well as knowledge about species and their habitat requirements and the role of biological diversity in life's history on Earth.

KEY TERMS

adaptive radiation *137*
biogeography *112*
biological diversity *118*
biological evolution *114*
biome *134*
biotic province *131*
competitive exclusion principle *120*

convergent evolution *135*
cosmopolitan species *120*
divergent evolution *136*
ecological niche *123*
endemic species *120*
exotic species *120*
genetic drift *115*

habitat *123*
industrial melanism *117*
migration *115*
mutation *115*
natural selection *115*
ubiquitous species *120*

STUDY QUESTIONS

1. Why do introduced species often become pests?

2. What is a geologic barrier, and why is this concept important in the geography of living things?

3. On which kind of planet would you expect a greater diversity of species:
   a. a planet with intense tectonic activities or
   b. a tectonically dead planet? (Remember that tectonics refers to the geologic processes that involve the movement of tectonic plates and continents, processes that lead to mountain building and so forth.)

4. You conduct a survey of national parks. What relationship would you expect to find between the number of species of trees and the size of the parks?

5. What is meant by the statement "Every nature preserve must be managed as if it were an island"?

6. What are the major factors that determine which species live in a particular location on a continent?

7. What are the consequences of geographic isolation?

8. In Jules Verne's classic novel *The Mysterious Is-*

land, a group of Americans find themselves on an isolated volcanic island inhabited by kangaroos and large rodents closely related to the agoutis of South America. Why is this situation unrealistic? What would make this co-occurrence possible?

9. What are three ways in which people have altered the distribution of living things?

10. From the perspective of biogeography, why do people attach so much importance to the conservation of tropical rain forests?

11. A city park manager has run out of money to buy new plants. How can the park labor force alone be used to increase the diversity of **(a)** trees and **(b)** birds in the parks?

12. A plague of locusts visits a farm field. Soon after, many kinds of birds arrive to feed on the locusts. What changes occur in animal dominance and diversity? Begin with the time before the locusts arrive and end with the time after the birds have been present for several days.

13. What is the difference between dominance and diversity in **(a)** a zoo and **(b)** a natural wildlife preserve?

14. What is the difference between habitat and niche?

15. A person suggests the following alternative hypothesis about the history of the peppered moth: The adult moths are able to change color, and the observed change from light to dark forms is simply a change in color of existing individuals, not a change owing to natural selection. What would a possible experiment be to test this hypothesis?

16. There are more than 600 species of trees in Costa Rica, most of which are in the tropical rain forests. What might account for the coexistence of so many species with similar resource needs?

17. Which of the following can lead to populations that are less adapted to the environment than their ancestors were?
   a. natural selection,
   b. migration,
   c. mutation, and
   d. genetic drift.

FURTHER READING

Darwin, C. A. 1859. *The Origin of Species by Means of Natural Selection, or the Preservation of Proved Races in the Struggle for Life*. London: Murray. Reprinted variously. This book marked a revolution in the study and understanding of biotic existence.

Dawkins, R. 1996. *Climbing Mount Improbable*. New York: Viking. A discussion of some implications of modern discoveries in genetics and evolution.

Margulis, L., and Sagan, D. 1995. *What Is Life?* New York: Simon and Schuster. A beautifully illustrated and well-written introduction to the major forms of life on Earth and the effect of life's diversity on the global environment.

Pimentel, D. 1992. "Conserving Biological Diversity in Agricultural/Forestry Systems," *BioScience* 42:354–362. The author argues that most biological diversity exists in human-managed ecosystems and identifies means to conserve that diversity.

Reid, W. V., and Miller, K. R. 1989. *Keeping Options Alive: The Scientific Basis for Conserving Biodiversity*. Washington, D.C.: World Resources Institute. This book provides an overview of the distribution of the world's species and genetic resources.

Strickberger, M. W. 1990. *Evolution*. Boston: Jones and Bartlett Publishers. This work provides a framework to explain current information and concepts in the study of evolution.

Wilson, E. O., ed. 1992. *The Diversity of Life*. New York: W. W. Norton. This book outlines the story of evolution of life on Earth, how species became diverse, and the scope of the current threat to that diversity.

INTERNET RESOURCES

Center for Conservation Biology Network: *http://www.conbio. rice.edu*—Access to the Virtual Library of Ecology and Biodiversity which contains information on endangered and extinct species, biodiversity issues and legislation, exotic introductions, habitats.

Nature Conservation Council of New South Wales, Australia: *http://www.nccnsw.org.au/index.html*—A community-based network of organizations acting to encourage community involvement in biodiversity conservation. Information and resources on biodiversity.

The Biodiversity Programs of the National Museum of Natural History (NMNH): *http://www.nmnh.si.edu/biodiversity*—The Smithsonian Institution's contribution to understanding the natural world and our place in it. Summaries of studies relating to biological diversity, systematics, evolution and ecology, and research initiatives are available on the site.

The Wild Screen Trust: *http://arkive.uwe.ac.uk*—Contains information on biodiversity, endangered species, and habitats. Species are categorized in kingdoms or, where kingdoms would be unwieldy (e.g., animals), subclassifications like birds and mammals are used.

A moose feeding in wetland flowers participates in ecosystem energy flow, as do the flowering plants as they use sunlight in photosynthesis.

CHAPTER 8

BIOLOGICAL PRODUCTIVITY AND ENERGY FLOW

CASE STUDY

Harvesting of Forests in Michigan and England

Today there is much concern with the sustainability of the planet's biological resources, including forests, wildlife, and fish. We seem to be in the process of destroying many of these resources by harvesting them faster than they can regrow, and we have been doing so for a long time. For example, in the United States, between 1840 and 1920, 7.7 million ha (19 million acres) of white pine forests were logged in the state of Michigan alone. Those who started the logging believed that they would never run out of wood, that by the time the last acre had been cut the first acre logged would have regrown. But those first loggers underestimated the power of technology. Only 20 ha (49 acres) of uncut white pine remain in Michigan, and only a small fraction of the original white pine forests have regenerated to anything like their original size and abundance.

To make matters worse, over much of the area unused limbs, branches, twigs, and leaves were left on the ground and provided fuel for fires that were often started accidentally at sawmills. The complete clearing of the land removed seed trees, and the fires destroyed the organic matter in the soil. As a result of these practices, stump barrens, areas where large trees have never regrown, are common in northern Michigan (Figure 8.1).

In contrast, in medieval England, some small forested areas were harvested carefully and slowly. For example, in 1356, a survey of the estates of Bishop Ely stated that a "certain wood called Heylewode" was 32.4 ha (80 acres) in size. Every year the "underwood" (the shrubs and young, small trees) was harvested in 4.5 of the 32.4 ha (11 of the 80 acres), leaving mature trees everywhere and young trees on the remaining 27.9 ha (69 acres). This practice was continued "without causing waste or destruction"; in modern terminology, the practice was sustainable. Similar forests were managed in much the same way in other locations

FIGURE 8.1 Sometimes forests that are clear-cut do not regenerate. This happened with some large areas of white pine forests in Michigan, which are called "stump barrens," including the area shown here, the Kingston Plains. These areas were logged and subject to repeated fires. The high fuel loading from downed branches and twigs led to damaging fires. Some believe that the logging practices of the time, along with climatic change, prevented regeneration.

FIGURE 8.2 English woodlands that have been managed since medieval times for sustainable forestry.

in England from the fourteenth century until World War I with little decline in the woodlands.[1] Today, we are seeing projects designed to redevelop these methods (Figure 8.2).

These two contrasting examples suggest that at certain levels of harvesting forest production can be sustained, but once these levels are exceeded, forests will decline and may not recover. If they do recover, the recovery may take an exceedingly long time. Throughout the history of civilization, people have cut trees faster than trees have regrown. Beginning with the earliest civilizations, forests were cleared to make way for agriculture and to provide fuel and structural materials, the basis of early civilization.[2]

The general practice of clearing the land and eliminating forests has continued into our own time. Clearly, the amount of timber harvested cannot exceed forever the amount of new timber that grows between harvests. Eventually the supply of timber will run out. With modern pressures for more and more use of wood and other biological resources, there is an increasing need to understand limits to the growth of timber—and to all biological production.

LEARNING OBJECTIVES

To conserve and manage our biological resources wisely, we must understand the basic concepts of energy, energy flow in ecosystems, and biological production. After reading this chapter, you should understand:

- That energy flow determines the upper limit on the production of biological resources, including forests, fisheries, wildlife, and endangered species.
- Where energy comes from and how it is transferred from one living thing to another.
- How the first and second laws of thermodynamics affect energy and production.
- That an ecosystem is an open system with re-

spect to energy and that the flow of energy is one-way through the ecosystem.

- That a basic quality of life is its ability to create order from energy on a local scale.

- Why little of the energy available to an organism is fixed in new organic matter and how little of the energy available moves on to the next trophic level.

8.1 HOW MUCH CAN WE GROW?

Many factors can limit the growth of trees and other forms of life, but the ultimate limit on production of organic matter is determined by energy flow. The determination of how much organic matter can be produced in any time period is important to many environmental topics, especially those that concern biological resources. How many bushels of wheat can a farmer produce in a field in 1 year? What is the upper limit of food that can be produced for all the people on Earth? What is the limit on the number of whales in the ocean? What is the maximum production of forests that we can expect?

Before we can estimate the actual and the maximum possible production of organic matter of any kind, we must understand the basic concepts of energy, energy flow in ecosystems, and biological production. These concepts are the subject of this chapter.

8.2 BIOLOGICAL PRODUCTION

The total amount of organic matter on Earth or in any particular ecosystem or area is called its **biomass.** This includes all living things plus all products of living things. Biomass is usually measured as the amount per unit surface area of Earth (e.g., as grams per square meter [g/m^2], or metric tons per hectare [MT/ha]). Biomass is increased through biological production (growth); the change in biomass over a given period of time is called **net production.** Biological production involves the capture of usable energy and the production of organic compounds in which the energy is stored. There are three measures of production, which we can think of as the currency of production: biomass, energy stored, and carbon stored. General relationships for calculating production are given in Working It Out 8.1 and 8.2.

WORKING IT OUT 8.1
EQUATIONS FOR PRODUCTION, BIOMASS, AND ENERGY FLOW

We can write a general relation between biomass (B) and net production (NP):

$$B_2 = B_1 + NP \qquad (8.1)$$

where B_2 is the biomass at the end of the time period, B_1 is the amount of biomass at the beginning of the time period, and NP is the change in biomass during the time period:

$$NP = B_2 - B_1 \qquad (8.2)$$

General production equations are given as:

$$GP = NP + R \qquad (8.3)$$
$$NP = GP - R \qquad (8.4)$$

where GP is gross production, NP is net production, and R is respiration.

The three currencies of energy flow are biomass, energy content, and carbon content. The average of the energy in vegetation is approximately 21 kJ/g. Energy content of organic matter varies. Ignoring bone and shells, woody tissue contains the least energy per gram, about 17 kJ/g; fat con-

tains the most, about 38 kJ/g; and muscle contains approximately 21 to 25 kJ/g. Leaves and shoots of green plants have 21 to 23 kJ/g; roots have about 19 kJ/g.[3]

The kilojoule (1 kJ = 1000 J = 0.24 kcal) is the International System unit (SI unit) preferred in scientific notation for energy and work. It replaces the calorie or kilocalorie of earlier studies of energy flow. The kilocalorie is the amount of energy required to heat a kilogram of water 1 degree Celsius (from 15.5° to 16.5°C). (The calorie, which is one-thousandth of a kilocalorie, is the amount of energy required to heat a gram of water the same 1 degree Celsius.) Be careful, however, as the "calorie" referred to in diet books is actually the kilocalorie, though for convenience it is referred to in popular literature as a calorie. To keep all this straight, just remember that almost nobody uses the "little" calorie, regardless of what they call it. To compare, an average apple contains about 419 kJ, or 100 kcal. The calorie is typically used in studies of diets; the joule is used in physics and engineering.

WORKING IT OUT 8.2
ENERGY EQUALITIES

For Those Who Make Their Own Food (Autotrophs)

Photosynthesis is defined as

$$6CO_2 + 6H_2O \rightarrow C_6H_{12}O_6 + 6O_2 \qquad (8.5)$$

Chemosynthesis also takes place in certain environments. In chemosynthesis, the energy in hydrogen sulfide (H_2S) is used by certain bacteria to make simple organic compounds. The reactions differ among species and depend on characteristics of the environment. Therefore, there is not one equation like that for photosynthesis.

Net production for autotrophs is given as:

$$NPP = GPP - R_a \qquad (8.6)$$

where NPP is net primary production, GPP is gross primary production, and R_a is the respiration of autotrophs.

For Those Who Do Not Make Their Own Food

Secondary production of a population is given as

$$NSP = B_2 - B_1 \qquad (8.7)$$

where NSP is net secondary production, B_2 is the biomass at time 2, and B_1 is the biomass at time 1. The change in biomass is the result of the addition of weight of living individuals, the addition of newborns and immigrants, and the loss through death and emigration. The biological use of energy is through respiration, most simply expressed as

$$C_6H_{12}O_6 + 6O_2 \rightarrow 6CO_2 + 6H_2O + energy \qquad (8.8)$$

Two Kinds of Biological Production

There are two different kinds of biological production. Some organisms make their own organic matter from a source of energy and inorganic compounds. These organisms, introduced in the discussion of trophic levels in Chapter 6, are called **autotrophs** (meaning self-nourishing) (Figure 8.3). The autotrophs include green plants (those containing chlorophyll), such as herbs, shrubs, and trees; algae, which are usually found in water but occasionally grow on land; and certain kinds of bacteria that grow in water.

The production carried out by autotrophs is called **primary production.** Most autotrophs make sugar from sunlight, carbon dioxide, and water in a process called **photosynthesis,** which releases free oxygen (see Working it Out 8.2, "Energy Equalities"). Some autotrophic bacteria can derive energy from inorganic sulfur compounds; these bacteria are referred to as **chemoautotrophs.** Such bacteria have been discovered in deep-ocean vents, where they provide the basis for a strange ecological community. Chemoautotrophs are also found in muds of marshes, where there is no free oxygen.

Other kinds of life cannot make their own organic compounds from inorganic ones and must feed on other living things. These are called **heterotrophs.** All animals, including human beings, are heterotrophs, as are fungi, many kinds of bacteria, and many other small forms of life. Production by heterotrophs is called **secondary production** because it is dependent on the production of autotrophic organisms. This dependency is the basis for the food web described in Chapter 6.

Once an organism has obtained new organic matter, it can use the energy in that organic matter to do things: to move, to make new kinds of compounds, to grow, to reproduce, and so forth. The use of energy in organic matter in both heterotrophic and autotrophic organisms is accomplished through **respiration.** In respiration, an or-

FIGURE 8.3 Energy pathways through an ecosystem. Usable energy flows from the external environment (the sun) to the plants, then to the herbivores, carnivores, and top carnivores. Death at each level transfers energy to decomposers. Energy lost as heat is returned to the external environment.

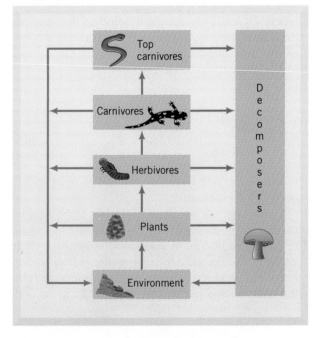

ganic compound is combined with oxygen to release energy and produce carbon dioxide and water (see Working it Out 8.2, "Energy Equalities"). The process is similar to the burning of organic compounds but takes place within cells at much lower temperatures through enzyme-mediated reactions. *Respiration is the use of biomass to release energy that can be used to do work.* Respiration returns to the environment the carbon dioxide that has been removed by photosynthesis.

Gross and Net Production

There are three steps in the production of biomass and its use as a source of energy by autotrophs. First, an organism produces organic matter within its body; next, it uses some of this new organic matter as a fuel in respiration; finally, some of the newly produced organic matter is stored for future use. The first step, production of organic matter before any use, is called **gross production.** This suggests another way to think about net production. Net production is what is left from gross production after use. In these terms,

Net production = gross production − respiration

This is a fundamental production relationship. The difference between gross and net production is like the difference between a person's gross and net income. Gross income is the amount you are paid. Net income is what you have left after money is deducted for taxes and other fixed costs. Respiration is like the necessary expenses that are required in order for you to do your work.

The gross production of a tree is the total amount of sugar it produces by photosynthesis before any is used. Within living cells in the tree some of the sugar is oxidized in respiration. Energy is used to convert sugars to carbohydrates, carbohydrates to amino acids, amino acids to proteins and new leaf tissue, and so forth. Energy is also used to transport material within the plant. What is not used in such ways may itself be transported to roots, stems, flowers, and fruits or stored within the leaf cell. Some energy is lost as heat in the transfer; other energy is used to make other organic compounds in other parts of the plant: cell walls, proteins, and so forth. Some is stored for later use. Net production of the tree is energy contained in what is left at the end of the year and includes new wood laid down in the trunk, new buds that will develop into leaves and flowers the next year, and new roots. Common units of measure of production are given in Table 8.1.

8.3 ENERGY FLOW

Most of the time energy is invisible to us, but with infrared film we can see the differences between warm and cold objects and we can see some factors about energy flow that affect life. In infrared film, warm objects appear red and cool objects blue. Figure 8.4 shows birch trees in a New Hampshire forest. In Figure 8.4*a* the trees are shown as we see

TABLE 8.1	Common Units of Measure for Biomass and Production	
Habitat	*Biomass*	*Production*
Units of Mass		
Land	g/cm^2	g/cm^2/d; g/cm^2/yr
	kg/ha	kg/ha/yr
	t/ha	t/ha/yr
Water	g/cm^3	g/cm^3/d
	kg/m^3	kg/m^3/yr
Units of Energy		
Land	kcal/cm^2	kcal/cm^2/d
	kcal/ha/yr	kcal/ha/yr
	kJ/cm^2	kJ/cm^2/d
	kJ/ha/yr	kJ/ha/yr
Water	kcal/m^3	kcal/m^3/yr
	kJ/m^3	kJ/m^3/yr

Note: Units of mass: 1.0 t (metric ton) = 1000 kg (kilogram) = 10^{6}g (gram). Units of area: 1.0 ha (hectare) = 10,000 square meters (m^2) = 10^8 square centimeters (cm^2). Units of time: yr (year, sometimes written as "a"); d (day). Units of energy: 1 kcal = 4.19 kJ.

(a)

(b)

FIGURE 8.4 (*a*) A birch forest in New Hampshire as we see it, using normal photographic film (left) and the same forest photographed with infrared film (right). Red color means warmer temperatures. This photograph shows that the leaves are warmer than the surroundings, because they are heated by sunlight. (*b*) A nearby rocky outcrop as we see it, using normal photographic film (left) and the same rocky outcrop photographed with infrared film (right). Blue means that a surface is cool. The rocks appear deep blue, meaning that they are much cooler than the surrounding trees.

them, using standard film; in Figure 8.4*b* infrared film shows the leaves bright red, indicating that they have been warmed by the sun and are absorbing and reflecting energy, whereas the white birch bark remains cooler. The ability of tree leaves to absorb energy is essential; it is this source of energy that ultimately supports all life in a forest. How energy is transferred from one living thing to another and the essential role of energy for life are the topics of the rest of this chapter.

Energy flow is the movement of energy through an ecosystem: from the external environment through a series of organisms and back to the external environment. It is one of the fundamental processes common to all ecosystems and one of the most profound and philosophical of all topics discussed in this book; it tells us about the very nature of life. The study of energy flow, however, poses difficult practical issues.

All life requires energy. Energy is the ability to do work, to move matter. As anyone who has dieted knows, our weight is a delicate balance between the energy we take in through our food and the energy we use. What we do not use and do not pass on, we store. Our use of energy, and whether we gain or lose weight, follows the laws of physics. This is not only true for people, it is also true of populations, communities, ecosystems, and the biosphere.

Energy is a difficult and abstract concept. When we buy electricity by the kilowatt-hour (1 kWh = 3.6 x 10^3 kJ), what are we buying? You can't see it or feel it, even if you have to pay for it. (The following discussion on the fundamental characteristics of energy is based on H. J. Morowitz, *Energy Flow in Biology*, 1979, Oxbow Press, Woodbridge, Conn., and is presented with his permission.)

At first glance energy flow seems simple enough: We take energy in and use it, just like a machine. But if we dig a little deeper into this subject, we discover its depth and philosophical importance. When we consider the role of energy in an ecosystem, we consider the heart of the matter of what it is

that distinguishes life and life-containing systems from the rest of the universe.

Ecosystem Energy

The equalities that link production, respiration, and biomass of individuals and communities can also be expressed for an ecosystem (see Working It Out 8.3, "Ecosystem Equalities"). In an ecosystem, the gross production is simply the gross production of all the autotrophs. The net ecosystem production is the amount of biomass added in some time period after all utilization, including the respiration of autotrophs and heterotrophs, has taken place.

There are two pathways by which energy enters an ecosystem. One is the pathway already discussed: Energy is fixed by organisms that can produce their own food from energy (through photosynthesis) or from inorganic minerals (Figure 8.3). In the second, heat energy transferred by the air or water currents or by convection through soils and sediments warms living things. For instance, when a warm air mass passes over a forest, heat energy is transferred from the air to the land and to the organisms.

8.4 THE ULTIMATE LIMIT ON THE ABUNDANCE OF LIFE

What ultimately limits the amount of organic matter that can be produced? How closely do ecosystems, species, populations, and individuals approach this maximum limit? The search for answers to these questions takes us to some of the most fundamental issues about life, issues that involve energy flow through ecosystems. The relationship between energy and life is explained by fundamental laws of physics.

Matter and energy are both subject to the *laws of conservation*. The *laws* are the laws of physics; the kind of *conservation* is this:

WORKING IT OUT 8.3
ECOSYSTEM EQUALITIES

$$GEP = GPP \qquad (8.9)$$

where GEP is gross ecosystem production and GPP is gross primary production.

$$R_e = R_a + R_h \qquad (8.10)$$

where R_e is net ecosystem respiration, R_a is respira-

tion of autotrophs, and R_h is respiration of heterotrophs.

$$NEP = GEP - R_e \qquad (8.11)$$

where NEP is net ecosystem production, GEP is gross ecosystem production, and R_e is net ecosystem respiration.

- In any physical or chemical change, matter is neither created nor destroyed but merely changed from one form to another.
- In any physical or chemical process, energy is neither created nor destroyed but merely changed from one form to another.

The law of conservation of energy is also called the *first law of thermodynamics* (discussed in

A Closer Look 3.1, "Electromagnetic Radiation"). It addresses the observation that energy changes form, not amount.

If the total amount of energy is always conserved (remains constant), why can't we just recycle energy inside our bodies? Similarly, why can't energy be recycled in ecosystems and in the biosphere? We are often told we eat to get particular chemical elements: calcium from milk, nitrogen from

 CLOSER LOOK 8.1

THE SECOND LAW OF THERMODYNAMICS

To better understand why we cannot recycle energy, imagine a closed system (i.e., it receives no other input) containing a pile of coal, a tank of water, air, a steam engine, and an engineer (Figure 8.5). Suppose that the engine runs a lathe that makes furniture. The engineer lights a fire to boil the water, creating steam to run the engine. As the engine runs, the heat from the fire gradually warms the entire system. When all the coal is completely burned, the engineer will not be able to boil any more water and the engine will stop. The average temperature of the system is now higher than the starting temperature.

The energy that was in the coal is now dispersed throughout the entire system, much of it as heat in the air. Why can't the engineer recover all that energy, recompact it, put it under the boiler, and run the engine? The answer is because of the second law of thermodynamics. Physicists have discovered that no real use of energy can ever be 100% efficient. Whenever useful work is done, some energy is inevitably converted to heat. Collecting all the energy dispersed in this closed system would require more energy than could be recovered.

Our imaginary system begins in a highly organized state, with energy

compacted in the coal. It ends in a less organized state, with the energy dispersed throughout the system as heat. The energy has been degraded, and the system is said to have undergone a decrease in order. The measure of the decrease in order (the disorganization of energy) is called **entropy.** The engineer did produce some furniture, converting a pile of lumber into nicely ordered tables and chairs. The system had a local increase of order (the furniture) at the cost of a general increase in disorder (the state of the entire system). All energy of all systems tends to flow toward states of increasing entropy.

FIGURE 8.5 Diagram of a system closed to the flow of energy, containing wood, water, and a steam-driven engine.

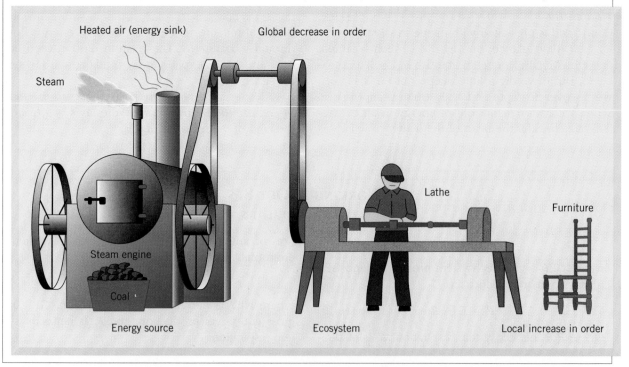

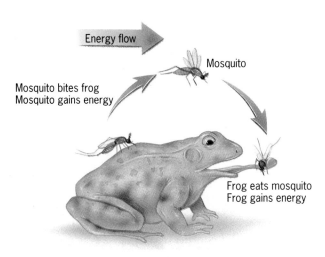

FIGURE 8.6 An impossible ecosystem.

meat. But if all matter must be conserved and one atom of an element is indistinguishable from another, why can't we recycle all our elements and only add new elements to replace those lost by accident or required for growth?[3]

Let us imagine how this might work, say, with frogs and mosquitoes. Frogs eat insects, including mosquitoes. Mosquitoes suck blood from vertebrates, including frogs. Consider an imaginary closed ecosystem consisting of water, air, a rock for frogs to sit on, frogs, and mosquitoes. In this system, the frogs get their energy from eating the mosquitoes and the mosquitoes get their energy from biting the frogs (Figure 8.6). Why can't this system maintain itself indefinitely? Such a closed system would be a biological perpetual-motion machine: It could continue indefinitely without an input of any new material or energy. This sounds nice, but unfortunately, it is impossible.

The general answer as to why this system could not persist is found in the *second law of thermodynamics*, which addresses the kind of change in form that energy undergoes.

- Energy always changes from a more useful, more highly organized form to a less useful, disorganized form.

That is, energy cannot be completely recycled to its original state of organized, high-quality usefulness. For this reason, the mosquito–frog system will eventually stop as there will not be enough useful energy left.

From the discussion presented in A Closer Look 8.1, "The Second Law of Thermodynamics" we reach a new understanding of a basic quality of life.[4] It is the ability to create order on a local scale that distinguishes life from its nonliving environment.

Obtaining this ability requires energy in a usable form, and this is the essence of what we eat. This is true for every ecological level: individual, population, community, ecosystem, and biosphere. Energy must continually be added to an ecological system in a usable form. Energy is inevitably degraded into heat, and this heat must be released from the system. If it is not released, the temperature of the system will increase indefinitely. This is what is meant by the statement that the net flow of energy through an ecosystem is one-way.

Now we have explained the general reason why the frog–mosquito system could not persist indefinitely without another source of energy. (There is also a specific reason: Only female mosquitoes require blood and then only in order to reproduce. Mosquitoes are otherwise herbivorous.)

From what we have just discussed, it is clear that an ecosystem must lie between a source of usable energy and a sink for degraded (heat) energy. The ecosystem is said to be intermediate between the energy source and the energy sink. The energy source, ecosystem, and energy sink form a thermodynamic system. The ecosystem can undergo an increase in order, called a local increase, as long as the entire system undergoes a decrease in order, called a global decrease. Creating local order involves the production of organic matter. Producing organic matter requires energy; organic matter stores energy.

Energy Efficiency and Transfer Efficiency

Energy efficiency is the ratio of output to input. How efficiently do living things use energy? This is an important question for the management and conservation of all biological resources. The laws of physics and chemistry state that no system can be 100% efficient. As energy flows through a food web, it is degraded and less and less is usable. Generally, the more energy an organism gets, the more it has for its own use. However, organisms differ in how efficiently they use the energy they obtain. A more efficient organism has an advantage over a less efficient one. Energy efficiency is usually defined as the amount of useful work obtained from some amount of available energy. Efficiency can therefore be defined for both artificial and natural systems: machines, individual organisms, populations, trophic levels, ecosystems, and the biosphere.[5]

Efficiency has different meanings to different users. A farmer thinks of an efficient corn crop as one that converts a great deal of solar energy to sugar and uses little of that sugar to produce stems, roots, and leaves. In other words, the most efficient crop is the one that has the most harvestable energy left at the end of the season.

A truck driver views an efficient truck as one that uses as much energy as possible from its fuel and leaves as little energy as possible to exit in the exhaust. When we view organisms as food, we define efficiency as the farmer does, in terms of energy storage (net production from available energy); when we are the users, we define efficiency as the truck driver does, in terms of how much useful work we accomplish with the available energy.

A common ecological measure of energy efficiency is called food chain efficiency, or **trophic-level efficiency,** which is the ratio of production of one trophic level to the production of the next lower trophic level. (See Chapter 6.) This efficiency is never very high. Green plants convert only 1 to 3% of the energy received from the sun during the year to new plant tissue. We would often like to increase trophic-level efficiency.

The efficiency with which herbivores convert the potentially available plant energy into herbivorous energy or the efficiency with which carnivores convert herbivores into carnivorous energy is usually less than 1%. It is frequently written in popular literature that the transfer is 10%, for example, that 10% of the energy in corn can be converted into energy in a cow. However, this is a managed ecological efficiency rather than the natural, trophic-level efficiency described here. In natural ecosystems, the organisms in one trophic level tend to take in much less energy than the potential maximum amount available to them, and they utilize more energy than they store for the next trophic level.

At Isle Royale National Park, an island in Lake Superior, wolves feed on moose in a natural wilderness. A pack of 18 wolves killed an average of one moose approximately every 2.5 days,[6] a trophic-level efficiency of wolves is about 0.01%. Wolves use most of the energy they take in from eating moose, especially for moving in the search for prey.[7] From the wolves' point of view, wolves are efficient, but from the point of view of someone who wants to feed on wolves, they appear inefficient.

The rule of thumb for ecological trophic energy efficiencies is that more than 90% (usually much more) of all energy transferred between trophic levels is lost as heat. Less than 10% (approximately 1% in natural ecosystems) is fixed as new tissue. In highly managed ecosystems, such as ranches, the efficiencies may be greater. Even in such highly managed systems, however, it takes an average of 3.2 kg (7 lb) of human edible vegetable matter eaten by livestock to produce 0.45 kg (1 lb) of edible meat. Cattle are by far the least efficient producers, requiring around 7.2 kg (16 lb) of vegetable matter to produce 0.45 kg (1 lb) of meat. At the lower end of the scale, it requires approximately 1.4 kg (3 lb) of vegetable matter to produce 0.45 kg (1 lb) of eggs or chicken meat. Much

attention has been paid to the idea of humans eating at a lower trophic level in order to use our resources more efficiently. (See Environmental Issue: Should People Eat Lower on the Food Chain?)

Other Measures of Energy Efficiency

Many other kinds of energetic efficiencies are widely used in ecological studies.[8] In Table 8.2 some values are given for the **growth efficiency,** or gross production efficiency (P/C), which is the ratio of the material produced (P is the net production) by an organism or population to the material ingested or consumed (C). The amount consumed is normally much less than the maximum amount available. Estimates for leaf-eating insects in forests and woodlands show that annually amounts of less than 1% to about 20% of the leaves available are consumed by leaf-eating insects.[9]

Table 8.2 also gives examples for the **net growth efficiency,** or net production efficiency (P/A), the ratio of the material produced (P) to the material assimilated (A), which is less than the material consumed because some food taken in is discharged as waste and never used by an organism.

TABLE 8.2 Ecological Efficiencies for Animal Populations

Trophic Types	Ecological Efficiencies (%)	
	Net Production Efficiency	Gross Production Efficiency
Terrestrial Animals		
Microorganisms[a]	~40	
Invertebrates		
Herbivores	20–40	8–27
Carnivores	10–37	~34
Saprophages	17–40	5–8
Vertebrates		
Herbivores[a]	2–10	
Carnivores[a]	2–10	
Aquatic Animals		
Fishes[a,b]	—	1–7

Sources: T. Penczak, 1992, *Comp. Biochem. Physiology,* 101(A4). pp. 791–798; D. E. Reichle, 1977, "The Role of Soil Invertebrates in Nutrient Cycling," in U. Lohm and T. Persson, eds., *Ecol. Bull.* (*Stockholm*), 25, pp. 145–156; and M. Schaefer, 1991, "Secondary Production and Decomposition," in E. Rohrig and B. Ulrich, eds., *Temperate Deciduous Forests,* Ecosystems of the World, vol. 7, Elsevier, Amsterdam.

[a]Data are based on characteristic values for trophic levels and populations.

[b]Populations in a tropical river.

8.5 SOME EXAMPLES OF ENERGY FLOW

Energy Flow in an Old Field Food Chain

In an old field in Michigan, meadow mice feed on grasses and herbs and least weasels feed on mice[10] (one of many food chains in this old field). The first step in the flow of energy is the fixation of light energy by photosynthesis in leaves of the grasses, herbs, and shrubs; in this step the energy in light is transferred to and stored in sugar or carbohydrates in the plants. Recall that this energy—the energy that is stored by autotrophs before any is used—is called gross primary production. Some of the energy is used immediately by the leaves to keep their own life processes going. As explained earlier, the amount stored by the autotrophs after using what is needed is net primary production.

Only a small fraction of the energy available to each trophic level (see Chapter 6) *is converted to net production of new tissue. A large fraction of the energy available to each trophic level is used in respiration.* In the old field in Michigan, about 15% of the vegetation's gross production was used in respiration; 68% of the energy taken up by the mice was used in respiration; and 93% of the energy taken up by the least weasel was used in respiration. Only a part of the energy flow in the old field moved through the food chain of the vegetation–mice–weasels. One reason for this is that mice eat only the seeds of the plant. Most of the energy remained in the vegetation until it was transferred from dead vegetation to animals, fungi, and bacteria by the decomposer food chain.

Energy Flow in a Stream or River

In most ecosystems, the original fixing of energy occurs within the ecosystem; however, some freshwater streams are an exception. In streams, photosynthesis of stream algae is often small compared to the amount of dead vegetation that falls in or flows in from the surrounding countryside (dead leaves and twigs fall into the stream or are washed into it during storms).[11] Detritivores (organisms that feed on dead organic material) are common in streams and feed mainly on this deposited vegetation. Some of the animals are shredders that tear up leaves; others feed on the smaller pieces.

Other grazing animals move along the surfaces of rocks and scrape off attached algae. Many stream predators are larvae of land-dwelling insects, such as dragonflies. Some animals capture prey from the land or air, as in the case of trout that catch flying insects.

An extreme case of a food chain based on external food input occurs in the floodplain of the Amazon River basin, in which fish feed on fruits and nuts carried into the streams during the rainy season. Here, the production of herbivorous fish exceeds what would be possible from aquatic primary production alone, yielding an abundant food supply for the human populations of the region.[12]

Energy Flow in Ocean Ecosystems

In the open ocean, the first trophic level is composed of many species of floating algae, collectively called phytoplankton.[13] These are fed on by many animals, including small, floating ones known as zooplankton, which in turn are fed on by larger organisms. A food chain is based on dead organic material that is produced at the surface and sinks toward the ocean floor, including cells of algae; dead bodies and excrement of heterotrophs; and some material, such as logs, transported from the land. These form the energy base for two groups of carnivores and detritivores: those that swim in the depths and catch the material as it sinks and those that live on the ocean bottom and hunt for material once it has been deposited. These food chains thus depend on an external input of energy as the base of energy flow.

Chemosynthetic Energy Flow in the Ocean

Earlier we mentioned organisms that make their own food from energy in sulfur compounds. This process creates a curious class of food chains recently discovered in the depths of the oceans. These food chains support previously unknown life-forms. The basis of the food chains is *chemosynthesis*, in which the source of energy is not sunlight, but hot, inorganic sulfur compounds emitted from vents in the ocean floor.

Sulfur-laden water is emitted from hot-water vents occurring at depths of 2500 to 2700 m (8200 to 8900 ft) associated with areas where flowing lava causes seafloor spreading. A rich biological community exists in and around the vents, including large white clams up to 20 cm (7.8 in.) in diameter, brown mussels, and white crabs. Clams and mussels filter chemoautotrophic bacteria and particles of dead organic matter from the water. Some vent communities contain limpets, pink fish, tube worms, and octopuses. Among the most curious creatures found in vents are giant worms, some 3.9 m (12.8 ft) long.[14]

Large areas of ocean have low productivity; combined, however, oceans account for a major portion of total energy fixed. Highly productive areas of the oceans occur in upwelling zones (see Earth's Biomes). These occur when deep ocean waters, rich in nutrients from dead organic material, flow upward, allowing abundant growth of algae and photosynthetic bacteria. Herbivores and carnivores move organic nutrients through the food chain. Although only

Should People Eat Lower on the Food Chain?

The energy content of a food chain is often represented by an *energy pyramid*, such as the one shown here for a hypothetical, idealized food chain. In an energy pyramid, each level of the food chain is represented by a rectangle whose area is more or less proportional to the energy content for that level. For the sake of simplicity, the food chain shown here assumes that each link in the chain has one and only one source of food. Assume that if a 75-kg (165-lb) person ate frogs (and some people do!), he would need ten a day, or 3,000 a year (approximately 300 kg or 660 lb). If each frog ate 10 grasshoppers a day, the 3,000 frogs would require 9,000,000 grasshoppers a year to supply their energy needs, or approximately 9,000 kg (19,800 lb) of grasshoppers. A horde of grasshoppers of that size would require 333,000 kg (732,600 lb) of wheat to sustain them for a year.

As the pyramid illustrates, energy content decreases at each higher level of the food chain. The result is that the amount of energy at the top of a pyramid is related to the number of layers the pyramid has. For example, if people fed on grasshoppers rather than frogs, each person could probably get by on 100 grasshoppers a day. The 9,000,000 grasshoppers could support 300 people for a year, rather than only one. If, instead of grasshoppers, people ate wheat, then 333,000 kg of wheat could support 666 people for a year.

This argument is often extended to suggest that people should become herbivores (vegetarians in human parlance) and eat directly from the lowest level of all food chains, the autotrophs. Consider, however, that humans can eat only parts of some plants. By eating herbivores that can eat the parts of plants that humans cannot eat, or those plants that humans cannot eat at all, more of the energy stored in plants becomes available for human consumption. The most dramatic example of this is in aquatic food chains. Because people cannot digest most kinds of algae, which are the base of most aquatic food chains, they depend on eating fish that eat algae, or on those fish that eat other fish. So, if people were to become entirely herbivorous, they would be excluded from many food chains. In addition, there are major areas of Earth where crop production damages the land, but

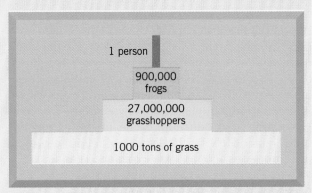

The total bulk (mass) of all the organisms making up each level of a food chain is less than that of the previous level. (*Source:* Nelson, 1970, *Fundamental Concepts of Biology*, 2nd ed., Wiley, New York, p. 299.)

grazing by herbivores does not. In those cases, conservation of soil and biological diversity leads to arguments that support the use of grazing animals for human food. This creates an environmental issue: How low on the food chain should people eat?

Critical Thinking Questions

1. Why does the energy content decrease at each level of a food chain? What happens to the energy that is lost at each level?

2. The pyramid diagram uses mass as an indirect measure of the energy value for each level of the pyramid. Why is it appropriate to use mass to represent energy content?

3. Using the average of 21 kJ of energy equals 1 g of completely dried vegetation (see Working It Out 8.3) and assuming that wheat is 80% water, what is the energy content of the 333,000 kg of wheat shown in the pyramid?

4. Make a list of the environmental arguments for and against people eating an entirely vegetarian diet. What might the consequences for U.S. agriculture be if everyone in the country began to eat lower on the food chain?

5. How low do you eat on the food chain? Would you be willing to eat lower? Explain.

1/1000 of the oceans' surface has natural upwellings, these zones account for more than 44% of the fish eaten by the world's human population.[15]

In every ecosystem, the energy flow provides a foundation for life and thus imposes a limit on the abundance and richness of life. The amount of energy available to each trophic level in a food chain depends not only on the strength of the energy source, but also on the efficiency with which the energy is transferred along the food chain.

SUMMARY

- In this chapter we have considered the process of one-way energy flow through an ecosystem. The study of energy flow is important to determine limits on food supply and on the production of all biological resources, such as wood and fiber.

- Energy is fixed by autotrophs, which are organisms that make their own food from energy and small inorganic compounds. There are two sources of the initial energy: light (mainly sunlight) and the energy in small sulfur compounds. Plants, algae, and some bacteria are autotrophs.

- All other organisms cannot make their own food and must feed on other organisms. Those that cannot make their own food are called heterotrophs.

- Biological production is the production of new organic matter, which we measure as change in biomass, change in stored energy, or change in stored carbon. Another way to think about biological production is that it is the change in biomass over time.

- There are several kinds of production. Gross production is the production before any utilization. Net production is the amount stored (not used) at the end of some time period. Respiration is the use of stored energy, so that net production equals gross production minus respiration.

- The laws of thermodynamics connect life to order in the universe. The second law of thermodynamics tells us that order always decreases when any real process occurs in the universe. However, life is more ordered than its environment. The ability to create order is the essence of what we get from our food.

- Energy efficiency is the ratio of output to input, or the amount of useful work obtained from some amount of available energy. Trophic-level efficiency is the ratio of production of one trophic level to the production of the next lower trophic level. This efficiency is never very high, often only about 1%.

REEXAMINING THEMES AND ISSUES

Human Population: The ultimate limit on the human population and its use of resources is set by available energy, although many other factors can limit our numbers well below the maximum.

Sustainability: Ecological energy flow sets an upper limit to sustainable biological production.

Global Perspective: From a cosmic perspective, Earth is a small planet where life uses the one-way flow of energy from the sun to create life's order: organisms, species, ecosystems, and the environment of the whole Earth.

Urban World: One of the fundamental characteristics of life is that it uses energy to create order on a local scale. From an ecological perspective, cities are an extreme example of this process.

Values and Knowledge: Until we know the upper limit on biological production, we cannot know the practical limits of our harvest and use of living resources. Understanding energy flow is fundamental to understanding the limits of our actions and what we can consider a good (i.e., sustainable) set of activities.

KEY TERMS

autotrophs *146*	gross production *147*	photosynthesis *146*
biomass *145*	growth efficiency *152*	primary production *146*
chemoautotrophs *146*	heterotrophs *146*	respiration *146*
energy flow *149*	net growth efficiency *152*	secondary production *146*
entropy *150*	net production *145*	trophic-level efficiency *152*

STUDY QUESTIONS

1. What is the difference among the following?
   a. ecological production, ecological productivity, and biomass;
   b. gross production and net production; and
   c. primary and secondary production.

2. What does the statement "Any living or life-containing system is always more ordered than its nonliving environment" mean?

3. Keep track of the food you eat during one day and make a food chain linking yourself with the sources of those foods. Using the diagram, what is the biomass (grams) and energy (kilocalories or kilojoules) you have eaten? Assuming that your net production is 10% efficient in terms of the energy intake, how much additional energy might you have stored during the day? Using an average of 5 kcal/g, what is your weight gain from the food you have eaten?

4. Referring to question 3, what is the amount of vegetation you eat during one day? If vegetation was 1% efficient in converting sunlight to organic matter stored as net primary production, how much sunlight was required to provide the vegetation you took in during the day?

FURTHER READING

Relative to many other topics discussed in this book, energy flow and productivity have a long history, and the most useful and interesting references tend to be classics published several decades ago. Although in most cases we try to provide the most up-to-date references, with this subject we believe some of the easiest to read or most important references are among the classical earlier works.

Blum, H. F. 1962. *Time's Arrow and Evolution*. New York: Harper & Row. How life is connected to the laws of thermodynamics and why this matters; a very readable book.

Gates, D. M. 1980. *Biophysical Ecology*. New York: Springer-Verlag. A discussion about how energy in the environment affects life.

Morowitz, H. J. 1979. *Energy Flow in Biology*. Woodbridge, Conn.: Oxbow. A book that provides the most thorough and complete discussion available about the connection between energy and life, at all levels, from cells to ecosystems to the biosphere.

Morowitz, H. J. 1981. "The Six Million Dollar Man," in *The Wine of Life and Other Essays on Societies, Energy and Living Things*. New York: Bantam Books. A fun essay about the second law of thermodynamics and life.

Peterson, R. O. 1995. *The Wolves of Isle Royale: A Broken Balance*. Minocqua, WI: Willow Creek Press. A first-hand account of Rolf Peterson's 25-year association with the long-running study of the wild wolves of Isle Royale National Park and their primary prey, the moose.

Schrödinger, E. 1942. *What Is Life?* Cambridge: Cambridge University Press. The original statement about how the use of energy differentiates life from other phenomena in the universe. Easy to read, yet a classic.

Sherman, K. 1990. *Large Marine Ecosystems: Patterns, Processes, and Yields*. Book News, Inc. Portland, OR. Based on an AAAS symposium, this book deals with the possible impacts of global change on ocean productivity and discusses managing large marine ecosystems as multinational units in order to sustain biomass yields of major coastal regions.

INTERNET RESOURCES

The Earth Council: *http://www.ecouncil.ac.cr*—An international non-governmental organization (NGO) created in September 1992 to promote and advance the implementation of the Earth Summit agreements. The site contains an on-line database, earth summit documents, and other sustainable development links.

National Marine Fisheries Service: *http://kingfish.ssp.nmfs.gov*—The National Marine Fisheries Service (NMFS) or "NOAA Fisheries" is a part of the National Oceanic and Atmospheric Administration (NOAA). NMFS administers NOAA's programs which support the domestic and international conservation and management of living marine resources. NMFS Web site provides access to NMFS publications and legislation.

Sea-viewing Wide Field-of-view Sensor Project (SeaWiFS) *http://seawifs.gsfc.nasa.gov*—Provides quantitative data on global ocean bio-optical properties. The SeaWiFS Project is developing and operating a research data system that processes and distributes data received from an Earth-orbiting ocean color sensor. The concentration of phytoplankton can be derived from satellite observation and quantification of ocean color. Technical reports, teacher's guide, and images are available on the Web site.

Workers in Nepal plant trees on a steep slope as part of a restoration project. Such projects are now recognized internationally as important in erosion control, water flow control, as well as providing benefits of the trees and their ecosystems. This project is funded by the US Agency for International Development.

CHAPTER 9

SUCCESSION AND RESTORATION: HOW ECOSYSTEMS RESPOND TO DISTURBANCE

◖ASE STUDY

Restoring Abandoned Mine Lands in Great Britain

Mining is a widespread cause of destruction and damage to land. In some areas, efforts are being made to undo mining damage and restore the landscape, if not to a natural ecosystem, then to some kind of vegetation cover. In Great Britain, where some mines have been used since medieval times, approximately 55,000 ha (136,000 acres) have been damaged by mining. Recently, programs have been initiated to restore these damaged lands to useful biological production, to remove toxic pollutants from the mines and mine tailings, and to restore the visual attractiveness of the landscape.[1]

One area damaged by a long history of mining lies within the British Peak District National Park, where lead has been mined since the Middle Ages and waste tailings are as much as 5 m (16.4 ft) deep (Figure 9.1). The first attempts to restore

this area used a modern agricultural approach: heavy application of fertilizers and planting of fast-growing agricultural grasses. It was hoped that rapidly growing grasses would be able to revegetate the site quickly. These grasses rapidly green on the good soil of a level farm field; the plan was that, with fertilizer, they would do the same in this situation. But after a short period of growth, the grasses died. On the poor soil, leached of its nutrients and lacking organic matter, erosion continued, and the fertilizers that had been added were soon leached away by water runoff. Without the fertilizers, the fast-growing grasses died. As a result, the areas quickly returned to a barren condition.

When the agriculturally oriented methods failed, an ecological approach was tried. This approach applied knowledge about **ecological suc-**

157

Figure 9.1 An old lead mining area in Great Britain, now undergoing restoration by plantings.

cession, the natural process of establishment or reestablishment of an ecosystem. Instead of planting fast-growing but vulnerable agricultural grasses, ecologists planted slow-growing, native grasses that were known to be adapted to minerally deficient soils and to the harsh conditions that exist in cleared areas. In choosing these plants, the ecologists relied on their observations of what vegetation first appeared in areas of Great Britain that had undergone succession naturally.[2] The result of the ecological approach has been successful restoration of once damaged lands.

The Peak District National Park provides an example of the application of ecological knowledge about succession to help restore a damaged landscape. Such information will be more and more important in the future, as landscapes continue to be damaged and as public interest in restoration continues to increase. In this chapter we consider the basic concepts of ecological succession, including what causes it, what can be done to promote it, and how we can use this knowledge to restore areas subjected to severe environmental damage to self-sustaining ecosystems.

Learning Objectives

Recent studies show that ecosystems, even when undisturbed by people, are not static but change over time. Studying the patterns of the changes in ecosystems in response to natural or human disturbance is an important tool for learning how to restore lands damaged by people. After reading this chapter, you should understand:

- How an ecosystem redevelops following a disturbance through a process known as ecological succession.
- The important role that disturbances play in the persistence of ecosystems.

- The patterns of succession that are involved in the return of life to disturbed land and the role played by various species, including how chemical cycling can lead to a loss of nutrients from ecosystems.
- The different effects on the land of physical forces and biological processes.
- How chemical cycling can result in a loss of nutrients in the soil from ecosystems.
- Why ecosystems do not maintain a steady-state condition.

9.1 RESTORATION OF DAMAGED LANDS

Restoration of ecologically damaged lands is an important practical problem. Throughout history, people have altered the landscape, often with undesirable effects; forests have been cleared and soil eroded.[3] These destructive effects on the land are not new. During the height of the ancient Greek civilization, for example, Plato wrote that the hills of Attica in Greece were a "skeleton" of their former selves.[4] As people sought more agricultural and pasture land, marginal areas of forests and grasslands were converted to crops and pasture and later abandoned when they were no longer productive.

A major problem in the twentieth century is that toxic pollutants have damaged ecosystems. For example, in an area known as the "black triangle," an industrial zone that spans the Czech Republic, Poland and eastern Germany, air pollution damage to forests is extensive. It is estimated that by the beginning of the 21st century, 76% of the forests in the Czech Republic will be changed by air pollutants.[5] Mining has ruined many areas. In eastern Germany, open-pit mining has destroyed about 60,000 ha (about 148,000 acres); in northern Bohemia, an estimated 155,000 ha (more than 380,000 acres) has been destroyed.[6] People are becoming increasingly active in trying to restore some of these damaged landscapes. However, restoration is not a simple task. For example, in 1991 the Black Triangle Environmental Programme was established and although air pollution emissions have decreased, contribution of emissions to forest damage in the region still remains considerable. We are finding that one of the important components of restoration efforts is to learn the patterns of succession in nature.[7]

9.2 ECOLOGICAL SUCCESSION

Recovery of ecosystems occurs naturally through a process called ecological succession. This natural recovery can occur if the damage is not too great. Sometimes, the rate of recovery is long in comparison to human desires. Natural areas are subject to disturbances of many kinds. These disturbances are not usually human induced; natural disturbances such as storms and fires have always been a part of the environment.[8] Such disturbances have existed for so long that animals and plants have adapted to them and benefit from their occurrence.[9]

If fundamental requirements for life are available, areas on Earth without life are soon filled with living things. Over time, ecosystems undergo patterns of development called ecological succession. There are two kinds of ecological succession: primary and secondary. **Primary succession** is initial establishment and development of an ecosystem; **secondary**

FIGURE 9.2 Revegetation on new lava flows on Hawaii (Big Island). Ferns are among the first plants to reestablish after a new volcanic flow.

succession is reestablishment of an ecosystem. In secondary succession, there are remnants of a previous biological community, including such things as organic matter and seeds. By contrast, in primary succession such remnants are nonexistent or negligible. Forests that develop on new lava flows (Figure 9.2) or at the edge of a retreating glacier (Figure 9.3) are

FIGURE 9.3 Revegetation of the land at the retreating edge of a New Zealand glacier.

examples of primary succession. A forest that develops on an abandoned pasture (Figure 9.4) or one that grows after a hurricane, flood, or fire is an example of secondary succession.

(a)

(b)

(c)

FIGURE 9.4 A series of photographs of succession on an abandoned pasture in the Poconos: (*a*) a second-year field; (*b*) young red cedar trees, grasses, and other plants some years after a field was abandoned; (*c*) mature forests developed along old stone farm walls.

One of the best documented examples of natural disturbance is the role of fire in the northern woods of North America. The Boundary Waters Canoe area is an example of nature relatively undisturbed by people. It consists of a million acres in northern Minnesota designated as wilderness under the U. S. Wilderness Act. With that protection, the area is no longer open to logging or other direct human-induced disturbances. In the early days of European exploration and settlement of North America, French voyagers traveled through this region hunting and trading for furs. In some places, logging and farming was common in the nineteenth and early twentieth centuries, but for the most part the land has been relatively untouched. In spite of the lack of human influence, the forests show a persistent history of fire. Fires occur somewhere in this forest almost every year, and on the average the entire area burns once in a century. Fires cover areas large enough to be visible by satellite remote sensing (Figure 9.5). When fires occur at natural rates and natural intensities, there are some beneficial effects. For example, trees in unburned forests appear more susceptible to insect outbreaks and disease. Thus recent ecological research suggests that wilderness depends on change and that succession and disturbance are continual processes. The landscape is dynamic.[10]

Succession is one of the most important ecological processes, and the patterns of succession have many management implications. We see examples of succession all around us. When a house lot is abandoned in a city, weeds begin to grow. After a few years, shrubs and trees can be found; secondary succession is taking place. A farmer weeding a crop and a homeowner weeding a lawn are both fighting against the natural processes of secondary succession. Succession may involve large areas, such as that affected by the eruption of Mount Saint Helens (see A Closer Look 9.1, "Reforestation of Mount Saint Helens"). Even more common is succession on a fairly local level. For example, one tree may be blown over, opening up a gap for early successional plants.

Stages in Succession

Succession involves recognizable, repeated patterns of change. The pattern is familiar in areas of the United States such as New England, Michigan, Wisconsin, and Minnesota, where much land was cleared for farming and logging in the nineteenth century and later abandoned. Redevelopment of forests after farm fields have been abandoned is a typical and classic pattern of secondary succession (see A Closer Look 9.2, "An Example of Succession in an Abandoned Farm Field").

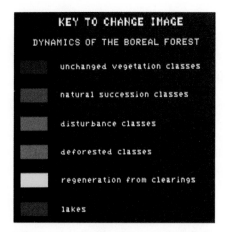

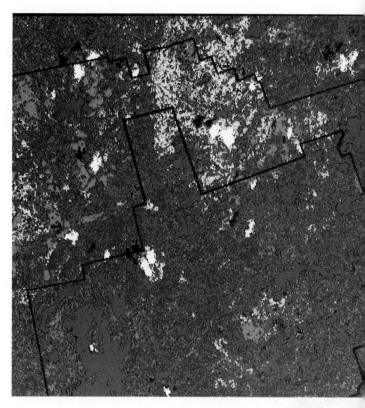

FIGURE 9.5 Forest fires can be natural or induced by human beings. This figure shows the change in a large area of the Superior National Forest in Minnesota between 1973 and 1983, as observed by the Landsat Satellite. The black boundaries show a central corridor where logging is permitted, surrounded by the Boundary Waters Canoe Area at the top and bottom. This is an area that is protected from all uses except certain kinds of recreation. The bright yellow shows areas that were clear of trees in 1973 but had regenerated to young forest by 1983. Most of this change is due to regrowth following a large fire that burned both inside and outside the wilderness. Red areas were forested in 1973 but cleared in 1983. Most of these are outside the wilderness, and some of these are due to logging (bright red) and some to fire or storms (dark red). Greens show areas that were forested both years.

General Features of Succession

Succession takes place not only in abandoned farm fields but also in areas of New England ravaged in 1938 by a hurricane that destroyed large areas of forest. Similar patterns occur elsewhere after fire, illustrated in the table in the Environmental Issue for lodgepole pine forests in the American West. In both cases, the forests grew back and recovered. From this we can describe several general features of succession.

First, the set of species that are present changes during succession.[11] In forested areas, plants that are rapid growing and short lived do well in the bright light and higher nutrient conditions that often occur after disturbances; these species tend to have widely and rapidly dispersing seeds. Such species are called pioneers, or **early successional species.** Plant species that dominate later stages of succession tend to be slower growing and longer lived. These plants do comparatively well in shade, and they have seeds that, while not as widely dispersing, can persist a rather long time. These are called **late successional species.**

In early stages of succession, biomass and biological diversity increase (Figure 9.7). In middle stages of succession, there are many species of trees and

many sizes of trees. Gross and net production (see Chapter 8) change during succession; gross production increases and net production decreases. Chemical cycling also changes. The organic material in the soil increases, as does the amount of chemical elements stored in the soils and the trees.[12] Some classic cases of succession are discussed in A Closer Look 9.3, "Some Classic Cases of Ecological Succession."

9.3 PATTERNS OF SPECIES CHANGE DURING SUCCESSION

During succession one species replaces another. In this section, we consider the causes of these changes. Do earlier species influence the timing of when other species are able to enter an ecosystem? The answer has practical consequences. White pine, for example, is an economically important tree in New England. White pine regenerates naturally following a disturbance; after the hurricane of 1938, white pine grew readily in abandoned pastures. But attempts to grow white pine commercially by planting it on bare soil did not succeed. Why is this so? Must grasses or other herbaceous plants be present before white pine can grow?

CLOSER LOOK 9.1

REFORESTATION OF MOUNT SAINT HELENS

Mount Saint Helens, a long-quiet volcanic mountain in Washington State, was a pleasantly wooded, peaceful area until early in the morning of May 18, 1980. Forests of Douglas-fir and western hemlock grew on the slopes below an elevation of 1200 m (4000 ft); mountain hemlock and firs grew above. Wildlife was abundant. On that May morning, life on the mountain was subject to a rapid series of devastating events when the volcano violently erupted. First, a large earthquake produced a massive avalanche of debris, moving the entire upper part of the mountain; then, superheated water and steam blasted the area. Mudflows spread downward to the valleys, and volcanic ash rained on other areas of the mountain. Finally, lava began to flow down the slopes. In all, 61,000 ha (151,000 acres) of land was damaged; on 21,000 ha (52,000 acres) forests were completely flattened; much of the soil was lost.[26]

The landscape was devastated and seemed devoid of life; it was believed that all animals above the ground at the time were killed, including 1600 elk. Stream habitats were destroyed, and fish almost disappeared from the streams. But to the surprise of almost everyone, including ecologists, life returned rapidly (Figure 9.6). Where the debris was not too thick above the original soil, fireweed and other plants common in clear-cut forest areas sprouted. Around fallen trees, spring flowers, such as trillium, appeared. Trees sprouted in areas protected from the blast and heat. Merely 4 years later, by 1984, the return of wildlife was well under way. More than 600 elk were living in the area,[27] and the streams were populated by 10 times as many steelhead trout as there had been before the eruption.[28] Life was returning, the establishment of an ecosystem and its biological community had begun on the mountain. [29]

FIGURE 9.6 Revegetation on Mt. St. Helens. Amid the standing and fallen trunks of trees killed by the volcanic explosion, herbs, shrubs, and trees have begun to sprout and revegetate the mountain slopes. This picture shows an early stage in forest ecological succession.

Patterns of Interaction

We find that there are at least three patterns of interaction among earlier and later species in succession (Figure 9.10).[13,14]

1. **Facilitation:** One species can prepare the way for the next (and may even be necessary for the occurrence of the next).

2. **Interference:** Early successional species may in some way prevent the entrance of later successional species.

3. **Life History Differences:** One species may not affect the time of entrance of another; two species may appear at different times during succession because of differences in transport, germination, growth, and longevity of seeds.

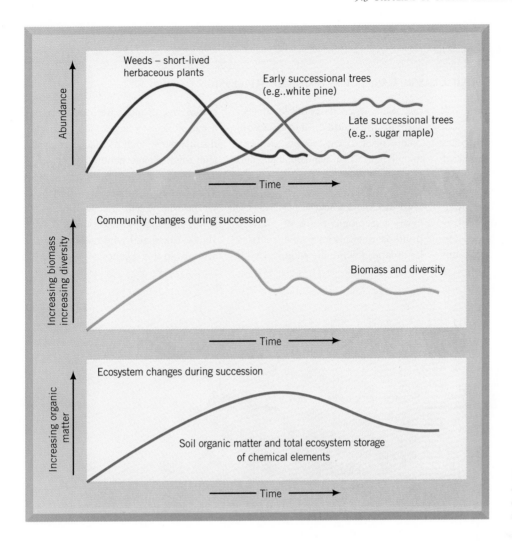

FIGURE 9.7 Graph of changes in biomass and diversity with succession.

CLOSER LOOK 9.2

AN EXAMPLE OF FOREST SUCCESSION

Within a few years after a field is abandoned, seeds of many kinds sprout, some of short-lived weedy plants and some of trees (Figure 9.4a). After a few years have passed, certain species, generally referred to as pioneer species, become established. In the Poconos of Pennsylvania, red cedar is such a species (Figure 9.4a). In New England, white pine, pin cherry, white birch, and yellow birch are especially abundant. These trees are fast growing in bright light, and have widely distributed seeds. For example, red cedar seeds are eaten by birds and dispersed; birches have very light seeds that are dispersed

widely by the wind. After several decades, forests of the pioneer species are well established, forming a dense stand of trees (Figure 9.6c).

Once the initial forest is established, other species begin to grow and become important. Typical dominants in the Northeastern United States are sugar maple and beech. These later successions of trees are slower growing than the ones that came into the forest first, but they have other characteristics that make them well adapted to the later stages in succession. These species are what foresters call shade tolerant; they grow relatively well in

the deep shade of the redeveloping forest.

After three or four decades, most of the short-lived species have matured, borne fruit, and died; these trees cannot grow in the shade of the forest and so do not regenerate in a forest that has been reestablished. For example, after five or six decades a New England forest is a rich mixture of birches, maples, beeches, and other species. The trees are a variety of sizes, although dominant trees are generally taller than those dominating earlier stages. After 1 or 2 centuries, such a forest will be composed mainly of shade-tolerant species.

CLOSER LOOK 9.3

SOME CLASSIC CASES OF ECOLOGICAL SUCCESSION

Succession Among the Ocean Tides

In the ocean, succession occurs where there is relative constancy of the environment such as between the high and low tidemarks along a rocky shore or in a coral reef. Where the ocean environment is constantly changing, succession does not appear to occur. For exam-ple, there is no perceptible succes-sion in the middle of the ocean, where open ocean waters are con-tinually stirred by winds, waves, and currents.

Wetland Succession

From a geologic point of view, a pond is a temporary feature of a landscape, eventually filling in with sediments. Ponds are common in areas subject to large-scale geologic disturbances, such as glaciation, that shift the drainage patterns of streams and create depressions and dams. This is why Minnesota, which lies within the heavily glaciated area of North America, is called the Land of 10,000 Lakes and why natural ponds are rare in the southern Great Plains

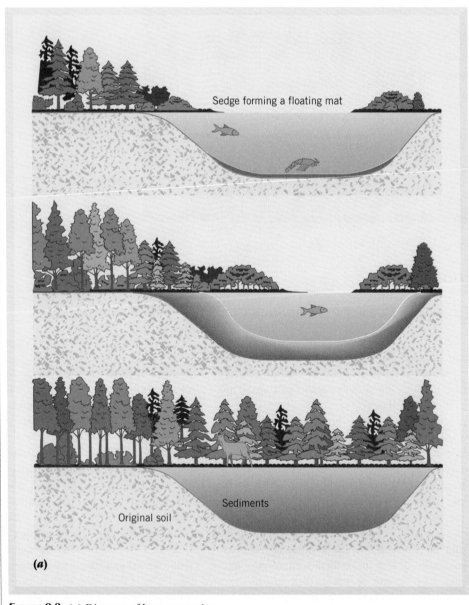

FIGURE 9.8 (*a*) Diagram of bog succession.

south of the glacial moraines, such as in Oklahoma.

When a pond is first created, it tends to have clear water with little sediment, low concentrations of chemical elements, and little organic matter. Over time these constituents increase. Streams carry sediments that are deposited in the comparatively quiet waters of the pond. The streams bring chemical elements to the pond, suspended in the particles and dissolved in the water. These enrich the pond, adding nutrients necessary for life. This increase in chemical elements is called the **eutrophication** of the pond. The young, nutrient-poor pond is called **oligotrophic;** the old, nutrient-rich pond is called **eutrophic.** The input of sediments and nutrients is sometimes referred to as the loading of the pond.

The increase in chemical elements in the pond allows a greater production of plants and animals, leading to an increase in live organic matter and an increase in the organic content of the sediments. This process is usually very gradual, and a pond may shift back and forth over a long period, from oligotrophic to eutrophic to oligotrophic, as climate and land vegetation upstream in the watershed supplying the pond's waters change.

Bog Succession

A bog is a body of water with acid waters and little if any surface outflow, so that the waters have little current. Often, a bog has sphagnum moss and inlets but no surface outlets. (This is a general description of typical bogs, as a technical definition may vary from expert to expert.) Succession in a bog is a process that begins with open water and ends with a forest (Figure 9.8).[30] Bog succession can be observed easily because the pond fills in from the edges toward the center. The center is successionally the youngest, and the bog's original edge is the oldest. In the quiet waters of the open part of a bog, sedge plants form floating mats that grow out over the water's surface. These short-lived shrubs are the pioneers. Their mat of thick, organic matter forms a primitive soil into which seeds of other plant species fall and germinate. Meanwhile, sediments build up on the bog bottom made up of dead organic matter from aquatic animals and plants as well as organic material that flows in from surface streams or is blown in by the wind.

The bog slowly fills in from the bottom to the top. Eventually the floating mat and sediments meet to form a base firm enough to support trees. The first trees that can survive under these conditions are adapted to wet grounds. For example, cedars and larches grow on these floating mats in northern forests, such as those of Minnesota, Michigan, and New England. These floating mats are called quaking bogs for a good reason. A visitor to a bog can easily demonstrate that the mat of sedges, shrubs, and trees is floating. If you jump up and down, the mat moves with you, like a water mattress.

If the process continues undisturbed, the entire bog fills in and a raised, heavily organic soil forms, in which other trees can survive. In some cases, the bog disappears and the area is taken over by tree species that are characteristic of mature forests on the well-drained soils of the region. In other cases, open water or a moss-covered wetland with some open water can persist. Wetlands can persist for very long periods without completely filling in. For example, in some Alaskan forests, a tree stage is replaced by a sphagnum moss stage, which can persist for a long time.

FIGURE 9.8 (*b*) Photograph of Livingstone Bog, Michigan.

Sand Dune Succession

Plant succession on sand dunes along beaches is a worldwide phenomenon, important along the shores of the Great Lakes in North America (Figure 9.9), in Cape Cod, Massachusetts, where Henry David Thoreau observed it in the nineteenth century, along the coasts of Australia, as well as in the famous Donana National Park in southern Spain. Dunes along the shores of Lake Michigan were among the first sites where succession was studied early in the twentieth century.[31]

Sand dunes are geologically unstable. They are continually formed, destroyed, and reformed by the action of winds, tides, and storms. The first pioneer plants that survive on a newly formed dune are dune grasses; they have long runners that anchor the plants in the sand. The runners have sharp ends that force their way through the sand as they grow (Figure 9.9). Dune grasses help stabilize the dune, making it possible for other plants to become established. Then seeds of shrubs and small trees germinate and grow on the stabilized dune. Eventually a small forest develops, including pines and oaks.

There are limits to the stability of the dune. Wind and water from a major storm can break through the dune, redistributing the sand and starting the process of succession over again. Often, as on the shores of Lake Michigan and the coast of Australia, series of dunes are visible,

FIGURE 9.9 Dune succession on the shores of Lake Michigan. Dune grass shoots appear scattered on the slope where they emerge from underground runners.

extending a considerable distance inland. The dunes closest to the water are in an early stage of succession, whereas the interior dunes were deposited earlier and have an older forest. It is possible to study the history of dune succession simply by walking inland.

From these examples, we can begin to see some general patterns of succession in the change in the physical structure of vegetation. Typically, succession begins with small plants (the dune's grasses, the bog's floating mats) and proceeds to larger and larger plants.

There is actually a fourth possibility: Succession never occurs and the species that enters first remains until the next disturbance. This fourth case is called **chronic patchiness.**

Each of these processes occurs in nature. Which occurs depends in part on the environmental conditions and in part on the pool of species that are available to take part in succession.

Facilitation

This pattern has been found to take place in tropical rain forests.[15] Early successional species speed the reappearance of the microclimatic condi-

tions that occur in a mature forest. In tropical forests, temperature, relative humidity, and light intensity at the soil surface can reach conditions similar to those of a mature rain forest after only 14 years.[16] Once these conditions are established, species that are adapted to deep forest shade can germinate and persist.

Sand dunes and bogs also illustrate facilitation. Dune grasses anchor the sandy soil so that seeds of plants that fall on the ground have a chance to germinate before they are buried too deep or blown away again (see A Closer Look 9.3). Sedges that form floating mats on the waters of a bog create a substrate where seeds of other species can lodge, germinate, and grow. (See "A Closer Look 9.3".)

Interference

Examples of interference can be found in certain tropical rain forests. When a rain forest is cleared, used for agriculture, and then abandoned, perennial grasses grow that form dense mats. For example, in parts of Asia, these include bamboo and *Imperata*, as well as thick-leaved small trees and shrubs. Together, these form stands so dense that seeds of other, later successional species cannot reach the ground, germinate, or obtain enough light, water, and nutrients to survive. *Imperata* either replaces itself or is replaced by bamboo, which then replaces itself.[17] Once established, *Imperata* and bamboo appear able to persist for a long time.

Life History Differences

An example of life history difference is illustrated by seed disbursal. When early successional species produce seeds, the seeds are readily transported by wind or animals, and so reach a clearing sooner and grow faster than seeds of late successional species (Figure 9.10). In many forested areas of eastern North America, birds eat the fruit of cherries and red cedar and their droppings contain the seeds that are spread widely. Species typical of old-growth forest, such as sugar maple, can grow in open areas, but these seeds take longer to get there. In this case the early succession of species might be neutral or interfere. Old-growth forest species typically have seeds with an adequate food supply for a seedling. Therefore, the seeds (maple, beechnut, acorns) tend to be heavy and not transported easily.

Chronic Patchiness

Whether a change in species occurs during succession depends on the complex interplay between life and its environment. Life tends to build up, to aggrade, whereas nonbiological processes in the environment tend to erode and degrade. In harsh environments, where energy or chemical elements required for life are limited and where disturbances are frequent, the physical, degrading environment dominates, and succession does not occur. Deserts exemplify the chronic patchiness that results. For example, in the warm deserts of California, Arizona, and Mexico, the major shrub species grow in patches that are often composed of mature individuals with few seedlings. These patches tend to persist for long periods until there is a disturbance.[18] Similarly, in highly polluted environments, a sequence of species replacement may not occur.

Knowledge of the causes of the succession of species can be useful in the restoration of damaged areas.[19] Plants that facilitate the presence of others

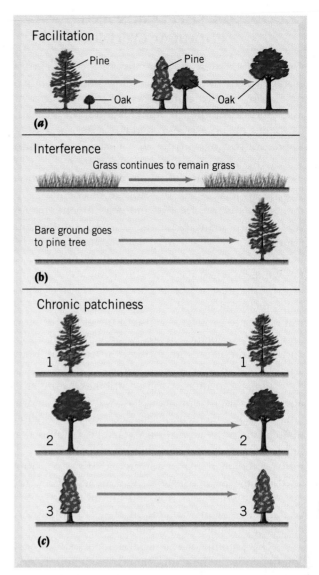

FIGURE 9.10 Several ways species might affect one another during ecological succession. (*a*) **Facilitation.** As Henry David Thoreau observed in Massachusetts, pines provide shade and act as "nurse trees" for oaks. Pines do well in openings. If there are no pines, few or no oaks will grow. The pines facilitate the entrance of oaks. (*b*) **Interference.** Some grasses that grow in open areas form dense mats that prevent seeds of trees from reaching the soil and germinating. The grasses' interference in the addition of tree species. (*c*) **Chronic patchiness.** The null condition, neither positive nor negative interactions. One species does not aid or hinder another. The physical environment tends to be dominant.

should be planted first, as on sandy areas, where dune grasses can help hold the soil before attempts are made to plant larger shrubs or trees.

This discussion suggests that species of plants show adaptations to specific stages in succession. When we attempt to restore landscapes, we can make use of our knowledge of these adaptations, just as the ecologists in our opening case study did when they used grasses adapted to early stages in succession in the restoration of mined lands.

9.4 SUCCESSION AND CHEMICAL CYCLING

On the land, there is generally an increase in the storage of chemical elements (nitrogen, phosphorus, potassium, and calcium, essential for plant growth and function) during the progression from the earliest stages of succession to middle or late succession. There are two reasons for this. First, organic matter stores chemical elements; as long as there is an increase in organic matter within the ecosystem, there will be an increase in the storage of chemical elements. This is true for live and dead organic matter. Additionally, many plants have root modules containing bacteria that can assimilate atmospheric nitrogen, which is then used by the plant in a process known as nitrogen fixation. The second reason is indirect: The presence of live and dead organic matter helps retard erosion. Both organic and inorganic soil can be lost to erosion because of the effects of wind and water. Vegetation tends to prevent such losses and therefore causes an increase in total stored material.

Organic matter in soil contributes to the storage of chemical elements in two ways. First, it contains chemical elements itself. Second, dead organic matter functions as an ion exchange column that holds onto metallic ions that would otherwise be transported in the groundwater as dissolved ions and lost to the ecosystem. As a general rule, the greater the volume of soil and the greater the percentage of organic matter in the soil, the more chemical elements will be retained.

The amount of chemical elements stored in a soil depends on the total volume of soil and its storage capacity for each element. Chemical storage capacity of soils varies with the average size of the soil particles. Large, coarse particles, like sand, have a smaller surface area per unit volume and can store a smaller quantity of chemical elements. Clay, which is made up of the smallest particles, stores the greatest quantity of chemical elements.

Soils contain greater quantities of chemical elements than do live organisms. However, much of what is stored in a soil may be relatively unavailable, or it may only become available slowly, because the elements are tied up in complex compounds that decay slowly. In contrast, the elements stored in living tissues are readily available to other organisms through food chains.

The rates of cycling and average storage times are system charcteristics (as discussed in Chapter 3). Soils store more elements than live tissue, but cycle them at a slower rate. The increase in chemical elements that occurs in the early and middle stages of succession does not continue indefinitely. If an ecosystem persists on the landscape for a very long period with no disturbance, there will be a slow but definite loss of stored chemical elements. Thus the ecosystem will slowly run downhill and become depauperate, that is, less able to support rapid growth, high biomass density, and high biological diversity (Figure 9.11).[20]

Changes in Chemical Cycling During a Disturbance

When an ecosystem is disturbed by fire, storms, or human actions, changes occur in chemical cycling. For example, when a forest is burned, complex organic compounds (such as wood) are converted to smaller inorganic compounds (including carbon dioxide, nitrogen oxides, and sulfur oxides). Some of the inorganic compounds from the wood are lost to the ecosystem during the fire, as particles of ash that are blown away or as vapors that escape into the atmosphere and are distributed widely. Other compounds are deposited on the soil surface as ash. These are comparatively highly soluble in water and

FIGURE 9.11 (*a*) Hypothesized changes in soil nitrogen during the course of soil development. (*b*) Change in total soil phosphorus over time with soil development. (*Source:* P. M. Vitousek and P. S. White, 1981, "Process Studies in Forest Succession," in D. C. West, H. H. Shugart, and D. B. Botkin, eds. *Forest Succession: Concepts and Applications,* Springer-Verlag, New York, Figure 17.1, p. 269).

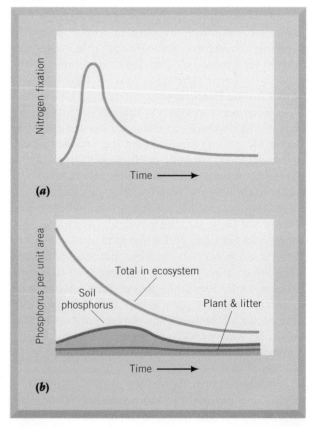

readily available for vegetation uptake. Therefore, immediately after a fire there is an increase in the availability of chemical elements. This is true even if the ecosystem as a whole has undergone a net loss in total stored chemical elements.

If sufficient live vegetation remains after a fire, then the sudden, temporary increase, or pulse, of newly available elements is taken up rapidly, especially if there is a moderate amount of rainfall (enough for good vegetation growth but not so much as to cause excessive erosion). The pulse of inorganic nutrients can then lead to a pulse in growth of vegetation and an increase in the amount of stored chemical elements in the vegetation. This in turn provides an increase in nutritious food for herbivores, which can subsequently undergo a population increase. The pulse in chemical elements in the soil can therefore have effects that extend through the food chain.

Other disturbances on the land have effects similar to those of fire. For example, severe storms, such as hurricanes and tornadoes, knock down and kill vegetation, which decays, increasing the concentration of chemical elements in the soil, which are then available for vegetation growth. Storms have another effect in forests; when trees are uprooted, the soil can be turned over, and chemical elements that were near the bottom of the root zone are brought to the surface, where they are more readily available.

Knowledge of the changes in chemical cycling and the availability of chemical elements from the soil during succession can be useful to us in restoring damaged lands. We know that nutrients must be available within the rooting depth of the vegetation. There must be sufficient organic matter in the soil to hold onto nutrients. Restoration will be more difficult on lands where the soil has lost its organic matter and has been leached than on lands that retain organic matter and have been little leached. Soils subject to acid rain and acid mine drainage pose special challenges to the process of land restoration.

9.5 SUCCESSION AND THE BALANCE OF NATURE

Succession on dunes and in wetlands and forests helps us think about one of the oldest and most intriguing questions people ask about wilderness: What are the characteristics of nature undisturbed? That is, what is the pristine, virgin state of nature unaffected by human beings? As we asked in Chapter 1: Is there a balance of nature, an ecological stability in an ecosystem undisturbed by humans? The answers to such questions tell us about the baseline state of nature, against which we can compare the

effects of our actions. Therefore these questions have much practical importance in the management of our natural resources and our entire planet and are also important to recreation, aesthetics, and philosophy.

As mentioned in Chapter 1, there has long been a belief that an undisturbed ecosystem has a kind of constancy and permanency. But recent evidence disputes this; the examples we have just discussed of very long successional patterns cast doubt on this idea. A forest in equilibrium must by definition have a zero net production: No addition of organic matter takes place. Such a forest can lose nutrients through geologic processes but has little means to accumulate them. Thus the ultimate fate of a never-disturbed forest is to go downhill biologically.

For much of the twentieth century, ecologists believed that succession proceeded to a fixed, classic **climax state,** defined as a steady-state stage that would persist indefinitely and have maximum organic matter, maximum storage of chemical elements, and maximum biological diversity.[21] This is an imaginary condition, however, never attained in nature. Evidence suggests that the greatest amount of organic matter and the greatest storage of chemical elements occur in the middle of succession (Figure 9.7).[22] There are two primary reasons for this situation. First, as organisms and populations grow, they must increase the amount of stored material (otherwise they could not grow). Thus, in a growing or aggrading ecosystem, inputs of chemical elements to the system will be greater than outputs, and there will be a net increase in ecosystem chemical storage. Second, at the same time that biological processes of growth are leading to an increase in stored chemical elements, nonbiological processes of erosion function to decrease this storage. Elements are dissolved in water and transported out of the ecosystem by ground and surface waters. Winds lift particles and move them away. In terms of chemical cycling, the process of succession is a continual conflict between aggrading forces of the biota and degrading forces of the physical environment.[23] Thus we cannot expect even those forests we protect from human interference to last forever.

In an old forest with a complex and diverse community and with trees of many species, the rate of loss from these physical processes may be small. In some ecosystems, such as certain tropical rain forests, there are many specific adaptations of vegetation to retain chemical elements. In such rain forests, only a small fraction of the chemical elements are stored in the soil; most chemical elements are held in the live biota and are recycled rapidly when trees die. In general, however, land ecosystems slowly lose some fraction of their stored chemi-

How Can We Evaluate Constructed Ecosystems?

What happens when restoring damaged ecosystems, as described for abandoned land mines in Great Britain at the beginning of this chapter, is not an option? In such cases, those responsible for the damage may be required to establish alternative ecosystems to replace the damaged ones, as in the case of saltwater wetlands on the coast of San Diego County, California. In 1984, construction of a flood control channel and two projects to improve interstate freeways damaged an area of saltwater marsh. The projects were of concern because California had lost 91% of its wetland area since 1943 and the few remaining coastal wetlands were badly fragmented. In addition, the damaged area provided habitat for three endangered species: the California least tern, the light-footed clapper rail, and a plant, the salt-marsh bird's beak.

The California Department of Transportation, with funding from the Army Corps of Engineers and the Federal Highway Administration, was required to compensate for the damage by constructing new areas of marsh in the Sweetwater Marsh National Wildlife Refuge. To meet these requirements, eight islands, known as the Connector Marsh, with a total area of 4.9 ha, were constructed in 1984. An additional 7 ha area, known as Marisma de Nacion, was established in 1990.

Goals for the constructed marsh, which were established by the U.S. Fish and Wildlife Service, included:

1. establishment of tide channels with sufficient fish to serve as food for the California least tern;
2. establishment of a stable or increasing population of salt-marsh bird's beak for three years;
3. establishment of habitat for the light-footed clapper rail with the elements needed for successful foraging and nesting of at least two breeding pairs.

The Pacific Estuarine Research Laboratory (PERL) at San Diego State University was selected to monitor progress on the goals and conduct research on the constructed marsh. In 1997, PERL reported that goals for the least tern and bird's beak had been met, but attempts to establish a habitat suitable for the rail had been only partially successful (see table).

During the past decade, PERL scientists have conducted extensive research on the constructed marsh to determine the reasons for its limited success. They found that rails live, forage, and nest in cordgrass more than 60 cm tall. Nests are built of dead cordgrass attached to stems of living cordgrass so that the nests can float above water as the level rises and falls. If cordgrass is too short, the nests cannot float high enough to prevent being washed out during high tides.

Researchers suggest that the coarse soil used to construct the marsh does not retain the amount of nitrogen

Species	Mitigation Goals	Progress in Meeting Requirements	Status as of 1997
California least tern	Tidal channels with 75% of the fish species and 75% of the number of fish found in natural channels	Met standards	In compliance
Salt marsh bird's beak	Through reintroduction, at least 5 patches (20 plants each) that remain stable or increase for 3 years	Did not succeed on constructed islands but an introduced population on natural Sweetwater Marsh thrived for 3 years (reached 140,000 plants); continue to monitor because plant is prone to dramatic fluctuations in population	In compliance
Tight-footed clapper rail	Seven home ranges (82 ha), each having tidal channels with:	Constructed	Not in compliance
	a. Forage species equal to 75% of the invertebrate species and 75% of the number of invertebrates in natural areas	Met standards	
	b. High marsh areas for rails to find refuge during high tides	Sufficient in 1996 but two home ranges fell short in 1997	
	c. Low marsh for nesting with 50% coverage by tall cordgrass	All home ranges met low marsh acreage requirement and all but one met cordgrass requirement and six lacked sufficient tall cordgrass	
	d. Population of tall cordgrass that is self-sustaining for 3 years	Plant height can be increased with continual use of fertilizer but tall cordgrass is not self-sustaining	

needed for cordgrass to grow tall. Adding nitrogen-rich fertilizer to the soil resulted in taller plants in the constructed marsh, but only if fertilizer were added on a continuing basis. PERL researchers also suspect that the lower diversity and numbers of large invertebrates, which are the major food source of the rails, in the constructed marsh compared with natural marshes is linked to low nitrogen levels. Because nitrogen stimulates the growth of algae and plants, which provide food for small invertebrates, and these in turn provide food for larger invertebrates, low nitrogen can affect the entire food chain.

Critical Thinking Questions

1. Make a diagram of the food web in the marsh showing how the clapper rail, cordgrass, invertebrates, and nitrogen are related.

2. The headline of an article about the Sweetwater Marsh project in the April 17, 1998 issue of *Science* declared "Restored Wetlands Flunk Real-World Test." Based on the information you have about the project, would you agree or disagree with this judgment? Explain your answer.

3. How do you think one can decide whether a constructed ecosystem is an adequate replacement for a natural ecosystem?

4. "Adaptive management" refers to the use of scientific research in ecosystem management. In what ways has adaptive management been used in the Sweetwater Marsh project? What lessons from the project could be used to improve similar projects in the future?

References

Malakoff, D. April 17, 1998. "Restored wetlands flunk real-world test." *Science* 280: 371–372.

Pacific Estuarine Research Laboratory (PERL). 1997 (November). "The status of Constructed Wetlands at Sweetwater marsh National Wildlife Refuge." Annual Report to the California Department of Transportation.

Zedler, J. B. 1997. "Adaptive management of coastal ecosystems designed to support endangered species." *Ecology Law Quarterly* 24: 735–743.

Zedler, J. B., and Powell, A. 1993. Problems in managing coastal wetlands: Complexities, compromises, and concerns. *Oceanus* 36(2): 19–28.

cal elements every year by the erosive effects of wind and water. This loss can be extremely slow in terms of human lifetimes. Whether physical erosion and loss dominate or whether biological aggradation and storage dominate depends on the relative rates of growth and erosion.

In an area of very heavy rainfall, erosion tends to override the biological effects. As an example, the west coast of New Zealand's southern island is famous for beautiful glaciers, glaciated valleys and harbors, and temperate rain forests. The glaciers are still melting and retreating as part of long-term trends that began at the end of the last ice age. Rainfall in this area of New Zealand is very high, as much as 700 cm/year (275 in./year). Erosional processes that might take much longer in areas with low rainfall occur rapidly here. The processes of succession and erosion can be observed on a walk from the present edge of a glacier toward the ocean shore. First, you pass over bare rock, then over rock covered by lichens, then over areas vegetated by grasses and lichens, then to a shrub and perennial grass stage, and on to several stages in the development of a temperate rain forest, which has great species diversity, high biomass, and large trees. This is the state that ecologists classically (throughout most of the twentieth century) have called the climax state. They assumed that this stage would persist indefinitely, as long as the forest was not disturbed.

But beyond the rain forest, farther from the glaciers and free of ice for thousands of years, is an even later stage of succession, composed of low shrubs and grasses with low species diversity and low biomass. Succession has not led to constancy or to maximum biomass and diversity.

Inspection of the soil shows a deeply leached layer. Over the centuries, rain has dissolved, transported, and deposited chemical elements from the upper soil, where plant roots can reach, to a depth below the reach of most plant roots. Without an abundance of chemical elements, the magnificent rain forests cannot persist.[24] Although the rate at which eroded elements are transported is unusually rapid on the west coast of New Zealand's south island, this process occurs in other places and is part of a general pattern.

Another example of the pattern occurs on a sequence of dunes in eastern Australia, which provides information about more than 100,000 years of succession.[25] Just as with the rain forests of New Zealand, the pattern of succession at first follows the classic idea. As one walks inland to older and older dunes, the stature of the forest first increases, as do total soil organic matter and the richness of the species. But then these factors decrease, and the very oldest dunes are depauperate. Gone are the large trees and the great diversity of species. On the oldest dunes one finds shrubs of low stature and a comparatively barren landscape. The soil has been slowly leached of its nutrients, which have been washed so far below the surface that tree roots can no longer reach them.

SUMMARY

- When damaged landscapes are restored, they undergo a process of recovery known as ecological succession, the process of establishment and development of an ecosystem. Knowledge of succession is important for the restoration of damaged lands.

- Understanding succession helps us answer some of the old questions about the characteristics of undisturbed wilderness; these questions are important to consider in the management of legally designated wilderness and natural areas.

- During succession, there is usually a clear, repeatable pattern of changes in species. Some species, called early successional, are adapted to the first stages, when the environment is harsh and variable, but necessary resources may be available in abundance. This contrasts with late stages in succession, when biological effects have modified the environment and reduced some of the variability but also have tied up some resources. Typically, early successional species are fast growing, whereas late successional species are slow growing and long lived.

- The causes of changes in the species found during succession can be due to facilitation, interference, or simply life-history differences. In facilitation, one species prepares the way for others. In interference, an early successional species prevents the entrance of later successional ones. Life-history characteristics of late successional species sometimes make them slow to enter an area.

- Biomass, production, diversity, and chemical cycling change during succession. Biomass and diversity peak in midsuccession, increasing at first to a maximum, and then declining and varying over time.

- Succession is one of the most important and useful ecological concepts. Succession is the key to restoration and recovery. We will encounter it throughout the rest of this text.

REEXAMINING THEMES AND ISSUES

Human Population: If we degrade ecosystems so that their recovery from disturbance is slowed or eliminated, then we reduce the local carrying capacity of an area for human beings. Therefore an understanding of the factors that determine ecosystem restoration is important to developing a sustainable human population.

Sustainability: Life tends to build up; nonbiological forces in the environment tend to tear down. By helping ecosystems to develop, we promote sustainability. Heavily degraded land, such as that significantly affected by pollution or overgrazing, loses the capacity to recover—to undergo ecological succession. Knowledge of the causes of succession can be useful in restoration of damaged ecosystems and therefore in achieving sustainability.

Global Perspective: Each degradation of land takes place locally, but such degradation has been happening around the world since the beginnings of civilization. Ecosystem degradation is therefore now a global issue.

Urban World: In cities, we generally eliminate or damage the processes of succession and the ability of ecosystems to recover. As our world becomes more and more urban, we must learn to maintain these processes within cities as well as in the countryside.

Values and Knowledge: Many people value natural ecosystems, the question is how can we maintain them. In our century, an important value has been placed on ecosystems completely undisturbed by human action; to maintain these, we must understand the natural role of disturbance and succession that takes place within them. Understanding succession is a key to maintaining these systems.

KEY TERMS

chronic patchiness 166
climax state 169
early successional species 161
ecological succession 157

eutrophic 165
eutrophication 165
facilitation 162
interference 162

late successional species 161
oligotrophic 165
primary succession 159
secondary succession 159

STUDY QUESTIONS

1. Farming has been described as managing land to keep it in an early stage of succession. Explain what this means and how it is achieved.

2. Redwood trees reproduce successfully only after disturbances (including fire and floods), yet individual redwood trees may live more than 1000 years. Is redwood an early or late successional species?

3. Why could it be said that succession does not take place in a desert shrub land (an area where rainfall is very low and the only plants are certain drought-adapted shrubs)?

4. What plan would restore an abandoned field in your town to natural vegetation for use as a park? The following materials are available: bales of hay; artificial fertilizer; seeds of annual flowers, grasses, shrubs, and trees.

5. Oil has leaked for many years from a gasoline tank of a gas station. Some of the oil has oozed to the surface. As a result, the gas station is abandoned and revegetation begins to occur.

   a. What effects would you expect this oil to have on the process of succession?

   b. What might you do to restore a different sequence of succession?

6. You are put in charge of a project to grow yellow birch, a species common in midsuccession, as a commercial crop. (Yellow birch is used to make dowels and hardwood plywood panels.) Considering the process of succession, how would you go about setting up plantations of yellow birch? How often would you log these plantations? Would the logging be complete or selective; that is, would you remove all the trees at one time or only some at any one time? Explain your answer.

FURTHER READING

Berger, J. J. 1990. *Environmental Restoration: Science and Strategies for Restoring the Earth.* Washington, DC: Island Press. Scientific and technical papers given at the first national conference on restoration, held in 1933, are presented in this informed and lively overview of the beginning of the restoration movement.

Cairns, J., Jr., ed. 1995. *Rehabilitating Damaged Ecosystems,* 2nd ed. Lewis Publishers. Contains discussions of natural and human-assisted restoration of different ecosystem types after either natural or human-caused disturbance.

Gorham, E., Vitousek, P. M., and Reiners, W. A. 1979. "The Regulation of Chemical Budgets over the Course of Terrestrial Ecosystem Succession," *Annual Review of Ecology and Systematics* 10:53–84. This paper examines the biogeochemical cycles of an ecosystem from the perspective of chemical inputs and outputs to that system.

Stevens, W. K. 1995. *Miracle Under the Oaks, the Revival of Nature in America.* Pocket Books. The story of several citizen action groups to restore damaged ecosystems.

West, D. C., Shugart, H. H., and Botkin, D. B., eds. 1981. *Forest Succession: Concepts and Application.* New York: Springer-Verlag. This volume of papers presents succession in forests from a variety of perspectives.

INTERNET RESOURCES

Conservation Ecology: *http://www.consecol.org/Journal*—*Conservation Ecology* is an electronic, peer-reviewed, scientific journal devoted to the rapid dissemination of current research. Content of the journal ranges from the applied to the theoretical. *Conservation Ecology* is published continuously on the Internet.

European Environment Agency (EEA): *http://www.eea.eu.int*—The EEA's goal is to help achieve significant and measurable improvement in Europe's environment through the provision of information. The Web site provides access to reports, a search engine, and links to other sites.

Organization for Economic Co-operation and Development (OECD): *http://www.oecd.org/env*—The OECD is an international organization that encourages the integration of environmental, economic and social policies. The Web site provides access to reports and publications.

Society for Ecological Restoration (SER): *http://nabalu.flas.ufl.edu/ser/SERhome.html*—SER is an international professional membership organization for people researching, practicing or interested in Ecological Restoration. An on-line library provides a browsable list of SER conference paper abstracts (progressively being added) and other SER publications related to restoration.

The Energy and Environmental Programme (EEP): *http://www.riia.org/Research/eep*—EEP is one of the seven research programs housed within the Royal Institute of International Affairs. Research is conducted on the political, strategic and economic aspects of domestic and international energy and environmental issues and policy. The Web site provides access to EEP research.

The Greater Yellowstone Coalition (GYC): *http://www.desktop.org/gyc*—The Greater Yellowstone Coalition's goals are to help shape a future where wildlife populations maintain their diversity and vitality, ecological processes function with minimal intervention, and outdoor recreational opportunities are available for residents and visitors alike. The Web site has news articles available on-line.

EARTH'S BIOMES

Similar environments lead to evolution of organisms similar in form and function but not necessarily in genetic heritage (see Chapter 7 for the distinction between *biotic realms*, and *biomes*). For example, deserts occur on all continents except Antarctica. Major desert plants of each continent often look similar but have different genetic heritage (see discussion of deserts later in this section). Deserts form a category of ecosystem. Ecologists call this category the desert biome. A biome is a kind of ecosystem. Major biomes of Earth are shown in Figure B1. The close relationship between environment and kinds of lifeforms is illustrated in Figure B2. The distribution of biomes gives Earth a distinctive pattern (Figure B1), as seen from space (Figure B8). Note the correspondence between this view from space

with the average summer temperature also measured from space (Figure B3). This pattern can be seen from space using satellite remote sensing (Figure B8). Note, for example, the correspondence between July average temperatures above 30°C (Figure B3) and deserts (Figure B1) in the Americas and in Africa. And note the correspondence between the location of boreal forests and July average temperature below 20°C. As Figure B2 suggests, a global map of precipitation, combined with the map of temperature, would lead to more refined correlations between climate and vegetation. Warm, wet climates favor high vegetation production, and biomes of these climatic zones tend to be highly productive—as long as other factors such as availability of chemical elements that plants require, are not limiting;

cold or dry climates do not permit high rates of vegetation growth. Therefore different biomes are characterized by different average rates of vegetation production (Figure B4). Knowledge of climate can be used to predict what kinds of vegetation are likely to be found in a region. Vegetation is the form of life most visible from space, but other forms of life have similar geographic relationships.

Knowledge of basic characteristics of the world's major biomes and features of life within each is important for planning, dealing with environmental issues, and determining beneficial introductions of new species. For example, we can make use of the correlation between geography of climate and geography of life to forecast possible effects of global warming on the geography of life. Biomes are usually

FIGURE B1 Global distribution of the major land biomes. (*Source:* H. J. deBlij, and P. O. Muller, 1996, *Physical Geography of the Global Environment,* New York, Wiley, Figure 27-1, p. 290.)

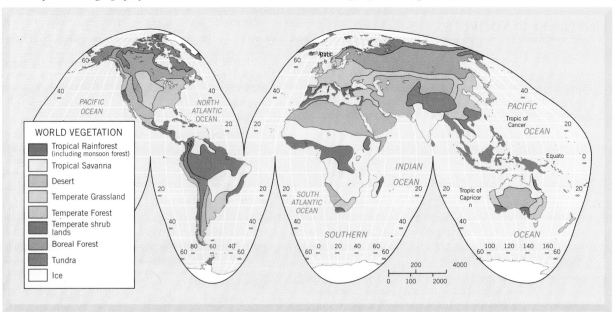

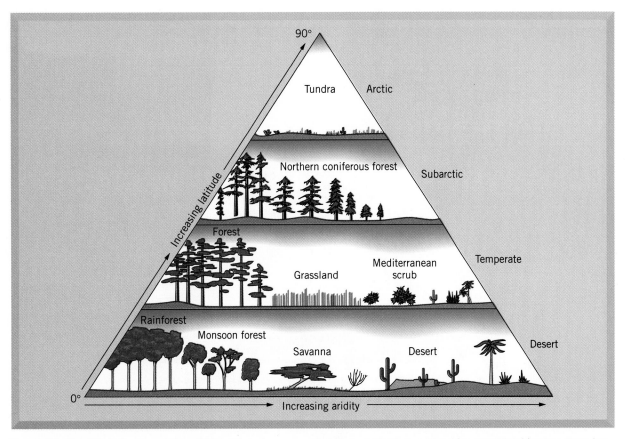

FIGURE B2 Simplified diagram of the relationship between precipitation and latitude and Earth's major land biomes. Here latitude serves as an index of average temperature, so "latitude" can be replaced by "average temperature" in this diagram. (*Source:* H. J. deBlij and P. O. Muller, 1996, *Physical Geography of the Global Environment,* Wiley, New York, Figure 27-4, p. 293.)

named for the dominant vegetation (e.g., coniferous forests as opposed to grasslands), for the dominant shape and form (called the physiognomy) of the dominant organisms (forest versus shrub land), or for the dominant climatic conditions (cold desert versus warm desert).

Tundra

Tundras are treeless plains that occur under the harsh climates of low rainfall and low average temperatures (Figure B5). (Figures B1 and B3 suggest that arctic tundra in the northern hemisphere occurs where the July average temperature is 15°C or lower.) The dominant vegetation includes grasses and their relatives (sedges), mosses, lichens, flowering dwarf shrubs, and mat-forming plants.

There are two kinds of tundra: arctic, which occurs at high

latitudes, and alpine, which occurs at high elevations. The vegetation of both is similar, but the dominant animals are different. Arctic tundras typically have important large mammals, as well as important small mammals, birds, and insects; in alpine tundras, the dominant animals are small rodents and insects. Alpine tundras occupy comparatively small, isolated areas, whereas arctic tundras cover the large territories required for populations of large mammals.

Parts of tundra have *permafrost,* which is permanently frozen ground. Such areas are extremely fragile ecologically; when disturbed by such activities as the development of roads, permafrost areas may be permanently changed or may take a very long time to recover. As the environment becomes more

harsh, the vegetation grades from dwarf shrubs and grasslike plants to mosses and lichens, and finally to bare rock surfaces with occasional lichens. The extreme tundra occurs in Antarctica, where the major land organism in some areas is a lichen that grows within rocks, just below the surface.

Taiga, or Boreal Forests

The taiga biome includes the forests of the cold climates of high latitudes and high altitudes. Taiga forests are dominated by conifers, especially spruces, firs, and larches, and certain kinds of pines. Aspens and birches are important flowering trees (Figure B6). Boreal forests are characterized by dense stands of relatively small trees, typically under 30 m, which form dense shade and make walking difficult. Although

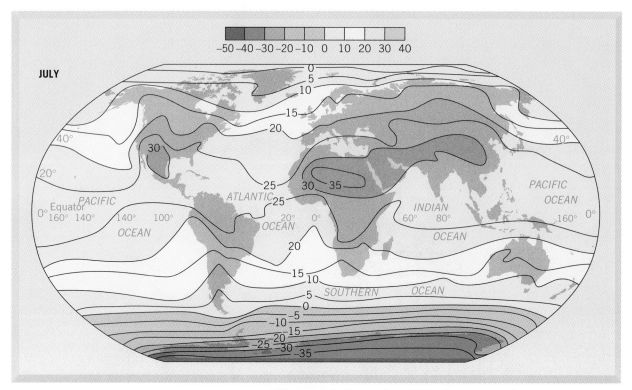

FIGURE B3 Mean sea-level air temperatures (°C) for July. (*Source:* H. J. deBlij and P. O Muller, 1966, *Physical Geography of the Global Environment,* Wiley, New York, Figure 8-11, p. 93.)

boreal forests cover very large areas, relatively few important species of trees occur. There are only about 20 major tree species in North American boreal forests, for example. Boreal forests are among the most economically important biomes as they are the source of much lumber and paper pulp. Because the northern areas of North America and Eurasia have been connected by land bridges during past ice ages, the animals and vegetation of this biome have been able to

FIGURE B4 Geography of vegetation production. The different biomes are characterized by different average rates of the production of vegetation. Tropical forests, temperate rain forests, and temperature deciduous forests are highly productive. This is another way that we can see the relationship between climate and biomes. (*Source:* H. J. deBlij and P. O. Muller, 1996, *Physical Geography of the Global Environment,* Wiley, New York, Figure 26-3, p. 281.)

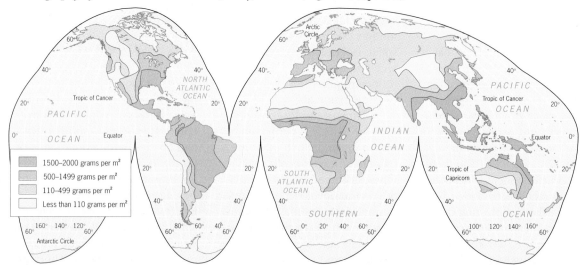

FIGURE B5 Tundra biome. Arctic tundra in Denali National Park, Alaska, with mountain avens in bloom.

fowl and carnivorous land birds, such as owls and eagles.

Disturbances—particularly fires, storms, and outbreaks of insects—are common in the boreal forests. For example, the entire million acres of the Boundary Waters Canoe Area of Minnesota burns over (through numerous small fires) an average of once each century, and individual forest stands are rarely more than 90 years old.

Temperate Forests

Temperate forests occur in climates somewhat warmer than those of the boreal forest. These forests occur throughout North America, Eurasia, and Japan and have many genera in common. Dominant vegetation includes tall deciduous trees; common species are maples, beeches, oaks, hickories, and chestnuts, typically larger in stature than trees of the boreal forest. These forests are important economically for their hardwood trees used for furniture. Temperate deciduous forests are among the biomes most changed by human beings because they occur in regions long dominated by civilization.

In the temperate deciduous forest, large mammals are less

common than in the taiga; herbivores include deer. The low density of large mammals results in part from the deep shade of the forest interior; there is less for ground-dwelling mammals to feed on. The dominant animals are small mammals, including those that live in trees (such as squirrels) and those that feed on soil organisms and small plants (such as mice). Birds and insects are abundant.

Temperate Rain Forests

Temperate rain forests occur under moderate temperature regimes where the rainfall exceeds 250 cm/year. Such rain forests are rare but spectacular; these are the giant forests (Figure B7). In the Northern Hemisphere, the dominant trees are evergreen conifers. This biome includes the redwood forests of California and Oregon, where the tallest trees in the world exist. It also includes forests of the state of Washington and adjacent Canada, dominated by such large trees as Douglas-firs and western cedars. Trees are taller than 70 m and are long lived; Douglas-firs live more than 400 years. Temperate rain forests also occur in the Southern Hemi-

spread widely. Moose, for example, are found in both continents. In this case, the boreal forests of North America and Eurasia share both genetic heritage and similar shapes and form.

The dominant animals of boreal forests include a few large mammals (moose, deer, wolves, and bears), small rodents (squirrels and rabbits), small carnivores (foxes), many insects, and migratory birds, especially water-

FIGURE B6 The boreal forest and a moose, a characteristic mammal of this biome.

FIGURE B7 Temperate rain forest biome. Moss covers Sitka spruce in Olympic National Park, a famous example of the temperate rain forest of the Pacific Northwest of North America.

sphere, the best known of which are the forests of western New Zealand.

Temperate rain forests have comparatively low diversity of plants and animals, in part because the climatic conditions tend to favor specialized species, in part because the abundant growth of the dominant vegetation produces a very deep shade in which few other plants can grow to provide food for herbivores.

This biome is important economically; redwoods, Douglas-firs, and western cedars are major North American timber crops.

Temperate Woodlands

Temperate woodlands occur where the temperature patterns are like those of deciduous forests, but the climate is slightly drier. Temperate woodlands are dominated by small trees such as pinyon pines and evergreen

oaks. The stands tend to be open, allowing considerable light to reach the ground. These generally pleasant areas are often used for recreation. Small pines are typical of many temperate woodlands and, in North America, occur from New England south to Georgia and the Caribbean islands. Fires are a common disturbance to which many species are adapted.

Temperate Shrub Lands

Under still drier climates temperate shrub lands occur. A distinctive feature of this biome is the *chaparral*, a miniature woodland dominated by dense stands of shrubs that rarely exceed a few meters in height. Chaparrals occur in Mediterranean climates, climates with low rainfall that is concentrated in the cool season. Chaparrals are found along the coast of California, in Chile, in South Africa, and in the Mediterranean

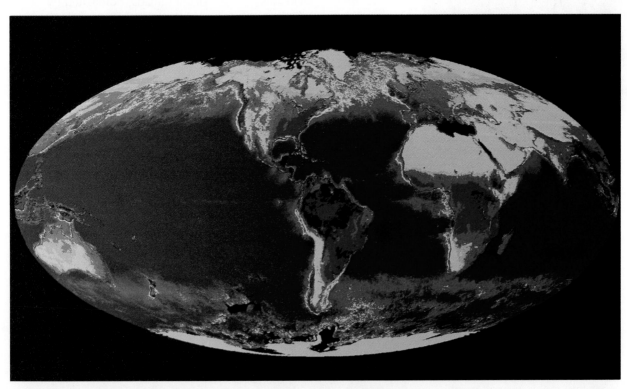

FIGURE B8 NASA Landsat imagery of the vegetation of Earth from space. (From "The Endangered Earth," a map supplement to the 100th anniversary issue of the National Geographic Magazine, 1988. The same map can be obtained from F. G. Hall, Goddard Space Flight Center, Code 623, Greenbelt, MD.)

region. There are few large mammals; reptiles and small mammals are characteristic.

Typically, the vegetation is distinctively aromatic, such as the scents of sage; some scientists believe the aromatic compounds are a means of competition among the plants. Its animals and plants have little economic value at present, but this biome is important for watersheds and erosion control. Because they occur in climates desirable for settlement, chaparrals are undergoing rapid change by human activities. The vegetation is adapted to fires; many species regenerate rapidly, and some promote fires by producing abundant fuel in the form of dead twigs and branches. As a result, stands are rarely more than 50 years old. When intense precipitation follows fire, erosion rates can increase to some of the highest known, until renewed vegetation again protects the slopes. Managing fires in chaparral is especially important because housing is often near the fire-prone vegetation.

Temperate Grasslands

Temperate grasslands occur in regions too dry for forests and too moist for deserts. A major biome in terms of area covered, these grasslands include the great North American prairies (Figure B9), the steppes of Eurasia, the plains of eastern and southern Africa, and the pampas of South America. Dominant species are grasses and other flowering plants, many of which are perennials with extensively developed roots.

Soils often have a deep organic layer. The highest abundance and the greatest diversity of large mammals are found in grasslands: the wild horses, asses, and antelope of Eurasia; the once-huge herds of bison that roamed the prairies of the

FIGURE B9 Temperate grassland biome. The Great American Prairie was one of the world's largest areas of temperate grasslands, prior to European settlement of North America.

American West; the kangaroos of Australia; and the antelope and other large herbivores of Africa. Temperatures are moderate, similar to those of the temperate deciduous forests, and rainfall is usually seasonal.

Tropical Rain Forests

Where the average temperature is high and relatively constant throughout the year, tropical biomes occur. In the wettest climates and in climates with rainfall well distributed throughout the year are found the tropical rain forests of South and Central America, northeastern Australia, Indonesia, the Philippines, Borneo, Hawaii, and parts of Malaysia. Species diversity is high, with hundreds of species of trees within a few square kilometers. Typically, some trees are very tall, but there are many kinds of plants; some, like palm trees remain relatively small; others, like bromeliads and some ferns, grow on trees (Figure B10). Many species of animals occur; mammals tend to live in trees but some are ground

dwellers. Insects and other invertebrates are abundant and show a high diversity. Rain forests occur in some of the most remote regions of Earth and remain poorly known; many undiscovered species are believed to exist there.

Except for dead organic matter at the surface, soils tend to be very low in nutrients. Most chemical elements (nutrients) are held in the living vegetation, which has evolved to survive in this environment; otherwise, rainfall would rapidly remove many chemical elements necessary for life.

Tropical Seasonal Forests and Savannas

Tropical seasonal forests occur where rainfall is high but very seasonal, in India and Southeast Asia, Africa, and South and Central America. In areas of even lower rainfall, tropical savannas—grasslands with scattered trees—are found. These include the savannas of Africa, which, along with the grasslands, have the greatest abundance of large

FIGURE B10 Tropical rain forest biome. The vegetation along the Segama River in Borneo illustrates lowland tropical rain forests.

mammals remaining anywhere in the world. The number of plant species is high.

Disturbances, including fires and the impact of herbivory on the vegetation, are common but may be necessary to maintain these areas as savannas; otherwise, they would revert to woodlands in wetter areas or to shrub lands in drier areas. Under still drier climates, savannas are replaced by shrub lands, characterized by small shrubs, a generally low abundance of vegetation, and a low density of vertebrate animals.

Deserts

Deserts occur where the rainfall is less than 50 cm/year. Although most deserts, such as the Sahara of North Africa and the deserts of the southwestern United States (Figure B11), Mexico, and Australia, occur at low latitudes, cold deserts occur in the basin and range area of Utah and Nevada and in parts of western

Asia. Most deserts have a considerable amount of specialized vegetation, as well as specialized vertebrate and invertebrate animals. Soils often have abundant nutrients but little or no organic matter and need only water to become very productive. Disturbances are common in the form

of occasional fires, occasional cold weather, and sudden, infrequent, and intense rains that cause flooding.

There are relatively few large mammals in deserts. The dominant animals of warm deserts are nonmammalian vertebrates (snakes and reptiles).

FIGURE B11 Desert biome. Desert lands at White Sands National Monument, New Mexico. Deserts vary greatly in the amount of vegetation they contain. This desert is relatively sparse, but some have no vegetation.

Mammals are usually small, like the kangaroo mice of North American deserts.

Wetlands

Wetlands include freshwater swamps, marshes, bogs and saltwater marshes. All have standing water: the water table is at the surface, and the ground is saturated with water (Figure B12). Standing water creates a special soil environment with little oxygen, so decay takes place slowly and only plants with specialized roots can survive. Bogs—wetlands with a stream input but no surface water outlet—are characterized by floating mats of vegetation. (See Figure 9.8(a) and (b).) Swamps and marshes are wetlands with surface inlets and outlets.

Dominant plants are small, ranging from small trees, such as mangroves of warm coastal wetlands to black spruces and larches of the north, to shrubs, sedges, and mosses. Small changes in elevation make a great difference; on slight rises, roots can obtain oxygen and small trees can grow; in lower areas are patches of open water with algae and mosses.

Although wetlands occupy a relatively small portion of Earth's land area, they are important in the biosphere. In the oxygenless soils, bacteria survive that cannot live in high-oxygen atmospheres. These bacteria carry out chemical reactions, such as the production of methane and hydrogen sulfide, that have important effects in the biosphere. Over geologic time, wetland environments produced the vegetation that today is coal. Saltwater marshes are important breeding areas for many oceanic animals and contain many invertebrates. Dominant animals include crabs and shellfish, such as clams. Saltwater marshes are therefore an important economic resource.

FIGURE B12 Wetland biome. Wetlands include areas of open standing water and areas of herbs, grasses, shrubs, and trees that can withstand persistent or frequent flooding.

Dominant animals in freshwater wetlands are many species of insects, birds, and amphibians; few mammals are exclusive inhabitants of this biome. The larger swamps of warm regions are famous for large reptiles and snakes, as well as for a relatively high diversity of mammals where topographic variation allows small upland areas.

Fresh Waters

Although freshwater lakes, ponds, rivers, and streams make up a very small portion of Earth's surface, they are critical for our water supply for homes, industry, and agriculture and play important ecological roles (Figure B13). They are major recreational resources but are easily polluted. Dominant plants

FIGURE B13 Freshwater biome. The Great Lakes of North America are among the world's most famous examples of the freshwater biome. In this picture, at Miners Castle Picture Rocks National Lakeshore, the sharp transition between the waters of the lake and the adjacent forests is clearly illustrated.

are floating algae, referred to as *phytoplankton*. Along shores and in shallow areas are rooted flowering plants, such as water lilies. Animal life is often abundant. Open waters have many small invertebrate animals (collectively called *zooplankton*)—both herbivores and carnivores. Many species of finfish and shellfish are found.

Rivers and streams are important in the biosphere as major transporters of materials from land to ocean. Fresh waters are economically important to people for their drinking water, production of energy, fish, bird life, and for recreation and transportation.

Estuaries—areas at mouths of rivers where river water mixes with ocean waters—are rich in nutrients and usually support an abundance of fish. They are important as breeding sites for many commercially significant fish.

Intertidal

The intertidal biome is made up of areas exposed alternately to air during low tide and ocean waters during high tide (Figure B14). Constant movement of waters transports nutrients into and out of these areas, which are usually rich in life and are major economic resources. Large algae are found here, from giant kelp of temperate and cold waters to algae of coral reefs in the tropics. Attached shellfish and birds are usually abundant and form a major part of economic resources. Near-shore areas are often important breeding grounds for many species of fish and shellfish, often of economic significance. This near-shore part of the oceanic environment is most susceptible to pollution from land sources; as a major recreational area, it is subject to considerable alteration by people. Some of the oldest environmental laws concern the rights to use resources of this biome, and today some major legal con-

flicts continue about access to intertidal areas and harvesting of biological resources.

Open Ocean

Called the *pelagic* region, the open ocean biome includes open waters of much of the oceans. These vast areas tend to be low in nitrogen and phosphorus—chemical deserts with low productivity and low diversity of algae. Many species of large animals occur but at low density.

Benthos

Benthos ("deeps") is the bottom portion of oceans. Primary input of food is dead organic matter that falls from above; the waters are too dark for photosynthesis, so no plants grow there.

Upwellings

Deep ocean waters are cold, dark and life is scarce. However, these waters are nutrient rich because of numerous creatures who die in surface waters and sink. (See Chapter 8 "Energy Flow in Ocean Ecosystems.") Upward flow or "upwelling" of

deep ocean waters brings nutrients to the surface, allowing abundant growth of algae and animals that depend on algae. Upwellings occur off the west coast of North America, South America, West Africa, and near the Arctic and Antarctic ice sheets. In some areas, deeper waters are brought to the surface by winds that push coastal waters away from shore. These fertile upwelling zones are among the most important regions for the production of commercial fish.

Hydrothermal Vents

This recently discovered biome (see Chapter 4) occurs in deep ocean, where plate tectonic processes create vents of hot water with a high concentration of sulfur compounds. These sulfur compounds provide an energy basis for chemosynthetic bacteria, which support giant clams, worms, and other unusual lifeforms. (See Chapter 8.) Water pressure is high, and temperatures range from boiling point in waters of vents to the frigid (about 4°C) waters of deep ocean.

FIGURE B14 Intertidal biome. Rocky intertidal biome on the Atlantic coast of North America.

PART IV

Kalila wa Dimna (book of the fables of Bidpai): The Ascetic and the Jar of Honey. India (Gujarat); mid-16th century.

Sustaining Living Resources

With one-fifth of the world's population, China requires a large food supply which begins simply as in this picture with a family meal.

CHAPTER 10

WORLD FOOD SUPPLY

CASE STUDY

Food for China

China, with one-fifth of the world's population, is the most populous country in the world. Between 1980 and 1995, China's population grew by 200 million people—about three-fourths of the population of the United States—to reach 1.2 billion. Although its growth rate is expected to slow somewhat in the coming decades, population experts predict that there will be 1.5 billion Chinese by 2025 (Figure 10.1). But can China's food production continue to keep pace with its growing population? Should China develop a food deficit, it may need to import more grain from other countries than those countries can spare from their own needs.

To give some idea of the potential impact of China on the world's food supply, consider the following examples. All of the grain produced by Norway would be needed to supply two more beers to each person in China. If the Chinese were to eat as much fish as the Japanese do, China would consume the entire world fish catch. Food for all the chickens required for China to reach its goal of 200 eggs per person per year by 2000 will equal all the grain exported by Canada—the world's second largest grain exporter.[1] Increased demand by China for world grain supplies could result in dramatic increases in food prices and precipitate famines in other areas of the world.

FIGURE 10.1 Population of China, 1950–1994, with projections to 2050. [Source: Brown, L. R. 1995, *Who Will Feed China? Wake-up Call for a Small Planet.* New York: W. W. Norton, p. 36.]

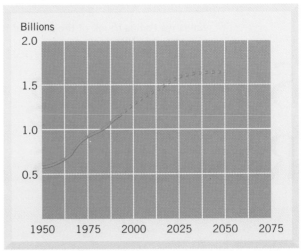

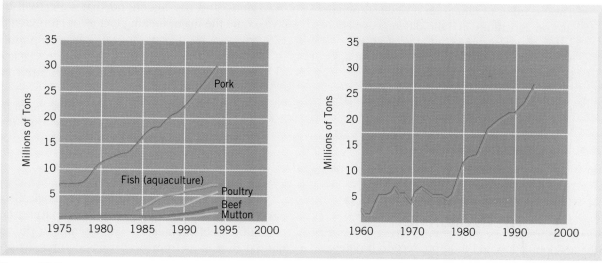

FIGURE 10.2 (*Left*) Meat consumption in China, 1975–1994. [Source: Brown, L. R., op. cit., p. 47.] (*Right*) Grain used for feed in China, 1960–1994. [Source: Brown, L. R., op. cit., p. 51.]

Between 1959 and 1961, 30 million Chinese starved to death during a famine brought on by a state policy of modernization that forced millions of farmers to work on large construction projects. In the wake of famine, China attempted to slow population growth by a "one couple–one child" policy and to increase food production through extensive agricultural reform. Between 1980 and 1995, total fertility rate in China fell from 4.8 children per woman to less than 2 children. In the same period, China achieved a remarkable increase in annual grain production from 200 kilograms per person needed to maintain a minimal level of physical activity to 300 kilograms.

Can China continue to produce enough food for its people? China has relatively little cropland—approximately 100 million hectares. By comparison, India has 170 million hectares of land with which to feed its population of 935 million. As a result of industrial development, China is losing agricultural land to roads, railroads, and manufacturing plants at the rate of 1.26 million hectares a year. At the same time, China, with half of its cropland under irrigation, is facing a severe water crisis due to seasonal variations in rainfall, periodic floods, and diversion of water to nonagricultural uses. Finally, with increased prosperity, the Chinese people are eating higher up the food chain; they are consuming more meat and dairy products—a diet that requires more grain production per person (Figure 10.2). Whether China's agricultural resources can sustain its population in coming decades is seen by many as a critical test of whether the world as a whole can find a sustainable balance between food production, consumer demand, and population growth. (See Environmental Issue "Should People Eat Lower on the Food Chain?" in Chapter 8.)

REFERENCES

A Food Crisis—or a Blip? 1996 (Feb.). *World Press Review* 43 (2): 34.

Brown, L. R. 1995. *Who Will Feed China? Wake-up Call for a Small Planet.* New York: W. W. Norton.

Cohen, J. E. 1995 (Nov./Dec.). How Many People Can the Earth Support? *The Sciences* 18–23.

Holmes, B. 1993 (Feb. 8). Feeding a World of 10 Billion. *U. S. News & World Report* 114 (5): 55.

Livernash, R. 1995 (July/Aug.). The Future of Populous Economies: China and India Shape Their Destinies. *Environment* 37 (6): 6–11, 25–34.

Ryan, M., and Flavin, C. 1995. Facing China's Limits. *State of the World, 1995.* New York: W. W. Norton.

Tyler, P. E. 1996 (May 23). China's Fickle Rivers: Dry Farms, Needy Industry Bring a Water Crisis. *The New York Times.*

LEARNING OBJECTIVES

The major agricultural challenge facing us today is to achieve sustainable production of crops and domestic animals. To do so requires that we take an ecological approach to food production. After reading this chapter, you should understand:

- The meaning of taking an ecological perspective on agriculture and the differences between agro-ecosystems and natural ecosystems.
- The major aspects of the relation between food supply and the environment.
- The importance of allocating various types of land for food production, rangeland, and so on.
- The role of limiting factors in determining crop yield, and the likelihood that water will become the major limiting factor for crop growth.
- The impact that the growing human population, the loss of fertile soils, and the lack of water for irrigation can have on future food shortages worldwide.
- The possibilities and limitations of some of the techniques of modern agriculture that may lead to increased food production.

10.1 WORLD FOOD SUPPLY AND THE ENVIRONMENT

China's population and food production problem, as well as similar food supply problems elsewhere, illustrate four important aspects of the relation between world food supply and the environment.

First, our modern food problem is the result of the great increase in human populations, which outstrips local food production in many areas, and an inadequate food distribution system for these growing populations.

Second, food production depends on the environment. When there is a drought, or a lack of any crucial resource, food production decreases. Because the environment is always varying, there are good years and bad years, and farmers must continually adjust to these environmental variations.

Third, agriculture changes the environment. When these changes are detrimental, large numbers of people can suffer. The history of agriculture can be viewed as a series of attempts to overcome environmental limitations and problems. But each new solution has created new environmental problems, which have in turn required their own solutions. In seeking to improve agricultural systems, we should expect some undesirable side effects and be ready to cope with them.

Fourth, world food supply is also greatly influenced by social disruptions and social attitudes, which affect the environment and in turn affect agriculture. In Africa, social disruptions since 1960 have included 12 wars, 70 coups, and 13 assassinations. This social instability makes sustained agricultural yields difficult.[2]

10.2 AN ECOLOGICAL PERSPECTIVE ON AGRICULTURE

When we farm, we create abnormal ecological conditions and novel ecosystems, known as **agro-ecosystems**. Agro-ecosystems differ from natural ecosystems in five ways (Figure 10.3).

First, in farming we try to stop ecological succession, to keep an agro-ecosystem in a constant condition (see Chapter 9). Any attempt to prevent natural processes from occurring requires actions, time, and effort on our part, and an input of energy. Most crops are planted on cleared land, which is then kept clear of other vegetation. By contrast, when a clearing is created by a natural disturbance, such as a fire or a storm, the vegetation returns—first, early successional species; then, later ones (see Chapter 9). Most crops are early successional species, which means that they do best when sunlight, water, and chemical nutrients in the soil are abundant. Under natural conditions, however, crop species would eventually be replaced by later successional plants.

Second, large areas are planted to a single species, or even to a single strain or subspecies, such as a single hybrid of corn. This practice, called **monoculture,** tends to reduce soil fertility because any species has its own set of requirements for specific chemical elements, and repeated planting of a single species can reduce the availability of certain essential chemical elements. This effect can be counteracted to a certain extent by crop rotation and with artificial fertilizers. In *crop rotation*, different crops are planted in turn in the same field, with the field occasionally left *fallow*. A fallow field is allowed to grow with a cover crop (sometimes planted, sometimes whatever plants germinate and grow) and is not harvested for at least one season; often the vegetation that grows on the fallow field is plowed under to add to soil fertility.

A third way in which agro-ecosystems differ from natural ecosystems is that crops are planted in neat, regular rows. In most natural ecosystems many species of plants grow mixed together in complex patterns, rarely forming regular rows. Regular patterns are important; a simple regular planting of one species makes it easy for pests; a complex pattern of many species makes it harder for the pests to find their victims.

Fourth, food chains are greatly simplified in agro-ecosystems. Pest control reduces the abun-

FIGURE 10.3 How farming changes an ecosystem.

dance and diversity of natural predators and parasites of crop plants; species that compete for nutrients with the desired crop are eliminated. The reduction in the number of species tends to make the ecosystem more susceptible to undesirable changes.

Fifth, plowing is unlike any natural disturbance of the soil. Nothing in nature repeatedly and regularly turns over the soil to a specific depth. Plowing exposes the soil to erosion and can damage its phys-ical structure, leading to a decline in organic matter and a loss of chemical elements to erosion.

10.3 SOURCES OF FOOD

Primitive societies obtained food through hunting and gathering. Some societies exist today that de-pend solely on these sources of food, but the great

FIGURE 10.4 Some of the world's major crops, including (*a*) wheat, (*b*) rice, (*c*) soybean, (*d*) bananas, and (*e*) coconuts. See text for a discussion of the relative importance of these five crops.

majority of people obtain food from cultivated plants and domesticated animals. Although some food is obtained from oceans and fresh waters, 95% of the human population's protein and most of its calories are obtained from traditional land-based agriculture of crops and livestock.[3]

Crops

Although there are 270,000 species of plants, only about 3,000 have been tried as agricultural crops, only 150 species have been cultivated on a large scale. Some crops provide food, others provide commercial products, including oils, drugs, pesticides, and fibers. Most of the world's food is provided by only 14 crop species. These are, in approximate order of importance, wheat, rice, maize, potatoes, sweet potatoes, manioc, sugarcane, sugarbeet, common beans, soybeans, barley, sorghum, coconuts, and bananas (Figure 10.4). Of these 14, six species provide more than 80% of the total calories consumed by human beings either directly or indirectly.[4] Wheat, rice, and maize alone make up the majority of food derived from food crops (Figure 10.5). Other crops, called forage, are important food for domestic animals; these include alfalfa, sorghum, and various species of grasses grown as hay. Alfalfa

is the most important such crop in the United States, where 14 million ha (about 30 million acres) are planted in alfalfa—one-half the world's total.

The bulk of human food is provided by eight grains (also called small grains): wheat, rice, maize, barley, oats, sorghum, millet, and rye. There is a large world trade in small grains, but only the USA, Canada, Australia, and New Zealand are major exporters; the rest of the world's nations are net importers. In 1998, the world cereal production was 1.9 billion tonnes, 1% below a 1997 record crop but still above trend. In the same year, food emergencies affected 40 countries worldwide due to the effect of both El Nino and La Nina weather patterns in Africa, Latin America, and Asia. Africa remains the continent with the most acute food shortages due to adverse weather and civil strife.[5]

It is useful to group farming into cash crops and subsistence crops. **Cash crops** are grown to be sold or traded in a large market. **Subsistence crops** are used directly for food by the farmer or sold locally where the food is used directly. Some cash crops may provide nonfood products. For example, rubber plantations produce latex from the sap of rubber trees; the latex is then sold to make rubber products. Other cash crops, such as coffee or tea, do not provide primary nutrition and are not necessary

FIGURE 10.5 Geographic distribution of world production of a few major crops.

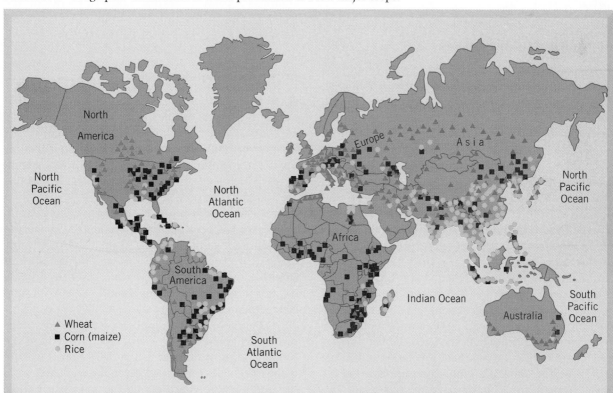

for survival, but in many countries such cash crops are a major source of international trade and foreign hard currencies. As a result, there is governmental and economic pressure to grow these cash crops instead of food crops.

Livestock

Domesticated animals are an important food source and have a major impact on the land. The major domesticated animals used as food are **ruminants** (see Figure 7.8). Ruminants have a four-chambered stomach within which bacteria convert the woody tissue of plants to proteins and fats that are, in turn, digested by the animal. In this way, ruminants convert cellulose, indigestible by people but Earth's most abundant organic compound, to human food. The major livestock are cattle, sheep, and goats, but there are other important species, including the water buffalo common in Southeast Asia, the camel of North Africa, and the llama of the South American Andes. Worldwide there are more than 1.7 billion sheep and goats and 1.3 billion cattle. There are 19.6 million camel and 188 million buffalo. These populations are projected to increase into the next century.[6] These animals are maintained on pasture and range. *Rangeland* is land that provides or can provide food for grazing and browsing animals without plowing and planting; *pasture* is land that is plowed and planted to provide forage for the animals.

Approaches to Agriculture

There are two major approaches to agriculture: (1) production based on highly mechanized technology, with a high demand for resources, including land, water, and fuel, and with little use of biologically based technologies and (2) production based on biological technology, with conservation of land, water, and energy. In the first kind, called **demand-based agriculture,** production is determined by economic demand and limited by that demand and not by resources. In the second approach, **resource-based agriculture,** production is limited by the availability of resources and economic demand usually exceeds production (Figure 10.6). The world as a whole is moving from demand-based to resource-based agriculture. The history of agriculture might be summarized most simply as (1) the introduction of resource-based agriculture about 10,000 years ago; (2) a shift to mechanized, demand-based agriculture during the industrial revolution of the last two centuries; and (3) a modern return to resource-based agriculture. Today's resource-based agriculture, however, differs from the original in that it in-

cludes many modern approaches and uses many forms of technology. (See A Closer Look 10.1, "New Genetic Strains and Hybrids" and 10.2 "Traditional Farming Methods.")

10.4 AQUACULTURE

Aquaculture is the production of food from aquatic habitats—marine and freshwater. Although aquaculture provides only a small amount of the world's food at present, it is important as a source of protein for many nations, especially in Asia and Europe, and offers a potentially important cash crop in other parts of the world. In China, growing fish in rice fields is an ancient practice that can be traced back to a treatise on fish culture written by Fan Li in 475 B.C.[7] In the Szechwan area alone there are more than 100,000 ha (about 250,000 acres) of rice fields; fish are farmed in the flooded fields with the rice.

Animals grown in aquaculture include oysters in many countries; shrimps and yellowtail in Japan; herbivorous crayfish in east Texas and Los Angeles; fish such as the carp, grown widely; tilapia in Africa, Israel, the United States, and many other countries; eels, Chinese carp, and minnows in China; catfish in the southern and midwestern United States; salmon in Norway and the United State; trout in the United States; the southeast Asian milkfish, plaice, and sole in Great Britain; mussels in France, Spain, and southeast Asian countries; and sturgeon in the Ukraine. Only a few species—trout and carp—have been subject to genetic breeding programs.[8]

Freshwater Aquaculture

Some fish growers in China and other Asian countries grow several species of fish in the same pond, exploiting their different ecological niches (see Chapter 6). For example, ponds developed mainly for carp, a bottom-feeding fish, also contain minnows, which feed at the surface on plant leaves that are added to the pond.

Sometimes, fish ponds are designed to use otherwise wasted resources. Some countries make use of fertilized water available from treated sewage. Other fish ponds make use of natural hot springs (Idaho) and of the warmed water used in cooling electric power plants (Long Island, New York, and Great Britain).[9]

An interesting example of an ecosystem approach to aquaculture occurs in China, where the production of freshwater fish and silkworms is combined. The silkworms feed on mulberry leaves, and the mulberry trees are planted around fish ponds. The leaves and the droppings from the silkworms

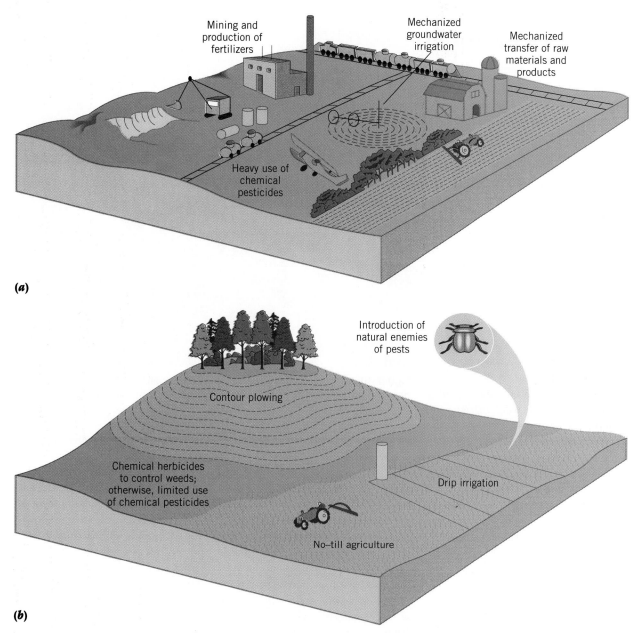

Mining and
production of
fertilizers

Mechanized
groundwater
irrigation

Mechanized
transfer of raw
materials and
products

Heavy use of
chemical
pesticides

(a)

Introduction of
natural enemies
of pests

Contour plowing

Chemical herbicides
to control weeds;
otherwise, limited use
of chemical pesticides

Drip irrigation

No–till agriculture

(b)

FIGURE 10.6 Agricultural technologies: (*a*) demand-based agriculture; (*b*) resource-based agriculture.

are added to the ponds to provide fertilizers for the growth of algae, which in turn provide food for the fish. The sediment from the pond is used to fertilize the trees. In this way the chemical elements are recycled (Figure 10.9).

In China, aquaculture includes a wide variety of organisms, from 50 species of fish, shrimp, crab, and shellfish to marine algae, sea turtles, sea cucumbers (a marine animal), and water chestnuts.[10] Aquaculture can be extremely productive on a per-area basis, especially because flowing water brings food into the pond or enclosure from outside. Although the area of Earth that can support freshwater aquaculture is small, we can expect this kind of aquacul-

ture to increase in the future and to become a more important source of protein.

Mariculture

For the most part, on the ocean, we are still only hunters and gatherers, not farmers. This is the great difference between traditional production of food from marine habitats and traditional agriculture. This situation is changing rapidly, but it remains more difficult for us to learn about methods of farming in the ocean in contrast to methods on land. However, **mariculture,** the farming of marine organisms, has grown rapidly in the last decade and can be ex-

A **CLOSER LOOK 10.1**

NEW GENETIC STRAINS AND HYBRIDS

FIGURE 10.7 Experimental rice plots at the International Rice Research Institute, Philippines, showing visual crop variation based on the use of fertilizer.

From the beginning, farming has affected the genetics of domesticated animals and plants, as people selected strains that were easy to grow and harvest. The act of farming makes certain species abundant where they were rare; thus farming selectively favors certain genotypes (populations with certain genetic characteristics).

Features that make a species or a genotype a weaker competitor under natural conditions sometimes make it more desirable as a crop. For example, wild relatives of wheat lose their seeds when the seeds are ripened and gently shaken by wind or animals. If you try to cut wild wheat stems and take them home, most of the seeds will be shaken off and there will be few left when you arrive. This is an adaptation that helps spread the seeds. Some individuals of wild wheat have a mutation that makes the seeds remain on the stalk. In the wild, these mutants leave fewer offspring and do not persist in the population; in the wild, the mutants are less fit genetically (see Chapter 7). Early farmers selected these mutants because they were easier to collect, transport, and use. In this way, people changed selective pressures on wheat and hastened the evolution of wheat strains that are useful to us but that could not survive on their own. One might also say that the result has been a symbiotic relationship between people and these otherwise less-competitive forms of wheat.

People also domesticated wheat by moving it to habitats to which it was not originally adapted, then developing new strains that could persist in these environments, and also changing the environmental conditions of the new habitat to better suit the wheat.[23] Corn (maize) went through a similar process of domestication. Likewise, domestic animals

have been bred to make them more docile and better producers of the meat and dairy products that we prefer. Modern agriculture has carried this process a step further with the frequent, intentional development of hybrids of different genotypes, bred to overcome new strains of disease and changes in climate. The process of intentional selection of new strains or strains with favorable characteristics for agriculture continues, as discussed in the next section.

The Green Revolution

The **green revolution** is the name attached to post–World War II programs that have led to the development of new strains of crops with higher yields, better resistance to disease, or better ability to grow under poor conditions. An advancement of the green revolution was the development of super strains of rice at the International Rice Research Institute in the Philippines. Although genetic hybridization of rice vastly increased rice production per acre, the new strains required a greater

use of fertilizers (Figure 4.7), as much as four to seven times the water, and in some cases produced a rice that was not considered desirable to eat. Another development in the green revolution was the formation of strains of maize with improved disease resistance at the International Maize and Wheat Improvement Center in Mexico.

Genetic Engineering

There is considerable interest in the potential for genetic engineering to develop strains of crops with entirely new characteristics. One focus of this research is the development of new crops that fix nitrogen, creating the same symbiotic relationship found in legumes (members of the pea family). Recall that legumes have bacteria that grow in root nodules. The bacteria live on products produced by the legumes; in turn, the bacteria fix nitrogen (meaning that they convert atmospheric gaseous nitrogen to a form that can be used by green plants). Legumes are often rotated with other crops so that the soil is enriched in nitrogen.

It may be possible to develop new strains of corn and other crops that, along with new strains of bacteria, can form a symbiotic nitrogen-fixing relationship. Such an accomplishment would lead to an increase in the production of these crops and a reduction in the utilization of fertilizers.

Another goal of agricultural genetic engineering is the development of strains with improved tolerance to drought, cold, heat, and toxic chemical elements. For example, one effort is aimed at developing wheat that is resistant to high levels of aluminum, an element that has negative effects on the growth of many plants.[24]

Although genetic modifications have proved to have great benefit, there are limitations and environmental concerns. Proponents of the approach argue that genetic engineering methods are necessary if we are to double or triple world food production in the next several decades to feed all the people of the world. Opponents argue that (1) the approach has too many environmental drawbacks and relies too heavily on mechanized agriculture, which is costly for the developing nations where the increase is most needed; (2) the approach does not pay enough attention to methods of nat-

ural pest control or to eliminating factors that lead to a loss in harvests; (3) some new strains require high amounts of fertilizers, draining resources; (4) the high production of certain crops requires intense cultivation methods, which may lead to greater erosion and to greater dependence on artificial chemical pest control; and (5) new superstrains might be too successful, that is, they could get away and dominate locations where they are not wanted, thus becoming pests.

New Crops

There is considerable potential for developing new crops by domesticating species that are currently wild. Although new crops are unlikely to replace current crop species as major food sources in this century, there is great interest in new crops to increase production in marginal areas and to increase the production of nonfood products, such as oils.

The development of new crops has been a continuing process in the history of agriculture. As people spread around the world, new crops were discovered and transported from one area to another. The process of introduction and increase in the production of crops continues. For example, the seed area har-

vested in the United States has increased from 17,000 ha in 1961 to 1.15 million ha in 1998 (42,000 acres to 2.8 million acres).[18]

Among the likely candidates for new crops are amaranth for seeds and leaves; *Leucaena*, a legume useful for animal feed; and triticale, a synthetic hybrid of wheat and rye. A promising source of new crops is the desert; none of the 20 major crops are plants of arid or semiarid regions, yet there are vast areas of desert and semidesert. In the United States there are 200,000 million ha (about 500,000 million acres) of arid and semiarid rangeland. In Africa, Australia, and South America the areas are even greater. Several species of plants can be grown commercially under arid conditions, allowing us to use a biome for agriculture that has been little used in this way in the past. Examples are guayule (a source of rubber), jojoba (for oil), bladderpod (for oil from seeds), and gumweed (for resin). Jojoba, a native shrub of the American Sonoran Desert, produces an extremely fine oil, remarkably resistant to bacterial degradation, which is useful in cosmetics and as a fine lubricant. Jojoba is now grown commercially in Australia, Egypt, Ghana, Iran, Israel, Jordan, Mexico, Saudi Arabia, and the United States.[26]

pected to increase in the future. Although mariculture provides only a small fraction of the world's protein, it is important in certain countries and can provide valuable cash crops.

Such delicacies as abalone and oysters, whose natural production is limited, are undergoing increased production as part of mariculture. In the United States and Canada, for example, there is research to learn how to attract the young, swimming stages of these shellfish to areas where they can be conveniently grown and harvested.

Oysters and mussels are grown on rafts that are lowered in the ocean, a common practice in Portugal, or on artificial pilings in the intertidal zone, a practice in the state of Washington (Figure 10.10). As filter feeders, these animals obtain food from water

that moves past them in currents. A small raft is exposed to a large volume of water and the food it contains, so that rafts can be extremely productive. For example, mussels grown on rafts in bays of Galicia, Spain, produce 300 metric tons/ha; by comparison, public harvesting grounds of wild shellfish in the United States yield about 10 kg/ha.[11]

10.5 SOIL AND AGRICULTURE

When a forest or prairie is cleared for agriculture, changes take place in the ecosystem. The original soil developed over a long period and is typically rich in organic matter, which is concentrated in its upper layers. Organic matter in soil is rich in chemi-

A CLOSER LOOK 10.2

TRADITIONAL FARMING METHODS

In industrialized countries of temperate zones there is a long history of plowing, but in less industrialized, tropical areas there is a history of agricultural methods that depend on clearing the vegetation without plowing. Where the loss of nutrients from the soil occurs rapidly following clearing, as in some tropical rain forests, the traditional practice is to cut the forest in small patches, but not cut it completely (Figure 10.8). Some shrubs and herbaceous plants are left. Several crops are planted together among existing vegetation. Crops are harvested for a few years. Then the land is allowed to grow back to a forest. The natural process of secondary succession—the redevelopment of the ecosystem—is allowed to occur; in fact, these farming practices promote this redevelopment and increase the conservation of chemical elements in the ecosystem.

After the forest has grown back, the process is repeated. This kind of agriculture has many names. It is sometimes called *cultivation with forest* or *bush fallow*. In Latin America, it is called *milpa* agriculture; in Great Britain, *swidden* agriculture; in west Africa, *fang* agriculture. In this type of agriculture, a mixture of crops is utilized, including root, stem, and fruit crops. For example, in western Africa, fang agriculture includes yams (a root crop) plus maize; in Southeast Asia, root crops are grown with rice and millet or with rice and maize.[26]

In theory, this method could be sustainable if human population density remained low. Erosional losses are minimized, and the soil eventually recovers its fertility. Uncut vegetation provides future seed sources. When human population pressures are low, a longer time occurs between the periods of use of any site. This is known as a *long rotation period*. Under high population pressures, such as occur in many places today, the rotation period is much shorter, and the land may not be able to recover sufficiently from previous usage. In such cases, production is not sustainable.

For many years agricultural experts from developed nations viewed this method of agriculture as a poor process with low, short-term productivity, used only by primitive peoples. Now it is understood that this kind of agriculture is well suited to high-rainfall lands where soils readily become impoverished when the land is completely cleared. The mixture of crops allows different species to contribute to soil fertility in different ways. Some perennial plants slow physical erosion; native legumes add nitrogen to the soil; and so forth.

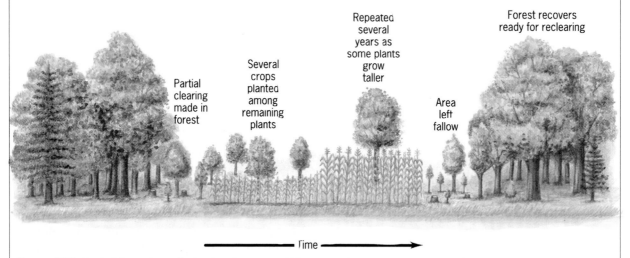

FIGURE 10.8 Bush fallow, also called milpa, fang, or swidden agriculture. Over time, secondary succession will take place on land partially cleared by slash and burn.

cal nutrients and provides a physical structure conducive to plant growth. When the original vegetation is cleared, there is less input of dead organic matter from the vegetation. The soil is exposed to sunlight, which warms it and increases the rate of

decomposition of its organic matter. The amount of organic matter declines.

Crop plants, which are typically early successional plants, grow well at first in the good soil under bright sun (see Chapter 9). But over the years,

FIGURE 10.9 Diagram of the traditional aquaculture practice in China. Leaves from mulberry trees and droppings from silkworms are added to ponds to provide fertilizers for algae, which in turn provide food for fish. The pond sediment is used to fertilize the trees. In this way, ecosystem chemical recycling is used to increase the production of both silk and fish.

the quality of the soil declines. Harvest of the crop leaves less dead vegetation to renew the organic matter in the soil. Plowing and harvesting open the soil to erosion and lead to compaction of the soil. The soil's physical structure becomes less conducive to plant growth.

Ironically, agriculture depends heavily on soil quality, but agriculture can lead to a decline in that quality. A high-quality agricultural soil has all the chemical elements required for plant growth and a physical structure that lets both air and water move freely through the soil, yet retains water well. Such a soil has a high organic matter content, which helps retain chemical elements, and a mixture of sediment particle sizes. Some sizes of particles help retain moisture and chemical elements, and other sizes help the flow of water. Physical qualities required vary with the crop. Lowland rice that grows in flooded ponds, for example, requires a heavy,

FIGURE 10.10 An oyster farm in Poulsbo, Washington. Oysters are grown on artificial pilings in the intertidal zone.

water-saturated soil, whereas watermelons grow best in very sandy soil.

Traditional demand-based agriculture has combated the decline in soil fertility through the use of organic fertilizers, such as animal manure. Organic fertilizers have an advantage in that they can improve both the chemical and physical characteristics of the soil. But organic fertilizers have drawbacks. They can be low in specific elements required by a particular soil.

One of the factors that has led to great increases in agricultural production is the development of modern, industrially produced, chemical fertilizers. For instance, industrial processes convert molecular nitrogen gas in the atmosphere to forms that can be used directly by plants. Phosphorus is mined. Then these and other elements are combined in proportions that are appropriate for specific crops in specific locations.

Limiting Factors

In many areas of the world, soils lack one or more of the chemical elements required by crops. Such an element is called a **limiting factor.** *An agricultural limiting factor is the single requirement for growth that is available in the least supply in comparison to the need of the crop.* In most cases, this lack is one of the *macronutrients*—a chemical element required in relatively large amounts by all living things. Macronutrients are sulfur, phosphorus, magnesium, calcium, potassium, nitrogen, oxygen, carbon, and hydrogen. In other cases, soils are lacking in *micronutrients*, which are specific trace elements, typically metals required only in small amounts. Micronutrients include molybdenum, copper, zinc, manganese, iron, boron, and chlorine. (Macronutrients and micronutrients are also discussed in Chapter 4.)

The older a soil, the more likely it is to lack trace elements because, as a soil ages, its supply of chemical elements tends to be leached by water from the upper layers to deeper layers (see Chapter 9). When crucial chemical elements are leached below the reach of crop roots, the soil becomes infertile.

Some of the most striking cases of soil nutrient limitations have been found in Australia, which has some of the oldest soils in the world. The Australian soils occur on land that has been above sea level for many millions of years, during which time severe leaching has taken place. Sometimes trace elements are required in extremely small amounts. For example, crops require very small amounts of molybdenum, and the addition of very small amounts has very great effects. It is estimated that adding an ounce of molybdenum to a field increased the yield of grass in sheep pastures by 1 ton/year.

Liebig's Law of the Minimum

The idea that some single factor determines the growth, and therefore the presence, of a species is known as *Liebig's law of the minimum.* Justus von Liebig was a nineteenth-century agriculturalist. He knew that crops required a number of nutrients in the soil and that crop yields could be increased by adding these as fertilizers. However, the factor that caused an increase varied. A general statement of this law is: *The abundance and distribution of a species is limited by the single factor in shortest supply or having the greatest impact.* Liebig wrote in 1840 that the yield of a crop was limited by the single nutrient that was in the shortest supply. Since then, ecologists have generalized Liebig's law beyond soil nutrients to include all life requirements. Although the law can be helpful, it should be used within limits and with caution.

If Liebig's law were always true, then environmental factors would always act one-by-one to limit the distribution of living things. For example, commercial fertilizers commonly contain nitrogen, phosphorus, and potassium. If a particular soil had nitrogen in least supply, phosphorus second, and potassium third, then, if Liebig's law were correct, the growth of plants would be increased by adding nitrogen up to some maximum amount. Adding phosphorus or potassium before adding nitrogen would have no effect. However, once the nitrogen content of the soil reached the maximum a particular plant required, then the plant's growth would be limited by phosphorus. Because phosphorus would now be the limiting factor, adding it would increase the yield. We could add phosphorus until the maximum was reached, at which point potassium would become the limiting factor.

This could continue until we had fertilized the soil to the point that the plant had all the nutrients it could use. Its growth might then be limited by water and light, which we could increase to a maximum. (Any of these factors can be limited at any time.) When all life requirements are available at a maximum amount, the yield of the plant is restricted by its own genetic characteristics and by the constraints imposed by fundamental laws of energy, the role of which was discussed in Chapter 8.

All this seems very reasonable, and Liebig's law seems to be true for the Australian soil that lacked molybdenum. But Liebig's law of the minimum as just stated is too restrictive, as it implies that there is no interaction between environmental factors, that each operates independently. This situation may not always be true. For example, nitrogen is a necessary part of every protein, and proteins are essential building blocks of cells. Enzymes, which make many cell reactions possible, contain nitrogen. A plant subject to little nitrogen and phosphorus might not make enough of the enzymes involved in taking up and using phosphorus. Increasing nitrogen to the plant might therefore increase the plant's uptake and use of phosphorus. If this were so, the two elements have a **synergistic effect.** In a synergistic effect, change in availability of one resource affects response of an organism to some other resource.

Sustainable Yields

A **maximum sustainable yield** of one crop is an amount produced per unit area that can be continued indefinitely without decreasing the ability of that crop species to sustain that yield. An **optimum sustainable yield** of a crop is the largest yield that can be obtained without decreasing the ability of the agro-ecosystem, as well as of the crop itself, to sustain that yield in the future. Little of modern agriculture is sustainable, and a major challenge in the future will be to create sustainable agriculture.

In terms of human beings and world food production, the major question is: How many people can be fed on a sustainable basis? In ecological terms, the major question is: What is the optimum sustainable population (or optimum sustainable carrying capacity) for human beings? The challenge before us is to produce adequate food to meet the demands of improved diets for an expanding population and an increasingly affluent population.

A major obstacle to answering this question is that technology for food production, processing, and distribution changes faster than we can understand the effects of the changes, and these changes have both positive and negative impacts on the environment. With modern technology, given enough energy and materials, we can grow food in a completely artificial environment, such as a spacecraft. But this is too expensive a solution in terms of money, resources, and effects on the environment to be a large-scale solution to food production.

Optimum Environmental Conditions

The concept of sustainable yield leads to the concept of optimum environmental conditions for growth of any single crop. Because every crop has its own set of optimum environmental conditions for growth, crops differ in where they grow best. Some crops, such as rye, grow best under relatively cool, moist conditions. Other crops, such as maize, do better in warmer, dryer climates.

Regions of Earth differ greatly in their capacity for crop production. Many factors influence which crops are produced in which areas, including tradition; access to technology and supplies; and local politics. For example, even within the United States, states differ greatly in their crop production (Figure 10.11). California has the greatest value of crop production in the United States, totaling $26.8 billion in 1997, and producing nearly 39 million tons of fruits, nuts, and vegetables in 1997. This accounts for more than half of U.S. production.[12] Much of California would be too dry to support the agriculture it does without irrigation. In the high plains east of the Rocky Mountains (where short-grass prairie originally grew) rangelands and irrigated farmlands are common. There, irrigated cropland is mainly planted in small grains (corn, wheat, etc.). Farther east in Nebraska are areas where winter wheat is most important; spring wheat is important in the Dakotas. The Midwest is the corn belt. In northern states from Minnesota to Maine, dairy and hay for cattle are important products. In the Southeast, major crops include cotton and tobacco, vegetables and fruits.

High crop production requires fertile soil, along with moderate rainfall distributed throughout the growing season, moderate to warm temperatures during growing season, and appropriate seasonal variations in climate. Some areas, such as far north, are unsuited to agriculture because of low temperatures. Other areas lack water or crucial chemical elements.

Only 11% of Earth's land surface is considered suitable for plant crops; about 19% of land in the United States is arable—144 million ha (about 357 million acres)—and of that, 80% was in production by the mid-1970s.[13] There is an additional 300 million ha (740 million acres) in pasture and rangeland. The best agricultural land is already in production. It is, however, constantly being lost to urbanization and soil erosion (about 1 million ha/year, or 2.5 million acres/year), as indicated by the following numbers. In 1982, 68.6 million acres of the 339 million acres of prime farmland in the United States were considered by the Census Bureau to be under urban influence. By 1992, prime farmland inside urban areas had increased to 82 million acres, while total acres nationwide had declined from 340 million acres in 1982 to 330 million in 1992.[14] About 79% of the yearly destruction of wetlands is attributed to agricultural practices. The drainage of wetlands adds about 117,000 acres per year, so the net loss to agricultural lands is 2.38 million acres per year.[15]

Irrigation

Plants require water. Much of modern agriculture involves *irrigation.* The importance of water is indicated by the fact that soil moisture in spring is a

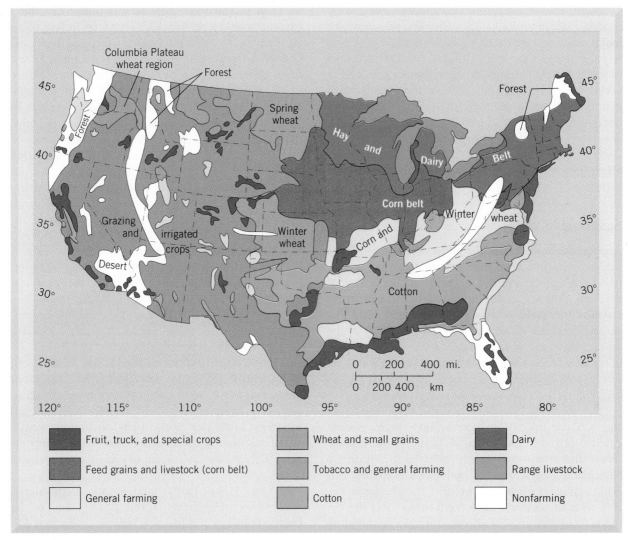

FIGURE 10.11 Major types of land use in the United States.

major determinant for agricultural production in the corn and wheat belts of the United States.

Like the use of fertilizers, techniques to add water to cropland are as old as agriculture (Figure 10.12). Sources of irrigation water include groundwater; nearby watercourses, such as rivers and streams; natural lakes and rivers; and artificial reservoirs. Large-scale irrigation projects cause environmental problems, as discussed in Chapter 11. Construction of reservoirs changes the local environment. Some habitats disappear. Stream patterns change, and erosion rates increase in the watershed of the reservoir (Figure 10.13).

10.6 WORLD FOOD SUPPLY

There are two kinds of food insufficiency: undernourishment and malnourishment. *Undernourish-*

ment is the lack of sufficient calories in available food, so that one has little or no ability to move or work. *Malnourishment* is the lack of specific components of food, such as proteins, vitamins, or essential chemical elements. Both are global problems.

Undernourishment leads to famines that are obvious, dramatic, and fast acting when they happen. Malnourishment is long-term and insidious. Although people may not die outright, they are less productive than normal and can suffer permanent brain damage. Among the major problems of malnourishment are *marasmus*, progressive emaciation caused by a lack of protein and calories; *kwashiorkor*, a lack of sufficient protein in the diet, which leads in infants to a failure of neural development and therefore to learning disabilities; and *chronic hunger*, which occurs when people have enough food to stay alive but not enough to lead satisfactory and productive lives.

(a)   *(b)*

FIGURE 10.12 Irrigation. (*a*) A United Nations Food and Agricultural Organization project helps farmers in the Sudan develop irrigation canals. (*b*) At a modern agricultural research station in Saudi Arabia, a grated pipe irrigation system brings water to a corn field.

Measures of Food Availability

A measure of the world's ability to deal with under-nourishment and famine is the *supply of grain on hand, measured in days*. This number has fluctuated since 1960, with a low of a 56-day supply in 1973 to a peak of a 104-day supply in 1987. Recent stocks have been declining; the 1995 estimate was a 62-day supply and the 1996 estimate was a 49-day supply—less than the low 1973 supply.[16]

The United States stores more than one-half the world's surplus grains and has 4 million metric tons of grain for emergency relief. Since World War II, the United States has been the leading food exporter and donor in the world.

Today, sufficient food is produced to provide adequate calories for the entire human population, but distribution is uneven and inequitable.[17] It is im-

FIGURE 10.13 Some ecological considerations in watershed development include: slope, elevation, floodplain, and river delta location.

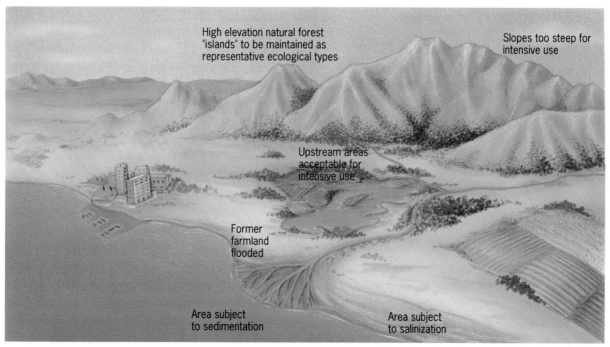

portant to have a system that includes stocks (reserves of food) and the ability to transport this food to where it is needed. For the entire world, such a system would require a standing stock of 10 million to 20 million metric tons; at a price of $137/ton, this would be about $2 billion worth of crops; maintenance of this stock would require $200 million/year.[18]

Another measure of the availability of food is **per-capita food production,** which is the amount of food produced per person. Today, population growth and food production are in a tug-of-war, and it is unclear in the short run which will increase faster, although we know in the long run there are limits to both. Both population and food supply have increased, but food production has varied more; on the average there has been no gain in the food available per person. This trend supports Malthus's 1798 prediction that at some point the growth of population will outstrip the capacity of Earth to supply food (see Chapter 5 and the Environmental Issue).

The *demand for food* is the amount of food that would be bought at a given price if it were available. The demand increases with population and with per-capita income. **Per-capita demand** is the economic demand per person. The per-capita demand increases as the standard of living and average income increase. The per-capita demand also increases with rising expectations, when people become aware that they could have more to eat and a greater variety of food. Modern communication has led to rising expectations and to growth in per-capita demand. As we saw in Chapter 5, an increase in either the environmental impact per person or the total number of people results in an increase in the total human environmental impact.

Limits to Food Production

Even given the great strides that can take place in food production from the use of new crops, genetic engineering, more fertilizer, and more water, there is a limit to the production of food on Earth. If populations continue to grow at present rates, the production of food must double or triple by the first decade of the next century for all people to be adequately fed. There are obstacles to this increase. Only some lands can be used for agriculture (Figures 10.13, 10.14). Second, for many crops we appear to have reached the limit of production gains realized for fertilizer applied. In the future, increases in yield per unit area must come from the development of new, higher producing crops or new superstrains of existing ones. Such increases can cause additional environmental problems.

Another limitation is that climatic change is more likely to decrease yield than to increase it be-cause areas with the best soils in the world also happen to have climates well suited to agriculture. Thus, most climatic changes are likely to make things worse. If global warming takes place as forecast by models of global climate, there will be major disruptions to agriculture.[19] The best climates for agriculture might shift from their present locations to locations farther north. For example, the climate of the midwestern corn belt of the United States, which has been so good for crop production, could move northward, favoring Canada. This move could have a real impact on the amount of grain produced because soils in Canada are generally not as suitable for grain production as are U.S. midwestern soils. Global warming could lead to an increase in evapotranspiration in many midlatitude areas.[20] Supplying water for irrigation would become an even greater problem than it is today. Those concerned about the future of agriculture must take the possibility of global warming into account in their planning.

Ways to Increase the Food Supply

As stated, most increases in agricultural production must come from the development of high-yield strains of crops. Several other methods of increasing the food supply may hold some promise, although always with limitations.

Improved Irrigation

A possible means to increase yield is with better irrigation techniques, which can improve yield and reduce overall water use (Figure 10.12). **Drip irrigation,** the application of water to the soil from tubes that drip water slowly, greatly reduces the loss of water from direct evaporation and increases yield. However, it is an expensive approach, most likely to be applied in developed nations or nations with a large surplus of hard currency, in other words, in few of the countries where hunger is most severe.

Increasing the Amount of Agricultural Land

What are the opportunities to add new lands to agriculture? Because most of the best land for agriculture is already so used, more and more marginal lands will be farmed. These lands are considered marginal because they lack water, have poor soil quality, or lie on slopes so steep that the soil readily erodes.

Some people suggest that in the future we will rely increasingly on artificial agriculture, such as **hydroponics.** Hydroponics is the growing of plants in a fertilized water solution on a completely artificial substrate in an artificial environment, such as a greenhouse. This approach is extremely expensive and is unlikely to be effective in the developing nations, where hunger is the greatest.

Eating Lower on the Food Chain

Some people believe that it is ecologically unsound to use domestic animals as food on the grounds that each step farther up a food chain leaves much less food to eat per acre. This argument is as follows. No organism is 100% efficient. Of the energy in food taken in, only a fraction is converted to new organic matter. Crop plants may convert 1–10% of sunlight to edible food, and cows may convert only 1–10% of hay and grain to meat. Thus, the same area could produce 10–100 times as much vegetation per year as meat (see the Environmental Issue in Chapter 5). This holds true for the best agricultural lands, which have deep, fertile soils on level ground.

However, as with so many things in environmental studies, this simple generalization does not apply to all cases. The fundamental question that faces us is how to achieve sustainable agriculture. Land too poor for crops can be excellent rangeland (Figure 10.14). If we could eat grass and hay, then the comparison would be valid. On poorer land— land on steeper slopes, with thinner soils, or with lower rainfall—crop production may not be sustainable (Figure 10.13). Plowing erodes the land; repeated harvesting of a crop reduces the fertility of the soil. Grazing animals may produce a sustainable product on such lands. Thus, from the point of view of sustainable agriculture, there is value in rangeland or pasture. Many lands have been ruined when they were converted unwisely from range to croplands. Similarly, many marginal rangelands have been ruined by overgrazing.

The wisest approach to sustainable agriculture involves a combination of different kinds of land use: using the best agricultural lands for crops and poorer lands for pastures and rangelands and avoiding the use of the best lands for grain production for animal feed.

Another problem with the argument that we should eat lower on the food chain is that food is more than just calories, and animals are important sources of proteins and minerals. Animals provide the major source of protein in human diets: 56 million metric tons of edible protein per year worldwide. In the United States, 75% of the protein, 33% of the energy, and most of the calcium and phosphorus in human nutrition come from animal products.

A third factor to keep in mind is that domestic animals are often used for other purposes—as beasts of burden; for plowing, carrying loads, and transportation; or as sources of clothing material (wool and leather)—as well as for food. The use of such animals for food represents an increase in efficiency.

Eating lower on the food chain makes sense from an ecological perspective mainly for prime agricultural land. Some people maintain vegetarian diets for religious or moral reasons or because of specific dietary problems, but these are independent of ecological arguments.

Modification of Food Distribution

If present trends continue, human population growth will eventually outstrip growth in world food production, and there will not be enough food to

FIGURE 10.14 Land unsuitable for crop production on a sustainable basis can be used for other purposes.

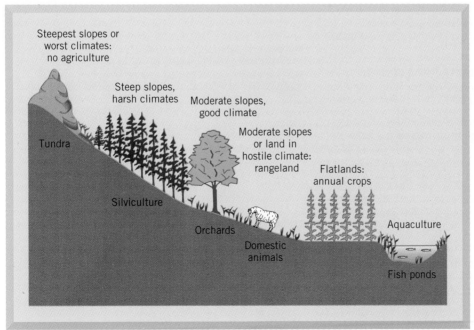

Will There Be Enough Water to Produce Food for a Growing Population?

By 2025, scientists estimate that the world population will have increased from 6 billion in 1999 to 8 billion, or approximately double what it was in 1974. To keep pace with the growing population, the United Nations Food and Agricultural Organization predicts that food production will have to double by 2025, and so will the amount of water consumed by food crops. Will the supply of freshwater be able to meet this increased demand or will water supply limit global food production?

Growing crops consume water through transpiration (loss of water from leaves as part of the photosynthetic process) and evaporation from plant and soil surfaces. The volume of water consumed by crops worldwide—including rainwater and irrigated water—is estimated at 3,200 billion m³ per year. An almost equal amount of water is used by other plants in and near agricultural fields; thus, it takes 7,500 billion m³ per year of water to supply globally crop ecosystems. Grazing and pasture land account for another 5,800 billion m³ and evaporation from irrigated water another 500 billion m³ for a total of 13,800 billion m³ of water per year for food production, or 20% of the water evaporated and transpired worldwide. By 2025, therefore, humans will be appropriating almost half of all the water available to life on land for growing food for their own use. Where will the additional water come from?

Although the amount of rainwater cannot be increased, it can be more efficiently used through farming methods such as terracing, mulching, and contouring. Forty percent of the global food harvest now comes from irrigated land, and some scientists estimate that the volume of irrigated water available to crops will have to triple by 2025—a volume equal to that of 24 Nile Rivers or 110 Colorado Rivers.[2] A significant saving of water can, therefore, come from more efficient irrigation methods, such as improved sprinkler systems, drip irrigation, night irrigation, and surge flow.

Additional water could be diverted from other uses to irrigation. But this may not be as easy as it sounds because of competing needs for water. For example, if water were provided to the 1 billion people in the world who currently lack drinking and household water, less would be available for growing crops. And, the new billions of people to be added to the world population in the next decades will also need water. Already, humans use 54% of the world's runoff and increasing this to more than 70%, as will be required to feed the growing population, may result in loss of freshwater ecosystems, decline in world fisheries, and extinction of aquatic species.

Irrigation projects are costly to construct and fewer sites than in the past are now available that are acceptable ecologically and socially. As a result, irrigation has been growing slower than population in recent years. If this trend continues, the irrigation gap will become even larger. Irrigation has also been responsible for salinization of agricultural areas, which makes them less suitable for growing crops and requires more water to flush the salts from the soil. In many places, groundwater and aquifers are being used faster than they are replaced—a process that is unsustainable in the long run. Many rivers are already so heavily used that they release little or no water to the ocean. These include the Ganges and most other rivers in India, the Huang He (Yellow River) in China, the Chao Phraya in Thailand, the Amu Dar'ya and Syr Dar'ya in the Aral Sea basin, and the Nile and Colorado Rivers.

Two hundred years ago, Thomas Malthus put forth the proposition that population grows more rapidly than the ability of the Earth to grow food and that at some time the human population will outstrip the food supply (see A Closer Look 5.1). Malthus might be surprised to know that, by applying science and technology to agriculture, food production has so far kept pace with population growth. For example, between 1950 and 1995, world population increased by 122% while grain productivity increased 141%. Since 1995, however, grain production has slowed down (see graph) and the question remains as to whether Malthus will be proven right in the next century. Will science and technology be able to solve the problem of water supply for growing food for

Estimated Water Requirements of Food and Forage Crops

Crop	Liters/kg
potatoes	500
wheat	900
alfalfa	900
sorghum	1110
corn	1400
rice	1912
soybeans	2000
broiler chicken	3500
beef	100,000*

* Includes water used to raise feed and forage. (*Source:* From Pimentel et al., 1997, p. 100.)

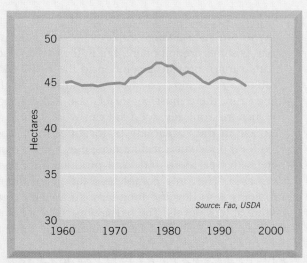

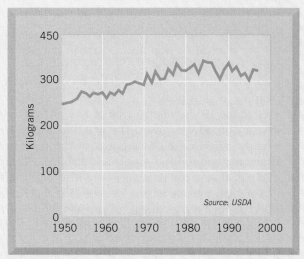

World irrigated area per thousand people, 1961–1995. (*Source:* From Brown et al., 1998, p. 47.)

World grain production per person, 1950–1997. (*Source:* From Brown et al., 1998, p. 29.)

people or will water prove a limiting factor in agricultural production?

Critical Thinking Questions

1. How might changes in diet in developed countries affect water availability?

2. How might global warming affect estimates of the amount of water needed to grow crops in the 21st century?

3. Why is withdrawing water from aquifers faster than the replacement rate sometimes referred to as "mining water."

4. Many countries in warm areas of the world are unable to raise enough food, such as wheat, to supply their populations. Consequently, they import wheat and other grains. How is this equivalent to importing water?

5. Malthusians are those who believe that sooner or later, unless population growth is checked, there will not be enough food for the world's people. Anti-Malthusians believe that technology will save the human race from a Malthusian fate. Analyze the issue of water supply for agriculture from both points of view.

References

Brown, L. R., Renner, M., and Flavin, C. 1998. *Vital Signs: 1998.* New York: W. W. Norton.

Food and Agricultural Organization of the United Nations (FAO). 1996. Food for All. Rome: FAO

Pimentel, D., Houser, J., Preiss, E., White, O., Fang, H., Mesnick, L., Barsky, T., Tariche, S., Schreck, J., and S. Alpert. 1997 (February). "Water Resources: Agriculture, the Environment, and Society." *Bioscience* 4 (2), 92–106.

Population Reference Bureau. 1998. *1998 World Population Data Sheet.* Washington, DC: Population Reference Bureau.

Postel, S. L. 1998 (Aug.). "Water for Food Production: Will There Be Enough in 2025?" *Bioscience* 48 (8), 629–637.

feed all the people of the world. However, the primary world food problem today is poor distribution. Today there is enough food produced in the world to feed all the people. Starvation and malnutrition occur because this food is not distributed where needed. Since the end of World War II, there has not been one year in which there was not a famine.[21] The distribution problem is clearly illustrated by the recent famines suffered in Brazil (1979–1984), Ethiopia (1984–1985), and Somalia (1991–1993) and the 1998 crises in Sudan. Food distribution fails because poor people cannot buy food and pay for its delivery or because transportation is lacking or too expensive or because food is withheld for political or military reasons. Although there is considerable international trade in food, most of the trade is among rich nations.

As long as food distribution is the primary problem, the best solution is to increase local production. Useful aid efforts will focus on the longer term and help people develop their own means of food production, improve the environment for agriculture, and make sure that agriculture is not damaging to the landscape in a way that harms future production.

Another solution, one that is commonly used, is food aid among nations, where one nation provides food to another or gives or lends money that is used to purchase food. This type of aid has become highly publicized in recent years, especially with the interest of pop music stars in the African famine. In the 1950s and 1960s, only a few industrialized countries provided food aid by using stocks of surplus food. A peak in international food aid occurred in the 1960s, when a total of 13.2 million tons per year of food was given. The world food crisis in the early 1970s raised awareness of the need for greater attention to food supply and stability. During the 1980s, donor commitments only reached a high of 7.5 million tons. A record level of 15 million tons of food aid in 1992/93 met less than 50% of the minimum caloric norms of the people fed. If food aid alone were to bring the world's malnourished people to a desired nutritional status, 55 million tons would be required by the year 2010—more than six times the level of food aid available in 1995.[22]

When a group of people starve, the world sorrows for them, and there has been much recent activity to provide starving people with food. Humanitarian gestures are important, but in themselves such efforts cannot provide a solution to the world food problem. Food aid is a short-term answer. In the longer run, there must be an increase in local food production. Ironically, food aid can work against an increase in the availability of locally grown food. Free food undercuts local farmers; they cannot compete with free food, and local production of food decreases. The only complete solution to famine problems is to develop long-term sustainable agriculture at local levels. The old saying, "Give a man a fish and feed him for a day; teach him to fish and feed him for life," is true. Agricultural techniques must be developed and taught that do not use up resources but that can be maintained over long periods. Availability of food grown locally also solves problems arising from disruptions in distribution and the need to transport food over long distances.

SUMMARY

- There are three major aspects to an environmental perspective on agriculture: (1) agriculture depends on the environment, and good agricultural practices require an understanding of the relationships between crop growth and the environment; (2) agriculture changes the environment, and the more intense the agriculture the greater the changes; and (3) as human population increases, there is a need to conserve our agricultural resources and to increase production, and at the same time to decrease the negative effects of agriculture on the environment.

- From an ecological perspective, agriculture is an attempt to keep an ecosystem in a specific stage, usually an early, highly productive, successional stage. Such stages favor early successional plants and animals, which become competitors and predators of crops.

- There are two major approaches to agriculture: (1) demand-based agriculture in which production is based on highly mechanized technology, with a high demand for resources, including land, water, and fuel, and with little use of biologically based technologies and (2) resource-based agriculture in which production is based on biological technology, with conservation and renewal of land, water, and energy.

- The history of agriculture can be viewed as a series of attempts to overcome environmental limitations and problems. Each new solution has created new environmental problems, which have in turn required their own solutions.

- Our modern food problem is the result of the great increase in human population growth, which outstrips local food production in many areas, and an inadequate food distribution system for this growing population. If the human population continues to grow, however, there will be limits on total food production in the future. One solution is to seek new sources of food.

- Most of the world's food is provided by only 14 crop species. Although some food is obtained from oceans and fresh waters, 95% of the protein and most of the calories are obtained from traditional land-based agriculture of crops and livestock. There is great potential for new crops from among the 270,000 species of plants.

- Farming greatly simplifies ecosystems, creating short, simple food chains, reducing species diversity, reducing the organic matter content and overall fertility of soils, and straining water supplies. Agro-ecosystems also differ from natural ecosystems in their spatial regularity. This simplicity makes these systems more vulnerable to outbreaks of diseases, undesired competitors, and herbivores.

- The major focus for the future must be on sustainable agriculture, which means that production per unit area is restricted to the amount that does not damage soils, eliminate water supplies, cause extinction of wild relatives of crops or extinction of potential new food species, and also does not lead to permanent pollution downstream (the topic of the next chapter).

REEXAMINING THEMES AND ISSUES

Human Population: Our modern food problem is the result of the great increase in human population growth, which outstrips local food production in many areas, and an inadequate food distribution system for this growing population. Human population growth will eventually be limited by total food production.

Sustainability: In the future, a major goal of agriculture must be to achieve sustainable food production in any location. This requires the development of farming methods that do not damage soils, eliminate water supplies, cause extinction of wild relatives of crops or of potential new food species, and do not lead to permanent pollution downstream.

The world as a whole is moving from demand-based to resource-based agriculture. The latter is more consistent with a sustainable approach to agriculture.

From an ecological perspective, agriculture is an attempt to keep an ecosystem in a specific stage, usually an early, highly productive, successional stage. In this way, agricultural works against natural mechanisms of sustainability, and we must compensate for this.

Global Perspective: Most of the world's food is provided by only 14 crop species. A challenge for the future is to search the world's diversity of life for additional food species. Food is a global resource in that it is traded internationally, and the availability of food in any area is the result of both local production and global agricultural markets.

Urban World: The artificiality of urban environments tends to lead us to believe that when we live in cities, we are independent of the environment. But when urban people in Brazil starved, they broke into food storages and caused social disruption, thus illustrating the intimate connection between urban life and agricultural production.

Values and Knowledge: The fundamental ethical problem we face concerning food supply is: Are we going to continue to attempt to feed more and more people? Do we attempt this at the risk of sacrificing habitats of noncrop species, natural ecosystems, and landscapes, reducing water supply for other purposes, and increasing the erosion of landscape? Or will we attempt to limit our numbers and also limit total food production?

KEY TERMS

agro-ecosystem *186*	hydroponics *200*	per-capita demand *200*
aquaculture *190*	limiting factor *196*	per-capita food production *200*
cash crops *189*	mariculture *191*	resource-based agriculture *190*
demand-based agriculture *190*	maximum sustainable yield *197*	ruminants *190*
drip irrigation *200*	monoculture *186*	subsistence crops *189*
green revolution *192*	optimum sustainable yield *197*	synergistic effect *197*

STUDY QUESTIONS

1. Explain differences between demand- and resource-based agriculture. Which characterizes traditional swidden agriculture (see A Closer Look 10.2)?

2. A city garbage dump is filled; it is suggested that the area be turned into a farm. What factors in the dump might make it a good area to farm, and what might make it a poor area to farm?

3. Most crops are characteristic of what stages in ecological succession? How might we use our knowledge of succession to make agriculture sustainable?

4. Ranching wild animals, that is, keeping them fenced but never tamed, has been suggested as a way to increase food production in Africa, where there is a great abundance of wildlife. Based on this chapter, what are the environmental advantages and disadvantages of such game ranching?

5. What is meant by the following statement: "The world food problem is one of distribution, not of production"? What are the major solutions to this world food problem?

6. You are sent into the Amazon rain forest to look for new crop species. In what kinds of habitats would you look? For what kinds of plants would you look?

7. How does agriculture simplify an ecosystem? In what ways is this simplification beneficial to people? In what ways does it pose problems for a sustainable food supply?

8. A vegetable garden is planted in a vacant lot in a city. Peas and beans grow well, but tomatoes and lettuce do poorly. What is the likely problem? How could it be corrected?

9. A second vegetable garden is planted in a vacant lot. Nothing grows well. Outside the city, in otherwise similar environments, vegetables grow vigorously. What might explain the difference?

FURTHER READING

Achebe, C., Hyden, G., Magadza, C., and Okeyo, A. P., eds. 1990. *Beyond Hunger in Africa: Conventional Wisdom and an Alternative Vision.* Portsmouth, NH: Heinemann.

Borgstrom, G. 1965. *The Hungry Planet: The Modern World at the Edge of Famine.* New York: Macmillan. The major book by one of the leaders of agricultural change.

Fisher, G., Frohberg, K., Parry, M. L., and Rosenzweig, C. 1994. "Climate Change and World Food Supply, Demand, and Trade: Who Benefits, Who Loses," *Global Environmental Change* 4 (1): 7–23. Uses computer simulation of climate change to examine implications for agriculture.

Pesek, J., ed. Board on Agriculture, National Research Council. 1989. *Alternative Agriculture.* Washington, DC: National Academy. This very readable report presents the scientific and economic soundness of alternative agricultural systems.

Wittwer, S., Yce, Y., Sun, H., and Wong, L. 1987. *Feeding a Billion: Agriculture in China.* Michigan State University Press, Lansing. The author looks at the modern and traditional agricultural technologies of China and discusses how China feeds almost one-quarter of the world's population on 7% of the world's arable land.

INTERNET RESOURCES

Aquaculture Information Center of the National Agricultural Library of the USDA: *http://www.nalusda.gov/answers/info-centers/aic/aic.html*—Bibliographies of publications on aquaculture in the U.S. and links to other aquaculture sites.

Aquaculture Magazine: *http://www.ioa.com/home/aquamag/index.html*—Web site of *Aquaculture Magazine,* a leading journal of aquaculture in the U.S. and abroad. Contains information about aquaculture from around the world.

Food and Agriculture Organization of the United Nations: *http://www.fao.org/*—Current information on the state of the world food situation, the world food summit, trends in food supplies, and environmental impacts of agriculture.

U.S. Department of Agriculture: *http://www.usda.gov/*—Contains statistics and information on agriculture in the U.S.

World Food Programme (WFP): *http://www.wfp.org*—WFP is the food aid organization of the United Nations. WFP became operational in 1963 and is now the world's largest international food aid organization. The Web site provides access to publications, press releases, and data regarding food aid.

Worldwatch Institute Online: *http://www.worldwatch.org*—Worldwatch is a nonprofit public policy research organization that provides information about emerging global problems, trends and links between the world economy and its environmental support systems. The Web site provides on-line access to some publications.

Mechanized irrigation of potato crops in Idaho illustrate the effect of modern agriculture on the environment.

CHAPTER 11

EFFECTS OF AGRICULTURE ON THE ENVIRONMENT

CASE STUDY

Clean Water Farms

Steve Burr, a farmer in the Salina, Kansas area, has 300 acres of crops and 400 acres of grassland on which he raises cattle. In 1994, Steve became concerned about the amount of erosion on his cropland and the effects of fertilizers, pesticides, and livestock wastes on water quality. He decided to make two major changes in the management of his farm. By converting some of the crop acreage to grass, he reduced the erosion and chemical pollution in the runoff from his land. At the same time, he divided his grazing land into sections, called paddocks, and rotated his animals through them so that they had a constant supply of fresh forage. Instead of the animal wastes piling up in feed lots and posing a major disposal problem, the cows dropped their wastes evenly as they moved through the series of paddocks. The rotation system prevented overgrazing and fertilized the pastures. In addition, Steve now can spend less money on dry feed for his cows, allowing him to increase the

size of his herd and generate more income. This method of raising livestock is known as intensive rotational grazing or management intensive grazing.

Steve Burr is not alone in his efforts to adopt farm management practices that reduce pollution and improve water quality, while improving his balance sheet. The Burr farm is one of 36 participants of the Kansas Rural Center's Clean Water Farms Project, which began in 1995. The goal of the Clean Water Farms Project is to farm in a manner that is beneficial to the environment as well as to the economics of farming.[1] And, farmers in many other states are using intensive rotational grazing systems and their numbers are increasing every year. But American farmers are not the only ones adopting this more environmentally benign and economically advantageous approach to raising livestock. In fact, the first system of rotational grazing was developed in France while an adaptation of it was pioneered in South Africa. New

(a) *(b)*

FIGURE 11.1 Grazing practices can affect the health of grasslands, as illustrated by longhorn cattle in (*a*) Colorado and (*b*) Nebraska.

Zealand ranchers are known for their expertise in intensive rotational grazing and the world-famous Argentinian beef is raised using similar methods.

According to Missouri farmer David Shafer, consumers also reap benefits from pasture-raised beef in higher quality products at lower prices. Shafer has found also that his animals have fewer parasites and that soil fertility, as well as water quality of runoff, has improved. On a small scale, farmers like David Shafer and Steve Burr are recreating the migration patterns of bison and elk as they move their cows in herds to ever new pastures. Shafer hopes that over time rotational grazing will rejuvenate the prairie ecosystem that once existed in the central United States.

Intensive rotational grazing is just one of many approaches to farming that are sustainable. Other methods include crop rotation, use of cover crops to reduce fertilizer needs and erosion, composting livestock wastes, no-till farming, integrated pest and weed management, and redesigning livestock waste management and watering systems.

This case history shows that practices that are environmentally benign can be economically advantageous. The possibilities for such alternative approaches to agriculture and the environmental effects of various forms of agriculture are the subjects of this chapter.

LEARNING OBJECTIVES

Agriculture changes the environment in many ways, both locally and globally. After reading this chapter, you should understand:

- How agriculture can lead to soil erosion, the severity of the problem, and what methods are available that can minimize erosion.

- How farming can deplete soil fertility, and why agriculture in most cases requires the use of fertilizers.
- Why some lands are most effectively used for grazing, and how overgrazing can damage land.
- What are the causes of desertification.

- How farming creates conditions that tend to promote pest species, the importance of effective pest (including weed) control, and the problems associated with chemical pesticides.
- How alternative agricultural methods, including integrated pest management, no-till agriculture, mixed cropping, and other methods of soil conservation can provide major environmental benefits.

11.1 HOW AGRICULTURE CHANGES THE ENVIRONMENT

Agriculture is the world's oldest and largest industry; more than one-half of all the people in the world still live on farms. Because the production, processing, and distribution of food all alter the environment, and because of the size of the industry, large effects on the environment are unavoidable. These effects can be both positive and negative. For example, modern pesticides have created a revolution in agriculture in the short term, but the long-term effects of some of these chemicals have proved extremely undesirable.

Agriculture has both *primary* and *secondary* environmental effects. A primary effect, also called an **on-site effect,** is an effect on the area where the agriculture takes place. A secondary effect, or **off-site effect,** is an effect on an environment away from the agricultural site, typically downstream and downwind (see Environmental Issue, *"Should Rice Be Grown in a Dry Climate?"*).

The effects of agriculture on the environment can be divided into three groups: local, regional, and global. Local effects are those that occur at or near the site of the farming; these effects include erosion, loss of soils, and increases in sedimentation downstream in local rivers. Regional effects are those that generally result from the combined effects of farming practices in the same large region. Regional effects include the creation of deserts, large-scale pollution, increases in sedimentation in major rivers and in the estuaries at the mouths of the rivers, and changes in the chemical fertility of soils over large areas. Global effects include climatic changes as well as potentially extensive changes in chemical cycles.

Major environmental problems that result from agriculture include deforestation, desertification, soil erosion, overgrazing, degradation of water resources, salinization, accumulation of toxic metals, accumulation of toxic organic compounds, and water pollution, including eutrophication (discussed in Chapters 4 and 20).

Crops, Soil Fertility, and Soil Erosion

Since the end of World War II, human food production activities have seriously damaged more than 1 billion ha (2.47 billion acres) of land (about 10.5% of the world's best soil), equal in area to China and India. Overgrazing, deforestation, and destructive crop practices have damaged approximately 9 million ha (22 million acres) to the point that recovery will be difficult; restoration of the rest will require serious actions.[2] How did this happen? How could one of Earth's most valuable resources be so badly damaged?

The simple answer is that farming easily damages soils (see A Closer Look 11.1, "Soils"). When land is cleared of its natural vegetation, such as forest or grassland, the soil begins to lose its fertility. Some of this occurs by physical erosion. Once the protection of the vegetative cover is lost, the soil is exposed directly to water and wind, which remove the loosened soil. In addition, the introduction of heavy, earth-moving machinery after World War II has led to a considerable increase in the compaction (packing down) of soil and the loss of the proper soil structure for crop production. Farmed soil also loses fertility when chemical elements are dissolved in water and transported away in streams and subsurface runoff. The rate of loss of fertility is sometimes measured as the time required for the soil to lose one-half of its original store of the chemical elements necessary for crops. The time over which a soil loses one-half of a chemical element varies. It is much faster in warmer and wetter climates, such as in tropical rain forests, than it is in colder or dryer climates, such as those where the natural vegetation is a temperate-zone grassland or forest.[3]

Plowing the Soil

In temperate climatic zones, there is a long history of use of the plow, which turns over the soil completely. There is nothing in nature like a plow or the action of a plow. Plants and soil organisms have not evolved or adapted to its effects. People must therefore be careful in using the plow. Typically, the same land is plowed and planted year after year (except for the occasional years when a field is allowed to lie fallow without plowing or harvesting). The practice of annual plowing and planting makes possible intense use of the land and high production of crops.

In some places, such as the farmlands of the province of Venice, Italy, plowing the soil and planting crops has been a way of life for several thou-

Closer Look 11.1

Soils

Soils may be defined as earth materials modified over time by physical, chemical, and biological processes such that, in addition to supporting rooted plant life, they are altered from the original *parent material* into a series of horizons that are sub-parallel to the surface (Figure 11.2). As soils develop, materials, such as clay, and minerals and nutrients such as iron, calcium, and magnesium, are leached from the upper horizons (*A* and *E*) and may be deposited in a lower horizon (*B*). *Leaching* is accomplished as water infiltrates from the surface, dissolves soil materials as part of chemical weathering processes, and transports the dissolved materials laterally or downward. These processes change the composition of horizons as well as the ability of the soil to retain water. The upper horizons (*O* and *A*) are complex microecosystems; each cubic centimeter of these horizons contains a multitude of microorganisms as well as insects, earthworms, and plant roots.

The type of soil at a particular site depends on such factors as climate, parent material, slope or topography, biological activity, and time. The soil horizons shown in Figure 11.2 are not necessarily all present in any one soil. Very young soils may have only an upper *A* horizon over a *C* hori-

Figure 11.2 Idealized diagram showing a soil profile with soil horizons.

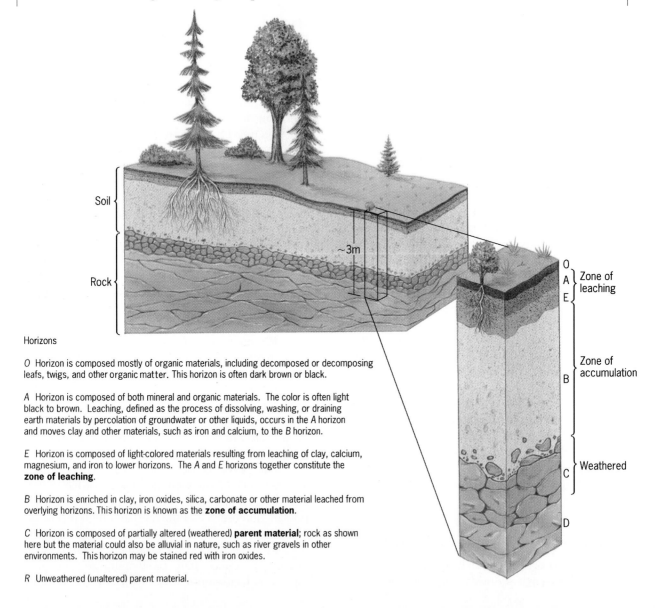

Horizons

O Horizon is composed mostly of organic materials, including decomposed or decomposing leafs, twigs, and other organic matter. This horizon is often dark brown or black.

A Horizon is composed of both mineral and organic materials. The color is often light black to brown. Leaching, defined as the process of dissolving, washing, or draining earth materials by percolation of groundwater or other liquids, occurs in the *A* horizon and moves clay and other materials, such as iron and calcium, to the *B* horizon.

E Horizon is composed of light-colored materials resulting from leaching of clay, calcium, magnesium, and iron to lower horizons. The *A* and *E* horizons together constitute the **zone of leaching**.

B Horizon is enriched in clay, iron oxides, silica, carbonate or other material leached from overlying horizons. This horizon is known as the **zone of accumulation**.

C Horizon is composed of partially altered (weathered) **parent material**; rock as shown here but the material could also be alluvial in nature, such as river gravels in other environments. This horizon may be stained red with iron oxides.

R Unweathered (unaltered) parent material.

zon, whereas older mature soils may have nearly all the horizons shown.

Soil fertility refers to the capacity of a soil to supply the nutrients necessary for plant growth when other factors, such as availability of water and climate, are also favorable. Soils that have formed on geologically young materials are nutrient rich, for example, the glacial deposits of northern Indiana and Illinois that form the famous midwestern corn belt. Soils in semiarid regions are often nutrient rich and need only water to become very productive for agriculture. Soils in humid areas and tropics may be heavily leached and relatively nutrient poor. Nutrients may be cycled through the organic-rich upper horizons, and, if forest cover is removed, reforestation may be very difficult (see Chapter 13).

Soils that accumulate certain clay minerals in semiarid regions may expand (swell) and contract (shrink) on wetting and drying, cracking roads, walls, buildings, and other structures. Expansion and contraction of soils in the United States causes billions of dollars in property damage each year.

Some soils, such as those composed of very small clay particles (less than 0.004 mm in diameter), retard the movement of fluids because the inner connecting space between soil particles is very small and tenaciously holds fluids. Other soils with coarser grains (greater than 0.06 mm in diameter), such as sand or gravel, have relatively large spaces between grains (pores) and fluids may move quickly through them. Thus the type of soil particles present is important in siting facilities such as landfills where retention of pollutants on site is an objective (see Chapter 27).

Some soils are more susceptible to erosion than are others. For example, soils consisting of unconsolidated sand particles (0.06–2 mm in diameter) are particularly susceptible to erosion by running water and wind. Soils composed of coarser (heavier) particles or finer particles that are usually more cohesive (held together by clay minerals) are more resistant to erosion.

In summary, soils are extremely important in many environmental considerations, such as agriculture (fertility), land use, siting of facilities, erosion potential, and shrink–swell potential. It is difficult to think of a human use of the near-surface land environment that does not involve consideration of the soils present. As a result, the study of soils continues to be an important part of environmental sciences.

---

sand years and continues today, with, however, major changes in soils. Plowing opens the land to erosion even more than removal of the original vegetation. Soil loosened by plowing can blow away when dry and wash away with rainwater. Plowed lands tend to lose the upper soil layers, where the most fertile organic matter is found. The less organic matter present in the soil, the more vulnerable the soil is to further erosion. Once erosion starts, the process can easily accelerate.

All forms of agriculture lead to soil loss, but the rate of loss varies with the crop and the methods of agriculture. In the United States, about one-third of the country's topsoil has been lost, resulting in 80 million ha (198 million acres) either totally ruined by soil erosion or made only marginally productive.[4] And soil loss in the United States continues. Today, 90% of the land used for row crops and small grains is farmed without soil conservation practices (some of which are shown in Figure 11.3). Erosion is estimated to be worse now than during the great Dust Bowl of the 1930s, when an estimated 40 million ha (100 million acres) was lost for agricultural production because of soil erosion (Figures 11.4 and 11.5).

Sediment Damage

Much of this eroded soil ends up in our waterways. Rivers carry about 3.6 billion metric tons/year (4 billion U.S. tons/year) of sediment in the United States, 75% of which is from agricultural lands. Of

(a)

(b)

Figure 11.3 Alternative agricultural methods: (*a*) contour strip crops in the midwest of the United States; (*b*) no-till soybean crop planted in wheat stubble on a Kansas farm.

(a)

(b)

(c)

FIGURE **11.4** (*a*) Gully soil erosion on cleared and plowed farmland in South Australia. (*b*) Agricultural runoff carrying heavy sediment load. (*c*) A farmer and his children flee the dust storm of the 1930s in Oklahoma.

this total, 2.7 billion metric tons/year (3 billion U.S. tons/year) is deposited in reservoirs, rivers, and lakes. Downstream sedimentation is a serious environmental effect of modern agriculture. Sediments fill in otherwise productive waters, destroying some fisheries. In tropical waters, sediments entering the ocean can destroy coral reefs that are near shore. Nitrates, ammonia, and other fertilizers carried by sediments can cause eutrophication in downstream waters; the resulting buildup of algae reduces fish production. Polluted sediments also can transport toxins. Sediment damage costs the United States about $500 million/year in dredging expenses and costs attributable to reduction in the useful life of reservoirs.

Making Soils Sustainable

Soil is continuously formed. Ideal farming would result in soil loss no greater than the production of new soil. On good lands soil forms at a rate that ranges from 1 mm/decade to 1 mm/40 years. With good management, however, the rate can be improved to 1 mm/year. Obviously, one way to counter soil erosion is to promote new soil formation. Another way to counter erosion from plowing is contour plowing (refer to Figure 11.3).

Contour Plowing

In traditional plowing, the plowed furrows make a path for water to flow, and if the furrows go downhill, then the water moves rapidly along them, increasing the erosion rate. In **contour plowing,** the land is plowed along the contours, perpendicular to the slope and as much in the horizontal plane as possible. Plowing along the contours can greatly reduce erosion loss owing to water runoff.

In the recent past, contour plowing has been the single most effective method for reducing soil erosion, as demonstrated by an experiment with sloping land planted in potatoes. Part of the land was plowed in uphill and downhill rows, and part was contour plowed. The uphill and downhill section lost 32 metric tons/ha (14.4 tons/acre) of topsoil, whereas the contour-plowed section lost only 0.22 metric ton/ha (0.1 ton/acre). In addition to drastically reducing soil erosion, contour plowing used less fuel and time. However, only a small fraction of the land receives this treatment. For example, of Minnesota's 4 million ha (10 million acres) of cropland, only 526,110 ha (1.3 million acres) are contour plowed.

Other Techniques for Sustaining Soil

Other practices that can aid in the sustainability of soils include fall plowing, which can be less harmful to the soil; multiculture (planting several

(a) *(b)*

Figure 11.5 (*a*) Soil erosion in Northern Natal caused by overgrazing and other land-use practices. The amount of such soil erosion has led some to say that soil is South Africa's greatest export. (*b*) A commercial cattle feedlot in Tulare, California. The cattle stand on mounds of their own manure. Runoff from these feedlots pollutes waterways.

crops intermixed in the same field); strip-cropping; terracing; and proper crop rotation.

No-Till Agriculture

An even more efficient technique to slow erosion is to avoid plowing altogether. **No-till agriculture,** also referred to as *conservation tillage,* is a recent term for a combination of farming practices that includes not plowing the land, using herbicides to keep down the weeds, and allowing some weeds to grow. In no-till agriculture the land is left unplowed most years. Stems and roots that are not part of the commercial crop are left in the fields and allowed to decay in place. In contrast to standard modern approaches, the goal in managing weeds is to suppress and control them but not to eliminate them at the expense of soil conservation. A variety of methods are used to control weeds in no-till agriculture, including integrated pest management (discussed later in this chapter) and chemical pesticides. These practices can greatly reduce soil loss, as well as the use of tractor fuel. Land in no-till agriculture increased from about 2 million ha (5 million acres) in 1962 to 30 million ha (74 million acres) in 1990 in the United States.

Carrying Capacity of Grazing Lands

The carrying capacity of land for cattle varies with the rainfall and the fertility of the soil. When the carrying capacity is exceeded, the land is overgrazed. **Overgrazing** reduces the diversity of plant species, leads to reduction in the growth of vegetation and dominance of plant species that are relatively undesirable to the cattle, increases the loss of soil by erosion as the plant cover is reduced, and results in

damage from the cattle trampling on the land (Figure 11.5). For example, paths made as the cattle travel to the same water hole or stream develop into gullies, which erode rapidly in the rain.

In areas with moderate to high rainfall evenly distributed throughout the year, cattle can be maintained at high densities, but in arid and semiarid regions the density drops greatly (Figure 11.6). In Arizona, for example, the rainfall is low, and cattle can be maintained only at low densities, one head of cattle for 7–10 ha (17–25 acres). Near Paso Robles, California, in an area where the rainfall is about 25 cm/year (9.8 in./year), that is, in desert to semiarid conditions, a ranch where cattle are grazed without artificial irrigation or fertilization supports about one head for 6 ha (15 acres).

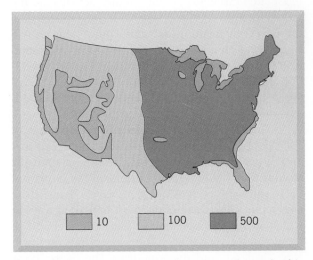

| 10 | 100 | 500 |

Figure 11.6 Carrying capacity of pasture and rangeland in the United States, in average number of cows per square mile. (*Source:* U.S. Department of Agriculture statistics.)

Rangeland is defined as land where native vegetation is mostly grasses or shrubs suitable for grazing or browsing. More than 99% of United States rangeland is west of the Mississippi River. Much of the world's rangeland is considered to be in poor condition from overgrazing. In the United States, rangeland conditions have improved since the 1930s, especially in upland areas. However, riparian areas are reported to be in poor condition. A 1994 study by the National Research Council advises that, instead of focusing on range condition only in terms of its capacity to produce food for livestock, the focus should be on rangeland health.

Traditional and Industrialized Use of Grazing and Rangelands

Traditional herding practices and industrialized production of domestic animals have different effects on the environment. In modern industrialized agriculture, cattle are initially raised on open range and then transported to feedlots, where they are fattened for market. Feedlots have become widely known in recent years as sources of local pollution. The cattle are often kept at extremely high densities and fed grain or forage that is transported to the feedlot. Manure builds up in large mounds. When it rains, manure pollutes local streams. Feedlots are popular with meat producers because they are economical for rapid production of good-quality meat. However, large feedlots require intense use of resources and have negative environmental effects.

Traditional herding practices, on the other hand, chiefly affect the environment through overgrazing. Goats are especially damaging to vegetation, but all domestic herbivores can destroy rangeland. As we have seen, the effect of domestic herbivores on the land varies greatly with their density (relative to rainfall and soil fertility). At low to moderate densities, the animals may actually aid the growth of above-ground vegetation by fertilizing soil through their manure and stimulating plant growth by clipping off ends in grazing, just as pruning stimulates plant growth. But at high densities the vegetation is eaten faster than it can grow, some species are lost, and the growth of others is greatly reduced.

The Biogeography of Agricultural Animals

People have distributed cattle, sheep, goats, and horses, as well as other domestic animals, around the world and then promoted the growth of these animals to densities that have changed the landscape. This is one of the most important ways in which agriculture has affected the environment. Domestic animals have been introduced into Australia, New Zealand, and the Americas. The horse, cow, sheep, and goat were brought to North America after the sixteenth century. The spread of cattle brought new animal diseases and new weeds, which arrived on their hooves and in their manure.

A recent important issue in cattle production is the opening up of tropical forest areas and their conversion to rangeland, for example, in the Brazilian Amazon basin. In a typical situation the forest is cleared by burning and crops are planted for about 4 years, after which time the soil has lost so much fertility that crops can no longer be grown economically. Ranchers then purchase the land, already cleared, and run cattle bred to survive in the hot, humid conditions. After about another 4 years, even grazing can no longer be supported and the land is abandoned. In such areas, the land's capability for many uses, including forest growth, has been greatly reduced by grazing.[6]

The spread of domestic herbivores around the world is one of the major ways we have changed the environment through agriculture. As the human population increases, and when income and expectations rise, the demand for meat increases. There will be an increased demand for rangeland and pastureland in the next decades. A major challenge in agriculture will be to develop ways to make the production of domestic animals sustainable.

Game Ranching

Maintaining wild herbivores, such as zebras and impalas in Africa and bison in North America, in their native habitat to be harvested for meat, leather, and other products is called **game ranching** (see Chapter 12 for a more detailed discussion of game ranching). The animals are not domesticated; they are not fed or cared for in barns or other enclosures. They are simply allowed to live wild on the range. Sometimes, this practice appears to do less damage to the vegetation and soils than would the grazing of domestic animals. Game may be sustainable at higher levels of production than domestic animals. Also, game ranching may utilize land unsuitable for the ranching of domestic animals, such as areas that are more easily eroded by cattle, sheep, and goats than by wild game. An interesting related recent development is the suggestion that some of the prairie states of the United States be returned to buffalo grounds, and the buffalo harvested as wild game. There are some ranches now raising buffalo and promoting the restoration of native prairie vegetation.

Desertification: A Regional Effect

Deserts occur naturally where there is too little water for substantial plant growth; the plants that do grow are too sparse and unproductive to create a soil rich in organic matter. Desert soils are mainly inorganic, that is, coarse and typically sandy. When rain does come, it often is heavy, and erosion is severe. The principal climatic condition that leads to desert is low or undependable precipitation. The warmer the climate, the greater the rainfall required to convert an area from desert to a productive area, such as grassland. But, even in cooler climates or at higher altitudes, deserts may form if precipitation is too low to support more than sparse plant life.

The crucial factor is the amount of water in the soil available for plants to use. Factors that destroy the ability of a soil to store water can create a desert.

Desert Regions

Earth has five natural, warm desert regions, all of which lie primarily between latitudes 15° and 30° north and south of the equator. These are the deserts of southwestern United States and Mexico; Pacific coast deserts of Chile and southern Ecuador; Kalahari Desert of southern Africa and its extensions into South Africa; Australian deserts that cover most of that continent; and the greatest desert region of all—the desert that extends from the Atlantic coast of North Africa (the Sahara) eastward to deserts of Arabia, Iran, Russia, Pakistan, India, and China.[7] Only Europe lacks a major warm desert; it lies north of the desert latitudinal band.

Based on climate, about one-third of Earth's land area should be desert, but estimates suggest that 43% of the land is desert. This additional desert area is believed to be a result of human activities.[8] **Desertification** is the deterioration of land in arid, semi-arid, and dry sub-humid areas due to changes in climate and human activities.[9]

Desertification is a serious global problem. It affects one-sixth of the world's population (about 1 billion people), 25% of the world's total land area, and 70% of all drylands (3.6 billion ha). Land degradation caused by people has altered 73% of drier rangelands (3.3 billion ha) and the soil fertility and structure of 47% of dryland areas with marginal rainfall for crops. Land degradation also affects 30% of dryland areas with high population density and agricultural potential.

A large part of desertification occurs in the poorest countries. These regions include Asia, Africa, and South America. Six million ha of land per year are lost to this process. Over $40 billion is lost annually, and the cost of recovery of these lands worldwide could reach $10 billion per year.[10]

Causes of Desertification

Some areas of Earth are *marginal* lands, easily converted to deserts when used for light grazing and crop production. In these areas, the rainfall is just barely enough to make the area more productive of vegetation than a desert, and then only if the soils are maintained within a certain range of conditions.

The leading causes of desertification are bad farming practices (such as failure to use contour plowing or simply too much farming; Figures 11.4 and 11.5); overgrazing; the conversion of rangelands to croplands in marginal areas where rainfall is not sufficient to support crops over the long term; and poor forestry practices, including cutting all the trees in an area marginal for tree growth.

In northern China, for example, areas that were once grasslands were overgrazed, then some of these rangelands were converted to croplands. Both practices led to the conversion of the lands to deserts. Between 1949 and 1980 an area of 65,000 km² (25,100 mi², an area larger than Denmark) became deserts, and an additional 160,000 km² (61,760 mi²) is in danger of becoming deserts. As a result of this desertification, the occurrence of sandstorms increased from about 3 days/year in the early 1950s to an average of 17 days/year in the next decade and to more than 25 days/year by the early 1980s.[11]

Desertlike areas can be created anywhere by the poisoning of soils. Poisoning can result from the application of persistent pesticides or other toxic organic chemicals or from industrial processes that lead to improper disposal of toxic chemicals. Airborne pollutant acidification, excessive manuring in feedlots, and oil or chemical spills can all lead to degradation of soils, forcing abandonment or reduced agricultural use of lands. Worldwide, chemical degradation accounts for about 12% of all soil degradation.

Irrigation of soils in arid areas can also lead to a desert. When irrigation water evaporates, a residue of salts is left behind. Although these may have been in very low concentrations in the irrigation water, over time the salts can build up in the soil to the point at which they become toxic. This effect can sometimes be reversed if irrigation is increased greatly; the larger volume of water then redissolves the salts and carries them with it as it percolates down into the water table. An example of other undesirable effects of irrigation is presented in A Closer Look 11.2, "Kesterson Wildlife Refuge."

Preventing Desertification

The major symptoms of desertification are:

- lowering of the water table (wells have to be dug deeper and deeper);
- increase in salt content of the soil;
- reduced surface water (streams and ponds dry up);
- increased soil erosion (the dry soil, losing its organic matter, begins to blow away and to be washed away in heavy rains); and
- loss of natural native vegetation (not adapted

CLOSER LOOK 11.2

KESTERSON WILDLIFE REFUGE: UNDESIRABLE EFFECTS OF IRRIGATION

An important example of chemical concentration as a result of heavy irrigation is found in the area near the Kesterson National Wildlife Refuge in California. In the spring of 1985, the U.S. Bureau of Reclamation announced that it was closing the Kesterson National Wildlife Refuge, a 17,000-ha (42,000-acre) preserve in the San Joaquin valley of California and that it was also closing a 132-km- (82-mi-) long irrigation drainage canal. The closing of the canal would take more than 20,000 ha (50,000 acres) of farmland out of production, with the loss of $45 million annually and 1200 jobs. The canal drained water that had been used in farm irrigation into the refuge, where it provided water for a wetland habitat for birds (Figure 11.7).

Why were the wildlife refuge and the canal closed? In 1983, biologists began to discover birth defects in water birds born in the refuge. These defects were due to a high concentration of the chemical element selenium, which, although harmless in the small concentrations normally found in soils and waters, causes genetic changes when present in high concentrations. The selenium was carried into the refuge in the irrigation water flowing from the drainage canal. Selenium concentration was low in the original irrigation water, but in the dry California valley, water used in irrigation evaporated quickly from the soil, concentrating the selenium.

In order to prevent severe buildup of salts within the soils, the farmers periodically had to use enough water to leach the soil of the salts. The water dissolved the salts, carried them down into the soil below the reach of the roots, and then transported them away from the farmland in subsurface water flow. This drainage water was high in many chemical elements, including selenium. According to plan, the saline water was transported by the canal into the wildlife refuge for two purposes: to provide a wetland habitat for waterfowl and to dispose of the water. An unplanned effect was the buildup in selenium.

The toxic levels of selenium in the refuge threaten life beyond the wildlife refuge. There is concern that selenium pollution might spread to thousands of acres of nearby marsh and farmlands, where it could poison livestock and enter the domestic water supply.

FIGURE 11.7 Kesterson National Wildlife Refuge, where agricultural runoff has polluted water and affected wildlife.

to desert conditions, native vegetation can no longer survive).[12]

Preventing desertification begins with monitoring these factors. Monitoring of aquifers and soils is an important activity in marginal agricultural lands. When undesirable changes are observed, activities producing these changes can be stopped, reduced, or changed.

Proper methods of soil conservation, forest management, and irrigation can help prevent the spread of deserts (see Chapters 10 and 19 for a background discussion of soil and farming and irrigation practices). In addition to the practices discussed earlier, good soil conservation includes the use of windbreaks (narrow lines of trees that help slow the wind) to prevent sandstorms and wind erosion of the soil. A landscape with trees is a landscape with a good chance of avoiding desertification. Practices that lead to deforestation in marginal areas should be avoided. Reforestation, including the planting of windbreaks, should be encouraged.

11.2 GLOBAL EFFECTS OF AGRICULTURE

Local and regional effects of agriculture on the environment are easily identified. Global effects are not as obvious, although they are potentially as serious. This section discusses a few of these global changes.

First, agriculture changes land cover, resulting in changes in reflection of light by the land surface; the evaporation of water; the roughness of the surface; and the rate of exchange of chemical compounds (such as carbon dioxide) produced and removed by living things. Each of these changes can have regional and global climatic effects.

Second, modern agriculture increases carbon dioxide in two ways. As a major user of fossil fuels, it contributes to the increase in carbon dioxide concentration in the atmosphere, adding to the buildup of greenhouse gases (discussed in detail in Chapter 21). Also, clearing land for agriculture leads to an increase in the decomposition of organic matter in the soil, transferring the carbon stored in the organic matter into carbon dioxide, which also increases the CO_2 concentration in the atmosphere.

Agriculture can also affect climate through fire. Fires associated with clearing land for agriculture, especially in tropical countries, may have significant effects on the climate because they add small particulates to the atmosphere.

Another global effect of agriculture results from the artificial production of nitrogen compounds for use in fertilizer, which may be leading to significant changes in global biogeochemical cycles. Finally, agriculture affects species diversity. The loss of competing ecosystems (because of agricultural land use) reduces biological diversity and increases the number of endangered species.

11.3 PEST CONTROL AND AGRICULTURAL CHEMICALS

All agriculture suffers from pests. From an ecological point of view, *pests* are undesirable competitors, parasites, or predators. Even today, with modern technology, the total losses from all pests are huge; in the United States, pests account for an estimated loss of one-third of the potential harvest and about one-tenth of the harvested crop. Preharvest losses are due to competition from weeds, diseases, and herbivores; postharvest losses are largely due to herbivores.[13]

Pests

The major agricultural pests are insects (feeding mainly on the live parts of plants, especially leaves and stems); nematodes (small worms that live mainly in the soil and feed on roots and other plant tissues); bacterial and viral diseases; weeds (flowering plants that compete with the crops); and vertebrates (mainly rodents and birds that feed on grain or fruit).

Weeds

Although we tend to think that the major pests are insects, in fact, weeds are the major problem in terms of potential crop loss. Farming produces special environmental and ecological conditions that tend to promote pests. To understand these, it is important to remember that the process of farming is an attempt to (1) hold back the natural processes of ecological succession; (2) prevent the normal entrance of migrating organisms into an area; and (3) prevent natural interactions (including competition, predation, and parasitism) between populations of different species.

As explained in Chapter 9, a farm is maintained in a very early stage of ecological succession. With the addition of fertilizers and water, it is a very good place not only for crops but for other early successional plants, such as weeds, to grow. Recall that early successional plants tend to be fast growing and to have seeds that are easily blown by the wind or spread by animals that have eaten their fruit. They spread and grow rapidly in the inviting habitat of open, early successional croplands.

There are about 30,000 species of weeds, and in any year a typical farm field is infested with between 10 and 50 weed species. Weeds compete with crops for all resources: light, water, and nutrients. The more weeds, the less crop. Some weeds can have a devastating effect on crops; for example, the production of soybeans can be reduced 60% if there is just one individual of the weed called cocklebur per row foot allowed to compete with the crop over the course of a growing season.[14] Agricultural losses in the United States as a result of weeds exceed $16 billion/year. About $3.6 billion is spent annually for chemical weed control, amounting to 60% of all pesticide sales in the United States.

Development of Pesticides

Before the industrial revolution farmers could do little to prevent pests except remove them when they appeared or use farming methods that tended to decrease their density. For example, swidden agriculture reduces the density of pests by allowing noncrop plants to grow and by maintaining a diversity of crops in a single field (see Chapter 10). Preindustrial farmers also could have planted vegetation, such as aromatic herbs, that tends to repel insects.

With the rise of modern agricultural sciences, chemical pesticides were developed. The use of pesticides has grown, reaching $31.25 billion worldwide in 1996. About 80% of the pesticides in use are applied in developing countries. Agricultural scientists are working to perfect pesticides. The most desirable pesticide would be *narrow spectrum,* that is, lethal to a single pest species but not harmful to other forms of life. Scientists refer to such an ideal, ultimate pesticide as a *magic bullet,* a chemical that rapidly seeks out individuals of a particular species and kills them, with no effect on any other form of life. No pesticide is that perfect.

Earlier chemical pesticides were *broad spectrum*, meaning that they affected a wide range of organisms. One of the earliest pesticides used was arsenic, a chemical element toxic to all life, including people. It was certainly effective in killing pests, but it killed beneficial organisms as well.

The second stage in the development of pesticides, which began in the 1930s, was the development of oil sprays and the use of natural plant chemicals. Many plants produce natural chemicals as a defense against disease and herbivores. Many of these natural chemicals are effective pesticides. Nicotine, from the tobacco plant, is one of these chemicals; it is the primary agent in some insecticides still in wide use today. However, although natural plant pesticides are comparatively safe, they are also not as effective as desired.

DDT

The real revolution in chemical pesticides—the development of more sophisticated pesticides—began with the end of World War II and the discovery of DDT and other chlorinated hydrocarbons, including aldrin and dieldrin. At first, DDT was thought to be the long-sought magic bullet: It appeared to have no short-term effect on people and seemed to kill only insects. At the time scientists believed that a chemical could not be readily transported from its original site of application unless it was water soluble. DDT was not very soluble in water and therefore did not appear to pose an environmental hazard.

Initially, DDT was used very widely until three things were discovered: (1) it has long-term effects on other, desirable organisms, including the ability of birds to produce eggs and a possible increased incidence of cancer in other organisms; (2) it is stored in oils and fats and is concentrated as it is passed up food chains, so that the higher an organism is on a food chain, the higher the concentration of DDT it contains, a process known as **food chain concentration** or **biomagnification** (discussed in detail in Chapter 14); and (3) the storage of DDT in fats and oils allows the chemical to be transferred biologically even though it is not very soluble in water.

In birds, intake of DDT and the products of its chemical breakdown (known as DDD and DDE) results in thin eggshells that break easily, so that the birds cannot reproduce successfully. This problem was especially severe in birds that are high on the food chain—predators that feed on other predators—such as the bald eagle or the osprey and the pelican, which feed on fish that may be predators of other fish.

As a result of these problems, DDT was banned in most developed nations. It was banned in the United States in 1971. Since then there has been a dramatic recovery in the populations of affected birds. For example, the brown pelican of the California coast, which had become rare and endangered and whose reproduction had been restricted to offshore islands where DDT had not been used, is common again. Another example, the bald eagle, has become abundant again in the north woods, where it can be seen in Voyageurs National Park and the Boundary Waters Canoe area of northern Minnesota.

However, DDT is still being produced in the United States for use in developing and less developed nations, especially as a control for malaria-spreading mosquitoes. There have been some benefits from the use of DDT. It was primarily responsible for eliminating malaria and yellow fever as major diseases, reducing incidence of malaria in the United States from an average of 250,000 cases a year prior to the spraying program to fewer than 10 per year in 1950. However, the fact remains that DDT is toxic to animals and its effectiveness has been decreased over the years because many species of insects have developed a resistance to it. Nevertheless, DDT continues to be used because it is cheap and effective and because people have become accustomed to using it. About 35,000 metric tons of DDT are produced annually in at least five countries, and it is legally imported and used in dozens, including Mexico.

Although people in developed nations believe they are free from the effects of DDT, in fact this chemical is transported back to industrial nations in agricultural products from nations still using the chemical. Also, migrating birds that spend part of the year in malarial regions are still subject to the effects of DDT on eggshells. Thus, in spite of its banning in the developed nations, DDT remains an important world issue in pest control. (The problem of developing nations using pesticides banned in other nations is not only an issue for DDT; the situation extends to other chemicals as well.)

Alternatives to DDT

With the banning of DDT in developed nations, other chemicals became more prominent. Chemicals were sought that were less persistent. Among the next generation of insecticides were organophosphates: phosphorus-containing chemicals that affect the nervous system. These chemicals are more specific and decay rapidly in the soil. Therefore they do not have the same persistence as DDT. However, they are toxic to people and must be handled extremely carefully by those who apply them.

Chemical pesticides have created a revolution in agriculture. However, in addition to the negative environmental effects of individual chemicals such as DDT, their use has other major drawbacks. One such problem is known as *secondary pest outbreaks*, which occur after extended use (and possibly because of extended use) of a pesticide. Secondary pest outbreaks can come about in two ways: (1) reduction in one target species reduces competition with a second, which then increases and becomes a pest or (2) the pest develops resistance to the pesticides through evolution and natural selection, which favor those in the population with a greater immunity to the chemical.[15] Developed resistance has occurred with many pesticides. For example, Dasanit, an organophosphate, was first introduced in 1970 to control maggots that attack onions in Michigan; although it was originally successful, it is now so ineffective that it can no longer be used for that crop.

11.4 INTEGRATED PEST MANAGEMENT

Modern approaches to pest control involve **integrated pest management** (IPM), an ecosystem approach to pest management that integrates a variety of techniques. These techniques include:

- the use of natural enemies of pests, including parasites, diseases, and predators;
- the planting of a greater diversity of crops to reduce the chance that pests will find a host plant;
- no-till or low-till agriculture, which helps natural enemies of some pests to build up in the soil; and
- the application of a set of highly specific chemicals, used much more sparingly than in earlier approaches.

Integrated pest management recognizes ecological communities and ecosystems. It takes into account the effect of one species on others—for instance, a decrease in one species may lead to an increase in another and decreases in still others.

Four principles govern IPM.[16]

1. The goal is control, not extinction. Pests are allowed to continue to exist at a low, tolerable level; the method is considered a success if pests are kept at these levels. (In contrast, advocates of older approaches involving heavy application of chemicals sought complete elimination of the pests.)
2. The use of natural control agents is maximized.
3. The ecosystem is the management unit.
4. Any control action can have unexpected and unwanted effects.

The components of IPM include chemicals, the development of genetically resistant stock, biological control, and land culture. *Land culture* refers to how the land is physically managed, including whether and how it is plowed, what kind of crop rotation is used, the dates of planting, and basic means of handling crop harvests to reduce the presence of pests in residues and products sold.

Biological Control

A part of IPM, **biological control** is a set of methods to control pest organisms by using natural ecological interactions including predation, parasitism, and competition. It includes the intentional introduction of predators, diseases, or other parasites of a pest. For example, ladybugs are common predators of many plant-eating pests. It is possible to buy quantities of ladybugs for release in home gardens or farms. The hope is that these ladybugs will feed on pests and reduce their abundance.

There are many specialized and effective biological controls. One of the most effective is a bacterial disease, *Bacillus thuringiensis*, which kills larval forms of many insect pests, including many caterpillars. It is used widely. Other effective biological control agents are small wasps that are parasites of caterpillars. The wasps lay their eggs in the caterpillars; the larval wasps then feed on the caterpillars and kill them. These wasps tend to have very specific relationships (one species of wasp will be a parasite of one species of pest) and so are both effective and narrow spectrum (Figure 11.8).

The control of the oriental fruit moth, which attacks a number of fruit crops, is an example of

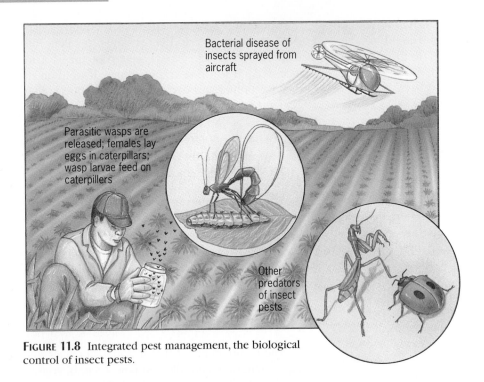

FIGURE 11.8 Integrated pest management, the biological control of insect pests.

(a)

(b)

FIGURE 11.9 The wasp that preys on the oriental fruit moth has been used in biological control (*a*) and the strawberry fields where they live (*b*).

IPM biological control. The moth was found to be a prey of a species of wasp. The introduction of the wasp into fields helped control the moth. Interestingly, the effectiveness of the wasp in peach fields was increased when there were nearby strawberry fields. These strawberry fields provided an alternative habitat for the wasp, especially important for overwintering (Figure 11.9).[17] As this example shows, spatial complexity and biological diversity also become part of the IPM strategy.

A study by the U.S. Office of Technology Assessment concluded that IPM could reduce the use of pesticides by as much as 75% and at the same time reduce preharvest pest-caused losses by 50%. This would also greatly reduce the costs to the farmers for pest control.[18]

Another technique to control insects involves the use of sex pheromones. In most species of adult insects, one sex (usually the female) releases a chemical called a sex pheromone, which acts as an attractant to members of the opposite sex. In some species it has been shown to be effective up to 4.3 km (2.7 mi) away. These chemicals have been identified and synthesized and used in insect control as bait in traps, in insect surveys, or simply to confuse the mating patterns of the insects involved.

Current agricultural practices in the United States involve a combination of approaches, but in most cases, they are more restricted than an IPM strategy. Biological control methods are used to a comparatively small extent. They are the primary tactic for the control of vertebrate pests (mice, voles,

Should Rice Be Grown in a Dry Climate?

Water is a precious resource, especially in California, where average rainfall is low (38–51 cm/year, or 15–20 in./year) and had dropped below normal in the latter part of the 1980s (see graph), population is large and growing (31 million, growing by about 2000 a day), and agricultural water use is high (farmers use 85% of the state's water to irrigate 3.2 million ha, or 7.8 million acres, more than in any other state). With cities and industries, not to mention fish and other wildlife, in need of water, the practice of growing crops with high water demand has come under intense criticism (see the accompanying table on water use). The conflict has been intensified because much of the water farmers receive is subsidized by the government, allowing many to sell it at a profit to municipal water districts. Although some farmers have responded by reducing the acreage of water-intensive crops and others have switched to crops such as fruits, vegetables, and nuts that require less water, rice farmers have tried to change agricultural practices without changing their crop.

California produces 18% of the nation's rice, making it the second largest rice-growing state. The rice industry, concentrated in the Sacramento valley, is worth about $500 million a year and uses enough water to supply one-fourth of the state's population for a year. Although rice growers are not the biggest water users in the state, they have been particular targets of attack because the flooded fields required for rice are a very visible re-minder of the amount of water used by agriculture. In addition to high water use, rice growing has had other adverse environmental effects: high pesticide and herbicide use, causing contamination of rivers and drinking supplies; and burning of stubble left after harvesting, contributing to air pollution in the valley.

Rice growers have responded to pressure to clean up their act in a number of ways. They have decreased water use by 32% and pesticide use by 98% in the past 10 years; they have converted to biodegradable pesticides; and they have decreased the burning of stubble by plowing it under, harvesting it, or flooding the fields in winter and allowing it to rot. In addition, the fields provide a wetland habitat for many migrating birds and other species. Winter flooding benefits an even wider diversity and greater number of species, particularly waterfowl, which declined from 10 to 12 million in 1967 to 4 to 5 million in 1990; California lost 90% of its wetlands to development and agriculture in that period.

The Sacramento valley, which lies along the Pacific flyway, a major migration route for waterfowl, is host to 20% of the ducks of the United States and 50% of all waterfowl in the flyway. Twenty-one wildlife species with special status (endangered, threatened, candidate species, or species of special concern) use the rice fields, attracted by the 114–136 kg (250–300 lb) of rice grains per acre left behind after harvesting and the 273–318 kg (600–700 lb) of invertebrates per acre that grow in the waters.

An experimental program was begun in 1992 to test the feasibility of winter flooding as an alternative to burning stubble, as a supplement to the state's water storage capacity, and as a way to create additional wetlands. Experts in the program are also trying to find ways to protect young salmon, which run in the rivers of the Sacramento valley, from being pumped into the channels leading to the rice fields. Although drawing water out of the rivers might have a negative impact on salmon, the release of water at the end of the winter, when the rivers are low, could help the spring run of salmon.

Critical Thinking Questions

1. Most of the areas where rice is grown have alkaline, hardpan soil, unsuited to other crops. If rice were not grown on the land, it probably would be developed for housing. Each acre of rice requires

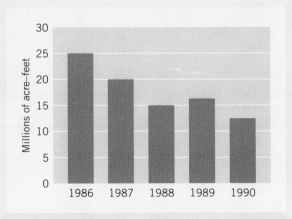

Dropping water levels in California reservoirs, 1986–1990: capacity, 38 million acre-feet; average, 22.3 million (1 acre-foot = 325.851 gallons). (*Source:* California Department of Water Resources, 1991.)

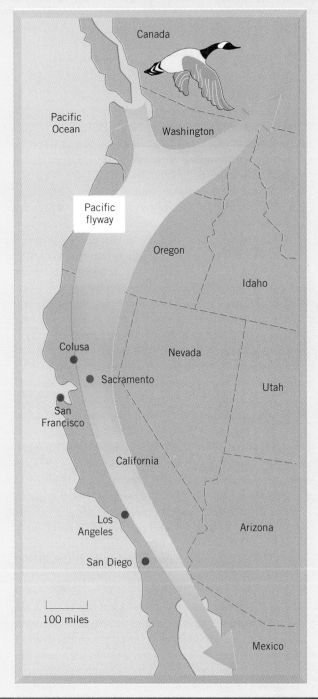

Pacific Ocean

Canada

Washington

Pacific flyway

Oregon

Idaho

Colusa

Nevada

Sacramento

Utah

San Francisco

California

Los Angeles

Arizona

San Diego

Mexico

100 miles

5 acre-feet of water. Less than 1 acre-foot could supply a family of four for a year. If housing lots were one-eighth acre and all housed families of four, how many acre-feet of water would be used in a year? Which uses more, an acre of rice or an acre of people? How would real estate development affect wildlife habitat?

2. Farmers find that the presence of waterfowl in flooded fields speeds up the rotting of stubble. What are at least two reasons for this?

3. Although birds can feed on rice grains in dry fields, why do they get a more balanced diet by feeding in flooded fields?

4. Two of the unknowns in this system are the long-term effects of flooding on the ability of the soil to support rice growing and on dryland species, such as rattlesnakes and rats. What is a scientific way of investigating one of these questions?

5. The new rice-growing practices are referred to as win–win. What is meant by the term in general, and how does this situation exemplify the term?

References

Martin, G. 1992 (June 29). "Rice Grower Proud of His Bird Habitat," *San Francisco Chronicle*, pp. A1, A6.

Vogel, N. 1992 (December 6). "Rice Farmers Change Ways, Reap Good Will," *Sacramento Bee*, pp. A1, A26.

Walker, S. L. 1992 (August 1). "Rice Growers Sow Good Will," *San Diego Union-Tribune*, pp. A1, A15.

Wood, D. B. 1992 (September 3). "California Rice Land Does Double Duty," *Christian Science Monitor*, p. 10.

World Resources Institute. 1992. *The 1992 Information Please Environmental Almanac.* Boston: Houghton Mifflin.

The Pacific Flyway, used by many birds that stop at agricultural wetlands.

and birds) of lettuce, tomatoes, and strawberries in California, but not a major technique for grains, cotton, potatoes, apples, or melons. Chemicals are the principal control methods for insect pests. For weed control, the principal controls are methods of land culture. The use of genetically resistant stock is important for disease control in wheat, corn, cotton, and some vegetable crops, such as lettuce and tomatoes.

Integrated pest management reduces the release of toxic chemicals into the environment while enabling economically viable production of crops. The more it can be employed, the better for the environment and for individual ecosystems.

SUMMARY

- The industrial revolution and the rise of agricultural sciences have led to a revolution in agriculture, with many benefits and some serious drawbacks. These developments have been accompanied by an increase in soil loss, erosion, and resulting downstream sedimentation, as well as the pollution of soil and water with pesticides, fertilizers, and heavy metals that are concentrated as a result of irrigation.

- Modern fertilizers have greatly increased the yield per unit area. Modern chemistry has led to the development of a wide variety of pesticides that have reduced, but not eliminated, the loss of crop production to weeds, diseases, and herbivores.

- Most twentieth-century agriculture has relied on machinery and the use of abundant energy, with relatively little attention paid to the loss of soils, the limits of groundwater, and the negative effects of chemical pesticides.

- Overgrazing has caused severe damage to lands. It is important to properly manage livestock, including using appropriate lands for grazing and keeping livestock at a sustainable density.

- Desertification is a serious problem that can be caused by poor farming practices as well as by the conversion of marginal grazing lands to croplands. Additional desertification can be avoided by improving farming practices, planting trees as windbreaks, and monitoring land for symptoms of desertification.

- The next revolution in agriculture, an ecological one, has begun. In this phase, pest control will be dominated by integrated pest management and a knowledge of agricultural lands as ecosystems, with a careful use of artificial chemicals. This phase will emphasize soil conservation through no-till agriculture and contour plowing. (It will also emphasize water conservation through methods discussed in Chapter 19.)

REEXAMINING THEMES AND ISSUES

Human Population: Agriculture is the world's oldest and largest industry; more than one-half of all the people in the world still live on farms. Because the production, processing, and distribution of food all alter the environment, and because of the size of the industry, large effects on the environment are unavoidable.

Sustainability: Alternative agricultural methods, such as those practiced by the Spray Brothers (see introductory case study) appear to offer the greatest hope of sustaining agricultural ecosystems and habitats over the long term, but more tests and development of methods are needed.

Global Perspective: Agriculture changes land cover, resulting in changes in: reflection of light by the land surface; the evaporation of water; the roughness of the land surface; and the rate of exchange of chemical compounds (such as carbon dioxide) produced and removed by living things. Each of these changes can have regional and global climatic effects. Modern agriculture increases carbon dioxide in two ways. As a major user of fossil fuels, it contributes to the increase in carbon dioxide concentration in the atmosphere, adding to the buildup of greenhouse gases (discussed in detail in Chapter 21).

Also, clearing land for agriculture leads to an increase in the decomposition of organic matter in the soil, transferring the carbon stored in the organic matter into carbon dioxide, which also increases the CO_2 concentration in the atmosphere. Agriculture can affect climate through fire. Fires associated with clearing land for agriculture, especially in tropical countries, may have significant effects on the climate because they add small particulates to the atmosphere.

Urban World: The agricultural revolution allows fewer and fewer people to produce more and more food and leads to greater productivity per acre. People, freed from dependence on farming, flood to cities. This leads to increased urban effects on the land. Thus agricultural effects on the environment indirectly extend to the cities.

Values and Knowledge: Human activities have seriously damaged one-fourth of the world's total land area, impacting one-sixth of the world's population (about 1 billion people). Six million ha of land per year are lost to desertification. A large part of desertification occurs in the poorest countries. Overgrazing, deforestation, and destructive crop practices have caused so much damage that recovery will be difficult; restoration of the rest will require serious actions. A major value judgment we must make in the future is whether our societies will allocate the funds required to restore these damaged lands. This process of restoration requires scientific knowledge, both about present conditions and actions required for restoration. Will we seek this knowledge and pay the costs for it?

KEY TERMS

biological control *219*
biomagnification *218*
contour plowing *212*
desertification *215*

food chain concentration *218*
game ranching *214*
integrated pest management *219*
no-till agriculture *213*

off-site effect *209*
on-site effect *209*
overgrazing *213*

STUDY QUESTIONS

1. How can an insect pest species become resistant to a pesticide?

2. What is meant by calling a pesticide a magic bullet? What would its characteristics be?

3. How can farming lead to the spread of deserts? What might be done to stop this desertification?

4. It has been said that farming can never be sustainable. What is meant by this statement? Do you agree or disagree? List your reasons.

5. What is meant by food chain concentration? How does this lead to differences in the effects of the same pesticide on different organisms?

6. Design an integrated pest management scheme for use in a small, city vegetable garden in a lot behind a house. How would this scheme differ from IPM used on a large farm? What aspects of IPM could not be employed? How might the artificial structures of a city be put to use to benefit IPM?

7. Arsenic is a potent toxic chemical. Why does it make a poor pesticide?

8. What are the off-site effects of an urban vegetable garden?

9. Under what conditions might grazing cattle be sustainable when growing wheat is not? Under what conditions might a herd of bison provide a sustainable supply of meat when cattle might not?

10. How can we avoid another dust bowl in the United States?

FURTHER READING

Grainger, A. 1982. *Desertification: How People Make Deserts, How People Can Stop and Why They Don't.* Earthscan Books, 2nd ed. London: Russell Press, Nottingham. This book provides examples and a discussion of the connections between desertification and human activity, particularly in nonindustrialized countries.

Lashof, J. C., ed. 1979. *Pest Management Strategies in Crop Protection*, vol. 1. Washington, DC: Office of Technology Assessment. U.S. Congress. Crop protection strategies are assessed and evaluated.

Matthews, A. 1992. *Where the Buffalo Roam.* Weidenfeld, NY: Grove Publishing.

Sheridan, D. 1981. *Desertification of the United States.* Washington, DC: Council on Environmental Quality. This report reviews scientific information on desertification in the United States, identifies policies and technologies that promote or control desertification, and provides an excellent overview of continuing conditions.

Young, M. D., 1991. *Towards a Sustainable Agricultural Development.* New York: John Wiley & Sons. N.Y. A broad survey of the use of agricultural chemicals, intensive animal production, soil erosion, land-use patterns and the impact on agriculture of pollution from other sources.

INTERNET RESOURCES

Agricultural Chemical Use (U.S. Geological Survey): *http://h2o. usgs.gov/public/pubs/bat/bat000.html*—Statistics and GIS maps of chemical and fertilizer use in the U.S.

National Integrated Pest Management Network: *http://ipmwww. ncsu.edu/nipmn/states/National.html*—Links to pest management resources in the Internet.

Not Just Cows: *http://www.snymor.edu/~drewwe/njc/*—An excellent starting point for agriculture resources on the Internet.

U.S. Department of Agriculture Natural Resources Conservation Service: *http://www.ncg.nrcs.usda.gov/Welcome.html*—Contains links to resources on agriculture, soil conservation, and other USDA sites on the Internet.

A Black Rhino and her calf, native to eastern and southern Africa, are among the most endangered of the world's large mammals.

CHAPTER 12

WILD LIVING RESOURCES: PLENTIFUL AND ENDANGERED

CASE STUDY

The American Whooping Crane and the California Condor

The American whooping crane and the California condor are two of North America's largest birds. The whooping crane is the tallest, measuring more than 1.5 m (approximately 5 ft) in height. The California condor has the greatest wingspan—almost 3 m (9 ft). Although both are rare and endangered, they are protected, and both species have large preserves set aside for them. The two species, however, seem to be responding differently to these conservation efforts.

In 1937 the whooping crane population was reduced to 14 individuals. It has since recovered; in 1991 it numbered 183 in the wild; an additional 49 birds were in captivity.[1] In the preservation of endangered species, the whooping crane is a success story.[2,3]

On the other hand, while the whooping crane population increased, the California condor population declined rapidly. Historical records suggest that after 1840 its numbers rapidly diminished, falling to about 60 in the 1940s and to about 25 in

the mid–1970s (Figure 12.1).[4] At that time, the U.S. Fish and Wildlife Service and the American Audubon Society began a controversial condor recovery program. They removed birds and eggs from the wild with the intent of hatching and rearing chicks in zoos and eventually returning them to the wild.

Removal of the remaining wild condors began in 1986 and the last bird was taken into captivity in 1987. In 1988 the first condor chick was hatched from an egg laid in captivity. Reintroductions of captive-bred condors began in 1992 and by 1995 there were 13 condors living in the wild, although they were being fed. Currently there are 90 condors still in captivity at four different locations: the Los Angeles Zoo, the San Diego Wild Animal Park, and breeding refuges in Ventura County and Boise, Idaho.[5] The goal of the condor recovery program is to establish three separate populations of 150 birds each: two geographically isolated populations (one in southern California and one in Ari-

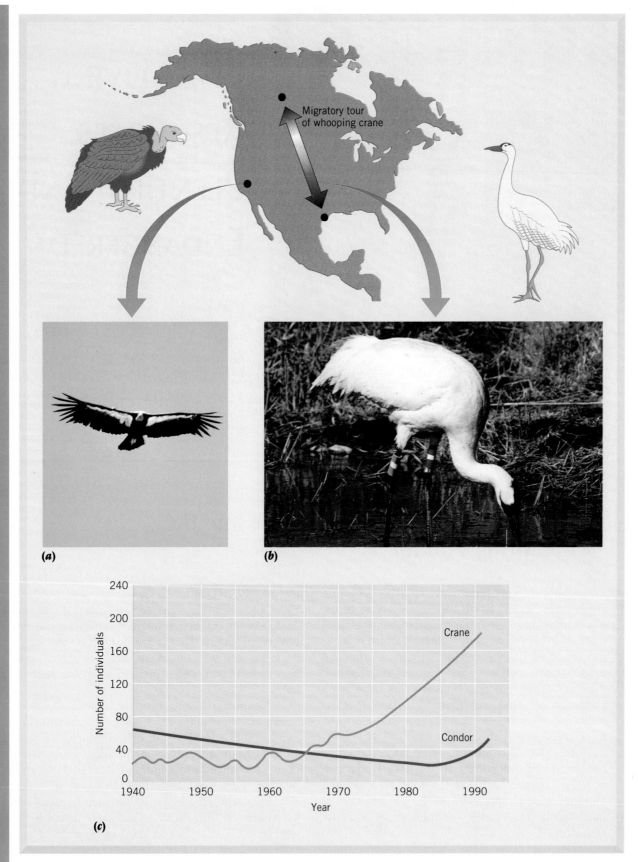

Figure 12.1 (*a*) The California condor. (*b*) The whooping crane in its habitat. (*c*) After a long period of decline, the condor population is showing a small recovery, while the whooping crane has been increasing generally along an exponential curve.

zona) and a third in captivity. The question now is whether the birds born in captivity can successfully return to and survive in the wild independently.

The first attempts to release two zoo-bred birds into their California habitat occurred in January 1992.[6] Ten months later, one of the birds was found dead from drinking a toxic antifreeze substance that had leaked from a car.[7] In mid–1993 two more of the total eight released were killed when they became entangled in high-voltage power lines. The release program continues, but it faces many difficulties and it remains controversial.

The extinction of species is not new; in a world of chance, the eventual fate of every species is extinction. Given that history, why is the whooping crane recovering while the California condor is having such difficulty? The differences are ecological, as affected by human history.

Although the whooping crane and the California condor are both large and long-lived, their habitat and food differ greatly. Unlike the crane, which migrates thousands of miles, the condor lives year-round in the Sierra foothills and coastal ranges of southern and central California. Its range was once much wider; Lewis and Clark shot a condor in 1805 on the Columbia River in what is now the state of Washington. Their feeding habits are different; the crane feeds in wetlands on small aquatic animals; the condor, a member of the vulture family, feeds on carrion, primarily of large mammals.

The recovery of the crane was possible because its habitat was in good condition; food and nesting sites were still sufficient and plentiful. The habitat of the condor, on the other hand, has been greatly altered (Figure 12.2). Food is scarce; the number of big game mammals, such as deer and elk, in the mountains has been greatly reduced over the years. What food exists is harder for the condor to find. Some areas that once burned often

FIGURE 12.2 The condor wilderness area in southern California.

and were relatively open grasslands have become more heavily grown with dense chaparral and woodlands. Even if a condor sees food and lands, without an open space it cannot take off and is trapped on the ground. Thus, although the condor was more abundant than the crane in the 1940s, the future is brighter for the crane than for the condor.

There is a lesson in the comparison of cranes and condors: For an endangered species, it is better to have an ecosystem in good condition and a small population than to have an abundant population in an ecosystem in poor condition.

Vigilance must still be maintained for the crane. It was found in the late 1980s that one of the major habitats of the whooping crane, the Aransas National Wildlife Refuge in Texas, is severely eroding. Between 0.4 and 1.6 ha (1 and 4 acres) of the 19,127-ha (47,261-acre) refuge are disappearing annually into the Gulf of Mexico as a result of digging a channel between the refuge and the barrier islands for the passage of commerce.[8] It is to be hoped that one of the few success stories in wildlife preservation will not become one of the failures.

LEARNING OBJECTIVES

There are many important reasons for preserving wildlife species and limiting both natural and human-induced risk to them. Much can be done to improve the ways in which we go about the conservation of species. After reading this chapter, you should understand:

- The major current causes of extinction.
- The utilitarian, ecological, aesthetic, and moral reasons for conserving wildlife and endangered species.
- The traits shared by species likely to become endangered through human activities.
- The risk factors for extinction, the human ef-

fect on extinction through history, and why modern technological civilization and the large number of people have greatly increased the rate of extinction.

- The concepts and terms related to conservation, such as carrying capacity, maximum sustainable yield, minimum viable population size, and minimum viable habitat.
- The differences between the goals and emphasis of modern wildlife management and those of traditional wildlife management.
- The necessary components of successful wildlife management.

12.1 THE GOALS OF CONSERVATION

The recovery plan for the condor discussed in the case study raises a philosophical question: When we say that we want to save a species, what is it that we really want to save? The possibilities include four options, one of which appears in Figure 12.3:

1. a wild creature in a wild habitat, as a symbol to us of wilderness;
2. a wild creature in a managed habitat, so that the species persists, feeds, and reproduces with little interference and so that we can see it in a naturalistic habitat;

FIGURE 12.3 A group of Przewalski horses in a protected park. These once-wild relatives of the domesticated horse now exist only in parks and zoos. Human activities have eliminated them from their original habitat.

3. preservation of a population in a zoo so that the genetic characteristics are maintained in live individuals; or
4. genetic material only—frozen cells containing DNA from a species for future scientific research.

The recovery plan for the condor accepts option 3 as a short-term option in the hope that option 2 will ultimately be achieved. Rarely are the specific goals of conservation spelled out in this fashion, but as the condor example illustrates, policies and actions differ widely depending on the specific goal.

If the goal were to maintain a wild creature in a wild habitat, then the condor recovery program would have emphasized habitat management, including controlled burning and other methods to clear the land of shrubs and open it to grasses, and the reintroduction of deer, elk, and other food species for the condor. At the other extreme, a DNA conservation program would have focused on catching individual condors just before they died of other causes and preserving some tissues.

This chapter discusses the conservation of biological diversity, introduced in Chapter 7. This topic includes the conservation of individual endangered species, a topic about which public interest has grown rapidly in recent years. We analyze the status of rare, threatened, and endangered species worldwide and review reasons for which we value endangered species and biological diversity. We explain the causes of extinction, both natural and human induced, to help understand what we can and cannot do to conserve endangered species. Finally, we discuss options for a constructive approach to conservation.

12.2 CATEGORIES OF THREATENED SPECIES

Species in danger of extinction are divided into several categories. The International Union for the Conservation of Nature (IUCN), a world-wide organization with headquarters in Switzerland, provides data on endangered species, including endangered, vulnerable, rare and indeterminate (see Table 12.1). A species that is rare is not necessarily in danger of becoming extinct; some species, like the whooping crane, are naturally rare. However, rarity does raise concerns about the possibility of extinction.

The number of species of animals listed in these categories worldwide has risen from 1672 in 1988 to 5876 in the most recent reports and the list continues to expand (see Table 12.1).[9] The 1996 IUCN *Red Book of Threatened Animals* reported that one-fourth of all known species of mammals are at risk of extinction, as well as 11% of known birds, 20% of known reptiles, 25% of amphibians, and 34% of fish, primarily freshwater fish. An especially large number of fish and mollusks are endangered or vulnerable (see Table 12.1). The *1997 IUCN Red List of Threatened Plants* estimates that 33,798 species of vascular plants (plants with a system for moving fluids, such as sap) have recently become extinct or endangered. This means that 12.5% of the world's vascular plants are under threat.[10] Assessment of the status of the world's tree species is difficult given the overall number of trees believed to exist. There are approximately 100,000 species of trees in the world. Over 8,700 tree species are listed as globally threatened by the IUCN. This means that nearly 9% of the world's tree flora is globally threatened.[11] Table 12.2 shows the percentages of species threatened in sev-

eral countries. The total number of species that are threatened is not known, because many areas, especially in the tropics, have been poorly explored for species diversity. In fact, only a small percentage of existing species in the tropics are believed to be identified.

12.3 WHY SAVE ENDANGERED SPECIES?

Chapter 1 introduced the concept of four justifications, or reasons, for saving endangered species. This chapter explores these four reasons—utilitarian, ecological, aesthetic, and ethical—in more depth. A strong case has been made for saving species for these reasons.[12]

Utilitarian Justification

Utilitarian justification is based on the consideration that many wild species might be useful to us and thus it is imprudent to destroy them before we have a chance to test their uses. The importance of maintaining genetic variation in general, in addition to conserving individual species, is also part of the utilitarian justification.

Genetic Characteristics

One example of utilitarian justification is the need to conserve wild strains of grains and other crops. Modern agricultural production of crops such as wheat and corn depends on the continued introduction of fresh genetic characteristics from wild strains to create new genetic hybrids. Disease organisms that attack crops evolve continually, changing their genetic characteristics. As new disease strains

TABLE 12.1 Status of Threatened and Endangered Animals Worldwide

Group	Endangered	Vulnerable	Rare	Indeterminate[a]	Total
Mammals	177	199	89	276	741
Birds	188	241	257	284	970
Reptiles	47	88	79	102	316
Amphibians	32	32	55	50	169
Fish	158	225	246	349	979
Mollusks	309	422	139	312	1182
Annelids	6	145	6	2	159
Spiders	3	2	4	9	18
Crustacea	10	6	28	114	158
Insects	252	119	241	572	1184
Totals	1182	1479	1144	2070	5876

[a]Includes those for which data are incomplete.

Source: IUCN 1996 Red List of Threatened Animals, International Union for the Conservation of Nature. Cambridge, England.

TABLE 12.2   Threatened Species as a Percentage of Species Known, Late 1980s

Country	Mammals	Birds	Fish	Reptiles	Amphibians	Vascular Plants
Australia	13.4	3.3	—	1.6	4.0	12.3
Austria	29.4	28.4	36.2	46.2	10.5	15.9
Belgium	21.5	29.0	—	75.0	100.0	24.0
Canada	7.3	3.8	1.2	2.4	2.4	0.8
Denmark	28.6	17.4	7.8	0.0	21.4	13.7
Finland	11.3	6.0	12.1	20.0	20.0	5.6
France	52.2	39.8	18.6	38.9	62.1	8.4
Ireland	16.1	23.7	—	0.0	33.3	—
Italy	13.4	14.3	13.9	52.2	46.4	10.0
Japan	7.4	8.1	10.6	3.5	6.3	10.2
Netherlands	48.3	33.1	22.4	85.7	66.7	—
New Zealand	20.3	5.7	0.4	17.9	—	4.8
Norway	7.4	10.2	1.2	20.0	40.0	4.5
Portugal	51.2	39.6	28.2	37.1	23.5	—
Spain	14.8	14.5	18.2	14.1	4.2	2.5
Sweden	15.4	6.8	4.6	0.0	38.5	8.2
Switzerland	46.3	50.9	—	80.0	78.9	25.8
Turkey	30.5	16.9	18.7	50.5	72.2	—
United Kingdom	31.2	15.0	3.4	45.5	33.3	9.6
United States	10.5	7.2	2.4	7.1	3.6	0.5
West Germany	46.8	32.1	70.0	75.0	57.9	28.2

Source: OECD Environmental-Indicators, 1991, *The State of the Environment,* OECD Publications Office, Paris, France. Reprinted by permission of Organization for Economic Cooperation and Development.

develop, crops become vulnerable. By introducing fresh genetic characteristics from the wild, new hybrid strains can be developed that are disease resistant. According to the U.S. Department of Agriculture, this increases farm production in the United States by $1 billion per year.

Chemical and Medical Uses

Another utilitarian justification for biological conservation is that many important chemical compounds come from wild organisms. Sales of plant-derived medicines are $14 billion a year in the United States and $40 billion worldwide.[15] Digitalis, an important drug in treating certain heart ailments, comes from purple foxglove. Aspirin is a derivative of willow bark. Well-known medicines derived from tropical forests include anticancer drugs from rosy periwinkles (Figure 12.4); steroids from Mexican yams; antihypertensive drugs from serpent-wood; and antibiotics from tropical fungi.[15] Some 25% of prescriptions dispensed in the United States contain ingredients extracted from vascular plants.[14] Only a small fraction of the estimated 270,000 plant species are utilized for the drugs used worldwide.[15] Other organisms may produce useful medical compounds that are as yet unknown. For example, scientists are testing marine organisms for use in pharmaceutical drugs. Coral reefs offer a promising area of study for such compounds, because many coral reef species produce toxins to defend themselves.[16]

New Crops and Products

A third utilitarian justification is potential for developing new crops or commercial products from wild plants and animals (see Chapter 10). Many horticultural crops and products have come from tropical rain forests, and hopes are high that new products will be found.[17] For example, of 275 species found in 1 ha (0.4 acre) in a Peruvian tropical forest, 72 yielded products with direct economic value. Of 842 individual trees, 350 yielded products with direct economic value. The average market price of fruit tree and palm products was estimated to be $650 per year, and net annual revenues were estimated to be $400/ha (about $160/acre).[18]

Indigenous Peoples

Fourth, biological diversity is of great importance to many indigenous peoples of less-developed nations, for whom diversity in forests and wildlife provides food, wood for shelter and tools, fuel, materials for clothing, and medicine. A reduction in biological diversity can severely increase the poverty of these people. For poor, indigenous people who depend on forests, there may be no reasonable replacement for these benefits except continual exter-

Figure 12.4 The rosy periwinkle from Madagascar produces chemicals used in the treatment of certain cancers and illustrates one of the utilitarian reasons for the conservation of the great biological diversity on Earth.

nal assistance, which development projects are supposed to eliminate. Urban residents share in the benefits of biological diversity, even if these benefits may not be apparent or may become apparent too late (see Chapter 26).

Pollution Control

Fifth, biological diversity provides pollution control. Aquatic filter-feeding animals remove excess primary production from eutrophic waters. Plants, fungi, and bacteria remove toxic substances from air, water, and soils. For example, carbon dioxide and sulfur dioxide are removed by vegetation, carbon monoxide is reduced and oxidized by soil fungi and bacteria, and nitric oxide is incorporated into the biological nitrogen cycle.[19] Because species have different capabilities for removal, a diversity of species can provide the best range of pollution control.

Tourism

A sixth utilitarian justification is tourism, a growing source of income for many developing countries. Over 125 nations consider tourism a major industry. In about one-third of them (i.e., Nepal, Bhutan, and until recently, Rwanda) it is the leading industry.[20] Although initial tourism was based on the larger forms of wildlife that inhabit open savanna lands, increasing attention is now being given to **ecotourism**, which includes interest in a whole array of plants and animals. For example, in Nepal there has been an explosion of trekking tourism over the last two decades. From 1980 to 1991, the number of trekkers increased 255%.[21]

Medical Research

Medical research provides a seventh utilitarian justification (see *A Closer Look 12.1, "Conflicting Goals"*). For example, the armadillo, one of only two animal species known to contract leprosy, is important to the study of this disease to find a cure.[22] Other animals such as the horseshoe crab, *Limulus,* and barnacles are important because of physiologically active compounds they make. Others may have similar uses as yet unknown to us.

Ecological Justification

When we reason that organisms are necessary to maintain the functions of ecosystems and the biosphere, we are using an ecological justification for their conservation. Individual species, entire ecosystems, and the biosphere provide public service functions essential or important to the persistence of life, and as such they are indirectly necessary for our survival. As Aldo Leopold wrote in *A Sand County Al-*

A Closer Look 12.1

Conflicting Goals

A new cancer drug, Taxol, was recently discovered in the Pacific yew, a conifer of the Pacific Northwest (Figure 12.5). Taxol has been found useful in treating ovarian cancer, which kills about 10,000 women a year in the United States. It takes about six 100-year-old trees to provide enough Taxol to treat the cancer in one patient, so its use would require harvesting many trees per year. The Pacific yew grows in habitats where the spotted owl is also found. In this case, a utilitarian argu-

ment and moral and aesthetic arguments are in conflict.

Those solely interested in the cancer drug would opt for harvesting the existing yews and developing plantations of yews that would be harvested at 100-year intervals. Those interested in the spotted owl would seek to keep intact areas of mature forest. Those interested in overall biological diversity would seek natural landscapes with a variety of stages in forest development. This example reveals the complexities of the issues

and the possibility for conflict among those who, on the surface, would seem to be in favor of the same goal.

The first thing we must do when we set out to conserve biological diversity is to clarify whether we seek to conserve one species or overall diversity and whether the primary justification is utilitarian, ecological, aesthetic, or moral. We could decide that both a specific endangered species and overall biological diversity are important, but it may be impossible to achieve both goals at the same time.

Figure 12.5 An important anticancer chemical, Taxol, has recently been discovered in the Pacific yew tree, *Taxus brevifolia,* which grows in the Pacific Northwest of North America. Like the rosy periwinkle, this species illustrates one of the utilitarian reasons for the conservation of the great biological diversity on Earth.

manac, "The outstanding scientific discovery of the 20th century is not TV or radio but the complexity of the land organism. . . . To keep every cog and wheel is the first precaution of intelligent tinkering."[23]

At the level of an individual species, the concept of a keystone species (one that carries out an essential function not performed by any other species) is an example of an ecological justification (see Chapter 7 for a discussion of keystone species). Although we might not use the species directly, it is

essential to the functioning of an ecosystem which can benefit us. Although ecological justifications have grown stronger in recent years, it is often difficult to determine whether a specific species is ecologically valuable or essential. Prudence suggests that we should assume that a species has ecological value unless there is concrete evidence that it does not.

Conserving Forests

At an ecosystem and regional level, forests illustrate several kinds of ecological justifications. Forests retard soil loss and erosion, especially in areas of high rainfall, high rates of tectonic uplift, and soft bedrock (conditions found in low-latitude mountainous regions, such as those of India and Nepal).[25] Forests stabilize water supply and runoff, a benefit known to the ancient Greeks, rediscovered in the nineteenth century, and quantified in the last 30 years through studies of forested watersheds in North America. It has been calculated, for example, that more than half the precipitation in the Amazon region is generated by the forests.[26,27] Some scientists suggest that deforestation of the Amazon basin in northern Brazil would so alter the climate in southern Brazil that agriculture would become impossible there.[28,29]

Global Perspective

As we learn more about the biosphere, we begin to understand how species affect one another around the world. Bacteria carry out chemical reactions that affect the chemical makeup of the atmosphere; for example, they convert molecular nitrogen in the atmosphere to chemical compounds that can be used by other living things. Wild vertebrates, invertebrates, and microorganisms play major roles in the pollination of wild and crop plants, germination, dispersal of seeds and other reproductive structures of plants, soil processes, and nutrient cycling. All these functions are vital not only to the maintenance of the ecosystems of which the organisms are a part but also to human welfare.

Aesthetic and Cultural Justification

An aesthetic justification asserts that biological diversity adds to the quality of life, providing some of the most beautiful and appealing aspects of our existence. Biological diversity is an important quality of landscape beauty. Many organisms—birds, large land mammals, and flowering plants, as well as many insects and ocean animals—are appreciated for their beauty (Figures 12.4 and 12.6).

Throughout human history people have emphasized the importance of the diversity of life to the purpose and meaning of our existence. Thousands of years of paintings and drawings, dating back to cave paintings drawn by Stone Age people, testify to the fundamental aesthetic role of nature and its diversity in human existence. Literature as well, from ancient epics to modern novels, films, and television, celebrates the beauty and diversity of life. The large sums spent on tourism to areas of lush vegetation, coral reef islands, and other areas of natural beauty underscore the economic implications of aesthetic and cultural values.

Today we continue to imbue certain animals with cultural significance; for instance, in the United States the bald eagle, which is endangered, is especially valued because it was adopted as the symbol of the new nation when the country was founded.

Moral Justification

Moral justification is based on the belief that species have a moral right to exist, independent of our need for them. Consequently, the argument follows that in our role as global stewards it is a human obligation to assist the continued existence of species and to conserve biological diversity. This right to exist was stated in the U.N. General Assembly World Charter

FIGURE 12.6 The American bald eagle, a national symbol, illustrates the aesthetic importance to people for the conservation of an endangered species.

for Nature, 1982. The U.S. Endangered Species Act also includes statements concerning the rights of organisms to exist. Thus a moral justification for the conservation of endangered species is part of the intent of the law.

Moral justification has deep roots within human culture, religion, and society. Those who focus on cost–benefit analyses tend to downplay moral justification. However, although moral justification may not seem to have economic ramifications, in fact it does. As more and more citizens of the world assert the validity of moral justification, more actions that have economic effects are taken to defend a moral position. For example, many Americans refuse to purchase tuna caught by fishing fleets using methods that kill the porpoises that swim with the tuna. The important economic effect of this action was made clear by the response of one major tuna company, which advertised that it would sell only tuna taken without harm to porpoises. There are other examples, including international boycotts of furs, teak, and ivory. Some carry this argument to another step. What has become known as the "deep ecology" movement argues that the biosphere ranks higher than people because the persistence of life depends on the biosphere. Therefore, from a moral perspective, choices should be made for the biosphere rather than for people. Because moral concerns about biological diversity are likely to increase in the future, we can expect more economic consequences to result.[30]

Are We Conserving Individual Species or Total Diversity?

The term *conservation of biological diversity* can mean several different things, as discussed in Chapter 7. Here we consider two primary meanings: the conservation of a specific endangered species in a specific place and the conservation of the total number of species in an area. Although these may seem to be the same thing, they are different and are sometimes in conflict.

For example, if we manage the forests of the northwestern United States solely for the conservation of the spotted owl that inhabits old-growth forests, we might lose species that depend on early stages in forest succession. If we focus on the overall conservation of biological diversity in the area, then the loss of any single species, such as the spotted owl, would be considered less important than the total number of species and the relative abundance of all species. In the latter case, we would attempt to conserve all stages in forest succession in conservation areas.

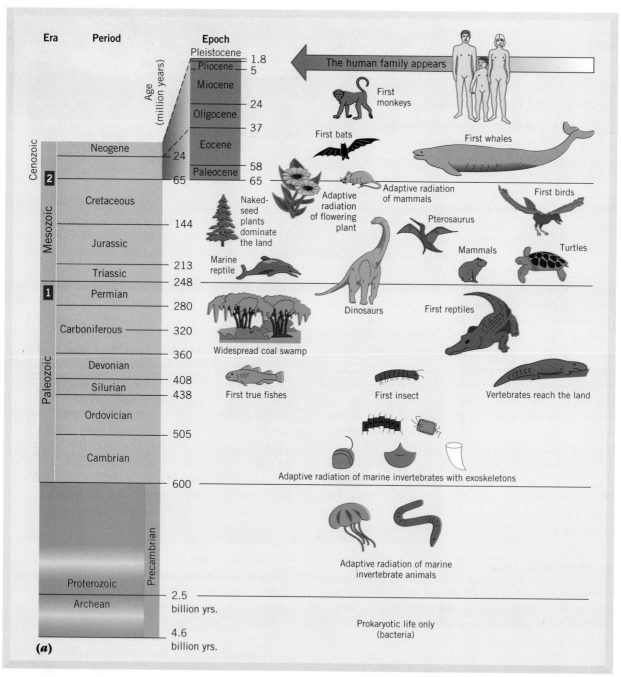

FIGURE 12.7 (*a*) See legend on facing page.

12.4 EXTINCTION

Extinction is the rule of nature. **Local extinction** occurs when a species disappears from a part of its range but persists elsewhere. **Global extinction** means that a species becomes extinct everywhere.

Although extinction is the ultimate fate of all species, the rate of extinctions has varied greatly over geologic time and has increased rapidly since the industrial revolution. From 580 million years ago until the beginning of the industrial revolution, on average, about one species per year became extinct. Over much of the history of life on Earth, the rate of evolution of new species equaled or slightly exceeded the rate of extinction. The average longevity of a species has been about 10 million years.[31] However, the fossil record suggests that there have been periods of catastrophic losses of species and other periods of rapid evolution of new species (Figure 12.7) which some refer to as "punctuated extinctions." About 250 million years ago a mass extinction occurred in which approximately 53% of marine animal species disappeared; about 65 million years ago, most of the dinosaurs became extinct. Inter-

spersed with the episodes of mass extinctions, there seem to have been periods of hundreds of thousands of years with comparatively low rates of extinction. Actually, episodes referred to as mass extinctions or punctuated extinction lasted a long time from our perspective—longer than the total time that human beings have had major effects on extinctions. The causes of these episodes of mass extinction are not well understood and remain the object of scientific controversy.

Natural extinctions often appear to follow understandable patterns, with the replacement of one form by a more successful one through the process of competition and evolution (see Chapter 7). This was not the case, however, about 10,000 years ago, at the end of the last great continental glaciation. At that time, massive extinctions of large birds and mammals occurred: 33 genera of large mammals (those weighing 50 kg [110 lb] or more) became extinct, whereas only 13 genera had become extinct in the preceding 1 or 2 million years. Smaller mammals were not so affected, nor were marine mammals. As early as 1876, Alfred Wallace, an English biological geographer, noted that "we live in a zoologically im-

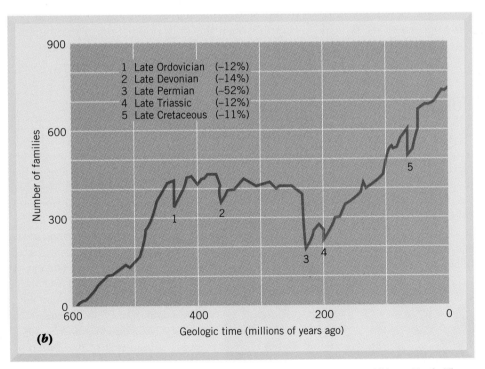

FIGURE 12.7 (*a*) Brief diagrammatic history of evolution and extinction of life on Earth. There have been periods of rapid evolution of new species and episodes of catastrophic losses of species. Two major catastrophes were the Permian loss of 52% of marine animals as well as losses of land plants and animals (1) and the Cretaceous loss of dinosaurs (2). (*b*) Graph of the number of families of marine animals in the fossil record, showing long periods of overall increase in the number of families punctuated by brief periods of major declines. [*Source: (a)* D. M. Raup, 1988, "Diversity Crises in the Geological Past," in E. O. Wilson, ed., *Biodiversity,* National Academy Press, Washington, DC, p. 53; derived from S. M. Stanley, 1986, *Earth and Life through Time,* W. H. Freeman, New York. Reprinted with permission. (*b*) D. M. Raup and J. J. Sepkoski, Jr., 1982, "Mass Extinctions in the Marine Fossil Record," *Science* 215, pp. 1501–1502.]

poverished world, from which all of the hugest, and fiercest, and strangest forms have recently disappeared."[32] It has been suggested that these sudden extinctions coincided with the arrival, on different continents, at different times, of Stone Age people and therefore may have been caused by hunting.[33]

Causes of Extinction

Causes of extinction are usually grouped into five risk categories: population risk, environmental risk, natural catastrophe, genetic risk, and human actions. Risk here means the chance that a species or population will become extinct owing to one of these causes.

Population Risk

Random variations in population rates (in birth rates and death rates) can cause a species in low abundance to become extinct. This is termed **population risk.** For example, blue whales swim over vast areas of ocean. Because whaling once reduced the total population to only several hundred individuals, there were probably year-to-year variations in the success of individual blue whales in finding mates. If in one year most whales were unsuccessful in finding a mate, then births could be dangerously low. Such random variation in populations, typical among many species, can occur without any change in the environment. It is a risk especially to species that consist of only a single population in one habitat. Mathematical models of population growth can help calculate the population risk and determine the minimum viable population size.[34]

Environmental Risk

Population size can be affected by changes in the environment that occur from day to day, month to month, and year to year, even though the changes are not severe enough to be considered environmental catastrophes. **Environmental risk** involves variation in the physical or biological environment, including variations in predator, prey, symbiotic, or competitor species. (See A Closer Look 12.2, "Extinction of the Heath Hen")

As an example, Paul and Anne Ehrlich described the local extinction of a population of butterflies in the Colorado mountains.[35] These butterflies lay their eggs in the unopened buds of a single species of lupine (a member of the legume family), and the hatched caterpillars feed on the flowers. One year, however, a very late snow and freeze killed all the lupine buds, leaving the caterpillars without food and causing local extinction of the butterflies. The plants survived, and their roots produced new stems, leaves, and flowers the next year. Had this been the only population of that butterfly, the entire species would have become extinct.

 ## CLOSER LOOK 12.2

EXTINCTION OF THE HEATH HEN

The difference between environmental risk and natural catastrophe is illustrated by the extinction of the heath hen. Once a common bird of the eastern United States, it was reduced to low numbers by human-induced habitat change and hunting.

By the last quarter of the nineteenth century, the heath hen existed only on the island of Martha's Vineyard, off Cape Cod, Massachusetts (Figure 12.8). In 1907, a part of the island was made into a preserve for this bird, whose numbers had been reduced to about 100. Predators of the heath hen were controlled. The population increased to 800 by 1916. Then, two things happened: A fire destroyed most of the nesting areas (a catastrophe) and the next winter an unusually heavy concentration of goshawks on the island (environmental risk) caused high mortality. The population fell to less than 200, and the bird became extinct by 1932, possibly the result of population or genetic risks.[69]

FIGURE 12.8 The heath hen, which once inhabited the island of Martha's Vineyard just offshore from Cape Cod, Massachusetts, became extinct as a result of the interplay of a complex set of factors.

In some cases, species are sufficiently rare and isolated that such normal variations can lead to their extinction. In other cases, species succumb to catastrophic variation in the environment.

Natural Catastrophe

A sudden change in the environment not the result of human action is a *natural catastrophe.* Fires, major storms, earthquakes, and floods are natural catastrophes on land; changes in currents and upwellings are ocean catastrophes. For example, the explosion of a volcano on the island of Krakatoa in Indonesia in 1883 caused one of recent history's worst natural catastrophes. Most of the island was blown to bits, bringing about local extinction of most life forms there. (See A Closer Look 9.1.)

Genetic Risk

Detrimental change in genetic characteristics not caused by external environmental changes is called **genetic risk**.[36] Genetic changes can occur in small populations from reduced genetic variation, genetic drift, and mutation (see Chapter 7). In a small population, only some of the possible inherited characteristics will be found. The species is vulnerable to extinction because it lacks variety or because a mutation can become fixed in the population.

Consider the last 20 condors that were in the wild in California. It stands to reason that this small number was likely to have less genetic variability than the much larger population that existed several centuries ago. This increased their vulnerability. Suppose that the last 20 condors, by chance, had inherited characteristics that made them less able to withstand lack of water. Then, if left in the wild, the condors would have been more vulnerable to extinction than a larger, more genetically variable population during the recent drought years in California.

Common Traits of Endangered Species

Species likely to become endangered through human activities tend to share certain traits. Knowledge of these common traits helps us protect and manage such species. Easily endangered species, particularly vertebrates, are generally long-lived and large. Such species tend to have low reproductive rates and recover slowly from lowered population levels (Figure 12.9*a*). It makes sense that the poten-

FIGURE 12.9 Certain characteristics tend to make some species more susceptible to extinction than others, especially from human activities. (*a*) Large, long-lived animals, like the African black rhinoceros, often are desirable for some commercial product. The horn of the rhinoceros is used as a medicinal and can be sold for a high value. (*b*) Species that are high on the food chain, such as the osprey, a fishfeeding predatory bird, concentrate chemical toxins such as DDT and are more susceptible to these chemicals than herbivores. (*c*) Species with highly specialized diets, like the giant panda of China, which feeds only on certain kinds of bamboo, are at risk due to human land clearing and modification. Large animals that require large habitats, like the panda, are among the most vulnerable to this effect.

(a)

(b)

(c)

tial growth rate of a pair of rats that can have several litters of multiple offspring each year is markedly greater than that of a pair of condors that raises one offspring once in several years. Another factor in their vulnerability is that the biggest and largest also require the largest territories and the most food per individual. Carnivores and large herbivores are particularly susceptible to extinction (see Chapters 8 and 11 and Figure 12.9*b*).

Specialist species—those adapted to very narrow sets of conditions and having highly specific habitat and behavior requirements—are also especially vulnerable (Figure 12.9*c*). As people clear land and modify the environment, the diversity of habitats is reduced, and specialist species are easily affected by these changes. Adaptable and generalist species, especially those that share foods and habitats with people, do well. Rats, sparrows, cockroaches, cats, and dogs fare well in a habitat dominated by humans, where there is an omnivore food base and a simplified environment.

12.5 HOW PEOPLE CAUSE EXTINCTIONS AND AFFECT BIOLOGICAL DIVERSITY

Human actions cause extinction of species through

1. intentional hunting or harvesting (for commercial purposes, for sport, or to control a species that is considered a pest);
2. disruption or elimination of habitats;
3. introduction of new parasites, predators, or competitors of a species (Figure 12.10); and
4. pollution of the environment.

People have caused extinctions over a long time, not just in recent years. The earliest people probably caused extinctions through hunting. With the knowledge of fire, people began to change habitats over large areas. With the development of agriculture and the rise of civilization, rapid deforestation and other habitat changes became important factors. Later, as people explored new areas, introductions of exotic species became an important cause of extinction (see Chapter 7), especially after Columbus's voyage to the New World and the spread of European civilization and technology. In the twentieth century, with the introduction of thousands of novel chemicals into the environment, pollution has become an increasingly significant cause of extinction.

The IUCN estimates that 75% of the extinctions of birds and mammals since 1600 were caused by human beings. Hunting is estimated to have caused 42% of the extinctions of birds and 33% of the extinctions of mammals. In the United States, of 47 species of wildlife that became extinct between 1700 and 1970, 25 were lost in the last 50 years of the period (Figure 12.11). Modern civilization has contributed greatly to the increased rate of extinction, and tropical deforestation accounts for much of those extinctions. In recent years the number of identified extinct species has increased significantly due to efforts to catalog species being lost. The current extinction rate among most groups of mammals is estimated to be 1000 times greater than the extinction rate at the end of the Pleistocene epoch.[37] Estimates today put the rate of extinctions worldwide at 50,000 species each year, resulting in a loss of 10% of all species alive today within 25 years.[38]

Introductions of Exotic Species

Since the beginning of the age of exploration, human beings have been conducting a gigantic experiment in biogeography, intentionally and unintentionally moving many species of animals, plants, and bacteria around the world (see A Closer Look 12.3, "The Asian Water Buffalo in Australia"). Changing the geography of life is one of the major ways in which we are changing the world. Since the beginning of agriculture, people have carried with them their favorite crops and domestic animals as they settled new territories (see the discussion of island biogeography in Chapter 7).

In Great Britain, for example, the only native foods that would have been found by the first settlers are nuts, berries, and wild game. Every other food grown in Great Britain has come from another part of the world. The discovery of new foods, including the New World's maize and potato, led to intentional introductions that greatly benefited human beings and were an important factor in nineteenth-century human population growth in parts of Europe.

Many major crops have been transported wherever they would grow. For example, citrus fruits, such as oranges, originated in southern China but are now grown in Mediterranean climates worldwide. Wheat's wild relatives are from the Middle East and northern Africa; corn's wild relatives are from Mexico and South America. Corn and wheat are now grown on every continent except Antarctica and in all temperate climates where people live.

People continue the process of introductions today. A recent example is the introduction of live turkeys from North America, where they are native, to mainland China. In 1979 several hundred fertile turkey eggs were sent from Canada to the Peking Animal Husbandry Bureau, and by 1984 turkeys had become a significant food item in China.

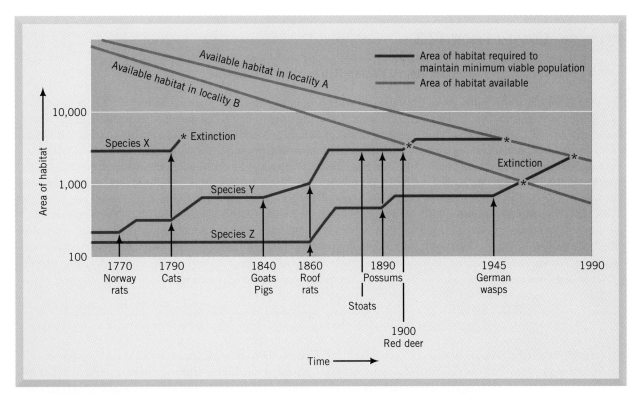

Figure 12.10 Hypothetical model showing the relationship between alien animal introductions and habitat loss in causing extinctions on islands. (*Source:* D. Western and M. C. Pearl, 1989, *Conservation for the Twenty-first Century,* Oxford University Press, New York, Figure 7.9, p. 65.)

Figure 12.11 Extinct vertebrate species and subspecies, 1760–1979. The number of species becoming extinct increased rapidly after 1860. Note that most of the increase is due to the extinction of birds. (*Source:* Council on Environmental Quality; additional data from B. Groombridge, ed., 1993, *1994 IUCN Red List of Threatened Animals,* IUCN, Gland, Switzerland, and Cambridge, UK.)

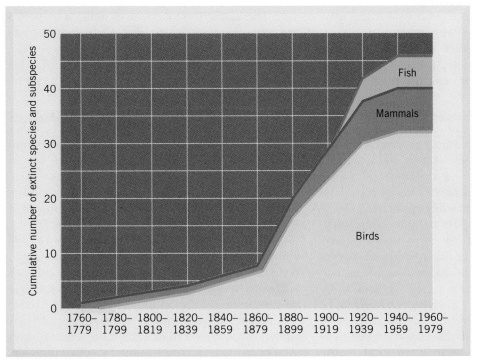

CLOSER LOOK 12.3

THE ASIAN WATER BUFFALO IN AUSTRALIA

In 1839 the Asian water buffalo was introduced as a beast of burden into northern Australia, an area with no native grazers on floodplain vegetation and no major predators of large mammals. The water buffalo arrived with only a few of its natural parasites; we know this because only 11 species of parasitic worms occur in the Australian water buffalo, whereas 77 species are found in Asia. Transported to a habitat lacking major disease organisms and predators, with a suitable climate and plentiful food, the water buffalo underwent a population explosion.

The population increased beyond the capacity of its food resources, resulting in several large die-offs, during which many animals died of slow starvation. In their native habitat in Asia, death would have come more quickly to the buffalo, as weakened individuals would have been easy targets for predators and parasites. A slow death from starva- tion is much worse for the Australian population's habitat, because slowly starving animals continue to eat all possible vegetation. The Australian vegetation was greatly affected by the water buffalo; during the dry season in some areas only bare soil was found.[70,71] The water buffalo population persists in Australia today but is less stable than the populations in its native Asia, fluctuating widely over time.

Inadvertent Introductions

Although many intentional introductions have been of great benefit to human beings, others have had disastrous effects, including the spread of human diseases. With the arrival of Europeans in North and South America, European diseases, particularly smallpox and chicken pox, were brought to the Americas, where they caused fatal epidemics among Native Americans. As Lewis and Clark traveled by boat up the Missouri River in 1804, they passed many abandoned villages; their native guides told them that the inhabitants had died of these new diseases.[39]

People have transported pests worldwide, usually unintentionally. Mice, rats, and many weeds have come along for the ride whenever people have migrated to new areas. Since the Renaissance, great human migrations have brought the house mouse and the European rat (known as the Norway rat) to every continent and many islands. Many familiar roadside flowers and weeds in North America were transported as seeds mixed in with crop seeds or carried in mud on the hooves of cows and horses.

Some recent inadvertent introductions of fish and shellfish have had negative effects. Introductions of European fish and shellfish into the Great Lakes of North America, brought by ships that passed through the Saint Lawrence Seaway, have disrupted major Great Lakes fisheries. These species include the sea lamprey and zebra mussel from Europe. The Baltic and Black Seas have suffered from inadvertent introductions of many marine and brackish-water species from North America and elsewhere, including a predator that has destroyed oyster beds along the Caucasian coast, and North American barnacles brought in on the outsides of ships. When freshwater or brackish-water species cause a problem, there is a comparatively simple control. Most are transported in ballast water, which is taken on board at the port of origin and then released at the destination. An exchange of ballast water in midocean solves the problem, since the freshwater and brackish-water species cannot survive in the higher salinities of the ocean and oceanic species taken on board as ballast cannot survive in the fresh or brackish water of the destination harbor.[40]

How Introduced Species Become Pests

Intentional introductions of animals as pets or for economic benefit or visual pleasure have often had disastrous consequences. The Asian walking catfish was recently introduced as a pet in Florida; now it has become wild and widespread; so far impossible to eradicate or control, it is becoming so abundant that it threatens many native freshwater fish species in Florida with extinction. The gypsy moth is another example of a well-intentioned introduction with undesirable effects. The moth was introduced into Cape Cod at the turn of the twentieth century to serve as the basis for a U.S. silk industry, but it escaped and is now a major pest of eastern U.S. forests and threatens to spread westward.

Why do some introduced species become abundant pests? If an introduced species is a superior competitor, then the competitive exclusion principle begins to operate to the detriment of native species (see Chapter 7). The introduced species increases rapidly and as a secondary consequence threatens native competitors with extinction. In general, organisms in one habitat have persisted together for a long time and have evolved together so that predator, parasite, prey, competitor, and symbiont (partner in a mutually beneficial relationship) have had time to adjust to one another. An introduced species rarely brings its own predators and usually brings some of its parasites. If it is a successful competitor, it undergoes a population explosion free from its natural enemies; therefore, the population increase is rapid.

Parasites

Often parasites transported by introduced species attack native species. An Asian fungus disease of elm trees (called Dutch elm disease because introductions came via Europe) was introduced into North America. Because trees had no resistance to it, it destroyed the American elm in most of its range.

Extinctions Caused by Introduced Species

Introductions of cats, rats, goats, sheep, dogs, insects and plants caused dramatic extinctions in the nineteenth century. Introductions of exotic species caused severe problems on islands, such as extinction of dodo birds, whose eggs, laid on ground, were easy prey for dogs. Norfolk Island, inhabited by descendants of the crew that mutinied on the British ship *Bounty*, provides one of the better documented cases of plant extinctions. Before the introduction of rabbits, there were over 34 species of vascular plants on the island. By 1967, only 21 indigenous plant species remained. (Figure 12.12).[41, 42]

New Zealand provides a good example of human-induced extinctions. Because New Zealand was isolated from continents for a long period that included the rise of mammals, native land vertebrates of New Zealand consist almost entirely of birds. Fossils suggest that more than 150 species of flightless birds lived there when the first Polynesians arrived approximately A.D. 950.[43] By the time Captain Cook reached New Zealand in 1769, 20 species of moas (large, long-necked, flightless birds) and several other birds had been driven to extinction. European settlers increased the extinction rate further. The last moas were killed in the late-eighteenth century, as well as some other native birds.[44] Such introductions and subsequent extinctions of species have occurred so widely they are a global problem.

Commercial Trade in Endangered Species

Despite legal protection in many countries, products from endangered species are widely traded. This trade can create additional threats to species already suffering at risk. According to the Fish and Wildlife Service, $3 billion worth of wild animals and wild-animal products are traded annually in the United States.[45] The World Wildlife Fund estimates that one-third of the wild animals, plants, or products of wildlife that enter the United States are imported illegally.

In Argentina alone, 5.4 million vertebrates from 25 species are exported each year, mostly as skins

FIGURE 12.12 Extinctions of plants on the Norfolk islands caused by rabbits. (*Source:* D. Western and M. C. Pearl, 1989, *Conservation for the Twenty-first Century,* Oxford University Press, New York, Figure 7.2, p. 55.)

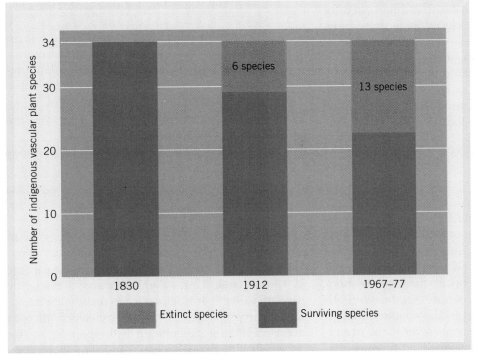

and leather. Although three-fourths of the more than $150 million produced by this trade comes from introduced species, native species are also threatened. The guanaco, a native wild relative of the llama, is one of the species exploited; of an estimated 500,000 remaining, about 10%, or 50,000, are exported each year.[46]

Wildlife is sold for fur, skin (for leather), and other products, such as ivory, antlers, and horns. Some wildlife is used for meat. Live animals are sold for medical and scientific research and as pets. More than 400,000 animals with an estimated value of $2 million were legally exported from the Philippines during the 1960s and 1970s. Most of these were monkeys used in medical research. There is legal trade in horns from nonendangered species, such as the red deer of New Zealand, and illegal trade in horns from endangered species, such as the rhinoceros.

12.6 WILDLIFE MANAGEMENT

Wildlife resources are renewable because populations and communities can be regrown, but as with all biological resources, populations of wildlife are vulnerable. In many parts of the world, people depend on wildlife for some of their essential food, clothing, and materials. Harvesting wildlife from forests is generally a less destructive use of land and is now being promoted as a possible solution to deforestation of tropical forests (see Chapter 13). There is a growing interest in game ranching, the practice in which, as discussed in Chapter 11, wild herbivores such as bison in North America and zebras in Africa are maintained on native habitats and harvested for meat, leather, and other products. Some studies suggest that wild game do less damage to vegetation and soils than cattle, sheep, and goats; this seems to be especially true for areas in which cattle, sheep, and goats are not native. Game ranching is also promoted as a means of increasing the economic value of endangered species to local human inhabitants, thereby increasing chances of survival of the endangered species, as discussed next.

Traditional Goals

In the past, goals of wildlife management seemed much simpler. Traditionally, wildlife management focused on terrestrial wild animals, primarily large mammals and birds, hunted for food, commercial products, or sport. Aldo Leopold, the father of modern wildlife management, whose *Sand County Almanac* was quoted earlier in the chapter, defined game (or wildlife) management earlier in the twentieth century as "the art of making land produce sus-

tained annual crops of wild game for recreational use."[47] When this was the major focus, specific goals were stated in terms of maintaining the maximum sustainable yield, later modified to be the optimum sustainable yield.

The **maximum sustainable yield (MSY)** of a population means the maximum production per unit time that can be sustained indefinitely without decreasing the capacity of the population to sustain that rate of production. According to Lee Talbot, one of the world's experts on wildlife management, the idea of MSY was considered virtually sacred to many wildlife managers until quite recently.[48] **Optimum sustainable yield (OSY)** means the maximum production that can be maintained without detriment to the population *or its ecosystem*. The MSY was traditionally estimated from the logistic growth curve as the population size exactly one-half of the carrying capacity (see Chapter 5 and Figure 12.13).

The Logistic Growth Curve

Recall that the logistic growth curve (see Chapter 5) is an S-shaped curve supposedly representing the growth of a population from a small number to its carrying capacity (Figure 12.13). It is calculated from the *logistic growth equation*, which is the simplest equation that includes the following ideas:

- A population that is small in relation to its resources grows at a nearly exponential rate.
- Competition among individuals in the population slows down the growth rate.
- The greater the number of individuals, the greater the competition and the slower the rate of growth.
- Eventually a point is reached, called the logistic carrying capacity, when the number of individuals is just sufficient for the available resources.
- At this level, the number of births in a unit time equals the number of deaths, and the population is constant.
- A population can be described simply by its total number.
- Therefore, the environment is assumed to be constant.

The **logistic carrying capacity** has been an important concept in wildlife management. However, it is a poor basis for realistic management (see A Closer Look 12.4, "Elephants and the Logistic Growth Curve"). It is important to realize that the logistic carrying capacity has an exact definition in terms of the logistic curve: It is the population size at which births equals deaths and thus there is no

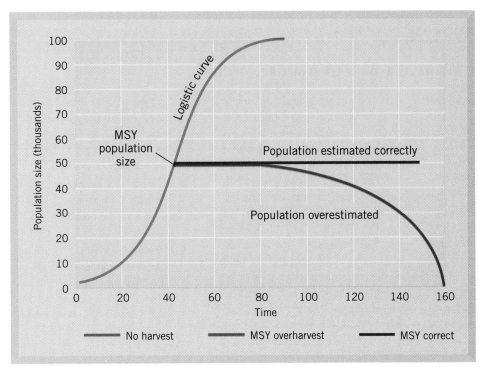

FIGURE 12.13 The logistic growth curve, showing the carrying capacity and the maximum sustainable yield population (where the population size is one-half the carrying capacity). The figure shows what happens to a population when we assume it is at MSY and it is not. Suppose that a population grows according to the logistic curve from a small number to a carrying capacity of 100,000 with an annual growth rate of 5%. The correct maximum sustainable yield would be 50,000. When the population reaches exactly that amount and we harvest at exactly the calculated maximum sustainable yield, the population continues to be constant. But if we make a mistake in estimating the size of the population and believe that it is 60,000 when, for example, it is really only about 50,000, then the harvest will always be too large and we will drive the population to extinction.

net change in the population. Note that the population is stable about the carrying capacity, which means that the population returns to the carrying capacity if it falls below or rises above it.

If immigration or any factor increases the population above the carrying capacity, deaths will exceed births and the population will return to the carrying capacity. If the population is temporarily harvested below the carrying capacity or reduced below the carrying capacity by any factor and the harvesting is stopped or the factor causing the reduction ceases to operate, births will exceed deaths until the carrying capacity is reached again.

Note that the term *carrying capacity*, without the qualifier *logistic*, is used much more broadly to mean the maximum population that can be sustained without detriment to the population or to the ecosystem of which the species is a part.

The Logistic Growth Curve and Endangered Species

In current practice, the logistic growth curve is often used as a basis to determine whether a species

is legally endangered. In such cases, a species is said to be endangered if its population falls below the MSY level. (see A Closer Look 12.4.) The problem with the logistic growth curve approach is discussed in A Closer Look 12.4 and in Chapter 5.

How Can We Better Manage Wildlife?

If the logistic growth curve fails as a basis for wildlife management (as illustrated in A Closer Look 12.4), what can replace it? A group of organizations, including the U.S. Council on Environmental Quality, the World Wildlife Fund United States, The Ecological Society of America, the Smithsonian Institution, and the IUCN, proposed four principles of wildlife conservation:

1. a safety factor in terms of population size, to allow for limitations of knowledge and the imperfections of procedures (an interest in harvesting a population should not allow the population to be depleted to some theoretical minimum size);

CLOSER LOOK 12.4

ELEPHANTS AND THE LOGISTIC GROWTH CURVE

Let's apply the logistic curve to elephants. Traditionally, this curve was thought to be the correct description of the growth of elephant populations and of all wildlife if nature were allowed to take its course. The logistic growth curve shows the growth of a hypothetical population plotted against time (Figure 12.13). We can also graph the relationship defined in the logistic equation between population growth and population size (Figure 12.14). In these graphs, we see that when the number of animals (such as elephants) is small and when the number is very high, the increase in population in a single time period is small. At the logistic carrying capacity, as explained earlier, net growth is zero. This must be true by definition; if the net growth were not zero, the population would be changing and thus would not be at its carrying capacity.

MSY and the Logistic Growth Curve

Figure 12.14 shows that the rate of growth changes between the smallest population and the carrying capacity and reaches a maximum between these extremes. *For a population growing exactly according to the logistic curve, the maximum rate of growth occurs when the population is exactly one-half its carrying capacity.* This point—a population size exactly one-half the carrying capacity—is therefore known as the point of maximum sustainable yield, or the MSY level. It is sustainable only in the narrow context of the logistic curve.

FIGURE 12.14 The logistic growth equation plotted to show population growth as a function of population size. The *horizontal* axis in this curve is the vertical axis in Figure 12.13. Maximum growth occurs when the population is exactly one-half the carrying capacity.

If an elephant population really did grow according to the logistic curve, then to achieve the traditional management goal of maximizing population growth, the strategy would be to harvest enough individuals every year to bring the population to exactly one-half its carrying capacity. Then the population growth in the second year would be the same as in the first, and it would be the maximum that could be added by that species. In the third year, the same number could be harvested as in the second and first years, and the population would grow again by exactly the same amount. The process could continue year after year. Harvests would be constant and sustainable forever (Figures 12.13 and 12.14).

Limitations of MSY and the Logistic Growth Curve

For the MSY concept to be true, and therefore for successful management to be based on it, three things must hold:

1. The population must have an exact and single carrying capacity, and its growth must be determined exactly by the classical logistic curve.

2. It must be possible to know precisely both the carrying capacity and the present population size.

3. It must be possible to obtain complete cooperation from all who harvest the animal involved (for elephants, this includes any leopards, lions, or hyenas that prey on young elephants) so that individual kills sum exactly the MSY level.

None of these conditions are met in the real world of wildlife management. If a population did grow according to the logistic, and measurements were subject to error, our management could easily lead to disaster for the population. Suppose we assume that an elephant population is at the MSY size but in fact it is slightly less. We have overestimated

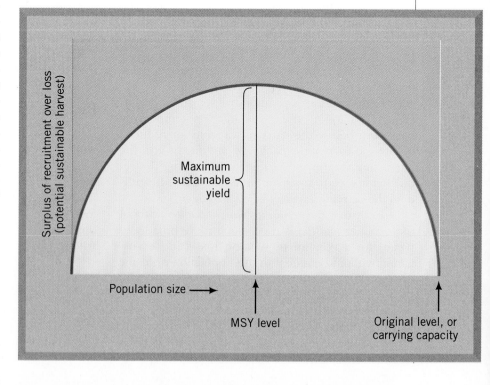

the population. Based on our over-estimate, we assume that we can harvest at the calculated MSY level and we proceed to do that. As illustrated in Figure 12.13, this leads to a disastrous situation; the population goes into a decline. Every year we harvest at the supposed MSY level,

believing the population has returned to the calculated MSY population size. But each year the population harvests exceed births and the population size declines.

Even the slightest mistake in overestimating the MSY will push the population to a smaller size the

next year. If we continue to maintain the same harvest levels, expecting the population to be at MSY, then we will drive the population to lower and lower levels and finally to extinction unless we recognize our error.

2. concern with the entire community of organisms and all the renewable resources, so that policies developed for one species are not wasteful of other resources;

3. maintenance of the ecosystem of which the wildlife are a part, minimizing risk of irreversible change and long-term adverse effects as a result of use; and

4. continual monitoring, analysis, and assessment; that is, the application of science and the pursuit of knowledge about the wildlife of interest and its ecosystem should be maintained and the results made available to the public.

These principles broaden the scope of wildlife management from the narrow focus on a single species to inclusion of the ecological community and ecosystem. They call for a safety net in terms of population size, meaning that no population should be held at exactly the MSY level or reduced to some theoretical minimum abundance. These new principles provide a starting point for an improved approach to wildlife management, to which we can add several other suggestions.

Age Structure

A key to successful management is knowledge about the age structure of a population. Sampling an age structure is one of our most powerful tools in managing populations. Recall from the discussion in Chapter 5 that age structure means the proportion of the population in each age group. Sometimes an age group represents the actual number of individuals; other times it represents the percentage or proportion of the population of each age.

Wildlife Management Techniques

Wildlife management techniques include direct regulation of the number of animals by setting limits on the sex, size, or age of individuals killed or removed from a habitat. They also include indirect methods that manipulate the habitat and ecosystem in a way that will affect population size. For example, changing the abundance of food supply, water, or physical shelter, changing the abundances of

predators, and controlling parasites are all techniques of indirect wildlife management.

One method of direct population management is varying the annual allowable take for a species to keep the population from falling below a certain number. In Maine, for example, wildlife managers count moose from aircraft each year to determine how many licenses to issue to hunters. One example of indirect management is habitat management in Wankie National Park, Zimbabwe, where artesian wells were dug to increase the water supply and the abundance of wildlife. Another example, involving ecosystem management, occurs in a number of parks, where fires are allowed to burn or controlled fires are started, in some cases to provide new forage for grazing animals.

Knowledge and Wildlife Management

To achieve a realistic approach to planning wildlife harvests, we must have the following minimum information: (1) sufficiently accurate estimates of population size, (2) growth rates, (3) sources of mortality other than hunting, (4) carrying capacity, and (5) an understanding of how the environment and the population of interest vary naturally over time. It is important to develop means to forecast future population sizes and to investigate the likely effects on the future of the population of various harvesting rates and other differences in harvesting policies. Accurate scientific observations and reliable mathematical or computer models are essential to managing wildlife at a sustainable and productive level.[31]

Management and Conservation of Migratory Species

Management and conservation of migratory species present problems because these species interact with many different habitats both during breeding and wintering times and during migration. Migratory birds of the Americas spend their breeding season in temperate zones in North America during the warmer months and then migrate to the tropics of Central and South America during the Northern Hemisphere winter. There has been a general decline in most species of these migratory birds over the last few decades. Loss of habitat or habitat fragmentation appears to be a major factor and, at present, loss of habitat through forest fragmentation in the temperate zones appears to be a greater problem than loss of habitat in the tropics.[51,52] Conservation of migratory birds should concentrate on either recovering or maintaining larger tracts of forest in breeding areas to increase survival success. Effective conservation should also identify habitats, areas, and species most in need of protection.[53]

12.7 WHAT IS BEING DONE TO HELP CONSERVE BIOLOGICAL DIVERSITY AND ENDANGERED SPECIES?

We can do a great deal to help conserve biological diversity and endangered species. Steps being taken in various places include the following:

- taking emergency actions for seriously endangered species;
- determining the minimum viable population size for a species (see next section);
- assisting a species in achieving that size, including artificial breeding in zoos and gardens;
- determining the characteristics of a minimum viable habitat and ecosystem for a species (see next section);
- restoring a damaged habitat and ecosystem or creating a new habitat appropriate for the species (see Chapter 9);
- taking an ecosystem approach to conservation;
- understanding and applying the concept of an ecological island (discussed later in this chapter) and acting to overcome the limitations of such islands through the transport of individuals among ecological islands and the development of a system of facilities ranging from wilderness preserves to zoos and gardens;
- monitoring the population and its habitat and ecosystem to maintain information about trends and conditions; and
- developing better means to project future population levels using mathematical models and computer simulation.

Immediate Emergency Actions

For species that are immediately endangered and likely to become extinct in the near future unless immediate action is taken, the appropriate immediate action is usually clear. Just as with an accident victim who is brought to the emergency ward of a hospital, a seriously endangered species requires

that life support be administered and that all negative activities cease. When a species is seriously endangered, it is essential to stop harvesting the species and to promote conservation of its habitat. Emergency actions may be temporary; once the population is reestablished and the habitat restored, some harvesting may be resumed.

An example of emergency action taken to save a species is the African black rhinoceros. Once totaling approximately 65,000 animals in 1970, the total number of black rhinos plummeted to fewer than 2000 in the early 1990s, largely due to illegal poaching. The black rhino survives today in a few isolated populations as a result of efforts to stem poaching, but in western Africa, a precarious population of only 30–50 animals struggles to expand. In Kenya, conservation efforts have resulted in the local black rhino population increasing from 366 animals in 1989 to 430 in 1994.[55] Another species, the white rhinoceros, faced extinction in southern Africa when its numbers dwindled to only 20 at the turn of the century. Placement on the endangered species list and efforts to recover the species have resulted in the population increasing to more than 7000 today.[56]

Minimum Viable Population Sizes

A **minimum viable population** is the smallest number of individuals that have a reasonable chance of persisting. In practice, the minimum viable population must be defined in terms of a specific chance of persisting (or a chance of avoiding extinction) for a specific time period. Based on genetic factors, it has been estimated that for a specific population a minimum of 50 breeding adults should be maintained to avoid the genetic problems of small populations, from inbreeding and genetic drift, and to keep the chance of extinction acceptably low.[57,58]

A population with 50 breeding adults will have many nonbreeding individuals, including newborns, young, and adolescents, as well as old, postbreeding adults. Thus several hundred individuals are, as a rule of thumb, a minimum safe number for a population even for a short period of time. *To maintain a population for a long period of time, 500 or more breeding individuals are preferable in order to avoid problems of low genetic variability.*[59]

Because few nature preserves can be large enough for a minimum viable population of some of the large animals, survival of such species requires active human manipulation. For example, because wolves require about 26 km² (10 mi²) each, a preserve that could maintain a minimum viable wolf population would have to be approximately 13,000 km² (5000 mi²) in area. In contrast, 500 field mice could persist in a preserve that is a fraction of 1 km². In the case of wolves and other large mammals, genetic drift can be avoided by occasionally moving individuals capable of breeding from one preserve to another (see Chapter 7).

Because larger organisms are more subject to extinction, we must be particularly careful in our management of them. Undoubtedly, choices will have to be made. One leading conservationist, Norman Myers, urges that we adopt the triage strategy used in emergency medical practice. The term originated with French medical procedure in World War I. When doctors found that there were more wounded than they could care for, they assigned each wounded soldier to one of three categories: those who would be helped by medical attention and might not survive without it; those who would survive without medical attention; and those who would die regardless. The doctors concentrated on the first group, which would benefit most. Myers suggests that we apply the same strategy to endangered species.[60]

Ecological Islands

Recall from our discussion in Chapter 7 that an **ecological island** is an area that is biologically isolated so that a species occurring within the area cannot mix (or only rarely mixes) with any other population of the same species (Figure 12.15). Mountaintops and isolated ponds are ecological islands. Geographic, or real, islands are also ecological islands. Insights gained from studies of the biogeography of islands have important implications for the conservation of endangered species and for the design of parks and preserves for biological conservation.[61]

Almost every park is a biological island for some species. A small city park occupying a square between streets may be an island for trees and squirrels. At the other extreme, even a large national park is an ecological island (Figure 12.15c). For example, the Masai Mara Game Reserve in the Serengeti Plain, which stretches from Tanzania to Kenya in East Africa, and other great wildlife parks of eastern and

(a)

(b)

(a)

(d)

FIGURE 12.15 Ecological islands: (*a*) Central Park in New York; (*b*) a mountaintop in Arizona where there are big-horn sheep; (*c*) an African wildlife park; (*d*) experimental ecological island in the Amazon rain forest as part of a Smithsonian Institution project.

southern Africa are becoming islands of natural landscape surrounded by human settlements.

Lions and other great cats exist in these parks as isolated populations, no longer able to roam completely freely and to mix over large areas. Other examples are islands of uncut forests left by logging operations and oceanic islands, where intense fishing has isolated parts of fish populations.

Size and Survival of Species on Ecological Islands

What is a sufficiently large ecological island to guarantee the survival of a species? The size varies with the species but can be estimated. Some islands that seem large to us are too small for species we wish to preserve. For example, a preserve was set aside in India in an attempt to reintroduce the Indian lion into an area where it had been eliminated by hunting and changing patterns of land use. In 1957, a male and two females were introduced into a 95-km^2 (36-mi^2) preserve in the Chakia forest, known as the Chandraprabha Sanctuary.[62] The introduction was carried out carefully and the population counted annually. There were 4 lions in 1958, 5 in 1960, 7 in 1962, and 11 in 1965, after which they disappeared and were never seen again.

Why did they disappear? Although 95 km^2 seems large to us, male Indian lions have territories of 130 km^2 (50 mi^2). Within such a territory, females and young also live. A population that could persist for a long time would need a number of such territories, so an adequate preserve would require 640–1300 km^2 (247–500 mi^2). Various other reasons were suggested for the disappearance of the lions, including poisoning and shooting by villagers, but regardless of the immediate cause, a much larger area than was set aside was required for long-term persistence of the lions.

Knowledge gained in the study of ecological islands has been put to use in a Brazilian project supported by the World Wildlife Fund and the Smithsonian Institution, known as the Minimum Critical Size Project. Logging operations in the Brazilian Amazon were carried out so as to leave uncut islands of many sizes, from a few hectares to hundreds of square kilometers, to learn the minimum size of a forest island adequate to protect the forest's species.[63]

Introducing Exotic Species to Minimize Effects on Native Species

Earlier in this chapter, we discussed the negative effects on biological diversity of artificial introductions of exotic species. If we are to shepherd our planet

successfully in the future, we must choose our introductions of new species much more carefully than we have in the past, and we must learn to reduce, if not eliminate, our unintentional transportation of organisms. How can introductions be made successfully, avoiding the negative effects? Several steps are needed. First, only those species that are truly useful or important should be introduced. Second, in a temporary, isolated, and controlled habitat, a population should be isolated and kept isolated for several generations to eliminate undesirable parasites. Third, the species' niche, population dynamics, habitat, and ecosystem requirements should be determined; some tests of competition between an exotic species and native species could be conducted in an intermediate, isolated habitat.

An example of a careful and (so far) successful introduction is that of dung beetles into Australia. Domestic cattle and sheep were brought to Australia by the early British settlers, and large areas of the country are devoted to grazing. There is no native equivalent of the dung beetle, which buries cattle droppings in the soil, speeding up their decomposition. A serious problem developed in Australia because cattle dung was not decaying rapidly enough and was accumulating over the pasturelands. Here was an unoccupied niche.

Australian government scientists obtained dung beetles from Africa and brought them to an intermediate location where they were raised through several generations to eliminate parasites that could spread to other animals. Then and only then were the beetles introduced into Australia.

Long-Term Planning

Earlier in this chapter we stated that short-term emergency measures are often clear. The more difficult set of decisions comes in formulating long-term plans for the maintenance of once-endangered species. For example, now that California condors have reproduced in zoos, an effort is underway to reestablish them in the wild. This has met with difficulties, and can be expected to take a long time. For the long term, there are more complex needs, and it is especially important to clarify specific policies and to obtain necessary scientific information.

Because the term *biological conservation* is often used vaguely, specific goals must be clarified. If our focus is simply on the maintenance of a genotype for utilitarian purposes, then actions can be restricted and artificial; in some cases the maintenance of species in zoos or botanic gardens will be sufficient. If the goal is simply the preservation of genetic material, it can even be maintained in a freezer. However, in most cases, the real concern of those interested in conservation is with the larger goal of maintaining wild creatures in their own habitats. This larger goal requires extensive scientific understanding. Long-term planning requires the following.

Establish Linkages among Facilities

Conservation of endangered species can involve many types of facilities, from zoos and botanic gardens to wilderness areas. In practice, successful programs often involve interconnections among different kinds of facilities, with zoos and gardens used as temporary, emergency facilities, much as we use hospitals for seriously ill people. We need to establish linkages, connecting zoos and gardens to parks, preserves, and wilderness areas.

When wild habitats are destroyed or seriously decreased and threatened, zoos provide the only remaining places where species can persist. In many cases, zoos are seen as holding areas where a species may be maintained for decades until suitable habitats can be restored and the species returned to the wild. For example, the only remaining truly wild horses live in zoos. The conservation of species within zoos has led to international programs to maintain proper genetic diversity in zoo populations by exchanging individuals of endangered species. Some zoos, such as the Jersey Island Trust on the Isle of Jersey, United Kingdom, are primarily devoted to the conservation of endangered species.

Obtain Necessary Information

In developing management strategies, we must have adequate information. We know surprisingly little about the basic population characteristics and habitat requirements of most endangered species. A prudent approach to conservation will involve continual scientific research programs. These can elucidate causes of decline and identify the requirements for persistence. Monitoring populations and ecosystems so that their status is known is an essential part of scientific programs.

Establish More Conservation Areas

In most of the world, human impacts are widespread and increasing, leading to the destruction of habitats of endangered species. Only by establishing more conservation areas can species be conserved in natural settings.

Endangered Species and Society

During the last 20 years, many countries have enacted laws protecting endangered species and international legal agreements among nations have increased. In the United States, laws affecting biological diversity were first passed early in the twentieth century. In the late 1960s and early 1970s several landmark acts, including the 1969 Endan-

gered Species Conservation Act, the 1972 Marine Mammal Protection Act, and the 1973 Endangered Species Act, were passed (see Table 12.3).

The 1972 Marine Mammal Protection Act and the 1973 Endangered Species Act represented a major change in the U.S. government's role in the protection of endangered species, in the legal attitude toward endangered species, and in the legal position of endangered species. The Endangered Species Act declared that endangered species of wildlife and plants "are of aesthetic, ecological, educational, historical, recreational, and scientific value to the nation and its people." The act provided a means to conserve ecosystems on which endangered species depend and established the need for an ecosystem approach to the management of endangered species. The act stated that all federal departments and agencies should seek to conserve endangered species and use their authorities to further the purpose of the act.

The Endangered Species Act of 1973 made unlawful the importation, exportation, or selling in interstate or international commerce any endangered species or any product of an endangered species. The law also made it illegal to harass, harm, or capture any such species within the United States or on the high seas. Since its inception, the Endangered Species Act has been reauthorized and strengthened three times, in 1978, in 1982, and again in 1988. Recently, the U.S. Congress introduced legislation, two bills in the Senate (S. 768 and S. 1364) and one in the House of Representatives (H.R. 2275), all of which seek to rewrite the Endangered Species Act. The proposed legislation would substantially weaken the scope and intent of the original act and would include such action as a moratorium on the naming of threatened and endangered species and elimination of habitat protection.[64,65] American scientific organizations have endorsed the current Endangered Species Act as sound policy. Along with a continued adherence to the Endangered Species Act, a sound scientific policy for protecting endangered species would emphasize the protection of certain hot spots of biodiversity, protection of umbrella species (species that are wide ranging and of low density), and protection of species and ecosystems before they become endangered; it would also supplement the Endangered Species Act regulatory requirements with economic incentives.[66,67]

Success of the Endangered Species Act

Although the intent of the Endangered Species Act was to identify and then recover threatened and endangered species, far more species have been added to the list of endangered species than removed from it. Since 1973, nearly 1000 species have been added

TABLE 12.3 Federal Laws Governing Biological Diversity

Common Name	Resource Affected	U.S. Code
On-site Diversity Mandates		
Lacey Act of 1900	Wild animals	16 U.S.C. 667, 701
Migratory Bird Treaty Act of 1918	Wild birds	16 U.S.C. 703 et seq.
Migratory Bird Conservation Act of 1929	Wild birds	16 U.S.C. 715 et seq.
Wildlife Restoration Act of 1937 (Pittman–Robertson Act)	Wild animals	16 U.S.C. 669 et seq.
Bald Eagle Protection Act of 1940	Wild birds	16 U.S.C. 668 et seq.
Whaling Convention Act of 1949	Wild animals	16 U.S.C. 916 et seq.
Fish Restoration and Management Act of 1950 (Dingell–Johnson Act)	Fisheries	16 U.S.C. 777 et seq.
Anadromous Fish Conservation Act of 1965 (Public Law 89-304)	Fisheries	16 U.S.C. 757 a–f
Fur Seal Act of 1966 (Public Law 89-702)	Wild animals	16 U.S.C. 1151 et seq.
Marine Mammal Protection Act of 1972	Wild animals	16 U.S.C. 1361 et seq.
Endangered Species Act of 1973 (Public Law 93-205)	Wild plants and animals	7 U.S.C. 136 16 U.S.C. 460, 668, 715, 1362, 1371, 1372, 1402, 1531 et seq.
Magnuson Fishery Conservation and Management Act of 1977 (Public Law 94-532)	Fisheries	16 U.S.C. 971, 1362, 1801 et seq.
Whale Conservation and Protection Study Act of 1976 (Public Law 94-532)	Wild animals	16 U.S.C. 915 et seq.
Fish and Wildlife Conservation Act of 1980 (Public Law 96-366)	Wild animals	16 U.S.C. 2901 et seq.
Salmon and Steelhead Conservation and Enhancement Act of 1980 (Public law 96-561)	Fisheries	16 U.S.C. 1823 et seq.
Fish and Wildlife Coordination Act of 1934	Terrestrial/aquatic habitats	16 U.S.C. 694
Fish and Game Sanctuary Act of 1934	Sanctuaries	16 U.S.C. 694
Historic Sites, Buildings, and Antiquities Act of 1935	Natural landmarks	16 U.S.C. 461–467
Fish and Wildlife Act of 1956	Wildlife sanctuaries	15 U.S.C. 713 et seq. 16 U.S.C. 742 et seq.
Wilderness Act of 1964 (Public Law 88-577)	Wilderness areas	16 U.S.C. 1131 et seq.
National Wildlife Refuge System Administration Act of 1966 (Public Law 91-135)	Refuges	16 U.S.C. 568dd et seq.
Wild and Scenic Rivers Act of 1968 (Public Law 90-542)	River segments	16 U.S.C. 1271–1287
Marine Protection, Research and Sanctuaries Act of 1972 (Public Law 92-532)	Coastal areas	16 U.S.C. 1431–1434 33 U.S.C. 1401, 1402, 1411–1421, 1441–1444
Federal Land Policy and Management Act of 1976 (Public Law 94-579)	Public domain lands	7 U.S.C. 1010–1012 16 U.S.C. 5, 79, 420, 460, 478, 522, 523, 551, 1339 30 U.S.C. 50, 51, 191 40 U.S.C. 319 43 U.S.C. 315, 661, 664, 665, 687, 869, 931, 934–939, 942–944, 946–959, 961–970, 1701, 1702, 1711–1722, 1731–1748, 1753, 1761–1771, 1781, 1782
National Forest Management Act of 1976 (Public Law 94-588)	National forest lands	16 U.S.C. 472, 500, 513, 515, 516, 518, 521, 576, 581, 1600, 1601–1614
Public Rangelands Improvement Act of 1978 (Public Law 95-514)	Public domain lands	16 U.S.C. 1332, 1333 43 U.S.C. 1739, 1751–1753, 1901–1908
Agricultural Marketing Act of 1946 (Research and Marketing Act)	Agricultural plants and animals	5 U.S.C. 5315 7 U.S.C. 1006, 1010, 1011, 1924–1927, 1929, 1939–1943, 1941–1943, 1947, 1981, 1983, 1985, 1991, 1992, 2201, 2204, 2212, 2651–2654, 2661–2668 16 U.S.C. 590, 1001–1005 42 U.S.C. 3122
Endangered Species Act of 1973 (Public Law 93–205)	Wild plants and animals	7 U.S.C. 136 16 U.S.C. 460, 668, 715, 1362, 1371, 1372, 1402, 1531 et seq.
Forest and Rangeland Renewable Resources Research Act of 1978 (Public Law 95-307)	Tree germ plasm	16 U.S.C. 1641–1647

Source: Office of Technology Assessment, 1986.

Note: Laws enacted prior to 1957 are cited by chapter and not public law number.

to the list, while only six species have been removed from it as a result of efforts to recover them; in some cases, other factors may have been more influential in the delisting than the recovery efforts alone. Opponents of the act point to its low success ratio. Supporters of the concept of an endangered species act counter that the current act is grossly underfunded and the agency responsible for carrying out its provisions, the Fish and Wildlife Service, has sometimes been at odds with the act's provisions, making for a difficult situation.[68]

More radical approaches to the conservation of endangered species include political activism. Members of the environmental organization Greenpeace have put themselves in small boats between whaling ships and whales, in a direct attempt to prevent the killing of the whales. This political radicalism for the conservation of endangered species is a novel development in the modern world and shows the great importance some people attach to conservation.

ENVIRONMENTAL ISSUE

Should Wolves Be Re-Established in the Adirondack Park?

With an area slightly over 24,000 km², the Adirondack Park in northern New York is the largest park in the lower 48 states. Unlike most parks, however, it is a mixture of private (60%) and public (40%) land and home to 130,000 people. When colonists from Europe first came to the area, it was, like the rest of North America, inhabited by gray wolves. By 1960 wolves had been exterminated in all of the lower 48 states, except for northern Minnesota. The last official sighting of a wolf in the Adirondacks was in 1890s.

Although the gray wolf was not endangered on a global level—there were more than 60,000 in Canada and Alaska—it was one of the first animals listed as endangered under the 1972 Endangered Species Act. As required, the United States Fish and Wildlife Service developed a plan for recovery that included protection of the existing population and reintroduction of wolves to wilderness areas. The recovery plan was to be considered a success if survival of the Minnesota wolf population was assured and at least one other population of more than 200 wolves had been established at least 320 km from the Minnesota population.

Under the plan the wolf population in Minnesota increased and some individuals, as well as others from southern Canada, dispersed into northern Michigan and Wisconsin, both of which had populations of approximately 100 wolves in 1998. Also, 31 wolves from Canada were introduced into Yellowstone National Park in 1995, where the population increased to over 100. By the end of 1998 it seemed fairly certain that the criteria for removing the wolf from the endangered species list would soon be met.

In 1992, when the fate of the recovery was still uncertain, the Fish and Wildlife Service proposed to investigate the possibility of re-introducing wolves to northern Maine and the Adirondack Park. A survey of New York State residents in 1996 funded by Defenders of Wildlife found that 76% of those living in the park supported re-introduction. However, many residents and organizations within the park vigorously opposed re-introduction and questioned the validity of the survey. Concerns expressed focused primarily on the potential dangers to humans, livestock, pets and the possible impact on the deer population. In response to the public outcry, Defenders established a citizen's advisory committee that initiated two studies by outside experts, one on the social and economic aspects and another on whether there was sufficient prey and suitable habitat for wolves.

The primary prey of wolves are moose, deer, and beaver. In recent years, moose have been returning to the Adirondacks and now number about 40, but this is far less than moose population in areas where wolves are successfully re-establishing. Beaver are abundant in the Adirondacks, with an estimated population of over 50,000. Because wolves feed on beaver primarily in the spring and the moose population is small, the main food source for Adirondack wolves would be deer.

Deer thrive in areas of early successional forest and edge habitats, both of which have declined in the Adirondacks as logging has decreased on private forest land and been eliminated all together on public lands. Furthermore, the Adirondacks is at the northern limit of the range for white-tailed deer, where harsh winters can result in significant mortality. Deer density in the Adirondacks, which is estimated at 3.25/km², is less than that found in wolf habitat in Minnesota, which also has 8,500 moose. If deer were the only prey available, wolves would kill between 2.5% and 6.5% of the deer population, while hunters take approximately 13% each year. Determining whether there is a sufficient prey base to support a population of wolves is complicated by the fact that coyotes have moved into the Adirondacks and occupy the niche once filled by wolves. Whether wolves would add to the deer kill or replace coyotes, with no net impact on the deer population, is difficult to predict.

An area of 14,000 km² in various parts of the Adirondack Park meets criteria established for suitable wolf habitat but this is about half of the area required to maintain a wolf population for the long term. Based on the average deer density and weight, as well as the food requirements of wolves, biologists estimate that this habitat could support about 155 wolves. Because human communities are scattered throughout much of the park, many residents are concerned that wolves would not remain on public land and threaten local residents as well as back country hikers and hunters. Also, private lands around the edges of the park, with their greater density of deer, dairy cows, and people, could attract wolves.

Critical Thinking Questions

1. Who should make decisions about wildlife management, such as returning wolves to the Adirondacks—scientists, government officials, or the public?

2. Some people advocate leaving the decision to the wolves, that is, waiting for them to disperse from southern Canada and Maine into the Adirondacks.

Study a map of the northeastern United States and southeastern Canada. What is the liklihood of natural recolonization of the Adirondacks by wolves?

3. Do you think that wolves should be re-introduced to the Adirondack Park? If you lived in the park, would that affect your opinion? How would removal of the wolf from the endangered species list affect your opinion?

4. Recently some biologists have concluded that wolves in Yellowstone and the Great Lakes region belong to a different subspecies (Rocky Mountain timber wolf) than those that formerly lived in the northeastern United States (eastern timber wolf). This means that the eastern timber wolf is still extinct inthe lower 48 states. Would this affect your opinion on re-introduction of wolves to the Adirondacks?

References

Hodgson, A. July 1997. "Wolf Restoration in the Adirondacks?," Wildlife Conservation Society, Working Paper No. 8.

Hosack, D. A. 1996. "Biological Potential for Eastern Timber Wolf Re-Establishment in the Adirondack Park," In Wolves of America Conference Proceedings, November 14–16, 1996, Albany, NY, Defenders of Wildlife, Washington, DC.

Stevens, W. K., March 4, 1997. "Wolves May Reintroduce Themselves to East," The New York Times.

SUMMARY

- Two main meanings of conservation of biological diversity are conservation of a single endangered species of special interest and conservation of all the species in an area.

- In deciding policies for conservation of endangered species, we must clarify specific goals and understand the reasons that we value biological conservation in a specific situation; these can be utilitarian, ecological, aesthetic, or moral justifications.

- Extinction is a fact of nature. Natural causes of extinction are usually grouped into four natural kinds of risks—genetic, population, environmental, and catastrophic—as well as human-induced risk.

- The bottom line in conservation of an endangered species is to achieve a minimum viable population in a minimum viable habitat.

- To save an endangered species, conservation must be at the level of habitat and ecosystem.

- People have greatly increased the rate of species extinctions; at present the primary threats to biological diversity are from human activities.

- Short-term actions required to save an endangered species are often clear, but for long-term conservation extensive scientific understanding is required. Unfortunately, the basic ecology of most endangered species is poorly known. New scientific research is an essential part of any conservation program.

- Traditional wildlife management was based on the logistic growth curve, which oversimplifies the dynamics of wildlife population and leads to overexploitation. Traditional wildlife management focused on carrying capacity and maximum sustainable yield as defined by the logistic. Today, wildlife conservation and management are undergoing fundamental changes.

- Keys to successful management include a broadened perspective to include the ecological community and ecosystem, a safety net in terms of population size, and knowledge of age structure of the population.

REEXAMINING THEMES AND ISSUES

Human Population: Human beings are a primary cause of the extinction of species. Nonindustrial societies have caused extinction by such methods as the introduction of exotic species into new habitats, including islands. With the age of exploration in the Renaissance and with the industrial revolution, the rate of extinctions increased greatly. People altered habitats more rapidly and over greater areas; hunting efficiency increased, as did the introduction of exotic species into new habitats. This chapter reviews the ways that our species has caused extinction and some of the solutions to these causes. Once again, we find that the human population problem underlies this environmental issue.

Sustainability: At the heart of issues concerning wild living resources is the question of sustainability of species and the ecosystems of which they are a part. One of the key questions is whether we can sustain these resources at a constant abundance. Or does sustainability require that populations vary over time without reaching zero? In general, it has been assumed that fisheries, wildlife hunted for recreation, such as deer, and forests could be maintained at some constant, highly productive level. Constant production is desirable economically because it would provide a reliable, easily forecast income each year. But in spite of direct management attempts, few wild living resources have remained at constant levels. New ideas about the intrinsic variability of ecosystems and populations lead us to question the assumption that such resources can or should be maintained at constant levels.

Global Perspective: Although the final extinction of a species takes place in one locale, the problem of biological diversity and the extinction of species is global because there has been a worldwide increase in the rate of extinction and because of the growth of the human population and its effects on wild living resources. Similarly, although fisheries and forests become overharvested in specific locations, the decline in fish populations and abundance of trees are now global problems because these have occurred worldwide.

Urban World: Although we have tended to think of wild living resources as existing outside of cities, there is a growing recognition that urban environments will be more and more important in the future in the conservation of biological diversity. This is in part because cities now occupy many sensitive habitats around the world, such as coastal and inland wetlands. It is also because appropriately designed parks and backyard plantings can provide habitats for some endangered species. As the world becomes increasing urbanized, this function of cities will increase.

Values and Knowledge: Among the most controversial environmental issues in terms of values is the conservation of biological diversity and the protection of endangered species. There are four kinds of values that we place on endangered species and other wild living resources: utilitarian, ecological, aesthetic, and moral. Ultimately our decisions about where to spend our efforts in sustaining wild living resources depend on our view of these values.

KEY TERMS

ecological island *247*
ecotourism *231*
environmental risk *236*
genetic risk *237*

global extinction *235*
local extinction *235*
logistic carrying capacity *242*
maximum sustainable yield (MSY) *242*

minimum viable population *246*
optimum sustainable yield (OSY) *242*
population risk *236*

STUDY QUESTIONS

1. Why are we so unsuccessful in making rats an endangered species?

2. What have been the major causes of extinction

(a) in recent times and (b) before people existed on Earth?

3. What would be a feasible plan to save the California condor for one of the two following goals?

   a. to maintain a wild species in the wilderness;

   b. to save the genetic characteristics of the condor.

4. What are some arguments for and against the statement "Eating meat helps preserve endangered species"?

5. Which of the four justifications for conservation of biological diversity apply to the following? (You can decide that none apply.)

   a. the black rhinoceros of Africa;

   b. the furbish lousewort, a rare, small flowering plant of New England, seen by few people;

   c. an unnamed and newly discovered beetle from the Amazon rain forest;

   d. smallpox;

   e. wild strains of potatoes in Peru;

   f. the North American bald eagle.

6. Locate an ecological island close to where you live and visit it. Which species are most vulnerable to local extinction?

7. Visit a local zoo or botanical garden. What activities are conducted there to promote biological conservation? Are these activities likely to be of much benefit?

8. How can a park or series of parks in your town promote biological conservation?

9. Using information available in libraries, what is the minimum area required for a minimum viable population of the following?

   a. domestic cats;

   b. the cheetah;

   c. the American bison;

   d. swallowtail butterflies;

   e. the peppered moth (see Chapter 7).

10. Monitoring populations is an important part of conservation. Using information available in libraries, what are the present best estimates of the populations of the species listed in question 9? How reliable are these estimates?

11. Both a ranch and a preserve will be established for the North American bison. The goal of the ranch owner is to show that bison can provide a better source of meat than introduced cattle and at the same time have a less detrimental effect on the land. The goal of the preserve is to maximize the abundance of the bison. How will the plans for the ranch and preserve differ and how will they be similar?

12. How can the new principles of wildlife management be applied to a plan for managing domestic cats that have no home and live in a large urban park?

FURTHER READING

Cadieux, C. L. 1991. *Wildlife Extinction*. Washington, DC: Stone Wall Press. An engaging overview of the issues surrounding endangered wildlife. Case studies are presented as well as appendixes, including lists of endangered wildlife and key players in their protection.

Caughley, G., and Sinclair, A. R. E. 1994. *Wildlife Ecology and Management*. London: Blackwell Scientific. A valuable textbook based on new ideas of wildlife management.

Ehrlich, A. 1981. *Extinction: The Causes and Consequences of the Disappearance of Species*. New York: Random House.

Ehrlich, P. R., and Wilson, E. O., eds. 1988. *Biodiversity*. Washington, DC: National Academy Press. An excellent compilation of papers, this book provides an overview of biological diversity and the threats that can lead to extinction.

Talbot, L. 1996. "Living Resource Conservation: An International Overview," *Ecological Applications*. How the conservation of wild living resources is actually being done today, based on interviews and conferences involving more than 400 scientists and policy makers.

McCullough, D. R. 1996. *Metapopulations and Wildlife Conservation*. Washington DC.: Island Press. Introduces a new geographically based approach to wildlife conservation, including the idea that isolated patches, such as parks and preserves, can help stabilize wildlife populations.

Soule, M. E., ed. 1980. *Conservation Biology: The Science of Scarcity and Diversity*. Sunderland, MA: Sinauer Associates. A major contribution to the developing discipline of conservation biology, this work defines the discipline and sets an agenda for continuing work.

INTERNET RESOURCES

Envirolink's Endangered Species Page: *http://envirolink.org/issues/endangered.html*—View distributions of endangered species, check out the U.S. position on international endangered species law and much more.

Environmental Educators (EE) Endangered Species: *http://www.nceet.snre.umich.edu/EndSpp/Endangered.html*—Designed for educators, it is a valuable source of information on endangered species, endangered species legislation, and related information.

Gap Analysis: *http://.www.nr.usu.edu/gap/*—The Utah State University, College of Natural Resources, Gap Analysis is a program of study in the management of biological diversity. Contains data, information, and links to other sites with information on wildlife management.

U.S. Fish and Wildlife Service: *http://www.fws.gov/*—Contains information on all USFWS activities including the Endangered Species Program.

World Conservation Union Red List of Threatened Animals: *http://www.wcmc.org.uk/data/database/rl_anml_combo.html* —Check out the most current list of endangered species from around the world. Check by country, species, family, order, or class.

World Wildlife Fund: *http://www.wwf.org/*—An environmental watchdog organization that monitors the status of endangered species, and trade in endangered species.

EE-Link: *http://eelink.nnet*—Environmental Education (EE) on the Internet. EE-Link develops and organizes Internet resources for environmental education. The EE-Link project has on-line publications, resources, and links to other environmental information.

The Wildlife Conservation Society (WCS): *http://wcs.org*—WCS is a nonprofit organization working to save wildlife and wild lands throughout the world. Among other efforts, WCS provides environmental education programs. The Web site contains access to news and scientific publications.

The coastal forests of Clayoquot Sound, British Columbia, Canada, are the subject of intense environmental debates.

CHAPTER 13

LANDSCAPES AND SEASCAPES

CASE STUDY

Conflict in Clayoquot Sound

In October 1993, over 800 people protesting logging in Clayoquot Sound, on the western side of Vancouver Island, British Columbia (Fig. 13.1), were arrested in the largest nonviolent mass protest in Canadian history. The protest climaxed a decade of controversy over how to manage the 250,000 ha (675,000 acres) of coastal temperate rainforest in the region. Temperate rainforests are now, and always have been, rarer than tropical rainforests. In ancient times, the temperate rainforests of the world covered an area of 30 million ha (75 million acres; approximately the size of Arizona), only 3% of that covered by tropical rainforests. In recent decades, a great deal of attention has been given to destruction of the tropical rainforests, although temperate rainforests are being lost just as rapidly. The largest remaining intact temperate rainforest in the Northern Hemisphere is on Vancouver Island, 17% of which is in Clayoquot Sound. The Sierra Club of Western Canada estimates that in 1954 two-thirds of the original rainforest of Vancouver Island—some 1.6 million ha (4 million acres)—remained, but by 1990, only 0.8 million ha (2 million acres) were left. At that rate,

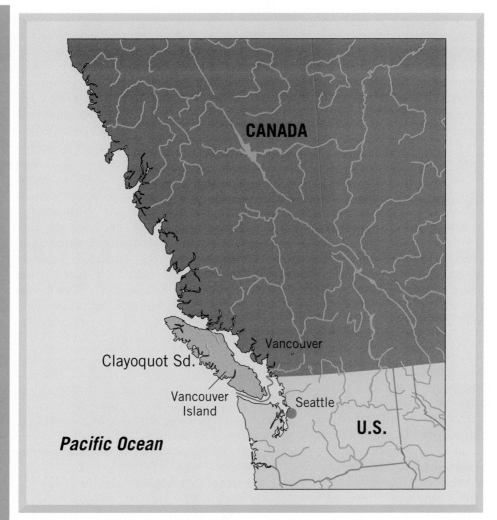

FIGURE 13.1 Map of western Canada showing Vancouver Island and indicating location of Clayoquot Sound.

according to the Sierra Club, the ancient temperate rainforests on southern Vancouver Island will be gone or severely fragmented by 2022.

The trees of this unique ecosystem—primarily western hemlock, western red cedar, amabilis fir, and sitka spruce—include some of the tallest and oldest trees in the world (Fig. 13.2), the most ancient of which are 1,500 years old, 6 m (20 ft.) wide, and 90 m (300 ft.) tall. A rich diversity of wildlife, including salmon, black bears, black-tailed deer, Roosevelt elk, cougars, wolves, and many species of birds abound in the forest of Clayoquot Sound. For 9,000 years it has also been the home of the nations of the Nuu-chah-nulth people. A total of 4,000 people, including residents of the nearby logging town of Ucluelet and the tourist town of Tofino, now live in the area. All Canadian forests are publicly owned and the provincial governments sell licenses to companies that give them exclusive logging rights in an area. In April 1993 the government of British Columbia released a

FIGURE 13.2 Picture of temperate rainforest on Vancouver Island.

plan that would allow clearcutting on 70% of Clay-oquot Sound. In addition to provoking protests by environmentalists, the plan angered the chiefs of the Nuu-chah-nulth First Nations, who expressed concern that they had not been consulted in its development. The following year the provincial government agreed to give the First Nations the right to review and veto proposals for development in Clayoquot and it established a panel of scientists to recommend how logging of Clayoquot Sound should be carried out. When the panel released its report in 1995, the government accepted all 128 recommendations and most environmental groups considered the new guidelines to be a step forward.

Early in 1996, a series of massive landslides occurred following a severe rainstorm. Environmental groups placed some of the blame for the landslides on clearcutting on steep slopes that left the ground relatively unprotected. They charged also that permits granted for logging were not consistent with the guidelines set forth by the Science Panel. Logging companies complain that the new procedures result in long delays in granting permits and threaten the economic stability of the logging community. They estimate that without new sources of wood, only three years of logging are left in Clayoquot Sound. In 1996, only 20 people were working in the woods of Clayoquot Sound, compared to more than 200 in 1991. Nevertheless, the largest timber company in the region pledged to carry out the panel's recommendations and to seek alternatives to clearcutting that would be sensitive to the ecology and to the traditions of the Nuu-chah-nulth. But when the major logging companies in the region were brought in to discuss alternatives to clearcutting, the Friends of Clayoquot Sound adopted a policy of no commercial cutting of ancient forests by any method. In the meantime,

residents of Tofino are concerned about the effects of clearcutting and controversy on their $14 million a year tourist-based economy.

Clayoquot Sound continues to be a battleground for competing interests in the region and, as yet, "the noble idea that sustainable forestry can co-exist with preservation in the Clayoquot remains an unfulfilled dream."[1] On the positive side, the case of Clayoquot Sound illustrates that people take the conservation of forests and other landscapes seriously and are willing to exert considerable pressure to save remaining old-growth stands and other wilderness areas. It also illustrates the complexity of such issues and the many different kinds of stakeholders and interest groups—from native peoples to local environmental organizations, from people who earn their living in the area, to people around the world who want to visit these areas and know that they still exist—who want to have a say in decisions that affect the future of these landscapes.

REFERENCES

A Crucial Test Founders in Clayoquot Sound. 1996 (Aug./Sept.). *Truck Logger,* 18–24.

Bulic, I. 1994 (Jan./Feb.). Clayoquot Summer. *Canadian Dimensions* 28:34–37.

Davey, R. 1995 (June). Rainforest Crunch. *E* 6:19–22.

Kerasote, T. 1994 (July). Canada: The Brazil of the North? *Sports Afield* 212:15–20.

Maxwell, J. 1994 (Jan./Feb.). The Last Best Rainforest. *Audubon* 98–103.

Sierra Club of Western Canada. 1993. Ancient Rainforests at Risk. Victoria, BC: Sierra Club of Western Canada.

Western Canada Wilderness Committee. 1996 (Summer). Beautiful Clayoquot Sound. Vancouver, BC: Western Canada Wilderness Committee, 8 pp.

The discussion of the conflict in Clayoquot Sound, British Columbia in the case study provides an illustration of the many societal and ecological interactions that make management of land and ocean resources difficult. This chapter explores problems in conserving and managing land and oceans and their biological resources, including forests, parks, wilderness, fisheries, and marine mammals. In the past, each was managed in isolation, as if each problem was independent of all others.

LEARNING OBJECTIVES

Forests and fisheries are among the biological resources that have commercial value. Our goal is to conserve these resources for sustained harvest as

well as for their recreational, aesthetic, and moral value. After reading this chapter, you should understand:

- The ecological services provided by forests and oceans.
- The ecosystem context for conservation and management of wild living resources.
- Although the land and the oceans differ greatly as habitats, conservation and management of their biological resources involve the same concepts, including sustainability, optimum populations and optimum yields, and an ecosystem and global perspective.
- The basic principles of forest management, including its historical context: Deforestation occurred in prehistory, has occurred throughout the history of civilization, and continues today.

- Conserving existing forests will be a difficult task in developing nations, which depend on firewood for fuel and have rapidly expanding populations.
- The roles of parks in the conservation of wilderness, representative natural areas, and wildlife habitats for outdoor recreation and scientific research.
- The basic principles of ocean productivity, including the geographic distribution of ocean life.
- What is being done to conserve marine mammals, many of which are endangered or seriously depleted because of human actions.

13.1 FORESTRY

Forests are areas where trees are the dominant vegetation. They include **closed canopy forests,** where leaves and twigs of adjacent trees touch, and **open woodlands,** where only some leaves and twigs of adjacent trees overlap (Figure 13.3). Estimates of forestland usually include the sum of closed forests and open woodlands as well as areas of scrubland and brushland that are neither forest nor open woodland and some deforested areas where forest regeneration is not taking place. Beyond the edges of forests the land gradually changes to **savannas,** grasslands with scattered trees.

Forests have always been important to people. In fact, there is a link between forests and civiliza-

FIGURE 13.3 (*a*) A closed canopy tropical rainforest. (*b*) Ponderosa pine open woodlands of the western United States. (*c*) Savannas of Africa.

(a)

(b)

(c)

tion. Forests are now and always have been a major economic resource. Wood provided one of the major building materials and the major source of fuel for early towns and cities and helped pave the way for the rise of civilization. Forests provided materials for the first boats and the first wagons. Even today, nearly half the people in the world depend on wood for cooking. In developing nations, wood remains the primary heating fuel. People also confer much symbolic meaning on forests or individual trees, perceiving them as places of beauty and grandeur or as fearsome, dark wildernesses.

Forests affect people indirectly as well. They retard erosion and moderate the availability of water (Figure 13.4), improving the water supply from major watersheds to cities. Forests are habitats for endangered species and other wildlife; and they are important for recreation, hiking, hunting, and bird and wildlife viewing. At regional and global levels, forests may be significant factors in the climate. They are also important in other global biospheric processes, such as the storage of carbon.

Forests are subject to change. As the famous nature writer Thoreau observed more than a century ago, change is natural in forests, as it is in other ecosystems (see the discussion of succession in Chapter 9). However, the pace of that change is slow compared with our life spans; forests seem to us relatively permanent aspects of the landscape.

Forest Resources

There are approximately 3.4 billion ha (8.4 billion acres) of forested area in the world today, covering approximately 27% of the earth.[2] Countries differ greatly in forest resources, depending on the potential of the land and climate for tree growth and on the history of land use and deforestation. About 61% of Indonesia is forest, 58% of Brazil, 23% of India, and less than 14% of China.[3] Some of the largest forest reserves are in the far north, in the boreal forests of the former United Soviet Socialist Republic (USSR), the United States, Canada, and the Scandinavian nations (see Table 13.1).

The United States has approximately 212 million ha (524 million acres) of *commercial-grade forest*, which is defined as forestland believed or known to be an economically profitable source of

Figure 13.4 Multiple-use concept applied to forests. A diagram of a forested watershed, showing the effects of trees in evaporating water, retarding erosion, and providing a wildlife habitat.

Trees intercept water, reducing erosive impact

Water evaporates

Roots anchor the soil

Riparian zone

TABLE 13.1 Forested Areas of the World by Region (millions of hectares)

	Forest land	Closed Forest	Open Woodland	Total Land Area	Closed Forest (% of land area)
North America	630	470	(176)	1,841	25
Central America	65	60	(2)	272	22
South America	730	530	(150)	1,760	30
Africa	800	190	(570)	2,970	6
Europe	170	140	29	474	30
Former USSR	915	785	115	2,144	35
Asia	530	400	(60)	2,700	15
Pacific area	190	80	105	842	10
World	4030	2655	(1200)	13,003	20

Source: Council on Environmental Quality and the Department of State, 1980.

Notes: Data on North American forests represent a mid-1970s estimate. Data on former USSR forests are a 1973 survey by the Soviet government. Other data are from R. Perrson, 1974, *World Forest Resources,* Royal College of Forestry of Sweden, Stockholm. They represent an early 1970s estimate. Forestland is not always the sum of closed forest plus open woodland; it also includes scrubland and brushland areas that are neither forest nor open woodland and includes deforested areas where forest regeneration is not taking place. In computation of total land area, Antarctica, Greenland, and Svalbard are not included: 19% of Arctic regions are included.

timber, capable of producing at least 1.4 m³/ha (20 ft³/acre) per year of wood. Commercial timberland occurs in many parts of the United States. Nearly 75% is in the eastern half of the country (about equally divided north and south); the rest is in the West (Oregon, Washington, California, Montana, Idaho, Colorado, and other Rocky Mountain states) and in Alaska. In the United States, 70% of forestland is privately owned, 15% is on U.S. Forest Service lands, and 15% is on other federal lands.[4] Publicly owned forests are primarily in the Rocky Mountain and Pacific Coast states on sites of poor quality and high elevation. In addition to these forests, about 300 million ha (741 million acres) in the United States are 10% stocked with trees and another 312 million ha (770 million acres) are rangeland, including natural grasslands, shrublands, most deserts, tundra, coastal marshes, and meadows (Figure 13.5).[5]

Commercial Forestry

Forestry is big business. The worldwide harvest of forest products, including that used in firewood, construction, paper, and industrial processes is about 3.4 billion m³ (120 billion ft³).Developed countries use timber nearly three times the per person rate of developing countries; the United States uses timber at a rate nearly twice the average of other countries. The United States is the world's leading importer of forest products and is second only to Canada as an exporter of forest products. The United States exported $13.4 billion of wood products in 1994 and imported $16.9 billion in the same period for a net trade deficit of $3.5 billion.

Sustainability as a Goal of Forestry

The key to forest management is sustainability. It is useful to distinguish between sustainability of timber harvest and sustainability of a forest ecosystem. A **sustainable timber harvest** is the amount of timber that can be removed periodically from a forest without decreasing the capacity of the forest ecosystem to sustain that level of harvest in the future. A **sustainable forest ecosystem** is one in which all the ecosystem properties of the forest are maintained. If the goal is to achieve a sustainable ecosystem, the amount of timber harvested will be less than the amount harvested when the goal is simply *sustainability of timber harvest.* Viewed in its simplest form, a sustainable timber harvest might seem to be identical with the concept of maximum sustainable yield discussed in Chapter 12. However, with forests the dynamics of stand growth and development and of ecological succession, are recognized so that the general interpretation is that a sustainable timber harvest involves much more complex ideas.

Although it is common to speak about sustainable forests, there are few demonstrated cases of successful sustainable forestry. (A historical example of sustained harvesting of trees occurred on an estate in medieval England where the understory was cut for firewood.) In more recent times, in the few cases where sustainability has been demonstrated, harvesting techniques have included coppicing woodlands—cutting limbs or branches from trees but not cutting the whole tree—and selective cutting of understory forests.

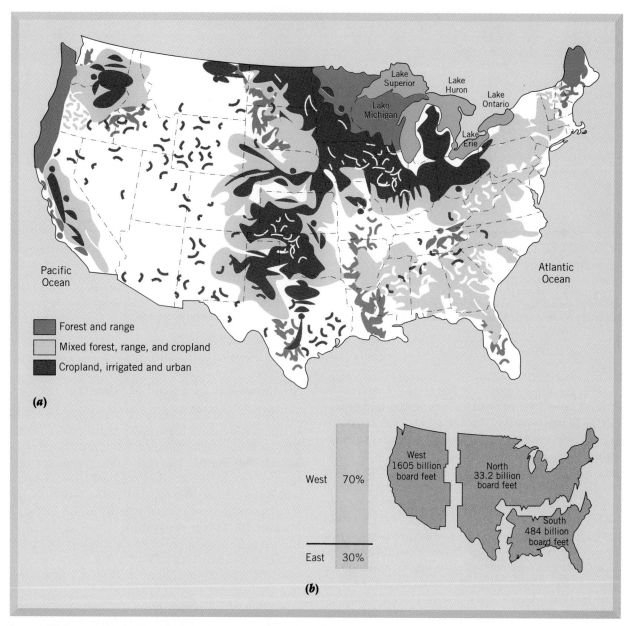

FIGURE 13.5 Forest and rangeland in the contiguous United States: (*a*) land use by area; (*b*) sawtimber volume. Three-fourths of this country's commercial forestland is in the east. However, in terms of sawtimber volume, almost the reverse is true: 70% is in the west and 30% in the east. [*Source:* (*a*) U.S. Forest Service, 1980; (*b*) U.S. Forest Service and C. H. Stoddard, 1978, *Essential of Forestry Practice,* 3rd ed., Wiley, New York.]

Certification of Forest Practices

Worldwide concern with the need for forest sustainability has led to international activities of two kinds: attempts to impose bans on imports of wood obtained from purportedly nonsustainable forest practices and the development of international programs for certification of forest practices. Some European nations have banned the import of certain tropical woods. Some environmental organizations have led demonstrations in support of such bans.

The original goal for formal certification was to certify to consumers that forests were in fact managed sustainably and provide certification for wood products that had been grown in those forests. The idea was to provide a "green seal" that appears on a product (a wood door, a table, a bookcase) stating that the wood used to produce that product was grown in a sustainable forest. However, the lack of proven sustainability means that it is only possible to certify that present practices are consistent with what is known to be sustainable. There is a gradual movement away from calling certified forest practices "sustainable" and instead referring to them as "well-managed forests" or "improved management."[6] A small industry has developed consisting of companies that review the management of a forest and provide a certification; however, no uniform criteria have been established. The search for acceptable uniform criteria has led to an ongoing series of international meetings, including major ones in Montreal and Helsinki. Canada is developing national standards for certification.

Certification programs began because of the concern of environmentalists that forests were being cut in an unsustainable fashion. However, the certification process has evolved into a means for producers of wood and wood products (private corporations and nations that depend on wood exports) to reach new markets, keep existing ones, or increase the value of the product (because people will pay more for a "certified" wood product) and therefore increase the selling price.

Recognizing factors that should be taken into account, some scientists have begun to call for a **new forestry** that includes a variety of practices to increase the likelihood of sustainability, including recognition of the dynamic characteristics of forests and of the need for an ecosystem and landscape context for management.[25] In developing certification criteria, it must be recognized that any new methods create an experiment and should be treated as such. Therefore any new programs that claim to provide sustainable practices must have control areas where no cutting is done, for comparative purposes, and must have adequate scientific monitoring of the status of the forest ecosystem.

13.2 DEFORESTATION: A GLOBAL DILEMMA

In spite of the importance of forests, it is difficult to determine the net rate of change in forest resources worldwide. Some experts have argued that there is a worldwide net increase in forests because large areas in the temperate zone, such as in the eastern and midwestern United States, were cleared in the nineteenth and early twentieth centuries and are now regenerating. Most experts believe that there is a worldwide, and perhaps rapid, net decrease in total forest area and biomass. Because few forests are actually successfully managed to achieve sustainability, it seems likely that forests of the world are undergoing a net decline.

The surprising and important point is that information is lacking on which to base an accurate evaluation. Because forests cover large, often remote areas that are little visited or studied, it is difficult to assess the total amount of forest area. Only recently have programs begun to obtain accurate estimates of the distribution and abundance of forests, and these suggest that past methods overestimated forest biomass by 100% to 400%.[7]

Recognizing limits of existing information, it is useful to review standard information about forests. It is estimated that forests covered one-quarter of Earth's land area in 1950 but only one-fifth in 1980.[8]

History of Deforestation

Forests were cut in the Near East, Greece, and Roman Empire before the modern era. Removal of forests continued northward in Europe as civilization advanced. Fossil records suggest that prehistoric farmers in Denmark cleared forests so much that early successional weeds occupied large areas. In medieval times Great Britain was cut over and many forested areas were removed. With colonization of the New World, much of North America was cleared.[9]

Deforestation is concentrated in the developing world, which lost approximately 200 million ha (494 million acres) from 1980 to 1995.[10] Many of these forests are in the tropics, mountain regions, or high latitudes, places difficult to exploit before the advent of modern transportation and machines.[11] The problem is especially severe in the tropics because they have high human population growth (Figure 13.6). For example, fire-related deforestation rose sharply in Indonesia in 1997. Estimates of forest area destroyed in 1997 range from 150,000 to

Figure 13.6 A satellite image showing clearings in the tropical rain forests in the Amazon in Brazil. The image is in false infrared. The bright red is reflection from the leaves of the rain forest. The straight lines of other colors are clearings by people extending from roads. Much of the clearing is for agriculture.

300,000 ha (370,500 to 741,000 acres).[10] Other areas where tropical deforestation is a major concern include the Amazon basin, western and central Africa, and Southeast Asia and the Pacific regions, including Borneo, the Philippines, and Malaysia. According to the World Resources Institute, Asia cleared 30%, Africa 18%, Latin America 18%, and the world as a whole cleared 20% of tropical forests between 1960 and 1990. It is important to note that estimates are, if anything, underestimates, because organizations must rely primarily on official statements from nations and losses resulting from illegal cutting or other unstated cutting are not necessarily included.[11]

Global and Regional Effects

Forests are a global resource, so cutting forests in one country affects other countries. For example, since 1950 erosion due to deforestation has caused the loss of 580 million hectares (1.4 billion acres) of soil worldwide. Heavier flooding in India's Ganges Valley has caused $1 billion a year in property damage and is blamed on the loss of large forested watersheds in other countries.[12] Nepal, one of the most mountainous countries in the world, lost more than half its forest cover between 1950 and 1980. This cutting destabilizes soil, increasing the rate of landslides, amount of runoff, and sediment load in streams. Many Nepalese streams feed rivers that flow into India. (Figure 13.7). The loss of forest cover in Nepal continues at a rate of about 100,000 ha

(247,000 acres) per year. Reforestation efforts replace less than 15,000 ha (37,050 acres) per year. If present trends continue, little forestland will remain in Nepal, thus permanently impacting India's flooding problems.[13]

Causes of Deforestation

The most common reason that people cut forests is to use or sell timber for lumber and paper products or fuel. Logging by large timber companies and local cutting by villagers are both major causes of deforestation. Another important cause is clearing of forestland for agriculture, including conversion to crops and pasture. Land clearing for agriculture is a principal cause of deforestation in Nepal and Brazil.

The World Firewood Shortage

In many parts of the world wood is a major energy source. Slightly more than half of all wood used in the world is used for firewood. In developed countries, firewood provides less than 1% of total commercial energy, but it provides one-quarter of the energy in developing countries and more than half the energy in Africa.[14] As the human population grows, firewood use increases. In this situation, intentional management practices need to be introduced, including management of woodland stands to improve growth. However, well-planned management of firewood stands has been the exception rather than the rule. Good management prac-

FIGURE 13.7 (*a*) Planting pine trees on the steep slopes in Nepal to replace forests that have been cut. The hazy dark green in the background is yet uncut forest, and the contrast between foreground and background suggests the intensity of clearing that is taking place. (*b*) The Indus River in Northern India carries a heavy load of sediment, as shown by the sediments deposited within and along the flowing water and by the color of the water itself. This scene, near the headwaters, shows that erosion takes place at the higher reaches of the river.

(a)   *(b)*

tices and some successful community-based reforestation projects are discussed later in this chapter.

Indirect Deforestation

A more subtle cause of the loss of forests is indirect deforestation: the death of trees from pollution or disease. Acid rain and other pollutants may be killing trees in many areas in and near industrial countries. In Germany, there is talk of *Waldsterben* ("forest death"). The German government has estimated that one-third of the country's forests has suffered damage: death of standing trees, yellowing of needles, or poorly formed shoots. The causes are unclear but appear to be the result of a number of influences, including acid rain, ozone, and other air pollutants that tend to weaken the trees and increase their susceptibility to disease. This problem extends throughout central Europe and is especially acute in Poland and the former Czechoslovakia. In the New England area of the United States, similar, curious damage to red spruce is occurring.

If global warming were to occur as projected by global models of climate, indirect forest damage could occur over vast regions, with major die-offs in many areas and major shifts in the potential areas for the growth of each species of trees.[15,16] The combination of temperature and rainfall required for different tree species could be altered by global warming, and some species might not continue to grow in their current locations. Such changes would dwarf other indirect effects and could occur by the turn of the twenty-first century, affecting the habitat for many endangered species.[17] However, even if a climate conducive to growth were to exist elsewhere, for trees to migrate northward, other combinations of factors would have to exist, such as capability of dispersal to favorable climates, beneficial soil conditions, and lack of preexisting, competitive species.

13.3 FOREST MANAGEMENT

Good management of forests is a major environmental priority. Many new plans and programs have been proposed, especially for tropical forests and other uncut forests, such as in the Pacific Northwest of North America.[18] The management of forests is called **silviculture.** As explained earlier, a major goal of forest management is sustained yield. Some forests are managed like mechanized farms: a single species is planted in straight rows and the land is fertilized, sometimes by helicopter. Modern machines make harvesting rapid; some remove the entire tree, root and all. Intensive management such as this is characteristic of Europe and parts of the northwestern United States. Other forests, such as those of New England, are managed less actively. In these, regeneration takes place from seeds from existing trees, and ecological succession follows (see Chapter 9). What approach is best depends on the type of forest, environment, and characteristics of the commercially valuable species.

Useful Forestry Terms

Foresters call an area of forest a *stand*, and they classify stands on the basis of tree composition. The two major kinds of commercial stands are **even-aged stands,** where all live trees began growth from seeds and roots germinating the same year, and **uneven-aged stands,** which have at least three distinct age classes.

In even-aged stands, trees are approximately the same height and differ in girth and vigor. A forest that has never been cut is called a *virgin forest*; one that has been cut and has regrown is called **second growth. Old growth,** a term that has gained popularity in several well-publicized disputes about forests, is not a scientific term and does not yet have an agreed-on, precise meaning. In popular usage it often refers to virgin forest. Another important management term is **rotation time,** which, as applied to forests, is the time between cuts of a stand. Trees are divided into the *dominants* (tallest, most common, and most vigorous), *codominants* (fairly common, sharing the canopy or top part of the forest), *intermediate* (forming a layer of growth below dominants), and *suppressed* (growing in the understory).

Management and Productivity

The productivity of a forest varies according to soil fertility, water supply, and local climate. Foresters classify sites by **site quality,** which is the maximum timber crop the land can produce in a given time. Site quality can decrease with poor management. For example, too frequent burning of forests decreases the potential for tree growth by lowering soil fertility. Traditionally, foresters developed site indices for types of forest lands and derived yield tables to estimate future production. Today, forecasts of forest production rely more and more on computer simulation methods. The management of forests can involve removing poorly formed and unproductive trees (or selected other trees) to permit larger trees to grow more rapidly, planting genetically controlled seedlings, and fertilizing the soil. Forest geneticists breed new strains of trees like agricultural geneticists have bred new strains of corn, wheat, tomatoes, and other crop plants. New "supertrees" are supposed to be able to maintain a high rate of growth and increase the total production of forests.

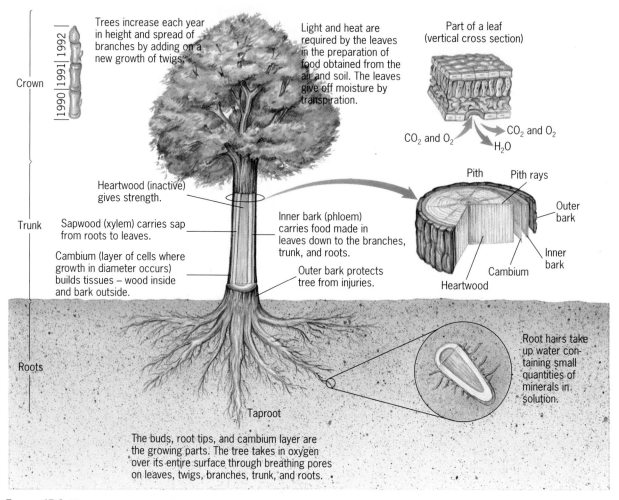

Trees increase each year in height and spread of branches by adding on a new growth of twigs.

Light and heat are required by the leaves in the preparation of food obtained from the air and soil. The leaves give off moisture by transpiration.

Part of a leaf (vertical cross section)

CO_2 and O_2 CO_2 and O_2

H_2O

Crown

Heartwood (inactive) gives strength.

Pith   Pith rays

Sapwood (xylem) carries sap from roots to leaves.

Inner bark (phloem) carries food made in leaves down to the branches, trunk, and roots.

Outer bark

Trunk

Cambium (layer of cells where growth in diameter occurs) builds tissues – wood inside and bark outside.

Inner bark

Outer bark protects tree from injuries.

Cambium

Heartwood

Roots

Taproot

Root hairs take up water containing small quantities of minerals in solution.

The buds, root tips, and cambium layer are the growing parts. The tree takes in oxygen over its entire surface through breathing pores on leaves, twigs, branches, trunk, and roots.

FIGURE 13.8 How a tree grows. (*Source:* C. H. Stoddard, 1978, *Essentials of Forestry Practice,* 3rd ed., Wiley, New York.)

There has been relatively little success in controlling diseases in forests. Insect outbreaks tend to occur infrequently, but when they do, they can have devastating results. Some insect problems are due to introductions of exotic species. For example, the gypsy moth was introduced intentionally into New England around the turn of the century as a source of silk, but it escaped and has spread through many eastern states. Other insect outbreaks appear to be naturally recurrent and to have existed for a long time. For example, a nineteenth-century New Hampshire gazetteer referred to a "plague of loathesome worms" that removed all the leaves from large areas of forest.* Insects affect trees by defoliating them, by eating the buds at the tops of the trees and destroying a straight form (destroying the main trunk of the tree, causing a forked growth), by eating fruits, and

*(*Gazetteer* of Cheshire County, N.H., 1736–1885. Compiled and published by Hamilton Child, Syracuse, N.Y. Printed at the Journal Office, 1885.)

by serving as carriers of diseases. Insecticides are sometimes used to combat these insects. Tree diseases are primarily fungal. Often, as with the Dutch elm disease, an insect spreads the fungus from tree to tree.

Forests are complex and difficult to manage. However, trees provide many kinds of easily obtained information that can be of great assistance in management. The age and growth rate of trees can be measured from tree rings (Figure 13.8); in temperate and boreal (northern) forests, trees produce one growth ring per year, and a tree can be aged by counting the number of rings.

It is important that we take an ecosystem approach to forest management for two reasons: (1) the success of trees depends on soils, climate, competition, and the abundance of parasites and herbivores on ecosystem and ecological community processes and (2) an ecosystem approach is necessary to biological diversity within forests. One way in which an ecosystem approach is taken in forest management is in the use of the concept of ecologi-

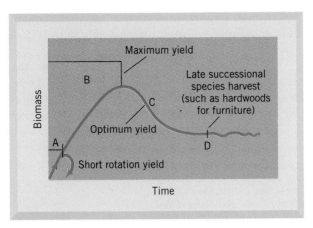

FIGURE 13.9 Use of an ecosystem approach in modern forestry: forest options for harvest, shown in relation to forest succession.

cal succession, the process of development of an ecosystem over time explained in Chapter 9. Some forests are managed for early successional species, and others are managed for late successional species (Figure 13.9). For example, in northern forests such as those in Canada, useful species include aspen, birch, and pine, all of which are early successional. For these, rotation times are comparatively short (less than a century). In contrast, oaks in Germany or maples in the eastern United States are late successional species. Forests of these species, which are used to make furniture, must be managed for a long rotation time or cutting must be done very selectively so that the forest is never opened up greatly.

Multiple Use

A forest can be managed for several uses at the same time. For example, the stated goal of the management of U.S. Forest Service lands is not only timber supply but also recreation, wildlife, and water supply. Because it is not possible to manage a resource so that more than one part is maximized at any time, **multiple use** involves compromises and trade-offs. In the United States, the Multi-Use Sustained Yield Act of the 1960s mandated that the Bureau of Land Management had to promote forest use by specific interest group, including: recreation, extraction, mining, water and other concessions, and the protection of biological diversity. The problem is that some of these uses conflict with others. Some balance must be struck between cutting for the most efficient production of trees and cutting at a level that facilitates other uses. In the past, multiple use has often been interpreted to mean that every parcel of land is managed for all uses. An alternative approach is to make use of the natural variation of different sites by identifying some areas as better for

one use and other areas as better for other uses. Then multiple use can be achieved by having each area managed for the needs it best fits, with the combination of areas meeting all needs.

Clear-cutting and Its Alternatives

Clear-cutting is the practice of cutting all trees in a stand at the same time. Alternatives to clear-cutting are **selective cutting, strip cutting, shelterwood cutting,** and **seed tree cutting.** Shelterwood cutting is the practice of cutting dead and less desirable trees first, and then later cutting mature trees. As a result, there are always young trees left in the forest. Seed tree cutting removes all but a few seed trees (trees with good genetic characteristics and with high seed production that are mature and seed producing), so as to promote regeneration of the forest. In selective cutting, individual trees are marked and cut. Sometimes, smaller, poorly formed trees are selectively removed; this practice is called *thinning*. At other times, trees of specific species and sizes are selectively removed. For example, some forestry companies in Costa Rica cut only some of the largest mahogany trees, leaving other, less valuable trees to help maintain the ecosystem and permitting some of the large mahogany trees to continue to provide seeds for future generations.

In strip cutting, narrow rows of forest are cut, leaving wooded corridors. Strip cutting offers several advantages. The uncut strips protect regenerating trees from wind and direct sunlight, and these remaining trees provide seeds. In addition, strip cutting can minimize the negative aesthetic effects of logging by leaving buffer zones and allowing the corridors of forest that remain to be used for recreation and as wildlife habitats.

Experimental Tests of Clear-cutting

Scientists have tested the effects of clear-cutting.[19] For example, in the U.S. Forest Service Hubbard Brook experimental forest in New Hampshire, an entire watershed was clear-cut and herbicides were applied to prevent regrowth for 2 years. The results were dramatic. Erosion increased, and the pattern of water runoff changed substantially. The exposed soil decayed more rapidly, and the concentrations of nitrates in the stream water exceeded public health standards.

In another experiment at the U.S. Forest Service H. J. Andrews experimental forest in Oregon, clear-cutting greatly increased the frequency of landslides, as did the construction of logging roads. In this forest, rainfall is high (about 240 cm, or 94 in., annually) and the trees (mainly Douglas-fir, western hemlock, and Pacific silver fir) grow very tall and live a long time.[20]

Clear-cutting also changes chemical cycling in forests and results in the loss from the soil of chemical elements necessary for life. When a forest is clear-cut, trees are no longer available to take up nutrients. Opened to the sun and rain, the ground becomes warmer. The process of decay is accelerated; chemical elements, such as nitrogen, are con-verted more rapidly to forms that are water soluble and can be readily lost in runoff during rains (Figure 13.10).[21]

The Forest Service experiments show that clear-cutting can be a poor practice on steep slopes in areas of moderate to heavy rainfall. In addition, the worst effects of clear-cutting were demonstrated

FIGURE 13.10 Effects of clear-cutting on forest chemical cycling. Chemical cycling (*a*) in an old-growth forest and (*b*) after clear-cutting. (*c*) Increase in nitrate concentration in streams following logging and the burning of slash: leaves, branches, and other tree debris. [*Source: (a, b)* Modified from R. L. Fredriksen, 1971, "Comparative Chemical Water Quality—Natural and Disturbed Streams Following Logging and Slash Burning," in *Forest Land Use and Stream Environment,* Oregon State University, Corvallis, OR, pp. 125–137.]

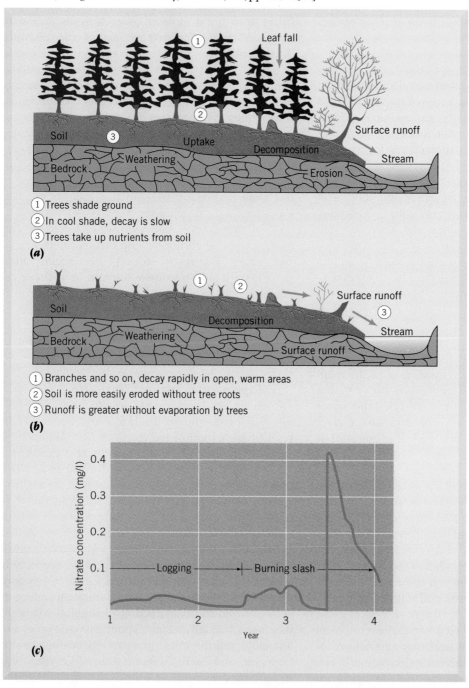

(1) Trees shade ground
(2) In cool shade, decay is slow
(3) Trees take up nutrients from soil

(a)

(1) Branches and so on, decay rapidly in open, warm areas
(2) Soil is more easily eroded without tree roots
(3) Runoff is greater without evaporation by trees

(b)

(c)

by the logging of vast areas of North America during the nineteenth and early twentieth centuries, when huge areas were clear-cut. Clear-cutting on such large scales is neither necessary nor desirable for the best timber production. However, where the ground is level or the slope slight, where rainfall is moderate, and where the desirable species require open areas for growth, clear-cutting at an appropriate spatial scale may be a useful method to regenerate desirable species. The key here is that clear-cutting is neither all good nor all bad for timber production or forest ecosystems. The use of clear-cutting must be evaluated on a case-by-case basis, taking into account the size of cuts, the environment, and available species of trees.

Shorter Rotation Times

In older, traditional forestry, rotation times were long (a century or more). In recent years, however, there has been an increasing emphasis on shorter and shorter rotation times. When seedlings are harvested for paper pulp, which requires wood fibers, not actual lumber, the rotation time is as short as 10 years. Such very short rotations can be hard on the soil and the forest ecosystem. Each cut results in some erosion and soil loss, especially when heavy machinery is used. Artificial fertilization also becomes necessary in this case.

Managing with and for Fire

For much of the twentieth century, it was the practice to try to suppress all fires. However, people have come to realize that some tree species and some forest animals depend on fire and grow only in areas that have burned (see Chapter 9). Areas with very great danger of forest fires, such as Yellowstone National Park, may best be managed through the intentional introduction of frequent light fires, which clear the ground of fuel and prevent conditions that lead to the fires that destroy homes, property, and life.

Prescribed fire, also called **controlled burning,** is becoming increasingly common. In the southeastern United States, prescribed fire is used on about 1 million ha (2.5 million acres) annually. In these forests, studies suggest that it has no significant effect on soils, nutrient cycling, or water flow from the forests but reduces risk from wildfires, controls certain tree diseases, increases food and habitat for wildlife, and can be used to manage the forests for greater production of desirable tree species.[22]

Old Growth and Biological Conservation

Some of the most hotly disputed forest issues concern the conservation of old-growth forests (in spite of the lack of precision about the term). This has become a major debate in the northwestern United States as well as in discussions of tropical rain forests. The issue is particularly acute in the Pacific Northwest. Trees there live a very long time and are famous for their large size, and forests can take centuries to develop. For example, the Douglas-fir, an important commercial timber species, lives more than 600 years and can grow more than 61 m (200 ft) high.

Coastal redwood is well known for its ability to live thousands of years and to reach more than 61 m (200 ft) in height. When a previously uncut stand is clear-cut, a forest stand is eliminated that may take 500 to 1000 years to regenerate. These unique forests of the Pacific Northwest are appreciated for their beauty and as habitats for many species of plants and animals. The old-growth stands, however, also provide the best timber. As the percentage of original forest declines, disputes over how to use and conserve remaining old growth increase. One difficulty in resolving this issue is the lack of direct legislation concerning virgin forest and old-growth stands. As a result, the disputes are reduced to surrogate issues, such as the conservation of specific endangered species whose habitats lie within these forests.

Plantation Forestry

An alternative solution to the pressure on natural forests is to emphasize forest plantations, where trees are planted as crops, on land totally devoted to that use. Much commercial timber is grown in this way. If plantations were used where forest production is high, than a comparatively small percentage of the world's forest land could provide all the world's timber. For example, high-yield forests produce 15 to 20 cubic meters per hectare per year. According to one estimate, if plantations were put on timber land that could produce at least 10 m³/ha/yr, then 10% of the world's forest land could provide the timber quantity for the world's timber trade.[23] This approach could reduce pressure on old-growth forests, on forests important for biological conservation, and on forest lands important for recreation.

Reforestation

As the original forests of the world are cut and the need for timber increases, it is important to plant new trees and develop programs in reforestation. There are many international and national efforts for reforestation. Most countries with a significant amount of economic forestry have such programs. Many private forestry corporations plant trees and reforest areas they have cut. One of the largest of these programs, Global ReLeaf, is conducted by

American Forests, the oldest conservation organization in the United States. The World Bank funds a reforestation program with 48 projects in different nations. In China, 700,000 farmers have cooperated to plant a green wall to protect crops, a shelter belt of trees 100 m (328 ft) wide and 2400 km (1500 miles) long. Although impressive, such programs are small compared to the rate of forest cutting.

Community Forestry

In many parts of the world, people cut nearby forests to meet the needs of small communities. This is particularly true in the developing nations, where, as noted, the use of firewood for fuel constitutes up to half or more of all energy used. In the past, most government forestry departments concentrated their efforts on government-owned forestland or acted merely to police a country's forests. Now there is a realization that this approach must change. In some countries, there is a new emphasis on *community forestry*, in which professional foresters help villagers develop woodlots with the goal of achieving some kind of sustainable local harvest to meet local needs. The FAO and the World Bank are supporting these programs. For example, in Malawi, Africa, a World Bank and FAO project sponsors reforestation in which almost 40% of the households have planted trees. In South Korea, villagers have been reforesting the country at the rate of 40,000 ha/year (98,840 acres/year).

In community forestry, good management practices include limiting access; cutting the slower growing and poorer burning species first to promote the growth of the better firewood species; making use of plantations; and supplementing firewood with other, more easily renewable fuels. However, some of these practices are in conflict with local, traditional activities or are difficult to implement for other reasons.

Such community efforts, like other reforestation projects, are impressive. Again, however, in total they have only a small effect on the problem of the worldwide shortage of firewood. It is not clear whether developing nations can implement a successful management policy in time to prevent serious damage to their forests and the land. If alternative fuels for developing nations cannot be found, the effects will be severe, not only for the land but also for the people.

13.4 PARKS AND PRESERVES

Parks are as old as civilization. The old French word *parc* referred to an enclosed area for keeping wildlife to be hunted. These areas were set aside for nobility, excluding the public. An example is the area now known as Donana National Park on the southern coast of Spain. Originally a country home of nobles, today it is one of the most important natural areas of Europe, used by 80% of birds migrating between Europe and Africa.

The first major public park of the modern era was Victoria Park in Great Britain, authorized in 1842. The concept of a national park, whose purposes include protection of nature as well as public access, originated in North America in the nineteenth century. In the twentieth century the purpose of national parks was broadened to emphasize biological conservation, and this idea was applied worldwide.[24] The first designated national park was Yosemite National Park in California, which was made a park by passage of a bill signed by President Lincoln in 1864 (Figure 13.11). The term *national park*, however, was first used with the establishment of Yellowstone in 1872.

In recent years the number of national parks throughout the world has increased rapidly. The first law establishing national parks in France was enacted in 1960. Taiwan had no national parks prior to 1980, but now has four. In the United States, the area in national and state parks has increased from less than 12 million ha (30 million acres) in 1950 to nearly 32 million ha (80 million acres) today, much of the increase resulting from the establishment of parks in Alaska.

Goals of Park Management

The goals of park and nature preserve management can be summarized as follows:

1. preservation of unique wonders of nature, such as Niagara Falls and the Grand Canyon;
2. preservation of nature without human interference (preserving wilderness for its own sake);
3. preservation of nature in a condition thought to be representative of some prior time period (e.g., the United States prior to European settlement);
4. wildlife conservation, including conservation of the required habitat and ecosystem of the wildlife;
5. maintenance of wildlife for hunting;
6. maintenance of uniquely or unusually beautiful landscapes for aesthetic reasons;
7. maintenance of **representative natural areas** for an entire country;
8. maintenance for outdoor recreation, including a range of activities from viewing scenery to wilderness recreation (hiking, cross-country skiing, rock climbing) to tourism (car and bus tours, swimming, downhill skiing, camping, etc.); and

FOREST 13.11 First national parks. Yosemite and Yellowstone were the first national parks established in the United States. El Capitan, rising above Yosemite Valley, shown here at sunrise, is one of the most famous scenes in the park.

9. maintenance of areas set aside for scientific research, both as a basis for park management and for the pursuit of answers to fundamental scientific questions.

The purpose of the earliest national parks established in the United States was to preserve the unique, awesome, and grand landscapes of the country, a purpose the historian Alford Runte refers to as *monumentalism*. In the nineteenth century, Americans saw their national parks as a contribution to civilization equivalent to the architectural treasures of the Old World and sought to preserve them as a basis of their pride in their country.[25]

In the second half of the twentieth century, the emphasis of park management has become more ecological, with parks established both for scientific research and to maintain examples of representative natural areas. For example, Zimbabwe established Sengwa National Park solely for scientific research; there are no tourist areas and tourists are not generally allowed. Its purpose is the study of the natural ecosystems with as little human interference as possible so that the principles of wildlife and wilderness management can be better formulated and understood. Other national parks in the countries of eastern and southern Africa, including those of Kenya, Uganda, Tanzania, Zimbabwe, and South Africa, have been established primarily for viewing wildlife and for biological conservation. The conservation of

representative natural areas of a country is an increasingly common goal of national parks. For example, the goal of New Zealand's park planning is to include in national parks at least one area representative of each major ecosystem of the nation, from seacoast to mountain peak.

Parks and Preserves as Islands

A *park* is an area set aside for use by people, whereas a *nature preserve* (although it may be used by people) has as its primary purpose the conservation of some resource, typically a biological one. Every park or preserve is an island of one kind of landscape surrounded by a different kind of land use. Islands have special ecological qualities (discussed in detail in Chapters 7 and 12), and concepts of island biogeography are used in the design and management of parks. Specifically, park and preserve planners know that the size of the park and the diversity of habitats affect the number of species that can be maintained there. Also, the farther the park is from other parks or sources of species, the fewer species are found (Figure 13.12). Even the shape of a park can determine what species can survive within it.

One of the important differences between a park and a truly natural wilderness area is that a park has definite boundaries. These boundaries are usually arbitrary from an ecological viewpoint and

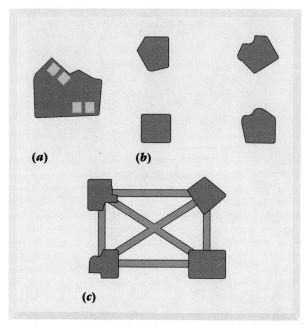

FIGURE 13.12 Park shapes and island biogeography. A large park (*a*) can maintain more species, but several small parks (*b*) provide a kind of insurance against catastrophe. For example, if a storm struck one park and killed all individuals of one species in it, other populations of that species could survive in the other parks. (*c*) A combination that provides the benefits of both a single large park and several small ones. Here the small parks are connected by **migration corridors** that allow occasional migration among the parks; the total area is equal to that of (*a*), and the distribution over the landscape provides greater insurance.

have been established for political, economic, or historical reasons not related to the natural ecosystem. In fact, many parks have been developed on what are otherwise considered wastelands, useless for any other purposes.

Even where parks or preserves have been set aside for the conservation of some species, the boundaries are usually arbitrary, and such arbitrariness has caused problems. For example, Lake Manyara National Park in Tanzania, famous for its elephants, was originally established with boundaries incorrect for elephant habits. Prior to the establishment of this national park, elephants would spend part of the year feeding along a steep incline above the lake. At other times of the year they would migrate down to the valley floor, depending on the availability of food and water. The annual movements of the elephants were necessary so that they could obtain food of sufficient nutritional quality throughout the year.

When the park was established, farms were laid out along its northern border. These farms crossed the traditional pathways of the elephants, creating two negative effects. First, elephants came into direct conflict with farmers. Elephants crashed

through farm fences, eating corn and other crops and disrupting the farms. Second, whenever the farmers were successful in preventing the movement of the elephants, the animals were cut off from reaching their feeding ground near the lake. It became clear that the park boundaries were arbitrary and inappropriate, and adjustments were made to extend the boundaries to cover those traditional migratory routes. This solution reduced the conflicts between elephants and farmers.

Because parks isolate populations genetically, they may provide too small a habitat for the maintenance of a minimum safe population size. If parks are to function as biological preserves, they must be adequate in size and habitat diversity to maintain a minimum population that will not undergo the serious genetic difficulties that can occur in small populations. An alternative, if necessary, is for a manager to move individuals of one species, say lions in African preserves, from one park to another to maintain genetic diversity.

When a park is first established and cut off from its surroundings, the number of species within the park decreases. This has been found to be true in a study of forest preserves left as islands when tropical forests in the Amazon basin of Brazil were cut for timber.[26] Once a forest island is formed, there is a rapid and immediate loss of some species. In addition, there is an **edge effect,** which means that many species escape from the cut area and seek refuge in the border of the forest. There is a negative effect of increasing borders on the interior of a habitat. Light and wind increase inside the habitat, near the border, affecting evaporation and moisture supply, and favoring early successional species of plants. Such effects seem to be important for a distance into the habitat equal to more than two tree heights. When the habitat becomes small enough, the edge effects may affect most of the area. Furthermore, where different habitat types come together, there may be more species than in either habitat alone. The net result of all this is that, when a patch is first produced, species may crowd in it, and the diversity may temporarily increase, but this may last only a short time. The longer term result is an exponential decline in the number of species following the initial creation of the island.

How Much Land Should Be in Parks?

Nations differ greatly in the percentage of their total area set aside as national parks. Costa Rica, a small country with great biological diversity, has more than 10% of its land in national parks, and Kenya, a larger nation also with great biological resources, has about 10% of its land in national parks. In France, an industrialized nation in which civilization

TABLE 13.2 Land Use Areas of the Great Lakes States

State	Total Area	State Forests	National Forests	State Parks	National Parks	Wilderness	Total Conservation Areas
Areas in km²							
Michigan	147,511	15,000	10,927	1,011	2,314	383	29,252
Minnesota	206,030	16,000	10,927	818	882	4,388	28,627
Illinois	144,120	1,113	1,033	—	—	—	2,146
Indiana	93,064	—	760	219	—	86	979
Ohio	106,201	699	769	801	—	—	2,270
Wisconsin	140,964	1,900	6,070	268	—	183	8,238
Total	837,890	34,712	30,487	3,117	3,196	5,040	71,512
Percentages of total							
Michigan		10.17	7.41	0.69	1.57	0.26	20
Minnesota		7.77	5.30	0.40	0.43	2.13	16
Illinois		0.77	0.72	0.00	0.00	0.00	1
Indiana		0.00	0.82	0.24	0.00	0.06	1
Ohio		0.66	0.72	0.75	0.00	0.00	2
Wisconsin		1.35	4.31	0.19	0.00	0.13	6
Total		4.14	3.64	0.37	0.38	0.60	9

Source: D. B. Botkin, 1992, "Global Warming and Forests of the Great Lakes States," in J. Schmandt, ed., *The Regions and Global Warming: Impacts and Response Strategies,* New York, Oxford University Press.

has altered the landscape for several thousand years, only 0.7% of the land is in the nation's six national parks. However, there are 24 regional parks in France that encompass 7% (3 million ha, or 7.4 million acres) of the nation's area.[27]

The total amount of protected natural areas in the United States is more than 98 million ha (about 240 million acres), approximately 10.5% of total U.S. land area. However, the percentage of land in parks, preserves, and other conservation areas differs greatly among the states. The West, for example, has vast parks, whereas the six Great Lakes states (Michigan, Minnesota, Illinois, Indiana, Ohio, and Wisconsin), covering an area approaching that of France and Germany combined, allocate less than 0.5% to parks and less than 1% to designated wilderness (see Table 13.2).[28] To manage our biological resources wisely, it seems only prudent to allocate a reasonable portion of the land to conservation uses. As the human population increases, pressures for more uses of the land will lead to more disputes about whether and where new parks should be established.

Conserving Wilderness

As a modern legal concept, **wilderness** is an area undisturbed by people. The only people in a wilderness area are visitors who do not remain. The conservation of such an area is a new idea introduced in the second half of the twentieth century and one that is likely to become more important as the human population increases and the effects of civilization become more pervasive throughout the world (Figure 13.13).

The United States Wilderness Act of 1964 was a landmark piece of legislation, marking the first time anywhere that wilderness was recognized by national law as a national treasure to be preserved. Under this law, wilderness includes "an area of undeveloped Federal land retaining its primeval character and influence, without permanent improvements or human habitation, which is protected and managed so as to preserve its natural conditions." Such lands are those in which (1) the imprint of human work is unnoticeable; (2) there are opportunities for solitude and for primitive and unconfined recreation; and (3) there are at least 5000 acres.

The law also recognizes that these areas are valuable for ecological processes, geology, education, scenery, and history. The Wilderness Act required certain maps and descriptions of wilderness areas, resulting in the U.S. Forest Service's *Roadless Area Review and Evaluation* (RARE I and RARE II), which evaluated lands for inclusion as legally designated wilderness.

Countries with a significant amount of wilderness include New Zealand, Canada, Sweden, Norway, Finland, Russia, and Australia; some countries of eastern and southern Africa; many countries of

FIGURE 13.13 In North America there are still large areas that have been little affected by human beings. Some of these are legally designated as wilderness areas, such as (*a*) the Boundary Waters Canoe Area in Minnesota, where human use is restricted to hiking and motorless boating. (*b*) Beautiful Swiss Alpine area famous for recreation, including skiing and hiking, and often thought of as wilderness, as praised by the Romantic poets of nineteenth-century England, are actually plantations of trees and land otherwise heavily used and much altered by people. (*c*) Hikers in Desolation Wilderness Area admire Lake Tahoe below.

South America, including parts of the Brazilian and Peruvian Amazon basin; the mountainous high-altitude areas of Chile and Argentina; some of the remaining interior tropical forests of Southeast Asia; and the Pacific Rim countries (parts of Borneo, the Philippines, Papua New Guinea, and Indonesia). In addition, wilderness can be found in the polar regions, including Antarctica, Greenland, and Iceland.

Many countries have no wilderness left to preserve. Switzerland is a country in which wilderness is not a part of preservation. In the Danish language even the word for wilderness has disappeared, although that word was important in the ancestral languages of the Danes.[29] The historian Roderick Nash contrasts a national park in Switzerland with a wilderness preserve. The Swiss park lies in view of the awesome scenery of the Alps, the scenery that inspired the English romantic poets of the early nineteenth century to praise what they saw as wilderness (Figure 13.13*b*). But the park is situated in an area that has been heavily exploited for such activities as mining and foundries since the Middle Ages. All the forests are planted.[30]

In another, perhaps deeper sense, wilderness is an idea and an ideal that can be experienced by some people in many places, such as in Japanese gardens. Such a spot may have more value as a place of solitude and beauty than some more traditional wilderness areas. In Japan, for instance, there are roadless recreation areas, but they are filled with people. There is a 2-day hiking circuit to a high-altitude marsh with huts. Trash is removed from the area by helicopter.

Conflicts in Managing Wilderness

The legal definition of wilderness has produced several controversies. The wilderness system in the United States began in 1964 with 3.7 million ha (9.2 million acres) designated in U.S. Forest Service control and by 1979 had grown to include nearly 8 million ha (19 million acres). There are 30 million ha (75 million acres) proposed, including 17 million ha (42 million acres) in Alaska. Those interested in developing the natural resources of an area, including mineral ores and timber, have argued that the rules are unnecessarily stringent, protecting too much land from exploitation when, they say, there is plenty of wilderness elsewhere. Those who wish to conserve additional wild areas have argued that the interpretation of the U.S. Wilderness Act is too stringent in the opposite direction, because the requirement for purity of the land from human influence has been taken too narrowly.

The management of wilderness may seem a paradox: A true wilderness would need no management. In fact, in modern life, with the great numbers of people in the world, even wilderness must be defined, legally set aside, and controlled. There are two distinct ways to view the goal of managing wilderness: one in terms of the wilderness itself and the other in terms of people. In the first, the goal is to preserve nature undisturbed by human beings. In the second, the purpose is to provide people with a wilderness experience.

Legally designated wilderness can be seen as one extreme in a spectrum of environments to manage.[31] Its management should involve as little direct action as possible, so as to minimize any human influence. Ironically, one of the necessities is to control human access and, therefore, to impose a carrying capacity for visitors so that any visitor has little, if any, sense that there are other people present. Access to wilderness for various kinds of activities, such as operation of motor vehicles, snowmobiles, and off-road vehicles, must also be restricted.

As an example of considerations in wilderness management, consider the Desolation Wilderness Area in California, an area of more than 24,200 ha (60,000 acres), which in one year had more than 250,000 visitors (Figure 13.13c).[32] Could each visitor really have a wilderness experience in that area or was the human carrying capacity exceeded? This is a subjective judgment. If all visitors saw only their own companions and believed they were alone, then the actual number of visitors did not matter for each visitor's wilderness experience. If, on the other

hand, every visitor found the solitude ruined by strangers, then the management failed, no matter how few people visited the area.

Wilderness must be managed with a knowledge of adjacent lands. A wilderness next to a garbage dump or a smoke-emitting power plant is a contradiction. Whether a wilderness can be adjacent to high-intensity campgrounds or near a city is a more subtle question that must be resolved by citizens.

Today those in wilderness management recognize that wild areas change over time and that these changes should be allowed to occur as long as they are natural. This is different from earlier views that nature undisturbed was unchanging and should be managed so that it did not change. It is generally argued now that, in choosing what activities can be allowed in a wilderness, emphasis should be given to those activities that depend on wilderness (the experience of solitude or the observation of shy and elusive wildlife) rather than to those that can be carried out elsewhere (such as downhill skiing).

A source of conflict is that wilderness areas frequently contain resources of economic importance, including timber, mineral ores, and sources of energy. Another controversy is between the need to study wilderness, to understand wilderness so that it can be better conserved, and the desire to leave the wilderness undisturbed. Those in favor of scientific research in the wilderness argue that it is necessary for the conservation of the wilderness. Those opposed argue that scientific research is contradictory to the purpose of a designated wilderness as an area undisturbed by people. One solution is to establish separate research preserves.

13.5 CONSERVING AND MANAGING LIFE IN THE OCEANS

Oceans and Climate

The oceans cover 70% of Earth's surface and play an important role in our global environment. Water has a high heat capacity, meaning that a gram of water can store more energy than a gram of many other compounds. Hence, oceans buffer and modulate climate; they reduce the extremes in temperature that would otherwise be experienced on Earth. Ocean currents can absorb heat from the atmosphere, thus delaying its warming; warm currents can give up heat to the atmosphere, thus delaying its cooling.

Oceans affect the global cycling of chemical elements (see Chapter 4). They are a major storehouse of carbon and exchange carbon dioxide rapidly with the atmosphere, and they can play a major role in the rate of global warming.

Ocean Food Webs

There are many different food webs in the oceans. Most of the major types were discussed in Chapters 6 and 8. Here we discuss the food webs from which we obtain most of the oceanic food we eat: planktonic, intertidal, and coral reef food webs.

Planktonic Food Webs

Planktonic food webs have small floating algae and photosynthetic blue-green bacteria as the primary producers. These are fed on by small, drifting herbivores. Other small, drifting, predatory animals, along with the floating herbivores, make up the zooplankton. Some large vertebrates, such as baleen whales and the whale shark, are also planktonic feeders. Planktonic food webs can exist only where there is light, oxygen, and an ample supply of all the other chemical elements required for life. Light penetrates only a relatively short distance into the column of water in the oceans, and surface waters in the center of large ocean basins are generally nutrient poor. Thus the most productive planktonic food chains are on continental shelves, which are discussed in more detail later in this chapter.

Intertidal Food Webs

The bases of intertidal food webs are attached, rather than floating or drifting, algae and photosynthetic blue-green bacteria familiar to anyone who has walked along an ocean shore. Kelp beds, such as those that occur along the west coast of North America, are examples of attached algae that form the base of an intertidal food chain. Intertidal ecosystems, especially salt marshes, are important habitats in the life cycles of many fish and shellfish that are sources of food, have commercial value, are endangered species, or are otherwise important for biological diversity.

Reef Food Webs

Coral reefs are structures that appear to be made of rocks but are actually formed of the hard material deposited by living organisms. These reef-building organisms include coralline algae and animals that have microscopic algae as symbionts. Corals produce a hard calcium carbonate material that forms the structure of the reef. The reefs provide a habitat for a great diversity of organisms and are among the most diverse of all kinds of ecosystems.[33]

13.6 FISHERIES

Fisheries are a major source of protein. People catch many species of fish and shellfish for food, but fewer than 20 species provide about 66% of the annual catch and 90% of the value. The worldwide harvest of fish is more than 118 million metric tons/year (130 million U.S. tons). The harvest from marine fisheries is approximately 82.7 million metric tons/year (91.2 million U.S. tons), and from fresh water it is 15.9 million metric tons/year (17.5 million U.S. tons).[34] Approximately 14.5 million metric tons (15.9 million U.S. tons) comes from aquaculture, the farming of fish in artificial ponds and tanks (see Chapter 10). the contribution of fish to the human diet is 6.6% in North America, 9.7% in Western Europe, 21.1% in Africa, 8.2% in Latin America, 27.8% in the Far East, and 21.7% in Central Asia so that fish provide about 16% of the world's protein. In developing countries, fish are an especially important part of the diet.[35]

Marine catch has changed dramatically over the last 45 years. Marine harvest peaked in 1989 at 86 million metric tons (94.8 million tons). In 1993, it dropped to 84 million metric tons (92.6 million tons)—although total global fish harvest continues to climb because of increased aquaculture production. The prices of most fish species continue to rise as harvests shrink.[35] Fish vary greatly in value, from tens of dollars to a thousand dollars per ton depending on the species and the port at which they are landed. Fishing is an international trade, but a few countries dominate. Japan, China, Russia, Chile, and the United States are among the major fisheries nations.[36]

Commercial fisheries are concentrated in relatively few areas of the world's oceans (Figure 13.14). Continental shelves, which make up only 10% of the oceans, provide more than 90% of the fishery harvest. Fish are abundant where their food is abundant and, ultimately, where there is a high production of algae at the base of the food chain. Algae are most abundant in areas with relatively high concentrations of the chemical elements necessary for life, particularly nitrogen and phosphorus. These areas occur most commonly along the continental shelf, particularly in regions of wind-induced upwellings and sometimes quite close to shore.

Upwelling areas are rich in chemical elements necessary for life because the rising currents carry these elements from several hundred meters deep in the ocean to the surface. Upwelling currents occur primarily near continents in locations where prevailing winds are offshore. These winds push the surface waters away from the coast. Water from the ocean bottom moves upward to replace the surface water, and this forms an upwelling current.

The Management of Fisheries

A fish population is called a *stock*. Fishermen are interested in obtaining as large a catch as possible, but to have a sustainable supply of fish, the harvest can-

(a)

FIGURE 13.14 Earth's major fishery regions. Red indicates regions of upwelling or high productivity. Pink indicates moderate fish production. The central ocean, which occupies the largest area, is a relative desert in terms of economically valuable fish. (*Source:* Mackintosh, 1965, *The Stocks of Whales,* Fishing News Books. Reproduced by permission of the Buckland Foundation.)

not be larger than the maximum annual growth of the population. Overfishing—catching a number of fish greater than the net growth in the population—leads in subsequent years to a decrease in the rate of growth and to a decline in the catch. The management of fisheries has concentrated on a single question: How rapidly can a fish stock grow? In the past, most projections of the growth of fish populations have been based on the logistic growth curve, and such projections suffer from all the limitations of this use discussed in Chapter 12. Remember that the logistic curve is S shaped and that a population growing according to this curve reaches a maximum size called the carrying capacity (see Chapters 1 and 5). The traditional goal of management of fisheries is to obtain a maximum sustainable yield, which is traditionally defined in terms of the logistic growth curve. In a population growing according to the logistic, the MSY is the point of maximum growth, which occurs when the number of individuals is exactly one-half the carrying capacity.

The Failures of Fisheries Management

During the last 200 years, one fishery after another has been developed and has failed (see A Closer Look 13.1, "The Peruvian Anchovy Fishery"). The failure of the world's fisheries continues today, even in those areas with the most conscientious and active management.[37,38] The repeated failure of fisheries around the world makes this a global environmental problem.

In the last 30 years, a number of other major fisheries have experienced failures similar to that described in A Closer Look 13.1 under intentional management regimes (see Table 13.3). Perhaps most well known of these in North America are the declines in the salmon fisheries of the west coast and in the fisheries in the Atlantic Ocean, particularly those of the Grand Banks off of Newfoundland, and the decline since 1950 by 95% of the oyster catch in the Chesapeake Bay (Figure 13.15*d*).[39]

How Can We Better Manage Fisheries?

In an attempt to avoid past pitfalls, fisheries managers have used optimum sustainable yield (OSY), another term introduced in Chapter 12. The OSY is the largest yield that can be sustained indefinitely, taking all factors into account: the growth rate of the population viewed in simplistic terms as well as the population as it really grows in a time-varying environment, in competition with other species, and as affected by predation and the populations of symbionts and of its food supply as well as nonbiological factors in the environment, such as changes in climate that affect ocean currents, including upwellings.

CLOSER LOOK 13.1

THE PERUVIAN ANCHOVY FISHERY

A classic case of failed management is the history of the Peruvian anchovy fishery, once the world's largest commercial fishery. Anchovies are found in upwelling regions, where nutrient-rich waters move upward. In 1970, 7.3 million metric tons were caught (8.0 million U.S. tons), but just 2 years later, by 1972, only 1.8 million metric tons (2.0 million U.S. tons) were caught—15% of the 1970 peak.

Why did the Peruvian anchovy fishery crash? Overfishing played a role, which may have been complicated by *El Niño*, a recurrent change in ocean upwellings and surface currents associated with climatic changes.

El Niño events illustrate the problems of managing fisheries from the logistic growth curve. In recent years El Niño events have been associated with changes in climate in California as well as other portions of the United States, and changes in the distribution of fisheries off the west coast of the United States. Key to the anchovy failure is the cessation in up-welling currents that normally bring chemical nutrients to the surface off the coast of Peru, providing nutrients for planktonic algae and creating a productive marine food chain that yields high anchovy production.

For reasons not entirely understood, there are periodic changes in climate. As explained earlier, upwellings occur when winds drive the surface waters offshore. During El Niño, these winds fail, the surface waters no longer move offshore, and the upwellings cease. Without the upward flow of nutrients, the algae no longer maintain high production rates. The production of the entire food chain crashes.

Historical records of the guano deposits from birds that eat anchovies on islands offshore of Peru show that El Niño–related failures in upwelling have occurred repeatedly in the past. Thus management policies based on the logistic MSY were bound to fail, because they assumed a constant environment. Harvest of anchovies must be set based on each year's expected production, which changes with upwellings. Successful management of anchovies would involve continued monitoring of the environment as well as the fish populations and would also include research aimed at better understanding the causes of El Niño and thereby improving the ability to predict its occurrence. With the necessary information and predictive capacity, a more flexible yield could be calculated, one that varies annually.

The failure of the anchovy fishery was a blow to the world's protein supply and to the management of fisheries. The anchovy fishery was actively managed by the Peruvian government with assistance from the FAO. They tried to manage the anchovies to keep the population at the level that would provide the MSY as defined by the logistic growth curve. Unintentionally, the population may have been brought to a level below the MSY, and as long as the catch was kept at the MSY level, the anchovy population could only decrease.

Although the idea of OSY makes sense, it is difficult to apply in practice because so many factors must be considered. As with so many environmental matters, information about fish stocks and their oceanic ecosystems is sparse and usually insufficient. In the future, sustainable fisheries can only be

---

TABLE 13.3 **Problems of Some Major Fisheries**

Anchovy: Reached peak in 1970 (10 million metric tons), then declined.

Atlantic herring: Exploitation so high that recruitment was decreased.

Arctonorwegian cod: High fishing level followed by 4 years of poor recruitment.

Downs' stock of herring in the North Sea: Managers failed to grasp stock and recruitment problems.

North Atlantic haddock: Catch averaged 50,000 metric tons for many years; increased to 155,000 in 1965, 127,000 in 1966, then fell to 12,000 in 1971–1974. In 1973, the International Commission for the Northwest Atlantic Fisheries (ICNAF) established a quota of 6000 metric tons. Apparently, haddock could sustain a 50,000-metric-ton catch, but when this was tripled, the population was so decreased that only a smaller catch could be sustained.

Atlantic menhaden: Peak catch was 712,000 metric tons in 1956; in 1969, it was 161,400. Fisheries experts believe the drop was due to overfishing.

Salmon: Loss has occurred worldwide wherever salmon and their relatives once used fresh waters to spawn and rear. This has been the result of construction of dams on rivers, channelization of rivers, pollution, overharvesting, and habitat alteration of many kinds.

Pacific sardines: Declined catastrophically from the 1950s through the 1970s.

Source: D. Cushing, 1975, *Fisheries Resources of the Sea and Their Management,* London, Oxford University Press.

achieved if the perspective is broadened to include ecosystem factors that vary over time.

An additional key to successful management of fisheries is monitoring the age structure of the population. Age structure provides many different kinds of information, as illustrated in Figure 13.15, which shows the age structure of the catch of salmon from the Columbia River in Washington for two different periods: 1941–1943 and 1961–1963. In the first period, most of the catch consisted of 4-year-olds (60% of the catch); 3-year-olds and 5-year-olds each made up about 15% of the population. Twenty years later, in 1961 and 1962, half the catch consisted of 3-year-olds, and the amount of 5-year-olds had declined to about 8%. The total catch had declined considerably also. For the combined period 1941–1943, 1.9 million fish were caught; for the second period, the total catch was 849,000, 49% of the total in the earlier period. The shift in catch toward younger ages, along with an overall decline in catch, suggests that the fish are exploited to the point where they are not reaching older ages. Such a shift in the age structure of a harvested population is a good early sign of overexploitation and of a need to alter allowable catches.

There have been two World Fisheries Congresses, one in 1992 in Athens and one in 1996 in Brisbane. At these, managers and scientists from every major country on Earth that harvest fish came together to discuss ownership, harvesting, and allocation of fishery resources. These have been the "Rio Conference" of fisheries. Some progress has been made as a result of these Congresses, in the adaptation of a "Fisheries Code of Conduct" for all nations.

FIGURE 13.15 The catch of salmon on the Columbia River by age for (*a*) 1941–1943 and (*b*) 1961–1963. The total catch is shown in (*c*). (*Source:* Botkin, D. et al., 1995, *Status and Future of Salmon of Western Oregon and Northern California: Findings and Options.* Santa Barbara, CA. Center for the Study of the Environment.)

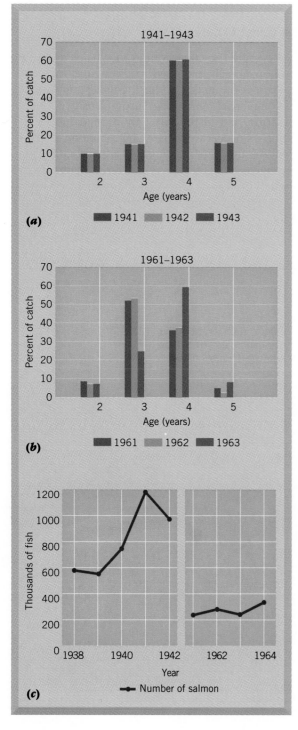

Darrell Jack, a Yakima Indian, pitches a salmon from boat to truck. The Yakima have lived along the Columbia and depended on salmon for many generations; alteration of the Columbia River and the surrounding environment threatens this way of life.

13.7 WHALES AND OTHER MARINE MAMMALS

Marine mammals have long fascinated people. Drawings of whales have been dated as early as 2200 B.C.[40] Eskimos used whales for food and clothing as long ago as 1500 B.C. In the ninth century A.D., whaling by Norwegians was reported by travelers whose accounts were written down in the court of the English king Alfred. Many people attach great significance to marine mammals. In the United States, whale watching, particularly of the grey whale off the southern California coast and of sperm, humpback, and other whales off Cape Cod, Massachusetts, has become quite popular. Few other issues in environmental sciences have raised as much public attention and debate as the harvesting of marine mammals.[41]

Evolution of Marine Mammals

All marine mammals were originally inhabitants of the land. During the last 80 million years, several separate groups of mammals returned to the oceans and underwent adaptations to marine life. Each group of marine mammals shows a different degree of transition to ocean life. Understandably, the adaptation is greatest for those who began the transition longest ago. Some marine mammals, such as dolphins, porpoises, and great whales, complete their entire life cycle in the oceans and have organs and limbs that are highly adapted to life in the water. They cannot move on the land. Others, such as seals and sea lions, spend part of their time on shore.

Whales fit into two major categories: baleen and toothed (Figure 13.16). The sperm whale is the only great whale that is toothed; the rest of the toothed group are smaller whales, dolphins, and porpoises. The other great whales are in the baleen group. These have highly modified teeth that look like giant combs and act as water filters. Baleen whales feed by filtering ocean plankton.

The Great Whales

In discussing the conservation of whales and other marine mammals, we need to distinguish two kinds of uses people make of these animals: subsistence and commercial. *Subsistence* use means that the animals form part of the basic food or materials of a culture, as occurs today for whales among Eskimos, Japanese, and inhabitants of Tonga and Greenland. All other uses of whales and marine mammals are *commercial* uses and involve harvesting for use as products not connected to tradition or culture. By tradition, whaling grounds are referred to as fisheries, even though whales are mammals.

The earliest whale hunters killed these huge mammals from the shore or from small boats near shore, but gradually whale hunters ventured farther

FIGURE 13.16 There are two kinds of great whales: (*a*) toothed whales, such as the killer whale, whose teeth appear to be much like those of other predatory mammals; (*b*) baleen whales, such as the southern right whale, whose teeth are modified into baleen, a comblike structure that filters plankton from the ocean.

(*a*)   (*b*)

out. In the eleventh and twelfth centuries, Basques hunted the Atlantic right whale from open boats in the Bay of Biscay, off the western coast of France. Later medieval whaling remained shore based. Whales were hunted from open boats and brought ashore for processing, and the boats returned to land once the search for a whale was finished.

Later, whaling became *pelagic:* Whalers took to the open ocean and searched for whales from ships that remained at sea for long periods; the whales were brought on board and processed there. **Pelagic whaling** became possible with the invention of furnaces and boilers (tryworks) for extracting whale oil at sea. Thus pelagic whaling was a product of the industrial revolution. With this invention, whaling grew as an industry. Pelagic whaling expanded outward from the Atlantic Ocean offshore from northern Europe. The English and the Dutch sought right and bowhead whales in the Atlantic in the seventeenth and eighteenth centuries, sailing from Spain to Hudson Bay. American fleets developed in the eighteenth century in New England, and by the nineteenth century the United States dominated the industry, providing most of the ships and even more of the crews for whaling.[42,43]

Sailors in these ships at first sought whales that were large, slow moving, and easy to process. The right whale got its name because it was the "right" whale to catch: It was relatively easy to catch, floated rather than sank after dying, and yielded great quantities of oil. Bowhead, gray, and sperm whales were the other species sought in the nineteenth century. Gray and sperm whales were more difficult to catch than right and bowhead whales.

Whales provided many nineteenth-century products. Whale oil was used for cooking, lubrication, and lamps. Whales provided the main ingredients for the base of perfumes. The elongated teeth (whalebone, or baleen) that enable baleen whales to strain the ocean waters for food are flexible and springy and were used for corset stays and other products before the invention of inexpensive steel springs.

The incidence of whaling increased in the twentieth century. Although the nineteenth-century whaling ships are more famous, made popular by such novels as *Moby Dick*, more whales have been killed in the twentieth century than were killed in the previous century. The worldwide decline in most species of whales as a result of human activities makes this a global environmental issue.

Commercial Value of Whaling and the Politics of Whale Conservation

Whaling was never a major world industry in terms of gross or net economic return. Even near its peak exploitation in the mid-twentieth century, the annual gross value of whaling production was approximately $500 million; the annual gross value of all other fisheries at the time was approximately $4 billion to $5 billion. However, some whaling products were crucial to nineteenth-century industry, especially whale oil, used as a lubricant, which continued to be an important product into the twentieth century.

Conservation of whales has been a concern among conservationists for many years. Attempts to control whaling began with the League of Nations in 1924. The first agreement, the Convention for the Regulation of Whaling, was signed by 21 countries in 1931. In 1946 a conference in Washington, D.C., initiated the International Whaling Commission (IWC). In 1982, IWC established a moratorium on commercial whaling. Currently 12 of approximately 80 species of whales are protected.[44]

The IWC has played a major role in the reduction (almost elimination) of the commercial harvest of whales. Since the formation of the IWC, no species has become extinct, the total take of whales has decreased, and harvesting of species considered endangered has ceased. Endangered species protected from hunting have had a mixed history (see Table 13.4). Blue whale appear to have recovered some but remain rare and endangered. Gray whale, have increased and are now relatively abundant, numbering about 23,000 individuals. However, global climate change, pollution, and ozone depletion have become greater risks to whale populations than whaling.[45]

The establishment of the IWC was a major landmark in wildlife conservation. It was one of the first major attempts by a group of nations to agree on a reasonable harvest of a biological resource. By comparison, no such agreement has ever been formed for the use of timber. The annual meeting of the IWC has become a forum for discussing international conservation, working out basic concepts of maximum and optimum sustainable yields, and formulating a scientific basis for commercial harvesting. The IWC demonstrates that even an informal commission whose decisions are accepted voluntarily by nations can function as a powerful force for conservation.

As with the management of fish populations, each marine mammal population has been treated as if it were isolated, had a constant supply of food, and was subject only to the effects of human harvesting. That is, it has been assumed that its growth followed the logistic curve. It is now realized that management policies for marine mammals must be expanded to include ecosystem concepts and the understanding that populations interact in complex ways.

The goal of marine mammal management, however, is different from that of fisheries management. *International agreements about whaling are*

TABLE 13.4 Recent Whale Population Estimates

Area/Region	Blue	Bowhead	Humpback	Fin	Right	Gray	Sperm	Sei	Minke
Western Arctic		8,000							
Pacific						23,109			
California–Mexico	2,134		482						
Central Pacific			1,407						
North Pacific				UNK	UNK		UNK	UNK	
California–Washington			597	575			756		526
East Tropical Pacific	1,400						22,700		
Atlantic									
Western North	308		5,543	2,700	295		337	870	2,650
Northern Gulf of Mexico							213		
Antarctica	610		15,000	18,000		UNK	290,000	17,000	700,000
IWC Estimates for South 30°S									

Source: National Marine Fisheries Service 1994 (unpublished); Barlow, J. et al. 1995. *U.S. Pacific Marine Mammal Stock Assessments.* NOAA Technical Memorandum NMFS. U.S. Department of Commerce, NOAA, National Marine Fisheries Service; and, Small, R. J. and DeMaster, D. P. 1995. *Alaska Marine Mammal Stock Assessments 1995.* NOAA Technical Memorandum NMFS-AFSC-57. U.S. Department of Commerce, NOAA, National Marine Fisheries Service. Antarctic estimates are from the International Whaling Commission.

concerned with preventing extinction and maintaining large population sizes rather than with maximizing production. For this reason, the Marine Mammal Protection Act enacted by the United States in 1972 has as its goal an **optimum sustainable population (OSP)** rather than a maximum or optimum sustainable yield. An OSP means the largest population that can be sustained indefinitely without deleterious effects on the ability of the population or its ecosystem to continue to support that same level.

Dolphins and Other Small Cetaceans

Among the many species of small "whales," or cetaceans, are dolphins and porpoises, more than 40 species of which have been hunted commercially or have been killed inadvertently as a part of other fishing efforts.[46] A classic case is the inadvertent catch of the spinner, spotted, and common dolphins of the eastern Pacific. Because these carnivorous, fish-eating mammals often feed with yellowfin tuna, a major commercial fish, more than 100,000 of these marine mammals have been netted and killed inadvertently in recent years.[47]

The U.S. Marine Mammal Commission and commercial fishermen have cooperated in seeking methods to reduce dolphin mortality. Research conducted on dolphin behavior helped in the design of new netting procedures that would be less likely to trap these mammals. When these new procedures were adopted by the fishermen, the trapping of dolphins was greatly reduced.

The attempt to reduce dolphin mortality illustrates cooperation among fishermen, conservationists, and government agencies and the role of scientific research in the management of renewable resources.

SUMMARY

- Conservation and management are important for biological resources that have, or have had, commercial value, such as forests, fisheries, and marine mammals, as well as for landscapes that have recreational, aesthetic, and moral interest, including parks and wilderness.

- In the past, these resources were managed as if their dynamics were simple and they could be considered in isolation from other factors, including the local environment, other species, and other parts of the landscape or seascape. Today, it is widely recognized that all biological resources must be viewed in a broad context, taking into account ecosystems and landscapes.

- Forests are among civilization's most important renewable resources. Forest management seeks a sustainable harvest that allows multiple

ENVIRONMENTAL ISSUE

How Does Fragmentation of Tropical Forests Contribute to Habitat Destruction?

Although tropical rain forests occupy only about 7% of the land area of the world, they provide the habitat for at least one-half the world's species of plants and animals. Approximately 100 million people live in rain forests or depend on them for their livelihood. Tropical plants provide useful products, such as chocolate, nuts, fruits, gums, coffee, wood, rubber, pesticides, fibers, and dyes. Drugs used to treat high blood pressure, Hodgkin's disease, leukemia, multiple sclerosis, and Parkinson's disease have been made from tropical plants, and medical scientists believe many more are yet to be discovered.

Most of the interest in tropical rain forests has focused on Brazil, whose forests are believed to have more species than any other habitat. Estimates of destruction in the Brazilian rain forest range from 6% to 12%. Numerous studies have shown, however, that deforested area alone does not adequately measure habitat destruction. For example, David Skole and Compton Tucker estimated that the clearing of 6% of the Brazilian rain forest by 1988 had affected 15% of the forest. To estimate the affected area, Skole and Tucker assumed an edge effect of 1 km, that is, that an area of forest bordering a deforested area would be affected for a distance of 1 km (0.6 miles) into the forest. The more fragmented a forest is, the more edges there are and the greater the impact on the living organisms.

Field studies have shown that the edge effect depends on the species, characteristics of the land surrounding the forest fragment, and distance between fragments. For example, a forest surrounded by farmland is more deeply affected than one surrounded by abandoned land in which secondary growth presents a more gradual transition between forest and deforested areas. Some insects, small mammals, and many birds find only 80 m (262.5 ft) a barrier to movement from one fragment to another, whereas one small marsupial has been found to cross distances of 250 m (820.2 ft). Corridors between

forested areas also help to offset the negative effects of deforestation on plants and animals of the forest.

Critical Thinking Questions

1. Assuming an edge effect of 1 km, what is the approximate area affected by deforestation of 100 km² (38.6 mi²) in the form of a square, 10 km (6.2 miles) on each side? If the 100-km² area were in the form of 10 rectangles, each 10 km long and 1 km wide, separated from each other by a distance of 5 km (3.1 miles), how large an area would be affected?

2. What environmental factors at the edge of a fragment would differ from those in the center? How might the differences affect plants and animals at the edge?

3. Why is a simple rule of thumb such as assuming an edge effect of 1 km too simplistic as a model of the effects of deforestation? Given that it is too simplistic, what advantages are there in using the rule?

4. Forest fragments are sometimes compared to islands. What are some ways in which this is an appropriate comparison and some ways in which it is not?

References

Bierregaard, R. O., Jr., Lovejoy, T. E., Kapos, V., Angusto dos Santos, A., and Hutchings, R. W. 1992. "The Biological Dynamics of Tropical Rainforest Fragments," *BioScience* 42(11): 859–866.

Holloway, M. 1993. "Sustaining the Amazon," *Scientific American* 269(1): 90–99.

Hosmer, E. 1987 (June). "Paradise Lost: The Ravaged Rainforest," *Multinational Monitor*, pp. 6–8, 13.

Skole, D., and Tucker, C. 1993. "Tropical Deforestation and Habitat Fragmentation in the Amazon: Satellite Data from 1978 to 1988," *Science* 260 (5116): 1905–1910.

Stevens, W. K. 1993 (June 29). "Loss of Species Is Worse Than Thought in Brazil's Amazon," *New York Times*, p. C4.

use of the landscape. The best practices vary with the kind of forest, the terrain and soil, and the climate.

• The continued use of firewood as an important fuel in developing nations is a major threat to forests, given the rapid population growth of those areas. It is doubtful whether developing

nations can implement a successful management program in time to prevent serious damage to their forests and land and severe effects on their people.

• The management of parks for biological conservation is a relatively new idea that began in the nineteenth century. The manager of a park must be concerned with its shape and size. Parks that are too small or the wrong shape or that have too small a population of the species for which the park was established may not be able to sustain the species.

- A special extreme in conservation of natural areas is the management of wilderness. In the United States, the 1964 Wilderness Act provided a legal basis for this conservation. The management of wilderness involves a basic contradiction: Trying to preserve an area undisturbed by people requires the interference of the manager to limit user access and to maintain the wilderness in a natural state.

- Most of the oceans are vast deserts with low biological productivity because the upper waters in these regions lack one or more of the chemical elements necessary for life. Most biological productivity in the oceans occurs near continents, in areas of upwelling currents, on continental shelves, or near river estuaries.

- Fisheries have been managed to achieve a maximum sustainable yield, as defined by the logistic growth curve. This practice has frequently resulted in overfishing.

- Marine mammals, heavily exploited in the past, are the focus of many conservation efforts. The goal of these efforts is an optimum sustainable population.

REEXAMINING THEMES AND ISSUES

Human Population: Forests provide essential resources for civilization; fisheries provide major sources of protein. As human population grows, there will be greater and greater demand for these resources but less area within which forests can grow and more pressure on fisheries.

Sustainability: This is the key to conservation and management of wild living resources. However, there are few cases where it has been demonstrated that sustainable harvests have been achieved and fewer (if any) where it has been shown that there is sustainability of the ecosystem within which the harvested resource lives. Sustainability must be the central focus for these resources in the future.

Global Perspective: Forests and fisheries are global resources. Decline in the availability of forest or fishery products in one region affects the rate of harvest and economic value of these products in other regions.

Urban World: We tend to think of cities as separated from living resources, but urban parks are important to making cities pleasant and livable. Also urban parks, if properly designed, help to conserve wild living resources.

Values and Knowledge: Throughout history people have shown a duality in the values they attach to forests. On the one hand, forests have provided products essential to civilization, large areas of majestic trees have been an inspiration for many people, and forested environments have been an important focus for biological conservation, especially since the nineteenth century. On the other hand, the dark interior of forests have often been viewed negatively. Rather than maintain forests, people have tended to clear and eliminate them. We have yet to come to terms with the various values we attach to forests.

KEY TERMS

clear-cutting *267*

closed canopy forests *259*

controlled burning *269*

edge effect *272*

even-aged stands *265*

migration corridor *272*

multiple use *267*

new forestry *263*

old growth *265*

open woodlands *259*

optimum sustainable population (OSP) *282*

pelagic whaling *281*

representative natural areas *270*

rotation time *265*

savannas *259*

second growth *265*

seed tree cutting *267*

selective cutting *267*

shelterwood cutting *267*

silviculture *265*

site quality *265*

strip cutting *267*

sustainable forest ecosystem *261*

sustainable timber harvest *261*

uneven-aged stands *265*

wilderness *273*

STUDY QUESTIONS

1. What environmental conflicts might arise when a forest is managed for the multiple uses of (a) commercial timber, (b) wildlife conservation, and (c) a watershed for a reservoir? In what ways would management for one use benefit another?

2. What are arguments for and against the statement "Clear-cutting is natural and necessary for forest management"?

3. Can a wilderness park be managed to supply water to a city? Discuss your answer.

4. A park is being planned in rugged mountains where there is high rainfall. What are environmental considerations if the purpose of the park is (a) to preserve a rare species of deer or (b) for recreation, including hiking and hunting?

5. What are the environmental effects of decreasing the rotation time in forests from an average of 60 years to 10 years? Compare these effects for (a) a woodland in a dry climate on a sandy soil and (b) a rain forest.

6. Based on the material discussed in this chapter, what is a definition of the term *old growth forest?* This term currently is used with a variety of meanings.

7. Develop a plan for the manatee and the sea otter (see the discussion of the sea otter and the community effect in Chapter 6) to attain an optimum sustainable population. How do these plans differ? What is the chance of success of each?

8. Salmon spawn in freshwater streams that flow through forested areas. How might (a) clear-cutting and (b) strip cutting of forests affect the reproduction of salmon?

FURTHER READING

Kimmins, J. P. 1996. *Forest Ecology: A Foundation for Sustainable Management.* New York: Prentice Hall. A new textbook that applies the recent developments in ecology to the practical problems of managing forests.

Kohm, K. A., and Franklin, J. F. 1996. *Creating a Forestry for the 21st Century: The Science of Ecosystem Management.* Washington, D.C.: Island Press. Applies the new understanding of ecological processes, including the role of disturbances, to management of forests. Views forest management within an economic and societal context.

Perlin, J. 1989. *A Forest Journey: The Role of Wood in the Development of Civilization.* New York: W. W. Norton. A fascinating presentation of the story of wood, forests, and people, covering a period of 5000 years and spanning five continents.

Runte, A. 1979. *National Parks: The American Experience.* Lincoln, NE: University of Nebraska Press. Select people, events, and legislation are interpreted to provide a history of the concept of national parks.

Sharma, N., ed., 1992. *Contemporary Issues in Forest Management: Policy Implications.* Washington, D.C.: The World Bank. A series of articles that analyze the reasons for conserving and sustaining forests.

Spurr, S. H., and Barnes, B. V. 1992. *Forest Ecology.* New York: Ronald. This most recent edition of a classic textbook provides an ecological basis for the management of forestland.

Yallee, S. L., Phillips, A. F., Frentz, I. C., Hardy, P. W., Maleki, S. M., and Thorpe, B. E. 1996. *Ecosystem Management in the United States: An Assessment of Current Experience.* Washington, D.C.: Island Press. A guide to where ecosystem management is being attempted. It reviews 105 representative sites and discusses the status of management efforts.

INTERNET RESOURCES

Consortium for International Earth Science Information Network (DIESIN): *http://www.ciesin.org/TG/LU/deforest.html*—This link within CIESIN includes sources on information on tropical deforestation.

Marine Mammal Center, The *http://www.tmmc.org/*—Find out more about efforts to protect marine mammals.

Marine Mammal Page: *http://www.phoenix.net/~mark/*—A great starting point for internet information on marine mammals, including the Marine Mammal Protection Act.

National Marine Fisheries Service: *http://kingfish.spp.nmfs.gov/*—Find information on the conservation and management of our marine resources.

National Park Service: *http://www.nps.gov/*—Contains information on our parks, park management, and other facts for students and educators.

Rainforest Action Network: *http://www.ran.org/ran/*—This RAN site tracks tropical deforestation and efforts to save tropical rainforests.

United Nations Food and Agriculture Organization: *http://www.fao.org/waicent/forestry.htm*—The FAO's forestry section contains information, programs, and documents of forestry, deforestation, and sustainable forestry.

World Resources Institute: *http://www.wri.org/*—Contains copies of publications produced by the WRI covering world resources including forests.

CHAPTER 14

ENVIRONMENTAL HEALTH AND TOXICOLOGY

Lead in the soil (legacy of using lead in gasoline and still concentrated in soils of open areas where children are likely to play), paint (in older homes) and general environment of inner cities such as New York shown here remains a hazard to children today.

CASE STUDY

Is Lead in the Urban Environment Contributing to Antisocial (Criminal) Behavior?

Lead is one of the most common toxic (harmful or poisonous) metals in our intercity environments. It is found to some extent in all parts of the urban environment (e.g., air, soil, and older pipes and paint) and in all biological systems, including people. There is no apparent biologic need for lead, but it is sufficiently concentrated in the blood and bones of children living in inner cities to cause health and behavior problems. In some populations over 20% of the children have levels of lead concentrated in their blood above that believed safe.[1] Lead affects nearly every system of the body. Acute lead toxicity may be characterized by a variety of symptoms, including anemia, mental retardation, palsy, coma, seizures, apathy, uncoordination, subtle loss of recently acquired skills, and bizarre behavior. Lead toxicity is particularly a problem for young children who tend to be exposed to higher

concentrations in some urban areas and apparently are more susceptible to lead poisoning than are adults. Following exposure to lead and having acute toxic response, some children manifest aggressive, difficult to manage behavior.[1–3]

The occurrence of lead toxicity or lead poisoning has cultural, political, and sociological implications. Over 2000 years ago the Roman Empire produced and used tremendous amounts of lead for a period of several hundred years. Production rates were as high as 55,000 metric tons per year. Romans had a wide variety of uses for lead, including pots in which grapes were crushed and processed into a syrup for making wine, cups and goblets from which the wine was drunk, as a base for cosmetics and medicines, and, finally, for the wealthy class of people who had running water in their homes, lead was used to make the pipes that carried the water. It has been argued by some historians that gradual lead poisoning among the upper class in Rome was partly responsible for Rome's eventual fall. Certainly it may have resulted in widespread stillbirths, deformities, and brain damage! Studies analyzing the lead content of bones of ancient Romans tend to support this hypothesis.[4] More recently, the occurrence of lead has been studied from ice cores from Greenland glaciers. Measurements of the concentration of lead from these cores show that during the period from approximately 500 B.C. to A.D. 300 lead concentrations in the glacial ice are about four times the norm. This suggests that the mining and smelting of lead during the Roman Empire added lead into the atmosphere that eventually settled out in the glaciers of Greenland.[5]

This discussion suggests that lead toxicity has been a problem for a long time. What is an emerging, interesting, and potentially significant hypothesis is that children in urban areas with lead concentrations in their bodies below the toxic levels are associated with an increased potential for antisocial, delinquent behavior. This is a testable hypothesis. (See Chapter 2 for a discussion of hypotheses.) If the hypothesis is correct, then some of our recent increases in urban crime may be traced to environmental pollution! A recent study[1] measured the relative amount of lead in bones and compared this with data concerning the behavior of children over a 4-year period from 7 years old to 11 years old. The study concluded that those children with above-average concentration of lead in their bones were associated with an increased risk for attention deficit, aggressive behavior, and delinquency. The study took into account factors such as maternal intelligence, socioeconomic status, and quality of child rearing.

What this study suggests is that environmental pollution of our urban areas may be in part responsible for some of our social problems. At another level the data and analysis may be used to argue that exposure of children to lead is in fact a preventable occurrence, and it should be included when considering the multivariant factors that contribute to antisocial and delinquent behavior.[1] The important environmental significance here is that although we have often considered that social unrest, antisocial behavior, and crime are related to social economic factors, they may in fact be due in part to environmental factors related to pollution of our inner city environments.

LEARNING OBJECTIVES

Serious environmental health problems and disease may arise from toxic elements in our water, air, soil, or even the rocks on which we build our homes. After reading this chapter, you should understand:

- How the terms *pollution, contamination, toxic, carcinogen, synergism,* and *biomagnification* are used in environmental health.
- What the classification and characteristics of major groups of pollutants in environmental toxicology are.
- Why there is controversy and concern about synthetic organic compounds such as dioxin.
- Why we should or should not be concerned with exposure to electromagnetic fields emitted by electric transmission lines and appliances such as computers.
- What the dose–response concept is and how it relates to LD–50, ED–50, ecological gradients, and tolerance.
- How the process of biomagnification works and why it is important in toxicology.
- Why the threshold effects of environmental toxins are important in affecting how we can control such toxins in the environment.
- How the process of risk assessment in toxicology is organized and what some related concerns and problem areas are.

14.1 SOME BASICS

Disease is often due to an imbalance resulting from a poor adjustment between the individual and the environment.[6] However, disease seldom has a one-cause, one-effect relationship with the environment. Rather, the incidence of a disease depends on several factors, including physical environment, biological environment, and life-style. Linkages between these factors are often related to other factors, such as local customs and the degree of industrialization. The more primitive societies that live directly off the local environment are usually plagued by a different variety of environmental health problems than is our urban society. Industrial societies have nearly eliminated environmental diseases, such as cholera, dysentery, and typhoid. However, we are more likely to suffer from chronic and acute diseases, such as respiratory problems and cancer.

People are often surprised to learn that the water we drink, the air we breathe, the soil in which we grow crops for food, and the rocks on which we build our homes and places of work may affect our chances of contracting serious environmental health problems and diseases. However, direct causative relationships between the environment and disease are difficult to establish.

On the other hand, the same environmental factors that contribute to disease—soil, rocks, water, and air—can also influence our chances of living a longer, more productive life. Surprisingly, many people still believe that soil, water, or air in a so-called natural, pure, or virgin state must be good

and that if human activities have changed or modified them, they have become contaminated, polluted, and therefore bad.[7] This belief is by no means the entire story; many natural processes, including dust storms, floods, and volcanic processes, can introduce materials harmful to humans and other living things into the soil, water, and air. A tragic example occurred on the night of August 21, 1986, when there was a massive natural release of carbon dioxide from Lake Nyos in Cameroon, Africa. The carbon dioxide was probably initially released from volcanic vents at the bottom of the lake and accumulated there with time as carbon dioxide saturated bottom water. Pressure of the overlying lake water normally keeps the dissolved gas at the bottom of the lake. The bottom water evidently became unstable and moved upward, and the gas was quickly released from the water to the atmosphere. The disturbance that caused the release was probably a subaqueous landslide or small earthquake that brought water from the bottom of the lake quickly to the surface. The gas is heavier than air and settled in nearby villages, killing many animals and more than 1700 people by asphyxiation (Figure 14.1). Assuming that carbon dioxide is still being released at the bottom of the lake from volcanic vents, the surface release is likely to occur again in the future, probably within about 20 years.[8]

What do we mean when we use the terms *pollution, contamination, toxic,* and *carcinogen*? A polluted environment is one that is impure, dirty, or otherwise unclean. The term **pollution** refers to the occurrence of an unwanted change in the environment caused by the introduction of harmful materials or production of harmful conditions (i.e., heat, cold, sound). **Contamination** has a meaning similar to that of pollution and implies making something unfit for a particular use through the introduction of undesirable materials, for example, contamination of water by hazardous waste. The term **toxic** refers to materials (pollutants) that are poisonous to people and other living things. **Toxicology** is the science that studies poisons (or toxins) and their effects as well as clinical, industrial, economic, or legal problems associated with toxic materials. A **carcinogen** is a particular kind of toxin, one that causes cancer. Because cancer is often fatal, carcinogens are among the most feared and regulated toxins in our society.

An important concept in considering pollution problems is **synergism,** which refers to the interaction of different substances where the combined effect is greater than the sum of the effects of the separate substances. For example, sulfur dioxide and particulates are both air pollutants. Either one taken separately may cause adverse health effects, but when they combine, as when sulfur dioxide is adsorbed onto small particles, they may be inhaled

FIGURE 14.1 In 1986, Lake Nyos in Cameroon, Africa, released carbon dioxide that moved down the slopes of the hills to settle in low places, asphyxiating animals and people.

_A_ CLOSER LOOK 14.1

SUDBURY SMELTERS: A POINT SOURCE

A famous example of a point source of pollution is provided by the smelters for the refining of nickel and copper ores at Sudbury, Ontario. Sudbury contains one of the world's major nickel and copper ore deposits. Within a small area, just a few kilometers, there are a number of mines, smelters, and refineries. The ores contain a high percentage of sulfur, and the smelter stacks emitted large amounts of sulfur dioxide (SO_2) as well as particulates containing nickel, copper, and other toxic metals. During its peak output in the 1960s, the complex was the largest single source of sulfur dioxide emissions in North America. In 1969, regulation with the objective to improve local air quality forced reduction of emissions or increased dispersal of emissions (by construction of tall smoke stacks). As a result, local concentration of sulfur dioxide was reduced by more than 50% after 1972. However, attempts to minimize the pollution problem in the immediate vicinity of the smelting operation by increasing smokestack height only spread the problem even more widely. In order to better control emissions from Sudbury, the Ontario government set standards to reduce emissions to less than 365,000 tons per year by 1994. This is about 14% of historical emissions of about 2,560,000 tons per year. The goal was achieved by reduced production from the smelters and treating the emissions to reduce pollution.[9]

Nickel has been found to contaminate soils 50 km (about 31 miles) from the stacks. As a result of decades of acid rain from the sulfur dioxide (produced when SO_2 combines with water) and the deposition of particulates containing heavy metals, the forests that once surrounded Sudbury were devastated. An area of approximately 250 km² (96 mi²) is nearly devoid of vegetation, and damage to forests in the region is visible over an area of approximately 3500 km² (1350 mi²; Figure 14.2). Secondary effects, in addition to loss of vegetation, include soil erosion and drastic changes in soil chemistry resulting from the influx of the heavy metals (acid rain is discussed in detail in Chapter 22).

Reductions in emissions from Sudbury have allowed areas denuded (polluted) by metal-contaminated soil to slowly begin to recover. Species of trees once eradicated from some areas have begun to grow again. Lakes that were sterilized due to acid precipitation in the area are rebounding and now support populations of plankton and fish.[9]

FIGURE 14.2 Lake St Charles, Sudbury, Ontario. Note high stacks in the background (smelters) and lack of vegetation in the foreground, resulting from air pollution (acid and heavy-metal deposition).

deeper than sulfur dioxide alone and cause greater damage to lungs. Another aspect of synergistic effects is that the body may be more sensitive to a toxin if it is simultaneously subjected to other toxins and stresses.

Pollutants are commonly introduced into the environment by way of **point sources,** such as smokestacks (A Closer Look 14.1, "Sudbury Smelters: A Point Source"), pipes, streams flowing into lakes or the ocean (Figure 14.3), or accidental spills. **Area sources** are more diffused sources, example include urban runoff or automobile exhaust. Area sources are difficult to isolate and correct because the problem is often widely dispersed over a region, as is agricultural runoff that contains pesticides (Chapter 19).

Measuring the Amount of Pollution

How we report the amount or concentration of a pollutant present in the environment varies widely, depending on the substance involved. For example, the amount of treated wastewater entering Santa Monica Bay in the Los Angeles area is reported in millions of gallons per day, whereas emission of nitrogen and sulfur oxides into the air is reported in millions of tons per year. However, we often use smaller measures, because even small amounts of toxic materials (such as dioxin) may cause environmental problems when released in soil or water.

The concentration of pollutants or toxins in the environment is often measured in units such as parts per million (ppm) or parts per billion (ppb). It is important to keep in mind that the concentration in ppm or ppb may be by either volume, mass, or weight. In many toxicology studies, the units used are milligrams of toxin per kilogram of body mass (1 mg/kg is equal to 1 ppm) reported as mass weight. Concentration may also be recorded as a percent [i.e., 100 ppm (100 mg/kg) is equal to 0.01%].

When dealing with water pollution, units of concentration for a pollutant may be milligrams per liter (mg/L) or micrograms per liter (μg/L). (A milligram is one-thousandth of a gram, a microgram is one-millionth of a gram.) For water pollutants that do not cause significant change in density of water (1 g/cm³), a concentration of pollution of 1 mg/L is approximately equivalent to 1 ppm. Air pollutants are commonly measured in units such as micrograms of pollutant per cubic meter of air (μg/m³).

Units such as ppm, ppb, or μg/m³ reflect very small concentrations. For example, if you were to use 3 g (one-tenth of an ounce) of salt to season

popcorn, to have that salt at a concentration of 1 ppm by weight you would have to pop approximately 3 metric tons of kernels!

14.2 CATEGORIES OF POLLUTANTS

A partial classification of pollutants follows. Some examples are discussed further in other parts of the book.

Infectious Agents

Environmentally transmitted infectious diseases, spread from interactions of individuals with food, water, air, or soil, constitute one of the oldest health problems that humans face. Specifically, diseases that can be controlled by manipulating the environment—such as improving sanitation and treating water—are classified as environmental health concerns. Although there is great concern about the toxins and carcinogens produced in industrial society today, the greatest mortality in developing countries is caused by environmentally transmitted infectious disease. In the United States, there are thousands of cases of water-borne illness and food poisoning each year. These diseases can be spread by mosquitoes or fleas, contact with contaminated food, water, or soil, or transmitted through ventilation systems in buildings. Some specific examples of environmentally transmitted infectious diseases are Legionellosis (Legionnaires' disease—often occurs where aerosolized water has been contaminated, i.e., air conditioning systems, shower heads), Giardiasis (protozoan infection of the small intestine spread via food, water, or person-to-person), Salmonella (food poisoning—bacterial infection spread via water or food), malaria (protozoan infection transmitted by mosquitoes), Lyme Borreliosis (Lyme disease—transmitted by ticks) and cryptosporidosis (protozoan infection transmitted via water or person-to-person, see Chapter 20).[25]

Toxic Heavy Metals

Among the major **heavy metals** that pose hazards are mercury, lead, cadmium, nickel, gold, platinum, silver, bismuth, arsenic, selenium, vanadium, chromium, and thallium. Each of these elements has uses in our modern industrial society, and each is also a by-product of the mining, refining, and use of other elements. Heavy elements often have direct physiological toxic effects. Some are stored or incorporated in living tissue, sometimes permanently. The content of heavy metals in our bodies is referred to as the *body burden*.

Figure 14.3 This southern California urban stream flows into the Pacific Ocean at a coastal park. The stream water often carries high counts of fecal coliform bacteria (see Chapter 20), as a result the stream is a point source of pollution for the beach which is sometimes closed to swimming following runoff events.

Notice that lead (for which we apparently have no biological need) has an average body burden of about twice that of the others combined. This problem reflects our heavy use of this toxic metal (see the chapter-opening Case Study).

Toxic Pathways

Many possible pathways (Figure 14.4) allow elements, initially released from rocks, eventually to become concentrated in humans (see chapter 4). These pathways often involve what is known as **biomagnification** (the accumulation or increase in concentration of a substance in living tissue as it moves through a foodweb, also known as bioaccumulation). For example, cadmium, which influences the risk of heart disease, may enter the environment via ash from burning coal. The cadmium in coal exists in very low concentrations (less than 0.05 ppm). Following the burning of coal in a power plant, the ash is collected in a solid form and disposed of in landfill-like basins, which, when full, are covered with soil and revegetated. The cadmium may become incorporated into the plants with concentrations of three to five times that found in the ash. As the cadmium moves through the food chain, it becomes more and more concentrated until, by the time it reaches people and other carnivores, it has increased in concentration by a factor of approximately 50–60. Many other potentially harmful materials, such as mercury, lead, and selenium, are also concentrated through the processes of biomagnification.

Mercury, thallium, and lead are very toxic to humans. They have long been mined and used, and their toxic properties are well known. Mercury, for example, is the Mad Hatter element (see A Closer Look 14.2, "Mercury and Minamata, Japan"). At one time, mercury was used in making felt hats stiff; because mercury damages the brain, hatters were known to act peculiarly in Victorian England. Thus, the Mad Hatter in Lewis Carroll's *Alice in Wonderland* had real antecedents in history.

Organic Compounds

Organic compounds are compounds of carbon produced either by living organisms or synthetically. It is difficult to generalize about the environmental and health effects of artificially produced organic compounds because there are so many of them, they have so many uses, and they can produce so many different kinds of effects (see A Closer Look 14.3, "Dioxin: The Big Unknown"). Synthetic organic compounds are used primarily in industrial processes, pest control, pharmaceuticals, food additives, and other consumer products. The production of synthetic organic compounds has grown rapidly in the twentieth century. A computer registry of chemical compounds maintained by the American Chemical Society has more than 4 million entries, and there are currently about 6000 additions per week.

Radiation

The topic of *nuclear radiation* is introduced here as a category of pollution. It is discussed in detail in Chapter 18 in conjunction with nuclear energy. We are concerned about nuclear radiation because exposure is linked to serious health problems, including cancer (see Chapter 23 for a discussion of radon gas).

Thermal Pollution

Thermal pollution, also called heat pollution, occurs when heat released into water or air produces undesirable effects. Heat pollution can occur as a sudden, acute event or as a long-term, chronic release. Sudden heat releases may result from natural events, such as brush or forest fires and volcanoes, or from human-induced events, such as fire storms. The major sources of chronic heat pollution are electric power plants that produce electricity in steam generators.

FIGURE 14.4 Potential pathways for toxic elements to become concentrated in plants, animals, and humans. (*Source:* J. J. Connor, 1990, *U.S. Geological Survey Circular* 1033, p. 61.)

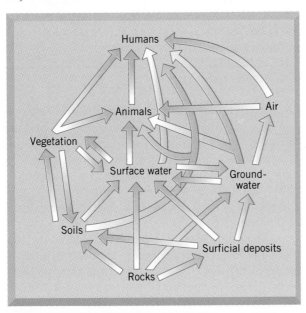

CLOSER LOOK 14.2

MERCURY AND MINAMATA, JAPAN

In the Japanese coastal town of Minamata, on the island of Kyushu, a strange illness began to occur in the middle of the twentieth century. It was first recognized in birds that lost their coordination and fell to the ground or flew into buildings and in cats that went mad, running in circles and foaming at the mouth.[10] The affliction, known by local fishermen as the "disease of the dancing cats," subsequently affected people, particularly families of fishermen. The first symptoms were subtle, fatigue, irritability, headaches, numbness in arms and legs, and difficulty in swallowing. More severe symptoms involved the sensory organs; vision was blurred and the visual field was restricted. Afflicted people became hard of hearing and lost muscular coordination. Some complained of a metallic taste in their mouths; their gums became inflamed, and they suffered from diarrhea. Eventually, 43 people died and 111 were severely disabled; in addition, 19 babies were born with congenital defects. Those affected lived in a small area, and much of the protein in their diet came from fish from the Minamata Bay.

A vinylchloride factory on the bay used mercury in an inorganic form in its production processes. The mercury was released in water effluent that flowed into the bay. Mercury forms few organic compounds, and it was believed that the mercury, although poisonous, would not get into food chains. But microbial action converted inorganic mercury into methyl mercury, an organic compound that turned out to be a much more harmful form. Inorganic mercury does not pass through cell membranes readily. However, methyl mercury readily passes through cell membranes and is transported by the red blood cells throughout the body; it enters and damages brain cells.[11] Fish absorb methyl mercury from water 100 times faster than they absorb inorganic mercury. This was not known before the epidemic in Japan. Once absorbed, methyl mercury is retained two to five times longer than is inorganic mercury.

Harmful effects of methyl mercury are dependent on a variety of factors, including the amount and route of intake, the duration of exposure, and the species affected. The effects of the mercury are delayed from 3 weeks to 2 months from the time of ingestion. If mercury intake ceases, some symptoms may gradually disappear, but others are difficult to reverse.[11]

The mercury episode at Minamata illustrates four major factors that must be considered in evaluating and treating environmental pollutants. First, *individuals vary in their response to exposure to the same dose, or amount, of a pollutant.* Not everyone in Minamata responded in the same way; there was variation even among those most heavily exposed. Because we cannot predict exactly how any single individual will respond, we need to find a way to state an average expected response of individuals in a population. Second, *pollutants may have a* **threshold,** that is, a level below which the effects are not observable and above which the effects become apparent. Symptoms appeared in individuals with concentrations of 500 ppb of mercury in their bodies; no measurable symptoms appeared in individuals with significantly lower concentrations. Third, *some effects are reversible.* Some people recovered when the mercury-filled seafood was eliminated from their diet. Fourth, *the chemical form of a pollutant, its activity, and potential to cause health problems are changed markedly by ecological and biological processes.* In the case of mercury, its chemical form and concentration changed as the mercury moved through the food webs.

The release of large amounts of heated water into a river changes the average water temperature and concentration of dissolved oxygen (warm water holds less oxygen than cooler water), changing the river's species composition (see the discussion of eutrophication in Chapter 20). Every species has both a range of temperature within which it can survive and an optimal temperature. For some species of fish, the range is small, and a slight change in water temperature can be a problem. Lake fish move away when the water temperature rises more than 1.5°C above normal; river fish can withstand a rise of about 3°C.[20] Heating the water can change the natural conditions and disturb the river ecosystem; fish spawning cycles may be disturbed, and the fish may have a heightened susceptibility to disease. If the warmer water causes physical stress in fish, they may be easier for predators to catch, and, finally, the warmer water can change the type and abundance of food available for fish at various times of the year.

There are several solutions to chronic thermal discharge into bodies of water. The heat can be released into the air by cooling towers (Figure 14.7), or the heated water can be temporarily stored in artificial lagoons until it is cooled to normal tempera-

A CLOSER LOOK 14.3

DIOXIN: THE BIG UNKNOWN

A colorless crystal made up of oxygen, hydrogen, carbon, and chlorine, dioxin is classified as an organic compound because it contains carbon. About 75 types of dioxin are known; they are distinguished from one another by the arrangement and number of chlorine atoms in the molecule.[12] Dioxin is not normally manufactured intentionally but is a by-product resulting from chemical reactions (burning of compounds containing chlorine) in the production of herbicides.[13] The dioxin in Agent Orange, used during the Vietnam War, is considered the most toxic component of the herbicides used in the war zone. A class-action lawsuit was filed for military personnel exposed to Agent Orange and the dioxin it contains. The lawsuit was settled before trial for $180 million. A fund was established to compensate about 250,000 veterans and their families, who have suffered from a variety of medical problems starting in the 1970s.

Although dioxin is known to be extremely toxic to mammals, its actions on the human body are not well known. What is known is that sufficient exposure to dioxin (usually from meat or milk containing the chemical) produces a skin condition (a form of acne) that may be accompanied by loss of weight, liver disorders, and nerve damage.[14]

Studies of animals exposed to dioxin suggest that some fish, birds, and other animals are sensitive to even small amounts of dioxin; as a result, it can cause widespread environmental damage to wildlife, including birth defects and death. However, the concentration necessary to cause human health hazards is still controversial. Studies suggest that workers exposed to high concentrations of dioxin for longer than 1 year have approximately a 50% increased risk of dying of cancer.[15] The

Environmental Protection Agency (EPA) has set an acceptable intake of dioxin at 0.006 pg (1 pg = 10^{-12} g; see appendix for prefixes and multiplication factors) per kilogram of body weight per day. This level is deemed by some scientists to be too low. They argue that the acceptable intake ought to be 100–1000 times (two to three orders of magnitude) higher, or approximately 1–10 pg/day.[15] The EPA believes that current human body burden levels of dioxin are within an order of magnitude of causing health effects. However, some scientists assert that lack of data still precludes establishment of a specific threshold concentration of dioxin at which health hazards begin.[16] Because of these uncertainties toxicity of dioxin will remain unknown until further studies better delineate the potential hazard.

Dioxin is a stable, long-lived chemical that is accumulating in the environment. Analysis of sediments taken from the bottom of Lake Superior suggest that the rate of deposition of dioxin increased eightfold from 1940 to 1970. However, since

then rates have slowly declined.[17] As yet we have not been able to determine a safe, reliable, and economically feasible way to clean up areas contaminated by dioxin (Figure 14.5). Many old waste disposal sites are contaminated by dioxin; it may also be found in soil and streams several kilometers around the sites. The dioxin problem became well known in 1983 when Times Beach, Missouri, a river town just west of Saint Louis with a population of 2400, was evacuated and purchased for $36 million by the government. The evacuation and purchase occurred after the discovery that oil sprayed on the town's roads to control dust contained dioxin, and the entire area had been contaminated. Times Beach was labeled a dioxin ghost town. Today the buildings have been bulldozed and all that is left is a grass and woody area enclosed by a barbed-wire-topped chain-link fence. The dioxin ghost town has disappeared. The evacuation has since been viewed by some scientists (including the person who ordered the evacuation) as an over-

FIGURE 14.5 Soil samples from Times Beach, Missouri, thought to be contaminated by dioxin.

reaction by the government to a perceived dioxin hazard.

In 1992 the EPA convened a panel of scientists to reevaluate the risk to the environment and to people from exposure to dioxin. A draft report released in 1994 concludes that:

- Dioxin is a probable human carcinogen, and exposure is known to cause disruption of the endocrine, immune, and reproductive systems. However, it is not a widespread significant cancer threat to people at ordinary (very low) levels of exposure.

- The cancer risk to workers handling chemicals containing dioxin and likely to be exposed to high levels of the chemical may be higher than previously thought.

- Very small levels of dioxin released into the environment can cause serious damage to wildlife sensitive to the chemical, potentially causing significant damage to ecosystems.[18]

The 20-year controversy concerning the toxicity of dioxin is not over.[19] Some environmental scientists argue that the regulation of dioxin must be tougher, whereas the industries producing the chemical argue that the dangers of exposure are overestimated. The EPA final report, when published, is not likely to put these issues to rest.

tures. Some attempts have been made to use the heated water to grow organisms of commercial value that require warmer water temperatures. Waste heat from a power plant may also be captured and used for a variety of purposes, such as warming buildings (see Chapter 15 for a discussion of cogeneration).

Particulates

Particulates are small particles of dust released into the atmosphere by many natural processes and human activities. Modern farming adds considerable amounts of particulates to the atmosphere, as do dust storms, fires (Figure 14.6), and volcanic eruptions.[11] The 1991 eruptions of Mount Pinatubo in the Philippines may be the largest of the century, explosively delivering huge amounts of volcanic ash, sulfur dioxide, and other volcanic material and gases into the atmosphere to elevations up to 30 km (18.6 miles). Such eruptions have significant impact on the global environment affecting, in conjunction with other processes, global climate and stratospheric ozone depletion (see Chapters 21 and 24). Many chemical toxins, such as heavy metals, enter the biosphere as particulates. Sometimes nontoxic particulates join forces with toxic substances, creating a dual (synergetic) threat (see discussion of particulates in Chapter 22).

Asbestos

As it is most commonly defined, **asbestos** refers to small, elongated mineral fragments (fibers) produced from amphibole (a group of minerals) and serpentine (a type of rock). Industrial use of as-

FIGURE 14.6 Fires in Indonesia in 1997 caused serious air pollution problems and people here are purchasing surgical masks in an attempt to breathe cleaner air.

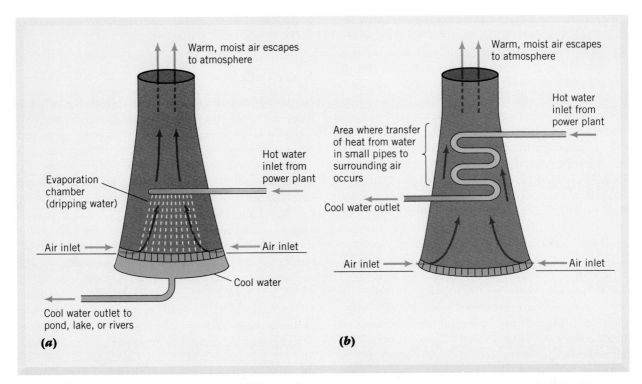

Warm, moist air escapes to atmosphere

Evaporation chamber (dripping water)

Hot water inlet from power plant

Air inlet →

← Air inlet

Cool water

← Cool water outlet to pond, lake, or rivers

(a)

Warm, moist air escapes to atmosphere

Hot water inlet from power plant

Area where transfer of heat from water in small pipes to surrounding air occurs

← Cool water outlet

Air inlet →

← Air inlet

(b)

FIGURE 14.7 Two types of cooling towers. (*a*) Wet cooling tower. Air circulates through tower; hot water drips down and evaporates, cooling the water. (*b*) Dry cooling tower. Heat from the water is transferred directly to the air, which rises and escapes the tower. (*c*) Cooling towers emitting steam at Didcot power plant, Oxfordshire, England. Red and white lines are vehicle lights resulting from long exposure time (photograph taken at dusk).

(c)

bestos has contributed to safety and convenience in fire prevention and has provided protection from the overheating of materials. Asbestos is used as insulation for a variety of purposes. Unfortunately, however, excessive contact with asbestos by some industrial workers has also led to asbestosis (a lung disease caused by the inhalation of asbestos) and to deaths from asbestosis and cancer. The hazard related to certain types of asbestos under certain conditions is thought to be so serious that extraordinary steps have been taken to reduce the presence of asbestos or to ban it outright. The expensive process of asbestos removal from old buildings (particularly schools) in the United States is one of those steps.

Recent experiments with animals have demonstrated that a wide variety of elongated mineral fragments can cause tumors to develop if the fibers are embedded in lung tissue. As a result, there is difficulty in defining just what asbestos is, because the crushing of almost any rock type produces some small, elongated mineral crystals.[21]

In addition, all types of asbestos are not equally hazardous. The most commonly utilized form of asbestos in the United States is white asbestos, which comes from the mineral chrysolite. It has been used as an insulation material around pipes and for brake linings of automobiles and other vehicles. It is thought that approximately 95% of the asbestos that is now in place in the United States is of the chrysolite type. Most of this asbestos was mined in Canada, and environmental health studies of Canadian miners show that exposure to chrysolite asbestos is not particularly harmful. Studies involving another type of asbestos, known as crocidolite asbestos (blue asbestos), suggest that exposure to this mineral can be very hazardous, and it evidently does cause lung disease. Several other types of asbestos have also been shown to be harmful.[21]

There has been a great deal of fear associated with nonoccupational exposure to chrysolite asbestos in the United States, and the health risk has been overstated. Tremendous amounts of money have been expended to remove chrysolite asbestos from homes, schools, public buildings, and other sites in spite of the fact that there has been no asbestos-related disease recorded among those exposed to chrysolite in nonoccupational circumstances. It is now thought that much of the removal was unnecessary because the existing chrysolite asbestos may not have presented a hazard. Obviously additional research into the types of asbestos and their health risks is necessary to better understand the potential problem and to outline strategies to minimize potential health problems.

Electromagnetic Fields

Electromagnetic fields (EMFs) are part of our everyday urban life. Electric motors, electric transmission lines for utilities, and appliances, such as toasters, electric blankets, and computers, all produce electromagnetic fields. There is currently a tremendous controversy over whether these fields produce a health risk. Early on, investigators did not believe that these small fields, which drop off quickly with distance from the source, could pose any health hazard. However, several recent health studies have gained the support of a number of scientists, and the controversy is not likely to subside quickly.

Magnetic and electric fields were not thought to be harmful because the EMFs that most people come into contact with are relatively weak. For example, the magnetic fields generated by power transmission lines or by a computer terminal are normally only about 1% of Earth's magnetic field; directly below power lines the electric field induced in the body is about what the body naturally produces within cells. Nevertheless, several studies have concluded that children exposed to EMFs from power lines have an increased risk of contracting leukemia, lymphomas, and nervous system cancers.[22] These children are about one and a half to three times as likely to develop cancer as children with very low exposure to EMFs. Unfortunately, all the studies are flawed to one degree or another, reflecting how difficult it is to perform this type of research. What most investigators now state is that additional work

is necessary to evaluate links between cancer and EMFs.

Noise Pollution

Noise pollution is unwanted sound. Sound is a form of energy that travels as waves (sound waves). We hear sound because our ears respond to sound waves through sympathetic vibrations of the eardrum stimulated by alternating compressions and rarefactions of the air molecules in the vicinity of the ear. The sensation of loudness is related to the intensity of the energy carried by the sound waves and is measured in units of decibels (dB). The threshold for human hearing is 0 dB; the average interior of a home is about 45 dB; automobiles, 70 dB; and jet aircraft taking off, about 120 dB (see Table 14.1). A 10-fold increase in the strength of a particular sound adds 10 dB units on the scale. An increase of 100 times would add 20 units.[10] The decibel scale is logarithmic; it increases exponentially as a power of 10. For example, 50 dB is 10 times louder than 40 dB and 100 times louder than 30 dB.

Environmental effects of noise depend not only on the total energy but also on the sound's pitch, frequency, time pattern, and length of exposure. Very loud noises (more than 140 dB) cause pain, and high levels can cause permanent hearing loss. Human ears can take sound up to about 60 dB without damage or hearing loss. Any sound above 80 dB is potentially dangerous. The noise of a lawn mower or motorcycle will begin to damage hearing after about 8 hours of exposure. In recent years,

TABLE 14.1 Examples of Sound Levels

Sound Source	Intensity of Sound (dB)	Human Perception
Threshold of hearing	0	
Rustling of leaf	10	Very quiet
Faint whisper	20	Very quiet
Average home	45	Quiet
Light traffic (30 m away)	55	Quiet
Normal conversation	65	Quiet
Chain saw (15 m away)	80	Moderately loud
Jet aircraft flyover at 300 m	100	Very loud
Rock music concert	110	Very loud
Thunderclap (close)	120	Uncomfortably loud
Jet aircraft takeoff at 100 m	125	Uncomfortably loud
	140	Threshold of pain
Rocket engine (close)	180	Traumatic injury

there has been concern for teenagers (and older people, for that matter) who have suffered some permanent loss of hearing following extended exposure to amplified rock music (110 dB). At a noise level of 110 dB, damage to hearing can occur after an exposure time of only half an hour. Loud sounds at the workplace are another hazard. Levels of noise that are below the hearing-loss level may still interfere with human communication, cause annoyance, and certainly be unpleasant. For example, noise in the range of 50–60 dB is sufficient to interfere with sleep and produce a feeling of fatigue on awakening.

Voluntary Exposures

Voluntary exposure to toxins and potentially harmful chemicals is sometimes referred to as exposure to personal pollutants. The most common of these are tobacco, alcohol, and other drugs. Use and abuse of these substances has led to a variety of human ills ranging from suffering and/or dying from chronic disease, criminal activity such as reckless driving or manslaughter, street crime, loss of careers, and straining of human relations at all levels.

Scientific evidence that use of tobacco, in all of its forms, is both habit forming and dangerous to human health has been demonstrated in many studies. Tobacco contains a variety of components that are either toxic, carcinogenic, radioactive, or additive. It has been estimated that 30% of all cancers in the United States are tied to smoking-related disorders. The American Cancer Society has stated that cigarette smoking is responsible for approximately 80% of lung cancers, and that second-hand smoke is also a hazard, as are so-called tobacco products such as chewing tobacco. Although the number of people who smoke in the United States as a percentage of adults has decreased in recent years, there is fear that more young people are becoming addicted to cigarettes and cigars.

A very large portion of people in our society use alcohol at social gatherings and celebrations. Approximately 70% of all American adults drink some alcohol, and moderate use of alcohol is legal and accepted in our society. However, when abused, alcohol is a very serious problem, and approximately one-half of all deaths in automobile accidents are related to alcohol use by drivers. Furthermore, a significant portion of violent crime and other criminal activity is committed by people under the influence of alcohol. Some people believe that alcohol is the most abused drug in our society today.

A variety of illegal drugs are commonly used in the United States and other parts of the world. These drugs have a variety of effects on their users, but the end result is often the degradation of the mind and/or body. Illegal drugs, because there is often little quality control concerning their strength, composition, and other chemical characteristics, are particularly dangerous to people. Of particular concern in recent years has been the development of synthetic (designer) drugs that are addictive and capable of causing significant health problems to people who use them.

14.3 GENERAL EFFECTS OF POLLUTANTS

Although there are many kinds of pollutants, each with its own method of action and environmental pathways, certain features are characteristic of most environmental toxins.

Almost every part of the human body is affected by one pollutant or another (Figure 14.8a). For example, lead and mercury affect the brain, arsenic the skin, carbon monoxide the lungs, and cadmium the heart. Similarly, the spectrum of effects of pollutants on wildlife has been documented for many organs and aspects of the life cycle (see Table 14.2 and Figure 14.8b). The list of potential toxins

TABLE 14.2 Effects of Pollutants on Wildlife

Effect on Population	*Pollutants*
Changes in abundance	Arsenic, asbestos, cadmium, fluoride, hydrogen sulfide, nitrogen oxides, particulates, sulfur oxides, vanadium
Changes in distribution	Fluoride, particulates, sulfur oxides
Changes in birth rates	Arsenic, lead, photochemicals, oxidants
Changes in death rates	Arsenic, asbestos, beryllium, boron, cadmium, fluoride, hydrogen sulfide, lead, particulates, photochemicals, oxidants, selenium, sulfur oxides
Changes in growth rate	Borium, fluoride, hydrochloric acid, lead, nitrogen oxides, sulfur oxides

Source: J. R. Newman, *Effects of Air Emissions on Wildlife*, U.S. Fish and Wildlife Service, 1980. Biological Services Program, National Power Plant Team, FWS/OBS-80/40, U.S. Fish and Wildlife Service, Washington, DC.

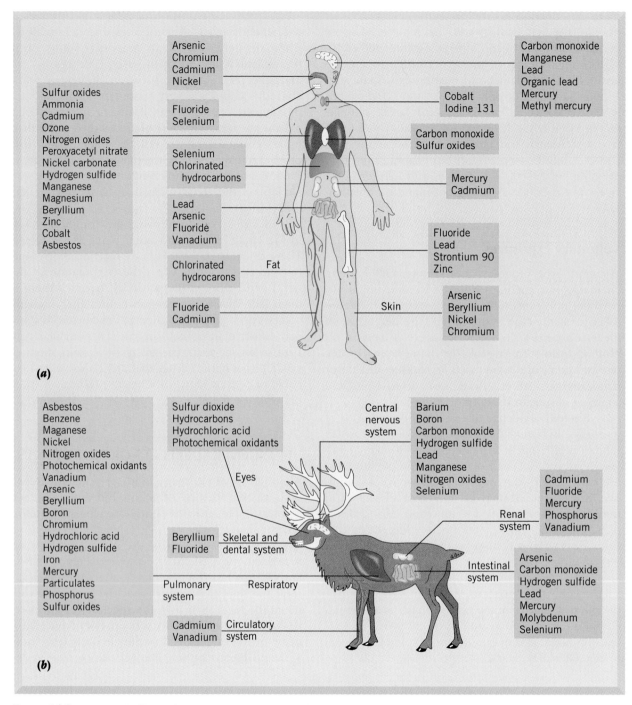

FIGURE 14.8 (*a*) Site of effects of some major pollutants in human beings. (*b*) Known sites of effects of some major pollutants in wildlife. [*Source:* (*a*) G. L. Waldbott, 1978, *Health Effects of Environmental Pollutants*, 2nd ed., Mosby, St. Louis. Copyright 1978 by C. V. Mosby; (*b*) J. R. Newman, 1980, *Effects of Air Emissions on Wildlife Resources*, U.S. Fish and Wildlife Services Program, National Power Plant Team, FWS/OBS–80/40, U.S. Fish and Wildlife Service, Washington, DC.]

and affected body sites for humans and other animals in Figure 14.8 is somewhat misleading. For example, chlorinated hydrocarbons, such as dioxin, are stored in the fat cells of animals, but they cause damage not only to fat cells but to the entire organism through disease, damaged skin, or birth defects.

Similarly, a toxin that affects the brain, such as mercury, causes a wide variety of problems and symptoms, as illustrated in the Minamata, Japan, example (A Closer Look 14.2). The real value of Figure 14.8 is in helping us to understand the adverse effects of excess exposure to chemicals.

Concept of Dose and Response

Five centuries ago Paracelsus wrote that "everything is poisonous, yet nothing is poisonous." By this he meant essentially that a substance in too great amounts can be dangerous, yet anything in extremely small amounts can be relatively harmless. Every chemical element has a spectrum of possible effects on a particular organism. For example, selenium is required in small amounts by living things, but may be toxic to or cause cancer in cattle and wildlife when it is in high concentrations in soil. Copper, chromium, and manganese are other exam-

ples of chemical elements required in small amounts but toxic in higher amounts.

It was recognized many years ago that the effect of a certain chemical on an individual depends on the dose or concentration of the toxic factor. This concept is termed **dose response.** Dose dependency can be represented by a generalized dose–response curve (Figure 14.9).

When various concentrations of a chemical or toxic factor present in a biological system are plotted against the effects on the organism, three things are apparent. First, although relatively large concentrations are toxic, injurious, and even lethal (points D, E, and F in Figure 14.9), trace concentrations may be beneficial or necessary for life (points A and B). The dose–response curve forms a plateau of optimal concentration and maximum benefit between two points (B and C). If there is a threshold concentration, then harmful effects to life do not start at zero concentration but at some concentration greater than zero. Effects of a toxin vary with concentrations less than those at point A (where there is too little of the factor) and greater than those at point D (where there is too much of the factor).

Points A, B, C, D, E, and F in Figure 14.9 represent significant concentrations. Unfortunately, points E and F are known only for a few substances for a few organisms, including people, and the very important point, D, is all but unknown. The width of the maximum benefit plateau (between points B and C) for a particular form of life depends on the organism's particular physiological state. The levels that are beneficial, harmful, or lethal may differ widely quantitatively and qualitatively and are difficult to characterize.

Fluorine illustrates the general dose–response curve. An important trace element, fluorine forms fluoride compounds that prevent tooth decay and promote development of a more perfect bone structure. Fluorine is fairly abundant in rocks and soil, and industrial activity and application of fertilizers have, on a limited basis, contributed locally to an increase in the concentration of fluorine in soils.

Relationships between the concentration of fluoride (in a compound of fluorine, such as sodium fluoride, NaF) and health indicate a specific dose–response curve, as shown in Figure 14.10. The plateau for optimal fluoride concentration (point B to point C) to reduce dental caries (cavities) is from about 1 ppm to just less than 5 ppm. Fluoride levels greater than 1.5 ppm do not significantly decrease tooth decay but do increase the occurrence of mottling (discoloration) of teeth. Concentrations of 4 to 6 ppm reduce the prevalence of osteoporosis, a disease characterized by loss of bone mass. Toxic effects begin between 6 and 7 ppm (point D in Figure 14.10).

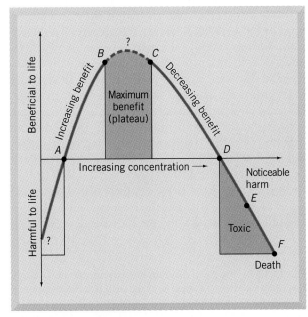

FIGURE 14.9 General dose-response curve.

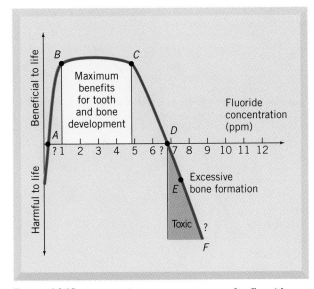

FIGURE 14.10 General dose-response curve for fluoride.

Dose–Response Curve (LD-50, ED-50, and TD-50)

The effect of an environmental pollutant is often described by a slightly different dose-response curve, which is actually a more detailed view of the generalized curve from point *D* to point *F* in Figure 14.9. A toxic dose–response curve shows a negative response, either death or injury, plotted against the increasing intensity of exposure to a pollutant. The upper limit of this curve represents 100% of the population affected.

Individuals differ in their response to environmental toxins, just as they do in response to all environmental conditions. It is difficult to estimate the exact dose that will cause a response in a particular individual. For this reason, it is practical to consider the average effect of a toxin on a population. The most commonly used measure, linked to the dose–response curve, is the amount of exposure required for 50% of a population or observed subjects to show a response. The dose at which 50% of the population dies is called the median lethal dose 50, or **LD-50.** The LD-50 is the first and most crude approximation of a chemical's toxicity. It is a gruesome index that does not adequately convey the sophistication of modern toxicology and is of little use in setting a standard for toxicity. However, the LD-50 determination is required for new synthetic chemicals as a way of estimating their toxic potential. Table 14.3 lists, as examples, LD-50 values in rodents for selected chemicals.

The **ED-50** (effective dose 50%) is the dose that causes an effect in 50% of the population or observed subjects. For example, the ED-50 of aspirin would be the dose that relieves headaches in 50% of the people.[23] The **TD-50** (toxic dose 50%) is defined as the dose that is toxic to 50% of the population. TD-50 is often used to indicate responses such as reduced enzyme activity, decreased reproductive success, or onset of specific symptoms, such as loss of hearing, nausea, or slurred speech. For a particular toxic agent there may be a whole family of dose-response curves as illustrated in Figure 14.11. Which dose is of interest depends on what is being evaluated. For example, for insecticides we may wish to know the dose that will kill 100% of the insects exposed and therefore the LD-95 (the dose that kills 95% of the insects) may be the minimum acceptable level. However, when considering human health and exposure to particular toxins, we are concerned with very small death rates at the LD-0 which is the maximum dose that does not cause any deaths.[23] That is, for potentially toxic compounds such as insecticides, which may form a residue on food or food additives, we want to ensure that the expected levels of exposure will have no known toxic effects. From an environmental perspective this is important due to concerns about cancers developing from exposure to some toxic agents. For example, lawsuits have been initiated in California regarding spraying malathion to eradicate the Mediterranean fruit flies (Medflies). Serious political and social consequences would have occurred had even 1% of the population become ill.[23]

For drugs used to treat a particular disease, the safety of the drug is of paramount importance. Thus, in addition to knowing what the therapeutic value (ED-

TABLE 14.3 Approximate LD-50 Values (for rodents) for Selected Agents	
Agent	*LD_{50}(mg/kg)**
Sodium chloride (table salt)	4,000
Ferrous sulfate (to treat anemia)	1,520
2,4-D (a weed killer)	368
DDT (an insecticide)	135
Caffeine (in coffee)	127
Nicotine (in tobacco)	24
Strychnine sulfate (still used to kill certain pests)	3
Botulinum toxin (in spoiled food)	0.00001

*Milligrams per kilogram of body mass (termed mass weight, although it really isn't a weight) administered by mouth to rodents.

Source: H. B. Schiefer, D. C. Irvine, and S. C. Buzik, 1997, *Understanding Toxicology,* New York, CRC Press.

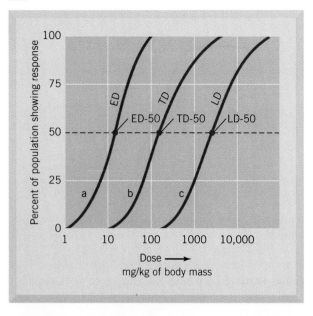

FIGURE 14.11 Idealized diagram illustrating a family of dose-response curves for a specific drug: (*a*) ED (effective dose); (*b*) TD (toxic dose) and (*c*) LD (lethal dose). Notice there is overlap for some parts of the curves. For example, at the ED-50, a few percent of the people exposed to that dose will suffer a toxic response, but none will die. At the TD-50 dose about 1% of the people experiencing that dose will die.

50) is, it is also important to know the relative safety of the drug. For example, there may be an overlap between the theraputic dose and the lethal dose (see Figure 14.11). That is, the dose that causes a positive theraputic response in some individuals might kill others. A quantitative measure of the relative safety of a particular drug is the theraputic index, defined as the ratio of the LD-50 to the ED-50. The greater the therapeutic index, the safer the drug is thought to be.[24]

Threshold Effects

Recall from A Closer Look 14.2 Mercury and Minamata, Japan that a threshold is a level below which no effect occurs and above which effects begin to occur. If a threshold exists (Figure 14.12), then any level in the environment below that threshold would be safe. Alternatively, it is possible that even the smallest amount of toxin has some negative effect. Whether there is a threshold effect in nature for environmental toxins is an important environmental issue. For example, the U.S. Federal Clean Water Act originally stated there could be zero discharge of a pollutant into a body of water, a statement that implies there is no such thing as a threshold effect, since no level of toxin is to be legally permitted. However, it is unrealistic to believe zero discharge of a water pollutant can be realized or to believe that we can reduce to zero the concentration of chemicals shown to be carcinogenic in animals on which we rely for food

A problem in evaluating thresholds for toxic pollutants is that it is difficult to account for syner-

gistic effects. Little is known about if or how thresholds might change if an organism is exposed to more than one toxin at the same time or to a combination of toxins and other chemicals, some of which are beneficial. Real-world exposures of people to chemicals is complex, and we are only beginning to understand and conduct research on the possible interactions and consequences of multiple exposure.

Ecological Gradients

Species differences in dose–response effects often reflect ecological conditions. The kinds of vegetation that persist nearest to a toxic source are often small plants with relatively short lifetimes (grasses, sedges, and weedy species usually regarded as pests) that are adapted to harsh and highly variable environments. Farther from the toxic source, trees may be able to survive. Changes in vegetation with distance, the **ecological gradient,** from toxins are similar to the kinds of changes one finds in walking down a mountain from the summit, the harshest environment, to the valley floor, the most benign environment.

Ecological gradients may be found around smelters and other industrial plants that discharge pollutants to the atmosphere from smokestacks. For example, these patterns can be seen in the area around smelters of Sudbury, Ontario, discussed earlier in this chapter (A Closer Look 14.1). Near the smelters, an area that was once forest is now a patchwork of bare rock and soil occupied by small plants.

Tolerance

An increase in resistance or ability to withstand stress that results from exposure to a pollutant or harmful condition is referred to as **tolerance.** The determination of a dose–response curve may be made more difficult because of the development of tolerance in individuals and in populations. Tolerance can develop for some pollutants in some populations, but not for all pollutants in all populations.

Tolerance may result from behavioral, physiological, or genetic adaptation. For example, mice learn to avoid traps. *Physiological tolerance* means the body of an individual adjusts to tolerate a higher level of pollutant. For example, people become tolerant of low oxygen at high altitudes because the body increases red blood cells. As another example, studies at the University of California Environmental Stress Laboratory involved exposing students to relatively high levels of ozone (triatomic oxygen, O_3), an air pollutant often present in large cities (Chapter 22). Students at first experienced symptoms that included irritation of eyes, throat and shortness of breath. However, after a few days their bodies adapted to the ozone, and students reported that

FIGURE 14.12 In this hypothetical toxic dose-response curve, toxin *A* has no threshold; even the smallest amount has some measurable effect on the population. The TD-50 for toxin *A* is the dose required to produce a response in 50% of the population.

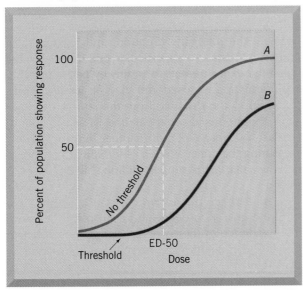

they were no longer breathing ozone-contaminated air, even though the concentration of O_3 stayed the same. This phenomenon explains why some people who regularly breathe polluted air report that they do not notice the pollution. Of course, this does not mean that the ozone is not damaging to the system; it is, especially to people with existing respiratory problems. There are many mechanisms for physiological tolerance, including *detoxification*, which is the conversion of the toxic chemical to a nontoxic form, and the internal transport of the toxin to a part of the body, such as fat cells, where it is not harmful.

Genetic tolerance, or adaptation, results when individuals who are more resistant survive an exposure to a toxin and have more offspring than do other individuals. These resistant individuals prevail in later generations. Adaptation has been observed among some insect pests following exposure to some chemical pesticides. For example, certain strains of malaria-causing mosquitoes are now resistant to DDT, which had been used for years to kill these mosquitoes (see the discussion of agriculture and pest control in Chapter 11), and some organisms that cause deadly infectious diseases have become resistant to common antibiotic drugs, such as penicillin, previously prescribed by physicians to cure those diseases.

Acute and Chronic Effects

Pollutants can have acute and chronic effects. An *acute effect* is one that occurs soon after exposure, usually to large amounts of a pollutant. A *chronic effect* takes place over a long period, often as a result of exposure to low levels of a pollutant. For example, a person exposed to a high dose of radiation at one time may be killed (an acute effect), but that same total dose, received slowly in small amounts over an entire lifetime, may instead cause mutations, lead to cancer, or affect the person's offspring.

14.4 RISK ASSESSMENT

Risk assessment may be defined as the process of determining potential adverse environmental health effects to people following exposure to pollutants and other toxic materials (recall the discussion of measurements and methods of science in Chapter 2). Such an assessment generally includes four steps[25]:

1. *Identification of the hazard.* This process consists of testing materials to determine whether exposure is likely to cause environmental health problems. One method used is the investigation of populations of people who have been previously exposed. For example, to understand the toxicity of radiation produced from radon gas, researchers studied workers in uranium mines. Another method is to perform experiments to test the effects of materials on animals. This method has drawn increasing criticism from groups of people who believe such experiments are unethical. Another approach is to try to understand how a particular material works at the molecular level on cells. For example, research has been done to determine how dioxin interacts with living cells to produce an adverse response. After quantifying the response, scientists can develop mathematical models to predict or estimate dioxin's risk.[15] This is a relatively new approach that might also be applicable to other potential carcinogens that work at the cellular level.

2. *Dose–response assessment.* This step involves identifying relationships between the dose of a pollutant or toxin and the nature and extent of adverse environmental health effects in a particular population. Sometimes, however, the studies involve fairly high doses administered to an animal population, with proportional effects inferred for lower doses on human populations. (This process of inferring results is called *extrapolation*, the projection of known data into an area not directly studied.) This step is often difficult and the results controversial. Part of the controversy arises because the dose that results in a particular response may be very small and subject to measurement errors. There may also be arguments over whether thresholds are present or absent. Finally, the assessment may rely on probability and statistical analysis. Although statistics may show strong relationships between, for example, a dose of a toxic substance and a response (symptom or disease), they may not establish that the substance tested *caused* the observed response. On the other hand, statistical results from experiments or observations are accepted as evidence to support an argument.

3. *Exposure assessment.* Attempts are made to estimate the intensity, duration, and frequency of human exposure to a particular pollutant or toxin. Often the exposure hazard is directly proportional to the population exposed and may be inversely proportional to the distance from the source of exposure. This step is also difficult and its results controversial because of difficulties in accurately measuring the concentration of a toxin in concentrations as small as parts per million, billion, or even trillion.

4. *Risk characterization.* During this final step, the goal is to delineate health risk in terms of the magnitude of the potential environmental health problem that might result from exposure to a particular pollutant or toxin. To do this it is necessary to identify the hazard, complete the dose–response as-

Are We Getting Too Much Fluoride?

Fluoride, at concentrations of 0.7 mg/L or higher, is in the drinking water of approximately 132 million Americans. More than 40 years after the introduction of fluoridation in the United States, there is still controversy concerning its health benefits and risks. Two questions dominate the debate: (1) Are parts of the population receiving more fluoride than necessary? (2) Does the amount of fluoride pose a health risk? What can we conclude if we follow the steps associated with risk assessment?

1. *Identification of the hazard.* Fluoride is harmful at concentrations above 6 ppm; overexposure affects teeth and bones, doses of 5 mg/kg of body weight can cause poisoning. At low doses, white areas may appear on the teeth, at moderate doses brown discolorations may occur; at high doses destruction of tooth enamel may occur. Mild or moderate effects on teeth are found in 1.3% of children in the United States.

In combination with other factors, fluoride may cause bones to become opaque and brittle, leading to abnormal bone growth and crippling. However, this is rare in the United States. Absorption of fluoride in the intestines and the percentage retained in the body vary with the person's age, the type of fluoride compound, the compound's pH (a measure of acidity or alkalinity), and the presence of other ions. Of three studies to determine whether fluoride causes mutations, two were negative and one was positive.

A 1993 study by the National Research Council found no evidence of fluoride (at concentrations to which humans are exposed) causing kidney toxicity, gastrointestinal problems, reproductive problems or genotoxicity. However, further studies need to be conducted to determine the role of fluoride in bone fractures and cancer occurence. In 1990, the National Toxicology Program concluded that there was no evidence relating fluoride to birth defects or to gastrointestinal, genital-urinary, or respiratory problems and that evidence was insufficient to determine whether fluoride had any effects on reproduction or development. The National Cancer Institute reviewed data on cancer from 1973 to 1987 and found no links between cancer and fluoride. Exposing animals to moderate and high doses of fluoride produced equivocal results: four were negative, but one found bone cancers in 5% of the rats, at least 10 times the normal rate.

2. *Dose-response assessment.* Although high doses of fluoride are associated with adverse, even toxic effects, too little fluoride intake is associated with increased incidence of dental caries (cavities). When first introduced, fluoridated water decreased the incidence of caries in children by 60%. Research in the 1980s, however, showed only a 20% to 40% difference. One interpretation of the change is that fluoride is now used in beverages, food, dental products, and dietary supplements, so that the control groups are actually receiving some fluoride.

The dose–response curve in Figure 14.10 shows the relationship between concentration of fluoride and human health.

3. *Exposure assessment.* Groundwater contains trace amounts to 25 ppm of fluoride. Optimal concentration for preventing dental caries and avoiding adverse effects has been set at 0.7–1.2 ppm for drinking water. People may ingest additional fluoride in food or in substances to which fluoride has been added, for example, toothpaste. Estimates of daily consumption of fluoride appear in the following table.

Critical Thinking Questions

1. Given the information provided in the accompanying table, what is the risk from fluoride to both children and adults in the general population?

Daily Fluoride Consumption

Source of Fluoride	Amount in Source	Total Intake (mg)		Concentration in Body[a] (mg/kg)	
		Child	Adult	Child	Adult
Drinking water	1.20 mg/L	1.68	2.40	0.084	0.034
Toothpaste, mouthwash	0.145–0.66 mg	0.66	0.145	0.033	0.002
Diet	0.2–0.4 mg	0.20	0.40	0.010	0.006
Total intake	—	2.54	2.95	0.127	0.042
Safe intake	—	0–1 year: 0.1–1, 1–2 years: 0.5–1.5	1.5–4.0		

[a]Assumes that a child weighs 20 kg and drinks 1.4 L/day; adult weighs 70 kg and drinks 2.0 L/day, and absorption in both cases is 100%.

2. Fluoridated toothpastes are not required to display information on the concentration of fluoride, but they average 1000 mg/L. Some dental experts have recommended half this amount for children's toothpaste. What would your recommendation be and why?

3. People living in areas in which groundwater levels exceed the recommended 1 ppm can remove fluoride, at considerable expense, or look for other sources of water. In an area with a fluoride concentration varying from 3 to 8 ppm, what recommendation would you make?

4. What recommendations would you make for the U.S. government concerning fluoride policy and research?

5. According to B. Walker, codirector of the Center for Epidemiologic Research and dean of public health at the University of Oklahoma, quantitative risk assessments are not science in the classical sense. What do you think he means by this?

References

December 1992. "Fluoride for Kids: Too Much, Too Soon?" *American Health* 9(10): 15.

Centers for Disease Control. 1991. "Public Health Service Report on Fluoride Benefits and Risks," *Journal of the American Medical Association* 266(8): 1061–1067.

Hamilton, M. 1992. "Water Fluoridation: A Risk Assessment Perspective," *Journal of Environmental Health* 54(6): 27–32.

Marshall, E. 1990. "The Fluoride Debate: One More Time," *Science* 247: 276–277.

National Research Council, 1993. *Health effects of ingested fluoride*. Washington, D. C. National Academy Press.

Walker, B. W., Jr. 1992. "Perspectives on Quantitative Risk Assessment," *Journal of Environmental Health* 55(1): 15–20.

sessment, and evaluate the exposure assessment as outlined above. This step involves all the uncertainties of the prior steps, and results are likely to be equivocal and controversial.

Risk assessment requires making scientific judgments and deriving actions to help minimize environmental health problems related to human exposure to pollutants and toxins. The process of *risk management* integrates the assessment of the risk with technical, legal, political, social, and economic issues.[15] The scientific arguments concerning the toxicity of a particular material are often open to debate. It is the role of the risk managers to make decisions based on available information. For example, there is debate concerning whether the risk from dioxin is linear, that is, effects start at minimum levels of exposure and gradually increase, or whether there is a threshold exposure beyond which environmental health problems arise (see A Closer Look 14.3, "Dioxin: The Big Unknown").[15,19] It is the task of people doing the risk management to make a judgment or decision about this issue and then set an acceptable intake level, sometimes termed the *action level*.

The risk presented by a particular suspected pollutant or toxin may also be evaluated in terms of the probability of an adverse environmental health effect and the consequences of exposure. Determining the probability also involves an assessment of the uncertainties in the experiments used to evaluate the risk to the human population.

SUMMARY

- Pollution produces an impure, dirty, or otherwise unclean state. Contamination means making something unfit for a particular use through the introduction of undesirable materials. The term *toxic* refers to materials that are poisonous to people and other living things; toxicology is the study of toxic materials, including the clinical, industrial, economic, and legal problems associated with them. A concept important in studying pollution problems is synergism, which refers to actions of different substances wherein the combined effect is greater than the sum of the effects of the separate substances.

- How we measure the amount of a particular pollutant introduced into the environment or the concentration of that pollutant varies widely depending on the substance involved. Common units for measuring the concentration of pollutants are parts per million (ppm) or parts per billion (ppb). Air pollutants are commonly measured in units such as micrograms of pollutant per cubic meter of air ($\mu g/m^3$).

- Categories of environmental pollutants include toxic chemical elements (particularly heavy metals), organic compounds, radiation, heat, particulates, electromagnetic fields, and noise.

- Organic compounds of carbon are produced by living organisms or synthetically. Artificially produced organic compounds may have physiological, genetic, or ecological effects when introduced into the environment. Some organic compounds are potentially more hazardous than others and some are more readily degraded in the environment than others. In particular, fat-soluble compounds are likely to undergo biomagnification, and others are extremely toxic even at very low concentrations. Some organic compounds that cause serious concern include pesticides and dioxin.

- The effect of certain chemicals or toxic materials on an individual depends on the dose of that material. With respect to pollutants and toxins, it is also important to determine potential tolerances of individuals as well as acute and chronic effects.
- Risk assessment involves hazard identification, assessment of dose response, assessment of exposure, and risk characterization.

REEXAMINING THEMES AND ISSUES

Human Population: Examination of environmental health and toxicology, as it relates to risk assessment, includes evaluation of exposure to people of natural and human-made toxins. As total population and, particularly, population density, increases, the probability of exposure of hazardous materials to a greater number of people increases. Finding acceptable sites for disposal of toxic chemicals also becomes more difficult as populations increase and encroach upon industrial areas and possible disposal sites.

Sustainability: Ensuring that future generations inherit a quality and relatively toxic-free environment remains a difficult problem. Sustainable development requires that our use of chemicals and other materials not damage the environment for us and future generations. This is one reason why there is research in toxicology. An objective is to understand the risk and take appropriate steps to minimize hazards associated with environmental exposure to toxic materials.

Global Perspective: Release of toxins into the environment may result in global patterns of contamination or pollution. This is particularly true when a toxin or contaminant enters the atmosphere, surface water, or oceans and becomes dispersed on a global basis. For example, some toxins, such as pesticides, herbicides, or heavy metals, may be emitted into the atmosphere in the midwestern United States and travel with the winds to be deposited on ice fields or lands in polar regions.

Urban World: Processes of industrial activity in urban areas concentrate potential toxic materials that may be inadvertently or deliberately released into the environment. Human exposure to a variety of toxins including materials such as lead, asbestos, particulates, organic chemicals, radiation, and noise are often greater in urban areas. Thus, we continue to develop regulations and standards in an attempt to minimize risk associated with these materials in urban areas.

Values and Knowledge: Because we value both human and nonhuman life we are interested in expanding our knowledge concerning risk of exposure of living things to chemicals and compounds in the natural and human-made environment. Unfortunately the state of our knowledge is fairly primitive, and the dose response for many chemicals is poorly understood.

KEY TERMS

STUDY QUESTIONS

1. What do you think of the hypothesis that social unrest and crime are in part caused by environmental pollution? How might the hypothesis be further tested? What are the social ramifications?

2. What kinds of life-forms would most likely survive in a highly polluted world? What would their general ecological characteristics be?

3. Some environmentalists argue that there is no such thing as a threshold for pollution effects. What is meant by this statement? How would you determine if it were true for a specific chemical on a specific species?

4. What is biomagnification, and why is it important in toxicology?

5. You are lost in Transylvania while trying to locate Dracula's castle. Your only clue is that the soil around the castle has an unusually high concentration of the heavy metal arsenic. You wander in a dense fog, able to see only the ground a few meters in front of you. What changes in vegetation warn you that you are nearing the castle?

6. Distinguish between acute and chronic effects of pollutants.

7. Design an experiment to test whether tomatoes or cucumbers are more sensitive to lead pollution.

8. Why is it difficult to establish standards for acceptable levels of pollution? In giving your answer, consider the physical, climatological, biological, social, and ethical reasons.

9. A new highway is built through a pine forest. Driving along the highway, you notice that the pines nearest the road have turned brown and are dying. You stop at a rest area and walk into the woods. One hundred meters away from the highway the trees seem undamaged. How could you make a crude dose–response curve from direct observations of the pine forest? What else would be necessary to devise a dose–response curve from direct observation of the pine forest? What else would be necessary to devise a dose–response curve that could be used in planning the route of another highway?

FURTHER READING

Amdur, M., Doull, J., and Klaasen, C. D., eds. 1991. *Casarett & Doulls Toxicology: The Basic Science of Poisons*, 4th ed. Tarrytown, N.Y.: Pergamon. Comprehensive and more advanced coverage of toxicology.

Carson, R. 1962. *Silent Spring*. Boston: Houghton Mifflin. A classic book on problems associated with toxins in the environment.

Gunn, J., ed. 1955. *Restoration and Recovery of an Industrial Region: Progress in Restoring the Smelter-Damaged Landscape Near Sudbury, Canada*. New York: Springer-Verlag. An informative review of past and current status of the area around Sudbury.

Schiefer, H. B., Irvine, D. G., and Buzik, S. C. 1997. *Understanding Toxicology: Chemicals, Their Benefits and Risks*. Boca Raton. CRC Press. This book provides a concise introduction to toxicology as it pertains to every day life. It includes information about pesticides, industrial chemicals, hazardous waste, and air pollution.

Travis, C. C., and Hattemer-Frey, H. A. 1991. "Human Exposure to Dioxin," *The Science of the Total Environment* 104: 97–127. An extensive technical review of dioxin accumulation and exposure.

INTERNET RESOURCES

Environmental Protection Agency, Office of Air and Radiation: *http://www.epa.gov.oar/oarhome.html*—Plenty of information on air pollution, air toxics, urban air quality and more.

Extension Toxicology Network (Extoxnet): *http://ace.orst.edu/info/extoxnet/*—A multi-university effort, this resource contains an abundance of both general information on toxicology, pesticides, and environmental health as well as specific information such as health effects of pesticides, current issues in toxicology, and environmental fates and consequences of toxins, and much more.

National Institute for Environmental Health Sciences: *http://www.niehs.nih.gov/*—Provides a wealth of information on environmental effects of pesticides, articles on environmental health topics, fact sheets on the toxicity of chemicals to humans, and details of research programs, as well as links to other sources of information.

National Toxicology Program: *http://ntp-server.niehs.nih.gov/*—Find information on chemical testing, chemical safety, and publications from the national program charged with determining chemical toxicities.

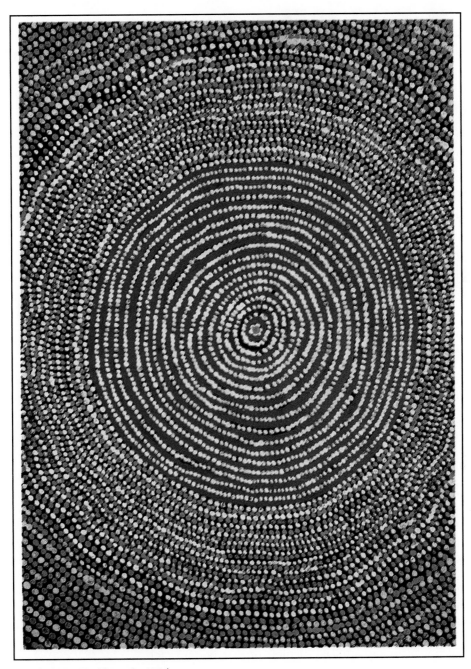

PART V

Untitled, Uta Uta Tjangala, 1984

Energy

Temple of Jupiter with Acropolis in background (Athens, Greece).

CHAPTER 15

ENERGY: SOME BASICS*

CASE STUDY

Energy Crises in Ancient Greece and Rome

It is fashionable today to think that an energy problem is something new. In fact, people have encountered energy problems for thousands of years, going back at least to the early Greek and Roman cultures. The climate in coastal areas of Greece 2500 years ago was characterized by warm summers and cool winters, much as it is today. However, at that time the Greeks did not have any artificial method of cooling their houses during the summer, and their small, charcoal-burning heaters (charcoal is produced from burning wood) were undoubtedly not very efficient in warming their homes during the winter. Wood was their primary source of energy, as it is today for half the world's people.

By the fifth century B.C. fuel shortages had become common, and much of the forest in many

*Written with assistance from Mel S. Manalis.

parts of Greece was depleted of firewood. As local supplies were depleted, it became necessary to import wood from farther and farther away. Olive groves became sources of fuel, which was made into charcoal for burning, reducing a valuable resource in the process. By the fourth century B.C. the city of Athens had banned the use of olive wood for fuel.

At about this time the Greeks began to build their houses facing the south, so that the low winter sun penetrated areas that needed heat and the high summer sun did not penetrate areas that needed to be cool. Recent excavations of ancient Greek cities suggest that large areas were planned so that individual homes could take maximum advantage of passive solar energy. The Greeks' use of solar energy in heating homes was a logical answer to their energy problem.[1]

The use of wood in ancient Rome is somewhat analogous to the use of oil and gas in the United States today. Wealthy Roman citizens about 2000 years ago had central heating in their large homes, burning as much as 125 kg (275 lb) of wood every hour. Not surprisingly, local wood supplies were exhausted quickly, and the Romans had to import wood from outlying areas, as had the Greeks. Eventually wood had to be imported from as far away as 1600 km (about 1000 miles).[1]

The Romans turned to solar energy for the same reasons that the Greeks sought it out but with much greater application and success. The Romans used glass windows to increase the effectiveness of solar heating, developed greenhouses to raise food during the winter, and oriented large public bathhouses (some of which accommodated up to 2000 people) to use passive solar energy (Figure 15.1). The Romans believed that sunlight in bathhouses was healthy, and it also saved greatly on fuel costs, making more wood available for heating the bath waters and steam rooms. The use of solar energy in ancient Rome was evidently quite widespread and resulted in the establishment of laws to protect a person's right to solar energy. In some areas it was illegal for one person to construct a building that shaded another's.[1]

FIGURE 15.1 Roman bathhouse (lower level) in the town of Bath, England.

This case history demonstrates that energy problems are not a new phenomenon. With this in mind we explore in this chapter some of the basic principles associated with what energy is, how much energy we consume, and how we might manage energy for the future.

LEARNING OBJECTIVES

Increased efficiency in the use of energy, conservation, and the expanded use of alternative energy sources are essential goals for effective future energy plans. After reading this chapter, you should understand:

- That energy is never created nor destroyed but is transformed from one kind to another.

- Why in all transformations energy tends to go from a more usable to a less usable form.

- What energy efficiency is and why it is always less than 100%.

- That people in industrialized countries consume a disproportionately large share of the world's total energy and how efficiency and conservation of energy can help make better use of global energy resources.

- Why some energy planners propose a hardpath approach to energy provision and others a soft path and why both of these have positive and negative points.

- Why it is an important goal to move toward global sustainable energy planning or integrated energy planning.

- What elements are needed to develop integrated energy planning.

15.1 OUTLOOK FOR ENERGY

The energy situation facing the United States and the world today is in some ways similar to that faced by the early Greeks and Romans. The use of wood in the United States peaked in the 1880s, when the use of coal became widespread. The use of coal in turn began to decline after 1920, when oil and gas started to become available. Today we are facing the peak of oil and gas utilization. Fossil fuel resources, which took millions of years to form, may be essentially exhausted in just a few hundred years. The decisions we make today will affect energy use for generations. Should we choose complex, centralized energy production methods, use simpler, widely dispersed energy production methods, or use a combination of the two? Which sources of energy should be emphasized? Which uses of energy should be emphasized for increased efficiency? How can we rely on current energy sources and provide for integrated energy planning? There are no easy answers.

The only thing certain about the energy picture for tomorrow is that it will involve living with a good deal of uncertainty when it comes to energy availability, energy cost, and the environmental effects of energy use. The sources and patterns of energy utilization will undoubtedly change. Supplies will probably continue to be regulated, and there will be a good deal of potential for disruption of supplies. In recent years the Organization of Petroleum Exporting Countries (OPEC) has attempted to control the flow of crude oil. In some instances this has been successful, but not nearly as successful as the exporting countries would have liked. In the 1980s and the mid 1990s there was a glut of petroleum on the world market; this situation may be short lived because petroleum is a nonrenewable energy source.[2]

The environmental effects of energy use at local, regional, and global levels will continue to be a topic of particular public concern. Serious shocks can be expected to occur in the future as they have in the past. Perhaps even a permanent shock will occur. For example, oil embargoes could cause significant economic impact in the United States and other countries or a war or revolution in a petroleum-producing country could cause exports of petroleum to drop significantly.

15.2 ENERGY BASICS

The concept of **energy** is somewhat abstract; you cannot see it or feel it, even though you have to pay for it.[3] To understand energy it is easiest to begin with the idea of a *force*. We all have had the experience of exerting force, of pushing or pulling. The strength of a force can be measured by how much it accelerates an object. Suppose that your car stalls while you are going up a hill and you get out to push it to the side of the road (Figure 15.2). You apply a force against gravity. If the brake is on, the brakes, tires, and bearings might heat up from friction. The longer the distance over which you exert the same force, the greater is the change in the car's speed, its position, or the heat in the brakes, tires, and bearings; in a physicist's terms, this is the work done. **Work** is exerting a force over a distance; that is, *work is the product of force times distance* (conversely, it is often said that energy is the ability to do

FIGURE 15.2 Some basic energy concepts, including potential energy, kinetic energy, and heat energy.

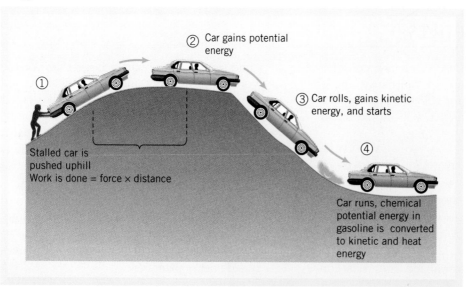

① Stalled car is pushed uphill
Work is done = force × distance

② Car gains potential energy

③ Car rolls, gains kinetic energy, and starts

④ Car runs, chemical potential energy in gasoline is converted to kinetic and heat energy

work). If you push hard, but the car does not move at all, you have exerted a force, but you have not done any work on the car (according to the definition), even if you feel tired and sweaty.[3]

In pushing your stalled car, you have done three things: changed its speed, moved it against gravity, and heated parts of it. These three things have something in common: They are all forms of energy. You have converted chemical energy in your body to the energy of motion of the car (kinetic energy), the gravitational (or potential) energy of the car, and heat energy. Energy can and often is converted or transformed from one kind to another, but the total energy is always conserved. The principle that energy may not be created or destroyed but is always conserved is known as the **first law of thermodynamics** (recall the discussion of energy conservation and laws introduced in Chapters 3 and 8 and the detailed discussion of energy flow in the biosphere in Chapter 8). Thermodynamics is the science that keeps track of energy as it undergoes various transformations from one type to another. We use the first law to keep track of the quantity of energy.[4]

The conservation and conversion of energy can be illustrated by the example of a clock pendulum (Figure 15.3). When the pendulum is held in its highest position, it is neither moving nor getting hotter. It does, however, contain energy (owing to its position). We refer to the stored energy as *potential*

energy, which is converted to other forms when it is released. Other examples of potential energy are the gravitational energy in water behind a dam; chemical energy in coal, fuel oil, and gasoline, as well as in the fat in your body; and nuclear energy, which is related to the forces binding the nuclei of atoms.[3]

When the pendulum is released, it moves downward. At the bottom of the swing the speed is greatest, and there is no potential energy. At this point all its energy is in the energy of motion, which is called *kinetic energy*. As the pendulum swings back and forth, the energy continuously changes between the two forms, potential and kinetic. But at each cycle the pendulum slows down because of the friction of the pendulum moving through air and the friction at its pivot. The friction generates heat. Eventually all the energy is converted to heat and the pendulum stops.[3]

This example illustrates another property of energy, that is, its tendency to dissipate and end up as heat. It is relatively easy to transform various forms of energy into low-grade heat but difficult to change heat into high-grade energy. Physicists have found that it is possible to change all the gravitational energy in a pendulum to heat, but not to change all the heat energy (energy of random motion of atoms and molecules) thus generated back into potential energy.

Energy is certainly conserved in the pendulum example; all the initial gravitational potential energy is transformed to heat when the pendulum finally stops. If the same amount of energy, in the form of heat, were returned to the pendulum, would you expect the pendulum to start again? The answer is NO! What, then, gets used up? It is not energy because energy is always conserved. What gets used up is the energy *quality*, or the availability of the energy to perform work. The higher the quality of the energy, the more easily it can be converted to work; the lower the energy quality, the more difficult it is to convert to work. This example illustrates another fundamental property of energy: Energy always tends to go from a more usable (higher quality) form to a less usable (lower quality) form. This statement, the **second law of thermodynamics,** means that, when you use energy, you lower its quality. Lower quality energy is more difficult to utilize. In Chapter 8 the term *entropy* was introduced as a measure of the energy unavailable to do useful work. When we use energy and it is converted to a less useful form, we say the entropy of the system has increased.

Let us return to the example of the stalled car, which you have now pushed to the side of the road. Having pushed the car a little way uphill, you have increased its potential energy. You can convert this to kinetic energy by letting it roll back downhill.

FIGURE 15.3 Idealized diagram of a grandfather (pendulum) clock illustrating the relation between potential and kinetic energy.

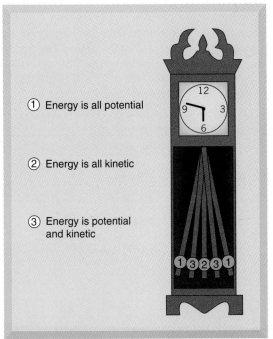

① Energy is all potential

② Energy is all kinetic

③ Energy is potential and kinetic

You engage the gears to restart the car. As the car idles, the potential chemical energy to move the car (from the gasoline) is converted to waste heat energy and various amounts of other energy forms, including sound and electricity to charge the battery and play the radio.

According to the first law of thermodynamics, the total amount of energy is always conserved. If this is true, why should there ever be any energy problem? Why could we not collect that wasted heat and use it to run the engine? Here we discover the importance of the second law of thermodynamics. According to that law, energy used to move the car is degraded to low-quality heat, which can never regain its original availability or energy grade. When we refer to heat energy as low grade, we mean that relatively little of it is available to do useful work. High-grade energy, such as gasoline, coal, or natural gas, has high potential to do useful work. The biosphere continuously receives high-grade energy from the sun and radiates low-grade heat to the depths of space.[3,4]

15.3 ENERGY EFFICIENCY

Two fundamental types of energy efficiencies are derived from the first and second laws of thermodynamics: the first-law efficiency and the second-law efficiency. The first-law efficiency deals with the amount of energy without any consideration of the quality or availability of the energy. The **first-law efficiency** is the ratio of the actual amount of energy delivered where it is needed to the amount of energy supplied in order to meet that need. Expressions for efficiencies are given as fractions; multiplying the fraction by 100 converts the fraction to a percentage. As an example, consider the use of a furnace system that keeps a home at a desired temperature of 18°C (65°F) when the outside tempera-

ture is 0°C (32°F). A furnace that burns natural gas or oil and delivers 1 unit of heat energy to the house for every 1.5 units of energy extracted from burning the fuel has a first-law efficiency of 1 divided by 1.5, or 67% (see Table 15.1 for other examples).[4]

In a sense, first-law efficiencies are misleading, because a high value suggests (often incorrectly) that little can be done to save energy through additional improvements in efficiency. This problem is overcome by the use of the **second-law efficiency,** which is defined as the ratio of the minimum available work needed to perform a particular task to the actual work used to perform the task. *Available work* is energy that is accessible to do work and is referred to simply as *availability.* Availability is higher for sources of high-quality energy; the available work of a fuel is close to its heat of combustion. Calculating the second-law efficiency for the home furnace yields a result of 5%, much lower than the first-law efficiency of 67%.[4] Table 15.1 also lists some second-law efficiencies for common uses of energy.

Processes with a low second-law efficiency are particularly important because they indicate where improvements in energy technology and planning may save significant amounts of high-quality energy. Notice that the definition of the second-law efficiency calls for the specification of the particular *task* needing energy. This task is referred to as the energy *end use.* The second-law efficiency monitors the energy quality or availability used in performing the defined task. For example, you could use a welder's acetylene blowtorch to light a candle, but a match is much more efficient (and safer as well).

We are now in a position to understand why the second-law efficiency is so low (5%) for the house heating example discussed earlier. This low efficiency implies that the furnace is consuming too much available work (i.e., high-quality energy) in carrying out the task of heating the house. Energy

TABLE 15.1 Examples of First- and Second-Law Efficiencies

Energy (End Use)	First-Law Efficiency (%)	Waste Heat (%)	Second-Law Efficiency (%)	Potential for Savings
Incandescent light bulb	5	95		
Fluorescent light	20	80		
Automobile	20–25	75–80	10	Moderate
Power plants (electric); fossil fuel and nuclear	30–40	60–70	30	Low to moderate
Burning fossil fuels (used directly for heat)	65	35		
Water heating			2	Very high
Space heating and cooling			6	Very high
All energy (U.S.)	50	50	10–15	High

with less availability (i.e., lower quality energy) could do the task with a higher second-law efficiency because there would be a better match between the required energy quality and the house-heating end use. For example, we could use a heat pump (a device that extracts heat from the outside environment and transfers it to the interior of the house) to heat the house. In other words, the task of heating the house requires heat at a relatively low temperature, near 18°C (65°F), not heat with temperatures in excess of 1000°C (1832°F), such as is generated inside the gas furnace. Through better energy planning, such as matching the quality of energy supplies to the end use, higher second-law efficiencies can be achieved, resulting in substantial savings of high-quality energy.

Examination of Table 15.1 indicates that electricity generating plants have nearly the same first-law and second-law efficiencies. These generating plants are examples of *heat engines*. A heat engine produces work from heat. Most of the electricity generated in the world today is from heat engines that use nuclear fuel, coal, gas, or other fuels. Our own bodies are examples of heat engines, operating with a capacity (power) of about 100 watts (W) and fueled indirectly by solar energy (see A Closer Look 15.1, "Energy Units," for an explanation of watts and other units of energy). The internal combustion engine (used in automobiles) and the steam engine are examples of heat engines. A great deal of the world's energy is used in heat engines, with profound environmental effects, such as thermal pollution, urban smog, acid rain, and potential global climate change.

The maximum possible efficiency of a heat engine, known as thermal efficiency, was discovered by the French engineer Sadi Carnot in 1824, before the first law of thermodynamics was formulated.[5] Modern heat engines have actual thermal efficiencies that range between 60% and 80% of their ideal Carnot efficiencies. Modern 1000-megawatt (MW) electrical generating plants have thermal efficiencies ranging between 30% and 40%; at least 60% to 70% of the energy input to the plant is rejected as waste heat. For example, assume that the electric power output from a large generating plant is 1 unit of power (typically 1000 MW). Producing that 1 unit of electric power requires 3 units of input (i.e., burning coal) at the power plant, and the entire process produces 2 units of waste heat, for a thermal efficiency of 33%. The astounding number here is the waste heat, 2 units, which amounts to twice the actual electric power produced.

Once electricity is produced from large power plants that burn coal or natural gas, from nuclear fuel, or from smaller producers, such as geothermal, solar, or wind sources (see Chapters 16 and 17), it is fed into the grid, which is the network of power lines, or the distribution system, eventually reaching homes, shops, farms, and factories, where it lights, heats, and drives motors and other machinery used by society. As electricity moves through the grid, there are losses[6]; the wires that transport electricity (power lines) have a natural resistance to electrical flow, known as *electrical resistivity*. This resistance converts some of the electric energy in the transmission lines to heat energy, which is lost to the environment.

15.4 ENERGY SOURCES AND CONSUMPTION

People living in industrialized countries are a relatively small percentage of the world's population, but they consume a disproportionate share of the total energy produced in the world. For example, the United States, with only 5% of the world's population, uses approximately 25% of the total energy consumed in the world. There is a direct relationship between a country's standard of living (as measured by gross national product) and energy consumption per capita. In the next several decades, as petroleum and natural gas become more scarce and expensive or their use is curtailed to reduce potential global climate change, both developed and developing countries will need to find innovative ways to obtain energy. The relationship between affluence and energy consumption needs to reflect more efficient use of energy resources, rather than a continuous increase in total energy consumed.

Today, approximately 90% of the energy consumed in the United States is produced by oil, natural gas, and coal. Because of their organic origin, these are called fossil fuels. They are produced from plant and animal material and are forms of stored solar energy that are part of our geologic resource base. They are essentially nonrenewable. The other sources of energy, which include geothermal, nuclear, hydropower, and solar, among others, are referred to as the alternative energy sources. The term *alternative* designates these as sources that might possibly replace fossil fuels in the future. Many of the alternative sources such as solar and wind are not depleted by consumption and are known as *renewable energy*.

The shift to alternative energy sources may be gradual, as fossil fuels continue to be utilized, or it could be accelerated as a result of concern over the potential environmental effects of burning fossil fuels. Regardless of which path we take, one thing is certain: Fossil fuel resources are finite. It has taken millions of years to form them, yet fossil fuels will be burned in only a few hundred years of human history. Using even the most optimistic predictions,

CLOSER LOOK 15.1

ENERGY UNITS

When we buy electricity by the kilo-watt-hour, what are we buying? We say we are buying energy, but what does that mean? Before we go deeper into the concepts of energy and its uses, we need to define some basic units.

The fundamental energy unit in the metric system is the *joule* (one joule is defined as a force of one newton* applied over a distance of one meter; recall the discussion of energy as calories in Chapter 8). To work with large quantities such as the amount of energy used in the United States in a given year, we use the unit *exajoule*, which is equivalent to 10^{18} (a billion billion) joules or roughly equivalent to 1 quad-

*A newton (N) is the force necessary to produce an acceleration of 1 m per sec per sec (m/s²) to a mass of 1 kg.

rillion, or 10^{15}, British thermal units (Btu), referred to as a *quad*.

To put these big numbers in perspective, the United States today consumes approximately 86 exa-joules (or quads) of energy per year, and world consumption is about 325 exajoules (quads) annually.

In many instances, we are particularly interested in the rate of energy use, or **power,** which is energy divided by time. In the metric system, power may be expressed as joules per second, or *watts,* W (1 joule per second is equal to 1 watt). When larger power units are required, we can use multipliers, such as kilo (thousand), mega (million), or giga (billion). For example, the rate of production of electrical energy in a modern nuclear power plant is 1000 megawatts (MW) or 1 gigawatt (GW).

Sometimes it is useful to use a hybrid energy unit, such as the watt-

hour, Wh (remember energy is power multiplied by time). Electrical energy is usually expressed and sold in *kilowatt-hours* (kWh, or 1000 Wh). This unit of energy is 1000 W applied for 1 hour (3600 s), the equivalent energy of 3,600,000 J (3.6 MJ).

Average estimated electrical energy in kilowatt-hours used by various household appliances over a period of 1 year is shown in Table 15.2. The total annual energy used is the power rating of the appliance multiplied by the time the appliance was actually used. The appliances that use most of the electrical energy are water heaters, refrigerators, clothes dryers, and washing machines. A list of common household appliances and the amounts of energy they consume is useful in identifying those appliances that might help save energy through conservation or improved efficiency.

TABLE 15.2 Average Estimated Electrical Energy Use per Year for Typical Household Appliances

Appliance	Power (W)	Average Hours Used per Year	Approximate Energy Used (kWh/year)
Clock	2	8760	17
Clothes dryer	4600	228	1049
Hair dryer	1000	60	60
Light bulb	100	1080	108
Compact fluorescent[a]	18	1080	19
Television	350	1440	504
Water heater (150 L)	4500	1044	4698
Energy-efficient model[a]	2800	1044	2900
Toaster	1150	48	552
Washing machine	700	144	1008
Refrigerator	360	6000	2160
Energy-efficient model[a]	180	6000	1100

Sources: Data from U.S. Department of Energy and D. G. Kaufman and C. M. Franz, 1993, *Biosphere 2000: Protecting Our Global Environment,* Harper Collins, New York.

[a]Newer, energy-efficient model.

the fossil fuel epoch that started with the industrial revolution will represent only about 500 years of human history. Therefore, although fossil fuels have been extremely significant in the development of modern civilization, their use will be an ephemeral event in the span of human history.

Energy Consumption in the United States Today

Energy consumption in the United States from 1950 to 1997 is shown in Figure 15.4, which dramatically illustrates our present dependence on the three major fossil fuels. Figure 15.4 shows that from approximately 1950 through the mid–1970s there was a tremendous increase in energy consumption, from approximately 30–70 exajoules. Since about 1980, energy consumption has increased by only about 20 exajoules. This situation is encouraging because it suggests that policies to improve energy conservation through efficiency improvements (such as requiring new automobiles to be more fuel efficient and buildings to be better insulated) have been at least partially successful.

What is not shown in these figures, however, is the tremendous energy loss. For example, energy consumption in the United States in 1965 was approximately 50 exajoules, and of this about half was effectively used. Energy losses were about 50% (the number shown in Table 15.1 for all energy). In 1997 energy consumption in the United States was about 95 exajoules, and again about 50% was lost in conversion processes. Energy losses in 1997 were nearly equivalent to total energy consumption in 1965! The largest energy losses are associated with the production of electricity and with transportation.

When electricity is generated, water is heated to produce steam that turns turbines, and heat energy is lost. Additional energy is lost through the transmission and distribution of the electricity throughout the grid system. Major losses of energy in the transportation sector of our economy are related to the burning of fossil fuels in the engines of our cars, buses, and trucks. Most of the losses related to the production of electricity and to transportation occur through the use of heat engines, which produce waste heat that enters the environment.

Another way to examine energy use is to look at the generalized energy flow of the United States for a particular year, as shown in Table 15.3 and Figure 15.5. These data show that for 1997 we imported considerably more oil than we produced and that consumption of energy is fairly evenly distributed in three sectors: residential/commercial, industrial, and transportation. It is clear that we remain dangerously vulnerable to changing world conditions affecting the production and delivery of crude oil. Evaluation of the entire spectrum of potential energy sources, as well as increased conservation through efficiency improvements and energy planning, is necessary to

FIGURE 15.4 Total energy consumption for the United States, 1950-1997. (*Source:* Energy Information Administration, 1998. *Annual Energy Review,* U.S. Department of Energy.)

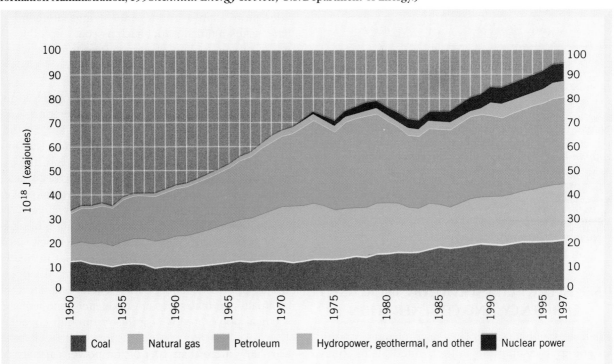

TABLE 15.3 Generalized Annual Energy Flow for the United States in 1997

Energy Source	Energy Production[a]	+	Net Imports (Imports–Exports)	±	Adjustments	=	Energy Consumed	Consumed by Sector
Coal	23.2		- 2.0					Residential/ commercial
Natural gas	21.9		+ 2.9					34.2
Oil	13.6		+ 19.2					Industrial
								34.2
Nuclear	6.7		0					
Hydropower	3.5		0					
Other	0.1		+ 0.3					Transportation 26.6
Total	69.0	+	20.4	±	5.6[b]	=	95.0	95.0

Source: OECD, Energy Information Administration, 1998, *Annual Energy Review,* U.S. Department of Energy.

[a]Exajoules (10^{18} J).

[b]Balancing to account for a variety of items including unaccounted-for supply, blending components, and changes in stock.

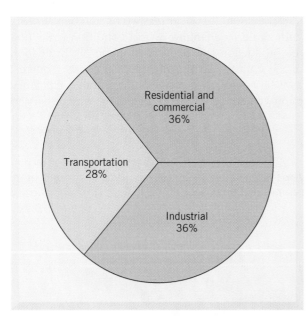

FIGURE 15.5 Energy consumption in the United States by sector. (*Source:* Energy Information Administration, 1998, *Annual Energy Review,* U.S. Department of Energy.)

ensure sufficient energy to maintain our society and a quality environment.

15.5 ENERGY CONSERVATION, INCREASED EFFICIENCY, AND COGENERATION

There is a significant movement to change patterns of energy consumption in the United States today through measures such as conservation, increased efficiency, and cogeneration. **Conservation** of energy refers to a moderation of energy use, simply getting by with less demand for energy. In a pragmatic sense this has to do with adjusting our energy needs and uses to minimize the amount of high-quality energy necessary to accomplish a given task. **Efficiency improvements** involve designing equipment to yield more energy output from a given amount of input energy.[2] **Cogeneration** refers to a number of processes designed to capture and use waste heat rather than simply release it into the atmosphere or water as thermal pollution. Energy conservation is particularly attractive because it provides more than a one-to-one savings. Remember that it takes 3 units of fuel such as coal to produce 1 unit of power such as electricity (two-thirds is waste heat). Therefore, not using 1 unit of power saves 3 units of fuel!

These three concepts—energy conservation, efficiency improvements, and cogeneration—are very much interrelated. For example, when electricity is produced at large, coal-burning power stations, large amounts of heat may be emitted into the atmosphere. Cogeneration, by using that waste heat, can increase the overall efficiency of a typical power plant from 33% to as much as 75%, effectively reducing losses from 67 to 25%.[2] Cogeneration of electricity also involves the production of electricity as a by-product from industrial processes that normally produce and use steam as part of their regular operations. Optimistic energy forecasters estimate that eventually we may meet approximately one-half the electrical power needs of industry through cogeneration.[2] Another source has estimated that, by the year 2000, more than 10% of the power capacity of the United States could be provided through cogeneration.

The average first-law efficiency of 50% (Table 15.1) emphasizes that, as energy from burning fuels, such as coal, oil, natural gas, and nuclear fuel, is converted to thermal and mechanical uses, large amounts of energy are lost in producing electricity and in transporting people and goods. Innovations in how we produce energy for a particular use can help prevent this loss. Of particular importance will be energy uses with applications below 100°C (212°F), because a large portion of the total U.S. energy consumption (for uses below 300°C, or 572°F) is for space heating and water heating (Figure 15.6).

In considering where to focus our efforts to develop more energy efficiency, we need to look at the total energy-use picture. In the United States, space heating and cooling of homes and offices, water heating, industrial processes (to produce steam), and automobiles account for nearly 60% of the total energy use. By comparison, transportation by train, bus, and airplane account for only about 5%. Therefore, the areas we should target for development of more energy efficiency are building design, industrial energy use, and automobile design.[2]

Building Design

A spectrum of possibilities exists for increasing energy efficiency and conservation in residential buildings. If we consider new homes, the answer may be *integral urban houses* (see the opening case history in Chapter 29 about Florida House), which are designed and constructed to minimize the amount of energy necessary to ensure comfortable living.[8] For example, buildings may be designed to take advantage of passive solar potential, that is, collecting solar heat using a system without moving parts. Windows and overhanging structures may be positioned so that the overhangs shade the windows from incoming summer solar energy, keeping the house cool. In the winter, the lower sun angle penetrates the windows, warming the house. A house designed for total environmental efficiency incorporates strong conservation principles for energy, water, and organic materials. The basic idea of the house is that the entire living area is connected in a unified way to provide the occupants with a comfortable living system that minimizes utilization of resources, maximizes the utility of the site, and is as efficient as possible.[8]

Many of our residential buildings were constructed a number of years ago; the best approach to conservation for them is insulation, caulking, weather-stripping, installation of window coverings and storm windows, and good maintenance. For these buildings and homes, the potential for energy savings through architectural design is extremely limited. The position of the building on the site is al-

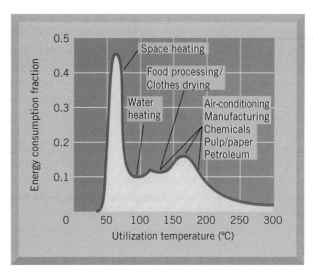

Figure 15.6 Energy use below 300°C in the United States. (*Source:* Los Alamos Scientific Laboratory, 1978, LASL 78-24.)

ready established, and reconstruction and modifications are often not cost effective.

Another thought concerning residential buildings is that with everything there is a price to pay. If we construct our buildings very tightly to conserve energy, there may be a higher potential for the development of indoor air pollution problems. (However, these potential problems may be reduced by using countercurrent heat exchangers that bring in fresh air; see Chapter 23.) Furthermore, construction of specific houses on specific sites may be difficult; fees to architects and engineers may be quite high. Initial costs may be higher for the features designed like those of the integral urban house. Nevertheless, moving toward better design of homes and residential buildings to conserve energy is certainly a worthwhile endeavor.

Industrial Energy

In looking at total energy consumption of the United States in the 1990s (Figure 15.4), it is apparent that the rate of increase in energy utilization leveled off in the early 1970s. Nevertheless, industrial production of goods (automobiles, appliances, etc.) is significantly higher! Today, U.S. industry consumes about one-third of the energy produced. The reason we have been able to have higher productivity with less energy use is that industries are more commonly using cogeneration and developing more energy-efficient machinery, such as motors and pumps designed to use less energy.[2,9]

Automobile Design

As a final note in our argument concerning conservation and efficiency, consider the automobile.

There have been steady improvements in the development of fuel-efficient automobiles during the last 15 years. In the early 1970s the average U.S. automobile burned approximately 1 gallon of gas for every 14 miles traveled. By 1996, the average miles per gallon (mpg) had risen to 28 for highway driving with some automobiles as high as 49 mpg.[10] Fuel consumption rates have not improved since 1991 due to lower oil prices and rise in concern for vehicle safety that resulted in heavier automobiles. However, vehicles in the future may reach a fuel consumption of nearly 100 mpg. This would result from increased efficiency and resulting conservation of fuel, and the construction of smaller cars that have smaller engines and are constructed of lighter materials.[2] Of course, there would be a price to pay for this change. Smaller cars may be more prone to damage on impact, and, as cars have gotten smaller, trucks have tended to stay the same size or even increased in size! As a result, the number of serious accidents between cars and trucks may increase as cars are made smaller.

15.6 ENERGY POLICY

Energy policy today is at or near a crossroads. One road leads to the development of so-called hard technologies, which involve finding ever greater amounts of fossil fuels and building larger power plants. Following the **hard path** means continuing as we have been for a number of years, that is, emphasizing energy quantity. In this respect, the hard path is more comfortable. That is, it requires no new thinking or realignment of political, economic, or social conditions. It also involves little anticipation of the inevitable depletion of the fossil fuel resources on which the hard path is built.

Hard Path versus Soft Path

Those in favor of the hard path argue that much environmental degradation has been caused by people in less developed countries who have been forced to utilize their local resources, such as wood, for energy. As a result, their lands have suffered from loss of plant and animal life and from soil erosion (loss of vegetation causes increased erosion because plants and their roots retard surface erosion, as discussed in Chapters 10 and 11). They argue that the way to solve the environmental problems in less developed countries is to provide them with cheap, high-quality energy, such as fossil fuels or nuclear energy, which utilizes more heavy industrialization and technology, not less.

In countries like the United States, with sizable energy resources of coal and petroleum, proponents of hard-path energy argue that we should exploit those resources to make certain that the environmental degradation now plaguing other countries of the world will not strike here. In other words, the argument is to let the energy industry develop the available resources, that industrial freedom is more likely than government regulation to ensure a steady supply of energy and less total environmental damage. Proponents of this view argue that the recent increase in the burning of firewood across the United States is an early indicator of the effects of strong governmental controls on energy supplies. They argue further that this is just the beginning and that the eventual depletion of forest resources will have an appreciably detrimental effect on the environment, as has occurred in so many less developed countries.

The other road is designated the **soft path.**[11] A champion of this strategy is Amory Lovins, who states that the soft path involves energy alternatives that emphasize energy quality and are renewable, flexible, and environmentally more benign than those of the hard path. As defined by Lovins, these alternatives have several characteristics:

- They rely heavily on renewable energy resources, such as sunlight, wind, and biomass (wood and other plant material).

- They are diverse and are tailored for maximum effectiveness under specific circumstances.

- They are flexible; that is, they are relatively low technologies that are accessible and understandable to many people.

- They are matched in both geographic distribution and scale to prominent end-use needs (the actual use of energy).

- There is good agreement, or matching, between energy quality and end use, increasing second-law efficiency.

This last point is of particular importance because, as Lovins points out, people are not particularly interested in having a certain amount of oil, gas, or electricity delivered to their homes; rather, they are interested in having comfortable homes, adequate lighting, food on the table, and energy for transportation.[11] According to Lovins, only about 5% of end uses require high-grade energy, such as electricity. Nevertheless, a lot of electricity is used to heat homes and water. For some purposes, continual use of electricity is very important; but for many others, it is not necessary. Lovins shows that there is an imbalance in using nuclear reactions at extremely high temperatures or in burning fossil fuels at high temperatures simply to meet needs where the temperature increase necessary may be only a few tens of degrees. Such large discrepancies are thought wasteful and a misallocation of high-quality energy.

The bridge, or transition, from the hard to the soft path will presumably take place through the utilization of present energy sources, such as coal and natural gas. Although there is sufficient coal to last hundreds of years, those in favor of the soft path would prefer to use this source as a transition, rather than an ultimate long-term, source.

Energy for Tomorrow

Energy supplies and demand for energy are difficult to predict because technical, economic, political, and social assumptions underlying predictions are constantly changing. Large annual and regional variations in energy consumption must also be considered. For example, in areas with cold winters and hot, humid summers, energy consumption peaks during the winter months with a secondary peak in the summer (the former resulting from heating and the latter from air conditioning). Regional variations are also significant. For example, in the United States the transportation sector uses approximately 23% of the energy consumed. In California, where population densities are high and people commute long distances to work, 48% of the energy is used for transportation. Energy sources vary by region as well. In California, which has a fair amount of oil and natural gas, coal is a minor energy source. However, in the eastern United States, the fuel of choice for power plants is still coal.

Future changes in population densities as well as intensive conservation measures will probably change existing patterns of energy use. This might be a shift that relies more heavily on alternative (particularly renewable) energy sources. One recent prediction holds that energy consumption in the United States in the year 2030 may be as high as 120 exajoules or as low as 60 exajoules (Figure 15.7). The high value assumes little change in energy policies, whereas the low value assumes very aggressive energy conservation policies. To bring about an environmental scenario that would also stabilize the climate in terms of global warming (as a result of emissions of carbon dioxide), energy use would need to be cut by an additional 50%.

Over the past decade or more, energy scenarios have consistently overestimated the energy demands of the future. It seems very unlikely that energy consumption in the United States in the year 2030 will be as low as 60 exajoules, and it is also unlikely that it will be as high as 120. Significant short-range reduction in the use of fossil fuels in the United States will be vigorously resisted in many segments of the society. One reason the change will not come quickly is that many policymakers are not

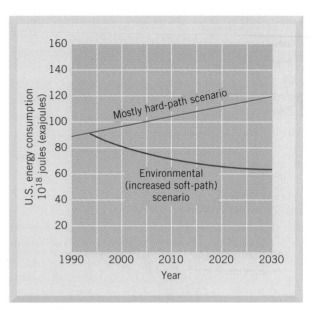

FIGURE 15.7 Two possible energy scenarios for the United States to the year 2030. (*Source:* Union of Concerned Scientists, 1991.)

convinced that serious environment degradation will actually occur as a result of burning fossil fuels. Thus we will continue on our global experiment and have to learn to live with potential climatic change.

A second low-energy scenario has been suggested by Lovins. This scenario involves a moderate decrease in energy consumption following a maximum of about 100 exajoules in the year 2000. The eventual reduction accompanies the shift from dependence on fossil fuels to the softer technologies (renewable, alternative energy resources). Authors of a similar low-energy scenario emphasize (as does Lovins) that reductions in energy use do not have to be associated with a lower quality of life, but rather with increased conservation of energy and a more energy-efficient distribution of urban population, agriculture, and industry.[7] These authors offer the following alternatives:

- New, more energy-efficient settlement patterns that maximize accessibility of services and minimize the need for transportation.
- New agricultural practices that emphasize locally grown and consumed foods and an emphasis on foods that require less total energy to produce than that required for diets more dependent on beef and pork.
- New industrial guidelines that promote energy conservation and minimize production of consumer waste.

To some extent these alternatives are already practiced in some areas; that is, in some instances highway construction has a lower priority than mass-transit systems; agricultural lands near urban centers are preserved; and industry is more receptive to recycling and decreased production of consumer waste (such as unnecessary packaging). Various low-energy scenarios differ in terms of the actual energy sources used. Some require more use of the fossil fuels, whereas others rely heavily on alternative sources. All projections of specific sources of energy use in the future must be considered speculative. Perhaps most speculative of all is the idea that we really can obtain most of our energy needs from alternative renewable energy sources in the next several decades. However, there are likely to be significant moves in that direction. From an energy viewpoint, the next 30 years will be crucial to the United States and to the rest of the industrialized world.

The transformation of energy use from a hard to a soft path has a long history, as seen in the case study of the early Greek and Roman cultures that opened this chapter. In the latter part of the twentieth century we experienced the shock of the 1973 oil embargo, which caused anxiety about our energy supply and the life-style that depends on it, as well as creating long lines at the gasoline pumps during a cold winter. That event in the United States spawned the research and development of alternative energy sources and significant government financial incentives to utilize sunlight, wind, and other alternative energy sources. With the return of abundant, inexpensive oil in the 1980s, there was much less government support of alternative energy. During the same period China, with one-fifth of the world's population, continued to fire its emerging industrial revolution by burning huge quantities of coal. Today the industrial countries are even more dependent on imported oil than they were in the 1970s, and it is argued that the Gulf War of 1991 was as much or more concerned with a steady, secure supply of Middle Eastern oil than with the human rights of a small country.

The inexpensive oil of the 1980s and mid 1990s contributed to a laissez-faire attitude toward the development of alternative energy sources. One proposal is to tax all forms of energy based on their heat content, the so-called BU tax, to encourage the conservation and development of alternative energy. For example, the price of gasoline in the United States remains very low compared to the price in other industrialized countries, which has reduced the incentive to conserve fuel and drive smaller automobiles. If the price of gasoline was more than $3/gal in the United States, as it is in countries such as Spain, France, Germany, and England, there certainly would be more incentive to change energy use patterns. Of course, the price of imported oil is about the same for all consuming countries. Most have very large taxes on oil.

There are strong arguments for raising the U.S. tax on oil. We might consider, for instance, raising the gasoline tax by $1/gal, with half the proceeds going toward building or improving rapid-transit systems and roads, energy conservation programs, and development of alternative energy sources and the other half used to help pay the national debt. In the United States today we use approximately 226 million gallons of gasoline per day, so the funds raised would be over $40 billion per year![12] Even with this increase, our gasoline prices would be moderate compared to those in the rest of the world. To minimize the shock to consumers, the tax could be put in place over a period of several years.

The energy decisions we make in the very near future will greatly affect both our standard of living and our quality of life. From an optimistic point of view, we have the necessary information and technology to ensure a bright, warm, lighted, and moving future—but time may be running out and we need action now. We can either continue to take things as they come and live in the year 2000 and beyond with the results of our present inaction or we can build for the future an energy picture based on careful planning.

Integrated Energy Management

The concept of **integrated energy management** recognizes that no single energy source can possibly provide all the energy required by both the developing and the developed countries of the world.[13] A range of options that vary from region to region will have to be employed. Furthermore, the mix of technologies and sources of energy will involve both fossil fuels and alternative, renewable sources. The basic objective of this integrated management is to obtain global **sustainable energy** that is realized at the local level. Sustainable energy development would have the following characteristics:

- It would provide reliable sources of energy.
- It would not cause destruction or serious harm to our global, regional, or local environments.
- It would help ensure that future generations inherit a quality environment with a fair share of the Earth's resources.

To implement integrated energy management, various regions of the world will have to develop carefully conceived energy plans based on local and regional conditions (such as sources of energy available, potential for conservation and efficiency, and desired end uses for energy). Such a plan recognizes that preservation of resources can be profitable and that the economy and the environment are intimately connected. There is a major philosophical

Is There Enough Energy to Go Around?

The developing countries have most of the world's population (4.7 billion of 5.9 billion people) and are growing at roughly twice the pace of the developed countries. The average rate of energy use for individuals in developing countries is 1.0 kW, whereas that for persons in developed countries is 7.5 kW. If the current annual growth rate of 1.3% is maintained, the world's population will double in 54 years, to 12 billion people. More people will mean more energy use, of course, but in addition, people in developing countries will likely consume more energy per capita if they are to achieve a higher standard of living.

With a worldwide average energy use rate of 2.6 kW per person, the 5.9 billion people on Earth use 15.3 terrawatt years annually (TW = trillion watts). A population of 12 billion with an average per-capita energy use rate of 6.0 kW would use 72 TW years annually, or about five times as much as is presently used. Can the world support this much energy use?

John Holdren, an energy expert, believes that a realistic goal is for annual per capita energy use to reach 3 kW with a world population to peak at 10 billion individuals by the year 2100. To achieve this goal, developing nations would not be able to increase their population by more than 60% and their energy use by more than 100%, and developed nations could increase their population by only 10% and would have to reduce their energy use by 2% each year.

Critical Thinking Questions

1. What would the energy-use rate be if Holdren's goals were realized? How much total energy would be required for all people on Earth to have a standard of living supported by 7.5 kW per person? How do these totals compare with the present energy-use rate worldwide?

2. How much greater efficiency would be required from today's mix of energy sources to provide an additional 45 TW years annually? How realistic do you think Holdren's assumption is? Explain.

3. In what specific ways could energy be used more efficiently in the United States? Compare your list with those of your classmates and compile a class list.

4. In addition to increasing efficiency, what other changes in energy consumption might be required to provide an average energy-use rate of 7.5 kW per person?

5. Would you consider Holdren's vision of the energy future an example of a hard or a soft path? Explain.

References

Ehrlich, P. R., and Ehrlich, A. H. 1991. *Healing the Planet*. Reading, MA: Addison-Wesley.

Fickett, A. P. 1990. "Efficient Use of Electricity," *Scientific American* 263(3): 65–74.

Holdren, J. P. 1990. "Energy in Transition," *Scientific American* 263(3): 157–163.

Lean, G. 1990. *Atlas of the Environment*. New York: Prentice-Hall.

U. S. Census Bureau, 1998. World population and growth rates. Accessed at http://www.census.gov/ipc/www/world.html.

recognition here that degradation of the environment and poor economic conditions go hand in hand.[14] That is, degradation of air, water, and land resources results in a depletion of assets that ultimately will result in a lowering of both the standard of living and the quality of life. A good energy plan is part of an aggressive environmental plan that has the goal of producing a quality environment for future generations. A good plan should[14]:

- provide for sustainable energy development;
- provide for aggressive energy efficiency and conservation;
- provide for a diversity of energy sources and competitive evaluation in determining future energy supplies, technologies, and fuels;
- provide for a balance between economic health and environmental quality; and
- use second-law efficiencies as an energy policy tool (i.e., strive to produce a good balance between quality of energy source and end uses for that energy).

Such a plan recognizes that new energy demands can be met in what is often the least expensive and most environmentally preferred way. An important element that the plan should address is the energy demand for transportation; development of new engines for vehicles and improvement of fuel technology to reduce both fuel consumption and air quality problems should be encouraged. Finally, the plan should address the economic marketplace through accurate pricing of fuels that reflects the economic cost of using the fuel as well as the cost to the environment. Of particular importance here is development of the true cost effectiveness of energy efficiency technology in terms of environ-mental costs and benefits. In summary, the plan should be an integrated energy management statement that moves toward sustainable development. Those who develop such plans recognize that a diversity of energy supplies will be necessary and that the key, or most prominent, components are (1) improvements in energy efficiency and conservation and (2) matching energy quality to end uses.[14]

SUMMARY

- The first law of thermodynamics tells us that energy is never created or destroyed, is always conserved, but is transformed from one kind to another. We use the first law to keep track of the quantity of energy.

- The second law of thermodynamics tells us that as energy is used it always goes from a more usable (higher quality) form to a less usable, lower quality.

- Two fundamental types of energy efficiency are derived from the first and second laws of thermodynamics. In the United States today first-law efficiencies average about 50%, which means about 50% of the energy produced is returned as waste heat. Second-law efficiencies average 10–15%, which means that there is a high potential for saving energy through better matching of quality of energy sources with end uses.

- Energy conservation and improvements in energy efficiency can have very significant impacts on energy consumption. It takes three units of a fuel such as oil to produce one unit of electricity. As a result each electicity unit conserved or saved through improved efficiency saves three units of fuel.

- There are arguments for and against both the hard and soft energy paths. The former is to some more comfortable, has a long history of success, and has produced the highest standard of living ever experienced. Those favoring the soft path argue that our present sources of energy are causing too much pollution and are not sustainable. They favor alternative energy sources that are renewable, decentralized, diverse, flexible, and that better match energy quality and end use, thus emphasizing second-law efficiencies.

- Global sustainable energy planning or integrated energy planning is necessary to make the eventual necessary transformation from the fossil fuels to other sources. Such planning must provide reliable sources of energy, not cause serious harm to the environment, and help ensure that future generations inherit a quality environment with sufficient natural resources.

REEXAMINING THEMES AND ISSUES

Human Population: The industrialized and urbanized countries of the world produce and use most of the world's energy. As societies change from rural to urban, energy demands generally increase. If this trend continues, there is fear that eventually our use of conventional sources of energy such as fossil fuels might lead to irreparable damage to many ecosystems. Controlling the growth of human population is an important factor in reducing future demand for energy.

Sustainability: It will be difficult to obtain the sustainability we are striving for if we continue with our present energy policies. To obtain a quality environment for future generations, it may be necessary to rethink the management and sources of energy used. This is at the heart of the controversy over the hard path versus the soft path.

Global Perspective: Understanding trends of energy production and consumption on a global basis is important if we are to directly address the global impact of energy policy related to burning of fossil fuels. Furthermore, the use of energy resources greatly influences global economic systems as these resources are transported around the world.

Urban World: A great deal of energy is used in our urban areas to fuel the urban systems for homes, industry, and critical facilities. How we choose to manage energy in our urban environments will greatly affect the quality of those environments in the future. For example, burning of cleaner fuels will result in less air pollution. This has already been observed in some areas.

Values and Knowledge: Society greatly values a quality environment. Public opinion polls consistently show this. As a result, intensive research is ongoing to better understand principles related to generation and use of energy (which often cause environmental degradation) at all levels of society. For example, we are evaluating how we may make more efficient use of our present energy resources, practice conservation of energy, and reduce adverse effects of energy consumption.

KEY TERMS

cogeneration *316*	first law of thermodynamics *311*	second law of thermodynamics *311*
conservation *316*	hard path *318*	soft path *318*
efficiency improvements *316*	integrated energy management *320*	sustainable energy *320*
energy *310*	power *314*	work *310*
first-law efficiency *312*	second-law efficiency *312*	

STUDY QUESTIONS

1. What evidence supports the notion that our present energy shortage is not the first in human history but may be unique in other ways?

2. What is the difference in meaning of the following terms: *energy, work,* and *power?*

3. Compare and contrast potential advantages and possible disadvantages of a major shift from hard-path to soft-path energy development.

4. You have just purchased a wooded island in Puget Sound. Your house is uninsulated and built of raw timber. Although the island receives some wind, trees over 40 m tall block most of it. You have a diesel generator for electric power, and hot water is produced by an electric heater run by the generator. Oil and gas can be brought in by ship. What steps would you take in the next 5 years to reduce the cost of energy with the least damage to the island?

5. Why is there a thermal bottleneck as chemical energy is converted to thermal and mechanical uses?

6. How might better matching of end uses for energy with potential sources yield improvements in energy efficiency?

7. Complete an energy audit of the building in which you live and develop recommendations that might lead to lower utility bills.

8. How might a plan that utilizes the concept of integrated energy management vary for the Los Angeles area and the New York City area? How might these differ from an energy plan for Mexico City, which is quickly becoming one of the largest urban areas in the world?

9. A recent energy scenario for the United States suggests that in the coming decades energy sources might be natural gas (10%), solar power (30%), hydropower (20%), wind power (20%), biomass (10%), and geothermal energy (10%). Do you think this is a likely scenario? What would be the major difficulties and points of resistance or controversy?

FURTHER READING

Alliance to Save Energy, American Council for an Energy-Efficient Economy, Natural Resources Defense Council, and Union of Concerned Scientists. 1991. *America's Energy Choices: Invest-ing in a Strong Economy and a Clean Environment.* Cambridge, MA: Union of Concerned Scientists. A detailed approach to altering America's energy use to achieve a cleaner environment and a stronger economy. Includes four scenarios that use improvements in energy efficiency and renewable energy technologies to substantially improve environmental quality and save money.

Committee for Economic Development, Research and Policy Committee. 1993. *What Price Clean Air? A Market Approach to Energy and Environmental Policy.* New York: Committee for Economic Development. A discussion of air quality, energy use, and environmental policy in the United States, including recommendations for policy change to improve environmental quality at a lower cost economically.

Gibbons, J. H., and Blair, P. D. 1991. "U. S. Energy Transition: On Getting from Here to There," *Physics Today* 44(7): 22–32. A good overview concerning our energy supplies, including dependence on Middle East oil, advances in the alternative energy systems, energy efficiency, and energy policy.

International Energy Agency. 1992. *Global Energy: The Changing Outlook.* Paris, France: Organization for Economic Co-operation and Development. A good discussion of the future shifts expected in the global energy balance, including how energy demand will change for each geographic region.

Lovins, A. B. 1979. *Soft Energy Path: Towards a Durable Peace.* New York: Harper & Row. This interesting book presents the argument for the soft path. Its message is more important today than when it was written.

Organization for Economic Co-operation and Development (OECD), International Energy Agency (IEA). 1994. *The Economics of Climate Change.* Proceedings of an OECD/IEA Conference. OECD, IEA. These proceedings discuss the economics of climate change, differences of opinion on the matter, method to link economic studies and climate change policy, and directions of international policy to deal with climate change.

INTERNET RESOURCES

Energy Information Administration: *http://www.eia.doe.gov/*—The Department of Energy's EIA has information on energy outlooks, consumption, and conservation for the U.S. and the world.

U.S. Department of Energy: *http://www.doe.gov/*—A great source of information on all aspects of energy in the U.S.

Slow-moving traffic during afternoon rush hour in San Francisco, California.

CHAPTER 16

FOSSIL FUELS AND THE ENVIRONMENT

CASE STUDY

Fuel Efficiency in U.S. Passenger Cars and Light–Duty Vehicles

There is good news and bad news about fuel efficiency and total use of gasoline. The good news is that fuel efficiency for U.S. passenger cars and light-duty vehicles (sports utility vehicles, minivans, sports cars, pick-up trucks) has nearly doubled since 1970. If fuel efficiency had not increased, we would need to import over twice as much oil as we currently do! The bad news is that we are burning more gasoline today than ever before (over 300 million gallons/day). This is because the number of motor vehicles in the United States has increased more than eight times in 40 years and more people are driving further and at greater speeds. Due to the strong U.S. economy in the 1990s, people have more money to spend on automobiles, gasoline, and traveling. Additionally, the price for gasoline in the United States is low and the price remains low due to an oversupply of gasoline. Some automobiles now have fuel efficiencies of more than 40 miles/gallon (17 km/liter), but there is little incentive for people to buy these vehicles.

In spite of the rosy picture for the supply side of gasoline, we are aware that the world's fossil fuel resources, which took millions of years to form, will be depleted in a period of about 500 years.[1] Sooner or later we will have to find alternatives to fossil fuels. We are faced with a dilemma— a situation in which a choice must be made between undesirable alternatives in the use of fossil fuels. Fossil fuels produce 90% of the energy we use but at a tremendous price, including urban air pollution, acid rain, and potential global warming. Continuing to rely on fossil fuels, with their known environmental consequences, is one option. The alternative is to use other sources of energy such as nuclear energy, hydropower, or solar energy. Unfortunately such a change might bring about undesirable economic consequences.

Regardless of the choices we make, eventually we will be forced to find alternatives to fossil fuels. In the long run there is no dilemma because there will be no choice. The short-term dilemma that we face is determining when the shift will take place. Should we burn fossil fuels as long as they are available, resulting in a degraded planet? Or should we shift to other fuels now and try to minimize the potential for environmental degradation? (See Chapters 15 and 17.)

The actual path we take for our energy future will undoubtedly be a compromise between the two extremes of the hard and soft paths, and we will need to develop intermediate policies and fuels to fill the gap during the conversion from fossil fuels to other energy sources. One fuel that may assist in this transitional period appears to be natural gas, which is relatively clean burning compared to coal. In this chapter we discuss fossil fuels in terms of origin, estimated time to depletion, and environmental concerns.

LEARNING OBJECTIVES

We rely almost completely on the fossil fuels (oil, natural gas, and coal) for our energy needs. However, these are nonrenewable resources, and their production and use have a variety of serious environmental impacts. After reading this chapter, you should understand:

- Why we are in an energy dilemma with respect to fossil fuels.

- Where oil, natural gas, and coal come from, that is, how they were formed.
- What the environmental effects of producing and using oil, natural gas, and coal are.
- How oil is produced from tar sands and oil shale.
- Why we will eventually have to switch from fossil fuels to other energy sources.

16.1 FOSSIL FUELS

The **fossil fuels** are all forms of stored solar energy. We say this because plants are solar energy collectors. They convert solar energy to chemical energy through photosynthesis (see Chapter 4). The fossil fuels were created from incomplete biological decomposition of dead organic matter (mostly plants and marine organisms). When organic matter is buried and escapes oxidation, it can be converted by complex chemical reactions in the geologic cycle to **hydrocarbons** and to fossil fuels (see Chapter 4).

The main fossil fuels (crude oil, natural gas, and coal) are considered our primary energy sources because on a worldwide basis they provide approximately 90% of the energy consumed (Figure 16.1). The biological and geologic processes in various parts of the geologic cycle produce the sedimentary rocks in which we find the fossil fuels. In this chapter we also briefly discuss two other forms of fossil fuels, oil shale and tar sands. However, these are really alternatives to the more readily available energy sources. Oil shale and tar sands, along with nonfossil energy sources, will become increasingly important as oil, gas, and coal reserves are depleted.

16.2 CRUDE OIL AND NATURAL GAS

Crude oil (petroleum) and **natural gas** are found primarily along tectonic belts at plate boundaries (see Chapter 4). Most geologists accept the hypothe-

sis that oil and gas are derived from organic materials that were buried with marine or lake sediments in what are known as depositional basins. The source material (or *source rock*) for oil and gas is fine-grained (less than $1/16$ mm, 0.0025 in., in diameter), organic-rich sediment buried to a depth of at least 500 m (1640 ft), where it is subjected to increased heat and pressure.

The elevated temperature and pressure initiate the chemical transformation of the organic material in the sediment into oil and gas. The elevated pressure causes the sediment to be compressed; this, along with the elevated temperature in the source rock, initiates upward migration of the hydrocarbons to a lower-pressure environment (known as the *reservoir rock*) that is coarser grained and relatively porous (has more and larger open spaces between the grains). Sandstone or a porous limestone, which has a lot of void space in which to store hydrocarbons, are common reservoir rocks. If there is a clear path to the surface, the oil and gas will escape to the atmosphere. This explains why oil and gas are not found in geologically old rocks; in such rocks the hydrocarbons have had ample time to migrate to the surface, where they either vaporized or were eroded away.[2]

The oil and gas fields from which we extract resources are places where the natural upward migration of the oil and gas to the surface is interrupted or blocked by what is known as a *trap* (Figure 16.2). The rock that helps form the trap, known as the *cap rock*, is usually a very fine grained sedi-

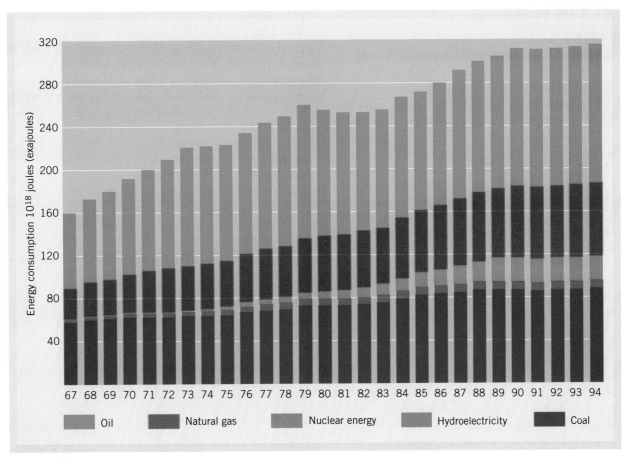

FIGURE 16.1 World energy consumption by primary sources, 1967–1994. (*Source:* British Petroleum Company, 1995, *BP Statistical Review of World Energy.*)

mentary rock, such as shale, which is composed of silt and clay-sized particles. Favorable geometry in the rock structure, such as an anticline (arch-shaped fold) or a fault (fracture in the rock along which displacement has occurred), may form traps, as shown in Figure 16.2. The important concept is that the combination of favorable rock structure and presence of a cap rock allow hydrocarbons to accumulate in oil and gas fields where they can be discovered and exploited as a valuable resource.[2]

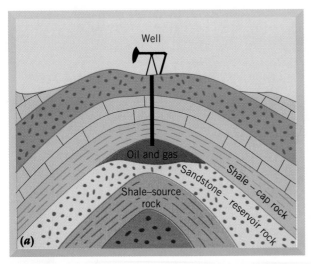

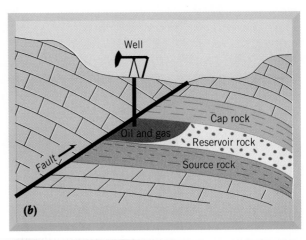

FIGURE 16.2 Two types of oil and gas traps: (*a*) anticline and (*b*) fault.

Petroleum Production

Production wells in an oil field recover oil through both primary and enhanced (secondary) production methods. *Primary production* involves pumping from wells under the natural pressure in the oil fields, which can be high and must be carefully controlled. This method can recover only about 25% of the available petroleum in the reservoir. To increase the amount of oil that can be recovered to 60%, *enhanced*, or *secondary recovery*, methods are used. In secondary recovery, the pressure in the reservoir is changed by injecting steam, water, or chemicals, such as carbon dioxide or nitrogen gases, into the reservoir to push the oil toward the wells where it can be recovered by pumping.

Next to water, oil is the most abundant fluid in the upper part of the Earth's crust. Most of the known, proven reserves, that part of the total resource that is identified and can be extracted now at a profit, are located in a few fields. Of a total reserve of 1000 billion barrels (bl), 65% are located in 1% of the fields, the largest of which are in the Middle East (Figure 16.3). Although new oil and gas fields have recently been and continue to be discovered in Alaska and other parts of the United States and in Mexico, South America, and other areas, the present world reserves will nonetheless be depleted in the next few decades.

However, the numbers suggest that we are not going to run out of crude oil immediately. At the present worldwide rate of consumption we have at least 45 years, if estimates of the resource and potential resource base are correct. There is, though, a good deal of uncertainty concerning the estimated size of the petroleum reserves and potential resources.

The total resource always exceeds known reserves; it includes petroleum that cannot be extracted at a profit and petroleum that is suspected but not proved to be present. Only 25 years ago the size of the proven reserves was estimated to be about 350 billion bl. Today that estimate is just over 1000 billion bl.[3] The increases in proven reserves of oil in the last 25 years have primarily been due to discoveries in the Middle East and Venezuela. Because so much of the world's oil is found in the Middle East, the tremendous oil revenues have resulted in huge trade imbalances. Figure 16.4 shows the major routes of trade for oil. Interestingly, in 1997 the United States imported more oil from Venezuela than from the Middle East. Most of the Middle Eastern oil goes to Europe, Japan, and Southeast Asia.

Natural Gas

The worldwide estimate of recoverable natural gas is about 140 trillion m³, which at the current rate of

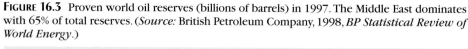

FIGURE 16.3 Proven world oil reserves (billions of barrels) in 1997. The Middle East dominates with 65% of total reserves. (*Source:* British Petroleum Company, 1998, *BP Statistical Review of World Energy.*)

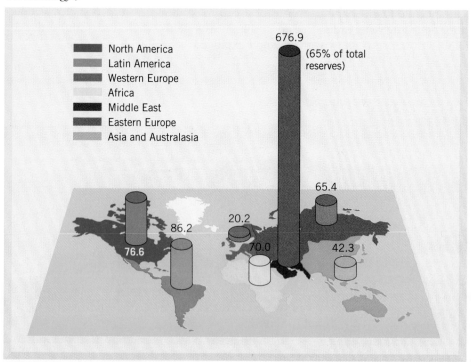

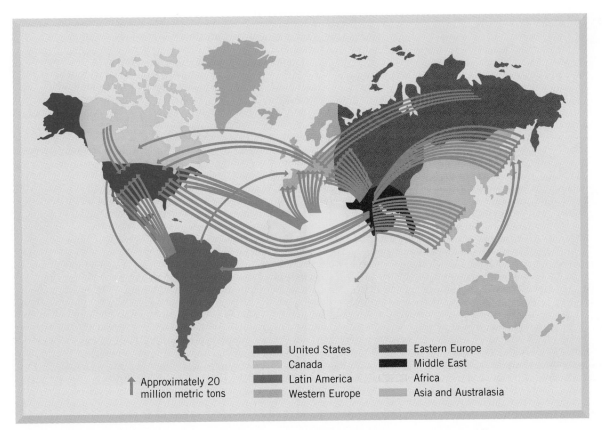

FIGURE 16.4 Major trade routes for the world's oil, emphasizing the countries that use Middle Eastern oil. (*Source:* British Petroleum Company, 1995, *BP Statistical Review of World Energy.*)

world consumption will last approximately 70 years.[3,4] Considerable natural gas has recently been discovered in the United States, and at present consumption levels the resource is expected to last at least 30 years.

We have only begun to look hard for natural gas and to utilize this resource to its potential. One reason for this slow start is that natural gas is transported primarily by pipelines, and only in the last few decades have these been constructed in large numbers. In fact, until recently, natural gas found with petroleum was often simply burned off as waste, and in some cases this practice continues.[5]

New supplies are being found in surprisingly large amounts, particularly at depths that are deeper than those at which oil is found. Optimistic estimates of the total resource suggest that at current rates of consumption the supply may last approximately 120 years.[1] This possibility has important implications because natural gas is considered a clean fuel (because burning it produces fewer pollutants than does burning oil or coal) that causes fewer environmental problems than do the other fossil fuels. As a result it is being considered as a possible transition fuel from the other fossil fuels (oil and coal) to alternative energy sources, such as solar power, wind power, and hydropower.

In spite of the new discoveries and the construction of pipelines, long-term projections for a steady supply of crude oil and natural gas carry many uncertainties. Everyone agrees that the supply is finite and that at present rates of consumption it is only a matter of time before the resources are depleted. Of course, we will never completely run out of all crude oil and natural gas, but the price will rise higher and higher as the resource becomes scarcer.

Environmental Effects of Oil and Natural Gas

Recovery

Development of oil and gas fields involves drilling wells on land or beneath the sea (Figure 16.5). A number of possible environmental impacts associated with the recovery of hydrocarbons at or near production sites are listed here:

- Disruption of the land to construct pads for wells, pipelines, or storage tanks and to build a network of roads and other production facilities.

- Pollution of surface waters and groundwater from runoff and infiltration or leaking from bro-

(a)

(b)

FIGURE 16.5 Drilling for oil in (*a*) the Sahara Desert of Algeria, and (*b*) the Cook Inlet of southern Alaska.

ken pipes or tanks of contaminated surface water, wastewater, or fluids used in secondary recovery. Production of petroleum brings to the surface with the oil a large but variable volume of salty water (brine) that must be disposed of. The brine is often toxic and may be disposed of in a number of ways: by evaporation in lined pits, by reinjection as part of secondary recovery, or by pumping into deep disposal wells outside the oil fields (see Chapter 27).

- Accidental release of air pollutants, such as hydrocarbons and hydrogen sulfide (in toxic gas).
- Land subsidence (sinking) as oil and gas are withdrawn.
- Loss or disruption and damage of fragile ecosystems, such as wetlands or other unique landscapes. This is the center of the controversy over the development of petroleum resources in pristine environments such as the Arctic National Wildlife Refuge in Alaska and the continent of Antarctica.

Additional environmental impacts associated with oil production in the marine environment include:

- oil seepages resulting from normal operations or large spills from accidents, such as blowouts or pipe ruptures;

- release of drilling muds (heavy liquids injected into the bore hole during drilling to keep the hole open) containing heavy metals, such as barium, that may be toxic to marine life; and
- aesthetic degradation from the presence of offshore oil-drilling platforms, which some people think are unsightly.

Refining

Refining crude oil and converting it to related products also have considerable environmental impacts. At refineries, crude oil is heated so that its components can be separated and collected (this process is called fractional distillation). Other industrial processes are then used to produce products such as gasoline and heating oil. Refineries may have accidental spills and slow leaks of gasoline and other products from storage tanks and pipes. Over years of operation, large amounts of liquid hydrocarbons may be released, polluting soil and groundwater resources below the site. Massive groundwater cleaning projects have been required at several West Coast refineries.

Crude oil and its distilled products are also used to make fine oil, plastics (such as Styrofoam), and organic chemicals. The various industrial processes involved have the potential for releasing a variety of pollutants into the environment.

Delivery and Use

The most extensive and significant environmental problems associated with oil and gas occur when the fuel is delivered and consumed. Crude oil is mostly transported on land in pipelines or across the ocean by tankers. Both methods have the potential to produce oil spills, however, marine oil spills are better known. Although the effects are usually relatively short lived, marine spills have killed thousands of seabirds, temporarily spoiled beaches, and caused loss of tourist revenue (see Chapter 20).

Air pollution is perhaps the most familiar and serious environmental impact associated with the use (burning) of oil. Combustion products from burning gasoline in automobiles produce primary and secondary pollutants that contribute to urban smog (adverse effects of smog on vegetation and human health are well documented and are discussed in detail in Chapter 22).

Oil Shale and Tar Sands

Oil shale is a fine-grained sedimentary rock containing organic matter (kerogen). When heated to 500°C (900°F) in a process known as destructive distillation, oil shale yields significant amounts of hydrocarbons, up to nearly 60 liters (14 gal) of oil per ton of shale. If not for the heating process, these hydrocarbons would remain locked in the rock. Shale oil is one of the so-called **synfuels** (from *synthetic* plus *fuel*), which are liquid or gaseous fuels derived from solid fossil fuels, such as oil from the kerogen in oil shale or oil and gas from coal. The best known oil shales in the United States are those in the Green River formation, which underlies approximately 44,000 km² (17,000 mi²) of Colorado, Utah, and Wyoming.

Total identified world oil shale resources are estimated to be about 3 trillion bl of oil. However, evaluation of the oil grade and the feasibility of economic recovery with today's technology is not complete. Oil shale resources in the United States amount to about 2 trillion bl of oil, or two-thirds of the total identified in the world. Of this, 90%, or 1.8 trillion bl, is located in the Green River oil shales.[6,7]

Environmental Impact

The environmental impact of developing oil shale resources varies with the recovery technique used. At present, surface and subsurface mining as well as in-place (in situ) techniques are being considered.

Surface mining, in either open-pit or strip mines, is attractive to developers because nearly 90% of the shale oil can be recovered, compared with less than 60% by underground mining. However, waste disposal will be a major problem with either surface or subsurface mining, which requires that the oil shale be processed, or *retorted* (crushed and heated) at the surface. The volume of waste will exceed the original volume of shale mined by 20% to 30%. Therefore, the mines from which the shale is removed will not be able to accommodate all the waste, and it will have to be piled up or otherwise disposed of.[8]

In the 1970s it was thought that there would be a tremendous rush to develop oil shale. This interest was fired by the oil embargo in 1973 and by concern that continued shortages of crude oil were likely. In the 1980s through the mid 1990s there was plenty of cheap oil, and oil shale development was put on the back burner; it is much more expensive to extract a barrel of oil from oil shale than to pump it from a well. However, it is likely that shortages will occur at some time in the future, and if an oil shortage occurs we may turn to oil shale. This would result in significant environmental, social, and economic impacts in the oil shale areas resulting from rapid urbanization to house a large work force, construction of industrial facilities, and increased demand on water resources.

Tar sands are sedimentary rocks or sands impregnated with tar oil, asphalt, or bitumen. Petroleum cannot be recovered from tar sands by pumping wells or other usual commercial methods because the oil is too viscous to flow easily. Seventy-five percent of the world's known tar sand deposits are in Canada. The total resource is about 1000 billion bl, but it is not known how much of this will eventually be recovered. Oil in tar sands is recovered by first mining the sands (which are very difficult to remove) and then washing the oil out with hot water.

In Alberta, Canada, tar sand is mined in a large open-pit mine. The mining process is complicated by the fragile native vegetation, a water-saturated mat known as a muskeg swamp, a kind of wetland that is difficult to remove except when frozen. Restoration of this fragile environment after mining is difficult. In addition, there is a waste disposal problem because the mined sand material occupies a greater volume than unmined material. The land surface after mining can be up to 20 m (66 ft) above the original ground surface.

16.3 COAL

When partially decomposed vegetation is buried deeply in a sedimentary environment, it may be slowly transformed into the solid, brittle, carbonaceous rock we call **coal.** This process is shown in

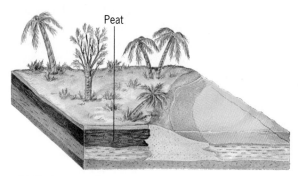

(*a*) Coal swamps form.

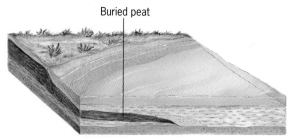

(*b*) Rise in sea level buries swamps with sediment.

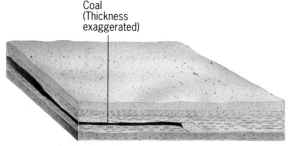

(*c*) Compression of peat forms coal.

FIGURE 16.6 Processes by which buried plant debris (peat) is transformed into coal.

Figure 16.6. Coal is by far the world's most abundant fossil fuel, with a total recoverable resource of about 1000 billion metric tons (Figure 16.7). Because the annual world consumption is about 4 billion metric tons, this resource should last about 250 years at current usage rates. However, if consumption of coal increases in the coming decades, the resource will not last nearly so long.[9]

Coal is classified according to its carbon and sulfur content (see Table 16.1). Energy content is greatest in anthracite coal, which has relatively few volatiles (oxygen, hydrogen, and nitrogen) and a low moisture content, and therefore the highest percentage of carbon. Energy content is lowest in lignite, which is high in moisture. The distribution of the common coals (*anthracite, bituminous, subbituminous,* and *lignite*) in the contiguous United States is shown in Figure 16.8. Most of the *low-sulfur coal* in the United States is relatively low-grade, low-energy lignite and subbituminous coal found west of the Mississippi River. The location of low-sulfur coal has environmental significance because low-sulfur coal causes less air pollution and is more desirable as a fuel for power plants. To avoid excessive air pollution, thermal power plants on the East Coast treat some of the local coal to lower its sulfur content, using methods discussed later on. Although it is expensive, treating the coal may be more economical than transporting low-sulfur coal from the western states.

Coal and the Environment

Most coal mining in the United States is done by *strip mining,* surface mining in which the overlying layer of soil and rock is stripped off to reach the coal. The practice of strip mining started in the late nineteenth century and has steadily increased because it tends to be cheaper and easier than underground mining. More than 40 billion metric tons of coal reserves are now accessible to surface-mining techniques. In addition, approximately another 90 billion metric tons of coal within 50 m (165 ft) of the surface is potentially available for strip mining. It is likely that more and larger strip mines will be developed as the demand for coal increases.

TABLE 16.1 U.S. Coal Resources

			Sulfur Content (%)		
Type of Coal	*Relative Rank*	*Energy Content (millions of joules/kg)*	*Low (0–1)*	*Medium (1.1–3.0)*	*High (3+)*
Anthracite	1	30–34	97.1	2.9	—
Bituminous coal	2	23–34	29.8	26.8	43.4
Subbituminous coal	3	16–23	99.6	0.4	—
Lignite	4	13–16	90.7	9.3	—

Sources: U.S. Bureau of Mines Circular 8312, 1966; P. Averitt, 1973, "Coal," in D. A. Brobst and W. P. Pratt, eds., United States Mineral Resources, *U.S. Geological Survey, Professional Paper 820,* pp. 133–142.

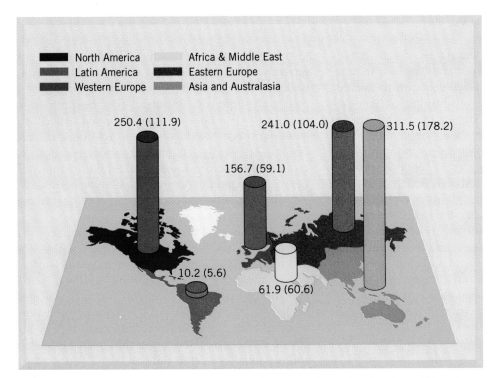

FIGURE 16.7 World coal reserves (billions of tons) in 1997 (share of anthracite and bituminous coal is given in parentheses). The United States has about 24% of the total reserves. (*Source:* British Petroleum Company, 1998, *BP Statistical Review of World Energy*.)

FIGURE 16.8 Coal areas of the conterminous United States. This is a highly generalized map and numerous relatively small occurrences of coal are not shown. (*Source:* S. Garbini and S. P. Schweinfurth, 1986, *U.S. Geological Survey Circular 979*.)

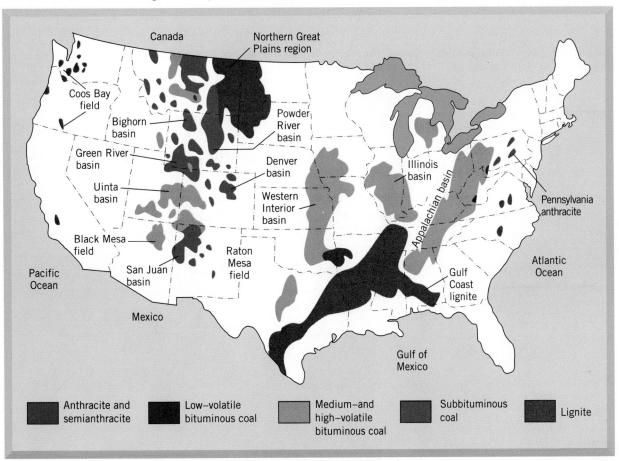

In the United States thousands of square kilometers of land have been disturbed by coal mining, and only about half this land has been reclaimed. Reclamation is the process of restoring and improving disturbed land often by reforming the surface and replanting vegetation (see Chapter 9). Unreclaimed coal dumps from underground mines and open-pit mines are numerous and continue to cause environmental problems. Abandoned coal dumps are so numerous in eastern Pennsylvania that they can be mapped using satellite images. Because little reclamation occurred prior to about 1960, abandoned mines are still common in the United States. One strip mine in Wyoming, abandoned more than 40 years ago, caused a disturbance so intense that vegetation has still not been reestablished on the waste dumps. Such barren, ruined landscapes emphasize the need for reclamation.[9]

Strip Mines

The impact of large strip mines varies from region to region depending on topography, climate, and reclamation practices. In areas of the eastern United States with abundant rainfall, drainage of acid water from mine sites is a serious problem (see Chapter 20). Surface water infiltrates the spoil banks (rock debris left after the coal or other materials are removed), where it reacts with sulfide minerals, such as pyrite (FeS_2),

to produce sulfuric acid, which pollutes streams and groundwater resources. Acid water also drains from underground mines and roads cut in areas where coal and pyrite are abundant, but the problem is magnified when large areas of disturbed material remain exposed to surface waters. Acid drainage can be minimized by channeling surface runoff or groundwater before it enters a mined area and diverting it around the potentially polluting materials.[10]

In arid and semiarid regions, water problems associated with mining are not as pronounced as in wetter regions, but the land may be more sensitive to activities related to mining, such as exploration and road building. In some arid areas of the western and southwestern United States, the land is so sensitive that tire tracks can remain for years. Wagon tracks from the early days of the westward migration reportedly have survived in some locations. To complicate matters, soils are often thin, water is scarce, and reclamation work is difficult.

All strip mining has the potential to pollute or destroy scenic, water, biological, or other land resources. However, good reclamation practices can minimize the damage (Figure 16.9). Although reclamation practices necessarily vary by site, the important principles of modern reclamation are emphasized in the case history of a modern coal mine in Colorado. (See a Closer Look 16.1, "The Trapper Mine.")

FIGURE 16.9 Open-pit coal mine in Wyoming. The land in the foreground is being mined and the green land in the background has been reclaimed following mining.

CLOSER LOOK 16.1

THE TRAPPER MINE

The Trapper Mine on the western slope of the Rocky Mountains in northern Colorado is a good example of a new generation of large coal strip mines. The operation is designed to minimize environmental degradation during the mining and to reclaim the land for dryland farming and grazing of livestock and big game.

Over a 35-year period the mine will produce 68 million metric tons of coal from the 20–24 km^3 (4.8–5.8 mi^3) site, to be delivered to an 800-MW power plant located adjacent to the mine. Four coal seams, varying from about 1–4 m (3.3–13.1 ft) thick, will be mined. The seams are separated by layers of rock, called overburden, and there is additional overburden above the top seam of coal. The depth of the overburden varies from 0 to about 50 m (165 ft).

A number of steps are involved in the actual mining. First, bulldozers and scrapers remove the vegetation and topsoil from an area up to 1.6 km long and 53 m wide (1 mile by 175 ft), and the soil is stockpiled for reuse. Then the overburden is removed with a 23-m^3 (800-ft^3) dragline bucket. Next, the exposed coal beds are drilled and blasted to fracture the coal, which is removed with a backhoe and loaded into trucks (Figure 16.10). Finally, the cut is filled, the topsoil replaced, and the land is either planted with a crop or returned to rangeland.

At the Trapper Mine the land is reclaimed without artificially applying water. The precipitation (mostly snow) is about 35 cm/year (about 14 in./year), which is sufficient to reestablish vegetation provided there is adequate topsoil. The fact that reclamation is possible at this site emphasizes an important point about reclamation: It is site specific. What works at one location may not be applicable to other areas.

Water and air quality are closely monitored at the Trapper Mine. Surface water is diverted around mine pits, and groundwater is intercepted while pits are open. Settling basins, constructed downslope from the pit, allow suspended solids in the water to settle before the water is discharged into local streams. Although air quality at the mine is degraded by dust produced from the blasting, hauling, and grading of the coal, dust is minimized by regularly watering or otherwise treating roads and other surfaces.

Reclamation at the Trapper Mine has been successful during the first years of operation. Although reclamation increases the cost of the coal by as much as 50%, it will pay off in long-range productivity of the land as it is returned to farming and grazing uses. It might be argued that the Trapper Mine is unique in its combination of geology, hydrology, and topography, which has allowed for successful reclamation. To some extent this is true, and perhaps the Trapper Mine presents an overly optimistic perspective on mine reclamation. On the other hand, the success of the mine operation demonstrates that, with careful site selection and planning, the development of energy resources can be compatible with other land uses.

(a)

(b)

FIGURE 16.10 (*a*) Mining and exposed coal bed at the Trapper Mine, Colorado, and (*b*) the land during restoration following mining. Top soil (lower right) is spread prior to planting vegetation.

Regulations

Since the adoption of the Surface Mining Control and Reclamation Act of 1977, the U.S. government has required that mined land be restored to support its premining use. The regulations also prohibit mining on prime agricultural land and give farmers and ranchers the opportunity to restrict or prohibit mining on their land, even if they do not own the mineral rights. Reclamation includes disposing of wastes, contouring the land, and replanting vegetation. Reclamation is often difficult, and it is unlikely that it will be completely successful. In fact, some environmentalists argue that success stories with reclamation are the exception and that strip mining should not be allowed in the semiarid southwestern states because reclamation is uncertain in that fragile environment.

Underground Mining

Underground mining still accounts for approximately 40% of the coal mined in the United States. There are many abandoned underground mines, particularly in the eastern U.S. coalfields of the Appalachian Mountains. Underground coal mining is a dangerous profession; there are always hazards of collapse, explosion, and fire. Respiratory illnesses are a risk, especially black lung disease, which is related to exposure to coal dust and has killed or disabled many miners.

Some of the environmental problems associated with underground mining are listed here:

FIGURE 16.11 Subsidence below coal mines in the Appalachian coal belt.

- Acid mine drainage from the mines and waste piles has polluted thousands of kilometers of streams (see Chapter 20).
- Land subsidence can occur over mines. Vertical subsidence occurs when the ground above coal mine tunnels collapses, often resulting in a crater-shaped pit at the surface that can develop quickly (Figure 16.11). Coal mining areas in Pennsylvania and West Virginia, for example, are well known for serious subsidence problems. In recent years a parking lot and crane collapsed into a hole over a coal mine in Scranton, Pennsylvania, and damage from subsidence caused condemnation of many buildings in Fairmont, West Virginia.
- Coal fires in underground mines may be either naturally caused or deliberately set. The fires may belch smoke and hazardous fumes, causing people exposed to them to suffer from a variety of respiratory diseases.

Transport of Coal

Getting the energy in coal from mining areas with low-energy demand to large population centers where the energy is needed is a significant environmental issue. Although coal may be converted at the production site to more easily transportable electricity, synthetic oil, or synthetic gas, these alternatives have their own problems. Power plants necessary to convert coal to electricity require water for cooling, and in semiarid coal regions of the western United States there may not be sufficient water. Furthermore, transmission of electricity over long distances is inefficient and expensive (see Chapter 15). The conversion of coal to synthetic oil or gas also requires a tremendous amount of water; the conversion process is expensive; and the technology is primitive.[11]

To transport the coal itself over long distances, freight trains and coal-slurry pipelines (designed to transport pulverized coal mixed with water) are options. Trains are typically used and will continue to be used because they provide relatively low-cost transportation compared with the cost of constructing pipelines. The economic advantages of slurry pipelines are tenuous, especially in the western United States, where large volumes of water are difficult to obtain.[11]

The Future of Coal

Coal is burned to produce nearly 60% of the electricity used, and about 25% of the total energy consumed in the United States today.[12] Coal accounts for nearly 90% of the fossil fuel reserves in the United States, and we have enough coal to last at least several hundred years. However, serious concern has been raised about burning that coal. The giant power plants that burn coal as a fuel to produce electricity in the United States are responsible for about 70% of the total emissions of sulfur dioxide, 30% of the nitrogen oxides, and 35% of the carbon dioxide (the effects of these pollutants are discussed in Chapter 22).

By contrast, the burning of natural gas produces only about half as much carbon dioxide and very little sulfur dioxide.[13] Recent legislation as part of the Clean Air Amendments of 1990 mandates that sulfur dioxide emissions be eventually cut by 70% to 90% depending upon the sulfur content of the coal and roughly in half by the year 2000. Nitrogen oxide emissions must be reduced by about 2 million metric tons per year. As a result of this legislation, the utility companies are struggling with various new technologies designed to reduce emissions of sulfur dioxide and nitrogen oxides from burning coal. Options being used or developed include the following list[13]:

- Chemical and/or physical cleaning of coal prior to combustion.
- New boiler designs that require a lower temperature of combustion, which reduces emissions of nitrogen oxides.
- Injection of material rich in calcium carbonate (such as pulverized limestone or lime) into the gases following burning of the coal. This is known as **scrubbing.** The carbonate reacts with sulfur dioxide, producing calcium sulfate as a sludge. The sludge is collected, thus removing the sulfur from the emissions.
- Finally, consumer education about energy conservation and efficiency, to reduce the demand for energy and thus the amount of coal burned and emissions released.

The real shortages of oil and gas are still a few years away, but when they do come, they will put tremendous pressure on the coal industry to open more and larger mines in both the eastern and western coal beds of the United States. Increased use of coal could have tremendous environmental impacts for several reasons. First, more and more land will be strip mined and thus will require careful restoration. Second, unlike oil and gas, burned coal produces large amounts of combustion products and by-products, including ash (5%–20% of the original amount of the coal becomes ash that is collected and disposed of); boiler slag (a vesicular rocklike cinder produced in the furnace); and sludge, produced from removing sulfur (scrubbing). Coal burning power plants in the United States today produce about 90 million tons of these materials. Calcium sulfate from scrubbing can be used to make wallboard and other products, and some slag can be used in construction projects. Nevertheless, about 75% of the combustion products of burning coal in the United States today end up on waste piles or disposed of in landfills.[14] Third, the handling of tremendous quantities of coal through all stages (mining, processing, shipping, combustion, and final disposal of ash) will have potentially adverse environmental effects. These include aesthetic degradation, noise, dust, and, most significant from a health standpoint, release of harmful or toxic trace elements into the water, soil, and air.[15]

It seems unlikely that coal will be abandoned in the near future in the United States because we have so much of it and we have spent so much time and money developing coal resources. It has been suggested that we should now promote the use of natural gas in preference to coal because it burns so much cleaner. However, there is a concern that we might then become dependent on imports of natural gas. Regardless, it remains a fact that coal is perhaps the most polluting of all the fossil fuels.

Allowance Trading

An innovative approach to managing our coal resources and reducing pollution is allowance trading. In this system, the Environmental Protection Agency grants utility companies tradable allowances for polluting. One allowance is good for up to 1 ton of sulfur dioxide emissions per year. In theory, some companies would not need all their allowances because they use low-sulfur coal or have installed equipment and methods that have reduced their emissions. Their extra allowances may then be traded to brokers and sold to other utilities that are unable to stay within the emission levels they have been allocated. The idea is to encourage competition in the utility industry and reduce overall pollution through economic market forces.[13]

Some environmentalists are not comfortable with the concept of allowance trading. They argue that, although buying and selling may be profitable to both parties in the transaction, it is less acceptable from an environmental viewpoint. That is, they believe that companies should not be able to buy their way out of pollution problems. Another negative factor is that, although the heavy polluters will be within the regulations because they have purchased sufficient allowances, people and other living things near these high-pollution sources will continue to suffer.

Should the Gasoline Tax Be Raised?

Fossil fuels are a limited and nonrenewable resource that must be conserved and used wisely during a period of transition to other energy sources. Petroleum, which occurs naturally in liquid form and is easily portable, is particularly well suited to transportation. The number of motor vehicles, most of which use petroleum as fuel, increased eight times between 1950 and 1989, and is expected to double in the next 20–30 years. In 1989 there were a half billion motor vehicles in the world; 34%, or 190 million, were in the United States alone. Each day, the United States uses about 11.4 million barrels of oil for transportation, 68% of the country's total oil consumption.

In addition to the drain on a limited resource, the use of petroleum in transportation contributes significantly to air pollution. Approximately 60% of the more than 170 million tons of air pollutants annually emitted into the atmosphere in the United States is attributed to automobiles. Gasoline accounts for 25% of the carbon dioxide emissions from burning fossil fuels in the United States. Each gallon of gasoline burned adds 8.6 kg (19 lb) of carbon dioxide to the air. A car that runs for 160,000 km (100,000 miles) with an average fuel consumption of 11.7 km/liter (27.5 miles/gal) emits 35 metric tons of carbon dioxide. Vehicles also emit carbon monoxide, nitrogen oxides, and organic chemicals called hydrocarbons. In the presence of sunlight, the hydrocarbons react with the nitrogen oxides to produce ozone. Carbon monoxide competes with oxygen in the blood, nitrogen oxides form nitric acid in the presence of water, some organic molecules are carcinogenic, and ozone is an irritant to the lungs.

Proposals for reducing the dependence on petroleum for transportation, thereby slowing the drain on oil resources and decreasing air pollution, include designing more efficient vehicles, using alternative fuels, shifting to other modes of transportation, such as mass transit and bicycles, improving roads, and encouraging consumers to conserve gasoline. Some energy policymakers have suggested that increasing the tax on gasoline would encourage Americans to purchase more fuel-efficient vehicles as well as discourage unnecessary use of motor transportation.

One of the main arguments advanced by proponents of higher gasoline taxes is that U.S. gasoline taxes are only 12% to 18% that of other industrialized countries (see graph) and per-capita use of gas is far greater. When adjusted for inflation, gasoline prices in 1993 were

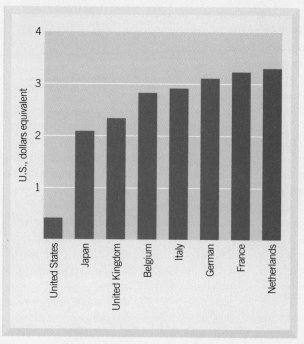

Taxes for 1 gal of gasoline in U.S. dollars equivalent for selected countries, 1995. (*Source:* Federal Highway Administration, U.S. Department of Transportation, August 1995, Monthly Motor Fuel Reports by States, Washington, DC.)

lower than at any other time since the mid–1960s. Every penny of tax would generate a billion dollars in revenue for the government. Opponents contend that higher taxes at the pump would have a greater impact on poorer families than on those who are better off and would unfairly punish those living in western states, which typically have lower populations and greater driving distances. As with many controversial issues, the facts themselves are in question. Proponents of the tax contend that the poor do not spend proportionally more on transportation and would not be adversely affected. Furthermore, they say that westerners spend only 9% more than easterners on gasoline.

Not surprisingly, higher gas taxes have been advocated by many environmentalists. But automobile companies have also supported the proposal, hoping that it would eliminate pressure for more stringent fuel-efficiency standards. Americans were divided on the issue in 1993, with 52% supporting a 15-cent tax increase, 46% opposed, and 2% not sure.

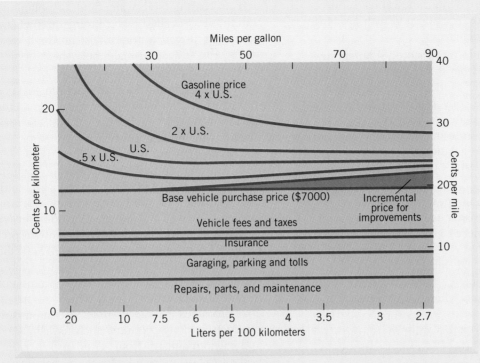

Estimating cost of driving with increasing fuel economy.

Critical Thinking Questions

1. Refer to the graph above to answer the following questions:
   a. What is the relationship of fuel efficiency (mpg) to gasoline price (cents per mile)?
   b. What incentive would doubling the price have on consumer choice of more fuel-efficient vehicles?

2. Make a table of the pros and cons of raising the price of gasoline through taxation in relation to environmental, economic, and social issues. What are both direct and indirect effects?

3. When you have completed the table, evaluate the arguments and decide where you stand on the issue. Write a paragraph explaining your position.

4. Some ethicists have suggested that one way of viewing issues such as the gasoline tax is to imagine that you are setting up rules for a card game and do not know which cards you hold. Most people, however, view social issues knowing what cards they hold and adopt a position in their own best interest. Examine your own position from the first point of view. Would this change your position? Explain.

References

Bleviss, D. L. 1988. *The New Oil Crisis and Fuel Economy Technologies.* New York: Quorum Books.

Bleviss, D. L., and Walzer, P. 1990. "Energy for Motor Vehicles," *Scientific American* 263(3): 103–109.

Corson, W. H., ed. 1990. *The Global Ecology Handbook.* Boston: Beacon. 1993 (January 18). "Driving Down the Deficit," *U.S. News and World Report* 114(2): 58–60.

Energy Information Administration. 1996 (January). *Monthly Energy Review.* Washington, DC: U. S. Department of Energy.

Greenwald, J. 1993. "Why Not a Gas Tax?" *Time* 141(7): 25–27.

Miller, E. W. and Miller, R. M. 1993. *Energy and American Society. A Reference Handbook.* Santa Barbara, CA: ABC-CLIO.

Nadis, S., and MacKenzie, J. J. 1993. *Car Trouble.* Boston: Beacon.

SUMMARY

- The United States has an energy dilemma with respect to fossil fuels. We have over 100 years of experience with fossil fuels; their use has allowed our society to industrialize and develop; and they provide 90% of the energy we consume to produce our high standard of living.

However, they are a nonrenewable resource and their utilization is associated with growing environmental problems.

- Fossil fuels are all forms of stored solar energy. They are created from the incomplete biological decomposition of dead organic material that is buried and converted to fossil fuels by complex chemical reactions in the geologic cycle. Coal forms from buried organic material such as swamp vegetation; oil and gas form from organic material buried with marine or lake sediments.

- Fossil fuels are nonrenewable; the entire period of their extensive use will only be a period of a few hundred years. We will eventually have to develop other sources to meet our energy demands. We must decide when the transition to alternative fuels will occur, and we must determine what the impacts of the transition will be.

- Environmental impacts related to oil and natural gas include those associated with exploration and development (damage to fragile ecosystems, water pollution, air pollution, and waste disposal); those associated with refining and processing (soil, water, and air pollution); and those associated with burning the fuels for energy to power automobiles, produce electricity, run industrial machinery, heat homes, and so on (air pollution).

- Coal is considered by many people to be a source of energy particularly damaging to the environment. The environmental impacts associated with mining, processing, transporting, and using coal are many. Problems associated with surface or subsurface mining include fires, subsidence, acid mine drainage, and difficulties related to land reclamation. Burning coal can release air pollutants, including heavy metals and sulfur dioxide (SO_2), a known precursor to acid rain. It also releases 35% of the total carbon dioxide emissions in the United States and is therefore a significant contributor to potential global warming. Finally, burning coal produces a large volume of combustion products and by-products such as ash, slag, and calcium sulfate (from scrubbing). It is important to find more uses for these materials to minimize the amount that must be disposed of as waste.

REEXAMINING THEMES AND ISSUES

Human Population: As human population has increased, particularly in industrial societies with high population densities, the use of fossil fuels and its potential pollution has also increased. As a result, we need to continually evaluate the *per-capita* human impact of burning coal, petroleum, and natural gas and reduce the pollution impact.

Sustainability: It has been argued that sustainable development and maintenance of a quality environment for future generations will be extremely difficult if we continue to increase our consumption of fossil fuels. Burning these fuels results in many known and potentially adverse effects on the environment. Obtaining sustainability will probably require wider use of a variety of energy sources with less dependency upon fossil fuels. We will also have to find more uses for the combustion products and by-products of burning coal, that is, turn wastes and pollutants into useful products.

Global Perspective: Various aspects of the global environment are affected by burning of fossil fuels. This is particularly true for the fast-moving atmospheric processes (see Chapters 21 and 22). As a result, an important issue remains as to how and when we will eventually switch from fossil fuels to other or alternative energy sources.

Urban World: Burning of fossil fuels in urban areas has a long history of problems. It was not too many years ago that black soot from burning coal covered the buildings of most of the major cities of the world. Today we are striving to improve our urban environments and as a result have taken a critical look at energy sources and urban areas to minimize urban environmental degradation.

Values and Knowledge: We have a lot of scientific information concerning the effects of burning fossil fuels. Much of the controversy relating to their use has to do with values. Do we value economic growth even though it means burning more fossil fuels and perhaps causing increased environmental problems? Or do we instigate further regulations to try to improve air quality? We know that air pollution resulting from burning of fossil fuels causes health problems to people and other living things; this is in conflict with our values of trying to improve the environment and the health of ecosystems.

KEY TERMS

coal *331*
crude oil *326*
fossil fuels *326*

hydrocarbons *326*
natural gas *326*
oil shale *331*

scrubbing *337*
synfuels *331*
tar sands *331*

STUDY QUESTIONS

1. Why are we in a dilemma with respect to fossil fuels? What can we do to get out of the dilemma?

2. What are the comparable potential environmental consequences of burning oil, burning natural gas, and burning coal?

3. What actions can you take at a personal level to reduce consumption of fossil fuels?

4. What potential environmental and eco-nomic problems could result from a rapid transition from fossil fuels to alternative sources?

5. What are some of the potential technical solutions to reducing air-pollutant emissions from burning coal. Which are best? Why?

6. What do you think about the idea of allowance trading as a potential solution to pollution resulting from burning coal?

FURTHER READING

Brookins, D. G. 1990. *Mineral and Energy Resources*. Columbus, OH: Merrill Publishing. This book was written by a geologist and has a good description of how fossil fuels form and how they are being used today. Also provided is information concerning the environmental effects of burning fossil fuels.

Fulkerson, W., Judkins, R. R., and Sanghvi, M. K. 1990. "Energy from Fossil Fuels," *Scientific American* 263(3): 128–135. A good overview of fossil fuels in terms of sources, consumption and environmental considerations.

Liu, P. I. 1993. *Introduction to Energy and the Environment*. New York: Van Nostrand Reinhold. A good summary of energy sources and issues with discussions of the environmental effects of various energy sources.

Miller, E. W., and Miller, R. M. 1993. *Energy and American Society. A Reference Handbook*. ABC-CLIO. Follows patterns of energy use from early American history through today and includes looks at future energy sources.

Vergara, W., Hay, N. E., and Hall, C. W. eds. 1990. *Natural Gas—Its Role and Potential in Economic Development*. Boulder, CO: Westview. A good discussion of the role of natural gas in providing future energy, including a chapter on how it compares with other energy sources in terms of environmental effects.

INTERNET RESOURCES

British Petroleum Company: *http://www.bpamoco.com*—Contains statistics about past, present, and future world fuel consumption, resources, and other information related to energy.

U.S. Department of Energy Energy Information Administration: *http://www.eia.doe.gov*—The EIA provides statistics and information on our oil and natural gas outlook, consumption, and supplies.

U.S. Department of Energy Office of Fossil Energy: *http://www.fe.doe.gov/*—A gateway to internet information on U.S. fossil fuel energy programs and projects. This site also supplies reports on those projects.

U.S. Geological Survey: *http://www.usgs.gov*—Contains statistics of present and predicted U.S. fossil fuel resources.

Interstate Oil and Gas Compact Commission: *http://www.iogcc.ok-lasf.state.ok.us/*—An excellent source for other internet resources in fossil fuel information.

National Petroleum Council: *http://npc.org/npc/index.html*—View summaries of reports from the group charged with advising the DOE on petroleum issues and matters.

Oil Online: *http://www.oilonline.com/*—Serving the oil industry, this site has an abundance of information on the U.S. oil industry.

Solar panels in California.

CHAPTER 17

ALTERNATIVE ENERGY AND THE ENVIRONMENT*

CASE STUDY

Is Large–Scale Centralized, Alternative Energy Likely?

On November 25, 1991, Luz International, the world's largest solar electric company and producer of over 95% of the world's solar energy, filed for bankruptcy. The company was founded in 1980 and was successful in developing nine solar thermal electric energy plants in the Mojave Desert of Southern California (Figure 17.1). The operation, which supplied electrical energy sufficient for approximately 540,000 people, was clearly demonstrating that solar power might be a financially viable enterprise. The cost of the electricity it generated was approximately 12 cents per kilowatt-hour from 1986 to 1988, and plants built between 1988 and 1990 reduced that price to 8 cents per kilowatt-hour. It was estimated that plants originally scheduled for construction in 1994–1995 would be able to produce electrical energy at a cost of approximately $6\frac{1}{2}$ cents per kilowatt-hour, which would be very competitive with nuclear en-

*Written with assistance from Mel S. Manalis.

ergy and electricity produced from burning fossil fuels.[1]

Advocates of solar energy plants argue that Luz's failure was due not to technological failures or mismanagement but to the lack of a coherent energy policy in the United States and the world today.[1] The solar thermal electric-generating industry is a young one, and to survive the crucial years of research and development, it needs support through tax incentives. Such incentives were present when Luz was started, but there were recurring problems in renewal of the state and federal legislation that provided the support. According to advocates for solar energy, the failure of the company resulted from the failure of the regulatory bodies and the government to recognize the environmental benefits and potential economic competitiveness of large-scale solar energy utilization; they also argue that there will not be sufficient incentives for investors to support solar energy until there is a shift in our energy policy.[1]

Power Block Assembly:
1. *solar collector assembly*
2. *natural gas boiler*
3. *turbine generator*
4. *steam generator and solar superheater*
5. *control building*
6. *cooling tower*
7. *Southern California Edison interconnect*

FIGURE 17.1 Luz International Solar Farm showing the system of solar collectors (curved mirrors that heat a synthetic oil that flows through a heated exchanger to drive steam turbine generators). The system is explained in greater detail in discussion of solar energy.

Although Luz's existing plants are still operating after a reorganization that split the energy-producing division of Luz into three separate companies, research and development efforts remain stagnant due to loss of federal and state tax incentives. Those incentives and indirect subsidies had enabled Luz to continue research and development and to produce solar power at a lower cost than previous attempts. Prior to the loss of tax incentives, Luz depleted its financial reserves and borrowed money, resulting in a $30 million debt. It has been suggested that the present energy policy and law make it financially impossible for producers of alternative energy to grow large. In the case of Luz, they became too big too fast.[2] Proponents of solar energy hope that this will not signal an end to large-scale solar energy plants but instead will motivate those responsible for developing energy policy to direct financial resources to alternative energy sources that have a real possibility of competing with fossil fuels.

It should be noted that one of the reasons for the Luz company's initial success was that it augmented its solar energy production by burning natural gas. The mix was approximately 75% solar and 25% natural gas. Because this ratio is governed by regulation, it could not be increased. Some studies have suggested that if more natural gas could be used and mixed with solar energy, the cost of the electricity might drop below that of both conventional fossil fuels and nuclear power stations.[1]

The Luz International story suggests that the path to produce more alternative energy sources (or mixtures of fossil fuels and alternatives) will not be easy. This chapter explores alternative energy in terms of sources, advantages and disadvantages, and environmental concerns.

LEARNING OBJECTIVES

Alternatives to fossil fuels include geothermal energy, solar energy, water power, wind power, and biomass fuels. Some of these alternatives are already used and efforts are underway to develop others. After reading this chapter, you should understand:

- What geothermal energy is and how developing and using it affects the environment.

- What passive and active solar systems are, the advantages and limitations of each, and the environmental effects of developing and using the various types of active solar systems.

- What photovoltaics are, their present and potential uses, and the environmental effects of their development and use.

- Why hydrogen may be an important fuel of the future.

- What the advantages, disadvantages, and environmental impacts of developing hydropower are.

- Why there is considerable potential for utilizing wind power and how its development and utilization could affect the environment.

- What the potential environmental consequences are of using biomass as an energy source.

17.1 ALTERNATIVE ENERGY SOURCES

As discussed in the previous chapter, the primary energy sources used today are fossil fuels, which supply approximately 90% of the energy consumed by people. All other sources are considered **alternative energy** and are subdivided into two groups: **renewable energy** and **nonrenewable energy.** Nonrenewable alternative energy sources include nuclear energy (see Chapter 18) and geothermal energy for the most part. The renewable sources are solar energy, water (hydro) power, wind power, and energy derived from biomass. Today these renewable and nonrenewable alternative sources are supplying a small but growing portion of our total energy needs. However, renewable energy sources have a difficult time competing with the fossil fuels because oil prices are low in the late 1990s and fossil fuels have enormous advantages of completed infrastructure, government support, and political power. Significant changes in energy policy, increased government support of renewable energy sources, and probably an increase in the price of oil are necessary before alternative energy can become a "big player" in national and world energy production.

The most significant sources of alternative energy are hydropower and nuclear power, which on a worldwide basis account for approximately 3% and 7%, respectively, of energy consumed. Although we get more energy from nuclear power, the gap is narrowing because hydropower is growing in usage while nuclear power is not. Not included in energy statistics are fuels such as wood and animal waste (biomass). Reliable statistics are difficult to obtain for biomass, even though it is an important energy source in many countries throughout the world.[2]

Our understanding of energy consumption is changing as we move to a more efficient use of energy sources. As efficiency increases, energy utilization per capita decreases. Energy efficiency and conservation are important concepts in the development of future energy management plans.

A discussion of alternative energy is important from an environmental viewpoint because, as previously stated, we will eventually run out of our primary fossil fuel energy sources and will need to make a transition to new sources. Those concerned with our urban, regional, and global environments are arguing that this transition should take place as soon as is technically and economically possible to help reduce environmental problems such as urban air pollution, acid rain, and global warming.

17.2 GEOTHERMAL ENERGY

Geothermal energy is the useful conversion of natural heat from the interior of Earth to heat buildings and generate electricity. The idea of harnessing Earth's internal heat is not new. As early as 1904, geothermal power was developed in Italy, and natural internal heat is now being used to generate electricity in Russia, Japan, New Zealand, Iceland, Mexico, Hawaii, and California.

Geothermal energy may be considered a nonrenewable energy source when it is used at rates of extraction greater than natural replenishment. Most geothermal energy has its origin in the natural heat production within Earth, and only a small fraction of the vast total resource base is being utilized today. Although most geothermal energy production involves the tapping of high heat sources, we are also experimenting with low-temperature geothermal energy. Groundwater, with a nearly constant temperature of about 13°C (55°F), is being used for both summer cooling and winter heating.

Geothermal Systems

Natural heat production within Earth is only partly understood. We do know that some areas have a

FIGURE 17.2 Geysers Geothermal Field located north of San Francisco, California. The Geysers is the largest geothermal power operation in the world and produces the energy directly from steam.

higher flow of heat from below than do others and that for the most part these locations are associated with plate tectonic boundaries (see Chapter 4). Oceanic ridge systems (divergent plate boundaries) and convergent plate boundaries, where mountains are being uplifted and volcanic island arcs are forming, are areas where this natural heat flow is anomalously high. A region of natural high heat flow is concentrated in the western United States, where there has been recent tectonic and volcanic activity.[3]

On the basis of geologic criteria, several types of hot geothermal systems (temperature greater than about 80°C, or 176°F) have been defined, and the resource base is larger than that of fossil fuels and nuclear energy combined. A common system for energy development is hydrothermal convection, characterized by the circulation of steam and/or hot water that transfers heat from depths to the surface. An example is The Geysers, a geothermal system 145 km (90 miles) north of San Francisco (Figure 17.2), where about 2000 MW of electrical energy are produced. The Geysers is the largest geothermal power operation in the world.

Yellowstone National Park is an area with a significant geothermal potential. The park contains about 10,000 hot springs and geysers (features that include eruptions of hot water and steam, including the famous "Old Faithful" geyser). Federal legislation has been passed to protect the hot springs and geysers of Yellowstone. What is controversial is—what is required to ensure protection? Certainly there must be an adequate buffer zone to be sure that geothermal energy development outside the Park doesn't degrade Yellowstone's geothermal features!

Groundwater Systems

Groundwater systems are present nearly everywhere, but only recently has groundwater at constant temperature been viewed as a geothermal energy source. The temperature of groundwater at a depth of about 100 m (320 ft) is typically about 13°C (55°F). In terms of a bath or a swim this is cold. However, compared to winter temperatures in much of the United States it is warm, and compared to summer temperatures it is cool. In recent years there have been significant improvements in the design of devices that transfer heat from one material (e.g., groundwater) to another (e.g., air in a building). In the summer, buildings are cooled by transferring heat from the warm air in the buildings to the cool groundwater. In the winter, when the outdoor temperature is below about 4°C (40°F), the groundwater is warm by comparison. Then the air in a building is warmed by transferring heat from the groundwater to the air.

Geothermal systems using constant-temperature groundwater are now helping to warm and cool buildings and homes in several states, including New Jersey, Indiana, Kentucky, Oklahoma, and Utah. Such systems have a greater initial expense because they require drilling wells. However, as energy prices for fossil fuels rise in the future, these systems will become more attractive.

Use of Geothermal Energy and the Environment

The environmental impact of geothermal energy may not be as extensive as that of other sources of energy, but it is nevertheless considerable. When geothermal energy is developed at a particular site, environmental problems include on-site noise, emissions of gas, and disturbance of the land at drilling sites and for roads and pipes. Fortunately, development of geothermal energy does not require the large-scale transportation of raw materials or the refining common with fossil fuels. Furthermore, geothermal energy does not produce the atmospheric particulate pollutants associated with burning fossil

fuels or the radioactive waste associated with nuclear energy. However, geothermal development often does produce considerable thermal pollution from hot wastewaters, which may be saline or highly corrosive, producing disposal and/or treatment problems.

Geothermal power is not very popular in some locations among some people. For instance, geothermal energy has been produced for years on the island of Hawaii, where active volcanic processes provide abundant near-surface heat. There is controversy, however, over further exploration and development. Native Hawaiians and others have argued that the exploration and development of geothermal energy degrade the tropical forest as developers construct roads, build facilities, and drill wells. There also are religious and cultural issues in Hawaii that are probably related to geothermal energy. Some

FIGURE 17.3 Routes of various types of renewable solar energy.

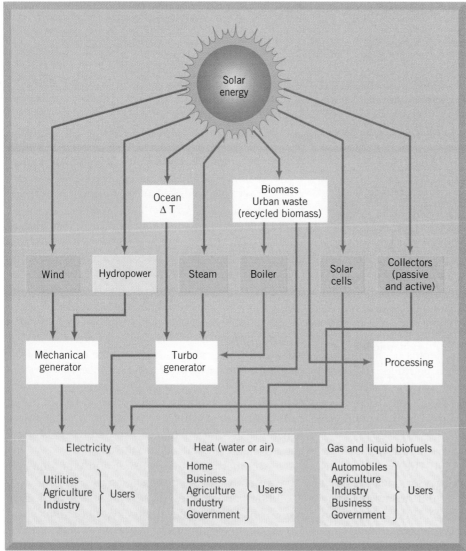

people may be offended by using the "breath and water of Pele" (the volcano goddess) to make electricity. This issue points out the importance of being sensitive to the values and cultures of people where development is planned.

At present, geothermal energy supplies only a small fraction of 1% of the electrical energy produced in the United States. With the exception of systems such as The Geysers in California, the production of electricity from geothermal reservoirs is still rather expensive compared to other sources of energy. Thus we cannot expect rapid commercial development of geothermal energy until the economic picture changes.

17.3 RENEWABLE ALTERNATIVE ENERGY SOURCES

The renewable energy sources—direct solar, water, wind, and biomass—are often discussed as a group because they are all derivatives of the sun's energy. That is, solar energy, broadly defined, comprises all the renewable energy sources, as shown in Figure 17.3. They are renewable because they are regenerated (renewed) by the sun within a time useful to humans.

These sources have the advantage of being inexhaustible and are generally associated with minimal environmental degradation. With the exception of burning biomass or its derivative, urban waste, solar energy sources (because no fuel is burned) do not pose a threat of increasing atmospheric carbon dioxide and thus modifying the climate. One major disadvantage is that many forms of solar energy are intermittent and spatially variable because they depend on sunshine. Furthermore, some of these sources are currently much more expensive than fossil fuels or nuclear energy. This is due in part to relatively large federal subsidies granted to domestic fossil fuel and nuclear energy compared to those for renewable energy.[4]

Another important aspect of renewable sources is the construction lead time necessary to implement the technology, which is often quite short relative to the development of new sources or the construction of power plants that utilize fossil or nuclear fuels.

17.4 DIRECT SOLAR ENERGY

The total amount of solar energy reaching Earth's surface is tremendous. For example, on a global scale, 10 weeks of solar energy is roughly equivalent to the energy stored in all known reserves of coal, oil, and natural gas on Earth. In the United States, on the average, 13% of the sun's original energy entering the atmosphere arrives at the surface (equivalent to approximately 177 W/m^2, or about 16 W/ft^2, on a continuous basis). However, the actual amount at any particular site is quite variable, depending on the time of the year and the cloud cover.[5]

Solar energy may be used directly through passive solar systems or active solar systems. **Passive solar energy systems** often involve architectural designs that enhance the absorption of solar energy and take advantage of the natural changes that occur throughout the year without requiring mechanical power (Figure 17.4). One simple technique is to design overhangs on buildings that block summer (high-angle) sunlight but allow winter (low-angle) sunlight to penetrate and warm rooms. Another technique is to build a wall that absorbs solar energy and emits it into a room, thus warming it. Many homes and other buildings in the southwestern United States as well as other parts of the country now use passive solar systems for at least part of their energy needs. **Active solar energy systems** require mechanical power, usually pumps and other apparatuses, to circulate air, water, or other fluids from solar collectors to a heat sink, where the heat is stored until used.

Solar Collectors

Solar collectors are usually flat panels consisting of a glass cover plate over a black background where water is circulated through tubes. Short-wave solar radiation enters the glass and is absorbed by the black background. Longer wave radiation is emitted from the black material, but it cannot escape through the glass, so it heats the water in the circulating tubes to 38° to 93°C (100–200°F).[5] Thus, solar collectors act as greenhouses. The number of systems using these collectors in the United States continues to grow.

Figure 17.5 shows a schematic of a flat-plate collector, with water as the heating fluid, and the basic plumbing used to run a solar water heater. Tax advantages in the United States until 1985 made systems such as this attractive to many people who lived in areas with consistent sunshine. The production of collectors used for water heating in homes and for space heating in homes and other buildings peaked in 1984 and then declined sharply. The decline is directly related to the fact that the federal energy tax credit for collectors for low- and medium-temperature uses expired in 1985.[6] Without such incentives the level of interest in solar energy is much lower, because present solar energy systems are expensive relative to the amount of hot water they produce. More

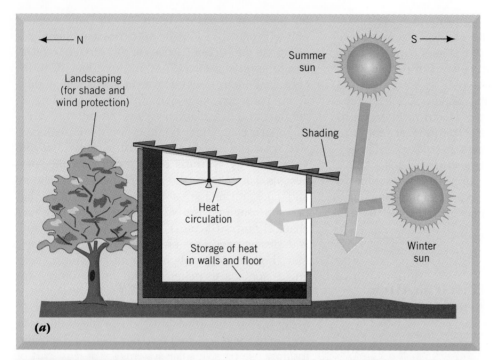

(a)

(b)

FIGURE 17.4 (*a*) Essential elements of passive solar design. High summer sunlight is blocked by the overhang, but low winter sunlight enters the south-facing window. Other features are designed to facilitate the storage and circulation of passive solar heat. (*b*) Design of this home utilizes passive solar energy. Sunlight enters through the windows and strikes a specially designed masonry wall that is painted black on the inside of the home. The masonry wall heats up and radiates this heat, warming the house during the day and into the evening. (*Source:* Moran, Morgan, and Wiersma, 1986, *Introduction to Environmental Science,* Freeman. Copyright by W. H. Freeman & Company. Reprinted with permission).

traditional methods of heating water are still preferred by most of the general public.

Photovoltaics

Photovoltaics is the term used for the technology that converts sunlight directly into electricity using a solid semiconductor material. The systems use solar cells made of silicon or other materials (such as gallium arsenide) and solid-state electronic components with few or no moving parts. The cells are constructed in standardized modules, which can be combined to produce systems of various sizes so that power output can be matched to the intended

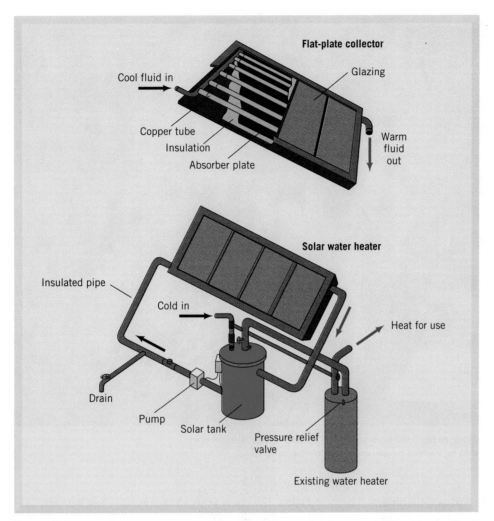

FIGURE 17.5 Detail of a flat-plate solar collector and pumped solar water heater. (*Source*: Farallones Institute, *The Integral Urban House*, 1979, Sierra Club Books. Copyright 1979 by Sierra Club Books. Reprinted with permission.)

use. Electricity is produced when sunlight strikes the cell, causing electrons to flow out of the cell through electrical wires. Photovoltaic cells can also work with light energy other than the sun, for example, the light of your desk lamp. The electricity in the circuit can be used to do work, such as lighting a bulb, running your calculator, or powering an electric appliance.

One of the major commercial uses for photovoltaics is in remote areas as power sources for equipment such as water-level sensors, meteorological stations, and space vehicles (Figure 17.6). Another use is in consumer products; many microsolar cells are used in hand-held calculators and in wristwatches.

Utility companies are interested in photovoltaics. Many experimental power plants, on a variety of scales ranging from solar cell panels on roofs of buildings to full power plants, are being evaluated. The largest photovoltaic power plant is in the Carrizo plain of central California, where about 5000 kW of power has been produced by large panels of solar cells.

There have been tremendous improvements in the technology of photovoltaics in the past few years. Currently there is an industrial race to produce the first low-cost systems that will be competitive with conventional fossil fuels as sources of energy. The efficiency of the conversion of solar energy in photovoltaics is now 10% to 25%, and the upper limit for efficiency is expected to be about 30%. Although this is not very high, what can improve the picture is the development of manufacturing techniques that would allow large-scale production at a lower cost. Some people believe that one day the price of solar cells will be so low that they might be built into the shingles that roof our homes and other buildings. The power produced by the cells could be stored in batteries (on site) or fed into the general electric grid (see Chapter 15). There are

FIGURE 17.6 Panels of solar cells are being used here to power a small refrigerator to keep vaccines cool. The unit is designed to be carried by camels to remote areas in Chad.

many benefits from the use of photovoltaics. They don't cause air pollution or release carbon dioxide into the environment, and the energy source will last as long as Earth does. Of course, photovoltaics do have some negative effects as well. Their production requires energy and chemicals and produces waste materials.

Power Towers

An interesting type of solar system is the **solar power tower** shown in Figure 17.7. The system works by collecting the heat of solar energy and delivering this energy in the form of steam to turbines that produce electric power. An experimental 10-MW (1 MW = 1,000,000 W) power tower near Barstow, California, is approximately 100 m (330 ft) high and is surrounded by approximately 2000 mirror modules, each with a reflective area of about 40 m² (430 ft²). The mirrors adjust continually to reflect as much sunlight into the tower as possible. However, because of their low efficiency and technical problems, power towers may prove not to be an economically viable technology for solar energy. Nevertheless, research to develop the technology continues, and a new generation of power towers that would use solar cells that turn sunlight directly

FIGURE 17.7 Solar power tower at Solar 1 in Barstow, California. Sunlight is reflected and concentrated at the central collector where the heat is used to produce steam to drive turbines and produce electric power.

to electricity without the intermediate step of making steam is being discussed.

Luz Solar Electric-Generating System

The system developed at Luz International is the most technically successful solar power experiment to date. The site in the Mojave Desert consists of nine solar farms, each comprising a power plant surrounded by hundreds of solar collectors. The system uses solar collectors (curved mirrors) to heat a synthetic oil that flows through heat exchangers that in turn drive steam turbine generators. The basic components of the system are shown in Figure 17.8. One advantage of this system is that it is modular, which allows for individual solar farms to be constructed in a relatively short period of time. The success of the system is based in part on the fact that it utilizes a natural gas burner backup system that ensures uninterrupted power generation both on cloudy days and during times of peak demand. Thus, the Luz system is really a combination of solar technology and conventional power generation. Although Luz International met financial difficulties, the system itself is very promising and illustrates that solar energy can be produced economically. Furthermore, if the prices of fossil fuels increase in the future, potential profits from the Luz system and others like it will undoubtedly increase substantially.[7]

Solar Ponds

The use of shallow, salt-gradient **solar ponds** to generate relatively low temperature water of about 40° to 70°C (104–158°F) is an interesting prospect for sources of commercial, industrial, and agricultural heat. The ponds are designed to collect incoming solar radiation, producing a bottom-water temperature of about 70°C. The heated water is kept on the bottom by the addition of salt, which makes it heavier. Circulation is restricted so that the dense bottom water does not mix with the water above. However, solar energy penetrates and warms the lower water, and heat may then be extracted from the bottom.[8]

In 1984 a pond in El Paso, Texas, with an area of 3200 m² (0.8 acre) was converted from a storage pond for fire protection to a salt-gradient solar pond. The El Paso Solar Pond Project was designed to sup-

FIGURE 17.8 Diagram illustrating how the Luz system works. (Courtesy of Luz International.)

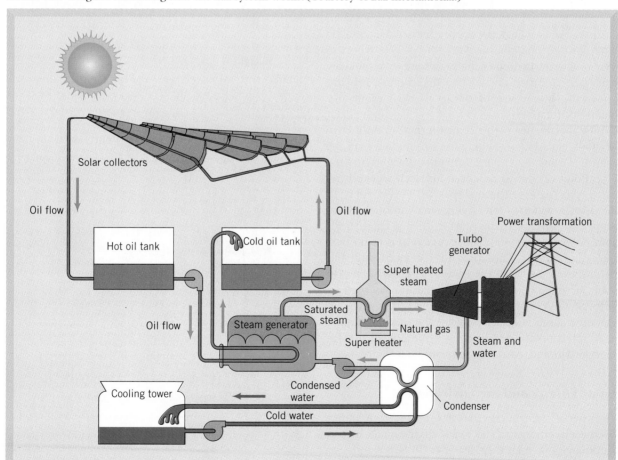

ply hot water to generate electricity for an adjacent food-processing plant. The water reached a temperature of 85°C (185°F) by the fall of 1986. During the summer it supplied electricity to the plant, and in the winter it supplied heat to the plant. Energy from the pond was also used to desalinate brackish (salty) water; in the late 1980s the desalination was producing about 19,000 liters (5000 gal) per day of drinking water.[8]

Ocean Thermal Conversion

A last example of direct utilization of solar energy involves using part of the natural oceanic environment as a gigantic solar collector. The process is called **ocean thermal conversion.** The surface temperature of ocean water in the tropics is often about 28°C (82°F). However, at the bottom of the ocean, at a depth as shallow as about 600 m (1970 ft), the temperature of the water may be 1° to 3°C (35–38°F). Equipment has been designed to exploit this temperature differential either by using the seawater directly or by using an appropriate heat exchange system in a closed cycle in which a fluid such as ammonia or propane is vaporized by the warm water. The expanding vapor propels a turbine and generates electricity. Following the generation of electricity, the vapor is cooled and condensed by the cold water. The construction of large-scale ocean thermal plants will depend on whether they can be built close to potential markets (the people who use the energy) and whether the plants are economically feasible.[5] Because answers to these questions are uncertain, the future of ocean thermal plants remains speculative.

Experiments with ocean thermal energy conversions have been ongoing for several years on the island of Hawaii. The experiments, located at Keahole Point on the Kona coast, are well situated because the seafloor there slopes off very steeply to deep, cold waters. Only a very small amount of energy has actually been produced, but there have also been experiments with aquaculture at the site (see Chapter 10). The cold, nutrient-rich water has been used by several companies to grow seafoods, such as salmon, oysters, and abalone, as well as kelp. Several of these organisms may be grown together in large tanks or ponds. When the laboratory was fully operating, it also used the cold water for air conditioning, reducing the utility bill.

Solar Energy and the Environment

The use of solar energy has a relatively low impact on the environment. The major disadvantage is that solar energy is relatively dispersed; a large land area is required to generate a large amount of energy. This problem is negligible when solar collectors can be combined with existing structures, as with the addition of solar hot-water heaters on the roofs of existing houses. Highly centralized and high-technology solar energy units, such as solar power towers, have a greater impact on the land. The impact of solar energy systems can be minimized by locating centralized systems in areas not used for other purposes and by making use of dispersed solar energy collectors on existing structures wherever possible.

The increased development of active solar energy technology may result in the widespread use of a variety of materials (metals, plastics, fluids, etc.) in the manufacture and use of solar equipment. This practice may cause environmental problems by producing toxic waste or by accidentally releasing toxic materials. Passive solar collectors, which use common materials such as water and rocks, pose negligible pollution problems.

Figure 17.9 shows in a very general way the estimated year-round availability of solar energy in the United States. However, solar energy is site specific, and detailed observation in the field is necessary to evaluate the solar energy potential in a given area. The energy that might be saved by converting to direct solar power is also quite variable. As recently as a few years ago, some optimistic people hoped that by the year 2000 direct solar energy

FIGURE 17.9 Estimated solar energy for the contiguous United States. (*Source:* Modified after Solar Energy Research Institute, 1978.)

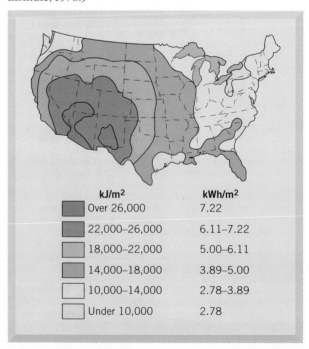

	kJ/m²	kWh/m²
	Over 26,000	7.22
	22,000–26,000	6.11–7.22
	18,000–22,000	5.00–6.11
	14,000–18,000	3.89–5.00
	10,000–14,000	2.78–3.89
	Under 10,000	2.78

would supply up to 20% of the total energy demand. Considering how slowly the solar energy industry is growing, these hopes have not proved to be realistic. Nevertheless, the future of solar energy is secure provided that incentives for developing the energy continue.

17.5 HYDROGEN

Hydrogen gas may be the fuel of the future. As a fuel, hydrogen can be used in any of the ways in which we normally use fossil fuels, such as to power automobile and truck engines or to heat water or buildings. When used in fuel cells (see A Closer Look 17.1, "Fuel Cells"), which are similar to batteries, it can also be used to produce electricity. Hydrogen, like natural gas, can be transported in pipelines and stored in tanks, and can be produced using solar and other renewable energy sources. It is a clean fuel; the combustion product of burning hydrogen is water, so it does not contribute to global warming, air pollution, or acid rain (see Chapters 21 and 22). It is expected that experimentation with hydrogen will continue and that the fuel produced may be substantially reduced in price. By the year 2000 the cost of hydrogen produced by solar sources may be equivalent to gasoline, at a cost of $3 to $5 per gallon.[9]

17.6 WATER POWER

Water power is a form of stored solar energy, as all weather and the flow of water on Earth are driven by the sun. Water power has been successfully harnessed since at least the time of the Roman Empire. Waterwheels that harness water power and convert it to mechanical energy were turning in western Europe in the seventeenth century; during the eighteenth and nineteenth centuries large waterwheels provided energy to power grain mills, sawmills, and other machinery in the United States. Today, hydroelectric power plants provide about 15% of the total electricity produced in the United States. Figure 17.11a shows the basic components of a hydroelectric power station.

Small-Scale Systems

Although the total amount of electrical power produced by running water from large dams will increase somewhat in the coming years in the United States, most of the acceptable sites are probably already being utilized. Small-scale hydropower systems, designed for individual homes, farms, or small industries, may be much more common in the future. Termed *micro-hydropower* systems, these typically have power output of less than 100 kW.[12]

Micro-hydropower is one of the world's oldest and most common energy sources. Numerous sites in many areas have the potential for producing small-scale electrical power. This is particularly true in mountainous areas, where potential energy from stream water is most available. Micro-hydropower development is by its nature site specific, depending on local regulations, economic situations, and hydrologic limitations. Hydropower can be used to generate either electrical power or mechanical power to run machinery; its use may help cut the high cost of importing energy and help small operations become more independent of local utility providers.[12]

An interesting aspect of hydropower energy management is the use of pump storage (Figures 17.11b and c), the objective of which is to make better use of the total electrical energy produced. During times when demand for power is low, electricity from oil, coal, or nuclear plants may be used to pump water up to a storage site or reservoir (high pool). Then, during times when demand for electricity is high, the stored water flows back down to the low pool through generators to supplement the power supply. It is important to keep in mind that pump storage systems are not as efficient as conventional hydroelectric plants. The bottom line is that about 3 units of energy from an oil, gas, or nuclear power plant are needed to produce 2 delivered units of energy from a pump storage facility. The advantage lies in the timing of energy production and use; the 2 units can be drawn during peak demand, and the 3 units are used to pump the water to the high pool when the demand is low.

Tidal Power

Another form of water power, known as **tidal power,** is derived from ocean tides in a few places with favorable topography, such as the north coast of France and the Bay of Fundy region of both the northeastern United States and Canada. The tides in the Bay of Fundy have a maximum range of about 15 m (49 ft). A minimum range of about 8 m (26 ft) is necessary to consider developing tidal power. To harness tidal power, dams are built across the entrance to a bay or estuary, creating a reservoir. As the tide rises (flood tide), water is initially prevented from entering the landward side of the dam. Then, when there is sufficient water (from the ocean-side high tide) to run the turbines, the water flows through the dam into the reservoir, turning the blades of the turbines and generating electricity.

CLOSER LOOK 17.1

FUEL CELLS—AN ATTRACTIVE ALTERNATIVE

Electrical power produced from burning fossil fuels, particularly coal and internal combustion engines (cars, trucks, ships, and locomotives), is associated with serious environmental problems. As a result, some people have concluded that we must seek an environmentally benign technology capable of generating power.[10] Fuel cells promise to be less polluting, less expensive, and have the potential to become an important source of high-quality energy.

Fuel cells are highly efficient power-generating systems that produce electricity by combining fuel and oxygen in an electrochemical reaction. Hydrogen and phosphoric acid are the most common fuel types, although fuel cells that run on methanol, ethanol, and natural gas are available. Traditional generating technologies require combustion of fuel in order to convert the resultant heat into mechanical energy (to drive pistons or turbines), and this mechanical energy is then converted into electricity. With fuel cells, however, chemical energy is converted directly into electricity, thus increasing second law efficiency (see Chapter 15) while reducing harmful emissions.

Basic components of a hydrogen-burning fuel cell are shown in Figure 17.10. Both hydrogen and oxygen are added to the fuel cell in an electrolyte solution. The reactants remain separated from one another and, upon ionization, migrate through the electrolyte solution from one electrode to another. The flow of electrons from the negative to the positive electrode is diverted along its path into an electrical motor, supplying current to keep the motor running. In order to maintain this reaction, hydrogen and oxygen are added as needed.

When hydrogen is used in a fuel cell, the only waste products are oxygen and water. Renewable energy sources, such as wind or solar power, may be used to split water molecules into hydrogen and oxygen, producing the necessary elements for the fuel cell. Burning natural gas in fuel cells produces some pollutants, but the amount is only about 1% of what would be produced by burning fossil fuels in an internal combustion engine or a conventional power plant.[10]

Not only are fuel cells efficient and clean, they may be arranged in series to produce energy for a particular task. Additionally, the efficiency of a fuel cell is largely independent of its size and output level. For these reasons, fuel cell technology is well suited to powering automobiles as well as large-scale power plants

Fuel cells are used to power buses at Los Angeles International Airport as well as to provide heat and power at Vandenberg Air Force Base in California.[11]

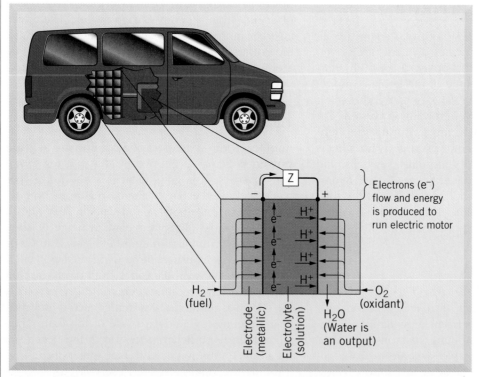

FIGURE 17.10 Idealized diagram showing how a fuel cell works and its application to power a vehicle.

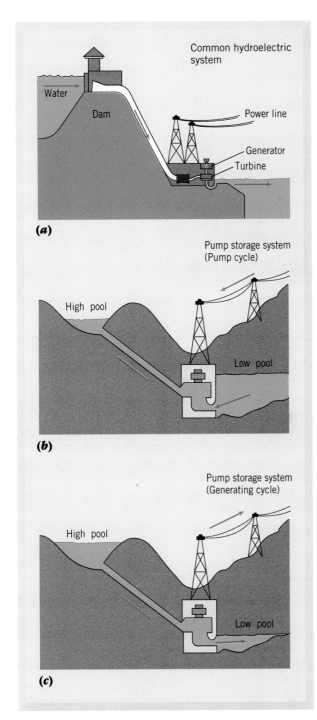

FIGURE 17.11 (*a*) Basic components of a hydroelectric power station. (*b*) During light power load water is pumped from low pool to high pool. (*c*) During peak power load water flow is from high pool to low pool through a generator. (*Source:* Modified from Council on Environmental Quality, 1975, *Energy Alternatives: A Comparative Analysis*, Science and Policy Program, University of Oklahoma, Norman.)

When the reservoir fills, the flow is stopped, holding the water in the reservoir. When the tide falls (ebb tide), the water in the reservoir is at a higher elevation. When there is sufficient water to run the turbines (which are reversible; water from either direction can turn them), the water is let out of the reservoir, producing power. Figure 17.12 shows the Rance tidal power plant on the north coast of France. The plant was constructed in the 1960s and at capacity produces over 200,000 kW from 24 power units spread out across the dam. At the Rance power plant, most electricity is produced from the ebb tide, which is easier to control.

Water Power and the Environment

Water power is clean power; it requires no burning of fuel, does not pollute the atmosphere, produces no radioactive or other waste, and is efficient. However, there is an environmental price to pay (see Chapter 19). Water falling over high dams may pick up nitrogen gas, which enters the blood of fish, expands, and kills them. This is similar to the condition known as the bends that occurs if divers using a compressed air source surface too rapidly from deep water. The fish respire nitrogen gas under pressure of the falling water. When the pressure is reduced quickly, gas bubbles in the blood expand and damage tissue. Nitrogen has killed many migrating game fish in the Pacific Northwest.

Furthermore, dams trap sediment that would otherwise reach the sea and replenish the sand on beaches. In addition, for a variety of reasons, many people do not want to turn all the wild rivers into a series of lakes. For these reasons, and because many good sites for dams are already utilized, the growth of large-scale water power in the future (with the exception of developing regions and countries, including Africa, South America, and China) appears limited.

There does seem to be an increase in interest in micro-hydropower to supply electricity or mechanical energy. The environmental impact of numerous micro-hydropower installations in an area could be considerable, as such installations change the natural stream flow, affecting the stream biota and productivity. Small dams and reservoirs also tend to fill more quickly with sediment, making their useful life much shorter. Because micro-hydropower development can adversely affect stream environments, careful consideration must be given to its development over a wide region. A few such sites may cause little environmental degradation, but if the number becomes excessive, the impact over a wider region may be appreciable. This consideration must be given to many forms of technology involving small sites. The impact of one single site may be nearly negligible in a broad region, but as the number of sites increases, the total impact may become significant.

FIGURE 17.12 Tidal power station on the River Rance near Saint-Malo, France.

17.7 WIND POWER

Wind power, like solar power, has evolved over a long period of time, from early Chinese and Persian civilizations to the present. Wind has propelled ships as well as driven windmills to grind grain or pump water. More recently, wind has been used to generate electricity. Winds are produced when differential heating of Earth's surface creates air masses with differing heat contents and densities. The potential for energy from the wind is tremendous, and yet there are problems with its use because wind tends to be highly variable in time, place, and intensity.[13]

Wind prospecting has become an important endeavor. On a national scale, regions with greatest potential for wind energy are the Pacific Northwest coastal area, the coastal region of northeastern United States, and a belt extending from northern Texas through the Rocky Mountain states and Dakotas. However, there are other good sites, such as mountain areas in North Carolina and northern Coachella Valley in southern California.

At a particular site the direction, velocity, and duration of wind may be quite variable, depending on local topography and on regional to local magnitude of temperature differences in the atmosphere.[13] For example, wind velocity often increases over hilltops or wind may be funneled through a mountain pass. The increase in wind velocity over a mountain is due to a vertical convergence of the wind, whereas in a pass the increase is partly due to a horizontal convergence. Because the shape of a mountain or a pass is often related to the local or regional geology, prospecting for wind energy is a geologic as well as a geographic and meteorological problem.

Significant improvements in the size of windmills and the amount of power they produce oc-curred from the late 1800s through approximately 1950, when many European countries and the United States became interested in large-scale generators driven by the wind. In the United States, thousands of small, wind-driven generators have been used on farms. Most of the small windmills generate approximately 1 kW of power, which is much too little to be considered for central power generation needs. Interest in wind power declined for several decades prior to the 1970s because of the abundance of cheap fossil fuels, but in recent years there has been a revival in interest in building windmills.

Small-Scale Producers

Small-scale power production from windmills in the United States was made more feasible by passage of the U.S. Public Utility Regulatory Policy Act (PURPA) in 1978. This act requires utilities to interconnect with small-scale independent producers of power and pay fair market price for the electricity produced. PURPA initiated an entrepreneurism by providing a market to small-scale power producers. However, the cost of producing the electricity from wind must be competitive with other sources to be economically viable. This is nowhere better exemplified than in California, where approximately 17,000 windmills, with a generating capacity of about 1400 MW, were installed in the 1980s. This is about 80% of the U.S. capacity of 1800 MW (megawatts of wind) [Figure 17.13(a)]. Individual windmills produce from 60 to 75 kW of power and are arranged in wind farms, consisting of clusters of windmills located in mountain passes. Electricity produced at these sites is connected to the general utility lines. In 1998, United States wind farms produced sufficient

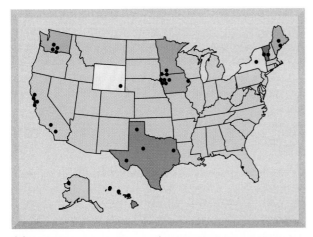

(a)

(b)

FIGURE 17.13 (a) Wind energy projects in the United States generate more than 3.5 billion kilowatt-hours of electricity each year. (*Source*: National Wind Technology Center, 1999. *http://www.nrel.gov/wind/deploy.html*). (*b*) Windmills on a wind farm near Altamont, California, a mountain pass region east of San Francisco.

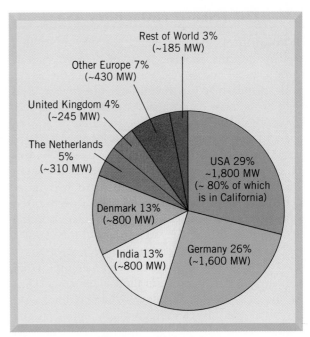

FIGURE 17.14 Worldwide installed wind power totaled 6170 MW at the end of 1996. (*Source:* International Energy Agency, February 1998. Http://www.afm.dtu.dk/wind/iea/status.htm#cap).

(this involves placing instruments at the site to measure the strength, duration, and direction of the wind) and installation of pilot wind power stations.[8]

Wind Power and the Environment

The use of wind power will not solve all our energy problems, but as one of the alternative energy sources, it can be used in particular sites to reduce our dependency on fossil fuel. Wind energy does have a few disadvantages:

- Demonstration projects have suggested that vibrations from windmills may produce objectionable noise.
- Windmills may interfere with radio and television broadcasts.
- Windmills kill birds (birds of prey, such as hawks and falcons, are vulnerable).
- Large windmill farms take up land for roads, pads for the windmills, and other equipment.
- Windmills may degrade an area's scenic resources.

However, everything considered, wind energy has a relatively low environmental impact, and its continued use should be carefully researched and evaluated. A number of large demonstration units are now being tested.

electricity to power, for example, both Washington, D.C. and San Francisco, CA, making a significant contribution to the modern utility grid. Tax incentives have helped establish the wind power industry. In California wind power may be the state's second least expensive source of power by the year 2010, second only to hydropower.

Wind energy is used in a variety of locations on Earth (Figure 17.14). Many locations have been evaluated, and many utility companies have experimented with wind power. The midsized wind turbines used in wind farms have been developed to the extent that many utility companies in the United States may now seriously consider wind power in their long-range energy planning. The steps will include research to evaluate the potential resource

17.8 ENERGY FROM BIOMASS

Biomass fuel is a new name for the oldest fuel used by humans. Biomass is organic matter, such as plant material and animal waste. Biomass fuel is organic matter that can be burned directly or converted to a more convenient form and then burned. For example, we can burn wood in a stove or convert it to charcoal and then burn it. Biomass has provided a major source of energy for human beings throughout most of the history of civilization. When North America was first settled, there was more wood fuel than could be used. The forests often were cleared for agriculture by girdling trees (cutting through the bark all the way around the base of a tree) to kill them and then burning the forests.

Until the end of the nineteenth century, wood was the major fuel source in the United States. During the mid-twentieth century, when coal, oil, and gas were plentiful and high-grade mines in the United States provided abundant cheap energy, burning wood became old-fashioned and quaint; it was something done for pleasure in an open fireplace that conducted more heat up the chimney than it provided for space heating. Now, with other fuels reaching a limit in abundance and production, there is renewed interest in the use of natural organic materials for fuel.

Firewood is the best known and most widely used biomass fuel, but there are many others. In India and other countries, cattle dung is burned for cooking. Peat, a form of compressed dead vegetation, provides heating and cooking fuel in northern countries such as Scotland, where it is abundant.

More than 1 billion people in the world today still use wood as their primary source of energy for heat and cooking. Energy from biomass may take several routes: direct burning of biomass either to produce electricity or to heat water and air, heating of biomass to form a gaseous fuel (gasification), or distillation or processing of biomass to produce biofuels such as ethanol, methanol, methane, or biocrude. Today in the United States various biomass sources supply nearly 4% of all energy consumed (approximately 3.0 exajoules per year), primarily in the form of wood for home heating, and cogeneration in the pulp and paper industry.[14]

Sources of Biomass Fuel

The primary sources of biomass fuels in North America are forest products, agricultural products, and combustible urban waste.

Today there are a number of facilities in the United States that process urban waste to generate electricity or to be used as a fuel. Presently, only about 1% of the nation's municipal solid wastes are being recovered for energy. However, if all the plants were operating at full capacity and if additional plants under construction were completed and operating, about 10% of the country's waste, or 24 million metric tons per year, could be used to extract energy. At processing plants such as those in Baltimore County, Maryland, Chicago, Milwaukee, Tacoma, Washington, Akron, Ohio, and Commerce, California, municipal waste is burned and the heat energy is used to make steam for a variety of purposes, from space heating to industrial generation of electricity. The United States has been slower than other countries to utilize urban waste as an energy source. For example, in Western Europe a number of countries now utilize from one-third to one-half their municipal waste for energy production. With the end of cheap, available fossil fuels, additional energy recovery systems utilizing urban waste will emerge.[15] However, burning urban waste is a controversial process with potentially adverse environmental impacts (see Chapter 27).

Net Energy Yield from Biomass

A problem with biomass fuels is their net yield of energy. Because the unit weight (weight per volume) of unprocessed vegetation is relatively low (i.e., it is bulky), considerable energy is required to collect biomass for fuel. Processing the biomass into more convenient fuels, such as methane and ethanol (alcohol), also requires energy. Biomass fuels provide the greatest net gain in energy when they are used locally. For example, in the production of sugar from sugarcane, unused parts of the cane can be burned to provide heat or electrical energy for local use. When fuels are consumed far from their source, and if much conversion is required, there may be no net gain in energy. For example, the energy it takes to convert cornstalks to alcohol and then transport the alcohol long distances to be used as fuel for cars and trucks may equal the energy available in the alcohol, thus providing no net gain in energy.

The efficiency of bioenergy production is improving. Gas turbine engines are now being introduced for use in biomass plants. The gas turbine engine uses the energy from burning a mixture of fuel and compressed air to spin a turbine to generate electricity; it has an efficiency of about 35%, similar to other forms of energy production, such as burning coal or natural gas.[16]

Energy from Biomass and the Environment

The use of biomass fuels can pollute the air and degrade the land. For most of us, the odor of smoke fumes from a single campfire is part of a pleasant

How Can We Evaluate Alternative Energy Sources?

The world is moving into a new era, one of transition between almost total dependence on fossil fuels to greater use of alternative, renewable sources of energy. Although each of the alternatives offers a way out of the energy dilemma created by population growth and technological development, each has advantages and disadvantages. How can we evaluate the alternatives and select the right mix of energy sources for the coming decades? We can begin by comparing them on the basis of those characteristics most important to us: cost, jobs lost or gained, environmental impact, and potential for supplying energy.

Critical Thinking Questions

1. Using what you have learned about alternative energy in this chapter and elsewhere, what would be the environmental impacts of the energy sources listed in the accompanying table? Complete the last column of the table. You may wish to subdivide the column into advantages and disadvantages.

2. Using the numbers 1–10, where 10 represents the best and 1 the worst, how would each column in the table be rated? For example, for carbon reduction, you might assign a rating of 10 to wind because it results in 100% reduction of carbon emis-

sions. Solar thermal energy would then receive a rating of 8.4. In rating environmental impact, you will have to use your judgment in assigning numerical values.

3. One way to evaluate the various alternatives would be to add up the rating scores for each energy source and see which ones receive the highest score. However, you may feel that some of the characteristics are more important than others and therefore should not be weighted equally. What would the weights in each column of the table be if you take into consideration the importance you believe each should have in decision making? For example, if you believe that costs are more important than land used, you would assign a higher value to costs. In order to be able to compare your evaluation with those of your classmates, use decimal fractions for the weights, such as 0.2. The total should add up to 1.0.

4. For each energy source, multiply its rating in each column by the weight you have assigned to the column. What is the total weighted score for each energy source? What are the sources in order of score, from highest to lowest?

5. Based on this analysis, what policy and research recommendations would you make to the U.S. government concerning alternative sources of energy?

Energy Source	U.S. Recoverable Resource[a] (exajoule/ year)	Costs in 1988 Cents[b] (per kilo-watt-hour) 1988	2000	Land Use[c] (m² per gigawatt-hour for 30 years)	Carbon Reduction[d] (%)	Carbon Avoidance Cost[d] ($$/ton)	Number of Jobs[c] (per thousand gigawatt-hours a year)	Environmental Impact
Wind	10–40	8	5	1335	100	95	542	
Geothermal	small	4	4	404	99	110	112	
Photovoltaic	35	30	10	3237	100	819	—	
Solar thermal	65	8	6	3561	84	180	248	
Biomass	13–26	5	na	—	100[f]	125	—	
Combined-cycle coal	—	6[g]	—	3642	10	954	116	
Nuclear	—	15[h]	—	—	86	535	100	

[a]*Recoverable resource* is a measure of how much of the energy can be captured or exploited. From M. Brower, 1990, *Cool Energy*, Union of Concerned Scientists, Washington, DC, p. 19.

[b]From L. R. Brown, C. Flavin, and S. Postel, 1991, *Saving the Planet*, W. W. Norton, New York, p. 27.

[c]From Brown et al., ibid., p. 60

[d]Based on comparison with existing coal-fired plants. From C. Flavin, 1990, "Slowing Global Warming," *State of the World*, W. W. Norton, New York, p. 27.

[e]From Brown et al., op. cit., p. 62.

[f]Assumes that the amount of carbon dioxide released in combustion with be consumed by replanted vegetation.

[g]From C. Flavin, 1992, "Building a Bridge to a Sustainable Future," *State of the World*, W. W. Norton, New York, p. 35.

[h]From A. K. Reddy and J. Goldemberg, 1990, "Energy for the Developing World," *Scientific American*, 263(3), p. 116.

References

Brown, L. R., Flavin, C., and Postel, S. 1991. *Saving the Planet.* New York: W. W. Norton.

Brower, M. 1990. *Cool Energy.* Washington, DC: Union of Concerned Scientists.

Flavin, C. 1990. "Slowing Global Warming," in L. R. Brown et al., eds., *State of the World.* New York: W. W. Norton.

Reddy, A. K. N. Reddy, and Goldemberg, J. 1990. "Energy for the Developing World," *Scientific American* 263(3): 110–118.

outdoor experience, but under certain weather conditions wood smoke from many chimneys in narrow valleys can lead to unpleasant and dangerous air pollution. In recent years, the renewed use of wood stoves in homes has led to reports of such air pollution in Vermont.

The use of biomass as fuel places another pressure on already heavily used resources. The worldwide shortage of firewood is adversely affecting natural areas and endangered species. For example, the need for firewood has threatened the Gir Forest in India, the last remaining habitat of the Indian lion. The world's forests will also decrease if our need for forest products and forest biomass fuel exceeds the productivity of the forests. If our forest and crop resources are managed properly (for sustainability), it may be possible to make biomass energy more attractive. The current estimates for the United States suggest that there are between 35 million and 200 million acres of land that are unsuitable for food production but with a potential for use for biomass plantations of trees and other plants using short rotation times (time between harvests). Forest plantations would have to be managed for sustainability. Deforestation accelerates the process of soil erosion (soils that are unprotected by vegetation erode more quickly). When fine-grained (silt and clay) particles enter the streams and rivers, the water quality is also degraded.

Advocates of using biomass for energy point out that, assuming areas harvested for bioenergy are replanted, the burning of biofuels contributes no net carbon dioxide emissions into the atmosphere. Replacing fossil fuels with sustainably produced biomass will actually reduce net CO_2 emissions, according to some estimates.[17] The burning of biomass produces significantly less CO_2 than does burning natural gas or oil. Finally, combustion of biomass-derived fuels releases much fewer pollutants such as sulfur dioxide and nitrogen oxides compared to combustion of coal and gasoline.

Biomass in its various forms appears to continue to have a future as an energy source. However, questions remain about the amount of energy it can provide and the rate of depletion. Any use of biomass fuel must be part of the general planning for all uses of the land's products.

As the prices of fossil fuels rise, particularly oil and gas, renewable energy sources certainly will become more attractive. Because we do not know when and how the transition from fossil fuels to renewable energy sources will occur, it is difficult to predict when, how long, and how complete the transition will be.

SUMMARY

- Geothermal energy is the use of the natural heat from Earth's interior as an energy source. The environmental effects of developing geothermal energy relate to specific site conditions and type of heat utilized (steam, hot water, or warm water). Geothermal energy may involve disposal of saline or corrosive waters as well as on-site noise, emission of gas, and industrial scars.

- Passive solar systems utilize architectural design to enhance or take advantage of solar energy at a site without requiring mechanical power or moving parts.

- Active solar systems include the use of solar collectors to heat water for homes or solar farms to produce heat for generating electricity. Other applications include the use of power towers or solar ponds to produce heat or electricity.

- Photovoltaics is the technology that converts sunlight directly into electricity. The system uses solar cells for a variety of purposes, such as powering remote equipment. Experiments are ongoing to evaluate photovoltaic power plants. This emerging technology remains expensive.

- Hydrogen gas may be an important fuel of the future, especially when used in fuel cells.

- Water power today provides about 15% of the total electricity produced in the United States. Most acceptable good sites for large dams are now utilized, but many small-scale systems are being evaluated. Water power is clean, but there is an environmental price to pay in terms of disturbance of ecosystems, sediment trapped in reservoirs, loss of wild rivers, and loss of productive land for water storage.

- Wind power has tremendous potential as a source of electrical energy in many parts of the world. Many utility companies are presently experimenting with or using wind power as part of their energy production or for long-term energy planning. Environmental impacts include loss of land to wind farms and killing of birds as well as degradation of scenic resources (some people think wind farms are unsightly) and noise pollution.
- Biomass as an energy source is the oldest

human fuel. More than 1 billion people still use wood as their primary source of energy. Recently there have been efforts to produce alcohol as a fuel from crops such as corn or sugarcane. The environmental impacts of burning wood include deforestation, soil erosion (soil unprotected by vegetation erodes easier), water pollution (eroded fine-grained sediment contaminates water), and air pollution. Burning organic urban waste (trash) can release toxins and cause air pollution as well. Conversion of crops to alcohol also produces air pollution and, considering total costs, may not be economically viable for many applications. That is, the costs of production, harvest, transport, and manufacturing of a fuel may exceed the value of the fuel.

REEXAMINING THEMES AND ISSUES

Human Population: As human population continues to increase, so does total world consumption of energy from all sources and the potential for continued environmental degradation from burning fossil fuels, especially coal. Environmental problems related to increased consumption of energy may be minimized through controlling human population, increased conservation efforts, and use of alternative energy sources that do not harm the environment.

Sustainability: In order to plan for sustainability from an energy perspective, we may need to rely more on alternative energy sources that will not cause potential global warming, pollute the atmosphere, or otherwise damage the environment. To do otherwise is antithetical to the concept of sustainability.

Global Perspective: Evaluation of the potential of alternative energy sources requires understanding of global Earth systems related to the flux of solar energy: wind and water. By this we may identify regions that are more

likely to produce higher quality alternative energy that could be utilized by nearby urban areas.

Urban World: Alternative energy sources have a future in our urban environment. For example, the roofs of buildings can be utilized for solar collectors or photovoltaic systems. Patterns of energy consumption can be regulated through use of innovative systems such as pump storage that allow production of electrical energy when demands are high in urban areas.

Values and Knowledge: We are seriously considering alternative energy today because we value a quality environment. Recognizing that present energy policies of burning fossil fuels have many serious environmental problems, we are trying to increase our scientific knowledge and technology to produce innovative alternative energy sources to meet our needs for the future while minimizing environmental damage. This is necessary to achieve a quality sustainable environment.

KEY TERMS

active solar energy systems *347*
alternative energy *344*
biomass fuel *358*
fuel cell *354*
geothermal energy *344*
nonrenewable energy *344*

ocean thermal conversion *352*
passive solar energy systems *347*
photovoltaics *348*
renewable energy *344*
solar collectors *347*
solar ponds *351*

solar power tower *350*
tidal power *353*
water power *353*
wind power *356*

STUDY QUESTIONS

1. What type of government incentives could be used to encourage use of alternative energy sources? Would their widespread use impact our economic and social environment?

2. Your town is near a large river that has a nearly constant water temperature of about 15°C (60°F). Could the water be used to cool buildings in the hot summers? How? What would be the environmental effects?

3. Which has greater future potential for energy production, wind or water power? Which has more environmental problems? Why?

4. What are some of the problems associated with producing energy from biomass?

5. It is the year 2500 and natural oil and gas are rare curiosities that people see in museums. Given the technologies available today, what would be the most sensible fuel for airplanes? How would this fuel be produced to minimize adverse environmental effects?

6. When do you think the transition from fossil fuels to other energy sources will (or should) occur? Defend your answer.

FURTHER READING

Ahmed, K. 1994. "Renewable Energy Technologies: A Review of the Status and Costs of Selected Technologies," World Bank Technical Paper No. 240. Discussion of the relative costs of energy production with biomass, solar, and photovoltaic energy sources compared to traditional sources and how these technologies have and will continue to become more competitive with oil and coal.

Brower, M. 1992. *Cool Energy. Renewable Solutions to Environmental Problems.* Cambridge: MA: MIT Press. Good coverage of alternative energy sources, showing fewer environmentally negative side effects than conventional sources.

Chartier, P., Beenackers, A. A. C. M., and Grassi, G., eds. 1995. *Biomass for Energy, Environment, Agriculture and Industry.* Proceedings of the 8th European Communities Conference, Vienna, Austria, October 3–5, 1994. London: Pergamon. A valuable, up-to-date summary of the status and promise of biomass use including papers covering policy issues, assessments of the resource base, economic assessments, environmental impacts, and commercialization issues.

Flavin, C., and Lenssen, N. 1994. *Power Surge: Guide to the Coming Energy Revolution.* The Worldwatch Environmental Alert Series. New York: W. W. Norton. An informative book that looks at problems of energy production and includes examples of alternative energy use and future paths for these alternative sources.

Sheffield, J. 1997. "The Role of Energy Efficiency and Renewable Energies in the Future and World Energy market." *Renewable Energy,* 10(2): 315–318.

U.S. Department of Energy, 1993. *Electricity from Biomass—Renewable Energy Today and Tomorrow.* Washington, D.C.: Solar Thermal Biomass Power Division, Office of Solar Energy Conversion. A brief introduction to the potential for biomass-generated electricity in the United States.

Van Koevering, T. E., and Sell, N. J. 1986. *Energy: A Conceptual Approach.* Englewood Cliffs, NJ: Prentice-Hall. This short book on energy has several chapters on alternative energy sources and related environmental concerns. It provides a good overview of a variety of alternative renewable and nonrenewable energy sources.

Weinberg, C. J., and Williams, R. H. 1990. "Energy from the Sun," *Scientific American* 263(3): 146–155. This review article discusses technologies in solar energy and argues that advances in solar alternative sources will become more important in the future.

INTERNET RESOURCES

The Center for Renewable Energy and Sustainable Technology: *http://solstice.crest.org/*—Find information on renewable energy, its status, sources and benefits, energy efficiency, and sustainable living. Read reports, letters, and articles on advances in renewable energy. Includes links to other internet sources on renewable energy.

U.S. Department of Energy Energy Efficiency and Renewable Energy Network: *http://www.eren.doe.gov/*—Find general information on renewable energy and many links to other internet sources.

U.S. Department of Energy National Renewable Energy Lab: *http://www.nrel.gov/*—See what current research and projects the NREL is involved with, and read their latest and past reports. Also find links to other internet resources.

International Energy Agency (IEA): *http://iea.org*—IEA was formed in 1974 as an autonomous body within the Organization for Economic Co-operation and Development (OECD). The Agency maintains an emergency system for dealing with oil supply disruptions and a permanent information system on the international oil market. Recently, attention has been paid to strengthening the link between environment and energy policies. The Web site contains IEA publications, statistics, and information about energy, economics, technology, and climate change.

Nuclear power plant at dusk in Cattenom, France.

CHAPTER 18

NUCLEAR ENERGY AND THE ENVIRONMENT*

CASE STUDY

Nuclear Energy and Public Opinion

In 1953, Dwight D. Eisenhower, a popular general of World War II and president after the war, described an optimistic vision of the future of atomic power, which he called "atoms for peace." He presented his ideas to the United Nations just ten years after the detonation of atomic bombs over Nagasaki and Hiroshima. President Eisenhower predicted that nuclear powered generators would provide such cheap, nearly unlimited clean power that we would never need to meter it. Optimism regarding nuclear power in the United States continued into the early 1980s. More than 100 nuclear power plants were constructed, and orders had been placed for more than 100 additional plants by 1978. However, no nuclear reactors have been ordered since, and polls suggest that the majority of Americans have been opposed to construction of nuclear power plants since the early 1980s. Reversal of

opinion occurred three years after the Three Mile island accident and four years before the Chernobyl accident (discussed later in this chapter). Anxiety of people concerning nuclear weapons and nuclear war peaked in the early 1980s at about the time attitudes changed toward constructing nuclear power plants, suggesting that worry over nuclear weapons had spilled over onto commercial nuclear power.[1]

Today nuclear power is at a crossroads. Polls suggest that many Americans believe nuclear power should and will be an important energy source in the future. However, people oppose construction of new nuclear power plants in the near future, and remain uncomfortable with issues of safety. Evidently the public is not willing to rule out the nuclear option as a potential alternative energy source. An important question is, if nuclear energy were phased out, would it make any signif-

363

icant difference? The answer is no! Nuclear energy is a small part of the U.S. energy package, and only minor substitution of other sources would be required to compensate for its loss. On the other hand, nuclear energy is one of several technologies that may eventually replace fossil fuels. Therefore, nuclear power, which does not contribute to global warming, may again be looked at as a major source of energy, particularly if issues of safety can be resolved.[2]

The discussion of public opinion about nuclear power suggests a change from the optimism of the 1950s but presents a dilemma because nuclear power remains an alternative to fossil fuels. This chapter explores nuclear power reactors, radiation, accidents, waste management, and the future of nuclear power.

Learning Objectives

As one of the alternatives to fossil fuels, nuclear energy generates much controversy. After reading this chapter, you should understand:

- What nuclear fission is and what the basic components of a nuclear power plant are.
- What nuclear radiation is and what the three major types are.
- Why it is important to know the type of radiation and half-life for a particular radioisotope.
- What the basic parts of the nuclear fuel cycle are and how each is related to our environment.

- What are the two major ways in which radioisotopes affect the environment and what are the major pathways of radioactive materials in the environment.
- What are the relationships between radiation doses and health.
- What we have learned from accidents at nuclear power plants.
- How we might safely dispose of high-level radioactive materials.
- What the future of nuclear power is likely to be.

18.1 NUCLEAR ENERGY

Nuclear energy is the energy of the atomic nucleus. Two nuclear processes can be used to release that energy to do work: fission and fusion. Nuclear **fission** is the splitting of an atom's nucleus into smaller fragments. Nuclear **fusion** is the combining of atomic nuclei to form heavier nuclei. A by-product of both reactions is the release of energy. Controlled nuclear fission reactions take place within commercial nuclear reactors to produce energy.

Nuclear energy from fission may be considered a nonrenewable alternative energy source. It is an alternative to fossil fuels, and it is nonrenewable because it requires uranium as a fuel. Uranium is a geologic resource in limited supply; it takes millions of years to form deposits of sufficient concentration to be mined for profit. However, worldwide uranium reserves are sufficient to supply nuclear energy at the present rate of consumption with present technology through the next century.

Fission Reactors

The first human-controlled nuclear fission, demonstrated in 1942, led to the development of the primary uses of uranium: in explosives and as a heat source to provide steam for generation of electricity. One kilogram of uranium oxide produces heat equivalent to approximately 16 metric tons of coal, making uranium an important source of energy in the United States and the world.

Three types of uranium occur in nature: uranium-238, which accounts for approximately 99.3% of all natural uranium; uranium-235, which makes up about 0.7%; and uranium-234, which makes up about 0.005%. Uranium-235 and uranium-238 are two naturally radioactive isotopes of uranium. Uranium-235 is the only naturally occurring fissionable material and therefore is essential to the production of nuclear energy. Processing uranium (called enrichment) to increase the concentration of uranium-235 from 0.7% to about 3% produces enriched uranium, which is used as fuel for the fission reaction. The process of radiation is explained and related terms are defined in A Closer Look 18.1, "Radiation."

Fission reactors split uranium-235 by neutron bombardment (Figure 18.1). The reaction produces neutrons, fission fragments, and heat. The released neutrons each strike other uranium-235 atoms, releasing more neutrons, fission products, and heat. The neutrons released are fast moving and must be slowed down, or *moderated*, to increase the probability of fission. Water is most commonly used as the moderator. As the process continues, a chain reaction develops as more and more uranium is split, releasing more neutrons and more heat.

Burner Reactors

Most reactors now in use consume more fissionable material than they produce and are known as **burner reactors.** The reactor itself is part of the nuclear steam supply system, which produces the steam to run the turbine generators that produce the electricity.[4] Therefore, the reactor has the same function as the boiler that produces the heat in coal-burning or oil-burning power plants (Figure 18.4).

The main components of the reactor shown in Figure 18.5 are the core (consisting of fuel and moderator), control rods, coolant, and reactor vessel.

FIGURE 18.1 Fission of uranium-235. A neutron strikes the U-235 nucleus, producing fission fragments and free neutrons and releasing heat. The released neutrons may then each strike another U-235 atom, releasing more neutrons, fission fragments, and energy. As the process continues, a chain reaction develops.

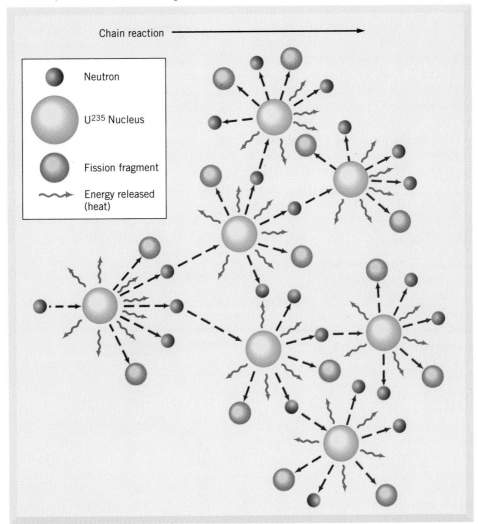

Closer Look 18.1

RADIATION

To many people, radiation is a subject shrouded in mystery. They feel uncomfortable with it. We learn from an early age that nuclear energy may be dangerous because of radiation and that nuclear fallout from the detonation of atomic bombs can cause widespread human suffering. What makes radiation scary is that we cannot see it, taste it, smell it, or feel it. However, we know if we are exposed to a lot of it, or a small amount over a long period of time, it causes health problems. In this chapter we try to demystify some aspects of radiation. In particular, we discuss the natural process of radiation, the units by which we measure radiation, radiation and human health, and the environmental effects of radiation.

First we need to understand that radiation is a natural process that has been going on since the creation of the universe. The process of radiation or radioactivity involves understanding the **radioisotope,** which is a form of a chemical element that spontaneously undergoes **radioactive decay.** During the decay process the radioisotope changes from one isotope to another and emits one or more forms of radiation. Isotopes are atoms of an element that have the same atomic number (the number of protons in the nucleus of the atom) but vary in atomic mass number (the number of protons plus neutrons in the nucleus of an atom). For example, two isotopes of uranium are $^{235}U_{92}$ and $^{238}U_{92}$. The atomic number for both isotopes of uranium is 92 (see Table 4.1); however, the atomic mass numbers are 235 and 238. The two different uranium isotopes may be written as uranium-235 and uranium-238 or U-235 and U-238.

An important characteristic of a radioisotope is its **half-life,** which may be defined as the time required for one-half of a given amount of the isotope to decay to another

form. For example, uranium-235 has a half-life of 700 million years, a very long time indeed! As a result, the change or radioactive decay of uranium-235 is a slow process compared to time frames we are used to dealing with, such as the human life span or even the historic record. Radioactive carbon-14 has a half-life of 5570 years, which is in the intermediate range, and radon-222 has a relatively short half-life of 3.8 days. Other radioactive isotopes have even shorter half-lives; for example, polonium-218 has a half-life of about 3 minutes. Still other radioactive isotopes have half-lives as short as a fraction of a second.

The main point here is that each radioactive isotope has its own unique and unchanging half-life. Those isotopes with very short half-lives are present for only a brief time, whereas those with long half-lives remain in the environment for long periods of time. Table 18.1 illustrates the general pattern for decay processes in terms of the elapsed half-lives and the fraction remaining. For example, if you start with 1 gram of polonium-218 with a half-life of approximately 3 minutes,

after an elapsed time of 3 minutes 50% of the polonium-218 remains. After five elapsed times, or 15 minutes, only 3% of the original gram of polonium is still present. After 10 half-lives have elapsed (30 minutes) only one-tenth of 1% of the original gram of polonium is still present. Where has the polonium gone? It has decayed to lead-214, another radioactive isotope, which has a half-life of about 27 minutes. The progression of changes is often known as a *radioactive decay chain* (see Figure 18.2). If we had started with 1 gram of uranium-235, with a half-life of 700 million years, then, following 10 elapsed half-lives, 0.1% of the uranium would be left, but this process would take 7 billion years.

Radioisotopes with short half-lives initially undergo a more rapid rate of change (nuclear transformations) than do radioisotopes with long half-lives. Conversely, radioisotopes with long half-lives have a less intense and slower initial rate of nuclear transformation but may be hazardous for a much longer time.[3]

There are three major kinds of nuclear radiation: **alpha particles, beta particles,** and **gamma rays.** Alpha

TABLE 18.1 Generalized Pattern of Radioactive Decay

Elapsed Half-life	Fraction Remaining	Percent Remaining
0	—	100
1	$1/2$	50
2	$1/4$	25
3	$1/8$	13
4	$1/16$	6
5	$1/32$	3
6	$1/64$	1.5
7	$1/128$	0.8
8	$1/256$	0.4
9	$1/512$	0.2
10	$1/1024$	0.1

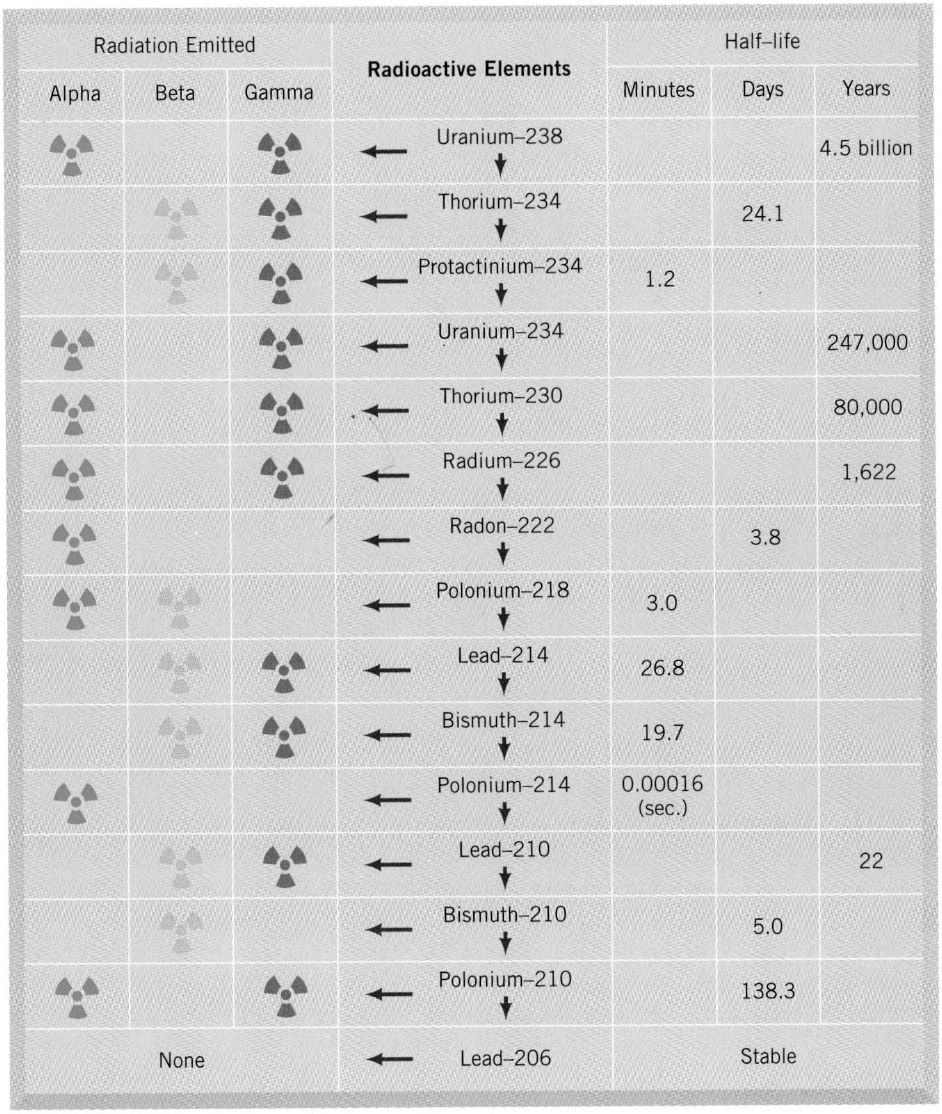

Radiation Emitted			Radioactive Elements	Half-life		
Alpha	Beta	Gamma		Minutes	Days	Years
☢		☢	← Uranium–238 ↓			4.5 billion
	☢	☢	← Thorium–234 ↓		24.1	
	☢	☢	← Protactinium–234 ↓	1.2		
☢		☢	← Uranium–234 ↓			247,000
☢		☢	← Thorium–230 ↓			80,000
☢		☢	← Radium–226 ↓			1,622
☢			← Radon–222 ↓		3.8	
☢	☢		← Polonium–218 ↓	3.0		
	☢	☢	← Lead–214 ↓	26.8		
	☢	☢	← Bismuth–214 ↓	19.7		
☢			← Polonium–214 ↓	0.00016 (sec.)		
	☢	☢	← Lead–210 ↓			22
	☢		← Bismuth–210 ↓		5.0	
☢		☢	← Polonium–210 ↓			138.3
None			← Lead–206	Stable		

FIGURE 18.2 Uranium-238 decay chain. (*Source:* F. Schroyer, ed., 1985, *Radioactive Waste*, 2nd printing, American Institute of Professional Geologists.)

particles consist of two protons and two neutrons (a helium nucleus) and have the greatest mass of the three types of radiation (Figure 18.3*a*). Because alpha particles have a relatively high mass, the particles do not travel far. For example, alpha particles can travel approximately 5–8 cm (about 2–3 in.) in air before they stop. However, in human tissue and that of other organisms, which is much denser than air, they can only travel about 0.005–0.008 cm (0.002–0.003 in.). Because this is a very short distance, to cause damage to living tissue and cells, alpha particles must originate very close to the cell.[4]

Beta particles are electrons and have a mass of 1/1840 of a proton. Beta decay occurs when one of the protons or neutrons in the nucleus of an isotope spontaneously changes. What happens is that a proton turns into a neutron or a neutron is transformed into a proton (Figure 18.3*b*). As a result of this process, another particle, known as a **neutrino,** is also ejected. A neutrino is a particle with no rest mass.[4] Beta particles travel farther through air than the more massive alpha particles but are blocked by even moderate shielding, such as a thin sheet of metal or a block of wood.

The third and most penetrating type of radiation is called *gamma decay*. When gamma decay occurs, a gamma ray, a type of elecromagnetic radiation, is emitted from the isotope. Gamma rays are similar to X rays but are more energetic and penetrating. Gamma rays travel the longest average distance of all types of radiation and can penetrate thick shielding.

Each radioisotope has its own characteristic emissions; some isotopes emit only one type of radiation, whereas others emit a mixture. In addition, the different types of radiation have different toxicities (degree or intensity of potential to harm

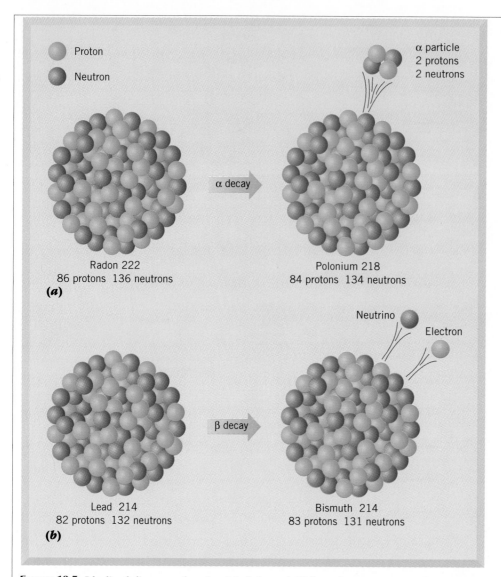

Figure 18.3 Idealized diagrams showing (*a*) alpha and (*b*) beta decay processes. (Source: D. J. Brenner, 1989, *Radon: Risk and Remedy,* Freeman. Copyright 1989 by W. H. Freeman and Company. Reprinted with permission.)

or poison). In terms of human health, or the health of other organisms, alpha radiation is most toxic or dangerous when inhaled or ingested. Because alpha radiation is stopped within a very short distance by living tissue, much of the damaging radiation is absorbed by the tissue. On the other hand, when alpha-emitting isotopes are stored in a container, they are relatively harmless. Beta radiation is intermediate in its toxicity, although most beta radiation is absorbed by the body when a beta emitter is ingested. Beta particles can be stopped by moderate shielding. Gamma emitters are toxic

and dangerous inside or outside the body, but when they are ingested, some of the radiation passes outside the body. Protection from gamma rays requires thick shielding.

Some radioisotopes, particularly those of very heavy elements such as uranium, undergo a series of radioactive decay steps (a decay chain) before finally reaching a stable nonradioactive isotope. For example, uranium decays through a series of steps to the stable nonradioactive isotope of lead. A decay chain for uranium-238 (with a half-life of 4 billion years) to stable lead-206 is shown in Figure 18.2. Also listed are the half-

lives and types of radiation that occur during the transformations. The symbol used to denote the half-life is $t_{1/2}$. Note that the simplified radioactive decay chain shown in Figure 18.2 involves 14 separate transformations from uranium-238 to lead-206. When considering radioactive decay, two important facts are the type of radiation emitted and the half-life.

The decay from one radioisotope to another is often stated in terms of parent and daughter products. For example, uranium-238, with a half-life of 4.5 billion years, is the parent of daughter product thorium-234, with a half-life of 24.1 days.

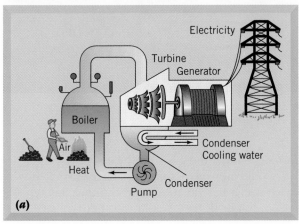 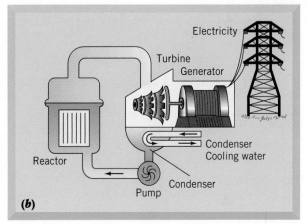

FIGURE 18.4 Comparison of (*a*) a fossil fuel power plant and (*b*) a nuclear power plant with a boiling water reactor. Notice that the nuclear reactor has exactly the same function as the boiler in the fossil fuel power plant. The coal-burning plant (*a*) is Ratclisse-on-Saw located in Nottinghamshire, England, and the nuclear power station (*b*) is located in Leibstadt, Switzerland. (*Source:* American Nuclear Society, 1973, *Nuclear Power and the Environment.*)

The core of the reactor is enclosed in a heavy, stainless steel reactor vessel; then, for extra safety and security, the entire reactor is contained in a reinforced concrete building.

Fuel pins, consisting of enriched uranium pellets placed into hollow tubes with a diameter less than 1 cm (0.4 in.) are packed together (40,000 or more in a reactor) into fuel subassemblies in the core. A minimum fuel concentration is necessary to keep the reactor *critical,* that is, to achieve a self-sustaining chain reaction. A stable fission chain reaction in the core is maintained by controlling the number of neutrons that cause fission. The control rods, which contain materials that capture neutrons, are used to regulate the chain reaction. As the con-

trol rods are moved out of the core, the chain reaction increases; as they are moved into the core, the reaction slows. Full insertion of the control rods into the core stops the fusion reaction.[5]

The function of the coolant is to remove the heat produced by the fission reactions. When water is used as the coolant, it acts as a moderator also, slowing the neutrons down, a process necessary for the fission of uranium-235.

Other parts of the nuclear steam supply system are the primary coolant loops and pumps, which circulate the coolant (usually water) through the reactor, extracting heat produced by fission, and heat exchangers or steam generators, which use the fission-heated coolant to make steam.

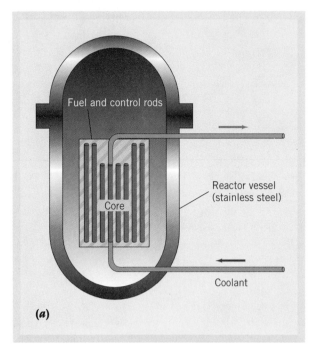

FIGURE 18.5 (*a*) Main components of a nuclear reactor. (*b*) Glowing spent fuel elements being stored in water at a nuclear power plant.

In recent years a new design philosophy has emerged in the nuclear industry. The concept is to build less complex, smaller reactors that are safer. For example, large nuclear power plants require an extensive set of pumps and backup equipment to ensure that adequate cooling is available to the reactor as necessary. Smaller reactors can be designed with cooling systems that work under the influence of gravity and as a result are not as vulnerable to failure of pumps or loss of electrical power and subsequent pump failure. The term for such cooling systems is *passive stability*.[6] Another approach being evaluated is the use of helium gas to cool reactors that have specially designed fuel capsules capable of withstanding temperatures as high as 1800°C (about 3300°F). The idea is to design the fuel assembly so that it cannot hold sufficient fuel to reach this temperature and is incapable of incurring a core meltdown.

A *meltdown* generally refers to a nuclear accident in which the nuclear fuel becomes so hot that it forms a molten mass that breaches the containment of the reactor and contaminates the outside environment with radioactivity. The gas-cooled reactors now being designed contain what are called fuel grains embedded in fuel pebbles about the size of a small orange. One fuel pebble contains thousands of fuel grains, each of which is coated with a ceramic material. In theory, such a reactor does not need a cooling system at all. The gas flowing through the core is the medium that transports the heat to the turbine. Should the gas supply be shut off, the reactor would not have sufficient heat to melt down. However, it has been pointed out that this type of design might be vulnerable to other types of accidents, such as ignition of the graphite that surrounds the fuel. It is important to note that considerable effort is going into the design of a new generation of nuclear reactors that are expected to be safer than those used in the past or currently in use.

Breeder Reactors

The nuclear burner reactors operating today use uranium very inefficiently. About 1% of uranium is actually utilized to produce steam for generating electricity. There is concern that if a large number of new burner reactors are built, the entire world's uranium supply could be used up in less than 100 years.

A nuclear reactor that can utilize between 40 and 70% of its nuclear fuel is called a **breeder reactor.** If these reactors are used in the future, global uranium supplies will last for many centuries. Breeder reactors convert the more abundant fertile isotopes uranium-238 or thorium-232 to fissile isotopes that can sustain a nuclear chain reaction, plutonium-239 or uranium-233. A breeder reactor is defined as a reactor that converts fertile nuclei to fissile nuclei faster than the rate of fission. The development and use of breeder reactors could greatly extend our nuclear fuel reserves as well as our uranium reserves.

Experiments with breeder reactors have not been very encouraging because the electricity produced has been considerably more expensive than that produced from burner reactors. The cost is high because it takes several decades for a breeder reactor to produce as much plutonium as it burns. Also, breeder reactors are more expensive to engineer and build. Concern has also arisen because these reactors produce plutonium-239, a major ingredient of nuclear weapons. It is feared that some of the plutonium might be directed to the manufacture of weapons and might even be obtained by terrorist organizations attempting to build nuclear bombs. On the positive side, breeder reactors produce less hazardous waste than do conventional burner reactors and require much less mining for fuel. Breeder reactors may indeed have a future, but that future is dependent on the increased use of conventional nuclear reactors to the point that fuel shortages become apparent. On the other hand, both burner and breeder reactors may become obsolete and be replaced by fusion reactors sometime in the next century.

Fusion

In contrast to fission, which involves splitting heavy nuclei (such as of uranium), fusion involves combining light elements (such as hydrogen) to form a heavier element (such as helium). As fusion occurs, heat energy is released (Figure 18.6). Nuclear fusion is the source of energy in our sun and other stars. In a hypothetical fusion reactor, two isotopes of hydrogen—deuterium and tritium—are injected into the reactor chamber, where the necessary conditions for fusion are maintained. Products of the deuterium–tritium (DT) fusion include helium, producing 20% of the energy released, and neutrons, producing 80% of the energy released (Figure 18.7).[7]

Several conditions are necessary for fusion to take place. First, there must be an extremely high temperature (approximately 100 million degrees Celsius for DT fusion). Second, the density of the fuel elements must be sufficiently high. At the temperature necessary for fusion, nearly all atoms are stripped of their electrons, forming a plasma. Plasma is an electrically neutral material consisting of positively charged nuclei, ions, and negatively charged electrons. Third, the plasma must be confined for a sufficient time to ensure that the energy released by the fusion reactions exceeds the energy supplied to maintain the plasma.[7,8]

The potential energy available when and if fusion reactor power plants are developed is nearly inexhaustible. One gram of DT fuel (from a water and lithium fuel supply) has the energy equivalent of 45 barrels of oil. Deuterium can be extracted economically from ocean water, and tritium can be produced in a reaction with lithium in a fusion reactor. Lithium can be extracted economically from abundant mineral supplies.

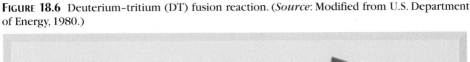

Figure 18.6 Deuterium–tritium (DT) fusion reaction. (*Source*: Modified from U.S. Department of Energy, 1980.)

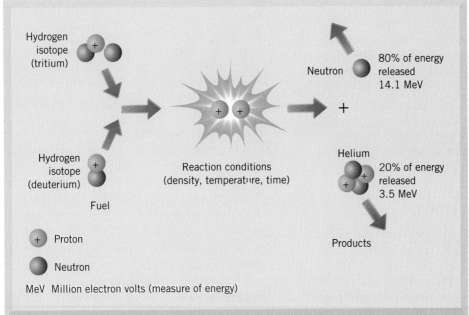

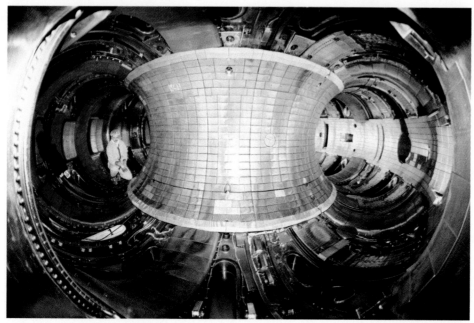

FIGURE 18.7 Experimental fusion nuclear reactor that magnetically confines plasma at very high temperatures.

Many problems remain to be solved before nuclear fusion can be used on a large scale. Research is still in the first stage, which involves basic physics, testing possible fuels (mostly DT), and magnetic confinement of plasma. Progress in fusion research has been steady in recent years, so there is optimism that useful power will eventually be produced from controlled fusion.[9]

18.2 NUCLEAR ENERGY AND THE ENVIRONMENT

Throughout the entire **nuclear cycle**—from the mining and processing of uranium to controlled fission, the reprocessing of spent nuclear fuel, the decommissioning of power plants, and the disposal of radioactive waste—various amounts of radiation may enter and affect the environment (Figure 18.8). In order to understand the environmental effects of radiation, it is useful to be acquainted with the units used to measure radiation and the amount or dose of radiation that may cause a health problem (see A Closer Look 18.2, "Radiation Units and Doses").

Problems with Nuclear Power

Let's look a little closer at the environmental effect of the nuclear fuel cycle (Figure 18.8):

- Uranium mines and mills produce radioactive waste material that can pollute the environment. There have been instances where ra-

dioactive mine tailings have been used for foundation and building materials and have contaminated dwellings.

- Uranium-235 enrichment and fabrication of fuel assemblies also produce waste materials that must be carefully handled and disposed of.

- Site selection and construction of nuclear power plants in the United States have been extremely controversial. The environmental review process is extensive and expensive, often centering on hazards related to the probability of such events as earthquakes damaging the plant.

- The power plant or reactor is the site most people are concerned about because it is the most visible part of the cycle. It is also the site of past accidents, including partial meltdowns that have released harmful radiation into the environment.

- The United States does not reprocess spent fuel from reactors to recover uranium and plutonium at this time, so there currently is no environmental impact from them. However, there are many problems associated with the handling and disposal of nuclear waste that are discussed later in this chapter.

- Waste disposal is a controversial part of the nuclear cycle because no one wants a nuclear waste disposal facility nearby.

- Decommissioning or modernization of a nuclear power plant (they all have a limited life-

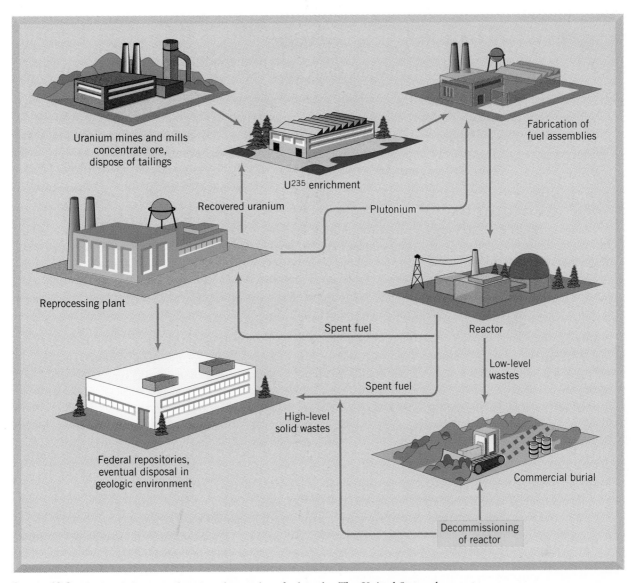

FIGURE 18.8 Idealized diagram showing the nuclear fuel cycle. The United States does not now reprocess spent fuel. Disposal of tailings, which because of their large volume may be more toxic than high-level waste, has been treated casually. (*Source*: Office of Industry Relations, 1974, *The Nuclear Industry*.)

time of between 20 and 30 years) is a controversial part of the uranium cycle with which we have little experience. Contaminated machinery needs to be disposed of or stored so that environmental damage will not occur. Decommissioning or refitting will be very expensive (perhaps $200 million to $500 million) and is an important aspect of planning for the use of nuclear power. It is possible that dismantling of old decommissioned reactors may become one of the highest costs for the nuclear industry.[10]

In addition to the hazards of transporting and disposing of nuclear material, there are potential hazards associated with supplying other nations with

reactors. Terrorist activity and the possibility of irresponsible persons in governments add a risk that is present in no other form of energy production. Nuclear energy may indeed be an answer to our energy problems, and perhaps someday it will provide unlimited cheap energy. However, with nuclear power comes responsibility. Many people share the value judgment that nuclear power should be used for, not against, people and that future generations should inherit a quality environment and be free from worrying about hazardous nuclear waste.

Fusion appears attractive from an environmental point of view. First, land-use and transportation impacts are small compared with those associated with fossil fuel or fission energy sources. Second, compared with fission breeders, fusion reactors pro-

Closer Look 18.2

Radiation Units and Doses

The units used to measure radioactivity are complex and somewhat confusing. Nevertheless, a modest acquaintance with them is useful in understanding and talking about the effects of radiation on the environment. A commonly used unit for radioactive decay is the **curie** (Ci), the amount of radioactivity from 1 gram of radium-226 that undergoes about 37 billion nuclear transformations per second. Radium-226 has a half-life of 1622 years. By the end of that time the initial gram of radium will be reduced to 0.5 gram, and the number of nuclear transformations will be reduced to about 18.5 billion per second. The amount of radioactivity present will be 0.5 curie.[3]

Radium was discovered in the 1890s by Marie Curie and her husband, Pierre. The unit still bears her name. Marie and Pierre also discovered polonium, which they named after Marie's homeland, Poland. The harmful effects of radiation were not known at that time, and both Marie Curie and her daughter died of radiation-induced cancer.[4] Her laboratory (Figure 18.9) is still contaminated today.

In the International System (SI) of measurement, the unit commonly used for radioactive decay is the **becquerel** (Bq), which is one radioactive decay per second. When considering radioactive isotopes, such as radon-222, the unit of measurement is often becquerels per cubic meter, or picocuries per liter (pCi\L). A picocurie is one-trillionth (10^{-12}) of a curie. Becquerels per cubic meter or picocuries per liter are, therefore, measures of the number of radioactive decays that occur each second in a cubic meter or liter of air.

When dealing with the environmental effects of radiation, we are most interested in the actual dose of radiation delivered by radioactivity. That dose is commonly measured in terms of **rads** (rd) and **rems.** In the International System (SI) the corresponding units are *grays* (Gy) and *sieverts* (Sv). Rads and grays are the units of absorbed dose of radiation, and 1 gray is equivalent to 100 rads. Rems and sieverts are units of equivalent dose, or effective equivalent dose, where 1 sievert is 100 rem.[3] The energy retained by living tissue that has been exposed to radiation is called the **radiation absorbed dose,** which is where the term *rad* comes from. Because different types of radiation have different penetrations and as a result do variable damage to living tissue, the rad is multiplied by a factor known as the relative biological effectiveness to produce the rem or sievert units. When very small doses of radioactivity are being considered, the millirem (mrem) or millisievert (mSv), that is, one-thousandth (0.001) of a rem or sievert, are used.[3,11,12] When considering gamma rays, the unit commonly used is the roentgen (R), or, in SI units, coulombs per kilogram (C/kg).

What is the naturally occurring, or background, radiation received by people? This question is commonly asked by people concerned with radiation. The average American receives about 1.5 mSv/year of natural radiation. The range, however, is from about 1.0 to 2.5 mSv/year.[3] The differences are primarily due to elevation and geology. More cosmic radiation is received at higher elevations, and some rock types (such as granite and organic shales) contain radioactive minerals. As a result, mountain states that also have an abundance of granitic rocks, such as Colorado, have a greater background radiation than do states that have a lot of limestone bedrock and low elevation, such as Florida.[11] In

Figure 18.9 Marie Curie in her laboratory.

spite of this general pattern, in locations in Florida where phosphate deposits occur, background radiation is above average because of a relatively high uranium concentration found in the phosphate rocks.

The precise amounts and sources of low background radiation received by people are not completely understood or agreed upon. However, some of the major sources include radioactive potassium-40 and carbon-14, which are present in our bodies and produce between 0.2 and 0.25 mSv/year. Potassium is an important electrolyte in our blood, and one isotope of potassium (potassium-40) has a very long half-life. Although potassium-40 is only a very small percentage of the total potassium in our bodies, it is present in all of us, so we are all slightly radioactive. As a result, if you choose to share your life with another person, you are also exposed to a little bit more radiation. Cosmic rays deliver between 0.35 and 1.5 mSv/year, depending on elevation, and radioactive materials in rocks, soils, and water deliver on average 0.35 mSv/year. However, the amount of radiation delivered from rocks, soils, and water may be much larger in some areas where radon gas (a naturally occurring radioactive gas) seeps into homes.

Sources of low-level radiation introduced by people include X rays for medical and dental purposes, which may deliver an average of 0.7

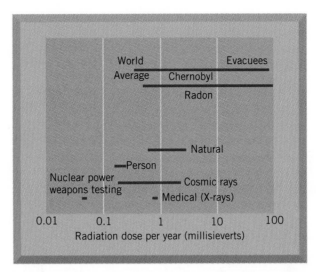

Figure 18.10 Annual radiation dose to people. (*Source:* Data in part from A. V. Nero, Jr., 1988, "Controlling Indoor Air Pollution," *Scientific American*, 258(5), pp. 42–48.)

to 0.8 mSv/year. Also, nuclear weapons testing and nuclear power plants may be responsible for approximately 0.04 mSv/year, and the burning of fossil fuels such as coal, oil, and natural gas may add another 0.03 mSv/year.[3,11]

A person's occupation and lifestyle can also affect the annual dose of radiation received. If you fly at high altitudes in jet aircraft, you receive an additional small dose of radiation. If you work at a nuclear power plant or a conventional coal-fired plant or a number of other industrial positions, you may also be exposed to low-level radiation. The amount of radiation received on the job site is closely monitored at obvi-

ous sites, such as nuclear power plants and laboratories where X rays are produced. At such locations personnel wear badges that indicate the dose of radiation received.

Figure 18.10 lists some of the common sources of radiation to which we are exposed. Notice that exposure to radon gas can equal what people were exposed to as a result of the Chernobyl nuclear power accident that occurred in the Soviet Union in 1986; in other words, in some homes people are exposed to about the same radiation as that experienced by the people evacuated from the Chernobyl area (radon gas is discussed in detail in Chapter 23).

duce no fission products and little radioactive waste and are less likely to be involved in an accident.[11] On the other hand, fusion power plants probably will use materials that are toxic to people; lithium, for example, is toxic in high concentrations if inhaled or ingested. Other potential hazards include strong magnetic fields and microwaves used in confining and heating plasma and short-lived radiation emitted from the reactor vessel.[7]

Effects of Radioisotopes

Radioisotopes affect the environment in two ways: by emitting radiation that affects other materials and

by entering the normal pathways of mineral cycling and ecological food chains.

The explosion of a nuclear atomic weapon does damage in both ways. At the time of the explosion, intense radiation of many kinds and energies is sent out, killing organisms directly. The explosion generates large amounts of radioactive isotopes, which are dispersed into the environment. Nuclear bombs exploded in the atmosphere produce a huge cloud that sends radioisotopes directly into the stratosphere, where the radioactive particles may be widely dispersed by winds. Atomic *fallout*—the deposit of these radioactive materials around the world—was an environmental problem in the 1950s

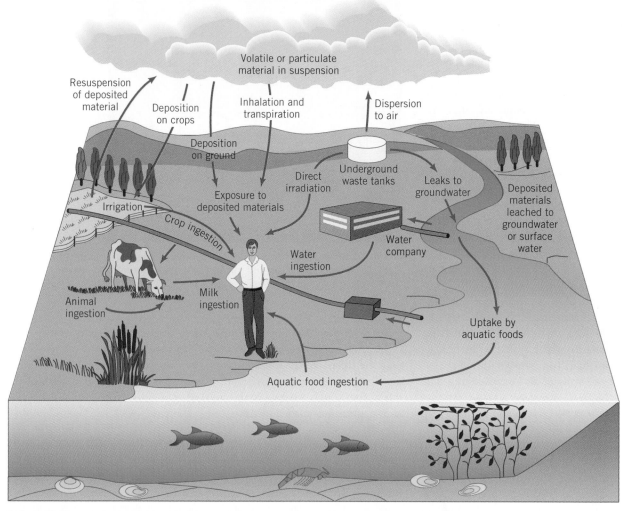

FIGURE 18.11 How radioactive substances reach people. (*Source*: F. Schroyer, ed., 1985, *Radioactive Waste*, 2nd printing, American Institute of Professional Geologists.)

and 1960s, when the United States, the former Soviet Union, China, France, and Great Britain were testing and exploding nuclear weapons in the atmosphere.

The pathways (Figure 18.11) of some of these isotopes illustrate the second way in which radioactive materials can be dangerous in the environment. One of the radioisotopes emitted and sent into the stratosphere by atomic explosions was cesium-137. This radioisotope was deposited in relatively small concentrations but was widely dispersed in the Arctic region of North America. It fell on reindeer moss, a lichen that is a primary winter food of the caribou. It was discovered that there was a strong seasonal trend in the levels of cesium-137 in caribou; the level was highest in the winter, when reindeer moss was the principal food, and lowest in the summer. Eskimo who obtained a high percentage of their protein from caribou ingested the radioisotope by eating the meat, and their bodies concentrated the cesium. The more that members of a group depended on

caribou as their primary source of food, the higher was the level of the isotope in their bodies.

The cesium was moved long distances by biospheric phenomena. After entering specific ecosystems through the vegetation, it underwent biomagnification, or ecological food chain concentration. That is, at each level in the food chain (moss to caribou to people), the concentration of the toxic material relative to concentrations of other materials in the bodies of organisms increased. Cesium-137 concentrations approximately doubled with each higher level in the food chain. This biomagnification was unknown until radioisotopes and toxic organic compounds were found to be occurring in higher and higher concentrations at higher and higher levels in the food chain. Biomagnification is one of the major ecological factors in environmental toxicology, as we saw in Chapter 14.

The actual body burden of cesium varied within the Eskimo population. Adult males between

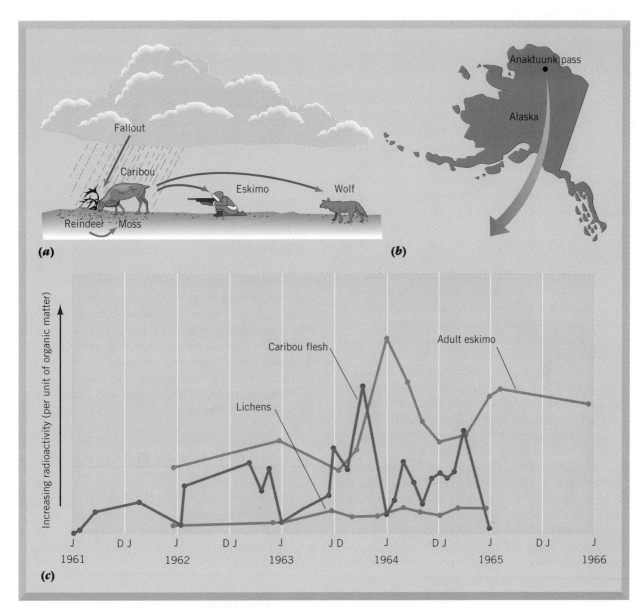

FIGURE 18.12 Cesium-137 released into the atmosphere by atomic bomb tests was part of the fallout deposited onto the soil and plants. (*a*) The cesium fell on lichens, which were eaten by caribou. The caribou were in turn eaten by Eskimo. (*b*) Measurements of cesium were taken in the lichens, caribou, and Eskimo in the Anaktuvuk Pass of Alaska. (*c*) The cesium was concentrated by the food chain. Peaks in concentrations occurred first in the lichens, then in the caribou, and last in the Eskimo. [*Source:* (*c*) W. G. Hanson, "Cesium-137 in Alaskan Lichens, Caribou, and Eskimos." *Health Physics*, 13, pp. 383–389. Copyright 1967 by Pergamon Press. Reprinted with permission.]

the ages of 20 and 50 had the highest levels, apparently because their diet contained the most caribou meat. Concentration varied seasonally and increased over the years of intensive atmospheric bomb testing (Figure 18.12).[13]

Another important factor in the toxicity of cesium-137 from fallout is the length of time the isotope remains in the body. This is measured by the biological half-life, the time for one-half the concentration to be lost. In the caribou, the average biolog-

ical half-life was 15 days. The biological half-life in the Eskimo, however, was approximately 65 days.

It is possible to predict the environmental pathways that radioisotopes will follow because we know the normal pathways of nonradioactive isotopes with the same chemical characteristics. Our knowledge of biomagnification and of large-scale air and oceanic movements that transport radioisotopes throughout the biosphere will also help us to understand the effects of radioisotopes.

Radiation Doses and Health

The most important factor in studying radiation exposure in people is determining at what point the exposure or dose becomes a hazard to health. Unfortunately, there are no simple answers to this seemingly simple question (see A Closer Look 18.2, "Radiation Units and Doses"). We do know that a dose of about 5000 mSv (5 sieverts) is considered lethal to 50% of people exposed to it. Exposure to 1000–2000 mSv is sufficient to cause health problems, including vomiting, fatigue, potential abortion of pregnancies of less than 2 months' duration, and temporary sterility in males. At 500 mSv, physiological damage is recorded. The maximum allowed dose of radiation per year for workers in industry is 50 mSv, which is approximately 30 times the average natural background radiation received by people.[3,11] For the general public, the maximum permissible annual dose is set in the United States at 5 mSv, which is about 3 times the annual natural background.[3]

Most information concerning the effects of high doses of radiation on people is from studies of the people who survived the atomic bomb detonations in Japan at the end of World War II. We also have information concerning people exposed to high levels of radiation in uranium mines, workers who painted watch dials with luminous paint containing radium, and people treated with radiation therapy for disease.[14] Workers in uranium mines who were exposed to high levels of radiation have been shown to suffer a significantly higher rate of lung cancer than the general population. Studies have shown that there is a delay of 10–25 years between the time of exposure and the onset of disease. Starting in about 1917 in New Jersey approximately 2000 young women were employed painting watch dials with luminous paint. To maintain a sharp point on their brushes, they licked them and as a result were swallowing radium, which was in the paint. By 1924 dentists in New Jersey were reporting cases of jaw rot, and within 5 years radium was known to be the cause. Many of the women died of anemia or bone cancer.[4]

Although there is a vigorous and ongoing debate about the nature and extent of the relationship between radiation exposure and cancer mortality, most scientists agree that radiation can cause cancer. Some scientists believe that there is a linear relationship, such that any increase in radiation beyond the background level will produce an additional hazard. Others believe that the body is able to successfully handle and recover from low levels of exposure to radiation but that beyond some threshold health effects (toxicity) become apparent. The verdict is still out on this subject, but it seems prudent to take a conservative viewpoint and accept that there may be a linear relationship. That is, any increase in radiation beyond the background level is likely to be accompanied by an increase in adverse health effects. Unfortunately, long-term chronic health problems related to low-level exposure to radiation are neither well known nor well understood.

Radiation has a long history in the field of medicine. Interestingly, drinking of waters that contain radioactive materials goes back to Roman times. By the year 1899 the adverse effects of radiation had been studied and were well known, and in that year the first lawsuit for malpractice in using X rays was filed. Because science had shown that radiation could destroy human cells, however, it was a logical step to conclude that drinking water containing radioactive material such as radon might help fight diseases such as stomach cancer. In the early 1900s it became popular to drink water containing radon, and the practice was supported by doctors who stated that there were no known toxic effects. Although we now know that statement to be false, radiotherapy, which uses radiation to kill cancer cells in humans, has been widely and successfully used for a number of years.[4]

18.3 NUCLEAR POWER PLANT ACCIDENTS

Although the chance of a disastrous nuclear accident is estimated to be very low, the probability of an accident occurring increases with every reactor put into operation. The two most well-known nuclear accidents are those of the Three Mile Island and Chernobyl reactors.

Three Mile Island

One of the most dramatic events in the history of U.S. radiation pollution occurred on March 28, 1979, at the Three Mile Island nuclear power plant near Harrisburg, Pennsylvania. Malfunctions of the nuclear plant resulted in a release of radioisotopes into the environment as well as intense radiation release within one of the nuclear facilities. Exposure from the plume emitted into the atmosphere has been estimated at 1 mSv, which is low in terms of the amount of radiation required to cause acute toxic effects. Average exposure to radiation in the surrounding area is estimated to be approximately 0.012 mSv, which is only about 1% of the natural background radiation received by people. However, radiation levels were much higher near the site. On the third day after the accident, 12 mSv/hour was measured at ground level near the site. By comparison, the average American receives about 1.5 mSv/year from natural radiation.

The Three Mile Island incident made it clear that there are many potential problems with the way in which our society deals with nuclear power. Historically, nuclear power had been relatively safe, and the state of Pennsylvania was somewhat unprepared to deal with the accident. For example, there was no state bureau for radiation help, and the state Department of Health did not have one book on radiation medicine (the medical library had been dismantled 2 years prior for budgetary reasons). One of the major impacts of the incident was fear, yet there was no state office of mental health and no staff member from the Department of Health was allowed to sit in on briefing sessions.

Because the long-term chronic effects of exposure to low levels of radiation are not well understood, the effects of Three Mile Island exposure, although apparently small, are difficult to estimate. This case illustrates that our society is in need of improvements in the way in which it handles crises arising from the sudden release of pollutants created by our modern technologies. It also illustrates our lack of preparedness and apparent willingness to treat nuclear power as an acceptable risk.[14]

Chernobyl

Lack of preparedness to deal with a serious nuclear power plant accident was dramatically illustrated by events that began unfolding on the morning of Monday, April 28, 1986 (Figure 18.13). Workers at a nuclear power plant in Sweden, frantically searching for the source of elevated levels of radiation near their plant, concluded that it was not their installation that was leaking radiation but that the radioactivity was coming from the Soviet Union by way of prevailing winds. Confronted, the Soviets announced that there had been an accident at their nuclear power plant at Chernobyl two days earlier, on April 26. This was the first notice to the world that the worst accident in the history of nuclear power generation had occurred.

It is speculated that the system that supplies cooling waters for the Chernobyl reactor failed, causing the temperature of the reactor core to rise to over 3000°C (about 5400°F), melting the uranium fuel. Explosions removed the top of the building over the reactor, and the graphite surrounding the fuel rods used to moderate the nuclear reactions in the core ignited. The fires produced a cloud of radioactive particles that rose high into the atmosphere. There were 237 confirmed cases of acute radiation sickness, and 31 people died of radiation sickness.[16]

In the days following the accident, nearly 3 billion people in the Northern Hemisphere received varying amounts of radiation from Chernobyl. With the exception of the 30-km (19-miles) zone surrounding Chernobyl, the world human exposure was relatively small. Even in Europe, where exposure was highest, it was considerably less than the natural radiation received during 1 year.[16]

In that 30-km zone approximately 115,000 people were evacuated and as many as 24,000 peo-

FIGURE 18.13 Guard halting entry of people into the forbidden zone evacuated in 1986 as a result of the Chernobyl nuclear accident.

ple were estimated to have received an average radiation dosage of 0.43 Sv (430 mSv). This group of people is being studied carefully, and it is expected, based on results from Japanese bomb survivors, that approximately 122 spontaneous leukemias are likely during the period from 1986 through 1998.[16] Surprisingly, as of late 1995, there was no significant increase in the incidence of leukemia, even among the most highly exposed people, but leukemia could still be expected in the future.[17] Studies have found that the number of childhood thyroid cancer cases per year has risen steadily in the three countries of Belarus, Ukraine, and the Russian Federation (those most affected by Chernobyl) since the accident. In 1994 there were a combined rate of 132 new thyroid cancer cases identified. Since the accident, there have been a total of 653 thyroid cancer cases diagnosed in children and adolescents. These cancer cases are believed to be linked to the released radiation from the accident, although other factors such as environmental pollution may also play a role. It is predicted that a few percent of the roughly 1 million children exposed to the radiation eventually will contract thyroid cancer.[18] Outside the 30-km zone the increased risk of contracting cancer is very small and probably not likely to be detected from an ecological evaluation.[16] Nevertheless, one estimate is that over the next 50 years Chernobyl will be responsible for approximately 16,000 deaths worldwide.[6]

Vegetation within 7 km of the power plant was either killed or severely damaged following the accident. Pine trees examined in 1990 around Chernobyl showed extensive tissue damage and still contained radioactivity. The distance between annual rings (a measure of tree growth) had decreased since 1986.[19] Scientists returning to the evacuated zone in the mid-1990s found, to their surprise, thriving and expanding animal populations. Species such as wild boar, moose, otters, waterfowl, and rodents seem to be enjoying a population boom in the absence of humans. The wild boar population has increased 10-fold since the evacuation. However, these animals may be paying a genetic price living within the contaminated zone. Study of gene mutations in voles (small mammals resembling rats or mice) within the contaminated zone has found that gene mutation rates are over five mutations per animal compared to a rate of only four-tenths per animal outside the zone. It is puzzling to scientists that the high mutation rate has not crippled the animal populations, but it appears so far that the benefit of excluding humans outweighs the negative factors associated with radioactive contamination.[20]

In the areas surrounding Chernobyl radioactive materials have and continue to contaminate soils, vegetation, surface water, and groundwater, presenting a hazard to plants and animals. The evacuation zone may be uninhabitable for a very long time unless some way is found to remove the radioactivity.[16] Estimates of the total cost of the Chernobyl accident vary widely but will probably exceed $200 billion.

Although the Soviets were accused of not giving attention to reactor safety and of using outdated equipment, people are still wondering if such an accident could happen again elsewhere. Because there are several hundred reactors producing power in the world today, the answer has to be yes. The Chernobyl accident follows a history of about 10 accidents that have released radioactive particles during the past 34 years. Therefore, although Chernobyl is the most serious nuclear accident to date, it certainly was not the first and is not likely to be the last. Although the probability of a serious accident is very small at a particular site, the consequences may be great, perhaps resulting in an unacceptable risk to society. This is really a political question more than a scientific one and a question of values. As a result of the Chernobyl accident, risk analysis in nuclear power is now a real-life experience rather than a computer simulation.

Advocates of nuclear power have argued that nuclear power is safer than other sources of energy. They say that the number of additional deaths caused by air pollution resulting from burning fossil fuels is much greater than the number of lives lost through nuclear accidents. For example, the 16,000 deaths that might eventually be attributed to Chernobyl are much fewer than the number of deaths caused each year by air pollution from burning coal.[6] Those arguing against nuclear power state that as long as people build nuclear power plants and manage them, there will be the possibility of accidents. It is argued that this is as much a problem of human nature as anything else. Because of the tremendous potential consequences of serious accidents such as that at Chernobyl, we may have to recognize that we are not ready as a people to handle nuclear power. We can build nuclear reactors that are safer, but people make mistakes, and as long as reactors are run by people, there will be chances of errors and of accidents.

18.4 RADIOACTIVE WASTE MANAGEMENT

Examination of the nuclear fuel cycle (Figure 18.8) illustrates some of the sources of waste that must be disposed of as a result of using nuclear energy to produce electricity. Radioactive wastes are by-products that must be expected when electricity is produced at nuclear reactors; they may be grouped into two general categories: low-level waste and high-level waste. In addition, the tailings (materials that are removed by mining activity but are not processed and remain at the site) from uranium mines and mills must also be considered hazardous. In the western United States more than 20 million metric tons of abandoned tailings will continue to produce radiation for at least 100,000 years.

Low-level radioactive waste contains sufficiently low concentrations or quantities of radioactivity that it does not present a significant environmental hazard if properly handled. Low-level waste includes a wide variety of items, such as residuals or solutions from chemical processing; solid or liquid plant waste, sludges, and acids; and slightly contaminated equipment, tools, plastic, glass, wood, and other materials.[21] Low-level waste has been buried and monitored in near-surface burial areas in which the hydrologic and geologic conditions were thought to severely limit the migration of radioactivity.[21] However, several of the U.S. sites for disposal of low-level radiation have not proved to provide adequate protection for the environment, and leaks of liquid waste have polluted groundwater. Of the original six burial sites, three had closed prematurely by 1979 due to unexpected leaks, financial problems, or loss of license. As of 1995, there were only two remaining low-level nuclear waste repositories, one in Washington and the other in South Carolina. Construction of new burial sites, such as the Ward Valley site in southeastern California, has been met with strong public opposition, and there remains controversy over whether low-level radioactive waste can be disposed of safely.[22]

High-level radioactive waste is extremely toxic, and a sense of urgency surrounds its disposal as the total volume of spent fuel accumulates. There is conservative optimism that the waste disposal problem will soon be solved. At present, spent fuel elements from commercial reactors are being stored and await disposal or eventual reprocessing to recover plutonium and unfissioned uranium. By the end of 1995, nuclear reactors in the United States had produced approximately 29,000 metric tons of spent fuel, with about 1900 more tons produced each year.[23]

Storage of high-level waste is at best a temporary solution, and serious problems with radioactive waste have occurred where they are being stored. Although improvements in storage tanks and other facilities will help, eventually some sort of disposal program must be initiated. There are some scientists who believe the geologic environment can best provide safe containment of high-level radioactive waste. Others disagree and have criticized proposals

for long-term disposal of radioactive waste underground. A comprehensive geologic disposal development program should have the following objectives:[24]

- identification of sites that meet broad geologic criteria of ground stability and slow movement of groundwater with long flow paths to the surface;
- intensive subsurface exploration of possible sites to positively determine geologic and hydrologic characteristics;
- predictions of behavior of potential sites based on present geologic and hydrologic situations and assumptions for future changes in variables such as climate, groundwater flow, erosion, and earth movements;
- evaluation of risk associated with various predictions; and
- the forming of political decisions based on risks acceptable to society.

The Nuclear Waste Policy Act of 1982 initiated a high-level nuclear waste disposal program. The Department of Energy was given the responsibility to investigate several potential sites and make a recommendation. The act was amended in 1987 specifically to state that the Yucca Mountain site in southern Nevada would be evaluated to determine if high-level radioactive waste could safely be disposed of there. Extensive scientific evaluations of the Yucca Mountain site are now ongoing, but use of this site is generating considerable resistance and controversy from the state and people of Nevada as well as those scientists not confident in the plan. Some of the scientific questions at Yucca Mountain have concerned natural processes and hazards that might allow radioactive materials to escape, such as surface erosion, groundwater movement, earthquakes, and volcanic eruptions.

One of the major problems with the disposal of high-level radioactive waste remains: How credible are long-range geologic predictions, those covering several thousand to a few million years?[24] Unfortunately there is no easy answer to this question, because geologic processes vary over both time and space. Climates change over long periods of time, as do areas of erosion, deposition, and groundwater activity. For example, large earthquakes even thousands of kilometers from a site may permanently change groundwater levels. The earthquake record for most of the United States extends back in time for only a few hundred years; therefore, estimates of future earthquake activity are tenuous at best. The bottom line is that geologists can suggest sites that have been relatively stable in the geologic past, but they cannot absolutely guarantee future stability. Therefore, decision makers (not geologists) need to evaluate the uncertainty of prediction in light of pressing political, economic, and social concerns.[24] These problems do not mean that the geologic environment is not suitable for safe containment of high-level radioactive waste, but care must be taken to ensure that the best possible decisions are made on this important and controversial issue.

18.5 THE FUTURE OF NUCLEAR ENERGY

Nuclear energy as a power source for electricity is now being seriously evaluated. Advocates for nuclear power have been arguing that nuclear power is good for the environment because:

- it does not produce potential global warming through release of carbon dioxide (see Chapter 21);
- it does not cause air pollution or emit precursors (sulfates and nitrates) that cause acid rain (see Chapter 22); and
- if breeder reactors are developed, it will greatly increase the amount of fuel available.

Those in favor of nuclear power argue that it is safer than other means of generating power and that we should build many more nuclear power plants in the future. This argument is predicated on the understanding that such power plants would be considerably safer than those being used today. That is, if we standardize nuclear reactors and make them safer and smaller, nuclear power could provide much of our electricity in the future.[6] With the exception of accidents, the safety of nuclear power under normal operating procedures is not disputed, although the disposal of spent fuel is a growing concern. The argument against reviving nuclear power is based more on political and economic considerations than on scientific ones. Those opposed to expanding nuclear power argue that if we were to convert from coal-burning plants to nuclear power plants for the purpose of reducing carbon dioxide

emissions, it would require an enormous investment in nuclear power to make a real impact. Furthermore, given the fact that safer nuclear reactors are only just being developed, there will be a time lag. As a result, nuclear power is not likely to have a real impact on environmental problems, such as air pollution, acid rain, and potential global warming, before at least the year 2050.[25]

Since its inception, the nuclear industry in the United States has benefited from significant governmental support. Between 1948 and 1992, it has been estimated that nuclear energy was the recipient of 65% of all federal energy research and development funds.[10] Critics of nuclear energy point out that if more money had been directed toward alternative and renewable energy sources rather than nuclear power, alternative energy would be supplying more of the nation's energy needs today.

In recent years the politics of nuclear energy has seemed to be losing ground. Nearly all energy scenarios are based on the expectation that nuclear power will continue to grow slowly or perhaps even decline in the coming years. Since the Chernobyl accident, many countries in Europe have been reevaluating the use of nuclear power, and in most instances the number of nuclear power plants being built has been significantly reduced. In the United States there have been no new orders for nuclear power plants since 1978. In spite of this situation, research and development into the smaller, safer nuclear power plants is going forward. Nuclear power produces about one-fifth of our electricity in the United States today, and U.S. consumption for the first time decreased slightly (about 7%) in 1997.[26] The nuclear option is again being evaluated in light of the environmental problems associated with fossil fuels. Producing electricity in nuclear power plants does not add carbon dioxide to the atmosphere or release air pollutants that can produce acid rain. However, these benefits of nuclear power must be balanced with consideration of the safety and waste disposal issues that have made nuclear energy an uncomfortable option for many people. The full impact of what began in 1942, when the atom was first split, is still to be determined.

Does Nuclear Power Have a Future in the United States?

Returning to the important question of future of nuclear power, in 1996 the United States had 109 nuclear power plants in operation in 32 states, producing 670 billion kilowatt-hours of electricity. This is almost twice as much as the next leading country, France, which generates 77% of its total electricity by nuclear power. Why, then, is there still a question as to the future of nuclear power in the United States? One reason is that, despite the impressive output of nuclear-generated electricity, the United States ranks 19th out of 33 countries with nuclear power plants in the percentage of its electricity generated by nuclear power (only 20%). Other reasons are that no new reactors have been ordered since 1978, and 61 plants will be 20 years old (more than half the expected lifetime for nuclear plants) by 2000. Finally, we are a long way from predictions made in the 1970s, that there would be 1000 large nuclear facilities in the United States by the beginning of the next century.

Why has the nuclear power industry been stalled in the United States? One factor has been that growth in energy demand has slowed since the 1970s as a reaction to the oil scares in that decade. Conservation and the use of alternative energy sources since then have exceeded expectations, so that predictions of growth in energy demand have had to be continuously revised downward (see graph) and may yet be inaccurate. Prior to 1970, energy use in this country was increasing parallel with the gross domestic product (GDP), but since then, growth in energy use has slowed while GDP has continued to increase.

At the same time, the cost of construction of nuclear power plants and the time required to complete them have increased at least two to three times, discouraging power utilities from becoming involved in new projects. Several factors may have contributed to these conditions: concern over safety in the wake of the accident at Three Mile Island; new safety regulations established by the Nuclear Regulatory Commission (NRC); public anger over the higher prices for electricity that would have to be charged to cover construction costs; poor management; and industry response to the decline in energy demand.

Public opinion about nuclear power in the United States is perceived as negative by many, primarily because of concerns about safety (low-level exposures as well as catastrophic accidents), disposal of radioactive wastes, and proliferation of raw materials that could be used to manufacture nuclear weapons. Certainly there have been outspoken critics of nuclear power and local opposition to plants and disposal sites. A 1992 poll found

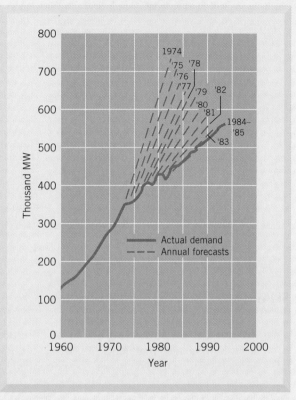

Growth in electric generating capacity in the United States compared to annual forecasts by the National Electricity Reliability Council (based on utilities' construction plans). (*Source:* J. F. Ahearne, 1993, "The Future of Nuclear Power," *American Scientist,* 81(1), pp. 24-35.

that 65% of the public opposes construction of new reactors but goes on to suggest that "opinion leaders" may exaggerate the opposition of the general public.

The possibility that carbon dioxide emissions from fossil fuel use are causing global warming has given new energy to the proponents of nuclear power. Developing countries are responsible for an increasing share of the world's carbon dioxide emissions—from 7% in 1970 to 28% in 1987—and the levels are increasing at 6% to 7% per year. In order for nuclear power to be economically competitive with coal (the most common energy source in developing countries), coal prices would have to double and the cost of nuclear power would have to decrease by half.

Critical Thinking Questions

1. If the United States is only 14th in the world in percentage of electrical energy generated by nu-

clear energy, how is it that it produces almost twice as much energy from nuclear power plants as its nearest competitor?

2. In the United States, about 45,000 deaths per year result from automobiles use, but no deaths have been attributed to nuclear industry. Why are Americans more willing to accept one risk than another?

3. What are some reasons for the decline in the nuclear power industry? How are they ranked in order of importance from the point of view of the utility companies and from the point of view of the public? Explain your rankings and the differences, if any, between the two points of view.

4. Under what conditions could nuclear power make a comeback in the United States? How likely do you think it is that these conditions will occur?

5. Alvin Weinberg, a pioneer in nuclear power, estimates that strict energy efficiency and 5000 nuclear plants worldwide could stabilize global warming in 100 years. Other experts estimate that it would take one new plant coming online every 2 days for 38 years to reduce carbon dioxide emissions. How would these estimates affect the argument that nuclear power could alleviate global warming?

References

Ahearne, J. F. 1993. "The Future of Nuclear Power," *The American Scientist* 81(1): 24–35.

Fox, M. R. 1987. "Perspectives in Risk: Compared to What?" *Vital Speeches of the Day* 53(23): 730–732.

Greenberg, P. A. 1993. "Dreams Die Hard," *Sierra* 78(6): 78.

Lanouette, W. April 1990. "Greenhouse Scare Reheats Nuclear Debate," *Bulletin of the Atomic Scientists* 46(3): 34–37.

Miller, A., and Mintzer, I. 1990. "Global Warming: No Nuclear Fix," *Bulletin of the Atomic Scientists* 46(5): 31–33.

United States Nuclear Regulatory Commission, 1997. *NRC Information Digest*, NUREG-1350, Vol. 9.

SUMMARY

- Nuclear fission is the process of splitting an atom into smaller fragments. As fission occurs, energy is released. The major components of a fission reactor are the core, control rods, coolant, and reactor vessel.

- Nuclear radiation occurs when a radioisotope spontaneously undergoes radioactive decay and changes into another isotope.

- The three major types of nuclear radiation are alpha, beta, and gamma.

- Each radioisotope has its own characteristic emissions. Different types of radiation have different toxicities, and in terms of human health and that of other organisms it is important to know the type of radiation emitted and the half-life.

- The nuclear cycle consists of uranium mining, enrichment of uranium-235, making fuel assemblies, generating nuclear power (burning the fuel), reprocessing uranium, disposal of nuclear waste, and decommissioning of power plants. Each part of the cycle is associated with characteristic processes, all with different potential environmental problems.

- The two major ways in which radioisotopes affect the environment are emitting radiation that affects other materials and entering ecological food chains. The major pathways for radiation to reach people through the environment include uptake by fish ingested by people, uptake by crops ingested by people, inhalation from air, and exposure to nuclear waste and the natural environment.

- The dose response for radiation is fairly well established. We know the dose response for the higher exposures, when symptoms such as illness or death of an organism occurs. However, there are vigorous debates concerning the health effects of low-level exposure to radiation and what relationships exist between exposure and cancer mortality. Most scientists believe that radiation can cause cancer. Ironically, radiation can be used to kill cancer cells, as in radiotherapy treatments.

- We have learned from accidents at nuclear power plants that it is difficult to plan for the human factor. People make mistakes. We have also learned that we are not as prepared for accidents as we would like to think. Some people believe that humans are not ready for the responsibility of nuclear power. Others believe that we can design much safer power plants where serious accidents are impossible.

- There is consensus that high-level nuclear waste may be safely disposed of in the geologic environment. The problem has been to locate a site that is safe and not objectionable to the people who make the decisions and to those who live in the region.

- Nuclear power is being seriously evaluated as

an alternative to the fossil fuels. It has advantages in that it emits no carbon dioxide and will not contribute to global warming or cause acid rain. On the other hand, people are uncomfortable with nuclear power because of possible accidents and waste disposal problems.

REEXAMINING THEMES AND ISSUES

Human Population: As human population has increased, so has demand for electrical power and as a result a number of countries have turned to nuclear energy. Although relatively rare, accidents at nuclear power plants such as Chernobyl in 1986 exposed people to an increase in radiation. There is considerable debate over potential adverse effects of that radiation. Nevertheless, the fact remains that as world population increases, the total number of people exposed to potential hazardous release of toxic radiation also increases.

Sustainability: It has been argued that a sustainable energy development will require return to nuclear energy because under normal operating conditions use of nuclear energy does not increase the possibility of potential global warming, release sulfur or precursors to acid rain into the atmosphere, or cause a variety of other environmental problems related to burning of fossil fuels.

Global Perspective: There are many potential pathways for radioactive substances to interact in the environment at a regional and global scale. Testing of nuclear weapons has spread radioactive isotopes around the entire planet, as have nuclear accidents. Radioactive isotopes that enter rivers and other waterways may eventually enter the oceans of the world, where oceanic circulation may further disperse and spread them. Use of nuclear energy also fits into our global management of the entire spectrum of energy sources. Building of smaller, safer nuclear reactors in the future may lead to their being more widely used.

Urban World: Development of nuclear energy is a product of our technology and urban world. In some respects it is near the pinnacle of our accomplishments in terms of technology.

Values and Knowledge: We have a great deal of knowledge concerning nuclear energy and nuclear processes. However, people remain very suspicious and in some cases, frightened by nuclear power. This results because of the value that people place on a quality environment and their perception that nuclear radiation is toxic to that environment. As a result, the future of nuclear energy will be related to political decisions based in part on the risk being acceptable by society.

KEY TERMS

alpha particles *366*	fusion *364*	nuclear energy *364*
becquerel *374*	gamma rays *366*	radiation absorbed dose *374*
beta particles *366*	half-life *366*	radioactive decay *366*
breeder reactors *370*	high-level radioactive waste *381*	radioisotope *366*
burner reactors *365*	low-level radioactive waste *381*	rads *374*
curie *374*	neutrino *367*	rems *374*
fission *364*	nuclear cycle *372*	

STUDY QUESTIONS

1. If exposure to radiation is a natural phenomenon, why are we worried about it?

2. What is a radioisotope and why is knowing its half-life important?

3. What is the normal background radiation that people receive? Why is it variable?

4. What are the possible relationships between exposure to radiation and adverse health effects?

5. Which processes in our environment may result in radioactive substances reaching people?

6. Suppose it is recommended that high-level nu-

clear waste be disposed of in the geologic environment of the region in which you live. How would you go about evaluating potential sites?

7. Are there good environmental reasons to develop and build new nuclear power plants? Discuss both sides of the issue.

FURTHER READING

Cohen, B. L. 1990. *The Nuclear Energy Option*. New York: Plenum. This book presents the case for nuclear energy and provides a complete discussion of the science concerning radiation and the potential health effects from exposure to radiation.

Murray, R. L. 1993. *Nuclear Energy: An Introduction to the Concepts, Systems, and Applications of Nuclear Processes* (4th ed). New York: Pergamon Press.

Nuclear Energy Agency (NEA), Organization for Economic Co-Operation and Development (OECD). 1993. *The Cost Of High-Level Waste Disposal In Geological Repositories: An Analysis of Factors Affecting Cost Estimates*. NEA, OECD. Analyses of management of spent fuel, methods of geological disposal, and costs associated with disposal of high-level radioactive waste.

Nuclear Energy Agency (NEA), Organization for Economic Co-Operation and Development (OECD). 1994. *Power Generation Choices: Costs, Risks, and Externalities*. Proceedings of an international symposium, Washington, D.C., September 23–24, 1993. NEA, OECD. Includes discussion of the economics of nuclear power versus other energy sources.

Nuclear Energy Agency (NEA), Organization for Economic Co-Operation and Development (OECD). 1995. *Environmental and Ethical Aspects of Long-Lived Radioactive Waste Disposal*. Proceedings of an international workshop, Paris, September 1–2, 1994. NEA, OECD. Essays covering topics of environmental policies, ethical and environmental considerations, cost–benefit analysis, and disposal issues of long-lived radioactive waste.

Thomas, S., and Berkhout, F. 1992. "The First 50 Years of Nuclear Power: Legacy and Lessons—Parts 1 and 2," *Energy Policy* 20(7, 8). This special issue of Energy Policy contains over 20 reviews, overviews, and case studies of the current state of nuclear energy supply, technology, and waste management and the political situation.

U.S. Department of Energy, Office of Environmental Management. 1995. *Closing the Circle on the Splitting of the Atom*. Washington, DC: U.S. Department of Energy. Descriptions of environmental, safety, and health problems associated with production of nuclear weapons and how the U.S. Department of Energy plans to deal with the problem.

World Health Organization. 1995. *Health Consequences of the Chernobyl Accident*. Geneva, Switzerland: World Health Organization. A short book covering the accident, response, health consequences, findings, and proposed future work.

Young, J. P. and Yalow, R. S., eds. 1995. *Radiation and Public Perception: Benefits and Risks*. Washington, DC: American Chemical Society. A comprehensive look at public perception of radiation risks and health effects of radiation through experimentation, occupational exposure, atomic detonation, and nuclear reactor accidents.

INTERNET RESOURCES

American Nuclear Society: *http://www.ans.neep.wisc.edu/*—From the nonprofit group advancing the study of nuclear research and engineering, find out facts about nuclear power related research, read reports and documents, and check out links to many web resources on nuclear power and engineering.

International Atomic Energy Agency: *http://www.iaea.org*—This site contains information about nuclear power worldwide, including statistics, public opinion, history, and technology.

U.S. Department of Energy Office of Civilian Radioactive Waste Management: *http://www.rw.doe.gov/*—The OCRWM is responsible for planning, constructing, and operating a system to dispose of radioactive waste generated in the U.S. Read facts about storage and disposal of spent fuel and see what is happening at the proposed Yucca Mountain long term disposal site.

Nuclear Energy Institute: *http://www.nei.org/*—The site of a trade association of the nuclear energy industry, it includes general (although slantedly pro-nuclear) information about nuclear power in the U.S. and abroad.

U.S. Nuclear Regulatory Commission: *http://www.nrc.gov/*—From the federal agency responsible for regulating the industry, this site contains useful facts about nuclear waste generation and disposal, performance of currently operating nuclear power plants, and other information about nuclear power's past, present, and future.

PART VI

Minoan vase from Palaikastro, c. 1,500 B.C.

Water Environment

Large rubber rafts docked on a beach (sandbar) on the Colorado River in the Grand Canyon, Arizona.

CHAPTER 19

WATER SUPPLY, USE, AND MANAGEMENT

CASE STUDY

The Colorado River

From its headwaters in Wyoming to its discharge point into the Gulf of California, the Colorado River drains much of the southwestern United States (Figure 19.1). The river basin is approximately 632,000 km² (244,000 mi²). Considering its size, the river has only a modest flow, but it is one of the most regulated and controversial bodies of water in the world. The total flow of water in the river has been appropriated among the various users, including seven states and Mexico. Today, Colorado River water only occasionally flows into the Gulf of California—it's all stored and used. As a result, ecosys-

tems of the lower river and delta, being deprived of water and nutrients, have been damaged.

The complex issues of water management for the Colorado River illustrate some of the major problems that are likely to be faced by other semi-arid regions of the world in the coming years: How are we going to appropriate our water resources when they are scarce? How can we best control water quality? How can we protect river ecosystems? There are no easy answers to these questions.

The Colorado River originates in the Wind River Mountains of Wyoming, and in its 2300-km

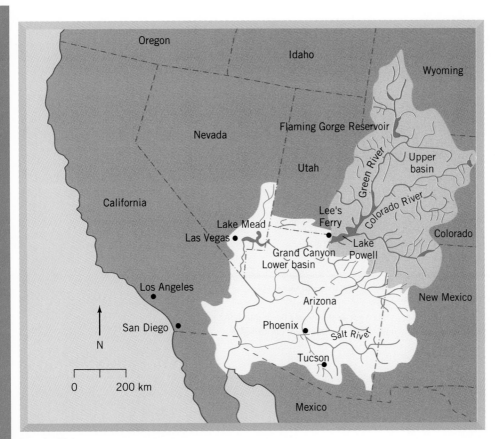

FIGURE 19.1 The Colorado River basin.

(1400-mile) journey to the Gulf of California flows through some of the most spectacular scenery in the world. Eight hundred years ago, Native Americans living in the Colorado River basin had a highly civilized culture and constructed a sophisticated water distribution system. In the 1860s settlers cleared the debris from these early canals and used them once again for irrigation.[1]

In spite of the Colorado River's long history of human occupation and use, it was not completely explored by the United States until 1869, when John Wesley Powell courageously navigated small wooden boats through the Grand Canyon. Today the water of the Colorado River basin is completely spoken for and distributed by canals and aqueducts to millions of urban people as well as to agricultural areas. However, the basin itself remains sparsely populated, and the largest city on the river is Yuma, Arizona, with a population of 42,000. Rod Nash, writing about his wilderness experience on the river, states that at the junction of the Green and Colorado Rivers, in the heart of a national park, it is 80 km (50 miles) to the nearest video game.[2]

Management of the waters of the Colorado River has been frustrating in part because of inherent uncertainties concerning how much water can be expected in a given year.[1] Water resources in the basin include snowmelt, long-term winter pre-

cipitation, and short-term summer thunderstorms; as a result the total water available on a year-to-year basis is tremendously variable. Table 19.1 compares the legal entitlements for water to the actual distribution to the users. Interestingly, the annual legal entitlements of 17.5 million acre-feet (21.5 km³) exceed the actual distribution of 14.5 million acre-feet (17.8 km³). (One acre-foot of water, equal to 1,238,000 liters, or 325,829 gal, is the volume of water necessary to cover one acre to a depth of one foot.[3])

Table 19.1 shows that all users do not take their full legal entitlement of water. If they did, serious problems would result, because the entitlements exceed the natural annual flow, which is about the amount currently being distributed. The reason the distribution can be maintained is that the Colorado River is one of the most managed rivers in the world. Figure 19.2 shows the profile of the river and some of the major dams and reservoirs in the system, which store approximately 86.3 km³ (70 million acre-feet) of water. The two largest reservoirs—Hoover and Glen Canyon dams (Figure 19.3)—store about 80% of the total in the basin. Total storage represents (with careful management) a buffer of several years' water supply. However, if there are several drought years in a row, maintaining a sufficient water supply for all users may not be possible.

TABLE 19.1 Legal and Actual Distribution of Colorado River Water

Area	Legal Entitlements (km³/yr)	Actual Distribution[a] (km³/yr)
Lower basin		
California	5.461[b]	5.461[c]
Arizona	4.674[b]	2.522[c]
Nevada	0.369[b]	0.369
Subtotal	10.455[d]	8.303
Upper basin		
Colorado	4.774[e]	2.959
Utah	2.122[e]	1.316
Wyoming	1.292[e]	0.801
New Mexico	1.038[e]	0.643
Subtotal	9.225[d]	5.720
Mexico	1.845[f]	1.845
Total	21.525	17.835

Source: W. L. Graf, 1985, *The Colorado River*, Resource Publications in Geography, Association of American Geographers, Washington, DC.

[a]Includes losses to evaporation of 0.6 million acre-feet/year in the upper basin and 0.9 million acre-feet/year in the lower basin and 0.9 million acre-feet/year inflow to lower basin from local streams.

[b]1928 Boulder Canyon Project Act.

[c]Agreement at time of Central Arizona Project authorization by Congress. Upper basin amounts agreed to by states as percentages.

[d]1922 Colorado River Compact.

[e]1948 Upper Colorado River Basin Compact.

[f]1944 Mexico–U.S. Treaty.

The Glen Canyon Dam, completed in 1963, is upstream from the Grand Canyon. From a hydrologic viewpoint the Colorado River is changed forever by the dam. The river is tamed; the higher flows have been reduced and the average flow has increased. The flow or discharge also changes often because of fluctuating needs to generate electrical power. Finally, changing the hydrology of the river also changed other aspects, including the rapids, distribution of sediments that form sandbars

FIGURE 19.2 Longitudinal profiles of the Colorado and Green Rivers, showing the major dams, reservoirs, and canyons. (*Source:* W. L. Graf, 1985, *The Colorado River*, Resource Publications in Geography, Association of American Geographers, Washington, DC.)

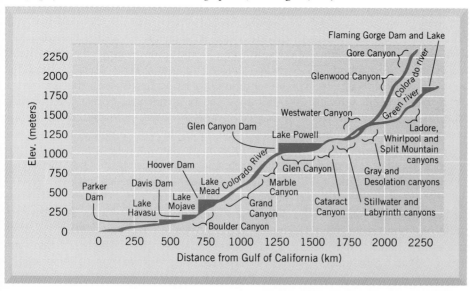

FIGURE 19.3 Glen Canyon Dam on the Colorado River in Arizona.

(called beaches by rafters, see photograph at beginning of chapter), and vegetation near the water's edge. The beaches are valuable wildlife habitats that were starved of sediment and shrank in size and number following construction of the dam.

A record snowmelt in the Rocky Mountains in June of 1983 forced the release of about 2500 m³ (88,000 ft³) of water per second, about three times the amount normally released and similar to a spring flood prior to the dam. Resulting floods scoured the river bed and banks, releasing stored sediment that replenished sandbars and broke off some vegetation that had taken root.[4] This release of water was beneficial to the river environment and highlighted the importance of floods in maintaining the system in a more natural state. Natural disturbances are a necessary part of the river ecosystem if it is to function on a sustainable basis. In late March of 1996 a flood with discharge about one-half the size of that in 1983 was deliberately released for a period of 1 week. Between March 26 and April 2 in 1996 water was allowed to flow at full flood, then flow was reduced for the last 2 days in order to redistribute the sand supply. The flood resulted in 55 new beaches and increased the size of 75% of existing beaches. It also helped rejuvenate marshes and backwaters, which are critical habitats to native fish and some endangered species.[5] This experiment is the first time that the U.S. government opened flood gates of a dam to improve a river's ecosystem. Some scientists were concerned that floodwaters would not be of sufficient volume and duration of 1 week too short. Nevertheless, experimental release of water marks a turning point in river management. The flood was hailed a great success, but it will take some time to see if the results will last. Hopefully, what was learned will be used to restore river environments and improve ecosystems of other rivers impacted by dams.

Conflicts over Colorado River water use and quality have spanned decades and extend beyond the river basin to urban centers and agricultural areas in California, Colorado, New Mexico, and Arizona. The crucial need for water in these semiarid regions has resulted in overuse of limited supplies of water as well as deterioration of water quality. This is a problem in the lower river particularly, where water is so salty that at times Mexican farmers are unable to use it. International and interstate agreements and court settlements have periodically intensified or eased tension among water users. These court decisions and laws, along with changes in water use, continue to significantly influence the lives of millions of people in the United States and Mexico.[6]

Water is a critical, limited, renewable resource in the Colorado River basin and many other regions on Earth. This chapter discusses our water resources in terms of supply, use, management, and sustainability and addresses important environmental concerns related to water: wetlands, reservoirs, channelization, and flooding.

LEARNING OBJECTIVES

Although water is one of the most abundant resources on Earth, there are many important issues and problems involved in water management. After reading this chapter, you should understand:

- Why the total abundance of water on Earth is not a problem, but making it available where and when it is needed *is* a problem.

- What characteristics of water make it unique as a liquid, critical for the existence of life, and an important factor in many physical and biochemical processes.

- Why the residence times of water in various parts of the hydrologic cycle are critical to water use and pollution potential.

- What a water budget is and why it is useful in analyzing water supply problems and potential solutions.

- What groundwater is and what environmental problems are associated with its use.

- How water may be conserved at home and in industrial and agricultural practice.

- Why the principles of water management will become more complex as the demand for water increases and why we must gather and use data concerning the natural flux of water resources.

- What the environmental impacts of water projects such as dams, reservoirs, canals, and channelization are.

19.1 WATER

To understand water as a necessity, as a resource, and as a factor in the pollution problem, we must understand its characteristics, its role in the biosphere, and its role in living things. Water is a unique liquid; without it, life as we know it is impossible. Compared with most other common liquids, water has a high capacity to absorb or store heat and is a good liquid solvent. Because many natural waters are slightly acidic, they can dissolve a great variety of compounds, from simple salts to minerals, including sodium chloride (common table salt) and calcium carbonate (calcite) in limestone rock. Water also reacts with complex organic compounds, including many amino acids found in the human body.

Compared with other common liquids, water has a high surface tension, a property that is extremely important in many physical and biological processes that involve moving water through or storing water in small openings or pore spaces. Among common compounds and molecules, water is the only one whose solid form is lighter than its liquid form (it expands by about 8% when it freezes, becoming less dense), which is why ice floats. If ice were heavier than liquid water, it would sink to the bottom of water bodies; the biosphere would be vastly different from what it is, and life, if it existed at all, would be greatly altered.[7] If bodies of water froze from the bottom up, they would freeze solid, killing all life in the water. Cells of living organisms are mostly water. As water freezes and expands, cell membranes and walls rupture. Finally, water is transparent to sunlight, allowing photosynthetic organisms to live below the surface.

A Brief Global Perspective

A short review of the concept of a global hydrologic cycle, introduced in Chapter 4, is worthwhile here. The main process of the cycle is the global transfer of water from the atmosphere to the land and oceans and back to the atmosphere (Figure 19.4). Table 19.2 lists the relative amounts of water in the major storage compartments of the cycle. Notice that more than 97% of Earth's water is in the oceans, and the next largest storage compartment, the ice caps and glaciers, accounts for another 2%. Together the oceans and ice tie up more than 99% of the total water, and both sources are generally unsuitable for human use because of salinity (seawater) and location (ice caps and glaciers). The total amount of water on Earth (from a human resource perspective) is constant (on a geologic time scale of millions of years new water is being added to Earth's supply by volcanic processes). It is a fascinating fact that only 0.001% of the total water on Earth is in the atmosphere at any one time. This relatively small amount of water in the global water cycle, with an average atmosphere residence time of only about 9 days, produces all our freshwater resources through the process of precipitation.

On a global scale, total water abundance is not the problem; the problem is water's availability in the right place at the right time in the right form. Water can be found in either liquid, solid, or gaseous form at a number of locations at or near Earth's surface. Depending on the specific location, the residence time may vary from a few days to many thousands of years (see Table 19.2). However, as mentioned previously, more than 99% of Earth's water in its natural state is unavailable or unsuitable

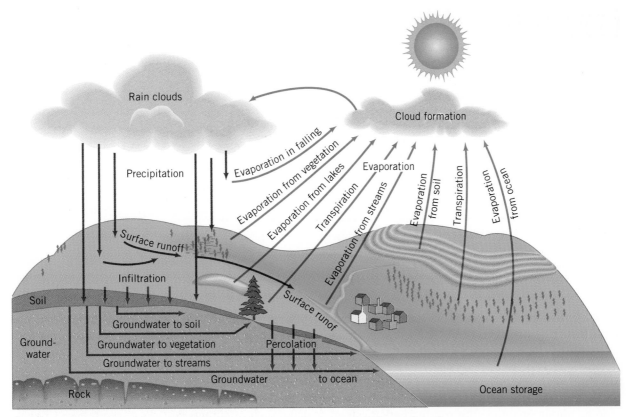

FIGURE 19.4 The hydrologic cycle showing important processes and transfer of water. (*Source:* Modified after Council on Environment Quality and the Department of State, 1980, *The Global 2000 Report to the President*, vol. 2, Washington, DC.)

for beneficial human use. Thus the amount of water for which all the people, plants, and animals on Earth compete is much less than 1% of the total.

As the world's population and industrial production of goods increase, the use of water will also accelerate. The world per-capita use of water in 1975 was about 700 m³/year, or 2000 liters/day (185,000 gal/year), and the total human use of water was about 3850 km³/year (about 10^{15} gal/year). It is estimated that by the year 2000 world use of water based on the projected population increase will ex-

ceed 6000 km³/year (about 1.58 x 10^{15} gal/year), which is a significant fraction of the naturally available fresh water.

The total average annual water yield (runoff) from Earth's rivers is approximately 47,000 km³ (1.2 x 10^{16} gal), but its distribution is far from uniform (see Table 19.3). Some occurs in relatively uninhabited regions, such as Antarctica, which produces about 5% of Earth's total runoff. South America, which includes the relatively uninhabited Amazon basin, provides about one-fourth of Earth's total

TABLE 19.2 The World's Water Supply (Selected Examples)

Location	Surface Area (km²)	Water Volume (km³)	Percentage of Total Water	Estimated Average Residence Time of Water
Oceans	361,000,000	1,230,000,000	97.2	Thousands of years
Atmosphere	510,000,000	12,700	0.001	9 days
Rivers and streams	—	1,200	0.0001	2 weeks
Groundwater (shallow to depth of 0.8 km)	130,000,000	4,000,000	0.31	Hundreds to many thousands of years
Lakes (fresh water)	855,000	123,000	0.009	Tens of years
Ice caps and glaciers	28,200,000	28,600,000	2.15	Tens of thousands of years and longer

Source: U.S. Geological Survey.

TABLE 19.3 Water Budgets for the Continents				
	Precipitation		Evaporation		Runoff
Continent	mm/yr	km³	mm/yr	km³	km³/yr
North America	756	18,300	418	10,000	8,180
South America	1600	28,400	910	16,200	12,200
Europe	790	8,290	507	5,320	2,970
Asia	740	32,200	416	18,100	14,100
Africa	740	22,300	587	17,700	4,600
Australia and Oceania	791	7,080	511	4,570	2,510
Antarctica	165	2,310	0	0	2,310
Earth (entire land area)	800	119,000	485	72,000	47,000[a]

[a]Surface runoff is 44,800; groundwater runoff is 2,200.

Source: I. A. Shiklomanov, 1993, "World Fresh Water Resources," in P. H. Gleick, ed., *Water in Crisis,* Oxford University Press, New York, pp. 3–12.

runoff. Total runoff in North America is about two-thirds that of South America. Unfortunately, much of the North American runoff occurs in sparsely settled or uninhabited regions, particularly in the northern parts of Canada and Alaska.

Compared with other resources, water is used in tremendous quantities. In recent years the total mass (or weight) of water used on Earth per year has been approximately 1000 times the world's total production of minerals, including petroleum, coal, metal ores, and nonmetals.[8] Because of its great abundance, water is generally a very inexpensive resource. However, in the southwestern United States the cost of water has been kept artificially low as a result of government subsidies and programs.

Federal water subsidies in the Southwest have been criticized by people in the midwestern and eastern United States, who have more water than the semiarid Southwest and yet may pay more for the water they use. Nevertheless, because the quantity and quality of water available at any particular time are highly variable, shortages of water have occurred and will probably continue to occur with increasing frequency. Such shortages can lead to serious economic disruption and human suffering.[9] The U.S. Water Resources Council estimates that water use in the United States by the year 2020 may exceed surface water resources by 13%.[9] Therefore, an important question is, how can we best manage our water resources, use, and treatment to maintain adequate supplies?

Groundwater and Streams

An introductory acquaintance with groundwater and surface water and the terms used in discussing them is important in understanding many environmental issues, problems, and solutions. Rain that falls on the land evaporates, runs off the surface, or moves below the surface and is transported underground. Locations where surface waters infiltrate the groundwater systems are known as *recharge zones*. Places where groundwater flows or seeps out at the surface, such as springs, are *discharge zones* or points. Water that moves into the ground from the surface (or infiltrates) first seeps through the soil and rock known as the *vadose zone*, which is seldom saturated, and then enters the groundwater system, the upper surface of which is called the *water table*. Below the groundwater table all the pore spaces (spaces between soil particles or in rock fractures) are filled with water (saturated).

The term **groundwater** usually refers to the water below the water table where saturated conditions exist. An *aquifer* is an underground zone or body of earth material from which groundwater can be obtained (from a well) at a useful rate. Unconsolidated gravel and sand with lots of space (openings) between grains and rocks or with many open fractures generally make good aquifers. Groundwater in aquifers usually moves slowly at rates of centimeters or meters per day. When water is pumped from an aquifer, the water table is depressed around the well, forming a *cone of depression*. Figure 19.5 shows the major features of a groundwater and surface water system.

Streams may be classified as effluent or influent (Figure 19.5). In an **effluent stream** flow is maintained during the dry season by groundwater seepage into the stream channel from the subsurface. An **influent stream** is entirely above the groundwater table and flows only in direct response to precipitation. Water from an influent stream seeps down into the subsurface.

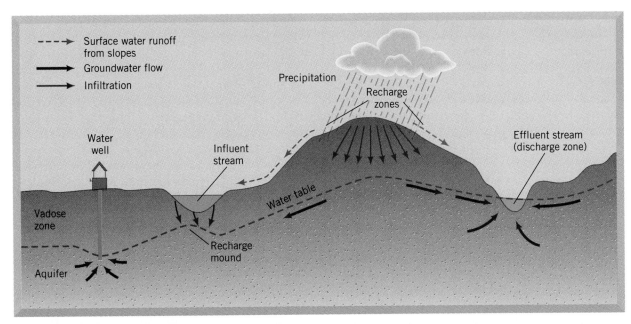

FIGURE 19.5 Groundwater and surface water flow system.

19.2 WATER SUPPLY: A U.S. EXAMPLE

The water supply at any particular point on the land surface depends on several factors in the hydrologic cycle, including the rates of precipitation, evaporation, transpiration (water in vapor form that directly enters the atmosphere from plants through pores in leaves and stems), stream flow, and subsurface flow. A concept useful in understanding water supply is the **water budget,** which is a model that balances the inputs, outputs, and storage of water in a system. On a continental scale, the water budget for the contiguous United States is shown in Figure 19.6. The amount of water vapor passing over the United

FIGURE 19.6 Water budget for the United States. (bgd = billion gallons per day). (*Source:* Water Resources Council, 1978, *The Nation's Water Resources 1975-2000,* Washington, DC.)

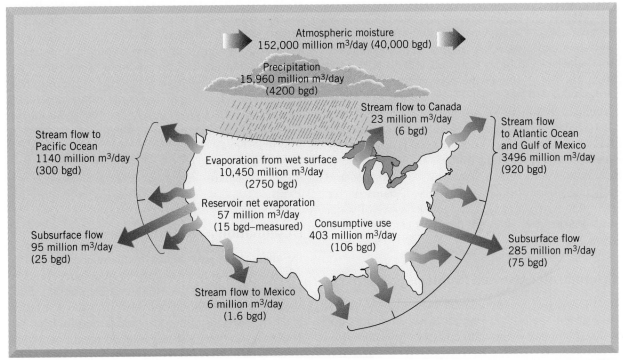

States every day is approximately 152,000 million m³ (40 trillion gal), and of this approximately 10% falls as precipitation in the form of rain, snow, hail, or sleet. Approximately 66% of the precipitation evaporates quickly or is transpired by vegetation. The remaining 34% enters the surface water or groundwater storage systems, flows to the oceans or across the nation's boundaries, is used by people, or evaporates from reservoirs. Owing to natural variations in precipitation that cause either floods or droughts, only a portion of this water can be developed for intensive uses (only about 50% is considered available 95% of the time).[9]

Precipitation and Runoff Patterns

In developing regional water budgets, it is necessary to consider annual precipitation and runoff patterns. Potential problems with water supply can be predicted in areas where average precipitation and runoff are relatively low, such as in the southwestern and Great Plains regions of the United States, as well as in some of the valleys of the Rocky Mountains. Although theoretically the surface water supply could be as high as the average annual runoff, this assumes all runoff could be successfully stored, which is not possible because of evaporative losses from large reservoirs and a limited number of suitable sites for reservoirs. As a result, shortages in the water supply are likely in areas with low precipitation and runoff, and strong conservation practices are necessary to ensure an adequate supply.[9]

Droughts

Because there are large annual and regional variations in stream flow, even areas with high precipitation and runoff may periodically suffer from droughts. For example, the dry years of 1985–1991 in the western United States produced serious water shortages. Fortunately for the more humid eastern United States, stream flow there tends to vary less than in other regions and drought is less likely.[9] Nevertheless, a drought in the summer of 1986 in the southeastern United States caused hardships and inflicted several billions of dollars of damage. We are often more prepared to handle excesses of water than deficiencies, although droughts can last several years.

Groundwater Use and Problems

Nearly one-half the population of the United States uses groundwater as a primary source for drinking water. Fortunately, the total amount of groundwater available in the United States is enormous, accounting for approximately 20% of all water used. In the contiguous United States the amount of groundwater within 0.8 km (about 0.5 miles) of the land surface is estimated to be between 125,000 and 224,000 km³ (3.3×10^{16}–5.9×10^{16} gal). To put this in perspective, the lower estimate is about equal to the total discharge of the Mississippi River during the last 200 years. However, the cost of pumping and exploration limits the total amount of groundwater that can feasibly be recovered.[9]

In many parts of the country, groundwater withdrawal from wells exceeds natural inflow. In such cases of **overdraft,** we can think of water as a nonrenewable resource that is being *mined*. This can lead to a variety of problems including damage to river ecosystems and land subsidence (see A Closer Look 19.1, "Loss of Riparian Vegetation and Land Subsidence"). Groundwater overdraft is a serious problem in the Texas–Oklahoma–High Plains area (which includes much of Kansas and Nebraska and parts of other states) as well as in California, Arizona, Nevada, New Mexico, and isolated areas of Louisiana, Mississippi, Arkansas, and the south Atlantic Gulf region. In the Texas–Oklahoma–High Plains area alone, the overdraft amount is approximately equal to the natural flow of the Colorado River.[9] This area contains the Ogallala aquifer, which is composed of water-bearing sands and gravels that underlie an area of about 400,000 km² from South Dakota into Texas. Although the aquifer holds a tremendous amount of groundwater, it is being used in some areas at a rate up to 20 times higher than the rate at which it is being naturally replaced. The water table in many parts of the aquifer has declined in recent years, causing yields from wells to decrease and energy costs for pumping the water to increase. There is concern that eventually a significant portion of land now being irrigated will be returned to dry-land farming as the resource is used up.

To date, only about 5% of the total groundwater resource has been depleted, but water levels have declined by as much as 30–60 m (about 100–200 ft) in parts of Kansas, Oklahoma, New Mexico, and Texas. The most severe problems in the High Plains and at the Ogallala aquifer today are in locations where irrigation was first used in the 1940s.

Desalination as a Water Source

Desalination of seawater, which contains about 3.5% salt (each cubic meter of seawater contains about 40 kg, or 88 lb, of salt) is an alternative source of usable water. This expensive form of water treatment is used at several hundred plants around the world. The salt content must be reduced to about 0.05%. Large desalination plants produce 20,000–30,000 m³ (about 5–8 million gallons) of water per day at a

CLOSER LOOK 19.1

LOSS OF RIPARIAN VEGETATION AND LAND SUBSIDENCE

In many areas, the pumping of groundwater has forever changed the character of the land. For instance, Tucson, Arizona, gets most of its water supply from groundwater sources, and overdraft has caused the water table to drop. Prior to the lowering of the water table through pumping, rivers in the Tucson area were perennial, with healthy populations of trout, beavers, and other animals. Today the rivers are dry much of the year and the native *riparian* trees (those on or adjacent to riverbanks) have died. Ironically, these processes have also increased the flood hazard in Tucson. Loss of riparian trees and the root strength they provided to stream banks have rendered the banks of the channels much more vulnerable to erosion. During the 1983 and 1993 floods in Tucson the effects of the problem became apparent as roads, bridges, and buildings were damaged by the shifting channels. Tree-lined channels are much more stable, but riparian trees need a groundwater table close to the surface. Unfortunately, restoration of trees along the streams is not possible, because the water table is now too deep.

When groundwater is pumped at a rate faster than it is naturally replaced, the water table or fluid pressure in an aquifer may be lowered or reduced. If this occurs, there may be effects at the surface. For example, withdrawal of groundwater in Mexico City over many years has resulted in surface **subsidence** (sinking or settling of the land) of about 5 m (16 ft) in some locations. The Palace of Fine Arts, constructed in 1939, originally had steps going up to the first floor. It now has steps going down to that floor. The subsi-

dence occurred because when the water is pumped out of the aquifer the natural fluid pressure that helps support a more open framework of the mineral particles in the aquifer is reduced, and the space between grains partially collapses or flattens. The cumulative effect is subsidence at the surface. Similar problems have been reported in Phoenix, Las Vegas, Houston, and the San Joaquin Valley in California.

Areas underlain by limestone aquifers may experience a sudden and spectacular form of subsidence owing in part to withdrawal of groundwater. Limestone is subject to chemical weathering that dissolves the stone, making a series of interconnected caves with large vertical or dome-shaped rooms that may come close to the surface. If the cave system is filled with groundwater, then the water helps support the

rocks above. The combination of a high water table and the strength of the overlying rocks helps prevent collapse. If the water table drops and a collapse does occur, a feature known as a collapsed sinkhole forms at the surface. The collapse can be considerable and it can happen quickly. Figure 19.7 shows the Winter Park, Florida, sinkhole that formed on May 8, 1981. The sinkhole grew rapidly for 3 days, swallowing a house, two businesses, part of a community swimming pool, a street, and several automobiles. The sinkhole developed during a drought, when groundwater levels were at a record low. Similar events have occurred in Alabama, Pennsylvania, Puerto Rico, the Yucatan Peninsula in Mexico, and many other areas around the world, where extensive limestone cave systems are part of groundwater systems.

FIGURE 19.7 The Winter Park, Florida, sinkhole that formed on May 8, 1981. Note damage to the municipal swimming pool, roads, and vehicles trapped in the depression.

cost of about 10 times that paid for traditional water supplies in the United States. Desalinated water has a *place value*, which means that the price increases quickly with the transport distance and the cost of moving water uphill from the plant at sea level. Because the various processes that remove the salt require large amounts of energy, the cost of the water is tied to ever-increasing energy costs. For these reasons, desalination will remain an expensive process used only when alternative water sources are not available.

19.3 WATER USE

In discussing water use, it is important to distinguish between off-stream and in-stream uses. **Off-stream use** refers to water removed from its source for use. Much of this water is returned to the source after use, for example, water used to cool industrial processes and water from sewage treatment plants that is returned to a river. **Consumptive use** is an off-stream use in which water does not return to the stream or groundwater resource immediately after use.[9] This water is consumed by plants and animals or in industrial processes and enters human tissue or products or evaporates during use but is not returned to its source.

In-stream use includes the use of rivers for navigation, hydroelectric power generation, fish and wildlife habitats, and recreation. These multiple uses usually create controversy because each in-stream use requires different conditions to prevent damage or detrimental effects. Fish and wildlife require certain water levels and flow rates for maximum biological productivity; these levels and rates may differ from those needed for hydroelectric power generation, which requires large fluctuations in discharges to match power needs.

Similarly, both these uses may conflict with requirements for shipping and boating. Figure 19.8 demonstrates some of these conflicting demands on a graph that shows optimal discharge for various uses throughout the year. Notice that navigation prefers constant discharge, and hydroelectric demands change daily, with a minimum in the spring, when spawning fish prefer higher flows. A major problem concerning in-stream use is how much water may be removed from a stream or river and transported to another location without damaging the stream system. This is an issue in the Pacific Northwest, where certain fish, such as steelhead trout and salmon, are on the decline partly because stream flow diversions (removal of water for agricultural, urban, and other uses) have damaged fish habitats. Another important factor in this situation is logging practices that increase soil erosion, leading

to sediment pollution in streams (Chapters 11 and 13 provide background on water use in agriculture, soil erosion, and logging practices).

Transport of Water

In many parts of the world, demands are being made on rivers to supply water to agricultural and urban areas. This is not a new trend. Ancient civilizations, including the Romans and Native Americans, constructed canals and aqueducts to transport water from distant rivers to where it was needed. In our modern civilization, as in the past, water is often moved long distances from areas with abundant rainfall or snow to areas of high usage (usually agricultural areas). For instance, in California, two-thirds of the state's runoff occurs north of San Francisco, where there is a surplus of water, and two-thirds of the water use occurs south of San Francisco, where there is a deficit. In recent years canals of the California Water Project have moved tremendous amounts of water from the northern to the southern part of the state, mostly for agricultural uses but increasingly for urban uses as well.

On the opposite coast, New York City has imported water from nearby areas for more than 100 years. Water use and supply in New York City shows a repeating pattern. Originally, local groundwater, streams, and the Hudson River itself were used. However, water needs exceeded local supply, so in 1842 the first large dam was built more than 48 km (30 miles) north of the city. As the city rapidly expanded from Manhattan to Long Island, water needs increased. The shallow aquifers of Long Island were at first a source of drinking water, but this water was removed faster than rainfall replenished it. In addition, the groundwater became contaminated with pollutants and with salt water intruding from the

FIGURE 19.8 In-stream water uses and optimal discharges for each use. Discharge is the amount of water passing by a particular location and is measured in cubic meters per second.

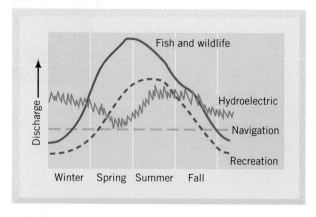

ocean. (The pollution of Long Island groundwater is explored in more depth in the next chapter.)

A larger dam was built at Croton in 1900. Further expansion of the population created the same pattern: initial use of groundwater; pollution, salinization, and exhaustion of this resource; and subsequent building of new, larger dams farther and farther upstate in forested areas. The western counties of Long Island (Queens and Brooklyn), have to import upstate water because their aquifers have been hopelessly polluted since the beginning of the twentieth century. The eastern counties of Long Island (Nassau and Suffolk) do not import upstate water and have enacted stringent regulations to conserve and protect their groundwater resources. Nevertheless, they also are experiencing groundwater pollution problems. The pattern continues with the development of tract housing in eastern Long Island, where there are now problems of pollution, salinization, and exhaustion of the resource. The cost of obtaining water from long distances and the competition for the available water from other users will place an upper limit on the water supply of the city.

Some Trends in Water Use

Trends in freshwater withdrawals and human population for the United States from 1950 to 1995 are shown in Figure 19.9. Examination of the data suggests, first, that withdrawal of surface water far exceeds groundwater and, second, that withdrawals of both surface water and groundwater increased from 1950 until 1980, reaching a total maximum of approximately 375 thousand million gallons per day. However, since 1980, water withdrawals have decreased and leveled off. It is particularly encouraging that water withdrawals have decreased after 1980 while the population of the United States has continued to increase. This suggests that there have been improvements in water management and water conservation during the last 20 years.[10]

Trends in freshwater withdrawals by water-use categories for the United States from 1960 to 1995 are shown in Figure 19.10. Examination of this graph suggests that (1) the big uses of water are for irrigation and the thermoelectric industry; (2) the use of water for irrigation by agriculture leveled off starting in about 1975; (3) water use by the thermoelectric industry and industry has decreased slightly since 1980; and (4) use of water for public and rural supplies continued to increase through the period from 1950 to 1995, presumably related to the increase in human population.[10]

19.4 WATER CONSERVATION

Water conservation is the careful use and protection of water resources and involves both the quantity of water used and its quality. Conservation is an important component of sustainable water use.

FIGURE 19.9 Trends in United States fresh groundwater and surface-water withdrawals and human population (1950–1995). (*Source:* Solley, W. B., Pierce, R. P., and Perlman, H. A. 1998. "Estimated Use of Water in the United States in 1995," *U.S. Geological Survey Circular 1200.*)

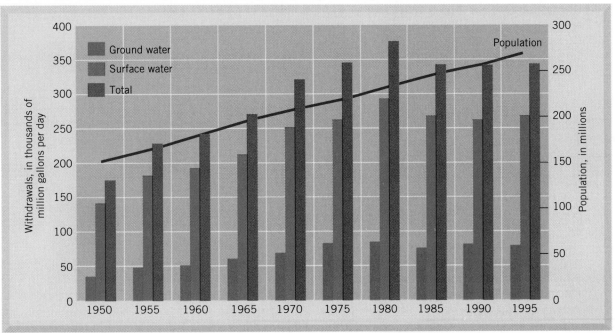

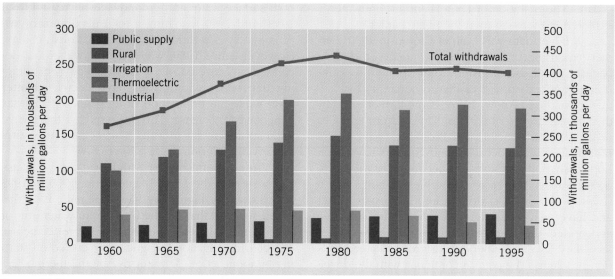

FIGURE 19.10 Trends in United States water withdrawals (fresh and saline) by water-use category and total (fresh and saline) withdrawals (1960–1995). (*Source:* Solley, W. B., Pierce, R. P., and Perlman, H. A. 1998. "Estimated Use of Water in the United States in 1995," *U. S. Geological Survey Circular 1200.*)

Improved agricultural irrigation (Figure 19.11) could reduce withdrawals by between 20 and 30%. Because agriculture is the biggest water user (80% of total fresh water consumptive use), this would be a tremendous saving. Suggestions for agricultural conservation include the following:

- Price agricultural water to encourage conservation (subsidizing water may encourage overuse).
- Use lined or covered canals that reduce seepage and evaporation.
- Use computer monitoring and schedule release of water for maximum efficiency.
- Integrate the use of surface water and groundwater to more effectively use the total resource. That is, irrigate with surplus surface water when it is abundant and also use surplus surface water to recharge groundwater aquifers by applying the surface water to specially designed infiltration ponds or injection wells. When surface water is in short supply, use more groundwater.
- Irrigate at times when evaporation is minimal, such as at night or in the early morning.
- Use improved irrigation systems, such as sprinklers or drip irrigation, that more effectively apply water to crops.
- Improve land preparation for water application; that is, improve the soil to increase infiltration and minimize runoff. Where applicable, use mulch to help retain water around plants.
- Encourage the development of crops that require less water or are more salt tolerant so that

less periodic flooding of irrigated land is necessary to remove accumulated salts in the soil.

Domestic Use

Domestic use of water accounts for only about 10% of total national water withdrawals. However, because domestic water use is concentrated in urban areas, it may pose major local problems in areas where water is periodically or often in short supply. Most water in homes is used in the bathroom and for washing clothing and dishes. Water use for domestic purposes may be substantially reduced at a relatively small cost by implementing the following measures:

- In semiarid regions, replace lawns with decorative gravels and native plants.
- Use more-efficient bathroom fixtures, such as low-flow toilets that use 1.5 gal per flush rather than the standard 5-gal-per-flush toilet and low-flow shower heads that deliver less but sufficient water.
- Turn off water when not absolutely needed for washing, brushing teeth, shaving, and so on.
- Flush the toilet only when really necessary.
- Fix all leaks quickly. Dripping pipes, faucets, toilets, or garden hoses can use a lot of water.
- Purchase dishwashers and washing machines that minimize water consumption.
- Take a long bath rather than a long shower.
- Don't wash sidewalks and driveways with water (sweep them).

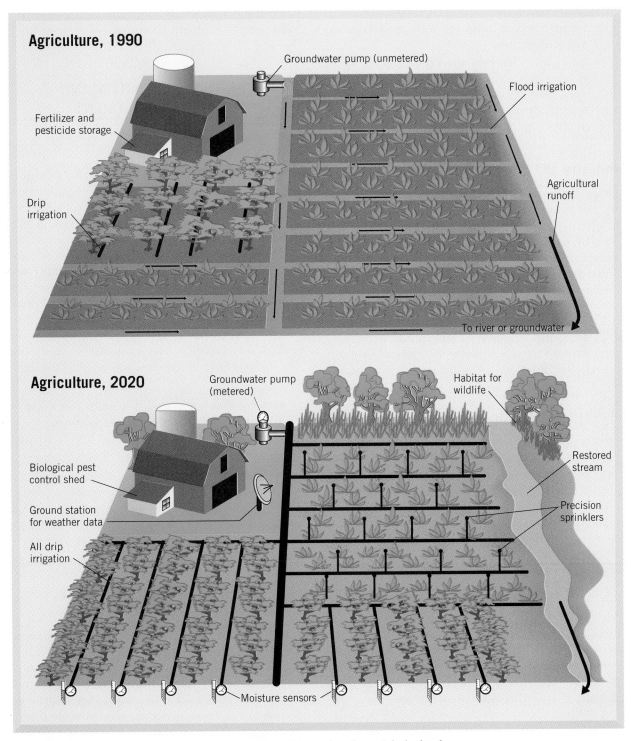

FIGURE 19.11 Comparison of agricultural practices in 1990 to what they might be by the year 2020. The improvements call for a variety of agricultural procedures from biological pest control to more efficient application of irrigation water to restoration of water resources and wildlife habitat. (*Source:* P. H.Gleick, P. Loh, S. V. Gomez, and J. Morrison. 1995, "California Water 2020, A Sustainable Vision," Pacific Institute for Studies in Development, Environment and Security, Oakland, CA.)

- Consider using gray water from washing machines to water vegetation.

- Water lawns and plants in the early morning, late afternoon, or at night to reduce evaporation.

- Use drip irrigation and place water-holding mulch around garden plants.

- Plant drought-resistant vegetation that requires less water.

- Learn how to read the water meter to monitor for unobserved leaks and record your conservation successes.
- Local water districts should encourage water pricing policies in which water is much more expensive beyond some baseline amount determined by the number of people in a home and the size of the property.

Industry and Manufacturing Use

Other conservation measures can be taken by industry. For instance, water removal for steam generation of electricity could be reduced as much as 25% to 30% by using cooling towers that use less or no water. Manufacturing and industry might curb water withdrawals by increasing in-plant treatment and recycling of water or by developing new equipment and processes that require less water.[9] Because the field of water conservation is changing so rapidly, it is expected that a number of innovations will reduce the total withdrawals of water for various purposes, even though consumption will continue to increase.[9]

Perception and Water Use

How water supply is perceived by people is also important in determining how much water is used. For example, people in Tucson, Arizona, perceive the area as a desert (which it is) and use a lot of native plants (cactus and other desert plants) in yards and gardens around homes and buildings there. As discussed earlier in the chapter, Tucson's water supply is mostly from groundwater, which is being mined (used faster than it is being naturally replenished). Water use in Tucson is about 605 liters (160 gal) per person per day. By way of comparison, remember that the world per-capita use of water is about 2000 liters/day. Not far from Tucson, the people of Phoenix use about 983 liters (260 gal) of water per person per day. In parts of Phoenix as much as 3780 liters (1000 gal) of water per person per day is used to water mulberry trees and high hedges!

Phoenix has been accused of having an oasis mentality concerning water use. Water rates also make a difference. The people in Tucson pay about 75% more for water than do the people in Phoenix, where the water supply is drawn from the Salt River rather than from groundwater. Water rates in Tucson are structured to encourage conservation, and some industries consider water conservation a cost control measure.[11] Whether Tucson can maintain its desert mentality following the planned importation of Colorado River water, however, remains to be seen. The message is that we could all do with a little of Tucson's desert mentality. This is particularly true for those in large urban areas, such as Los Angeles and San Diego, in Southern California.

19.5 SUSTAINABILITY AND WATER MANAGEMENT

Sustainable Water Use

Water is essential to sustain life. We also recognize that water is essential for maintaining ecological systems necessary for the survival of humans. As a result water plays important roles in ecosystem support, economic development, cultural values, and community well-being. From a water supply use and management perspective, **sustainable water use** may be defined as the use of water resources by people that allows our society to develop and flourish into an indefinite future without degrading the various components of the hydrologic cycle or the ecological systems that depend on it.[12]

Some criteria for water use sustainability are as follows[12]:

- Develop water resources in sufficient volumes to maintain human health and well-being.
- Provide sufficient water resources to guarantee the health and maintenance of ecosystems.
- Ensure minimum standards of water quality for the various users of water resources.
- Ensure that actions of humans do not damage or reduce long-term renewability of water resources.
- Promote the use of water-efficient technology and practice.
- Gradually eliminate water pricing policies that subsidize the inefficient use of water.

To work toward this sustainability, we must be sure that we have institutional processes and mechanisms to prevent and resolve conflicts over water use. There is fear that if we continue our present policies concerning water resources, there will be an increase in incidents of social, economic, and environmental conflicts over water, which are antithetical to sustainable water use. We will now consider some aspects of water management.

Water Management

Management of water resources for water supply is a complex issue that will become more difficult as demand for water increases in the coming years. This difficulty will be especially apparent in the southwestern United States and other semiarid and arid parts of the world where water is or soon will be in short supply. Options for minimizing potential water supply problems include locating alternative water supplies and managing existing supplies better. In some areas location of new supplies is unlikely, and serious consideration is being given to ideas as original as towing icebergs to coastal regions where fresh water is needed. It seems apparent that water will become much more expensive in the future, and if the price is right, many innovative programs are possible.

A method of water management utilized by a number of municipalities is known as the *variable-source approach*. The city of Santa Barbara has developed a variable-source approach that involves several interrelated measures to meet present and future water demands. Details of the plan (shown in Figure 19.12) include features such as enlarging reservoirs, developing new sources, using reclaimed water, and instituting a permanent conservation program. In essence, this seaside community has developed a master water plan.

A Master Plan for Water Management

Luna Leopold[13] suggests that a new philosophy of water management is needed, one based on geo-logic, geographic, and climatic factors as well as on the traditional economic, social, and political factors. He argues that the management of water resources cannot be successful as long as it is naively perceived from an economic and political standpoint. The essence of Leopold's water management philosophy is that surface water and groundwater are both subject to natural flux with time. In wet years there is plenty of surface water and the near-surface groundwater resources are replenished. During dry years, which must be expected even though they may not be accurately predicted, we should have specific strategies to minimize hardships. For example, there are subsurface waters in various locations in the western United States that are too deep to be economically extracted or have marginal water quality. These waters may be isolated from the present hydrologic cycle and therefore are not subject to natural recharge. Such water might be used when the need is great. However, advance planning to drill the wells and connect them to existing water lines is necessary if they are to be ready when the need arises. Another possible emergency plan might involve the treatment of wastewater. Reuse of water on a regular basis might be too expensive, but advance planning to reuse treated water during emer-

FIGURE 19.12 Schematic drawing of a variable source model (present and future) for water supply for the city of Santa Barbara, California. (*Source:* Santa Barbara City Council, 1991.)

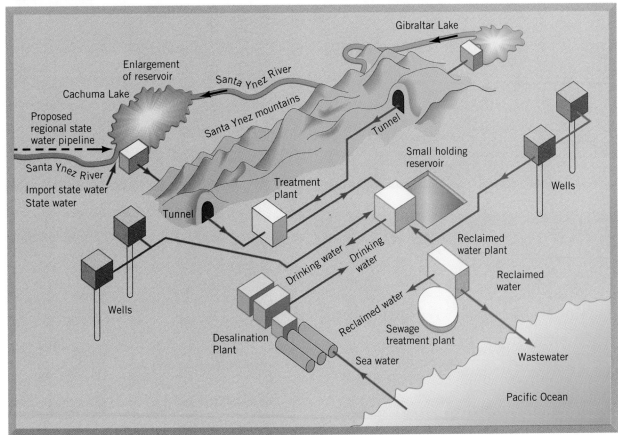

gencies could be a wise decision. Finally, we should develop plans to use surface water when available and we should not be afraid to use groundwater as needed in dry years. During wet years natural recharge as well as artificial recharge (pumping excess surface water into the ground) will replenish the groundwater resources. This water management plan recognizes that excesses and deficiencies in water are natural and can be planned for.[13]

Water Management and the Environment

Many agricultural and urban areas require water delivered from nearby (and in some cases not-so-nearby) sources. To deliver the water, a system is needed for water storage and routing by way of canals and aqueducts from reservoirs. We also build dams, modify or fill wetlands, and channelize rivers to help control flooding. There often is a good deal of controversy concerning water development, and the day of development of large projects in the United States without careful environmental review has passed. The resolution of development issues now involves input from a variety of groups that may have very different needs and concerns. These range from agricultural groups who see water development as critical for their livelihood to groups whose primary concerns are with wildlife and wilderness preservation. It is a positive sign that the various parties on water issues are now at least able to meet and communicate their desires and concerns.

(a)

FIGURE 19.13 Several types of wetlands: (*a*) aerial view of part of the Florida Everglades at a coastal site; (*b*) marsh environment at the edge of Chesapeake Bay; (*c*) cypress swamp, water surface covered with a floating mat of duckweed, northeast Texas; and (*d*) aerial view of farmlands encroaching on prairie potholes, North Dakota.

(b)

(c)

(d)

Wetlands

Wetlands is a comprehensive term for landforms such as salt marshes, swamps, bogs, prairie potholes, and vernal pools (shallow depressions that seasonally hold water). Their common feature is that they are wet at least part of the year and as a result have a particular type of vegetation and soil. Figure 19.13 shows several types of wetlands.

Wetlands may be defined as areas that are inundated by water or where the land is saturated to a depth of a few centimeters for at least a few days per year. Three major components used to determine the presence of wetlands are hydrology or wetness, type of vegetation, and type of soil. Of these, hydrology is often the most difficult to define, because some freshwater wetlands may be wet for only a few days a year. The duration of inundation or saturation must be sufficient for the development of wetland soils, which are characterized by poor drainage and lack of oxygen, and for the growth of specially adapted vegetation.[14]

Functions of Wetlands

Wetland ecosystems may serve a variety of natural service functions for other ecosystems and for people, including the following:

- Freshwater wetlands are a natural sponge for water. During high river flow they store water, reducing downstream flooding. Following a flood they slowly release the stored water, nourishing low flows.
- Many freshwater wetlands are important as areas of groundwater recharge (water seeps into the ground from a prairie pothole, for instance) or discharge (water seeps out of the ground in a marsh that is fed by springs).
- Wetlands are one of the primary nursery grounds for fish, shellfish, aquatic birds, and other animals. It has been estimated that as many as 45% of endangered animals and 26% of endangered plants either live in wetlands or depend on them for their continued existence.[14]
- Wetlands are natural filters that help purify water; plants in wetlands trap sediment and toxins.
- Wetlands are often highly productive and are a place where many nutrients and chemicals are naturally cycled.
- Coastal wetlands provide a buffer for inland areas from storms and high waves.
- Wetlands are an important storage site for organic carbon; storage is in living plants, animals, and rich organic soils.
- Wetlands are aesthetically pleasing places for people.

Although most coastal marshes are now protected in the United States, freshwater wetlands are still threatened in many areas. One percent of the nation's total wetlands is lost every 2 years, and freshwater wetlands account for 95% of this loss. Wetlands such as prairie potholes in the midwestern United States and vernal pools in Southern California are particularly vulnerable because their hydrology is poorly understood and establishing their wetland status is more difficult.[15] Over the past 200 years, over 50% of the wetlands in the United States have disappeared because they have been diked or drained for agricultural purposes or filled for urban or industrial development. Perhaps as much as 90% of the freshwater wetlands have disappeared.

The extensive salt marshes at many of the nation's major estuaries, where rivers entering the ocean widen and are influenced by tides, have been modified or lost. These include estuaries of major rivers such as the Potomac, Susquehanna (Chesapeake Bay), Delaware, and Hudson, as well as San Francisco Bay.[16] The San Francisco Bay estuary, considered the estuary most modified by human activity in the United States today, has lost nearly all its marshlands to leveeing and filling (Figure 19.14).[16] Modifications result not only from filling and diking

FIGURE 19.14 Loss of marshlands in the San Francisco Bay and estuary from about 1850 to the present. (*Source:* T. J. Conomos, ed., 1979, *San Francisco, The Urbanized Estuary*, American Association for the Advancement of Science, San Francisco, CA; F. H. Nichols, J. E. Cloern, S. N. Luoma, and D. H. Peterson, 1986, "The Modification of an Estuary," *Science*, 231, pp. 567–573. Copyright 1986 by the American Association for the Advancement of Science.)

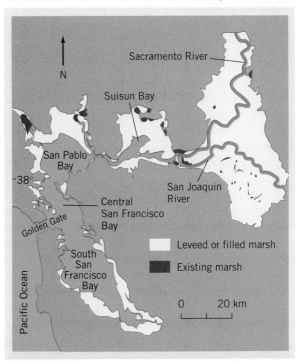

but also from loss of water. The freshwater inflow has been reduced by more than 50%, dramatically changing the hydrology of the bay in terms of flow characteristics and water quality. As a result of the modifications, the plants and animals in the bay have changed as habitats for fish and wildfowl have been eliminated.[16]

Most people are in agreement that wetlands are valuable and productive lands for fish and wildlife. But wetlands are also valued as potential lands for agricultural activity, mineral exploitation, and building sites. Wetland management is drastically in need of new incentives for private land owners (who own the majority of several types of wetlands in the United States) to preserve wetlands rather than fill them in and develop the land.[15] Management strategies must also include careful planning to maintain the water quantity and quality necessary for wetlands to flourish or at least survive. Unfortunately a national wetland policy for the United States is not in place. There is debate as to what constitutes a wetland and how property owners should be compensated for preserving wetlands.[17]

Restoration

A related management issue is restoration of wetlands. A number of projects have attempted to restore wetlands with varied success. The most important factor to be considered in most freshwater marsh restoration projects is the availability of water. If water is present, wetland soils and vegetation will likely develop. The restoration of salt marshes is more difficult because of the complex interactions among the hydrology, sediment supply, and vegetation that allow salt marshes to develop. Careful studies of relationships between the movement of sediment and the flow of water in salt marshes is providing information crucial to restoration, which makes successful reestablishment of salt marsh vegetation more likely. The restoration of wetlands has become an important topic because of the mitigation requirement related to environmental impact analysis. That is, if wetlands are destroyed or damaged by a particular project, the developer must obtain or create additional wetlands at another site to compensate.[14] Unfortunately, the state of the art of restoration is not adequate to ensure that specific restoration projects will be successful.

Construction of human-made wetlands for the purpose of cleaning up agricultural runoff is a new idea being attempted in areas with extensive agricultural runoff. Using a wetland's natural ability to remove excess nutrients, break down pollutants, and cleanse water, water managers in Florida are creating a series of wetlands to remove nutrients (especially phosphorus). The purpose is to help restore the Everglades of south Florida to a more natural state. The Everglades are a huge wetland system that functions as a wide shallow river flowing south through southern Florida to the ocean. Fertilizers applied to farm fields north of the Everglades make their way directly into the Everglades by way of agricultural runoff, disrupting the ecosystem (phosphorus enrichment causes undesired changes in water quality and aquatic vegetation; see discussion of eutrophication in the next chapter). The human-made wetlands are designed to intercept and hold the nutrients, so they do not enter and damage the Everglades.[18]

Dams and the Environment

Dams and their accompanying reservoirs generally are designed to be multifunctional structures. People who propose the construction of dams and reservoirs point out that reservoirs may be used for recreational activities as well as to provide flood control and ensure a more stable water supply. However, it is often difficult to reconcile these various uses at a given site. For example, water demands for agriculture might be high during the summer, resulting in a drawdown of the reservoir and the production of extensive mudflats or an exposed bank area subject to increased erosion. Those interested in recreation find the low water level and the mudflats aesthetically degrading. Also, high water demand may cause quick changes in lake levels, which may interfere with wildlife (particularly fish) by damaging or limiting spawning opportunities. Finally, dams and reservoirs tend to provide a false sense of security to those living below the water retention structures, because dams cannot fully protect us against great floods.

The environmental effects of dams are considerable and include the following:

- Loss of land and biological resources in the reservoir area.
- Storage behind the dam of sediment that would otherwise move downstream to coastal areas where it would supply sand to beaches. The trapped sediment also reduces water storage capacity, limiting the life of the reservoir.
- Downstream changes in hydrology and in sediment transport that change the entire river environment and the organisms that live there.

There is little doubt that if our present practices of water use continue we will need additional dams and reservoirs and some existing dams will be heightened to increase water storage. Conflicts over the construction of additional dams and reservoirs are bound to occur. Water developers may view a canyon dam site as a resource for water storage, whereas other people view it as a wilderness area and recreation site for future generations. The conflict is particularly pointed, because good dam sites are often sites of high-quality scenic landscape.

There is also an economic aspect to dams: They are expensive to build and operate. Some people complain that they are often constructed with federal tax dollars in the western United States, where they provide inexpensive subsidized water for agriculture. This has been a point of concern to some taxpayers in the eastern United States, who do not have the benefit of federally subsidized water. Perhaps a different pricing structure for water would encourage conservation and fewer new dams and reservoirs would be needed.

Canals

Water from upstream reservoirs may be routed downstream by way of natural watercourses or by canals and aqueducts. Canals, whether lined or unlined, are often a hazard to people and animals who attempt to swim in them or drink the water. Where they flow, drownings are an ever-present threat.

The construction of canal systems, especially in developing countries, has led to serious unanticipated environmental problems. For example, when the High Dam at Aswan was completed in 1964, a system of canals was built to convey the water to agricultural sites. The canals became infested with snails that carry the disease schistosomiasis (snail fever). This disease has always been a problem in Egypt, but the swift currents of the Nile floodwaters flushed the snails out each year. The tremendous expanse of waters in irrigation canals now provides happy homes for these snails. The disease is debilitating and so prevalent in parts of Egypt that virtually the entire population of some areas may be affected by it.

Channelization and the Environment

Channelization of streams consists of straightening, deepening, widening, clearing, or lining existing stream channels. It is an engineering technique that has been used to control floods, improve drainage, control erosion, and improve navigation.[19] Of these four objectives, flood control and drainage improvement are cited most often in channel improvement projects.

Thousands of kilometers of streams in the United States have been modified by channelization that all too often has produced adverse environmental effects, including:

- degradation of the stream's hydrologic qualities, turning a meandering stream with pools (deep, slow flow) and riffles (faster, shallow flow) into straight channels that are nearly all riffle flow, resulting in loss of important fish habitats;
- removal of riparian vegetation, which removes wildlife habitats and shading of the water;
- downstream flooding where the channelized flow ends, because the channelized section has a larger channel area and carries a greater amount of floodwater than the downstream natural channel can carry without overbank flow or flooding;
- damage or loss of wetlands (because their source of water is removed by the channelization, which often lowers the groundwater table and thus drains the wetlands); and
- aesthetic degradation (channelized streams are much less attractive than natural streams).

Figure 19.15 compares selected environmental features of natural and channelized streams.

Not all channelization causes serious environmental degradation, and in many cases drainage projects are beneficial. Moreover, channel design is being improved on the basis of past experience. Currently more consideration is being given to the environmental aspects of channelization, and some projects are being designed with modified channels that behave more like natural streams than do straight ditches.

Channelization of the Kissimmee River in Florida is an example of potential problems and conflict in river management. Channelization of the river started in 1962 and, after 9 years and $24 million of construction, the sinuous river was converted into an 83-km-long (about 50-mile) straight ditch (Figure 19.16). Unfortunately, not only did the channelization fail to provide the expected flood protection, but it degraded a valuable wildlife habitat, contributed to water quality problems associated with the land drainage, and caused aesthetic degradation. Now, in the 1990s, at a cost that may exceed the original cost of channelization, the river may be returned to its original sinuous path! The restoration of the Kissimmee River may become the most ambitious restoration project ever attempted in the United States. Work has begun on a 16-km (10-mile) experimental stretch of river and, if successful, will be expanded to other parts of the channelized river.

Natural stream

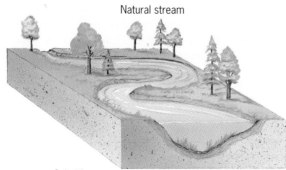

Channelized stream

Suitable water temperatures:
adequate shading; good cover for fish life; minimal
temperature variation; abundant leaf material output.

Increased water temperatures:
no shading; no cover for fish life; rapid daily and
seasonal temperature fluctuations;
reduced leaf material input.

Pool—riffle sequence

Pool—silt, sand,
and fine gravel

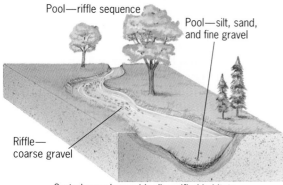

Riffle—
coarse gravel

Sorted gravels provide diversified habitats
for many stream organisms.

Mostly riffle

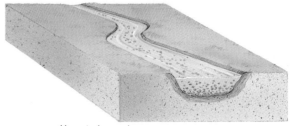

Unsorted gravels:
reduction in habitats; few organisms.

Pool environment

High flow

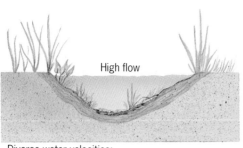

Diverse water velocities:
high in pools, low in riffles. Resting areas
abundant beneath banks, behind large rooks, etc.

Low flow

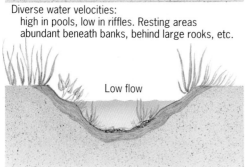

Sufficient water depth to support fish and
other aquatic life during dry season.

High flow

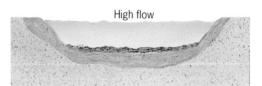

May have stream velocity higher than some aquatic
life can stand; fewer or no resting places.

Low flow

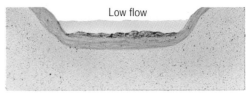

Insufficient depth of flow during dry season
to support fish and other aquatic life;
few if any pools (all riffle).

Figure 19.15 A natural stream compared with a channelized stream in terms of general characteristics and pool environments. (*Source:* Modified after R. V. Corning, February 1975, *Virginia Wildlife*, pp. 6–8.)

(a)

(b)

FIGURE 19.16 (*a*) Channelized Kissimmee River and (*b*) restoration work in progress on the Kissimmee River.

Flooding

A river and the flatland adjacent to it, known as the **floodplain,** together constitute a natural system. In most natural rivers the water flows over the riverbanks and onto the floodplain every year or so. This natural process has many benefits for the environment:

- water and nutrients are stored on the floodplain;
- deposits on the floodplain contribute to the formation of nutrient-rich soils;
- wetlands on the floodplain provide an important habitat for many birds, animals, plants, and other living things; and

- the floodplain functions as a natural greenbelt that is distinctly different from adjacent environments and produces environmental diversity.

Natural flooding is not a problem until people choose to build homes and other structures on floodplains. These structures are subject to damage and loss when inundated by floodwaters. We have chosen to build on so many floodplains that flooding is the most universal natural hazard in the world. In the United States about 100 persons lose their lives in floods every year and accompanying damages exceed $3 billion. The 1993 flood of the Mississippi River took about 50 lives and caused over $10 billion in damages (Figure 19.17)! This loss of life,

(a)

(b)

FIGURE 19.17 (*a*) Flood damage to farmlands and homes as a result of the 1993 Mississippi River floods. (*b*) Failure of a levee in Illinois caused flooding in the town of Valmeyer.

How Wet Is a Wetland?

Areas where land meets water, whether fresh or salt, are known generally as *wetlands*. Characteristically they are covered by surface water or have saturated soils. Landscape features such as swamps, bogs, marshes, potholes, and sloughs are wetlands. For most of our history, wetlands were considered to be wastelands and were destroyed by filling, draining, or polluting. Only 230 of the 530 million ha (568.3 of the 1309.6 million acres) of wetlands that existed in the lower 48 states at the time of European colonization remain, and 40% of them suffer from some degree of pollution.

Today, however, it is generally recognized that wetlands have many values. They provide food, water, and cover for fish, shellfish, waterfowl, game animals, and many amphibians and reptiles. In addition, soil and plants in wetlands purify water by absorbing or destroying pollutants. Wetlands also help to recharge groundwater stores and, by holding water, control flooding and erosion. One-third of endangered and threatened species, two-thirds of the commercial saltwater fish and shellfish, one-third of the birds, and almost all the amphibians in our country depend on wetlands. Wetlands are among the most productive ecological communities in the world, many times more productive than a heavily fertilized cornfield.

Protection of these unique communities began with the goal of preserving wetlands used by wildlife, particularly ducks, but federal protection has since been broad-

ened to include most of the remaining wetlands. Still, 81,000–162,000 ha (200,151–400,302 acres) of wetlands are lost in the United States each year. Of particular concern are the freshwater wetlands, many of which occur on private land. A recent policy of no net loss of wetlands has been praised by some environmental scientists but attacked by farmers and developers.

Particularly controversial are the small, seasonal potholes in the agricultural areas of the Midwest and northern Great Plains and other seasonal wet areas of the West that may not appear wet to the casual observer. They do, however, provide habitats for many species, including about half the 10 to 31 million waterfowl that nest in the lower 48 states. Critics of applying strict regulations to potholes say that if an area is not wet enough for a duck to land in and splash, it is not wet enough to qualify as a wetland.

Critical Thinking Questions

1. Results of a study comparing wildlife use of seasonal wetlands with those that are permanently inundated or saturated in the San Francisco estuary are shown in the accompanying table. What conclusions can you draw from the data about the importance of seasonal wetlands to wildlife in the estuary? What additional data would you need to extend your conclusions to potholes in the Midwest?

Area and Wildlife Use of Wetlands in the San Francisco Estuary

| | Area | | Wildlife Use | |
| | | | Number of | |
Type of Wetland	ha	%	Species	%[a]
Permanent	58,765	23		
Mudflat	25,949	10.2	49	11.9
Marsh	17,964	7.1		
Tidal salt			95	23.1
Brackish			192	46.6
Tidal fresh			174	42.2
Salt pond	14,852	5.8	82	19.9
Seasonal	195,709	77		
Diked marsh and other	34,467	13.5		
Diked marsh			72	17.5
Other			—	
Farmed wetlands	156,176	61.3	92	22.3
Riparian woodlands	5066	1.9	207	50.2
Total	*254,474*	*100[b]*	*412*	*100*

[a]Percent of total number of species found in all wetlands.

[b]Permanent + seasonal = 100%.

2. Some people have proposed defining wetlands so as to exclude many of the seasonal ones, some 4.5 million ha (11.1 million acres). What position would you take on such a proposal and why? How would you reconcile the conflicting needs of the farmers and developers with the need to preserve wildlife habitats?

3. What effects would you expect a substantial decrease in wetlands to have on populations of migratory birds?

References

Baldwin, M. F. 1987. "Wetlands: Fortifying Federal and Regional Cooperation," *Environment* 29(7): 16–20, 39.

Leidy, R. A., Fiedler, P. L., and Micheli, E. R. 1992. "Is Wetter Better?" *Bioscience* 40(9): 58–61, 65.

National Institute for Urban Wildlife. *Wetlands Conservation and Use* (Issue Pak). Columbia, MD: Author.

Stevens, W. K. 1990 (March 13). "Efforts to Halt Wetland Loss are Shifting to Inland Areas," *New York Times*, p. C1.

World Resources Institute. 1992. *The 1992 Information Please Environmental Almanac*. Boston: Houghton Mifflin.

although terrible, is relatively low compared to areas of the world that lack the sophisticated monitoring and warning systems that we have for our rivers. For example, flooding associated with two cyclones that struck Bangladesh in 1970 and 1991 killed more than half a million people (see Chapter 29).

Urbanization and Flooding

Many small drainage basins (the land area that contributes water to a particular stream system) with an area less than 20 km² or so have been intensively urbanized. In these drainage basins it may be hard to even find the streams that have been forced to flow through underground pipes or concrete lined channels. Urbanization has caused significant environmental impacts on these basins and streams. With urbanization, much of the land is paved over or covered with buildings and rendered impervious to water. As a result, rainwater quickly runs off the artificial surfaces to drainage systems (storm sewers) and then to streams. Because both the amount of runoff and the speed at which it reaches streams increases, there is an increase in the flood hazard. Thus, urban areas experience more frequent and larger floods than do natural systems of similar size. Because less water infiltrates soils in the drainage basin (pavement and buildings do not allow for infiltration), the groundwater is not replenished as it otherwise would be. Groundwater seepage into streams supplies water to streams during dry seasons, so if the groundwater table falls below the stream channel, the stream stops flowing. Many urban streams have experienced lower flow or even dried up.

Although runoff from washing cars, watering vegetation, and treating sewage may augment flow, it has a cost, because water quality is greatly reduced. Urban runoff is usually polluted water, especially following dry periods. As a result, urban streams have a much-reduced water quality. This ad-versely affects all living things in the stream system. Finally, urban streams are often neglected streams used for dumping all types of garbage. It is ironic that we allow this aesthetic degradation of urban streams yet travel long distances on weekends and holidays to visit unpolluted, free-flowing streams.

The environmental impacts of urbanization can be minimized or removed if we change our consciousness about urban streams. These streams can be perceived as a valuable resource rather than as an eyesore. Runoff from urban areas during storms can be delayed with retention ponds, storage in parking lots, and other creative means. Stream channels can be cleaned of trash, and development on floodplains can be controlled. Wastewater can be better treated before being added to streams. Urban dwellers may then enjoy a unique greenbelt and environment of higher quality.

Adjustments to the Flood Hazard

By far the most appropriate adjustment to flooding is land-use planning that avoids developing and building on floodplains. This option is also the most desirable from an environmental perspective. The construction of upstream dams, levees (raised embankments along a river), and floodwalls often leads to a false sense of security and cannot protect us from floods greater than these structures are designed to control. Also, dams and levees can fail! During the 1993 flood of the Mississippi about 70% of the levees failed (Figure 19.17). They simply were not designed to withstand a flood that lasted over 2 months. Following the flood some communities are considering relocation rather than rebuilding on the floodplain. Urban land presently developed on floodplains will in most cases continue to be protected, but we should not repeat past mistakes with future development. Not developing on any floodplain in the future would be a good start.

SUMMARY

- Water is a liquid with unique characteristics that has made life on Earth possible.

- Although it is one of the most abundant and important renewable resources on Earth, more than 99% of Earth's water is unavailable or unsuitable for beneficial human use because of its salinity or location.

- The pattern of water supply and use on Earth at any particular point on the land surface involves interactions between the biological, hydrologic, and rock cycles. To evaluate a region's water resources and use patterns, a water budget is developed to define the natural variability and availability of water.

- During the next several decades it is expected that the total water withdrawn from streams and groundwater in the United States will decrease slightly, but the consumptive use will increase because of greater demands from a growing population and industry.

- Water withdrawn from streams competes with in-stream needs, such as maintaining fish and wildlife habitats and navigation, and may therefore cause conflicts.

- Groundwater use has resulted in a variety of environmental problems, including overdraft, loss of riparian vegetation, and land subsidence.

- Because agriculture is the big user of water, conservation of water in agriculture has the most significant effect when considering sustainable water use. However, it is also important to practice water conservation at the personal level in our homes and to price water to encourage conservation and sustainability.

- There is a need for a new philosophy in water resource management that considers sustainability and uses creative alternatives and variable sources. Development of a master plan involves inclusion of normal sources of surface water and groundwater, conservation programs, and use of reclaimed water.

- Development of water supplies and facilities to more efficiently move water may cause considerable environmental degradation; construction of reservoirs and canals and channelization of rivers should be considered carefully in light of potential environmental impacts.

- Wetlands serve a variety of natural service functions at the ecosystem level that benefit other ecosystems and people.

- Flooding is perhaps the most universal natural hazard in the world; both the frequency and the severity of flooding of small streams are increased by urbanization.

- The preferable and most environmentally sound adjustment to flooding is land-use planning that avoids building on floodplains.

REEXAMINING THEMES AND ISSUES

Human Population: As human population has increased, so has the demand for water resources. As a result we must be more careful in managing Earth's water resources, particularly near urban centers.

Sustainability: The water resources of the planet are sustainable provided we manage them properly and they are not overused, polluted, or wasted. This requires good water management strategies. We believe that the move toward sustainable water use must be facilitated now to avoid conflicts in the future.

Global Perspective: The water cycle is one of the grandest geochemical cycles we have. It is responsible for the transfer and storage of water on a global scale. Fortunately, on a global scale the total abundance of water on Earth is not a problem. However, ensuring that

it is available when and where it is needed in a sustainable way *is* a problem.

Urban World: Although urban areas consume only a small portion of the water resources used by people, it is in urban areas where shortages are often most apparent. Thus, the concepts of water management and water conservation are critical in urban areas.

Values and Knowledge: Conflicts result because of varying values related to water resources and our knowledge of hydrology. We value natural areas such as wetlands and free-running rivers but we want the water resources and protection from hazards such as flooding. As a result, we must learn to plan more effectively with nature to minimize natural hazards, maintain a quality water resource, and provide the water resources necessary for the ecosystems of our planet. This remains a major task.

KEY TERMS

channelization *407*

consumptive use *398*

effluent stream *394*

floodplain *409*

groundwater *394*

influent stream *394*

in-stream use *398*

off-stream use *398*

overdraft *396*

subsidence *397*

sustainable water use *402*

water budget *395*

water conservation *399*

wetlands *405*

STUDY QUESTIONS

1. If water is one of our most abundant resources, why are we concerned about its availability in the future?

2. Which is more important from a national point of view, conservation of water use in agriculture or in urban areas? Why?

3. Distinguish between in-stream and off-stream uses of water. Why is in-stream use controversial?

4. What are some important environmental problems related to groundwater use?

5. How might your community better manage its water resources?

6. What are some of the major environmental impacts associated with the construction of dams and canals? How might these be minimized?

7. What are some of the environmental problems associated with channelizing rivers? How might the potential adverse effects of channelization be minimized?

8. How does urbanization affect the flood hazard and how can this hazard best be managed?

FURTHER READING

Brookes, A. 1988. *Channelized Rivers: Perspectives for Environmental Management.* New York: Wiley. Covers consequences of river channelization and methods to improve river modification.

Graf, W. L. 1985. *The Colorado River.* Resource Publications in Geography. Washington, DC: Association of American Geographers. A good summary of the Colorado River water situation.

James, W., and Neimczynowicz, J., eds. 1992. *Water, Development and the Environment.* Lewis Publishers. Covers problems with water supplies imposed by a growing population, including urban runoff, pollution and water quality, and management of water resources. Florida: CRC Press.

La Riviere, J. W. M. 1989. "Threats to the World's Water," *Scientific American* 261(3): 80–84. Summary of supply and demand for water and future threats to continued supply.

Leopold, L. B. 1974. *Water: A Primer.* San Francisco: W. H. Freeman. An excellent introduction to water processes.

Reisner, M. 1990. *Overtapped Oasis: Reform or Revolution for Western Water.* Washington, DC: Island. An overarching review of water supply, water resource development, and government water policy in the American West.

Spulber, N., and Sabbaghi, A. 1994. *Economics of Water Resources: From Regulation to Privatization.* Kluwer Academic. Authors discuss water supply and demand, pollution and its ecological consequences, and water on the open market.

Twort, A. C., Law, F. M., Crowley, F. W., and Ratnayaka, D. D. 1994. *Water Supply,* 4th ed. Edward Arnold. Contains good coverage of water topics from basic hydrology to water chemistry, its use, management, and treatment.

Van der Leeden, F. 1990. *The Water Encyclopedia,* 2nd ed. Lewis Publishers. A great source of all sorts of water facts.

Water Resources Council. 1978. *The Nation's Water Resources, 1975–2000,* vol. 1. Washington, DC: Author. A good summary of water use in the United States.

Wheeler, B. D., Shaw, S. C., Fojt, W. J., and Robertson, R. A. 1995. *Restoration of Temperate Wetlands.* New York: Wiley. Discussions of wetland restoration around the world.

INTERNET RESOURCES

Environmental Protection Agency's Office of Water: *http://www.epa.gov/watrhome*—This site provides information about water topics including laws and regulations, drinking water, pollution, and current water news.

U.S. Bureau of Reclamation, Water Conservation: *http://ogee.byd-lab.do.usbr.gov/rwc/rwc.html*—From the federal agency responsible for management, use, supply, and conservation of water in the western U.S., check out current water conservation plans and programs.

U.S. Geological Survey Water Resources Division: *http://h2o.usgs.gov/*—Look at data, graphs, and maps of water use, river flows, and reservoir levels throughout the U.S.

CHAPTER 20

WATER
POLLUTION
AND
TREATMENT

Pollution control officer measuring oxygen content of the River Severn, near Shrewsbury, England.

CASE STUDY

Outbreak

The occurrence of a disease may quickly and surprisingly occur. The term for this occurrence is **"outbreak"** and has been the subject of several popular science fiction movies and novels. Recent outbreaks of diseases that are not waterborne that are now familiar include acquired immunodeficiency syndrome (AIDS) in America and Ebola virus in Africa.

Today the primary water pollution problem in the world is the lack of clean disease-free drinking water. In the past, epidemics (outbreaks) of

waterborne diseases have been responsible for the deaths of thousands of people in the United States. Fortunately, epidemics of diseases such as cholera have been largely eliminated in the United States as a result of treating drinking water prior to consumption. This certainly is not the case worldwide. Every year several billion people are exposed to waterborne diseases. For example, an epidemic of cholera occurred in South America in the early 1990s, and outbreaks of waterborne diseases continue to be a threat, even in developed countries.

The largest waterborne outbreak in the history of the United States occurred in April of 1993 in Milwaukee, Wisconsin. The disease, which causes flulike symptoms, is a gastrointestinal illness that is carried by a microorganism (a parasite) known as *Cryptosporidium* (the disease is known as cryptosporidiosis). Between March 11 and April 9, approximately 400,000 people of a total of 1.6 million people in a five-county area became ill following exposure to *Cryptosporidium* in the drinking water. Most people who contracted the illness were sick for approximately nine days, but the disease can be fatal to people with a depressed immune system, such as cancer patients or AIDS patients. The parasite is resistant to chlorination and evidently passed through one of the city's water treatment plants. The source of the parasite remains largely unknown, but possible sources include cattle grazing along rivers that flow into Milwaukee Harbor, slaughter houses, or human sewage. It is possible that runoff from spring rains and snowmelt transported the parasites to Lake Michigan where they entered the intake to water treatment plants.[1,2]

The outbreak in Milwaukee is a "wake-up call" concerning the quality of our drinking water.

Many other cities in the United States that utilize surface water resources are just as vulnerable as Milwaukee.[1] In fact, recent tests suggest that *Cryptosporidium* is present in 65% to 97% of the surface waters tested in the United States.[3] Although it is very resistant to disinfectants, it may be removed by filtration. The outbreak in Milwaukee happened even though the water treatment plants met all existing federal and state quality standards. Federal guidelines are even stricter now, but there still is concern as to whether waterborne parasites such as *Cryptosporidium* are being effectively removed prior to consumption of water by people. The 1993 Milwaukee outbreak killed approximately 100 people, and in May of 1994 another outbreak of the same disease, caused by the same parasite, killed approximately 19 people in Las Vegas.[4]

Upgrading water treatment plants is a cost-effective mechanism for reducing the threat of waterborne disease. The price of inaction is very high. For example, in 1994, people in the United States spent approximately $4.2 billion for bottled water. Considering the many costs of illness and death associated with drinking contaminated water, future investments in technology and facilities to treat our water should probably be considered a bargain.[4]

Outbreak of a waterborne disease is a particularly visible type of water pollution. Other types, such as pesticides in groundwater, are often more difficult to identify without careful sampling and testing. This chapter discusses major categories of water pollution and traditional as well as innovative options for wastewater treatment.

LEARNING OBJECTIVES

Degradation of our surface water and groundwater resources is a serious problem, the effects of which may not be fully known for some time. There are a number of steps we can take to treat water and to minimize pollution. After reading this chapter, you should understand:

- What constitutes water pollution and the major categories of pollutants.
- That the primary water pollution problem present in many locations around the world is lack of disease-free drinking water.
- What the difference between point and non-point sources of water pollution is.
- How biochemical oxygen demand works and why it is important.

- What eutrophication is, why it is an ecosystem effect, and how human activity may cause cultural eutrophication.
- What the twofold effect of sediment pollution is.
- What acid mine drainage is and why it is a problem.
- How urban processes may cause shallow aquifer pollution.
- What the various methods of wastewater treatment are and why some are more environmentally preferable than others.
- What environmental laws protect water resources and ecosystems.

20.1 WATER POLLUTION

Water pollution refers to degradation of water quality. In defining pollution, we generally look at the intended use of the water, how far it departs from the norm, its effects on public health, or its ecological impacts. From a public health or ecological view, a pollutant is any biological, physical, or chemical substance that in identifiable excess is known to be harmful to other desirable living organisms. Water pollutants include excessive amounts of heavy metals, certain radioactive isotopes, fecal coliform bacteria, phosphorus, nitrogen, sodium, and other useful (even necessary) elements, as well as certain pathogenic bacteria and viruses. In some instances, a material may be considered a pollutant to a particular segment of the population although not harmful to other segments. For example, excessive sodium as a salt is not generally harmful, but it is to some people who must restrict salt intake for medical reasons.

It is a fundamental principle that the quality of water determines its potential uses. The major uses for water today are for agriculture, industrial processes, and domestic (household) supply. Water for domestic supply must be free from constituents harmful to health, such as insecticides, pesticides, pathogens, and heavy-metal concentrations; it should taste and smell good; and it should not damage plumbing or household appliances. The quality of water required for industrial purposes varies widely depending on the process involved. Some processes may require distilled water; others simply need water that is not highly corrosive or that is free

of particles that could clog or otherwise damage equipment. Because most vegetation is tolerant of a wide range of water quality, agricultural waters may vary widely in physical, chemical, and biological properties.[5]

Many different processes and materials may pollute surface water or groundwater. Some of these are listed in Table 20.1. All segments of our society (urban, rural, industrial, agricultural, and military) may contribute to the problem of water pollution. Most of the sources result from runoff and leaks or seepage of pollutants into surface water or groundwater (see A Closer Look 20.1, "Pollution from Leaking Buried Gasoline Tanks"). Pollutants are also transported by air and deposited in water bodies. The following sections present a few selected water pollutants to emphasize the principles applicable to many other pollutants (see Table 20.2). Other water pollutants are discussed elsewhere in this book (heavy metals, organic chemicals, and thermal pollution in Chapter 14 and radioactive materials in Chapter 18).

Increasing population often results in more pollutants being introduced into the environment as well as placing more demands on our finite water resource.[6] As a result, it can be expected that sources of drinking water in some locations will degrade in the near future. More than one-quarter of drinking water systems in the United States reported at least one violation of federal health standards during 1993 and 1994.[7] During the period 1992 to 1994, approximately 36 million people in the United States were supplied with water from systems that violated (at least once) federal drinking water standards.[4]

TABLE 20.1 Some Sources of Water Pollution

Surface Water	*Groundwater*
Urban runoff (oil, chemicals, organic matter, etc.) (U, I, M)	Leaks from waste disposal sites (chemicals, radioactive materials, etc.) (I, M)
Agricultural runoff (oil, metals, fertilizers, pesticides, etc.) (A)	Leaks from buried tanks and pipes (gasoline, oil, etc.) (I, A, M)
Accidental spills of chemicals including oil (U, R, I, A, M)	Seepage from agricultural activities (nitrates, heavy metals, pesticides, herbicides, etc.) (A)
Radioactive materials (often involving truck or train accidents) (I, M)	Saltwater intrusion into coastal aquifers (U, R, I, M)
Runoff (solvents, chemicals, etc.) from industrial sites (factories, refineries, mines, etc.) (I, M)	Seepage from cesspools and septic systems (R)
Leaks from surface storage tanks or pipelines (gasoline, oil, etc.) (I, A, M)	Seepage of acid-rich water from mines (I)
	Seepage from mine waste piles (I)
Sediment from a variety of sources, including agricultural lands and construction sites (U, R, I, A, M)	Seepage of pesticides, herbicide nutrients, and so on from urban areas (U)
Air fallout (particles, pesticides, metals, etc.) into rivers, lakes, oceans (U, R, I, A, M)	Seepage from accidental spills (train or truck accidents, for example) (I, M)
	Inadvertent seepage of solvents and other chemicals including radioactive materials from industrial sites or small businesses (I, M)

Key: U = urban; R = rural; I = industrial; A = agricultural; M = military.

TABLE 20.2 Categories of Water Pollutants

Pollutant Category	Examples of Sources	Comments
Dead organic matter	Raw sewage, agricultural waste, urban garbage	Produces biochemical oxygen demand and diseases.
Pathogens	Human and animal excrement and urine	Examples: Recent cholera epidemics in South America and Africa; 1993 epidemic of cryptosporidiosis in Milwaukee, Wisconsin. See discussion of fecal coliform bacteria under "Waterborne Disease."
Organic chemicals	Agricultural use of pesticides and herbicides (see Chapter 11); industrial processes that produce dioxin (Chapter 14)	Potential to cause significant ecological damage and human health problems. Many of these chemicals pose hazardous waste problems (see Chapter 27).
Nutrients	Phosphorus and nitrogen from agricultural and urban land use (fertilizers) and wastewater from sewage treatment	Major cause of artificial eutrophication. Nitrates in groundwater and surface waters can cause pollution and damage to ecosystems and people.
Heavy metals	Agricultural, urban, and industrial use of mercury, lead, selenium, cadmium, and so on (see Chapter 14)	Can cause significant ecosystem damage and human health problems. For example, mercury from industrial processes that is discharged into water (see Chapter 14).
Acids	Sulfuric acid (H_2SO_4) from coal and some metal mines; industrial processes that dispose of acids improperly	Acid mine drainage is a major water pollution problem in many coal mining areas, damaging ecosystems and spoiling water resources.
Sediment	Runoff from construction sites, agricultural runoff, and natural erosion	Reduces water quality and results in loss of soil resources.
Heat (thermal pollution)	Warm to hot water from power plants and other industrial facilities	Causes ecosystem disruption (see Chapter 14).
Radioactivity	Contamination by nuclear power industry, military, and natural sources (see Chapter 18)	Often related to storage of radioactive waste. Health effects vigorously debated (see Chapters 14 and 18).

20.2 WATER POLLUTANTS

The Environmental Protection Agency has set thresholds, or limits, of water pollution levels for some (but not all) pollutants. As a result of difficulties in determining effects of exposure to low levels of pollutants, only a small fraction of the more than 700 identified drinking water contaminants have had maximum concentration standards set for them. If the pollutant is in a concentration greater than an established threshold, the water is unsatisfactory for a particular use. A list of selected pollutants or indications of pollution included in the national drinking water standards for the United States can be found in Table 20.3. The following sections focus on several pollutants as well as on biochemical oxygen demand and dissolved oxygen, which is not a pollutant but is needed for healthy aquatic ecosystems.

Biochemical Oxygen Demand (BOD)

Dead organic matter in streams decays. Bacteria carrying out this decay require oxygen. If there is enough bacterial activity, the oxygen in the water available to fish and organisms can be reduced to levels so low that they may die. A stream without oxygen is a dead stream for fish and for many organisms we value; a stream with an inadequate oxygen level is considered polluted for those organisms that require dissolved oxygen above the existing

level. The amount of oxygen required for such biochemical decomposition is called the **biochemical oxygen demand (BOD).**

TABLE 20.3 National Drinking Water Standards

Contaminant	Maximum Contaminant Level (mg/L)
Inorganics	
Arsenic	0.05
Cadmium	0.01
Lead	0.015 action level[a]
Mercury	0.002
Selenium	0.01
Organic chemicals	
Pesticides	
Endrin	0.0002
Lindane	0.004
Methoxychlor	0.1
Herbicides	
2,4-D	0.1
2,4,S-TP	0.01
Silvex	0.01
Volatile organic chemicals	
Benzene	0.005
Carbon tetrachloride	0.005
Trichloroethylene	0.005
Vinyl chloride	0.002
Microbiological organisms	
Fecal coliform bacteria	1 cell/100 mL

Source: U.S. Environmental Protection Agency.
[a]Action level is related to the treatment of water to reduce lead to a safe level. There is no maximum contaminant level for lead.

CLOSER LOOK 20.1

POLLUTION FROM LEAKING BURIED GASOLINE TANKS

Pollution from leaking gasoline tanks belonging to automobile service stations is a widespread environmental problem that no one thought very much about until only a few years ago. Underground tanks are now strictly regulated. Many thousands of old, leaking tanks have been removed, and the surrounding soil and groundwater have been treated to remove the gasoline. Cleanup can be a very expensive process, involving removal and disposal of soil (as a hazardous waste) and treatment of the water using a process known as vapor extraction (Figure 20.1). On the other hand, treatment may be accomplished under the ground by micoorganisms that consume the gasoline. This is known as **bioremediation** and is much less expensive than removal, disposal, or vapor extraction. Notice in Figure 20.1 that the liquid gasoline and vapor from the gasoline are above the water table; a small amount dissolves into the water. Nevertheless, all three phases of the pollutant (liquid, vapor, and dissolved) float on the denser groundwater. The extraction well (shown in the figure) takes advantage of this situation, and the function of the dewatering wells (also shown) is to pull the pollutants in where the extraction is most effective. Pollution from leaking buried gasoline tanks emphasizes some important points about water pollutants:

- Some are lighter than water and thus float on the water.
- Some pollutants have multiple phases: liquid, vapor, and dissolved. Dissolved phases chemically combine with the water (e.g., salt dissolves into water).
- Some pollutants are heavier than water and sink or move downward through surface water or groundwater. Examples of sinkers include some particulates and cleaning solvents. Pollutants that sink may become concentrated in sediment at the bottom of lakes or rivers.
- The method used to treat or eliminate a water pollutant must take into account the physical and chemical properties of the pollutant and how these interact with surface water or groundwater. For example, the extraction well for removing gasoline from a groundwater resource (Figure 20.1) takes advantage of the fact that gasoline floats on water.
- Because cleanup or treatment of water pollutants is very expensive and undetected or untreated pollutants may cause environmental damage, the emphasis on water pollution should be on preventing pollutants from entering water in the first place.

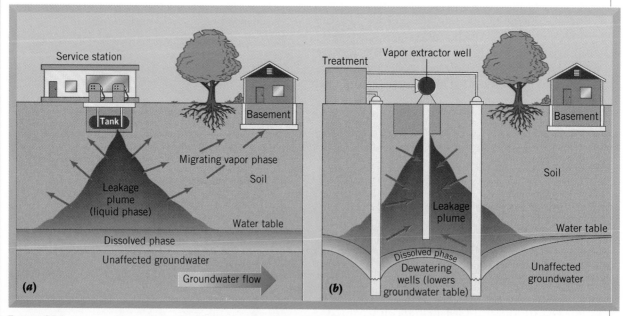

FIGURE 20.1 Diagram illustrating (*a*) leak from a buried gasoline tank, and (*b*) possible remediation using vapor extractor system. See text for explanation. (*Source:* Courtesy of University of California Santa Barbara Vadose Zone Laboratory and David Springer.)

BOD, a commonly used measure in water quality management, is a measure of the amount of oxygen consumed by microorganisms as they break down (consume) organic matter within small water samples analyzed in a laboratory. The BOD is routinely measured as part of water quality testing, particularly at discharge points into surface water, such as at wastewater treatment plants. At treatment plants, the BOD of the incoming sewage water from sewer lines is measured, as is the water both upstream and downstream of the plant. This practice allows comparison of the upstream, or background, BOD to the BOD of the water being discharged by the plant.

Dead organic matter is contributed to streams and rivers from natural sources (such as dead leaves from a forest) as well as from agricultural runoff and urban sewage. Approximately 33% of all BOD in streams results from agricultural activities, but urban areas, particularly those with combined sewer systems (storm-water runoff and urban sewage share the same line), also considerably increase the BOD in streams. This results because during times of high flow, sewage treatment plants are unable to handle the total volume, and raw sewage mixed with storm runoff is discharged untreated into streams and rivers.

When the BOD is too high, the *dissolved oxygen content* of the water becomes too low to support all the life in the water. The Council on Environmental Quality defines the threshold for a water pollution alert as a dissolved oxygen content of less than 5 mg/L of water. The diagram in Figure 20.2 illustrates the effect of BOD on dissolved oxygen content in a stream when raw sewage is introduced as a result of an accidental spill. Three zones are recognized:

1. *a pollution zone,* with a high BOD and reduced dissolved oxygen content as initial decomposition of the waste begins;

2. *an active decomposition zone,* where the dissolved oxygen content is at a minimum owing to biochemical decomposition as the organic waste is transported downstream; and

3. *a recovery zone,* in which the dissolved oxygen increases and the BOD is reduced, because most oxygen-demanding organic waste from the input of sewage has decomposed and the natural stream processes are replenishing the water with dissolved oxygen. For example, quickly moving water causes a mixing of the water surface with the air, and oxygen enters the water.

All streams have some capability to degrade organic waste after it enters the stream. Problems result when the stream is overloaded with biochemical oxygen-demanding waste, overpowering the stream's natural cleansing function.

Waterborne Disease

As emphasized in the opening of this chapter, the primary water pollution problem in the world today is the lack of clean, disease-free drinking water. We tend not to think of this much, because the water supply in the United States, Europe, and most places we are likely to visit is treated to remove disease-carrying microorganisms. Historic epidemics of waterborne diseases that killed thousands of people in cities such as Chicago, Illinois, were eliminated by providing a safe source of drinking water and by not allowing sewage water to contaminate drinking

FIGURE 20.2 Relationship between dissolved oxygen and biochemical oxygen demand (BOD) for a stream following the input of sewage.

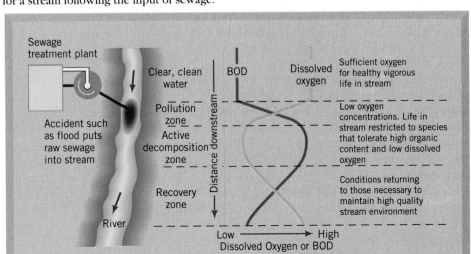

water supplies. However, this is not the case everywhere, and each year several billion people, particularly in poor countries, are exposed to waterborne diseases whose effects vary in severity from an upset stomach to death. Historically, outbreaks of serious waterborne diseases, such as cholera and typhoid fever, have caused tremendous hardships on society and killed thousands of people. As recently as the early 1990s, epidemics of cholera in South America caused much suffering and death.

Because it is difficult to monitor disease-carrying organisms directly, we use the count of **fecal coliform bacteria** as a standard measure and indicator of disease potential. These mostly common, usually (but not always) harmless bacteria are normal constituents of human and animal intestines and in all human and animal waste. Escherichia coli (E. Coli), a type of fecal coliform bacteria, has been responsible for causing human illness and death. Outbreaks have resulted from eating contaminated meat and drinking contaminated juices or water. There was an outbreak due to contaminated meat at a popular fast food chain in 1993. In 1998, 26 children became ill and one died after visiting a Georgia water park. As recently as July 1998, the community of Alpine, Wyoming suffered a major outbreak of illness due to the presence of E. coli in the town's drinking water supply.[8] It is clear that fecal coliform bacteria present a real threat to human health and must be carefully regulated. The threshold used by the Council on Environmental Quality for swimming water is not more than 200 cells of fecal coliform bacteria per 100 milliliters of water; if fecal coliform is above the threshold level, the water is considered unfit for swimming. Most state agencies consider water with any fecal coliform unsuitable for drinking.

Nutrients

Nutrients released by human activity lead to water pollution. Two important nutrients that cause pollution problems are phosphorus and nitrogen, both are released from sources related to land use. Forested land has the lowest concentrations of phosphorus and nitrogen in stream waters. In urban streams, these nutrients are greater because of fertilizers, detergents, and products of sewage treatment plants. However, the highest concentrations of phosphorus and nitrogen are found in agricultural areas, where the sources are fertilized farm fields and feedlots (Figure 20.3). Over 90% of total nitrogen added to the environment by human activity is from agriculture. Therefore, nitrogen entering the Gulf of Mexico from the Mississippi may have originated from fertilizer applied to fields in Minnesota or

FIGURE 20.3 Cattle feed lot located in Colorado. High concentrations of cattle have the potential for both surface and groundwater pollution.

Ohio.[9] One aspect of this problem is illustrated in A Closer Look 20.2, "Eutrophication of Medical Lake."

Oil

Oil discharged into surface water (usually the ocean) has caused major pollution problems. Several large oil spills from underwater oil-drilling have occurred in recent years. However, although spills make headlines, normal shipping activities probably release more oil over a period of years than is released by the occasional spill. The cumulative impacts of these releases are not well known.

The best known oil spills are caused by tanker accidents. On March 24, 1989, supertanker *Exxon Valdez* ran aground on Bligh Reef south of Valdez in Prince William Sound. Alaskan crude oil that had been delivered to Valdez through Trans-Alaska Pipeline poured out of ruptured tanks of the tanker at about 20,000 barrels per hour. The tanker was loaded with about 1.2 million barrels of oil, and about 250,000 barrels (11 million gallons) entered the sound (Figures 20.6 and 20.7). An even bigger spill was avoided when the remainder of oil was offloaded onto another vessel. The *Exxon Valdez* spill produced an environmental shock that resulted in passage of the Oil Pollution Act of 1990 and in renewed evaluation of cleanup technology.[15,16]

A CLOSER LOOK 20.2

EUTROPHICATION OF MEDICAL LAKE

In the summer of 1971, the waters of Medical Lake in Washington became clogged with algae and bacteria and turned a dark, turbid green. Fish, algae, and bacteria died, and the wind blew them into stinking masses on the lee shore (Figure 20.4).[10] It was then apparent that Medical Lake was dying.

Fish in the lake did not die from a direct effect of pollution: from disease organisms in the polluted water or from something in the polluted water that was directly toxic to them. They died as the result of an ecosystem effect: the complex effect produced by a community of interacting organisms exerted through the local environment and involving that environment (recall the discussion of ecosystems and the community effect in Chapter 6).

This is how the process worked. Phosphorus, a chemical element necessary for all living things and one that is often in short supply, became available in large concentrations in treated sewage water that entered the lake. Just as when phosphorus is added as a lawn fertilizer and causes increased growth of grass, the phosphorus added to the lake caused a population explosion of photosynthetic blue-green bacteria and algae. Mats of the algae and bacteria became so thick that those at the top shaded those at the bottom; not receiving enough light, those at the bottom died. The dead algae and bacteria became food for other bacteria, and as these bacteria increased, they used more and more of the oxygen dissolved in the lake water. (Bacteria require oxygen for respiration, just as we do.) The supply of oxygen in the water was soon depleted. Algae, bacteria and fish began to die. Because fish require a higher concentration of oxygen in water than do bacteria, most of the

FIGURE 20.4 Mats of dying green algae in a pond undergoing eutrophication.

fish died, whereas some bacteria continued to live (Figure 20.5).

The fish did not die in Medical Lake from phosphorus poisoning. If you added phosphorus to water in an aquarium where there were only fish and no algae or bacteria, so that the phosphorus concentration was the same as that in Medical Lake, the phosphorus would not affect the fish. The fish died from a lack of oxygen resulting from a chain of events that started with the input of phosphorus and affected the whole ecosystem. The unpleasant effects resulted from the interactions among different species, the effects of the species on chemical elements in their environment, and the condition of the environment (the lake and the air above it). This is what we call an **ecosystem effect.** The increase in the concentration of chemical elements required for living things (such as phosphorus) is called the **eutrophication** of the ecosystem.

Eutrophication can be a natural process. What happened to Medical Lake is artificial eutrophication (sometimes also called *cultural eu-*

trophication). A lake that has a high concentration of chemical elements required for life is called a *eutrophic lake;* one that has a relatively low concentration of these elements is called an *oligotrophic lake.* Oligotrophic lakes are relatively sterile and therefore have clear waters that are pleasant for swimmers and boaters. Eutrophic lakes have an abundance of life, often mats of algae and bacteria that are unpleasant. In contrast to the typical idea of what is desirable in nature, low nutrient conditions are generally preferred by people when it comes to lakes for swimming and boating.

Problems associated with the artificial eutrophication of bodies of water are not restricted to freshwater lakes. In recent years concern has grown about the outflow of sewage from urban areas into tropical coastal waters and the possible effects of eutrophication on coral reefs.[11,12] Such concern has grown for the famous Great Barrier Reef of Australia, an important area for biological diversity and recreation as well as some reefs fringing the Hawaiian Islands.[13,14]

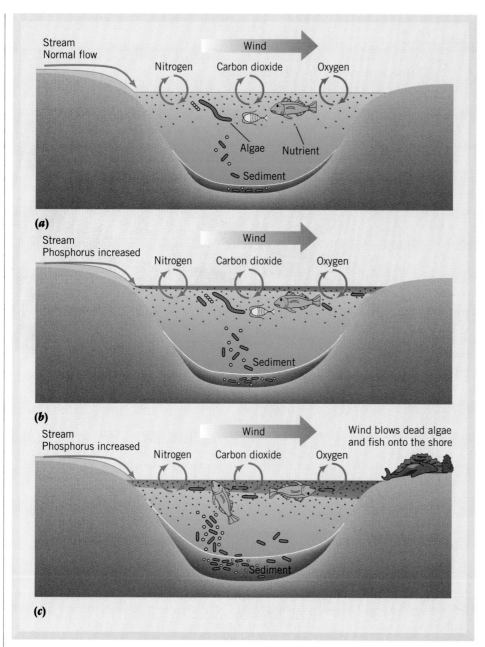

FIGURE 20.5 The eutrophication of a lake: (*a*) an oligotrophic, or low nutrient, lake; the algae (green) abundance is low, the water clear; (*b*) phosphorus is added to streams and enters the lake; algae growth is stimulated; a dense layer is formed; and (*c*) the algae layer becomes so dense that the algae at the bottom die; bacteria feed on the dead algae and use up the oxygen; fish die from lack of oxygen.

The solution to artificial eutrophication is fairly straightforward and involves ensuring that high concentrations of nutrients from human sources do not enter lakes and other water bodies. This goal can be accomplished by taking care in disposing of treated wastewater and by using more advanced water treatment methods, such as special filters and chemical treatment that remove more of the nutrients.

Environmental shock resulted because oil was spilled into what is considered one of the most pristine and ecologically rich marine environments of the world.[15] Many species of fish, birds, and marine mammals are present in the sound. The effects of the spill included the death of 13% of harbor seals, 28% of sea otters, and 100,000–645,000 seabirds.[16]

FIGURE 20.6 *Exxon Valdez* tanker accident in Prince William Sound (1989). Oil is being offloaded from leaking *Exxon Valdez* (left) to the smaller *Exxon Baton Rouge* (right).

Within three days of the spill, winds began blowing the slick beyond any hope of containment. Of the 11 million gallons of spilled oil, about 20% evaporated and 50% was deposited on the shoreline. Only 14% was collected by skimming and waste recovery. The extent of oil sheens, tar balls, and mousse (a thick, weathered patch of oil with the consistency of soft pudding) is shown in Figure 20.7.

FIGURE 20.7 Extent of Alaskan oil spill of 1989. (*Source: Alaska Department of Fish and Game, 21(4), 1989.*)

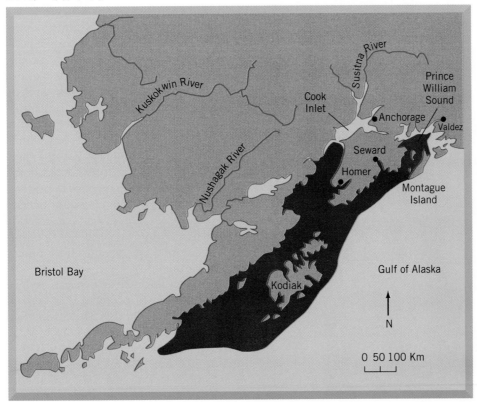

Before the *Exxon Valdez* spill it was generally believed that the oil industry was capable of dealing with oil spills. However, more than $3 billion has been spent to clean the spill, and few people are satisfied with the results. Some scientists argue that the recovery might have been faster if some of the cleanup methods, such as spraying rocks and beaches with high-pressure hot water, had not been used. They argue that the high pressure and heat killed coastal organisms that live under the rocks and had survived the initial impact of the spill.[16] There is no doubt that the cleanup work posed enormous problems (Figure 20.8); photographs and videotapes of workers attempting to clean individual pebbles on beaches are a vivid reminder of the difficulty and virtual futility of achieving an effective cleanup after an event of this magnitude. In addition, the oil spill disrupted the lives of the people who live and work in the vicinity of Prince William Sound.

The long-term effects of large oil spills are uncertain. We know that the effects can last several decades; toxic levels of oil have been identified in salt marshes 20 years following a spill.[16] The ocean abounds with natural processes capable of eventually consuming the oil. Therefore, compared to some other water pollutants (such as pathogens, heavy metals, and organic chemicals), oil toxicity is relatively low and recovery can occur fairly quickly. Nevertheless, there are significant short-term effects of spills, including disruption of fisheries, curtailment of tourism, and loss of seabirds and mammals.

FIGURE 20.8 Attempt to clean oil from the shoreline of Eleanor Island, Alaska, four months after the oil spill from the *Exxon Valdez.*

The *Exxon Valdez* spill demonstrated that the technology for dealing with oil spills is inadequate. The first step is to avoid large spills; a primary method for doing so is to use supertankers with double hulls designed to minimize the release of oil on collision and rupture of tanks. Once a spill occurs, the collection of oil at sea using floating barriers and skimmers (oil is lighter than water and so floats on water) is a worthwhile endeavor, but if conditions include high winds and rough seas, it is nearly impossible. Oil on beaches may be collected by spreading absorbent material, such as straw, allowing the oil to soak in, and then collecting and disposing of the oily straw.

Sediment

Sediment consisting of rock and mineral fragments ranging from gravel particles greater than 2 mm in diameter to finer sand, silt, clay, and even finer colloidal particles can produce a sediment pollution problem. By volume and mass, sediment is our greatest water pollutant. In many areas it chokes streams; fills lakes, reservoirs, ponds, canals, drainage ditches, and harbors; buries vegetation; and generally creates a nuisance that is difficult to remove. It is truly a resource out of place. *Sediment pollution* is a twofold problem: It results from erosion, which depletes a land resource (soil) at its site of origin (Figure 20.9), and it reduces the quality of the water resource it enters.[17]

Land Use and Sediment

Many human activities affect the pattern, amount, and intensity of surface-water runoff, erosion, and sedimentation. Streams in naturally forested or wooded areas may be nearly stable; that is, there is relatively little excessive erosion or sedimentation. Converting forested land to agriculture generally increases the runoff and sediment yield or erosion of the land. Application of soil conservation procedures to farmland can minimize but not eliminate soil loss. The change from agricultural, forested, or rural land to highly urbanized land has even more dramatic effects. Large quantities of sediment may be produced during the construction phase of urbanization. Fortunately sediment production and soil erosion may be minimized by on-site erosion control measures.

That sediment control measures can reduce sediment pollution in an urbanizing area is demonstrated by a study in Maryland.[18] The suspended sediment transported by the northwest branch of the Anacostia River near Colesville, Maryland, with a drainage area of 54.6 km² (21 mi²), was measured over a 10-year period. During that time, urban construction within the basin involved about 3% of the

FIGURE 20.9 Erosion of a highway embankment in Georgia. The erosion damages and removes vegetation, and the sediment may be transported offsite to degrade water resources.

area each year. The total urban land area in the basin was about 20% at the end of the 10-year study. Sediment pollution was a problem because of the amount of rainfall and the type of soil, which is highly susceptible to erosion when not protected by a vegetative cover. Most of the sediment was transported during spring and summer rainstorms.

A sediment-control program reduced sediment yield by an estimated 35%. The program's basic principles were to tailor the development to the natural topography, expose a minimum amount of land, provide protection for exposed soil, minimize surface runoff from critical areas, and trap eroded sediment on the construction site.[18]

20.3 ACID MINE DRAINAGE

The term **acid mine drainage** is perhaps a poor one because it does not refer to an acid mine at all. Rather, it refers to water that drains from mostly coal but also metal mines (copper, lead, and zinc) with a high concentration of sulfuric acid (H_2SO_4). Coal and the rocks containing coal are often associated with a mineral known as fool's gold or pyrite (FeS_2), which is iron sulfide. When the pyrite, which may be finely disseminated in the rock and coal, comes into contact with oxygen and water, it weathers; a product of the weathering is sulfuric acid. Pyrite is also associated with metallic sulfide deposits, which,

when weathered, also produce sulfuric acid. The acid is produced when surface water or shallow groundwater runs through or moves into and out of mines or tailings (waste piles composed of worthless material removed from a mine) (Figure 20.10). If the acid-rich water runs off to a natural stream, pond, or lake, significant pollution and ecological damage may result. The acidic water is toxic to the plants and animals of an aquatic ecosystem; it damages biological productivity, and fish and other aquatic life may die. Acid-rich water can also seep into and pollute groundwater.

Acid mine drainage is a significant water pollution problem in Wyoming, Indiana, Illinois, Kentucky, Tennessee, Missouri, Kansas, and Oklahoma and is probably the most significant water pollution problem in West Virginia, Maryland, Pennsylvania, Ohio, and Colorado. The total impact is significant, because thousands of kilometers of streams have been damaged.

Even abandoned mines can cause serious problems. For example, subsurface mining in the tristate area of Kansas, Oklahoma, and Missouri for sulfide deposits containing lead and zinc began in the late nineteenth century and ended in some areas in the 1960s. When the mines were operating, they were kept dry by pumping out groundwater that seeped in. However, since the termination of mining, some of them have flooded and overflowed into nearby creeks, polluting the creeks with acid-rich water. The problem was so severe in the Tar Creek area of Oklahoma that it was at one time designated by the U.S. Environmental Protection Agency as the nation's worst hazardous waste site.

FIGURE 20.10 Acid drainage from an abandoned mine site is entering a small stream channel, polluting surface waters. This site is located in the mountains of southwestern Colorado.

20.4 SURFACE-WATER POLLUTION

Pollution of surface waters occurs when too much of an undesirable or harmful substance flows into a body of water, exceeding the natural ability of that water body to remove the undesirable material, dilute it to a harmless concentration, or convert it to a harmless form.

Water pollutants, like pollutants in general (see Chapter 14), are categorized as emitted from point or nonpoint sources. **Point sources** are distinct and confined, such as pipes from industrial or municipal sites that empty into streams or rivers. In general, point-source pollutants from industries are controlled through on-site treatment or disposal and are regulated by permit. Municipal point sources are also regulated by permit. In older cities in the northeastern and Great Lakes areas of the United States, most point sources are outflows from combined sewer systems, which combine storm-water flow with municipal wastewater. During heavy rains, urban storm runoff may exceed the capacity of the sewer system, causing it to overflow and deliver pollutants to nearby surface waters.

Nonpoint sources, such as runoff, are diffused and intermittent and are influenced by factors such as land use, climate, hydrology, topography, native vegetation, and geology. Common urban nonpoint sources include urban runoff from streets or fields; such runoff contains all sorts of pollutants, from heavy metals to chemicals and sediment. Rural sources of nonpoint pollution are generally associated with agriculture, mining, or forestry. Nonpoint sources are difficult to monitor and control.

From an environmental view two approaches to deal with water pollution are (1) reduce the sources or (2) treat the water to remove or convert the pollutant to a harmless form. Which of the options is used depends on the specific circumstances of the pollution problem. Reduction at the source is the most environmentally preferable way of dealing with pollutants. For example, air-cooling towers, rather than water-cooling towers, may be utilized to dispose of waste heat from power plants (see Chapter 14). The second method (water treatment) is used for a variety of pollution problems. These include chlorination and filtering of water to remove contaminants and chemical treatment with materials such as ozone, which oxidizes certain pollutants.

There is a growing list of success stories in the treatment of water pollution. One of the most notable is the cleanup of the Thames River in Great Britain. For centuries London's sewage had been dumped into that river, and there were few fish to be found downstream in the estuary. In recent decades, however, improvement in water treatment has led to the return of a great number of species of fish, some not seen in the river in centuries.

Large cities in the United States such as Boston, Miami, Cleveland, Detroit, Chicago, Portland, and Los Angeles grew on the banks of rivers, nearly destroying them with pollution and concrete. Today there are grass roots movements all around the country to restore our urban rivers and adjacent lands as greenbelts, parks, or other environmentally sensitive developments. For example, the Cuyahoga River in Cleveland was so polluted by 1969 that sparks from a train ignited oil-soaked wood in the river, setting the surface of the river on fire! Burning of an American river became a symbol for a growing environmental consciousness. The Cuyahoga River today is much cleaner and no longer flammable. In fact, it is being transformed into a 140-km- (87-mile-) long greenbelt and trail system, changed from a sewer into a valuable public resource and focal point for economic and environmental renewal.[19]

20.5 GROUNDWATER POLLUTION

Approximately one-half of all people in the United States today depend on groundwater as their source of drinking water. We have long believed that groundwater is in general pure and safe to drink. Therefore, it may be alarming for some people to learn that groundwater may in fact be easily polluted by any one of several sources (see Table 20.1) and that the pollutants, even though they are very toxic, may be difficult to recognize.

In the United States today, only a small portion of the groundwater is known to be seriously contaminated, but (as mentioned earlier) the problem may become worse as human population pressure on water resources increase. Nevertheless, several million people have used that water, and the extent of the problem is growing as the testing of groundwater becomes more common. For example, Atlantic City and Miami are two eastern cities threatened by polluted groundwater that is slowly migrating toward their wells.

It is estimated that 75% of the 175,000 known waste disposal sites in the country may be producing plumes of hazardous chemicals that are migrating into groundwater resources. Because many of the chemicals are toxic or are suspected carcinogens, it appears that we have inadvertently been conducting a large-scale experiment on the effects on people of chronic low-level exposure to potentially harmful chemicals. The final results of the experiments will not be known for many years.[20] Preliminary results suggest that we had better act now before a hidden time bomb of health problems explodes.

The hazard presented by a particular groundwater pollutant depends on several factors, including the concentration or toxicity of the pollutant in the environment and the degree of exposure to people or other organisms (see the section on risk assessment in Chapter 14).[21]

Groundwater pollution differs in several ways from surface-water pollution. Groundwater often lacks oxygen, a situation that is helpful in killing aerobic types of microorganisms but may provide a happy home for anaerobic varieties. Bacterial breakdown of pollutants, generally confined to the soil or to material a meter or so below the surface, does not occur readily in groundwater. Furthermore, the channels through which groundwater moves are often very small and variable. Thus, the rate of movement is low in most cases, and the opportunity for dispersion and dilution is limited. (See A Closer Look 20.3, "Long Island, New York".)

CLOSER LOOK 20.3

LONG ISLAND, NEW YORK

Two counties on Long Island, New York (Nassau and Suffolk), with a population of several million people, are entirely dependent on groundwater for their water supply. Two major problems associated with the groundwater in Nassau County are intrusion of salt water and shallow-aquifer contamination.[22] Saltwater intrusion is a problem in many coastal areas of the world (see Figure 20.11).

The general movement of groundwater under natural conditions for Nassau County are illus-

FIGURE 20.11 How saltwater intrusion might occur. The upper drawing shows the groundwater system near the coast under natural conditions and the lower drawing shows a well with both a cone of depression and a cone of ascension. If pumping is intensive, the cone of ascension may be drawn upward, delivering salt water to the well. H and $40H$ represent the height of the freshwater table above sea level and the depth of salt water below sea level, respectively.

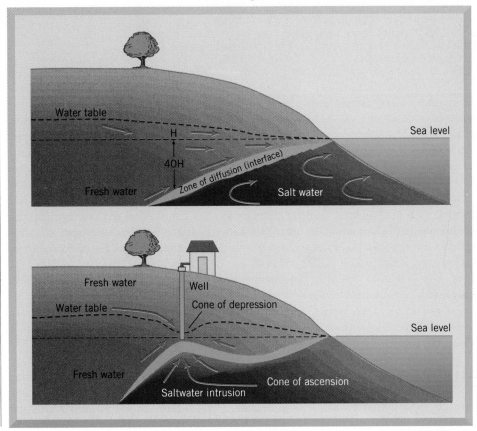

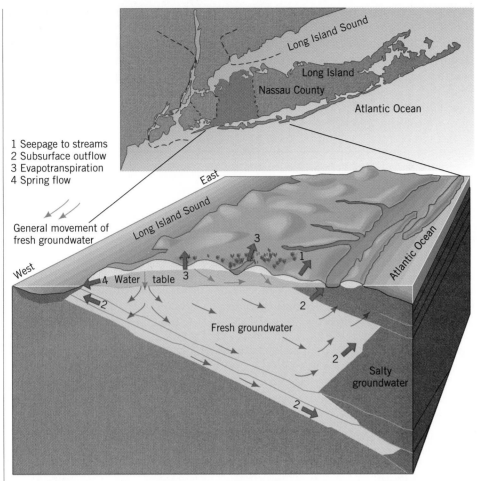

1 Seepage to streams
2 Subsurface outflow
3 Evapotranspiration
4 Spring flow

General movement of
fresh groundwater

FIGURE 20.12 The general movement of fresh groundwater for Nassau County, Long Island. (*Source:* From G. L. Foxworth, 1978, "Nassau County, Long Island, New York—Water Problems in Humid Country," in G. D. Robinson and A. M. Spieke, eds., *Nature to be Commended,* U.S. Geological Survey Professional Paper 950, pp. 55–68.)

trated in Figure 20.12. Salty groundwater is restricted from migrating inland by the large wedge of fresh water moving beneath the island. Notice also that the aquifers are layered, with those closest to the surface the most salty.

In spite of the huge quantities of water in Nassau County's groundwater system, intensive pumping in recent years has caused water levels to decline as much as 15 m (50 ft) in some areas. As groundwater is removed near coastal areas, the subsurface outflow to the ocean decreases, allowing salt water to migrate inland. Saltwater intrusion has become a problem for south shore communities, which now must pump from a deeper aquifer, below and isolated from saltwater intrusion problems.

The most serious groundwater problem on Long Island is shallow-aquifer pollution associated with urbanization. Sources of pollution in Nassau County include urban runoff, household sewage from cesspools and septic tanks, salt used to deice highways, and industrial and solid waste. These pollutants enter surface waters and then migrate downward, especially in areas of intensive pumping and declining groundwater levels.[22] Landfills for municipal solid waste have been a significant source of shallow-aquifer pollution on Long Island because pollutants (garbage) placed on sandy soil over shallow groundwater quickly pollute the water. For this reason, most Long Island landfills were closed in the last decade.

20.6 WASTEWATER TREATMENT

The water used for industrial and municipal purposes is often degraded during use by the addition of suspended solids, salts, nutrients, bacteria, and oxygen-demanding material. By law, these waters must be treated before being released back into the environment. Treatment in the United States now costs approximately $15 billion per year, and the price is expected to double during the next 10 years. Wastewater treatment is big business. Conventional methods include disposal and treatment of household wastewater by way of septic-tank disposal systems in rural areas and centralized water-treatment plants that collect wastewater from sewer systems in cities. Recent innovative approaches to wastewater treatment include the application of wastewater to the land, aquaculture, and wastewater renovation and reuse.

Septic-Tank Disposal Systems

In recent years in the United States, people have been moving in great numbers from rural to urban or urbanizing areas. In many instances, construction of city sewage systems and wastewater treatment facilities has not kept pace with growth. As a result, the individual septic-tank disposal system, long used in rural areas not connected to sewer systems, continues to be an important method of sewage disposal in outlying areas of cities. Because not all land is suitable for the installation of a septic-tank disposal system, evaluation of individual sites is usually required by law before a permit can be issued. An alert buyer will check to make sure that the site is satisfactory for septic-tank disposal before purchasing property on the fringe of an urban area where such a system is necessary. Failure to do so has made many buyers unhappy, and they may pass the property on to another unsuspecting person.

The basic parts of a septic-tank disposal system are shown in Figure 20.13. The sewer line from the house leads to an underground septic tank in the yard. The tank is designed to separate solids from liquid, digest (biochemically change) and store organic matter through a period of detention, and allow the clarified liquid to discharge into the drain field (absorption field), a system of piping through which the treated sewage may seep into the surrounding soil. As the wastewater moves through the soil, it is further treated by the natural processes of oxidation and filtering. By the time the water reaches any freshwater supply, it should be safe for other uses.

Sewage absorption fields may fail for several reasons. The most common causes are failure to pump out the septic tank when it is full of solids and poor soil drainage, which allows the effluent to rise to the surface in wet weather. When a septic-tank absorption field does fail, serious pollution of groundwater and surface water may result.

Wastewater Treatment Plants

Wastewater treatment, or sewage treatment, occurs at specially designed plants that accept municipal sewage from homes, businesses, and industrial sites. The raw sewage is delivered to the plant through a network of sewer pipes. Following treatment the wastewater is discharged into the surface-water environment (river, lake, or ocean) or in some limited cases may be used for another purpose, such as crop irrigation.

Wastewater treatment methods are usually divided into three categories: **primary treatment, secondary treatment,** and **advanced wastewater treatment.** Primary and secondary treatment are required by federal law for all municipal plants in the United States. However, treatment plants may qualify for a waiver to be exempt from secondary treatment if installment of secondary treatment facilities would pose an excess financial burden. Where secondary treatment is not sufficient to protect the water quality of the surface water into which the

FIGURE 20.13 Septic-tank sewage disposal system and location of the absorption field with respect to the house and well. (*Source:* After Indiana State Board of Health.)

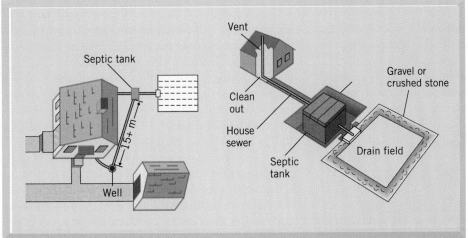

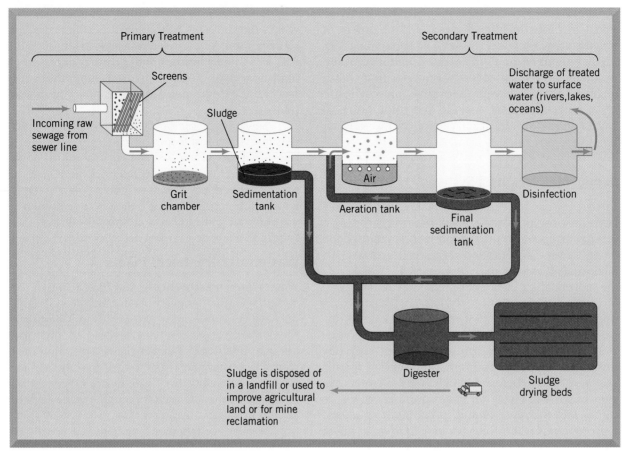

FIGURE 20.14 Diagram of sewage treatment processes. Note: the use of digesters are relatively new and many older treatment plants do not have them.

treated water is discharged, advanced treatment may be required.[23]

A simplified diagram of a wastewater treatment plant is shown in Figure 20.14, and a description follows.[23]

Primary Treatment

Incoming raw sewage enters the plant from the municipal sewer line and is first passed through a series of screens, the purpose of which is to remove large floating material. The sewage next enters the grit chamber, where sand, small stones, and grit are removed and disposed of. The sewage then enters the primary sedimentation tank, where particulate matter settles out to form a sludge. Sometimes chemicals are used to help the settling process. The sludge is removed and transported to the digester for further processing. Primary treatment removes approximately 30% to 40% of the pollutant volume from the wastewater, mainly in the form of suspended solids and organic matter.[23]

Secondary Treatment

There are several methods of secondary treatment. What is described here is known as activated sludge, the most common treatment. The wastewater from the primary sedimentation tank enters the aeration tank (Figure 20.14), where the wastewater is mixed with air (pumped in) and some of the sludge from the final sedimentation tank, which contains aerobic (an oxygen-rich environment) bacteria that consume (break down) organic material (pollutants) in the waste. After several hours the wastewater enters the final sedimentation tank, where sludge settles out. Some of this activated sludge rich in bacteria is recycled and mixed again in the aeration tank with air and new, incoming wastewater, repeating the process. The bacteria are used again and again. Most of the sludge from the final sedimentation tank, however, is transported to the sludge digester. There, along with sludge from the primary sedimentation tank, it is treated by anaerobic (an oxygen-deficient environment) bacteria, which further degrade the sludge by microbial digestion.

Methane gas is a product of the anaerobic digestion and may be used at the plant as a fuel to run equipment or heat and cool buildings. In some cases it is burned off. Wastewater from the final sedimentation tank is then disinfected, usually by chlorination to eliminate disease-causing organisms. The treated wastewater is then discharged into a river, lake, or ocean or, in some limited cases, used to irrigate farmland. Secondary treatment removes about 90% of the pollutants that enter the plant in the sewage.[23] The sludge from the digester is dried and disposed of in a landfill or may be applied to improve soil.

Advanced Wastewater Treatment

Primary and secondary treatment do not remove all pollutants from incoming sewage. More pollutants can be removed by adding steps of treatment similar to secondary treatment. Removal of nutrients such as phosphates and nitrates, organic chemicals, or heavy metals require treatments designed specifically for those contaminants. Treatments may include sand filters, carbon filters, or application of chemicals that assist in the removal process. Advanced treatment may remove more than 95% of the pollutants in wastewater.[23] Treated water is then discharged into surface water or may be used for irrigation of agricultural lands or municipal properties, such as golf courses, city parks, or grounds surrounding wastewater treatment plants. Advanced wastewater treatment is used when it is particularly important to maintain good water quality. For example, if a treatment plant discharges treated wastewater into a river and there is concern that nutrients remaining after secondary treatment may cause damage to the river ecosystem (eutrophication), advanced treatment may be used to greatly reduce nutrients in discharge from the treatment plant.

20.7 WATER TREATMENT FOR DOMESTIC USE

Water for domestic use in the United States is drawn from surface waters and groundwater. Although some groundwater sources have high water quality and need little or no treatment, most sources are treated to conform to national drinking water standards (see Table 20.3). Prior to treatment and purification water is usually stored in reservoirs or special ponds. Storage allows for solids, such as fine sediment and organic matter, to settle out, improving clarity of water. Water is then run through a treatment plant, where it is filtered and purified by chemical treatment (usually chlorination) before it is distributed to individual homes. At the homesite, water may be further treated by additional filtering before being used for drinking and cooking, but most people use it directly from the tap. Some people are suspicious of tap water that runs through metal pipes and contains chlorine. In recent years there has been a growing market for high-quality bottled water for personal consumption. Although drinking water in the United States is thought to be safe, local varia-

tions in clarity, hardness (concentration of calcium and magnesium), and taste or fear of contamination by minute concentrations of pollutants has led many people to prefer bottled water, now a multi–billion dollar industry.[4] The major complaint about tap water is a chlorine taste that may occur with chlorine concentrations as low as 0.2–0.4 mg/L.

There is no doubt that water purification at treatment plants has nearly eliminated waterborne diseases, such as typhoid and cholera, which previously caused widespread suffering and death in the developed world and still do in many parts of the world. However, we need to know much more about the long-term effects of exposure to low concentrations of toxins in our drinking water. How safe is the water in the United States? It's much safer than it was 100 years ago, but low-level contamination (below what is thought dangerous) of materials such as organic chemicals and heavy metals is a concern that requires continued research and evaluation.

A recently discovered potential problem with treatment of domestic water sources is that the chemicals used to disinfect the water, such as chlorine and ozone, give rise to a variety of by-products, some of which are identified as potentially hazardous to humans and other animals. For example, a recent study involving riverine fish in Britain revealed that in some rivers, male fish sampled downstream from waste treatment plants had testes containing both eggs and sperm, a phenomenon called *intersex*. This is likely related to the concentration of sewage effluent and the treatment method used.[24] Evidence also suggests that these by-products may pose risks of cancer and other human health effects. However, the degree of risks posed by the actual concentrations of by-products in drinking water is controversial and is currently being hotly debated.[25]

20.8 LAND APPLICATION OF WASTEWATER

The innovative practice of applying wastewater to the land involves the fundamental belief that waste is simply a resource out of place. The idea is sometimes expressed as the **wastewater renovation and conservation cycle,** as shown schematically in Figure 20.15. The major steps in the cycle are the following: (1) return of treated wastewater to crops via a sprinkler or other irrigation system; (2) *renovation*, or natural purification by slow percolation of the wastewater through the soil, to eventually recharge the groundwater resource with clean water; and (3) reuse of the water by pumping it out of the ground for municipal, industrial, institutional, or agricultural purposes.

Recycling of wastewater is now being practiced at a number of sites around the United States.

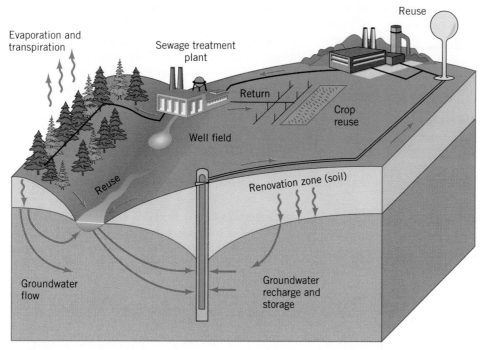

FIGURE 20.15 The wastewater renovation and conservation cycle. (Source: From R. R. Parizek, L. T. Kardos, W. E. Sopper, E. A. Myers, D. E. Davis, M. A. Farrel, J. B. Nesbitt, Pennsylvania State Studies: Waste Water Renovation and Conservation. University Studies: 23. Copyright 1967 by the Pennsylvania State University. Reproduced by permission of the Pennsylvania State University Press.

In a large-scale wastewater recycling program near Muskegon and Whitehall, Michigan, raw sewage from homes and industry is transported by sewers to the treatment plant, where it receives primary and secondary treatment. The wastes are then chlorinated and pumped into a network that transports the effluent to a series of spray irrigation rigs that apply the treated water to the land. After the wastewater trickles down through the soil, it is collected in a network of tile drains and transported to the Muskegon River for final disposal. This last step is an indirect advanced treatment using the natural physical and biological environment as a filter. Data collected to date indicate that this system removes most of the potential pollutants as well as heavy metals, color, and viruses.

Technology for wastewater treatment is rapidly evolving. An important question being asked is: Can we develop environmentally preferred, economically viable wastewater treatment plants that are fundamentally different from those in use today? One such idea is shown on Figure 20.16, illustrating a resource recovery wastewater treatment plant. The term *resource recovery* here refers to the production of resources including methane gas (which can be burned as a fuel) as well as the production of plants that have commercial value. The processes in the treatment are as follows: first, the wastewater is run through filters to remove large objects; second, the water undergoes an anaerobic processing (this

process produces the methane gas); and third, nutrient-rich water flows over an incline surface containing plants (the plants use the nutrients and further purify the water). This process is thought to clean the water to standards obtained from secondary treatment in conventional wastewater treatment plants. If further purification is necessary, then the water may be utilized by still other plants before being discharged into the environment.

Wastewater treatment plants that utilize the resource recovery concept shown in Figure 20.16 are only in the experimental stage in small pilot plants. This technology faces several problems before it is likely to be more widely used: first, there has been a tremendous investment in traditional wastewater treatment plants and engineers and other technicians are familiar with how to build and operate them. Second, economic incentives to provide for new technologies are not sufficient; third (and perhaps most important) there are not sufficient personnel trained to design and operate new types of wastewater treatment plants. This may be changing, however, because more universities are developing environmental engineering programs with a broader view of technological development.[26]

Wastewater is also being applied experimentally to freshwater marshes and forestlands. In the future we can expect more advanced treatment of wastewater through the use of biological systems (see A Closer Look 20.4, "Wastewater and Wetlands").

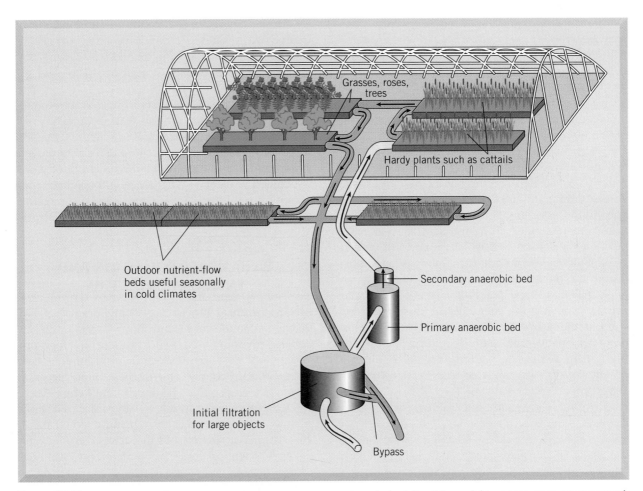

FIGURE 20.16 Components of a resource-recovery wastewater treatment plant. For this model two resources are recovered: methane, with can be burned to produce energy from the anaerobic beds; and ornamental plants that may be sold. After W. J. Jewell, 1994. Resource-recovery wastewater treatment. *American Scientist* 82:366–375.

20.9 WATER REUSE

Water reuse generally refers to the use of wastewater following some sort of treatment. It may be discussed in terms of an emergency water supply, a long-term solution to a local water shortage, or a fringe benefit to a reduction in water pollution. Data have been and are being collected in many locations in the United States as part of water-reuse research and implementation.[29]

 CLOSER LOOK 20.4

WASTEWATER AND WETLANDS

The wastewater treatment system utilized by the city of Arcata in northern California services more than 20,000 people. Wastewater comes mostly from homes, with minor inputs from lumber and plywood plants. It is treated by standard primary and secondary methods, then chlorinated and dechlorinated before being discharged into Humboldt Bay.

Oxidation ponds (constructed wetlands), part of a large wetland system in the bay, are utilized for some secondary and advanced treatment. Aquaculture experiments have also been conducted as part of the Arcata experience. Pacific salmon fingerlings were raised in water drawn from the oxidation ponds.

The experiments at Arcata certainly indicate that the use of artificial wetlands or constructed wetlands (ponds) in part of a larger natural wetland is a worthwhile endeavor in wastewater treatment. The concept of a more natural water treatment in constructed or artificial wetlands has led to the development of emergent treatment plant systems in which the treatment of sewage occurs in wetland conditions that encourage a more natural breakdown of waste.[27,28]

Water reuse can be inadvertent, indirect, or direct. *Inadvertent water reuse* results when water is withdrawn, treated, used, treated, and returned to the environment without specific plans for further withdrawals and use, which nevertheless take place. Such use patterns occur along many rivers and are accepted as a common and necessary procedure for obtaining a water supply. Several risks are associated with inadvertent reuse. Inadequate treatment facilities may deliver contaminated or poor-quality water to users. Because the fate of disease-causing viruses during and after treatment is not completely known, the environmental health hazards of treated water are uncertain. In addition, each year many new chemicals, some of which cause birth defects, genetic damage, or cancer in humans, are introduced into the environment. Harmful chemicals are often difficult to detect, and their effects on humans may be hidden if the chemicals are ingested in low concentrations over many years.[27] In spite of these problems, inadvertent reuse of water will by necessity remain a common pattern. If we recognize the potential risks, we can plan to minimize them by using proper water treatment.

Indirect water reuse is a planned endeavor. An example is the wastewater renovation and conservation cycle previously discussed and illustrated in Figure 20.15. Similar plans have been used in southern California, where several thousand cubic meters of treated wastewater per day have been applied to surface recharge areas. The treated water eventually enters into groundwater storage to be reused for agricultural and municipal purposes.

Direct water reuse refers to treated wastewater that is piped directly to the next user. In most cases the user is industry, agricultural activity, or irrigation of golf courses, institutional grounds (e.g., a university campus), or parks. Very little direct use of water is planned (except in emergencies) for human consumption because of perceived risks and negative cultural attitudes toward using treated wastewater.

20.10 WATER POLLUTION AND ENVIRONMENTAL LAW

Environmental law is a growing branch of law that is becoming more and more important as we debate environmental issues and make decisions about how best to protect our environment (see Chapter 29). Federal laws in the United States to protect water resources go back to the Refuse Act of 1899, the purpose of which was to protect navigable streams, rivers, and lakes from being polluted. Table 20.4 lists major federal laws that have a strong water resource/pollution component. State and local laws

TABLE 20.4 Federal Water Legislation

Date	Law	Comment
1899	Refuse Act	Protects navigable water from pollution.
1956	Federal Water and Pollution Control Act	Enhances the quality of water resources and prevents, controls, and abates water pollution.
1958	Fish and Wildlife Coordination Act	Water resources projects such as dams, power plants, and flood control must coordinate with U.S. Fish and Wildlife Service to enact wildlife conservation.
1969	National Environmental Policy Act	Requires environmental impact statement prior to federal actions (development) that significantly affect the quality of the environment. Included are dams and reservoirs, channelization, power plants, bridges, and so on.
1970	Water Quality Improvement Act	Expands power of 1956 act through control of oil pollution, and hazardous pollutants and research and development to eliminate pollution in Great Lakes and acid mine drainage.
1972 (amended in 1977)	Federal Water Pollution Control Act (Clean Water Act)	Purpose is to clean up nation's waters. Provides billions of dollars in federal grants for sewage treatment plants. Encourages innovative technology, including alternative water treatment methods and aquifer recharge of wastewater.
1974	Federal Safe Drinking Water Act	Aims to provide all Americans with safe drinking water. Sets contaminant levels for dangerous substances and pathogens.
1980	Comprehensive Environmental Response, Compensation, and Liability Act	Established revolving fund (Superfund) to clean up hazardous-waste disposal sites, reducing groundwater pollution.
1984	Hazardous and Solid Waste Amendments to the Resource Conservation and Recovery Act	Regulates underground gasoline storage tanks. Reduces potential for gasoline to pollute groundwater.
1987	Water Quality Act	Established national policy to control nonpoint sources of water pollution. Important in development of state management plants to control nonpoint water pollution sources.

How Can Polluted Waters Be Restored?

The Illinois River begins in the northeast region of the state and flows west and south, draining parts of Indiana and Wisconsin (see map). From Chicago's Lake Michigan, which is connected to the river by a canal, to the confluence with the Mississippi, is a distance of 526 km (327 miles). The surrounding floodplains, once a mixture of prairie and oak–hickory forest, are now primarily used for raising crops. Formerly, the river was highly productive, especially in the lower 320 km (200 miles); it produced 10% of the U.S. freshwater fish catch in 1908 (11 million kg, or 24 million lb; 200 kg/ha, or 178 lb/acre). By the 1970s, the same stretch of river produced a mere 0.32% of the total freshwater fish harvest (4.5 kg/ha or 4 lb/acre). Two major factors are responsible for the change in the productivity of the Illinois River: diversion of Chicago's sewage from Lake Michigan to the river, and agriculture. A brief history of events related to water quality in the Illinois River is given in the accompanying table.

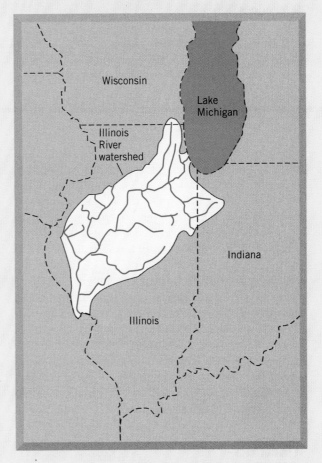

Critical Thinking Questions

1. Develop a hypothesis to explain why the fish population peaked in 1908, after the construction of the Chicago Sanitary and Ship Canal, and declined after that. Your hypothesis should also be able to explain the recovery of the fish in the 1920s and 1930s and the causes of the environmental problems of the 1940s and 1950s. Design a controlled experiment to test your hypothesis.

2. Why did water quality show some improvement by 1990, although the TARP Project was not yet completed? (Hint: See Table 20.4.)

3. The most important variables that affect the life of a river or stream are energy source (the amount of organic material entering the stream from sources outside it), water quality, habitat quality, water flow, and interactions among living things. In the case of the Illinois River what are those variables affected by human activities? For each variable, cite examples of specific activities, their environmental effects, and what could be done to improve water quality in the river even more.

Year	Critical Event	Environmental Impact
1854–1855	Heavy rains resulted in untreated sewage from Chicago entering Lake Michigan and then the city's drinking water	Cholera and typhoid epidemic in Chicago
1900	Chicago Sanitary and Ship Canal built to convey sewage away from Lake Michigan and into Illinois River	Waste entered Illinois River; commercial fish yield from river reached peak in 1908
1920		Fish populations declined in river
1920–1940	Most cities along river built sewage treatment plants	Some recovery in fish population
1940–1960	Rapid population growth in Chicago and other cities along the river; increase in agricultural acreage	Lower oxygen levels in river; further declines in fish populations; sport fish and ducks declined in backwaters and lakes of the river
1977–199?	Construction of Chicago Tunnel and Reservoir Plan (TARP) to capture and treat sewage overflows initiated	Some improvement in water quality by 1990, but no change in turbidity or total phosphorus; sodium increased

4. Why is there a conflict between managing the Illinois River for waterfowl versus managing it for fish? How could the conflict be resolved?

References

Allan, J. D., and Flecker, A. S. 1993. "Biodiversity Conservation in Running Waters," *BioScience* 43(1): 32–43.

Armour, C. L., Duff, D. A., and Elmore, W. 1991. "The Effects of Livestock Grazing on Riparian and Stream Ecosystems," *Fisheries* 16(1): 7–11.

Karr, J. R., Toth, L. A., and Dudley, D. R. 1985. "Fish Communities of Midwestern Rivers: A History of Degradation," *BioScience* 35(2): 90–95.

Sparks, R. 1992. "The Illinois River Floodplain Ecosystem," in National Research Council, *Restoration of Aquatic Ecosystems*. Washington, DC: National Academy Press, pp. 412–432.

Stevens, W. K. 1993 (January 26). "River Life through U.S. Broadly Degraded," *New York Times*, pp. B5, B8.

also regulate and control water resources and pollution at a variety of levels, from urban runoff to agricultural and industrial practices. In addition, many federal laws have been passed with the purpose of cleaning up or treating pollution problems or treating wastewater. However, there is also a focus on preventing pollutants from entering water. Prevention has the advantage of avoiding environmental damage and costly cleanup and treatment.

From a water pollution viewpoint, the mid-1990s in the United States is a time of debate and controversy. Republicans, taking control of the House of Representatives and the Senate in 1994, attempted to rewrite major environmental laws, including the Clean Water Act. The purpose was to provide industry greater flexibility in choosing how to comply with environmental regulations concerning water pollution.

Industry is in favor of proposed new regulations that in their estimation would be more cost-effective, without causing an increase in environmental degradation. Environmentalists, on the other hand, view the attempts to rewrite the Clean Water Act as a giant step backward in the nation's twenty-year fight to clean up our water resources. Apparently, the Republican majority read the public wrong on this issue. Survey after survey have established that there is strong support for a clean environment in the United States today. People are willing to pay to have clean air and clean water for this and future generations. There has been a strong backlash and criticism of steps taken to weaken environmental laws, particularly as they relate to important resources such as water. Thus the debate goes on.[30]

SUMMARY

- The primary water pollution problem, present in many locations on the globe, is lack of disease-free drinking water.

- Water pollution is degradation of quality that renders water unusable for its intended purpose.

- Major categories of water pollutants include disease-causing organisms, dead organic material, heavy metals, organic chemicals, acids, sediment, heat, and radioactivity.

- Sources of pollutants may be point sources, such as pipes that discharge into a body of water, or nonpoint sources that are diffuse, such as organic chemicals applied to crops, which enter the soil and pollute groundwater.

- Eutrophication is the natural or human-induced (artificial or cultural eutrophication) increase in the concentration of nutrients required for living things. Water with a high concentration of human-induced nutrients, such as phosphorus and nitrogen, may experience damaging ecosystem effects, including a population explosion of photosynthetic bacteria, which lowers the concentration of dissolved oxygen in the water and leads to the death of fish.

- Sediment pollution has the twofold effect of loss of soil where sediment is produced (soil erosion) and reduction of water quality when sediment enters a body of water.

- Acid mine drainage is a serious water pollution problem that results when water and oxygen react with sulfide minerals (e.g., pyrite, FeS_2), often associated with coal or metal sulfide deposits such as copper and zinc, forming sulfuric acid. Acidic water draining from mines or tailings pollutes streams and other bodies of water, damaging aquatic ecosystems and degrading water quality.

- Urban processes, such as waste disposal in landfills, application of fertilizers, and dumping chemicals, such as motor oil and paint, all may contribute to shallow-aquifer contamination. Overpumping of aquifers near the ocean may draw in salt water, contaminating a water resource.

- Wastewater treatment at conventional treatment plants includes primary, secondary, and perhaps advanced treatment. In some locations natural ecosystems, such as wetlands and soils, are being used as part of the treatment process.

- Cleanup and treatment of water pollution for both surface-water and groundwater resources is expensive and may not be completely successful. Furthermore, environmental damage may result before a pollution problem is identified and treated. Therefore, we should continue to focus our attention on preventing pollutants from entering water, which is a purpose of water quality legislation.

REEXAMINING THEMES AND ISSUES

Human Population: We state in this chapter that the number one water pollution problem in the world today is the lack of disease-free drinking water. This problem is only likely to get worse in the future as the number of people, particularly in developing countries, continues to increase. As population increases, so does the possibility of continued water pollution from a variety of sources relating to agricultural, industrial, and urban activities.

Sustainability: Any activity by people that leads to water pollution is antithetical to sustainability. Groundwater resources are fairly easy to pollute, and once degraded, these waters may remain polluted for long periods of time. Therefore, if we wish to leave a fair share of groundwater resources to future generations, we must ensure that these resources are not polluted, degraded, or made unacceptable for human use and use of other living organisms on Earth.

Global Perspective: Several aspects of water pollution have global implications. For example, some pollutants may enter the atmosphere and be transported on a global basis prior to deposition, where they may degrade water quality. Examples include radioactive fallout from nuclear reactor accidents or experimental detonation of nuclear devices. Also included are waterborne pollutants from rivers and streams that enter the ocean and circulate with marine waters around the ocean basins of the world.

Urban World: Urban areas are centers of activities that may result in serious water pollution. A broad range of chemicals and disease-causing organisms are present in large urban areas, and these may enter surface and groundwaters to pollute them. Many large cities have grown along the banks of streams and rivers, and the water quality of those streams and rivers is often degraded as a result. There are positive signs here, however; many large U.S. cities are viewing their rivers as valuable resources, the careful management of which can be the focus of environmental and economic renewal. Thus rivers through some cities are designated as greenbelts, with parks and trail systems being created along the river corridor.

Values and Knowledge: It is clear that the people of the United States are placing a high value on the environment and in particular on protection of resources such as water. Attempts to weaken water quality standards are viewed critically by the public at large. There is also considerable concern for protection of water resources necessary for the variety of ecosystems found on Earth. This has led to research and development to find new technologies to reduce, control, and treat water pollution. Examples include development of new methods of wastewater treatment and support of laws and regulations that protect water resources.

KEY TERMS

acid mine drainage *425*
advanced wastewater treatment *429*
biochemical oxygen demand
  (BOD) *417*
bioremediation *418*
ecosystem effect *421*

environmental law *434*
eutrophication *421*
fecal coliform bacteria *420*
nonpoint sources *426*
outbreak *414*
point sources *426*

primary treatment *429*
secondary treatment *429*
wastewater renovation and
  conservation cycle *431*
wastewater treatment *429*
water reuse *433*

STUDY QUESTIONS

1. Will outbreaks of waterborne diseases be more common or less common in the future? Where? Why?

2. What was learned from the *Exxon Valdez* oil spill that might help reduce the number of future spills and their environmental impact?

3. What is meant by the term *water pollution* and what are several major processes that contribute to water pollution?

4. Compare and contrast point and nonpoint sources of water pollution. Which is easier to treat and why?

5. What is the twofold effect of sediment pollution?

6. In the summer you buy a house with a septic system that appears to function properly. In the winter, effluent discharges at the surface. What could be the possible environmental cause of the problem? How could the problem be alleviated?

7. Describe the major steps in wastewater treatment (primary, secondary, advanced). Can natural ecosystems do any of these functions? Which ones?

8. In a city along an ocean coast, rare water birds inhabit a pond that is part of a sewage treatment plant. How could this have happened? Is the water in the sewage pond polluted? Consider this question from the birds' and from your point of view.

9. How does water that drains from coal mines become contaminated with sulfuric acid? Why is this an important environmental problem?

10. What is *eutrophication* and why is it an ecosystem effect?

11. How safe do you believe the drinking water is in your home? How did you reach your conclusion? Are you worried about low-level contamination by toxins in your water? What might sources of contamination be?

FURTHER READING

Borner, H. (ed.). 1994. Pesticides in Ground and Surface Water. Vol. 9 of *Chemistry of Plant Protection*. New York: Springer-Verlag. Essays on the fate and effects of pesticides in surface and groundwater, including methods to minimize water pollution from pesticides.

Dinar, A. and Loehman, E. T. (eds.). 1995. *Water Quantity/Quality Management and Conflict Resolution*. Praeger. Includes essays and case studies on water pollution problems and resolutions.

Dunne, T., and Leopold, L. B. 1978. *Water and Environmental Planning*. San Francisco: W. H. Freeman. A great summary and detailed examination of water resources and problems.

Hester, R. E., Harrison, R. M. (eds.) 1996. *Agricultural Chemicals and the Environment*. Cambridge: Royal Society of Chemistry, Information Services. A good source for information about the impact of agriculture on the environment, including eutrophication and the impact of chemicals on water quality.

Manahan, S. E. 1991. *Environmental Chemistry*. Chelsea, MI: Lewis. A detailed primer on the chemical processes pertinent to a broad array of environmental problems, including water pollution and treatment.

Newman, M. C. 1995. *Quantitative Methods in Aquatic Ecotoxicology*. Chelsea, MI: Lewis. Up-to-date text on fate, effects, and measurement of pollutants in aquatic ecosystems.

Nichols, C. 1989. "Trouble at the Waterworks," *The Progressive* 53: 33–35. A concise report on the problem of tainted water supplies in the United States.

Rao, S. S. (ed.). 1993. *Particulate Matter and Aquatic Contaminants*. Chelsea, MI: Lewis. Coverage of the biological, microbiological, and ecotoxicological principles associated with interaction between suspended particulate matter and contaminants in aquatic environments.

INTERNET RESOURCES

National Water Resources Association: *http://www.nwra.org*—Provides information about water legislation, as well as water links and publications.

U.S. Environmental Protection Agency Office of Water: *http://www.epa.gov/OW/*—This site contains considerable information including programs and activities for protecting our drinking water, groundwater, and other resources; water quality; environmental indicators of water quality; and the Clean Water Act and congressional efforts to change it.

U.S. Geological Survey Water Resources Division: *http://h2o.usgs.gov/*—Contains plentiful information on federal water resources programs, water use, water quality assessments for the U.S. and more.

PART VII

Six Birds, c. 1969, Kingmeata

Air Environment

Clouds show the dynamic nature of atmosphere, especially to the south, west, and east of the southern tip (Cape of Good Hope) of Africa in this image.

CHAPTER 21

THE ATMOSPHERE, CLIMATE, AND GLOBAL WARMING*

CASE STUDY

El Niño, 1997–1998

"El Niño" became a household word during the winter of 1997–1998 when it was blamed for everything from tornadoes and thunderstorms in Florida to catastrophic fires in Indonesia (see Chapter 22). The term El Niño refers to the child, in particular the Christ Child, because the event often begins off the coast of South America near Christmas time. El Niño events cause a disruption of the ocean-atmosphere system in the tropical Pacific and are, in part, responsible for weather phenomena that can cause billions of dollars in property damage and the loss of thousands of human lives. Occuring at intervals of two to seven years, El Niño events typically last for 12 to 18 months. They start with the weakening of east to west trade winds and warming of eastern Pacific Ocean waters. This results in tropical rainfall shifting from Indonesia to South America as shown in Figure 21.1. During non-El Niño conditions, trade winds blow west across the tropical Pacific and the warm surface water in the western Pacific tends to pile up so that the sea surface can be as much as

0.5 m higher at Indonesia than at Peru (Figure 21.1). In contrast, during El Niño the trade winds weaken and may even reverse. As a result, the eastern equatorial Pacific Ocean becomes unusually warm and the westward moving equatorial ocean current weakens or reverses. The rise in temperature of sea surface waters off the South American coast has significant ecological consequences. Warm surface waters inhibit the upwelling of nutrient-rich cold water from deeper levels which normally support a diverse marine ecosystem and major fisheries (see A Closer Look 13.2, "The Peruvian Anchovy Fishery"). Because rainfall follows warm water eastward during El Niño years, there are high rates of precipitation and flooding in Peru while droughts and fires are commonly observed in Australia and Indonesia. Because warm ocean water provides an atmospheric heat source, El Niño changes global atmospheric circulation which causes changes in weather in regions that are far removed from the tropical Pacific.

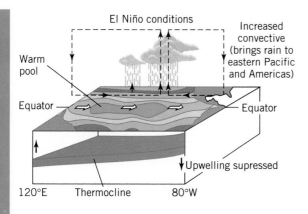
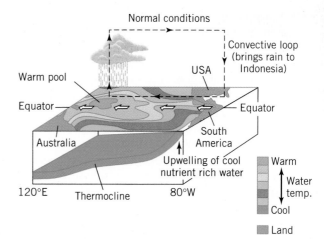

FIGURE 21.1 Idealized diagram comparing selected normal and El Niño conditions. (*Source:* NOAA.)

It is important to remember that El Niño events are natural phenomena and part of the dynamic system involving the coupling of Earth's atmosphere and ocean. El Niño events temporarily alter the weather that humans are accustomed to, sometimes with devastating effects. El Niño events can alternate with La Niña events. In marked contrast to the warm waters of El Niño, La Niña events are characterized by unusually cool ocean water temperatures and the effects are just the opposite. For example, during La Niña, winter (for example, 1998–1999) temperatures are cooler than normal in the northwest. La Nina events do not necessarily follow every El Niño event, but together they constitute a natural cycle of ocean-atmospheric disruption with global consequences.[1]

Our discussion of El Niño suggests that global change in the ocean-atmospheric system can be rapid and we should strive to better understand the processes responsible for the maintenance of important atmospheric, hydrologic, and lithospheric systems. This chapter addresses the global issues related to weather, climate change, and global warming with an emphasis on the role of humans on climate change.

LEARNING OBJECTIVES

Earth's atmosphere is a dynamic system that is changing continuously while undergoing complex physical and chemical processes. After reading this chapter, you should understand:

- How urban areas affect local climate and produce an urban dust dome or heat island.
- How the climate has changed during the last million years.
- Why there is controversy concerning whether we are now in or entering a period of human-induced global warming.
- How human activity has resulted in increased emissions of greenhouse gases.
- How positive- and negative-feedback cycles in the atmosphere might affect global temperature change.
- What the potential effects of and adjustments to global warming are.

21.1 THE ATMOSPHERE

The **atmosphere,** the thin layer of gases that envelops Earth, is a great resource to all living things on the planet. It is composed of gas molecules held close to Earth's surface by a balance between gravitation and thermal movement of air molecules. Major gases in the atmosphere are nitrogen (78%), oxygen

(21%), argon (0.9%), and carbon dioxide (0.03%). The atmosphere also contains trace amounts of numerous elements and compounds, including methane, ozone, hydrogen sulfide, carbon monoxide, oxides of nitrogen and sulfur, hydrocarbons, chlorofluorocarbons (CFCs), and various particulates or aerosols (small particles). Water vapor is also present in the lower few kilometers of atmosphere.

The atmosphere is a dynamic system, changing continuously. Physical movement of air masses, each with a different temperature, pressure, moisture, and aerosol content, produces weather and climate. A vast chemical-reacting system, the atmosphere is fueled by sunlight, high-energy compounds emitted by living things, and human industrial and agricultural activities. Many complex chemical reactions take place in the atmosphere; changing from day to night and with chemical elements available.

Two important measurable quantities are pressure and temperature. *Pressure* is force per unit area. Atmospheric pressure is the weight of overlying atmosphere (air) per unit area; it decreases as altitude increases because there is less weight from

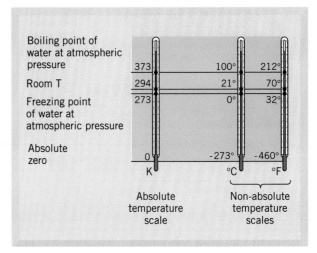

Figure 21.2 Temperature scales. °C = Celsius; °F = Fahrenheit; K = kelvin. °C = 5/9 (°F −32); °C = K −273.

overlying air. At sea level atmospheric pressure is 10^5 N/m² (newtons per square meter), which is equivalent to 14.7 lb/in.² *Temperature* refers to relative hotness or coldness of materials, such as air, water, soil, and living organisms. In a quantitative

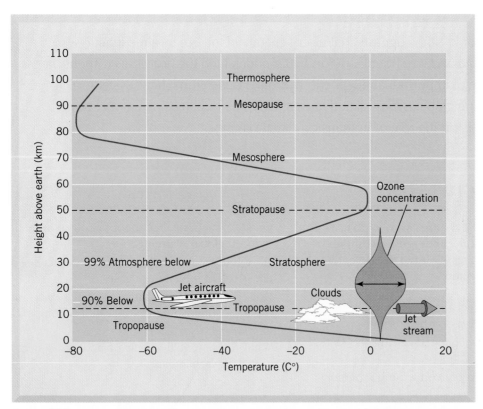

Figure 21.3a The structure of the atmosphere. Temerature profile and ozone layer of the atmosphere to an altitude of 110 km. Note 99% of the atmosphere by weight is below 30 km, the ozone layer is its thickest at about 25 km with a concentration of about 200 ppm, and the weather occurs below about 11 km, about the elevation of the jet stream. (*Source:* Duxbury, A. C. and Duxbury, A. B. *An Introduction to the World's Oceans,* ©1997. Wm. C. Brown Publishers, 5th ed.).

sense, temperature is a measure of the thermal energy (heat) of a substance and is measured with a thermometer. Figure 21.2 shows the three common temperature scales. This book uses the Kelvin (K) and Celsius (°C) scales.

In the lower atmosphere water vapor varies from approximately 1% to 4% by volume. The amount of water vapor present in the atmosphere at a particular location is dependent on many factors, including air temperature, pressure and availability of water vapor from processes such as evaporation from water bodies and soil and transpiration from vegetation (loss of water from plants to the air). When air holds the maximum amount of water that it can, given its particular temperature, it is said to be *saturated,* which means that no more water vapor may be added to the air. The term *relative humidity* is a measure of how close the air is to saturation. For example, a relative humidity of 100% means that the air is completely saturated; a relative humidity of only 5% means that if additional water vapor were available the air could hold more vapor. At high relative humidity people and other animals may feel uncomfortable and perceive the air as sticky. In desert regions with low relative humidity

the air is dry; even with a warm temperature the air to many people feels more comfortable than air with a lower temperature but a higher relative humidity.[2]

Winds, the movement of clouds, and transitions from stormy to clear skies show us that the atmosphere changes rapidly and continuously. This atmospheric circulation takes place on a variety of scales. On a local or regional scale, the lower atmosphere is heated by Earth's surface. Heat energy is also transferred to the air from the land and from bodies of water by the evaporation of water. As warm air rises, it cools by expansion; winds are produced as cooler surface air is drawn in to replace rising warm air. The lower atmosphere is therefore said to be unstable because it tends to circulate and mix, particularly in the lowest 4 km (2.5 miles) or so.

An idealized diagram showing the structure of the atmosphere is shown in Figure 21.3a. The relationship between altitude and air temperature for the lower atmosphere is shown in Figure 21.3a. The lower part of the atmosphere (lower 10–12 km, or about 6–8 miles) is known as the *troposphere,* and it is here that weather occurs. In the troposphere the temperature of the atmosphere decreases systematically with elevation at a global average of approxi-

FIGURE 21.3b Generalized circulation of the atmosphere. (*Source:* From Williamson, *Fundamentals of Air Pollution,* © 1973. Figure 5.5. Reprinted with permission of Addison-Wesley, Reading, Mass.)

mately 6.5°C/km (7.3°F/miles). At the top of the troposphere, or *tropopause* (about 10–12 km above sea level), a cold trap causes condensation of most of the water vapor. For this reason there is very little water vapor in the stratosphere (about 1% relative humidity). Condensation of water in the troposphere produces clouds. The role of clouds, including how they develop and move, is becoming an important area of research to better understand the global processes that operate in the atmosphere. Figure 21.3a also shows the *stratospheric ozone layer,* which extends from the tropopause to an elevation of approximately 50 km (30 miles), with a maximum concentration of ozone at about 20 to 25 km (12–16 miles). Stratospheric ozone (O_3) protects life in the lower atmosphere from receiving harmful doses of ultraviolet radiation and is discussed in detail in Chapter 24.

On a global scale, atmospheric circulation results primarily from Earth's rotation and the differential heating of Earth's surface and atmosphere. These processes produce global patterns that include prevailing winds and latitudinal belts of low or high air pressure from the equator to the poles (Figure 21.3b). In general, belts of low pressure develop at the equator and at 50° to 60° north and south latitude as a result of rising columns of air; belts of high pressure resulting from descending air develop at 25° to 30° north and south latitude. These belts have names such as doldrums, areas at the equator with little air movement; trade winds, northeast and southeast winds important in the early days of international trade when clipper ships moved the world's goods; and horse latitudes, two belts centered at about 30° north and south of the equator. Earth's major deserts occur in the horse latitudes as a result of pervasive high pressure sandwiched between the equatorial and midlatitudinal lows.

Except for molecular argon and other noble or inert gases, all the compounds in Earth's atmosphere either are produced primarily by biological activity or are greatly affected by it. Although the atmosphere has been greatly modulated by life during the last 3.5 billion years, most of these alterations have been natural; that is, they have produced an atmosphere whose makeup is relatively constant and essential to our own survival.

Ever since life began on Earth, the atmosphere has been an important resource for chemical elements and a medium for the deposition of wastes. The earliest plants that carried out photosynthesis released oxygen (the element that was their waste) into the atmosphere. The long-term increase in atmospheric oxygen, in turn, made possible the development and survival of higher life-forms that required high rates of metabolism and rapid use of

energy. For our biological ancestors and for ourselves, oxygen became a necessary resource for respiration, the process by which we burn our internal biological fuels and provide the energy to sustain our life processes.

21.2 CLIMATE

Climate refers to the representative or characteristic atmospheric conditions (what we call weather) at a place or places on Earth. Whereas climate refers to long time periods, such as seasons, years, or decades, *weather conditions* refer to short periods of time, such as days or weeks. Because the climate of a particular location may depend on extreme or infrequent conditions, it is more than just the average temperature and precipitation. The simplest classification of climate is by latitude—tropical, subtropical, midlatitudinal (continental), subarctic (continental), and arctic—but several other categories are necessary, including humid continental, Mediterranean, monsoon, desert, and tropical wet–dry, among others (Figure 21.4). Both precipitation and temperature show tremendous variability on a global scale. Although detailed discussion of climatic types is beyond the scope of this book, it is important to recognize the significance of potential climatic variability in determining what kinds of organisms live where. Recall from the discussion of biogeography in Chapter 7 that similar climates produce similar kinds of ecosystems. This concept is important and useful for environmental science. Knowing the climate, we can predict a great deal about what kinds of life we will find in an area and what kinds could survive there if introduced.

On a regional scale, air masses that cross oceans and continents may have a profound influence on seasonal patterns of precipitation and temperature. On a local scale, climatic conditions can also vary considerably and produce a local effect referred to as a **microclimate.** Microclimate may vary even from one side of a small rock to another or from one side of a tree to another. Organisms often take advantage of these different conditions. Furthermore, urban areas produce a characteristic microclimate with important environmental consequences. The very presence of a city affects the local climate, and as the city changes, so does its climate. For example, in the middle of the eighteenth century, according to Peter Kam, then traveling in North America, (Manhattan Island) was "generally reckoned very healthy," perhaps because of its nearness to the ocean and its relatively unobstructed ocean breezes. Today, air pollution and the effects of tall buildings on air flow lead the average visitor to

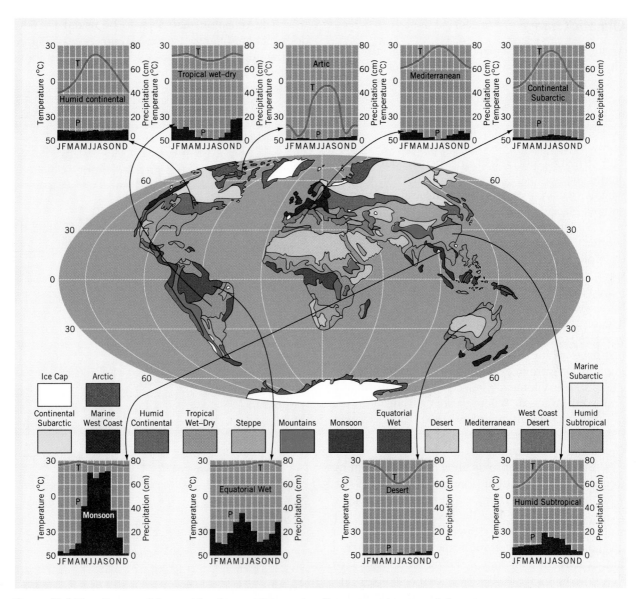

FIGURE 21.4 The climates of the world and some of the major climate types in terms of characteristic precipitation and temperature conditions. (*Source:* Modified from W. M. Marsh and J. Dozer, *Landscape,* © 1981, John Wiley & Sons, reprinted with permission of John Wiley & Sons, Inc.)

Manhattan to a quite different conclusion. Although air quality in urban areas is in part a function of the amount of pollutants present or produced, it is affected also by the city's ability to ventilate and flush out pollutants. The amount of ventilation depends on several aspects of the urban microclimate. Compared to surrounding rural areas, cities in midlatitude regions tend to: (1) be warmer (0.5–2°C, due to heat sources from burning fuel for transportation, industry, and homes); (2) be less humid (6–10%, due to less evaporation of water from standing water and soil); (3) have up to ten times as many dust particles (due to pollution, forming the **urban dust dome**); and (4) have 30–100% more fog and 5–10% precipitation (due to the presence of small particles in city air that upon which water may condense).[3,4]

The abundance of particulates above a city reduces incoming solar radiation by as much as 30% and thus cools the city, but the cooling effects of particles are relatively small compared to the processes that heat the city.[4] As a result, the combination of lingering air, abundance of particles, and other pollutants produces a heat island effect (see Chapter 26) that initiaters a general circulation pattern of air; moving cooler, cleaner air from rural or suburban areas toward the inner city where heat rises and then flows out to the sides completing the pattern.

Climatic Change

Another important aspect of climate is **climatic change.** The mean annual temperature of the Earth has swung up and down by several degrees Celsius over the past million years (Figure 21.5). Times of high temperature reflect relatively ice-free periods (interglacial periods) over much of the planet; times of low temperature reflect the glacial events (Figure 21.5a,b). It is not yet clear whether our current warm climate marks the end of the ice ages or whether we are merely in an interglacial period with another glacial age due.

Global climate also changes over time scales shorter than that of glacial–interglacial periods. For example, continental glaciation ended about 12,500 years ago with very rapid warming, perhaps over a period as short as a few decades.[5] This was followed by a short global cooling about 11,500 years ago [Figure 21.5(c)]. Climatic change over the last 18,000 years reflects several warming and cooling trends that have greatly affected people. For example, during a major warming trend from 800 to 1300 (Figure 21.5c,d), the Vikings colonized Iceland, Greenland, and North America. When glaciers advanced during the cold period starting around 1400 (Little Ice Age), the Viking settlements in North America and parts of Greenland were abandoned.

Starting in approximately 1850 a warming trend became apparent, lasting until the 1940s, when temperatures began again to cool. Figure 21.5e shows change over the last 130 years. At this scale the 1940s event is clearer and is followed by a leveling off of temperature in the 1950s and then a further drop in temperature during the 1960s. Since that time temperature has increased steadily to approximately the mid 1990s. The line that shows the temperatures for the last 100 years is a running average of the annual temperature (a running average is a statistical technique to help smooth out data that tend to be quite variable). What is evident from the record of the last 100 years is that global mean annual temperature has increased by approximately $0.5 \pm 1.2°C$. Furthermore, the period from 1986 through 1998 have been the warmest in the 138 years that global temperatures have been monitored. This period includes the warmest years of the twentieth century.[6,7]

FIGURE 21.5 Change in Earth's temperature over varying time periods during the past 1 million years (*Sources:* UCAR/DIES, 1991. "Science Capsule, Changes in the Temperature of the Earth." *Earth Quest,* Spring vol. 5, no. 1; Houghton, J. T., Jenkins, G. L. and Ephranns, J. J. (eds.) 1996. *Climate Change, the Science of Climate Change.* Cambridge University Press, Cambridge, U.K.)

The question that begs to be asked is: Why does climate change occur? Examination of Figure 21.5(*a*) suggests there are cycles of change about

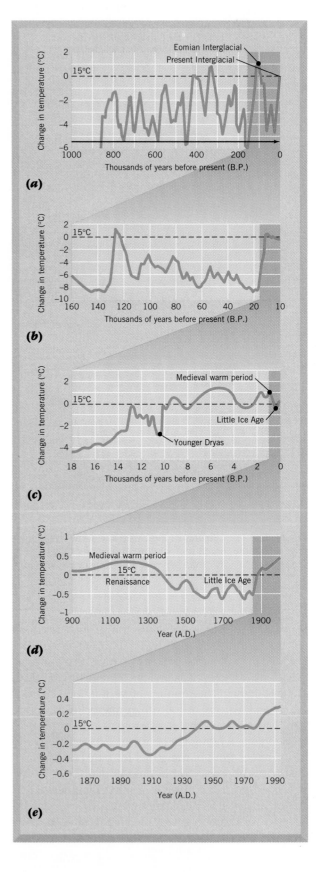

100,000 years long separated by shorter cycles of 20,000 to 40,000 years in duration. These cycles have not gone unnoticed and were first identified by Milutin Milankovitch in the 1920s as a hypothesis to explain climate change. Milankovitch realized that the spinning Earth is like a wobbling top unable to keep a constant position in relationship to the Sun, and that position determines (in part) the amount of sunlight reaching and warming Earth. He discovered that variations in Earth's orbit around the Sun follow an approximate 100,000 year cycle which correlates with the major glacial and inner-glacial of Figure 21.5(*a*) (unfortunately the magnitude of the changes of the amount of sunlight change resulting from these cycles is not nearly enough to produce the observed climatic effect). Cycles of approximately 40,000 and 20,000 years are the result of changes in the tilt of the Earth's axis and wobble of the Earth's axis, respectively. While Milankovitch cycles faithfully reproduce, most of the long-term cycles we see in the climate only have a small effect on the amount of sunlight reaching Earth and are sufficient, by themselves, to produce the large-scale climatic variations in the geologic record. Therefore, these cycles, along with other processes, must be invoked to explain global climatic change. Thus, the Milankovitch cycles can be looked at as natural forcing mechanisms that, along with other processes, may produce climatic change. Shorter cycles have also been suggested, and in fact one study suggests that during the past 4,000 years, there has been an approximately 1,500 year cycle, perhaps explaining the medieval warming period in the Little Ice Age as well as the present warming trend which is predicted to continue naturally until approximately 2400 A.D.[8] If this is correct, then any warming caused by human activity would be superimposed on a system that is already slowly warming. However, the human component of warming may be much greater than that warming not associated with human activity. There is considerable debate about whether we are entering a period of increased global warming or whether the recent trends are part of normal climatic cycles. Because of the complexities of climate and climatic change, we are unable to answer this question at present. What we do think is that our climate system may be inherently unstable and capable of changing quickly from one state (cold) to another (hot) in as short as a few decades.[9] Part of what may drive the climate system and its potential to change is the "ocean conveyor belt"—a global scale circulation of ocean waters characterized by strong northward movement of upper waters in the Atlantic Ocean that are approximately 12–13°C when they arrive near Greenland where they are cooled to 2–4°C (Figure 21.6).[9] As

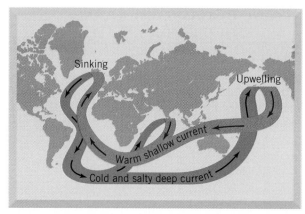

FIGURE 21.6 Idealized diagram of the oceanic conveyor belt. The actual system is more complex, but, in general, warm surface water (red) is transported eastward and northward (increasing in salinity due to evaporation) to near Greenland where it cools from contact with cold Canadian air. As the water increases in density, it sinks to the bottom and flows south, then east to the Pacific where upwelling occurs. The mass of sinking and upwelling waters balance, and the total flow rate is about 20 million m^3/sec. The heat released to the atmosphere from the warm water keeps northern Europe 5°C to 10°C warmer than if the oceanic conveyor belt were not present. (*Source*: Modified after Broker, W., 1997. "Will Our Ride Into The Greenhouse Future Be A Smooth One?," *Geology Today*. vol. 7 no. 5, pp. 2–6.)

the water is cooled it becomes more salty, increasing in density and causing it to sink to the bottom. The current then flows southward around Africa adjoining the global pattern of ocean currents. The flow in this conveyor belt current is huge—about equal to 100 Amazon rivers.[9] The amount of warm water and heat released is sufficient to keep northern Europe 5–10°C warmer than it would be if the conveyor belt were not present.[9] If the conveyor belt were to shut down, global cooling could result and northern Europe would become much less habitable. If this were to happen in the future when there are a few more billion people to feed, global catastrophe might result![9]

A recent report from an international panel of climate experts concluded that, based upon the balance of scientific evidence, there is a discernible human influence on global climate. However, panel members could not unequivocally state that human-induced global warming is in fact occurring.[10] Nevertheless, serious consideration is being given to the possibility that global warming is in fact occurring as a result of increased emissions of gases that tend to trap heat in the atmosphere or that global cooling is occurring as a result of increased particulate emissions from burning coal that reflect incoming solar radiation back into space.

21.3 GLOBAL WARMING: THE GREENHOUSE EFFECT

Global warming is defined as a natural or human-induced increase in the average global temperature of the atmosphere near Earth's surface. The temperature at or near the surface of Earth is determined by four main factors[11]:

- the amount of sunlight Earth receives;
- the amount of sunlight Earth reflects;
- retention of heat by the atmosphere; and
- evaporation and condensation of water vapor.

Sunlight that reaches Earth warms both the atmosphere and the surface. Earth's atmospheric system then reradiates heat as infrared radiation.[11] Water vapor and several other gases, including carbon dioxide, methane, and chlorofluoro carbons (CFCs), warm Earth's atmosphere because they absorb and reemit radiation. They trap some of the heat energy radiating from Earth's atmospheric system. The trapping or warming is somewhat analogous to a greenhouse, which also traps heat; thus the process has been called the **greenhouse effect.** Actually, trapping heat in the atmosphere might better be called the atmospheric effect, because the dominant process responsible for heating the air in a greenhouse is quite different from that which heats the lower atmosphere.

Although some infrared radiation is trapped in a greenhouse, the dominant process responsible for warming the air is the restriction of cooling by air circulation (wind) because of the glass enclosure. Nevertheless, *greenhouse effect* has become the accepted term for the trapping of heat by the atmosphere. It is important to understand that the effect is in fact a natural phenomenon that has been occurring for millions of years on Earth as well as on other planets in our solar system. The majority of natural greenhouse warming is due to water in the atmosphere. On a global level, water vapor and small particles of water in the atmosphere produce about 85% and 12%, respectively, of our total greenhouse warming. We are not particularly worried about water vapor in the future because it is not significantly increasing in the atmosphere as a result of human-induced processes. The gases we are concerned with are those that result in part from *anthropogenic* processes, that is, those that result from human activities. These include carbon dioxide, CFCs, methane, nitrous oxides, and ozone, all of which have increased significantly in the atmosphere in recent years. The CFCs are a group of inert, stable, human-made chemicals that are used as propellants in spray cans (deodorants, paints, etc.)

FIGURE 21.7 Idealized diagram showing greenhouse effect. Incoming visible solar radiation is absorbed by the Earth's surface to be reemitted in the infrared region of the electromagnetic spectrum. Most of this reemitted infrared radiation is absorbed by the atmosphere maintaining the greenhouse effect. (*Source:* Developed by M. S. Manalis and E. A. Keller, 1990.)

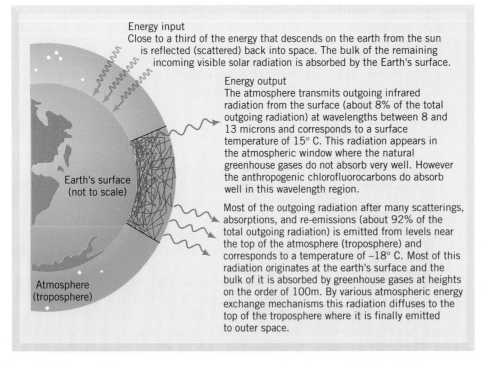

Energy input
Close to a third of the energy that descends on the earth from the sun is reflected (scattered) back into space. The bulk of the remaining incoming visible solar radiation is absorbed by the Earth's surface.

Energy output
The atmosphere transmits outgoing infrared radiation from the surface (about 8% of the total outgoing radiation) at wavelengths between 8 and 13 microns and corresponds to a surface temperature of 15° C. This radiation appears in the atmospheric window where the natural greenhouse gases do not absorb very well. However the anthropogenic chlorofluorocarbons do absorb well in this wavelength region.

Most of the outgoing radiation after many scatterings, absorptions, and re-emissions (about 92% of the total outgoing radiation) is emitted from levels near the top of the atmosphere (troposphere) and corresponds to a temperature of −18° C. Most of this radiation originates at the earth's surface and the bulk of it is absorbed by greenhouse gases at heights on the order of 100m. By various atmospheric energy exchange mechanisms this radiation diffuses to the top of the troposphere where it is finally emitted to outer space.

Earth's surface
(not to scale)

Atmosphere
(troposphere)

and as the working fluid in appliances such as air conditioners and refrigerators. Use of CFCs as propellants for spray cans was banned in the United States in 1978 (CFCs are discussed in detail in Chapter 24, which addresses stratospheric ozone depletion).

Because carbon dioxide, CFCs, methane, nitrous oxides, and ozone absorb infrared radiation emitted from Earth, it has been hypothesized that Earth may be warming because of the increases of these anthropogenic greenhouse gases. Approximately 200 billion metric tons of carbon in the form of carbon dioxide enters and leaves Earth's atmosphere each year as a result of a number of biological and physical processes. What we are concerned about in the following discussion is the anthropogenic greenhouse effect as it relates to two factors:

- the burning of fossil fuels, which adds about 5.4 billion metric tons of carbon each year to the atmosphere, and deforestation, which adds another 1.6 billion metric tons per year, increasing the concentration of atmospheric CO_2; and

- human activities that emit other greenhouse gases, such as CFCs, ozone, methane, and nitrous oxides.

A highly idealized diagram showing some of the important aspects of the greenhouse effect is presented in Figure 21.7. The three arrows labeled "energy input" represent the energy from the sun absorbed at or near the surface of Earth. The four arrows labeled energy output represent energy emitted from the upper atmosphere and the surface of the earth, which balances the input consistent with Earth's energy balance. The highly contorted lines near the surface of Earth represent the absorption of infrared radiation occurring there and providing the 15°C (59°F) near-surface temperature. Following many scatterings and absorptions and reemissions, the infrared radiation emitted from levels near the top of the atmosphere (troposphere) corresponds to a temperature of approximately –18°C (0°F). The one output arrow that goes directly through Earth's atmosphere and is emitted represents the amount of outgoing radiation through what is called the *atmospheric window,* that is, wavelengths where natural greenhouse gases (water vapor and carbon dioxide) do not absorb very well (Figure 21.8). However, anthropogenic CFCs do absorb in this region. In other words, the atmospheric window (8–12 μm) is centered on a wavelength of 10 μm and denotes the region where outgoing radiation from Earth is not absorbed by water vapor or carbon dioxide but is

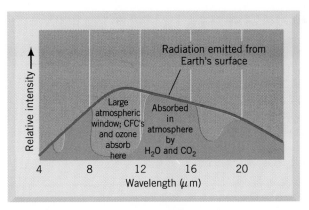

FIGURE 21.8 Absorption spectra of greenhouse gases, water vapor, and carbon dioxide in the long-wave infrared radiation region. The atmospheric window is the region where neither water vapor nor carbon dioxide absorb, but where CFCs do absorb. (*Source:* Modified after T. G. Spiro, and W. M. Stigliani, 1980, *Environmental Science in Perspective,* State Univ. of New York Press, Albany.)

absorbed by CFCs. Therefore, CFCs significantly contribute to the greenhouse effect.

Changes in Greenhouse Gases

The major anthropogenic greenhouse gases are listed in Table 21.1. The table also lists the recent growth rate in percent per year and the relative contribution in percent to the anthropogenic greenhouse effect.

Carbon Dioxide

Carbon dioxide has received a lot of attention with respect to global warming; 50% to 60% of the anthropogenic greenhouse effect is attributed to this gas. In order to evaluate recent changes in atmospheric carbon dioxide, we need to have a broader perspective of Earth's history. Ancient air may be sampled in glacial ice, which contains air bubbles, small samples of the atmosphere when the glacial ice formed. Measurements of carbon dioxide trapped in such air bubbles in the Antarctic ice sheet suggest that during the 160,000 years prior to the industrial revolution the atmospheric concentration of carbon dioxide varied from approximately 200 to 300 ppm.[12] The highest level or concentrations of carbon dioxide in the atmosphere other than at present occurred during the major interglacial period about 125,000 years ago.

About 130 years ago, at the beginning of the industrial revolution, the atmospheric concentration of carbon dioxide was approximately 280 ppm, a level apparently constant for the previous 700 years.[11] Beginning in about 1860, the concentration

TABLE 21.1 Relative Contribution of Trace Gases to the Anthropogenic Greenhouse Effect

Trace Gases	Relative Contribution (%)	Growth Rate (%/yr)
CFC	15[a]–25[b]	5
CH$_4$	12[a]–20[b]	1[c]
O$_3$ (troposphere)	8[d]	0.5
N$_2$O	5[d]	0.2
Total	40–50	
Contribution of CO$_2$	50–60	0.3[e]–0.5[d,f]

[a]W. A. Nierenberg, August 1989, "Atmospheric CO$_2$: Causes, Effects, and Options," *Chemical Engineering Progress* 85(8), p. 27.

[b]J. Hansen, A. Lacis, and M. Prather, November 20, 1989, "Greenhouse Effect of Chlorofluorocarbons and Other Trace Gases," *Journal of Geophysical Research* 94(D13), pp. 16, 417.

[c]Growth stopped in 1991–1992 possibly due to control of leaks in Russian natural gas production systems.

[d]H. Rodhe, 1990, "A Comparison of the Contribution of Various Gases to the Greenhouse Effect," *Science* 248, p. 1218, Table 2.

[e]W. W. Kellogg, "Economic and Political Implications of Climate Change," Presented at the Conference on Technology-based Confidence Building: Energy and Environment, University of California, Los Alamos National Laboratory, July 9–14, 1989.

[f]P. H. Abelson, March 30, 1990, "Uncertainties about Global Warming," *Science* 247(4950), p. 1529.

of carbon dioxide in the atmosphere has grown exponentially. Currently, the rate of increase of carbon dioxide in the atmosphere is about 0.5% per year; if it continues to grow at this rate, we will see a doubling of the concentration in approximately 140 years (or before the end of the 22nd century, recall the rule of thumb in Chapter 3). Data prior to the mid-twentieth century are from measurements of air trapped in glacial ice. The remaining data are from direct measurement from the monitoring station at Mauna Loa, Hawaii (see A Closer Look 21.1, "Monitoring of Atmospheric Carbon Dioxide Concentrations").

Today the concentration of carbon dioxide in the atmosphere is approaching 370 ppm, and it is predicted that the level may rise to approximately 450 ppm by the year 2050, more than 1.5 times the preindustrial level.[13] It is interesting to note that the rate of increase of carbon emissions (not carbon dioxide) from the burning of fossil fuels, deforestation, and other anthropogenic processes has been approximately 4.3% per year since the industrial revolution began, more than 8 times the 0.5% per year rate of increase in the concentration of carbon dioxide in the atmosphere.

The high rate of carbon dioxide emissions and the high growth rate of the emissions would seem to suggest that the increase in carbon dioxide in the atmosphere is a direct result of the anthropogenic input of carbon dioxide from sources such as burning fossil fuels and deforestation. Establishing this seemingly simple relationship as a fact has been difficult, however, because the global carbon cycle is complex; all the linkages and flows of carbon from the various sources to sinks are not yet well understood. What is apparent is that if all the carbon dioxide produced by human activities remained in the atmosphere, the concentration of that gas should be even higher than it is today! Therefore, we must hypothesize that there are sinks for carbon dioxide in the oceans or on the land that are not identified or well understood (see chapter 4). In spite of all these cautions and shortcomings, it is clear that carbon dioxide concentrations in the atmosphere have increased significantly since the industrial revolution. Furthermore, it is a reasonable hypothesis (something we assume without proof) that these increases will continue to contribute to global warming via the greenhouse effect.

Accepting that approximately 50% to 60% of the anthropogenic greenhouse effect is due to carbon dioxide, we can conclude that the remaining greenhouse gases must account for approximately 40% to 50% of the effect (see Table 21.1).

Methane

Until 1991, methane (CH$_4$), was increasing in the atmosphere at a rate of approximately 1% per year, and it is thought to contribute approximately 12% to 20% of the anthropogenic greenhouse effect

CLOSER LOOK 21.1

MONITORING OF ATMOSPHERIC CARBON DIOXIDE CONCENTRATIONS*

Human activity and other life affect characteristics of Earth's surface, waters, and atmosphere, even in areas removed from human activity and living things. For example, air pollutants and other artifacts of our society are found in glacial ice, and pesticides are found in lakes far from agriculture. Air samples taken near the summit of Mauna Loa, Hawaii (one of the largest active volcanoes and the highest mountain in the world based on elevation change from base to top), have helped show another dimension of how life and human activity affect the atmosphere. Samples are taken at Mauna Loa because it is far from direct effects of human life and other biological activity. Because carbon dioxide is taken up by green plants during photosynthesis and released in respiration of all oxygen-breathing organisms, a measure of carbon dioxide in the atmosphere is analogous to a measure of breathing in and out of all life on Earth.

The Mauna Loa measurements of carbon dioxide are truly remarkable observations. Two important aspects of the record are shown in Figure 21.9: a strong upward trend over the years and an annual cycle, which is obvious and regular. Carbon dioxide concentration reaches a peak in winter and a trough in summer. The annual curve is indeed a measure of life activities of the entire Northern Hemisphere. In summer, green plants are most active and total amount of photosynthesis exceeds total amount of respiration. As a result, carbon dioxide is removed from the atmosphere in summer. In winter, photosynthesis decreases and becomes less than total respiration, so the carbon dioxide concentration of the atmosphere increases.

Similar observations have been made in Antarctica. Because Antarctica is less accessible, measurements are more sporadic; nevertheless, the same trends are observed: a strong upward trend in concentration of carbon dioxide in the atmosphere and an annual cycle. The annual cycle from Antarctic is smaller in amplitude than the Mauna Loa cycle because of relatively smaller land area in the Southern Hemisphere and smaller amount of woody vegetation that stores carbon and exchanges carbon dioxide with the atmosphere.

The upward trend (increase) in carbon dioxide concentration of the atmosphere from both Mauna Loa and Antarctica is believed to be due to addition of carbon dioxide from burning fossil fuels and other human activities, such as cutting forests and burning wood (burning releases the stored carbon in wood, which combines with oxygen-producing carbon dioxide). The curves were some of the first data to show directly that life touches the entire Earth and that human activities have begun to affect the entire atmosphere of our planet.

The Mauna Loa data clearly demonstrate the benefits of long-term collection of information. Funding for long-term projects is often difficult to maintain because funding agencies may prefer to sponsor new projects rather than long-term monitoring. Nevertheless, understanding global change depends on collection and maintenance of supportive data. To that extent, the Mauna Loa CO_2 measurement project is unique, and its effectiveness is a tribute to the people who initiated and nurtured it for so many years.

* Material for this section is from: C. A. Edahl and C. D. Keeling, 1973. "Carbon and The Biosphere" Oak Ridge, Tenn.: Technical Information Service and C. D. Keeling, T. P. Worf, and J. Van der Plicht, 1995. "Nature" 375: 666–670.

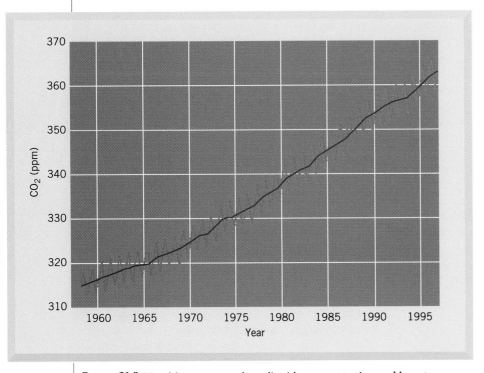

FIGURE 21.9 Monthly average carbon dioxide concentration and long-term trend. (*Source*: Mauna Loa Observatory, http://mloserv.mlo.hawaii.gov/mloinfo/programs/gases/CO2graph.htm—accessed 10/14/98, P. Tans and C. D. Keeling.)

(see Table 21.1).[14] As with carbon dioxide, there are important uncertainties in our understanding of the sources and sinks of methane in the atmosphere. Natural environments release methane into the atmosphere. Major contributors are termites, which produce methane as they process wood, and freshwater wetlands, where decomposing plants in oxygen-poor environments produce and release methane as a decay product. The several anthropogenic sources of methane include the burning of biomass (organic material such as logs), the production of coal and natural gas, and agricultural activities, such as the cultivation of rice and the raising of cattle. (Methane is released by anaerobic activity in flooded lands where rice is grown, and cattle expel methane gas as part of their digestive processes.) For unknown reasons the increase in atmospheric methane stopped in 1991 and 1992, most likely related to control of leaks in Russian natural gas systems, with a smaller contribution from atmospheric changes resulting from the 1991 eruption of Mount Pinatubo.[15,16]

Chlorofluorocarbons

The CFCs are highly stable compounds that have been or are being used in spray cans as aerosol propellants and in refrigeration units—the gas that is compressed and expanded in the cooling unit. Although CFCs are no longer used in spray cans in the United States and many other countries, they are not yet banned worldwide. Deliberate release and accidental leaks of CFCs into the atmosphere in recent years have been considerable. The rate of growth of CFCs in the atmosphere in recent years has been about 5% per year (see Table 21.1). It has been estimated that approximately 15% to 25% of the anthropogenic greenhouse effect may be related to CFCs in the atmosphere.[17] The potential global warming from CFCs is considerable, because they absorb in the atmospheric window (explained earlier), and each CFC molecule may absorb hundreds or even several thousand times more infrared radiation emitted from Earth than is absorbed by a molecule of carbon dioxide. Furthermore, because CFCs are highly stable compounds, their residence time in the atmosphere is long. Even if production of these chemicals is drastically reduced or eliminated within the next few years, their concentrations in the atmosphere will remain significant for many years, perhaps for as long as a century.[17]

Nitrous Oxide

Nitrous oxide (N_2O) is also increasing in the atmosphere and is probably contributing as much as 5% of the anthropogenic greenhouse effect.[14] Anthropogenic sources of nitrous oxide include agricultural activities (application of fertilizers) and the burning of fossil fuels. Reductions in the use of fertilizers and the burning of fossil fuels would reduce emissions of nitrous oxide. However, this gas also has a long residence time; even if emissions were stabilized or reduced, elevated concentrations of nitrous oxide would persist for at least several decades.[14]

In summary, carbon dioxide contributes between 50% and 60% of the anthropogenic greenhouse effect. The rest of the human-made effect comes from trace gases, the most important of which are the CFCs and methane. These trace gases contribute between 27% and 45% of the anthropogenic greenhouse effect (see Table 21.1), and they have accumulated in the atmosphere at much faster rates than carbon dioxide. The Montreal Protocol is an international treaty signed in 1987 by 24 countries to reduce and eventually eliminate the production of CFCs and to accelerate the development of alternative chemicals. The treaty calls for production of CFCs to be phased out by the year 2000. The worldwide production of CFCs has fallen since the treaty was signed. Many countries expect to achieve this goal prior to the deadline, because development and production of alternative chemicals has been more rapid than expected. However, not all countries have signed the treaty and illegal production and use of CFCs continues in some countries. If CFCs had not been regulated by the Montreal Protocol, by the early 1990s they would have become the major contributor to the anthropogenic greenhouse effect.[18] The reduced emissions are evidently responsible for the recent decrease in growth rates of atmospheric CFCs commonly used in refrigeration, air conditioning, and aerosol propellants. It is hypothesized that if the growth continues to decrease, atmospheric concentrations of these CFCs will peak before the year 2000 and then decline.

21.4 THE GLOBAL WARMING CONTROVERSY

As stated earlier, there were a number of warm years in the 1980's and 1990's, and 1998 broke these records month-by-month. Furthermore, it appears that in the past several decades the mean global temperature has increased approximately 0.5°C (0.9°F). However, these two observations are not sufficient to conclude that global warming is occurring as a result of anthropogenic increases in greenhouse gases. The controversy is not about whether there is a greenhouse effect. The greenhouse effect is not controversial; its existence is one of the best-established principles of atmospheric science. And there is no doubt that anthropogenic processes have resulted in increased emissions of greenhouse gases, such as car-

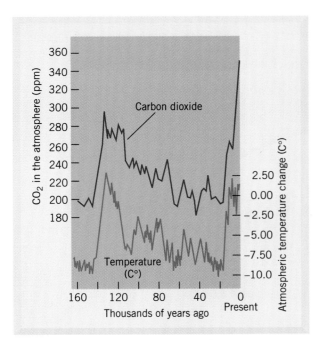

FIGURE 21.10 Inferred concentration of atmospheric carbon dioxide and temperature change for the past 160,00 years. The relation is based on evidence from Antarctica and indicates a high correlation between temperature and CO_2. (*Source:* Adapted from S. H. Schneider "The Changing Climate", © September 1989, *Scientific American, Inc.* All rights reserved. Vol. 261, p. 74.)

bon dioxide, CFCs, methane, and nitrous oxide. Finally, there is good reason to argue that increases in carbon dioxide and other greenhouse gases are related to an increase in the mean global temperature of Earth.[8,19] Over the past 160,000 years there has been a strong correlation between the concentration of atmospheric CO_2 and global temperature (Figure 21.10). When CO_2 has been high, temperature has also been high, and, conversely, low concentrations of CO_2 have correlated with a low global temperature. However, in order to further evaluate the issue of global warming, we need to consider both the positive and negative feedbacks that occur on Earth, on the sun, and in the atmospheric system.

Negative and Positive Feedbacks

Greenhouse warming is very complex. The warming effect initiates both negative- and positive-feedback loops that can offset any temperature increase or raise it more. Negative-feedback loops are self-regulating and result in global cooling in response to a warming circumstance. Positive feedbacks are self-enhancing; thus a perturbation that causes an increase in global temperature leads to further increases in that temperature. We first discussed positive and negative feedbacks with respect to

Earth systems and changes in Chapter 3; you may wish to review those concepts.

Several of the potential negative and positive feedbacks concerning global warming are shown in Figure 21.11.[19] It is important to remember that if the negative feedbacks are strong and persistent, global warming may not occur. On the other hand, if the negative-feedback systems are weak relative to positive feedback, warming is likely to occur more readily.

As shown in Figure 21.11a, negative feedbacks can be fairly complex. For example, negative-feedback cycle 1 is based on the hypothesis that as global warming occurs there will be an increase in the algae populations in the warming ocean. Algae will absorb more carbon dioxide, reducing the concentration of CO_2 in the atmosphere and causing cooling. Negative-feedback cycle 2 is related to terrestrial vegetation. Here it is hypothesized that an increase in carbon dioxide concentration will stimulate plant growth (as it does in laboratory experiments), and the increased amount of vegetation will absorb more carbon dioxide from the atmosphere, facilitating cooling. Finally, negative-feedback cycle 3 is related to cloud cover, which is poorly understood but extremely important. The idea is that as the global temperature increases, more water will evaporate from the ocean, leading to more water vapor in the atmosphere and thus more clouds. Because clouds tend to reflect incoming solar radiation, Earth will be cooled by the increased cloud cover.

Positive-feedback processes are idealized in Figure 21.11b. The warming Earth causes an increase in the evaporation of water from the oceans (cycle 4), which adds additional water vapor to the atmosphere. But here we look at a different effect of the water vapor: It causes additional warming (water vapor is an important greenhouse gas). The warming Earth causes increased melting of permafrost (areas where the soil, below an active zone that thaws in the summer, remains frozen from year to year) at high latitudes, which may result in additional release of the greenhouse gas methane (cycle 5), a by-product of decomposition of organic material in the melted permafrost layer. Another effect of the warming trend would be a reduction in the summer snowpack, which would reduce the amount of solar energy reflected from Earth (cycle 6). Collectively these processes would result in additional warming and thus are part of a positive-feedback loop. In cycle 7 the warming of Earth is felt by people in urban areas, who use additional air conditioning, thus increasing the burning of fossil fuels, which in turn releases additional carbon dioxide into the environment and results in additional global warming.

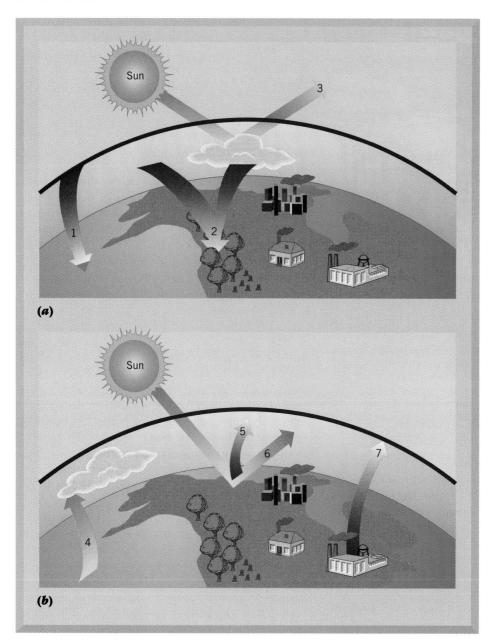

Figure 21.11 (*a*) Negative and (*b*) positive feedback mechanism (cycles) associated with the greenhouse effect. See text for explanation. (*Source:* J. R. Luoma, 1991, "Gazing into Our Greenhouse Future," *Audubon* 93 (2); p. 57.)

Both negative- and positive-feedback processes occur simultaneously in the atmosphere. Which are more important? At present, the answer is not known. A great deal of research is currently being carried out on cloud negative-feedback processes. Many discussions on the greenhouse effect state that if Earth's atmosphere did not trap heat, our planet would be approximately 33°C (60°F) cooler at the surface, and as a result all water would be frozen. However, since water vapor is the major greenhouse gas in the atmosphere, no greenhouse effect implies no (or very little) water vapor in the atmosphere. Further, this implies no clouds, which would lead to

a substantial reduction in the atmospheric reflection of incoming sunlight, which would result in warmer surface temperatures on Earth. The dual role of atmospheric water vapor as both a negative and a positive feedback with respect to global warming is extremely important to understanding possible climatic modifications created by an anthropogenic greenhouse effect.

There is vigorous debate among scientists as to whether global warming caused by human activities is in fact occurring. Some scientists believe that warming has indeed already started, whereas others suggest that the negative-feedback cycles will be suf-

CLOSER LOOK 21.2

EARTH SYSTEM SCIENCE AND GLOBAL CHANGE

Until very recently it was generally thought that human activity was only capable of causing local, or at most regional, environmental changes. We now know differently! The main goal of the emerging science known as **Earth system science** is to obtain a fundamental understanding of how our planet works as a system.

From a pragmatic point of view, the research priorities of Earth system science and global change can be summarized as follows[20]:

- establishment of worldwide measurement stations to better understand physical, hydrologic, chemical, and biological processes that are significant in the evolution of Earth on a variety of time scales;
- documentation of global changes, especially those that occur in a time period of several decades, that are of particular interest to the human environment;
- development of quantitative models useful in the predic-

tion of future global change; and

- assisting in the gathering of the essential information that decision makers need at the regional (national and international) and global levels.

The major tools for studying global change are:

- evaluation of the geologic record;
- monitoring; and
- mathematical models.

ficient to moderate the effect in the future and that anthropogenic global warming will not be a serious problem.[19] With each year, additional evidence suggests that humans are probably responsible for changes in atmospheric temperature, although not enough evidence and data exist to claim unequivocally that this is true.[8] Until data are collected for many more years it may be difficult to prove whether global warming has occurred or is occurring. Nevertheless, all the global modeling experiments suggest that warming as a result of anthropogenic increases in greenhouse gases will in fact occur (but, to be even-handed, some scientists find faults with the models). Although some model studies suggest that the average global temperature could rise as much as 5°C (9°F) by the middle of the next century, most predict a smaller rise of 2° to 4°C (3.6°–7.2°F). In the most optimistic case, if there are large reductions in emissions of greenhouse gases, the global warming may be less than 1°C.[6] The models on which global climate change is predicted are mentioned in A Closer Look 21.2, "Earth System Science and Global Change" and are discussed below. Despite the existence of such models, uncertainties related to sunspots, aerosols, nocturnal and daytime temperatures, volcanic eruptions, and El Niño events continue to cloud the issue of the anthropogenic greenhouse effect.

Geologic Record

The sediments deposited on floodplains, in bogs, and on lake or ocean bottoms are analogous to the pages of a history book in that they can be read to provide a history. Organic material, such as skeletal

material, shells, pollen, and bits of wood, leaves, and other plant parts, is often deposited with sediments and can yield valuable information concerning Earth's history. In addition, organic material may be dated to provide the necessary chronology to establish past changes. Finally, both sediments and organic material can be used to evaluate the past climate—what lived where and the nature and extent of changes that have occurred.

One of the more interesting uses of the geologic record has been the examination of glacial ice. The process of transformation of snow to glacial ice involves recrystallization and an increase in the density of the ice. The process also traps air bubbles that may be analyzed to provide information concerning the concentration of carbon dioxide in the atmosphere at the time the ice formed. Thus, glacial ice may be thought of as a time capsule that can be analyzed to provide information concerning the atmosphere in the past. To study the ice, long cores of glacial ice are extracted, and trapped air bubbles are carefully sampled. This method has been used to analyze the carbon dioxide content of the atmosphere up to 160,000 years ago. Figure 21.12 shows the record from 1500 to 2000.[10]

Real-Time Monitoring

Monitoring may be defined as the regular collection of data for specific purposes. For example, we monitor rainfall and the flow of water in rivers to evaluate water resources or flood hazards. We collect the data to provide baseline conditions from which to evaluate changes in the future. Similarly, samples of

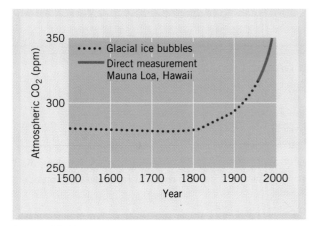

FIGURE 21.12 Average concentration of atmospheric carbon dioxide 1500–1900. (*Source:* Data from W. M. Post, T. Peng, W. R. Emanuel, A. W. King, V. H. Dale, and D. L. De Angelis, 1990, *American Scientist* 78(4) p. 310–326. Reprinted by permission of *American Scientist,* Journal of Sigma Xi, The Scientific Research Society.)

atmospheric gases, particulates, and chemicals are helping establish trends, or changes in the composition of the atmosphere (see A Closer Look 21.1, "Monitoring of Atmospheric Carbon Dioxide Concentration"). Finally, measurements of the temperature, composition, and chemistry of ocean waters may be used to help evaluate potential changes in the marine environment.

Mathematical Models

Mathematical models are developed as an attempt to represent, through numerical means, real-world phenomena, linkages, and interactions among physical, chemical, and biological processes. The models use equations that describe the phenomena and linkages being considered. Models have been developed to predict the flow of surface water and groundwater, ocean circulation, and atmospheric circulation. The models that have gained the most attention from the perspective of global change are the **global circulation models (GCMs).** The objective of GCMs is to predict atmospheric changes (circulation) at the global scale.[11]

The variables utilized in GCMs include temperature, relative humidity, and wind conditions. Values for many of these variables are estimated for the past based on data derived from such sources as tree-ring records. Annual growth rings of trees can be evaluated to provide a time scale (dendrochronology). They can also be studied to infer climatic information, such as precipitation and runoff, which are useful in calibrating results from mathematical models. To organize the data necessary for the calculations in GCMs, the surface of Earth is divided into large cells measuring several

degrees of latitude and longitude. The typical cell is a large part of the surface of Earth the size of, for example, Oregon or Indiana and Ohio together. Several layers of data are necessary. Most GCMs use 6 to 20 levels of vertical data throughout the lower atmosphere (Figure 21.13). Mathematical relationships are used in the calculations to make predictions of future atmospheric circulation. The GCMs are very complex and require supercomputers for their operation.[11]

Results from these models are relatively crude and so may not accurately predict future conditions.[11,18] Furthermore, it is difficult to estimate interactions of other factors, such as cloud cover, which may significantly affect atmospheric energy relations. As a result, GCMs can only be considered a first approach to solving very complex atmospheric problems. In spite of their limitations, mathematical models are providing information necessary for evaluating global change and Earth as a system. The GCMs are also helping to pinpoint what additional data are necessary to better predict future change. The models do predict, in a relative sense, areas or regions that are likely to be wetter or drier if certain changes in the atmosphere occur; thus predictions from mathematical models are being taken seriously, and their importance as a tool for studying global change will continue to increase.

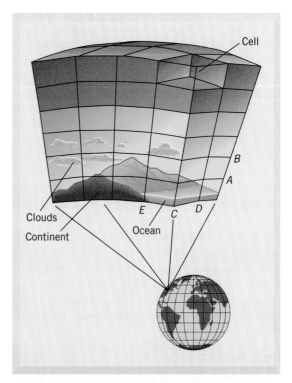

FIGURE 21.13 Idealized diagram showing how cells are arranged for a global circulation model. *A–B* is a few kilometers in elevation; *E–C* is a few hundred kilometers in longitude; *C–D* is a few hundred kilometers in latitude.

Sunspots

It is a simple fact that the sun is responsible for heating Earth. Therefore, when looking at climatic change, it would seem to make sense to first look to solar variation as a possible cause. Recent attention has been focused on the relation between sunspot cycles and temperature variations near the surface of Earth. Sunspots appear as small, black dots when viewed through a solar telescope. They are created when strong solar magnetic fields protrude through the sun's surface. There exists a startling correlation between Earth's Northern Hemisphere temperature over the past 130 years and changes in the length of the sunspot cycle, which is the time interval between peaks in sunspot abundance. The data show that during the past 130 years, as the cycle length decreased, Earth's temperature increased; conversely, when the sunspot cycle length increased, temperature dropped.

A problem that has bothered atmospheric scientists and climatologists for some time is that carbon dioxide concentrations have increased exponentially since the industrial revolution, yet there has been a fair amount of variability in Earth's temperature during that period.[21] What if the primary changes in Earth's temperature are caused by what the sun is doing rather than emission of carbon dioxide? Temperature increased from approximately 1880 to 1940, then decreased to 1970, and has been increasing again. These changes in temperature are highly correlated with changes in the length of the sunspot cycles. It is speculated that the solar cycle length may be a possible indicator of long-term variability in the total energy output of the sun.[21] Recent studies suggest that 11 year climate change cycles may be linked with solar variability[22], and the Little Ice Age (Figure 21.5d) occurred during a period of very low solar activity. It may be possible that anthropogenic emission of greenhouse gasses and solar radiative forcing may both play a role in contributing to the warming trend of the past 100 yrs.[23] However, atmospheric scientists have so far been unable to define a mechanism related to sunspot activity that would produce sufficient solar variability to account for the observed changes in global temperature during the past 130 years. If a mechanism is discovered, the correlation between the sunspot cycle and global temperature would imply that solar variation is the most important variable in determining Earth's climate over time periods as short as 130 years. What this would mean is that we would not have to call upon an anthropogenic greenhouse effect to account for temperature variations in the atmosphere. However, even this discovery would not discount the possibility of a future anthropogenic greenhouse effect that would interact with solar vari-

ation. Finally, correlations do not necessarily indicate a cause-and-effect relationship, so the good fit between sunspot cycles and global temperature change may be due to chance. Whatever the outcome, the work on sunspot activity certainly emphasizes the importance of gaining a better understanding of Earth's climate as it relates to solar activity before drawing firm conclusions about the anthropogenic greenhouse effect.[21]

Aerosols

Emission of aerosols to the atmosphere has increased significantly since the industrial revolution began. An aerosol is a particle with a diameter less than 10 µm. Because the effects of collisions with air molecules dominate over gravity, the smaller aerosol particles tend to remain in the atmosphere for a long time. Bigger particles (diameters greater than 10 µm) drop out of the atmosphere faster because of gravity. Recent research has indicated that aerosols emitted from coal (sulfates) may contribute to global cooling because sulfates act as seeding for clouds; the aerosol particles provide surfaces for water to condense on, forming clouds that reflect incoming solar energy. For example, the 1991 eruption of Mount Pinatubo in the Phillapines hurled 20 million tons of sulfer dioxide into the stratosphere that circled the glcbe and cooled the planet's surface for two years. The aerosol particles also reflect a significant amount of sunlight directly. The net cooling owing to sulfate aerosols may offset much of the global warming expected from the anthropogenic greenhouse effect.[24] For example, over the United States, increased atmospheric aerosols have probably produced mean temperatures roughly 1°C cooler than it would be otherwise. Aerosol cooling may thus help explain the disparity between model simulations of global warming and actual recorded temperatures that are lower than those predicted by models.[25]

Nocturnal and Daytime Temperatures and the Greenhouse Effect

During the past 40 years, average annual maximum daytime temperatures in the United States, the former Soviet Union, and China (together, about one-quarter of Earth's land area) have remained essentially unchanged, although the minimum nocturnal temperatures in the three countries have increased substantially. According to climate models, day and night temperatures should be equally influenced by the greenhouse effect. The greenhouse effect cannot function just at night and not during the daytime! Additional research is necessary to resolve this greenhouse inconsistency. Perhaps aerosol particles in the form of sulfates may reflect enough sunlight

FIGURE 21.14 Eruption column from Mount Pinatubo in the Philippines during 1991 eruptions. Such eruptions injected vast amounts of dust and sulfur dioxide as high as 30 km into the atmosphere.

during daylight hours to offset the greenhouse effect. This occurrence would partially explain the inconsistency, because the aerosols would have no effect at night.[26]

Volcanic Eruptions

Another uncertainty in predicting global temperatures is the effect of volcanic eruptions. How much cooling from one year to the next was related to the eruption of Mount Pinatubo in the Philippines (Figure 21.14) in the summer of 1991? Tremendous explosions sent volcanic ash to elevations of 30 km (19 miles) into the stratosphere, and as with similar past events, the aerosol cloud of ash and sulfur dioxide remained in the atmosphere, circling Earth for several years. The particles of ash and sulfur dioxide scattered incoming solar radiation, resulting in a slight cooling of the global climate during 1991 and 1992. Calculations suggest that aerosol additions to the atmosphere from the Mount Pinatubo eruption counterbalanced the warming effects of greenhouse gas additions through 1992. However, by 1994, most aerosols from the eruption had fallen out of the atmosphere and global temperatures returned to previous higher levels.

El Niño

Another natural perturbation in the physical Earth system that affects global climate is the occurrence of El Niño events (See the Case Study at beginning of this chapter.) During an El Niño, the normal conditions of equatorial upwelling of deep oceanic waters are diminished or eliminated. Upwelling re-

leases carbon dioxide to the atmosphere as carbon dioxide–rich deep water reaches the surface. El Niño events reduce the amount of oceanic carbon dioxide outgassing and thus perturb the global carbon dioxide cycle.

The 1982–1983 El Niño event was particularly strong. It is thought to have produced climatic events such as floods and droughts that killed several thousand people and caused billions of dollars in damage to crops, structures, and other facilities. El Niño events may change the patterns of the upper troposphere (jet streams) to produce wetter winters and larger storms in the United States. El Niño conditions also developed in 1991–1992 and again in 1993 and, although they were not as strong as the 1982–1983 event, contributed to a slight global warming. Nevertheless, the effects of volcanic eruptions, especially that of Mount Pinatubo, will more than offset the warming through global cooling (produced by the eruption of sulfur dioxide–rich aerosol clouds described earlier).

Recognition of El Niño events is important for understanding potential perturbations that affect the global climate. The events also provide data for global models that predict change. Because the volcanic eruptions and El Niño events are relatively short term, they provide important information from which models can be calibrated and tested.

21.5 POTENTIAL EFFECTS OF GLOBAL WARMING

If we continue emitting carbon dioxide into the atmosphere as we have in the past, it is estimated that by the year 2030 the concentration of carbon dioxide in the atmosphere will have doubled from pre–industrial revolution concentrations. The average global temperature (according to mathematical models) will have risen approximately 1° to 2°C (2°–4°F), with significantly greater temperature change at the polar regions.[9] Although the specific effects of this assumed temperature rise are difficult to predict, those that are being taken seriously are changes in global climate patterns and a rise in the sea level.

Changes in Climate Pattern

Estimates have been made of climatic change likely to occur by the year 2030 owing to the doubling of carbon dioxide in the atmosphere since the industrial revolution.[27] Figure 21.15 shows the potential changes for selected parts of the world. In central North America warming is expected to vary from approximately 2° to 4°C (4°–7°F), with an increase in winter precipitation but a decrease in summer rains. As a result, soil moisture is expected to decrease in the summer by as much as 20%. Clearly this could have a significant effect on the grain-growing areas of the United States. Similar changes will also occur in southern Europe if this scenario is correct. It is important to keep in mind that these projections are based on global circulation models that are controversial and subject to variability. Nevertheless, most of the models predict changes in the directions indicated and as a result are being taken seriously by both scientists and policymakers.

Global rise in temperature is expected to significantly change patterns of rainfall, soil moisture, and other climatic factors related to agricultural productivity. Studies utilizing global circulation models to predict patterns for the Northern Hemisphere suggest that some of the more northern areas, such as Canada and Russia, may become productive. However, it is emphasized that although global warming might move North America's prime farming climate north from the midwestern United States to the region of Saskatchewan, Canada, the U.S. loss would not simply be translated into a gain for Canada. Saskatchewan would have the optimum climate for growing, but the Canadian soils are thinner and less

fertile than the prairie-formed soils of the Midwest. Therefore, a climate shift could have serious effects on world food production. On the other hand, lands in the southern part of the Northern Hemisphere may become more arid. As a result, soil moisture relationships will change.

Soil moisture predicted by computer modeling generally supports these statements concerning changes in productivity. However, we emphasize that such predictions are extremely difficult to make because of uncertainties surrounding global warming. People are anxious when uncertainty is present, and particularly anxious when that uncertainty involves our food supply. There is real concern that hydrologic changes associated with climatic change resulting from global warming may seriously affect the global food supply.

There is also concern that global warming will alter normal weather and climatic patterns, including change in frequency or intensity of violent storms. This possibility may be more important than changes in climate. The hypothesis is that warming ocean waters could feed more energy into high-magnitude storms, such as cyclones and hurricanes, causing a significant increase in their frequency or intensity. Approximately half Earth's human popula-

FIGURE 21.15 Estimated climatic change if CO_2 concentration doubles industrial revolution levels. (*Source:* V. A. Mohnen, W. Goldstein, and W. Wang, 1991, "The Conflict over Global Warming," *Global and Environmental Change* 1(2), pp. 109–123.)

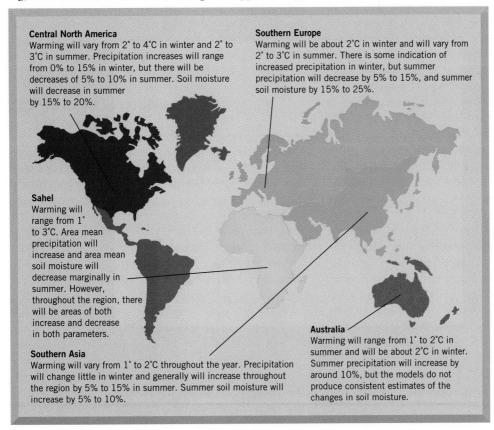

tion lives in coastal areas; potential problems would be exacerbated by the fact that many of these areas are low lying and are experiencing rapid population growth. Greenhouse warming is expected to result in wetter winters, hotter and drier summers, an increased frequency of large storm events, and an increased possibility of droughts in the northern temperate latitudes. Study of these changes in climate have already suggested a possible climate shift in the United States.[28]

In summary, global warming could affect Earth in various ways, often for the worse. These changes may affect crop production, forestry, and human health. Midlatitude climate zones could shift northward by as much as 550 km (330 miles) over the next century. At this rapid rate, some tree species may be nearly eliminated. Furthermore, the expansion of tropical climate zones expected in global warming will lead to an increase in tropical diseases such as malaria, dergue fever, yellow fever, and viral encephalitis.[29]

Rise in Sea Level

A rise in the sea level is a potentially serious problem as it relates to global warming. Although a precise estimate of the total potential rise in sea level is not possible at this time, there is a consensus that the level of the sea will in fact rise. In fact, sea level along much of the U.S. coast is already rising at a rate of 2.5 to 3.0 mm per year, or about 10 to 12 inches per century.[30] The causes for the rise are thought to be twofold: thermal expansion of warming ocean water (the primary cause) and melting of glacial ice (a secondary cause). The various models predict that the rise may be anywhere from 20 cm to approximately 2 m (8–80 in.) in the next century; the most likely rise is probably 20 to 40 cm (8–16 in.). One estimate is that sea level will likely rise 15 cm by the year 2050 and 34 cm by the year 2100. When other factors such as land subsidence and compaction, groundwater depletion, and natural climate variation are considered, some coastal regions could experience a sea level rise of 45 to 55 cm by the year 2100.[30] Such a change will have significant environmental impacts; it could easily cause increased coastal erosion on open beaches of up to 50 to 100 m (165–230 ft), making buildings and other structures in the coastal zone more vulnerable to damage from waves generated by high-magnitude storms. It could also cause a landward migration of estuaries and salt marshes, putting additional pressure on human structures in the coastal zone.[9] Finally, groundwater supplies for coastal communities may be threatened by saltwater intrusion should sea levels rise.

A rise in the sea level of approximately 1 m (3.3 ft) would have serious consequences. People would have to make significant alterations in the coastal environment to protect investments, and communities would be forced to choose between making very heavy financial investments in controlling coastal erosion and allowing for considerable loss of property.[9]

Considering the amount of coastal defenses present in the world today, it seems inevitable that a rise in sea level will lead to further investment for protecting cities in the coastal zone. Construction of seawalls, dikes, and other erosion-controlling structures will become more common as coastal erosion threatens urban property. In more rural areas, where development is set well back from the coastal zone, the most likely response to a rising sea level will be simply to adjust to the erosion that occurs. Coastal erosion is a difficult problem that is very expensive to deal with; it is prudent to allow erosion to naturally take place where feasible rather than to try to control it.

Finally, when considering a rise in the sea level, we must be concerned with the hundreds of millions of people who live in low-lying areas of developing countries. For example, two cyclones that hit Bangladesh in the last 25 years killed more than 400,000 people and caused over $1.6 billion in property damage (see Chapter 5). The double impact of a rising sea level and more frequent and powerful cyclones and other tropical disturbances (owing to warmer oceans, as discussed earlier) would have a devastating effect on people in developing countries.

21.6 ADJUSTMENTS TO POTENTIAL GLOBAL WARMING

There is considerable debate over climatic change and potential warming related to human activities that release greenhouse gases into the atmosphere. Scientists do agree that we need to develop better models and gain a better understanding of both climatic change and variability and short-term trends in weather. Three important questions related to global warming and climate change are being vigorously debated[27]:

- Has the global climate changed in the past 100 to 150 years and is that change likely to continue into the twenty-first century? There seems little doubt that the global temperature in the last few decades has in fact increased approximately 0.5°C (0.9°F). Whether that rise is natural or caused by human activity is being debated, but some component of the rise is undoubtedly human induced, and this will increase as more fossil fuels are burned in the future.

- Do combustion of fossil fuels and massive emissions of carbon dioxide into the atmosphere cause climatic changes such as global warming? Although we are unable to completely answer this question at present, many scientists believe and computer models predict that global warming will continue to occur as a result of increased greenhouse gases.

- Given that we do not completely understand the consequences of emitting greenhouse gases and that computer models are in early stages of development, should we spend vast sums of money to reduce emissions? There is no easy answer, and the answer is not likely to come soon from scientific information but will require a policy decision based on incomplete information. Nevertheless, we frequently make decisions based on inadequate information because potential consequences of inaction may result in an unacceptable risk to society.

What should we do about potential global warming? There are two basic adjustments:

- Mitigation (reducing the severity of the problem) through reduction of emissions of greenhouse gases or
- Do nothing to combat it and live with future global climatic change.

Living with Global Change

With our present understanding of global warming and trends in energy use and deforestation, the most likely adjustment will be learning to live with the changes. These include a warmer climate and new variability in weather patterns as well as a higher sea level. If changes are relatively slow, learning to live with new conditions may be feasible; in fact, in some cases the changes will offer opportunities. However, it is emphasized that this may not be the best adjustment, because there will probably be many unexpected surprises and problems.

If the hypothesis is accepted that global warming is due in part to increases in emissions of greenhouse gases, then reduction of these gases is a primary management strategy. To stabilize carbon dioxide concentrations at present levels would require a 60% reduction of global carbon dioxide emissions. To address this problem, the United Nations Framework Convention on Climate Change (FCCC) was adopted in 1992. This commits partici-

pants to produce a national program to reduce carbon dioxide to 1990 levels by 2000, and develop methods for protection of carbon dioxide sinks such as forests. Reducing carbon dioxide emissions was revisited in 1997 at the Summit on Global Warming in Kyoto, Japan. Congress, however, refused to acknowledge the problem, and prohibited studying global warming unless developing countries also agreed to commit to reductions in carbon dioxide emissions. This inactivity reduces chances of real progress. The meetings did result in bringing the issue of global warming to the forefront for debate. Approximately 70% of anthropogenic carbon dioxide emissions are related to burning of fossil fuels. Therefore energy planning that relies heavily on energy conservation and efficiency and use of alternative energy sources, will reduce emissions of carbon dioxide. Increased use of nuclear power would also reduce emissions of carbon dioxide. Other ideas to reduce emissions of carbon dioxide include increasing tax for using fossil fuels; providing economic incentives to increase use of mass transit and decrease use of automobiles; providing greater economic incentives to improve development of energy-efficient technology; requiring higher fuel economy standards for cars, trucks, and buses; and requiring higher standards of energy efficiency. Another important source of carbon dioxide emissions is **deforestation.** Burning forests to convert lands to agricultural purposes accounts for about 20% of anthropogenic carbon loading into the atmosphere. Management plans that seek to minimize burning and protect the world's forests would help reduce the threat of global warming, as would plans to plant trees (reforestation).

In summary, if global warming occurs, our most likely adjustment will be to live with it. However, if we are prudent, we will plan to reduce emissions of carbon dioxide and other greenhouse gases. Doing so will require changes in land management and energy use. Of particular importance will be energy conservation, using energy more efficiently, and emphasizing alternative energy sources. Essentially, we should reduce emissions of greenhouse gases as much as is feasible and have contingency plans for greater reductions in emissions should they become necessary. Thus, the recommended strategy is somewhere between full mitigation and learning to adapt to change.

Some people argue that there is consensus among scientists that global warming is now occurring. They might argue that the position taken in this chapter is too conservative, that the problem should be addressed now before it is too late. They might further argue that steps taken to abate global warming would have tremendous environmental benefits

Will Planting Trees Offset Global Warming?

Some scientists, foresters, environmentalists, and government officials have suggested that planting trees may ease the threat of global warming. Trees take up carbon dioxide during photosynthesis and can store carbon for long periods of time. It is an enticing proposal, since it does not require that we stop doing anything, such as driving large cars, and requires only a simple biological solution, which would have other advantages, such as shading buildings, improving the aesthetics of urban areas, and providing more wood, a renewable resource.

Who could argue against planting a tree? Or even millions or billions of trees? But if we are to ask millions of people to plant billions of trees, we should do so only if we are confident that they will have a substantial effect on reducing carbon dioxide in the atmosphere. According to current estimates, humans are adding 7 gigatons (1 gigaton equals 1 billion metric tons, or 1.1 billion U.S. tons) of carbon, as carbon dioxide, to the environment each year, most as a result of burning fossil fuels. About 75% of that, or 5.2 gigatons, remains in the atmosphere or is absorbed in the oceans. What happens to the remaining 1.8 gigatons is controversial; some scientists think it is absorbed by the oceans; others think it is absorbed by the forests; still others think that both the oceans and the forests are involved.

Living trees contain on the order of 500 gigatons of carbon, the atmosphere 700 gigatons as carbon dioxide, and soils an estimated 2000 to 4000 gigatons of carbon. Each year, the trees of the world remove about 5 gigatons of carbon through photosynthesis and return an equal amount to the atmosphere through respiration and decay. The forests of the world occupy an area of 4 Gha (1 Gha, a gigahectare, equals 1 billion ha, or 2.5 billion acres). Annual rates of deforestation are estimated at 0.2% (0.008 Gha, or 8 million ha, or 19.8 billion acres), a rate of 1 ha every 4 seconds.

Critical Thinking Questions

1. If we plant trees equal to 1% of our present forests, how much carbon dioxide would be removed from the atmosphere and why?

2. How many hectares of trees would we need to plant to offset the present rate of deforestation? Using an atlas, compare this area to some known area of the world, such as Peru, the United States, or Australia.

3. The numbers used to estimate how much carbon could be taken up by trees are crude estimates of maximum net growth, but trees do not always grow at their maximum rate. What factors could interfere with trees growing at their maximum? How would this affect the calculations you have done?

4. Would you recommend a program of tree planting? If so, why, and what type of program? If not, why not?

References

Botkin, D. B. 1989. "Can We Plant Enough Trees to Absorb All the Greenhouse Gases?" Paper delivered at the University of California Workshop on Energy Policies to Address Global Warming, September 6–8, 1989, Davis, CA.

Rosenfeld, A. H., and Botkin, D. B. 1989 (Oct. 27). *Trees Can Sequester Carbon, or Die, Decay, and Amplify the Threat of Global Warming: A Primer on Forestation, Deforestation, and Possible Feedback between Rising Temperature, Stressed Forests, and CO_2. Physics and Society* 19:4 pp.

Stevens, W. K. 1989 (July 18). "To Halt Climate Change, Scientists Try Trees," *New York Times*, pp. C1, C4.

even if global warming does not occur. We would respond that the environmental need for energy conservation, reduction in air pollutants, and use of alternative energy sources is clear and need not be tied to the global warming issue.

Consensus among scientists is not scientific consensus based on irrefutable data. However, this does not mean that we should not make policy decisions based on potential risks to society and our planet. Risks from potential global warming are sufficiently significant and grave that they need to be addressed. To this extent the Earth Summit meeting sponsored by the United Nations in 1992 at Rio de Janeiro established the principle of reducing greenhouse gas emissions by the year 2000 to 1990 levels. Meeting this goal will require the cooperation of the developed countries, where most of the carbon dioxide is emitted. For example, the United States, with 5% of the world's population, emits about 20% of the atmospheric carbon dioxide. It is encouraging to note that the U.S. government has acknowledged that reduction of emissions of carbon dioxide into the atmosphere is an important goal. Providing economic incentives to improve and design new energy-efficient technologies and alternative energy sources will be a positive step.

SUMMARY

- The atmosphere, a layer of gases that envelops Earth, is a dynamic system that is constantly changing. A great number of complex chemical reactions take place in the atmosphere, and atmospheric circulation takes place on a variety of scales, producing the world's weather and climates.

- Nearly all the compounds found in the atmosphere either are produced primarily by biological activity or are greatly affected by life.

- On a local scale climatic conditions vary considerably and may produce what is referred to as a microclimate. For example, the urban microclimate results because the presence of a city significantly affects local climate. Thus our midlatitude subhumid cities are generally cloudier, warmer, rainier, and less humid than are the surrounding areas.

- During the past 2 million years there have been major climatic changes with periodic appearances and retreats of large masses of land-bound ice known as glaciers. During the past 1500 years, several warming and cooling trends have affected people. During the past 100 years the mean global annual temperature has apparently increased by about 0.5°C.

- Water vapor and several other gases, including carbon dioxide, methane, and CFCs, tend to warm Earth's atmosphere because they absorb and trap some of the heat energy radiating from Earth's atmospheric system. This trapping or warming is referred to as the greenhouse effect. The vast majority of the greenhouse effect is produced by water vapor, which is a natural constituent of the atmosphere, as are carbon dioxide and other gases. Because some of the greenhouse gases, such as carbon dioxide, methane, and CFCs, are being emitted into the atmosphere by human activity, there is concern that the mean global temperature of Earth may rise. Some climatic models suggest that when carbon dioxide has doubled from its preindustrial levels in the next few decades the mean global temperature may rise by 1 or 2°C.

- Because there are many complex positive- and negative-feedback cycles affecting the atmosphere, the hypothesis of global warming resulting from increases in the greenhouse gases caused by human activity remains controversial. Furthermore, natural cycles, solar cycles, aerosols, volcanic eruptions, and El Niño events to a lesser or greater extent also affect the temperature of Earth. Solar output is receiving a lot of attention because it is obviously an important variable in possible climatic change.

- The two major effects of global warming would be (1) changes in climatic patterns and frequency and intensity of storms and (2) a rise in sea level. Changes in climatic patterns and storms are worrisome because they may adversely affect our agriculture and thus our food supply. A rise in sea level is a potentially serious problem because a significant portion of people live in or near coastal areas and coastal erosion is a difficult problem. Finally, if sea level does rise and more heat energy is fed into the atmosphere, the occurrence of large cyclones and tropical disturbances will be a greater hazard for people living in vulnerable areas.

- Adjustments to global warming include learning to live with the changes and attempting to mitigate warming through reduction of emissions of greenhouse gases. Given the uncertainty concerning global warming and the great expense involved in changing energy policies and land use it seems inevitable that the adjustment chosen will be to learn to live with change. A danger of this path is that climatic change may be rapid, resulting in problems with food production that will be difficult to address. The evidence suggests that some component of global warming as a result of anthropogenic processes is occuring, and thus it is prudent to reduce emissions of greenhouse gases. For example, carbon dioxide emissions may be reduced through energy conservation and the use of alternative energy sources. The actual strategy we take will likely be somewhere between full mitigation and adapting to the changes as they occur.

REEXAMINING THEMES AND ISSUES

Human Population: Much of the discussion in this chapter centers on human-induced climatic change at regional and global levels.

Burning of fossil fuels and deforestation, among other processes, have resulted in increased emissions of carbon dioxide into the atmosphere, potentially forcing climate change. Continued increase in the world's human population will place further demands on resources that result in continued increases in the emissions of greenhouse gases. Thus, as stated many times in this book, we need to control the increase in human population to maximize our chances of solving many environmental problems—such as global warming.

Sustainability: Through our emissions of greenhouse gases we are conducting global experiments, the results of which are difficult to predict. Because of this, obtaining sustainability in the future will be more difficult. If we do not know what the consequences or magnitude of human-induced climatic change will be, then it is difficult to predict how we might achieve sustainable development for future generations. Hopefully, some of the uncertainties concerning global climatic change will become more apparent in the coming years and we will be better prepared to develop plans that will work in addressing variables such as providing a sustainable food supply for the world's population.

Global Perspective: One of the major topics of this chapter is global warming; in particular we are interested in the potential effects that human activities may have on the subject. There seems little doubt that global warming is in fact occurring, but whether this is a natural process or human induced is still largely unknown. Our research on global warming directly centers on developing ways to look at what Earth is doing as a system with respect to climate; this is the essence of having a global perspective.

Urban World: In this chapter we have explored the effects of urbanization on the local climate in the urban environment. As large urban areas continue to expand, in the future we can expect that urban microclimate effects will likewise become more apparent. As a result it is important to incorporate the development of green zones and other such areas in urban planning to partially offset potentially adverse urban microclimate effects.

Values and Knowledge: It is becoming increasingly clear that people in industrialized countries as well as other areas of the globe are placing more value on a quality environment. Numerous research programs have been established to try to understand what factors effect global climate change and in particular if human activities are having a significant impact. It is expected that high priority will continue to be placed on understanding the global climate system and how it is likely to change in the future.

KEY TERMS

atmosphere *441*	Earth system science *455*	greenhouse effect *448*
climate *444*	global circulation models (GCMs) *456*	microclimate *444*
climatic change *446*	global warming *448*	urban dust dome *445*
deforestation *461*		

STUDY QUESTIONS

1. What is the composition of our atmosphere and how has life affected the atmosphere during the past several billion years?

2. What is meant by the urban dust dome and an urban microclimate? Compare a midlatitude city with a subhumid climate and with a lower latitude semiarid climate.

3. Why is there so much controversy concerning the greenhouse effect? What could be done to reduce the uncertainty?

4. What is meant by *anthropogenic greenhouse gas?* Discuss the different anthropogenic greenhouse gases in terms of their potential to cause global warming.

5. What are some of the major negative- and positive-feedback cycles present in the atmosphere that may increase or decrease potential global warming?

6. In terms of the effects of global warming, do you think that a change in climate patterns and

storm frequency and intensity is likely to be more serious than a global rise in sea level? Illustrate your answer with specific problems and areas where problems are likely to occur.

7. How would you criticize or defend the statement that the most likely adjustment to potential global warming is to do little or nothing and learn to live with change?

FURTHER READING

Anthes, A. R. 1992. *Meteorology,* 6th ed. New York: Macmillan. This short text provides a good overview of basic meteorology and atmospheric processes.

Goodess, C. M., Palutikof, J. P., and Davies, T. D. 1992. *The Nature and Causes of Climate Change: Assessing the Long-Term Future.* Lewis. A good text discussing short- and long-term climate changes and anthropogenic versus natural forces in global warming.

Houghton, J. 1994. *Global Warming: The Complete Briefing.* Lion. A review for the nonscientist of the current state of knowledge on a possible human-induced climate change, including discussions of ancient climates, weather forecasting, global freshwater use, and technological advances that could reduce the use of fossil fuels.

IPCC. 1991. *The Intergovernmental Panel on Climate Change Scientific Assessment.* New York: Oxford University Press. This is a detailed scientific review and assessment of global warming.

Mohnen, V. A., Goldstein, W., and Wang, C. W. 1991. "The Conflict over Global Warming," *Global and Environmental Change* 1(2): 109–123. This paper is a good one for detailed information on the global warming controversy.

Organisation for Economic Co-operation and Development, International Energy Agency. 1994. *The Economics of Climate Change.* Proceedings of an OECD/IEA Conference, OECD, IEA. These proceedings discuss the economics of climate change, differences of opinion on the matter, methods to link economic studies and climate change policy, and directions of international policy to deal with climate change.

Titus, J. G., and Narayanan, V. K. 1995. *The Probability of Sea Level Rise.* U.S. Environmental Protection Agency, Washington, DC. A close look at future sea-level rise due to global warming using mathematical models to predict sea-level changes.

INTERNET RESOURCES

U.S. Global Change Research Program: *http://www.usgcrp.gov/*— This interagency program conducts research to monitor and understand factors involved with a changing climate. Read findings from the Intergovernmental Panel of Climate Change, see what data are being collected and how data are changing policy, and check out what type of research is being conducted and why. Also find many good links to other global climate change internet sites.

Global Climate Information Programme: *http://www.doc.mmu. ac.uk/aric/gcciphm.html*—This program of the Atmospheric Research and Information Centre out of Manchester Metropolitan University, UK, has a number of good fact sheets on subjects of global warming, climate change, and the greenhouse effect.

Goddard Distributed Active Archive Center: *http://daac.gsfc.nasa. gov/*—From the Goddard Space Flight Center at NASA, this active archive has substantial resources. View data and satellite images, read about the FSFC research into atmospheric study,

and check out interesting facts on general atmospheric dynamics and chemistry.

Halogen Occultation Experiment home page: *http://haloedata. larc.nasa.gov/home.html*—From the Mission to Planet Earth program of NASA, HALOE uses satellites and sunlight to measure ozone depletion and the contribution of various chemicals to ozone depletion. This site describes current research findings and presents data.

Mauna Loa Observatory: *http://www.mloserv.mlo.hawaii.gov*— Provides access to sites for climate monitoring and change.

National Oceanic and Atmospheric Administration Climatic Data Center: *http://www.ncdc.noaa.gov/*—View current data on atmospheric ozone trends, temperature trends, and other atmospheric trends and find out what climate research NOAA is conducting.

National Oceanic and Atmospheric Administration El Niño Page: *http://www.elnino.noaa.gov*—Everything about El Niño.

Indonesian fires of 1997 at this city in Borneo caused very hazardous air pollution with the Pollution Standard Index over 800 (see Table 22.2). An APSI of 800 is off the scale! This photo taken during early evening hours showing vehicles with lights on and diminished visibility due to air pollution.

CHAPTER 22

AIR POLLUTION

CASE STUDY

London Smog and Indonesian Fires

In December 1952, air in London became stagnant and cloud cover blocked incoming solar radiation. The temperature dropped rapidly until at noontime it was about -1°C (30°F) and humidity climbed to 80%. Thick fog developed; cold and dampness increased demand for home heating. Because the primary fuel used in homes was coal, emissions of ash, sulfer oxides, and soot increased quickly. Stagnant air was filled with pollutants from home heating fuels and automobile exhaust. At the height of the crisis, visability was so reduced automobiles used headlights at midday. Between December 4 and 10, about 4,000 Londoners died from the pollution. The siege of smog ended when weather changed and air pollution was dispersed. Environment, not human activities, finally solved this problem. Since the beginning of the industrial revolution and before, people had survived in London in spite of the weather and pollution. What had gone wrong?

During the London smog crisis, stagnant weather conditions, combined with emissions from homes burning coal and from cars burning gasoline, exceeded the atmosphere's ability to remove or transform pollutants; even rapid natural mechanisms for removing sulfur dioxide were beyond their limit. As a result, sulfur dioxide remained in the air and

fog became acidic, adversely affecting people and other organisms, particularly vegetation. Human health effects were especially destructive because small acid droplets became fixed on larger particulates, which were easily drawn into the lungs.

The 1952 London smog crisis was a landmark event. Finally, the natural abilities of the atmosphere to serve as a sink for removal of wastes had been exceeded by human activities. The crisis was due, in part, to a reinforcing feedback situation. Burning fossil fuels added particulates to the air, increasing the formation of fog and decreasing visibility and light transmission; the dense, smoggy layer increased the dampness and cold and accelerated use of home heating fuels. As the weather and pollution worsened, more people burned fuel, which further worsened weather and pollution. Before 1952, London was known for its fogs; what was little known was the role of coal burning in exacerbating the fog conditions. The good news is, since 1952, London fogs associated with air pollution have been greatly reduced because coal has been replaced by cleaner natural gas as the primary home heating fuel.

Fifty-five years after the London smog event and following the most severe drought in 50 years, a serious air pollution event occurred in Indonesia

466

that dwarfed the London event. As a result of a strong El Niño (see Chapter 21), huge fires ravaged Indonesia for months, producing a thick, toxic haze of smoke. What went wrong?

Slash and burn has been part of farming in tropical rain forests for centuries. Each family burns a few hectares of rainforest, plants it, and after harvest moves on to a new area. Fire is the preferred choice of land clearing by farmers because the cost of clearing land by bulldozer is too expensive. It is generally thought that the process of slash and burn farming has not been particularly disturbing to the ecology of rain forests, as long as it is not too widespread and fires are carefully controlled.[1] Unfortunately in 1997 the combination of events related to El Niño and controlled burns (that burned out of control) resulted in what some think is one of the world's greatest environmental disasters. Because monsoon rains were late, people took advantage of dry weather to burn more forest. Although much of the blame was placed on small farmers, agriculture at an industrial scale also was responsible for clear-cut logging and burning. During the 1997 dry season at least 20,000 hectares of land burned. Vast amounts of smoke and particulate matter entered the atmosphere causing a severe pollution problem that resulted in 20 million Indonesians being treated for a variety of illnesses directly caused or aggravated by the fires. Smoke was so dense that a passenger airline crashed in Sumatra, killing 234 people. The Pollution Standards Index (PSI)—a measure of air quality—was at about 800 due to the fire. A PSI of 500 is considered very hazardous, and it is recommended that all people remain indoors. At a PSI of 500, premature death of ill and elderly people can be expected (see PSI discussion near the end of the chapter). A PSI of 800 is considered equivalent to smoking four packs of cigarettes a day! Burning land produced a dense, blinding haze, greatly reducing visibility. At one point people were so desperate for relief that they attempted to use surgical masks as filters, but this provided little protection. Near the end of summer, haze began to drift across the South China Sea to Malaysia and Singapore. Although many of the fires were extinguished with the arrival of monsoons, when the rains stopped in February 1998 more fires were again spotted. There is worry that there will be continued cycles of fire that will have a cumulative disastrous environmental impact. Certainly the damage is not limited to humans. Wildlife, including endangered species such as rhinoceros, and wilderness reserves also suffered damage.[1]

The London smog of 1952 and the Indonesian fires of 1997–98 are examples of serious, human induced air pollution. This chapter discusses the major air pollutants, urban air, acid rain, and control of air pollution.

LEARNING OBJECTIVES

The atmosphere has always been a sink (place for deposition and storage) for gaseous or particulate wastes. When the amount of waste entering the atmosphere in an area exceeds the ability of the atmosphere to disperse or degrade the pollutants, problems result. After reading this chapter, you should understand:

- Why human activities that pollute the air, combined with meteorological conditions, may exceed the natural abilities of the atmosphere to remove wastes.

- What the major categories and sources of air pollutants are.

- Why air pollution problems are different in different regions.

- What acid rain is and how it is produced.

- What the environmental impacts of acid rain are and how they might be minimized.

- Why, from an environmental standpoint, the best strategies for controlling air pollution are energy efficiency and conservation.

- What methods are useful in the collection, capture, or retention of pollutants before they enter the atmosphere.

- What air quality standards are and why they are important.

- Why the economics of air pollution is controversial and difficult.

22.1 POLLUTION OF THE ATMOSPHERE

As the fastest moving fluid medium in the environment, the atmosphere has always been one of the most convenient places to dispose of unwanted materials. Ever since fire was first used by people, the atmosphere has been a sink for waste disposal.

People have long recognized the existence of atmospheric pollutants, both natural pollutants and those induced by humans. Leonardo da Vinci wrote in 1550 that a blue haze formed from materials emitted into the atmosphere from plants; he had observed a natural photochemical smog whose cause is still not completely understood. The phenomenon of acid rain was first described in the seventeenth century, and by the eighteenth century it was known that smog and acid rain damaged plants in London. Beginning with the industrial revolution in the eighteenth century, air pollution became more noticeable. The word *smog* was probably introduced by a physician at a public health conference in 1905 to denote poor air quality resulting from a mixture of smoke and fog.

A major event in Donora, Pennsylvania, in 1948, was responsible for increasing research on air pollution. This event caused 20 deaths and 14,000 illnesses, and people recognized that meteorological conditions were part of the production of dangerous smog. This was reinforced by the 1952 London smog crisis, after which regulations to control air quality began. Today, in the United States and other countries, legislation to reduce emission of air pollutants has been successful, but more needs to be done. Chronic exposure to high levels of air pollutants such as sulfur dioxide, ozone, carbon monoxide, particulates, and nitrogen oxides continues to contribute to illnesses that kill people around the world.

What are the chances that another killing smog pollution event may occur again somewhere in the world? Unfortunately, the chances seem all too likely, given the tremendous amount of air pollution that is occurring in some large cities today. For example, Beijing might be a candidate; the city uses an immense amount of coal, and a cough is so pervasive among its residents that it is often called the Beijing cough. Another likely candidate is Mexico City, which has one of the worst air pollution problems found anywhere in the world today.

General Effects of Air Pollution

Air pollution has considerable effects on many aspects of our environment: visually aesthetic resources, vegetation, animals, soils, water quality, natural and artificial structures, and human health. Air pollutants affect visual resources by discoloring the atmosphere and by reducing visual range and atmospheric clarity so that the visual contrast of distant objects is decreased. We cannot see as far in polluted air, and what we do see has less color contrast. These effects were once limited to cities, but they now extend to some wide-open spaces of the United States. For example, near the area where the borders of New Mexico, Arizona, Colorado, and Utah meet, emissions from the Four Corners fossil fuel–burning power plant are altering the visibility in a region where in the past visibility was normally 80 km (50 miles) from a mountaintop on a clear day.[2]

The effects of air pollution on vegetation include damage to leaf tissue, needles, or fruit; reduction in growth rates or suppression of growth; increased susceptibility to a variety of diseases, pests, and adverse weather; and the disruption of reproductive processes.[2]

In some parts of the world, air pollution is thought to be a significant factor in the human death rate. For example, it has been estimated that in Athens the number of deaths is several times higher on days that are heavily polluted and in Hungary, where air pollution has been a horrendous problem in recent years, it may contribute to as many as 1 in 17 deaths. The United States is certainly not immune to health problems related to air pollution. The most polluted air in the country is found in the Los Angeles urban area, where millions of people are exposed to unhealthy air. It is estimated that as many as 150 million people live in areas of the United States where exposure to air pollution contributes to lung disease, which causes more than 300,000 deaths per year. Air pollution is directly responsible for annual health costs of about $50 billion.[3]

Air pollutants can affect human health in several ways (Figure 22.1). The effects on an individual depend on the dose or concentration of exposure (see the discussion of dose response in Chapter 14) and other factors, including individual susceptibility. Some of the primary effects of air pollutants include toxic poisoning, causing cancer, birth defects, eye irritation, and irritation of the respiratory system; an increased susceptibility to viral infections, causing pneumonia and bronchitis; an increased susceptibility to heart disease; and aggravation of chronic diseases, such as asthma and emphysema. In urban areas people suffering from respiratory diseases are most likely to be affected by air pollutants; healthy people tend to acclimate to pollutants in a relatively short period of time. However, this is a physiological tolerance (see Chapter 14). Urban air pollution is a serious health problem. Many of the pollutants have *synergistic effects* (in which, as noted in Chapter 10, the combined effects are greater than the sum of separate effects). For example, sulfate and nitrate may attach to small particles in the air, facilitating their in-

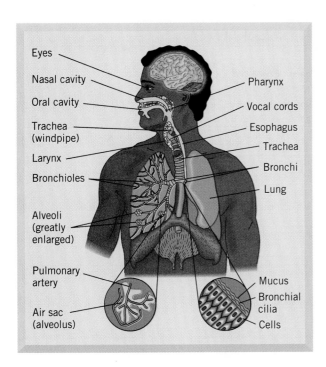

FIGURE 22.1 Idealized diagram showing some of the parts (brain and cardiovascular–lung systems) of the human body that may be damaged by common air pollutants. The most severe health risks to normal exposures are related to particulates. Others of concern include carbon monoxide, photochemical oxidants, sulfur dioxide, and nitrogen oxides. Toxic chemicals and tobacco smoke also can cause chronic or acute health problems.

halation deep into lung tissue, where greater damage to the lungs may occur than is attributable to either of these pollutants alone (this is a synergistic human health effect). This phenomenon has obvious health consequences; consider joggers breathing deeply of particulates as they run along the streets of a city.

The effects of air pollutants on vertebrate animals include impairment of the respiratory system; damage to eyes, teeth, and bones; increased susceptibility to disease, pests, or other stress-related environmental hazards; decreased availability of food sources (such as vegetation impacted by air pollutants); and reduced ability to reproduce.[2]

Air pollution can degrade soil and water resources when pollutants from the air are deposited. Soils and water may become toxic from the deposition of various pollutants. Soils may also be leached of nutrients by pollutants that form acids. The effects of air pollution on human-made structures include discoloration, erosion, and decomposition of building materials; these effects are discussed when we explore the topic of acid rain later in this chapter.

Sources of Air Pollution

Many of the pollutants in our atmosphere have natural as well as human-related origins. Examples of natural emissions of air pollutants include the release of gases, such as sulfur dioxide, from volcanic eruptions; the release of hydrogen sulfide from geyser and hot spring activities and by biological decay from bogs and marshes; an increased concentration of ozone in the lower atmosphere as a result of unstable meteorological conditions, such as violent thunderstorms; and the emission of a variety of particles from wildfires and windstorms.[2]

Air pollutants and sources are shown in Table 22.1. The data suggest that, with the exception of sulfur and nitrogen oxides, natural emissions of air pollutants exceed human-produced input. Nevertheless, it is the human component that is most abundant in urban areas and that leads to the most severe air pollution problems for human health.

The two major kinds of air pollution sources are stationary sources and mobile sources. **Stationary sources** are those that have a relatively fixed location. These include point sources, fugitive sources, and area sources. **Point sources,** as discussed in Chapter 14, are stationary sources that emit air pollutants from one or more controllable sites, such as smokestacks of power plants at industrial sites. **Fugitive sources** are types of stationary sources that generate air pollutants from open areas exposed to wind processes. Examples include dirt roads, construction sites, farmlands, storage piles, surface mines, and other exposed areas from which particulates may be removed by wind. **Area sources,** also discussed in Chapter 14, are locations from which air pollutants are emitted from a well-defined area within which are several sources, for example, small urban communities or areas of intense industrialization within urban complexes or agricultural areas sprayed with herbicides and pesticides. **Mobile sources** are emitters of air pollutants that move

TABLE 22.1 Major Natural and Human-Produced Components of Air Pollutants

Air Pollutants	Natural	Emissions (% of total) Human-Produced	Human-Produced Component of Major Sources	Percent
Particulates	89	11	Industrial processes	51
			Combustion of fuels (stationary sources)	26
Sulfur oxides (SO$_x$)	50	50	Combustion of fuels (stationary sources, mostly coal)	78
			Industrial processes	18
Carbon monoxide (CO)	91	9	Transportation (mostly automobiles)	75
			Agricultural burning	9
Nitrogen dioxide (NO$_2$)		Nearly all	Transportation (mostly automobiles)	52
			Combustion of fuels (stationary sources, mostly natural gas and coal)	44
Ozone (O$_3$)	A secondary pollutant derived from reactions with sunlight, NO$_2$, and oxygen (O$_2$)		Concentration present depends on reaction in lower atmosphere involving hydrocarbons and thus automobile exhaust	
Hydrocarbons (HC)	84	16	Transportation (mostly automobiles)	56
			Industrial processes	16
			Evaporation of organic solvents	9
			Agricultural burning	8

from place to place while yielding emissions. These include automobiles, trucks, buses, aircraft, ships, and trains.[2]

22.2 AIR POLLUTANTS*

There are two main groups of air pollutants: primary and secondary. **Primary pollutants** are those emitted directly into the air. They include particulates, sulfur dioxide, carbon monoxide, nitrogen oxides, and hydrocarbons. **Secondary pollutants** are pollutants produced through reactions between primary pollutants and normal atmospheric compounds. For example, ozone forms over urban areas through reactions of primary pollutants, sunlight, and natural atmospheric gases. Thus ozone is a secondary pollutant that is produced on bright, sunny days in areas where there is much primary pollution. This phenomenon has been particularly well documented for cities in southern California, such as Los Angeles, but occurs worldwide under appropriate conditions.

Again, the primary pollutants that account for nearly all air pollution problems are carbon monoxide, particulates, hydrocarbons, nitrogen oxides, and sulfur oxides. Each year well over a billion metric tons of these materials enters the atmosphere from human-related processes. About half is carbon monoxide; the other four pollutants listed account for a few percent each. At first glance this amount of pollutants appears to be very large.

If the air pollutants were uniformly distributed in the atmosphere, the concentration would be only a few parts per million by weight. Unfortunately, pollutants are not uniformly distributed but tend to be released, produced, and concentrated locally or regionally. For example, in large cities weather and climatic conditions combine with the high density of people and industry to produce air pollution problems.

The major air pollutants occur either in gaseous forms or as particulate matter (PM). The gaseous pollutants include sulfur dioxide (SO$_2$), nitrogen oxides (NO$_x$), carbon monoxide (CO), ozone (O$_3$), hydrocarbons (HC), hydrogen sulfide (H$_2$S), and hydrogen fluoride (HF). Particulate matter pollutants are particles of solid or liquid substances and may be organic or inorganic.

Sulfur Dioxide

Sulfur dioxide (SO$_2$) is a colorless and odorless gas normally present at Earth's surface at low concentrations. One of the significant features of SO$_2$ is that once it is emitted into the atmosphere it may be converted through complex reactions to fine particulate sulfate (SO$_4$). The major anthropogenic source of sulfur dioxide is the burning of fossil fuels, mostly coal in power plants. Another major source comprises a variety of industrial processes, ranging from petroleum refining to the production of paper, cement, and aluminum.

Adverse effects associated with sulfur dioxide depend on the dose or concentration present (see Chapter 14) and include corrosion of paint and met-

*This section is summarized in part from Air Resources Management Manual, National Park Service, 1984.

als and injury or death to animals and plants. Crops such as alfalfa, cotton, and barley are especially susceptible. Sulfur dioxide is capable of causing severe damage to human and other animal lungs, particularly in the sulfate form. It is also an important precursor to acid rain, as are nitrogen oxides.

Nitrogen Oxides

Nitrogen oxides (NO_x) are emitted largely in two forms: nitric oxide (NO) and nitrogen dioxide (NO_2) (the x in NO_x refers to the number of oxygen atoms present in the gas molecule). Although nitrogen oxides occur in many forms in the atmosphere, only NO and NO_2 are subject to emission regulations. The most important of these is NO_2, which is a visible yellow brown to reddish brown gas. A major concern with nitrogen dioxide is that it may be converted by complex reactions in the atmosphere to the ion (NO_3^{2-}) within small water particles, impairing visibility. Additionally, nitrogen dioxide is one of the main pollutants that contribute to the development of smog (as is NO) and is a major contributor to acid rain (both topics are discussed later in this chapter). Nearly all NO_2 is emitted from anthropogenic sources; the two major contributors are automobiles and power plants that burn fossil fuels.

The environmental effects of nitrogen oxides on humans are variable but include the irritation of eyes, nose, throat, and lungs and increased susceptibility to viral infections, including influenza (which can cause bronchitis and pneumonia).[2] Nitrogen oxides suppress plant growth and damage leaf tissue. When the oxides are converted to their nitrate form in the atmosphere, they impair visibility. However, when nitrate is deposited on the soil, it can promote plant growth through nitrogen fertilization.

Carbon Monoxide

Carbon monoxide (CO) is a colorless, odorless gas that at very low concentrations is extremely toxic to humans and other animals. The high toxicity results from a striking physiological effect, namely, that carbon monoxide and hemoglobin in blood have a strong natural attraction for one another. Hemoglobin in our blood will take up carbon monoxide nearly 250 times more rapidly than it will oxygen. Therefore, if there is any carbon monoxide in the vicinity, a person will take it in very readily, with potentially dire effects. Many people have been accidentally asphyxiated by carbon monoxide produced from incomplete combustion of fuels in campers, tents, and houses. The effects depend on the dose or concentration of exposure and range from dizziness and headaches to death. Carbon monoxide is particularly hazardous to people with known heart disease, anemia, or respiratory disease. In addition, it may cause birth defects, including mental retardation and impairment of growth of the fetus.[2] Finally, the effects of carbon monoxide tend to be worse at higher altitudes, where oxygen levels are naturally lower.

Approximately 90% of the carbon monoxide in the atmosphere comes from natural sources, and the other 10% comes mainly from fires, automobiles, and other sources of incomplete burning of organic compounds. Concentrations of carbon monoxide can build up and cause serious health effects in a localized area.

Photochemical Oxidants

Photochemical oxidants result from atmospheric interactions of nitrogen dioxide and sunlight. The most common photochemical oxidant is ozone (O_3), a colorless gas with a slightly sweet odor. In addition to ozone, a number of photochemical oxidants known as PANs (peroxyacyl nitrates) occur with photochemical smog.

Ozone is a form of oxygen in which three atoms of oxygen occur together rather than the normal two. Ozone is relatively unstable and releases its third oxygen atom readily, so that it oxidizes or burns things more readily and at lower concentrations than does normal oxygen. Ozone is sometimes used to sterilize; for example, bubbling ozone gas through water is a method used to purify water. The ozone is toxic to and kills bacteria and other organisms in the water. When it is released into the air or produced in the air, ozone may injure living things.

Ozone is very active chemically, and it has a short average lifetime in the air. Because of the effect of sunlight on normal oxygen, ozone forms a natural layer high in the atmosphere (stratosphere). This ozone layer protects us from harmful ultraviolet radiation from the sun. Ozone is considered a pollutant when present above the National Air Quality Standard threshold concentration in the lower atmosphere but is beneficial in the stratosphere. The important topic of stratospheric ozone depletion is discussed in Chapter 24.

The major sources of the chemicals that produce oxidants, and particularly ozone, are automobiles, fossil fuel burning, and industrial processes that produce nitrogen dioxide. Due to the nature of its formation (see discussion of smog in Section 22.3), ozone is a difficult pollutant to regulate and is the pollutant whose health standard is most frequently exceeded in urban areas of the United States.[4] The adverse environmental effects of ozone and other oxidants, as with other pollutants, depend in part on the dose or concentration of exposure and include damage to plants and animals as well as to materials such as rubber, paint, and textiles.

Effects of ozone on plants can be subtle. At very low concentrations, ozone can reduce growth rates while not producing any visible injury. At higher concentrations, ozone kills leaf tissue and, if pollutant levels remain high, killing whole plants. The death of white pine trees along highways in New England is believed due in part to ozone pollution. Ozone's effect on animals, including people, involves various kinds of damage, especially to eyes and the respiratory system.

Hydrocarbons

Hydrocarbons are compounds composed of hydrogen and carbon. There are thousands of such compounds, including natural gas or methane (CH_4), butane (C_4H_{10}), and propane (C_3H_8). Analysis of urban air has identified many different hydrocarbons, some of which are more reactive with sunlight (producing photochemical smog) than others. Potential adverse effects of hydrocarbons are numerous; many at a specific dose or concentration are toxic to plants and animals or may be converted to harmful compounds through chemical changes that occur in the atmosphere. Over 80% of hydrocarbons (which are primary pollutants) that enter the atmosphere are emitted from natural sources. The most important anthropogenic source is the automobile (see Table 22.1). Hydrocarbons also escape to the atmosphere when a car's tank is filled with gasoline or gasoline is spilled and it evaporates. Vapor recovery systems on hoses that feed gasoline to the tank are now required in many urban areas and are helping to reduce the problem of hydrocarbons (vapors) escaping.

Hydrogen Sulfide

Hydrogen sulfide (H_2S) is a highly toxic and corrosive gas, easily identified by its rotten-egg odor. Hydrogen sulfide is produced from natural sources, such as geysers, swamps, and bogs, as well as from human sources, such as petroleum processing–refining and metal smelting. The potential effects of hydrogen sulfide include functional damage to plants and health problems ranging from toxicity to death for humans and other animals.

Hydrogen Fluoride

Hydrogen fluoride (HF) is a gaseous pollutant that is released primarily by aluminum production, coal gasification, and the burning of coal in power plants. Hydrogen fluoride is extremely toxic, and even a small concentration (as low as 1 ppb) may cause problems for plants and animals.

Other Hazardous Gases

It is a rare month when the newspapers do not carry a story of an accident that releases toxic chemicals in a gaseous form into the atmosphere. In these incidents it is often necessary to evacuate people from the area until the leak is stopped or the gas dispersed to a nontoxic level. Chlorine gases are often the culprit, but a variety of other materials used in chemical and agricultural processes may be involved.

Some chemicals are so toxic that extreme care must be taken to ensure they do not enter the environment. The danger of such chemicals was demonstrated on December 3, 1984, when a toxic chemical (stored in liquid form) at a pesticide plant leaked, vaporized, and formed a toxic cloud over a 64-km^2 (about 25-mi^2) area of Bhopal, India. The gas leak lasted less than 1 hour, yet over 2000 people were killed and more than 15,000 injured by the gas, which causes severe irritation (burns on contact) to eyes, nose, throat, and lungs (Figure 22.2).

Particulate Matter

Particulate matter is the term used for the varying mixtures of particles suspended in the air we breathe. Particles are present everywhere, but high concentrations and/or specific types of particles have been found to present a serious danger to human health. Farming adds considerable particulate matter to the atmosphere, as do desertification and volcanic eruptions. Nearly all industrial processes, as well as burning fossil fuels, release particulates into the atmosphere. Much particulate matter is easily visible as smoke, soot, or dust; other particulate matter is not easily visible. Particulates include materials such as airborne asbestos particles and small particles of heavy metals, such as arsenic, copper, lead, and zinc, which are usually emitted from industrial facilities such as smelters.

Of particular importance are very fine particle pollutants less than 2.5 μm in diameter (2.5 millionths of a meter) (Figure 22.3). For comparison, the diameter of human hair is about 60 μm to 150 μm. Fine particles are easily inhaled into the lungs where they can be absorbed by the bloodstream or remain embedded for a long period of time. Among the most significant of fine particulate pollutants are sulfates and nitrates. These are mostly secondary pollutants produced in the atmosphere through chemical reactions between normal atmospheric constituents and sulfur dioxide and nitrogen oxides. These reactions are important in the formation of sulfuric and nitric acids in the atmosphere and are

FIGURE 22.2 Evacuation from Bhopal, India, in 1984, following a leak of toxic gas from a pesticide plant.

further discussed when we consider acid rain.[2] When measured, particulate matter is often referred to as *total suspended particulates* (TSPs).

Particulates affect human health, ecosystems, and the biosphere. In the United States, particulate air pollution contributes to the death of 60,000 people annually.[5] Recent studies estimate that 2% to 9% of human mortality in cities is associated with particulate pollution; risk of mortality is about 15% to 25% higher in cities with the highest levels of fine particulate pollution.[6] Particulates that enter lungs may lodge there and have chronic effects on respiration. Certain materials, such as asbestos, are particularly dangerous. Dust raised by road building and plowing and deposited on surfaces of green plants may interfere with absorption of carbon dioxide and oxygen and release of water. Heavy dust may affect breathing of animals. Particulate matter is particularly hazardous to the elderly, and those with respiratory problems such as asthma. There is a direct relationship between particulate pollution and increased hospital admissions for respiratory distress (see case study opening this chapter). Particulates associated with large construction projects may kill organisms and damage large areas, changing species composition, altering food chains, and affecting ecosystems. Modern industrial processes have greatly increased the total suspended particulates in Earth's atmosphere. Particulates block sunlight and may cause changes in climate. Such changes have lasting effects on the biosphere.

FIGURE 22.3 Size of selected particulates. Shaded area shows size range that produces the greatest lung damage. (*Source:* Modified from Fig. 7-8, p. 244 from *Chemistry, Man and Environmental Change: An Integrated Approach* by J. Calvin Giddings. Copyright © 1973 by J. Calvin Giddings. Reprinted by permission of HarperCollins Publishers, Inc.)

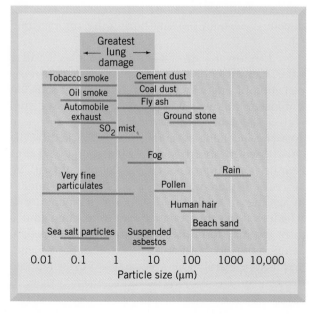

Asbestos

Asbestos is the term for several minerals that have the form of small elongated particles. In the past, asbestos was treated casually, and people working in asbestos plants were not protected. Asbestos was used in building and electrical insulation, roofing material, and brake pads for vehicles. As a result, a considerable amount of asbestos fibers have been spread throughout industrialized countries, especially in urban environments of Europe and North America. In one case, the products containing asbestos were sold in burlap bags that were recycled by nurseries and other secondary businesses, thus further spreading the pollutant. Some types of asbestos particles are believed to be carcinogenic, or carry with them carcinogenic materials, and so must be carefully controlled (see the discussion of asbestos in Chapter 14).

Lead

Lead is an important constituent of automobile batteries and other industrial products. When lead is added to gasolines, it helps protect the engine and promotes more even fuel consumption. The lead in gasoline is emitted into the environment in the exhaust. In this way, lead has been spread widely around the world and has reached high levels in soils and waters along roadways.

Once released, lead can be transported through the air as particulates to be taken up by plants through the soil or deposited directly on plant leaves. Thus it enters terrestrial food chains. When lead is carried by streams and rivers, deposited in quiet waters, or transported to oceans or lakes, it is taken up by aquatic organisms and thus enters aquatic food chains.

The concentration of lead measured in Greenland glaciers was essentially zero in A.D. 800 and reached measurable levels with the beginning of the industrial revolution in Europe. The lead content of the glacial ice increased steadily from 1750 until the mid-twentieth century (about 1950), when the rate of accumulation by the glaciers began to increase rapidly. This sudden upsurge reflects the rapid growth in the use of lead additives in gasoline. Lead reaches Greenland as airborne particulates and via seawater. The accumulation of lead in the Greenland ice illustrates that our use of heavy metals in this century has reached a point where the entire biosphere is affected. The reduction and eventual elimination of lead in gasoline is a good start. Lead has been removed from nearly all gasoline in the United States and Canada and is being phased out in much of Europe.

22.3 URBAN AREAS AND AIR POLLUTION

Wherever there are many sources emitting air pollutants over a wide area (whether we are talking about automobile emissions in Los Angeles or smoke from wood-burning stoves in Vermont), there is a potential for the development of smog. Whether air pollution develops depends on the topography and on weather conditions, because these factors determine the rate at which pollutants are transported away from their sources and converted to harmless compounds in the air. When the rate of production exceeds the rate of degradation and of transport, dangerous conditions may develop, as illustrated in the case study that opened this chapter.

Influences of Meteorology and Topography

Meteorological conditions can determine whether air pollution is a nuisance or a major health problem. The primary adverse effects of air pollution are damage to green plants and aggravation of chronic illnesses in people; most of these effects are due to relatively low level concentrations of toxins over a long period of time. Periods of pollution in the Los Angeles basin or other areas generally do not directly cause large numbers of deaths. However, as with the London and Pennsylvania cases mentioned earlier, serious pollution events (disasters) can develop over a period of days and lead to increases in illnesses and deaths.

In the lower atmosphere, restricted circulation associated with inversion layers may lead to pollution events. An **atmospheric inversion** occurs when warmer air is found above cooler air, and it poses a particular problem when there is a stagnated air mass. Figure 22.4 shows two types of developing inversions that may worsen air pollution problems. In the upper diagram, which is somewhat analogous to the situation in the Los Angeles area, descending warm air forms a semipermanent inversion layer. Because the mountains act as a barrier to the pollution, polluted air moving in response to the sea breeze and other processes tends to move up canyons, where it is trapped. The air pollution that develops occurs primarily during the summer and fall.

The lower part of Figure 22.4 shows a valley with relatively cool air overlain by warm air. An explanation of one way in which this type of situation can occur follows. When cloud cover associated with a stagnant air mass develops over an urban area, the incoming solar radiation is blocked by the clouds, which absorb some of the energy and thus heat up. On the ground, or near Earth's surface, the air cools. If there has been a fair amount of humidity, as the air cools the dew point (temperature at which water vapor condenses) is reached and a thick fog may form. Because the air is cold, people living in the city burn more fuel to heat their homes and factories, and more pollutants are delivered into the atmosphere. As long as the stagnant conditions

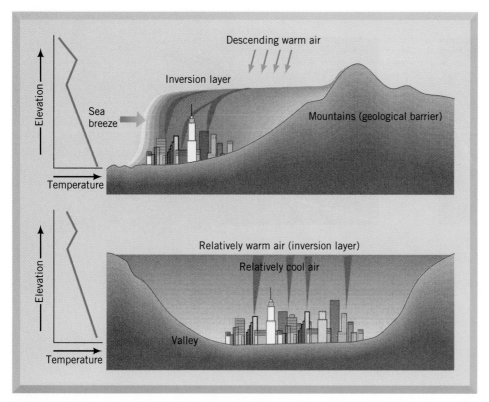

FIGURE 22.4 Two causes for the development of a temperature inversion, which may aggravate air pollution problems.

exist, the pollutants will build up. It was this mechanism that caused the deadly 1952 London smog.

Cities situated in a valley or topographic bowl surrounded by mountains are more susceptible to smog problems than are cities in open plains. Surrounding mountains and the occurrence of temperature inversions prevent the pollutants from being transported by winds and weather systems. The production of air pollution is particularly well documented for Los Angeles, which has mountains surrounding part of the urban area and lies within a region that tends to have stagnating air conditions that promote air pollution (Figure 22.5).

Potential for Urban Air Pollution

The potential for air pollution in urban areas is determined by the following factors: the rate of emission of pollutants per unit area; the distance downwind that a mass of air may move through an urban area; the average speed of the wind; and finally, the height to which potential pollutants may be thoroughly mixed in the lower atmosphere (Figure 22.6).[7] The concentration of pollutants in the air is directly proportional to the first two factors. That is, as either the emission rate or downwind travel distance increases, so will the concentration of pollutants in the air. A good example is provided by the Los Angeles basin. If there is a wind from the ocean, the coastal side of cities such as Santa Monica or Malibu will experience much less air pollution than will the inland side of those cities. Conversely, if there is a wind coming off the inland desert and down from the mountains, the air will be more polluted at the coast.

City air pollution decreases with increases in the other two factors, which are meteorological: the wind velocity and the height of mixing. The stronger

FIGURE 22.5 Part of southern California showing the Los Angeles basin (south coast air basin). (*Source:* Modified after S. J. Williamson, *Fundamentals of Air Pollution,* © 1973, by Addison-Wesley, Reading, Mass.)

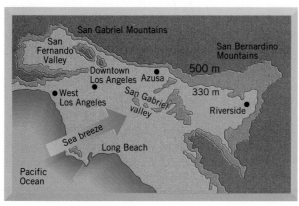

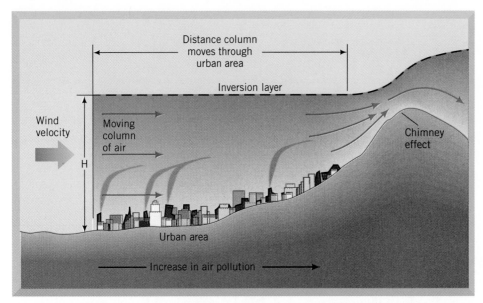

FIGURE 22.6 The higher the wind velocity and the thicker the mixing layer (shown here as H), the less is the potential air pollution. The greater the emission rate and the longer the downwind length of the city, the greater the air pollution. The chimney effect allows polluted air to move over a mountain down into an adjacent valley.

the wind and the higher the mixing layer, the lower the pollution. Assuming a constant rate of emission of air pollutants as the column of air moves through the urban area, it will collect more and more pollutants. The inversion layer acts as a lid for the pollutants, but near a geologic barrier, such as a mountain, there may be a chimney effect, in which the pollutants spill over the top of the mountain (see Figures 22.5 and 22.6). This effect has been noticed in the Los Angeles basin, where pollutants may climb several thousand meters, damaging mountain pine trees and other vegetation and spoiling the air of mountain valleys.

Smog

There are two major types of smog: **photochemical smog,** which is sometimes called *L.A.-type smog,* or *brown air;* and **sulfurous smog,** which is sometimes referred to as *London-type smog, gray air,* or *industrial smog.* Solar radiation is particularly important in the formation of photochemical smog (Figure 22.7). The reactions that occur in the development of photochemical smog are complex and involve both nitrogen oxides (NO_x) and organic compounds (hydrocarbons).

The development of photochemical smog is directly related to automobile use. Figure 22.8 shows a characteristic pattern in terms of how the nitrogen oxides, hydrocarbons, and oxidants (mostly ozone) vary throughout a typically smoggy day in southern California. Early in the morning, when commuter traffic begins to build up, the concentrations of ni-

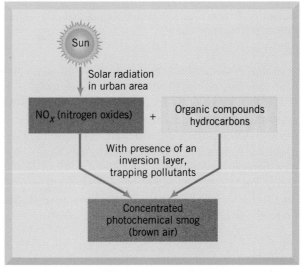

FIGURE 22.7 How photochemical smog may be produced.

trogen oxide (NO) and hydrocarbons begin to increase. At the same time, the amount of nitrogen dioxide NO_2 may decrease, because sunlight breaks it down to produce NO plus atomic oxygen (NO + O). The atomic oxygen (O) is then free to combine with molecular oxygen (O_2) to form ozone (O_3), so the concentration of ozone also increases after sunrise. Shortly thereafter, oxidized hydrocarbons react with NO to increase the concentration of NO_2 by midmorning. This reaction causes the NO concentration to decrease and allows ozone to build up, producing the midday peak in ozone and minimum in NO. As the smog matures, visibility may be greatly

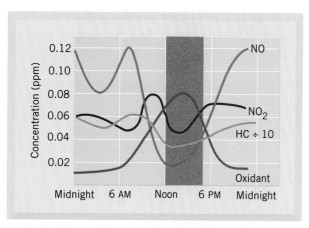

FIGURE 22.8 Development of photochemical smog over the Los Angeles area on a typical warm day.

(a)

(b)

FIGURE 22.9 The city of Los Angeles, California, on (*a*) a clear day and (*b*) a smoggy day.

reduced (Figure 22.9) owing to light scattering by aerosols.

Sulfurous smog is produced primarily by burning coal or oil at large power plants. Sulfur oxides and particulates combine under certain meteorological conditions to produce a concentrated sulfurous smog (Figure 22.10).

Future Trends for Urban Areas

The optimistic view concerning future air pollution in urban areas is that air quality will improve because we know so much about the sources of air pollution and have developed effective ways to reduce pollution. The pessimistic view, however, is that even though we know a lot about the sources and how to reduce pollution, population pressures and economics will dictate what is likely to happen in many parts of the world, and the result will be poorer air quality (more air pollution) in many locations. The actual situation in the 1990s and into the next century is likely to be a mixture of the optimistic and pessimistic points of view. Large urban areas in developing countries will probably experience a reduction in air quality even as they attempt to improve the situation, because the population and economic factors will likely outweigh pollution abatement.

Large urban areas in developed and more affluent countries, however, may well experience improved air quality in the coming years. For example, the Los Angeles urban area, with the worst air quality in the United States, is coming to grips with its air pollution problem. The people studying air pollution in the Los Angeles region now understand that pollution abatement will take massive efforts much different from past strategies, which have been more of a Band-Aid approach. A new, controversial multifaceted air quality plan being discussed involves the

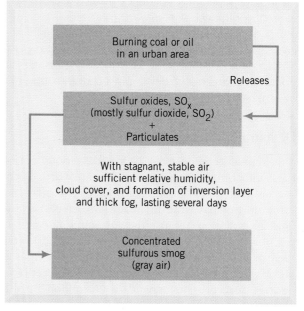

FIGURE 22.10 How concentrated sulfurous smog and smoke might develop.

entire urban region and includes the following aspects[2]:

- strategies to discourage automobile use and reduce the number of cars;
- stricter emission controls for automobiles;
- a requirement for a certain number of zero-pollutant automobiles (electric cars);
- a requirement for gasoline to be reformulated to burn cleaner;
- improvements in public transportation and incentives for people to use it;
- mandatory carpooling; and
- increased controls on industrial activities and household activities that are known to contribute to air pollution.

At the household level, for example, common materials such as paints and solvents will be reformulated so that their fumes will cause less air pollution, and eventually there may be a ban on certain equipment, such as gasoline-powered lawn mowers.

Even though southern California's air quality is the worst in the United States, there are encouraging signs that improvements have occurred. For example, from 1955 to 1992 the peak level of ozone (considered one of the best indicators of air pollution) declined from 680 to 300 ppb. This occurred in spite of the fact that during the period the population nearly tripled and the number of motor vehicles quadrupled![8] Nevertheless, air quality in southern California remains the nation's worst, and even if all the aforementioned controls in urban areas are implemented, air quality may still be a significant problem in coming decades. This is particularly true if the urban population continues to increase because even if pollution *per capita* (per person) is decreased, the overall amount of pollution may increase if the number of people continue to grow rapidly. With this in mind we will now consider air pollution problems of developing countries that are experiencing rapid population growth.

Developing Countries

As discussed earlier, cities in less developed countries with burgeoning populations are particularly susceptible to air pollution now and in the future. They do not have the financial base necessary to fight air pollution because they are more concerned with basic survival and finding ways to house and feed their growing populations. A good example is Mexico City, with a present population of 23 million people, projected to expand to 26 million by the end of the century, perhaps making it the largest urban area in the world. Cars, buses, industry, and power plants in Mexico City emit hundreds of thou-

sands of metric tons of pollutants into the atmosphere each year. The city is at an elevation of about 2255 m (7400 ft) in a natural basin surrounded by mountains, a perfect situation for a severe air pollution problem. It is becoming a rare day in Mexico City when the mountains can be seen, and physicians report that there has been a steady increase in respiratory diseases. Headaches, irritated eyes, and sore throats are common when the pollution settles in. Doctors advise parents to take their children out of the city permanently!

The people in Mexico City do not need to be told they have an air pollution problem; it is all too apparent. Developing a successful strategy to improve the quality of the air in the urban area is difficult, however. Two novel ideas have been suggested: using hundreds of large fans to blow the pollution away and excavating huge cuts in the mountains south of the city to allow the air pollutants to escape. Neither of these has been taken seriously.

A major source of pollutants in Mexico City is motor vehicles. It has been estimated that the 50,000 buses and taxis and several million automobiles in the city are responsible for much of the air pollution. Most of these vehicles are old and in poor running condition, and they pump immense amounts of pollutants into the atmosphere. Another major source of air pollution is leaks of liquefied petroleum gas (LPG) used in homes for cooking and heating water. The leaking LPGs (hydrocarbons) produce atmospheric precursors to formation of ozone, a major component of urban photochemical smog. It is possible that LPG leaks in Mexico City are responsible for a significant portion of the city's ozone pollution.[9] In an attempt to reduce the air pollution in the urban area, a large oil refinery was shut down. For nearly 60 years the refinery had emitted nearly 90,000 metric tons of air pollutants into the atmosphere annually. Although shutting down the refinery and ordering (as was recently done) thousands of other industrial plants to relocate will help the air quality of the urban area, these plants are not the primary source of pollutants. If the city is unable to control the increase in population and the use of buses, taxis, automobiles, and leaks of liquefied petroleum gas, air pollution will continue to be a serious problem for many years. To help address the problem, the license plate number on automobiles dictates on what days an automobile may be driven in the city. This regulation has resulted in more affluent people purchasing an additional automobile to avoid restriction of their driving. If the pollution problem is not controlled, it is quite possible that Mexico City will eventually have a pollution event of catastrophic proportions that causes thousands of deaths.

CLOSER LOOK 22.1

AIR QUALITY PROBLEMS IN REMOTE AREAS

The North Slope of Alaska is a vast strip of land approximately 200 km (125 miles) wide that is considered by many to be one of the last unspoiled wilderness areas left on Earth. It seems logical to assume that air in the Arctic environments of Alaska would have pristine quality, except perhaps near areas where petroleum is being vigorously developed. However, ongoing studies suggest that the North Slope has an air pollution problem that originates from sources in Eastern Europe and Eurasia.[10] It is suspected that pollutants from the burning of fossil fuels in Eurasia are transported via the jet streams, moving with speeds that may exceed 400 km (250 miles) an hour, northeast from Eurasia over the North Pole and eventually to the North Slope of Alaska. There the air mass slows, stagnates, and produces what is known as the Arctic haze. Concentrations of air pollutants, which include oxides of sulfur and nitrogen, are sufficient that the air quality is being compared with that of some eastern cities, such as Boston. Air quality problems in remote areas, such as Alaska, have significance as we try to understand air pollution at the global level.

Given our discussions concerning developed countries and less developed countries, what can be said about future trends for air pollution in urban areas or for that matter even once pristine remote areas (see A Closer Look 22.1, "Air Quality Problems In Remote Areas")? In summary, it will be a mixture of success stories and potential or actual tragedies. What is apparent is that air pollution is an important issue at all levels; ambitious air pollution control plans are being drawn up in many urban areas. Whether these plans are put into action will depend on factors related to the global, regional, and local economies, on population growth, or on international cooperation and the priority given to pollution abatement relative to other concerns. With these thoughts in mind, we turn to a discussion of acid deposition, or, as it is more commonly called, acid rain.

22.4 ACID RAIN

Acid rain encompasses both *wet* (rain, snow, fog) and dry (particulate) acidic depositions that occur near and downwind of areas where major emissions of sulfur dioxide (SO_2) and nitrogen oxides (NO_x) result from burning fossil fuels. Although the oxides (sulfur and nitrogen) are the primary contributors, other acids are also involved in the acid rain problem. For example, hydrochloric acid is emitted from coal-fired power plants.

The term *acid rain* is fairly new, even though the problem probably extends at least as far back as the beginning of the industrial revolution. In recent decades the problem of acid rain has gained more and more attention; today it is considered a major global environmental problem.

Many people are surprised to learn that all rainfall is slightly acidic; water reacts with atmospheric carbon dioxide to produce weak carbonic acid. Thus pure rainfall has a *pH* (a numeric value to describe the strength of an acid) of about 5.6, where 1 is highly acid and 7 is neutral (see Figure 22.11). Acid rain is defined as precipitation in which the pH is below 5.6. However, in some instances natural rainfall in tropical rain forests has been observed to have a pH of less than 5.6; this is probably related to acid precursors emitted by the trees. Because the pH scale is logarithmic, a pH value of 3 is 10 times more acidic than a pH value of 4, and 100 times more acidic than a pH value of 5. Automobile battery acid has a pH value of 1. It is alarming that in Wheeling, West Virginia, rainfall has been measured with a pH value of 1.5, nearly as acidic as stomach acid and far more acidic than lemon juice or vinegar.

Perhaps more important than isolated cases of very acid rain is the apparent growth of the prob-

FIGURE 22.11 The pH scale. (*Source:* Modified after U.S. Environmental Protection Agency, 1980.)

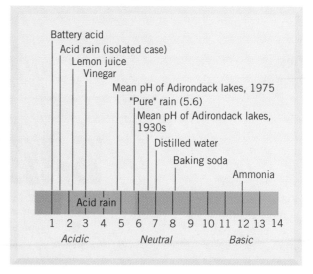

lem. Until quite recently it was believed that acid rain was primarily a European problem. It is now recognized that acid rain affects all industrial countries. In the United States it was thought that acid rain affected only a relatively small area in the northeastern United States; now it is believed to affect nearly all of eastern North America, and West Coast urban centers, such as Seattle, San Francisco, and Los Angeles, are now beginning to record acid rainfall and acid fog events. The problem is also of great concern in Canada, Germany, Scandinavia, and Great Britain. Finally, developing countries that are expected to rely heavily on coal in the future (such as China) will face serious acid rain problems.

Causes of Acid Rain

Amounts of sulfur dioxide and nitrogen oxides released by the United States into the environment

from 1940 to 1994 are shown on Figure 22.12. Emissions peaked in the 1970s and have generally declined since then, leveling off at about 20 million metric tons per year each in the 1980s. In the early 1990s sulfur dioxide emissions have decreased, while those of nitrogen have increased slightly. In the atmosphere, sulfur dioxide and nitrogen oxide are transformed by reactions with oxygen and water vapor to form sulfuric and nitric acids. These acids may travel long distances with prevailing winds to be deposited as acid precipitation (Figure 22.13). Such precipitation may be in the form of rainfall, snow, or fog. Sulfate and nitrate particles may also be deposited directly on the surface of the land as dry deposition. These particles may later be activated by moisture to become sulfuric and nitric acids.

Sulfur dioxide is emitted primarily from stationary sources, such as power plants that burn fossil

FIGURE 22.12 U.S. emissions (1940–1994) of (*a*) sulfur dioxide and (*b*) nitrogen oxides. (*Source:* Council on Environmental Quality, 1991; U.S. Environmental Protection Agency, 1995. National Air Quality and Emission Trends Report, 1994.)

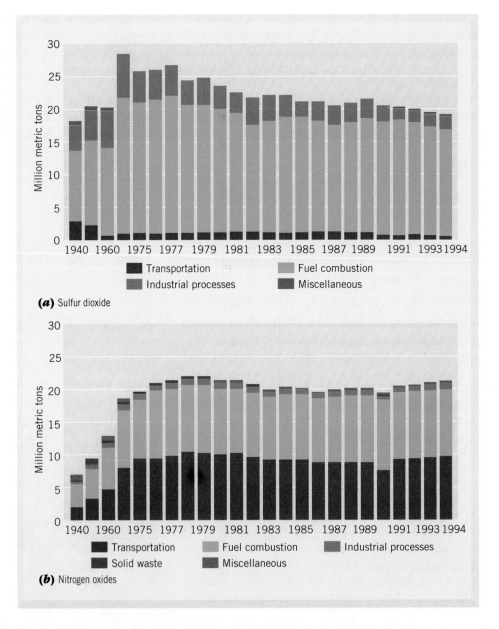

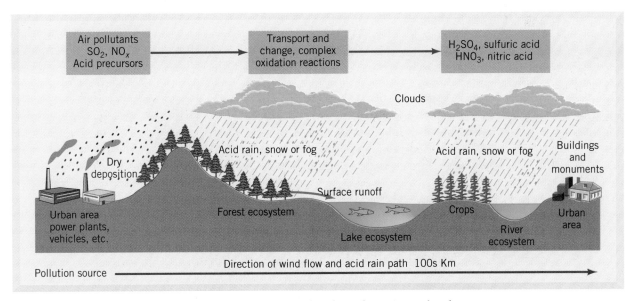

Figure 22.13 Idealized diagram showing selected aspects of acid rain formation and paths.

fuels, whereas nitrogen oxides are emitted from both stationary and transport-related sources, such as automobiles. Approximately 80% of sulfur dioxide and 65% of nitrogen oxides in the United States come from states east of the Mississippi River. In some instances, taller emission stacks have been constructed. Taller stacks reduced local concentrations of air pollutants but increased regional effects by spreading pollution more widely. Tall stacks increase average residence time of pollutants emitted into the atmosphere from 1 or 2 days to 10 to 14 days, because pollutants enter the atmosphere at a greater altitude, where mixing and transport by wind are more effective. In this case, dumping waste into someone else's backyard has created more rather than fewer problems. For example, problems associated with acid precipitation in Canada may be traced to emissions of sulfur dioxide and other pollutants in the Ohio Valley.

Analysis of the distances that sulfur compounds may be transported before deposition suggests that approximately one-third of the total amount deposited over the eastern United States originates from sources greater than 500 km (300 miles) away. Another one-third comes from sources between 200 and 500 km (about 125–300 miles) away, and the remainder comes from sources less than 200 km away.[11]

Effects of Acid Rain

Geology and climatic patterns as well as types of vegetation and soil composition affect the potential impact of acid rain. Figure 22.14, showing areas of the United States and Canada sensitive to acid rain and pH of precipitation for 1996, is based on some of these factors. Sensitive areas are those in which

bedrock or soil cannot *buffer* acid input. Materials (chemicals) that have the ability to neutralize acids are called **buffers.** Calcium carbonate ($CaCO_3$), the mineral calcite, present in many soils and rock (limestone), is an important natural buffer to acid rain. Hydrogen in acid reacts with calcium carbonate, and the reaction neutralizes acid. Areas with abundant granitic rocks, and those in which soils have little buffering action, are sensitive to acid rain. Studies of soils exposed to acid deposition in the last half century suggest that those with high pH (basic) became more acidic, whereas naturally acidic soils remained acidic. Areas likely to suffer less damage are those in which bedrock contains an abundance of limestone or other carbonate material or where soils contain a thick horizon rich in calcium carbonate. Soils in sensitive areas lose fertility, either because nutrients are leached out by acid or because acid releases elements into soil that are toxic to plants.

It has long been suspected that acid precipitation, whether as snow, rain, fog, or dry deposition, adversely affects trees. Studies in Germany led scientists to cite acid rain and other air pollution as the cause of death for thousands of acres of evergreen trees in Bavaria. Similar studies in the Appalachian Mountains of Vermont (where many soils are naturally acidic) suggest that in some locations half the red spruce trees have died in recent years. Damage is attributed in part to acid rain and fog with pH levels of 4.1 and 3.1, respectively. The reason trees died is not well understood, and some tree deaths have been erroneously attributed to acid rain. There is speculation that sulfate and nitrate deposition (acid rain) affects the ability of trees to tolerate cold temperatures, and weakened trees are killed by cold conditions or become susceptible to diseases.

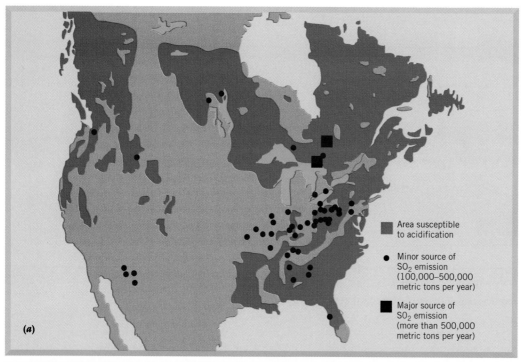

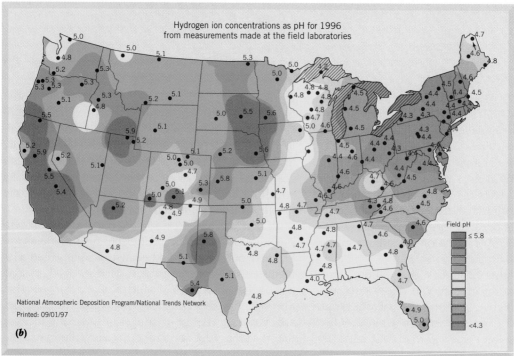

FIGURE 22.14 (*a*) Areas in Canada and the United States that are sensitive to acid rain. (*Source:* "How Many More Lakes Have to Die?" *Canada Today* vol. 12, no. 2, 1981.) (*b*) pH of precipitation over the United States in 1996. Notice strong relationship between sources of SO₂ and pH values. (*Source:* National Atmospheric Deposition Program/National Trends Network, 1997.)

Lake Ecosystems

In recent years fish have disappeared from lakes in Scandinavia. Records from Scandinavian lakes show an increase in acidity accompanied by a decrease in fish (Figure 22.15). The death of fish has been traced to acid rain, the result of industrial processes in other countries, particularly Germany and Great Britain.

Acid rain affects a lake ecosystem by dissolving chemical elements necessary for life and keeping them in solution so they leave the lake with water outflow. Elements that once cycled in the lake are lost. Without these nutrients, algae do not grow, and animals

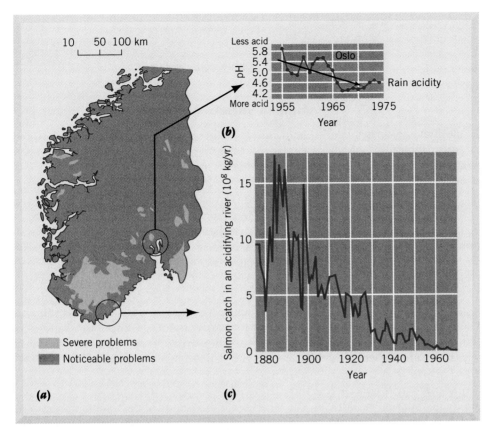

FIGURE 22.15 (*a*) In Norway, many lakes in the south have severe problems with acid rain. (*b*) The rain has become more acidic during the last 20 years, as measured at Oslo. Measurements at five other sites in southern Norway show the same trend. (*c*) The catch of fish, as illustrated by the catch of salmon in the Tovdalselva River of southern Norway, has decreased dramatically. (*Sources:* I. P. Muniz, H. Leivestad, E. Gjessing, E. Joranger, and D. Svalastog, 1976, *Acid Precipitation: Effects on Forest and Fish,* SNSF Project, IR 13/75, Government of Norway, As, Norway; S. Oden, 1976; "The Acidity Problem—An Outline of Concepts," *Water, Air, and Soil Pollution,* 6, pp. 137–166; E. Snekvik, 1970, *Norwegian Directorate for Game and Freshwater Fish,* unpublished report; R. F. Wright, T. Dale, E. G. Gjessing, G. R. Hendrey, A. Henriksen, M. Johannessen, and I. P. Muniz, 1976, "Impact of Acid Precipitation on Freshwater Ecosystems in Norway," *Water, Air, and Soil Pollution,* 6, pp. 438–499. Reprinted by permission of Kluwer Academic Publishers.)

that feed on the algae have little to eat. The fish that are typically predators of the small invertebrate animals also lack food. The acid water has other adverse effects on living organisms and their reproduction. For example, crayfish produce fewer eggs in acid water, and the eggs produced often grow into malformed larvae.

To better study the effects of acidification on lakes, scientists in Canada added sulfuric acid to a lake in northwest Ontario over a period of years and observed the effects. When the experiment started, the pH of the lake was 6.8. The following year, owing to addition of the acid, the pH had dropped to 6.1. The initial drop in pH was not harmful to the lake, but as more and more acid was added, the pH dropped first to 5.8, then to 5.6, then to 5.4, and finally, 5 years after the project started, to 5.1. The problems started when the pH was lowered to 5.8;

some species disappeared and others experienced reproductive failure. At a pH of 5.6, the death rate among lake trout embryos increased. When the pH was lowered to 5.4, lake trout reproduction failed.[12]

These experiments have proved valuable in pointing out what we can expect in thousands of other lakes that are now becoming acidified. The precise processes involved in the toxicity and damage to the lake are poorly understood. However, it is known that acid rain leaches metals, such as aluminum, lead, mercury, and calcium, from the soils and rocks in a drainage basin and discharges them into rivers and lakes. The elevated concentrations of aluminum are particularly damaging to fish, because the metal can clog the gills and cause suffocation. The heavy metals may pose health hazards to humans, because they may become concentrated in fish and then passed on to people, mammals, and birds when the fish are eaten. Drinking water taken from acidic lakes may also have high concentrations of toxic metals.

Not all lakes are as vulnerable to acidification as was the lake in the Ontario experiment. The acid is neutralized in waters with a high calcium or magnesium content, and lakes on limestone or other rocks rich in calcium or magnesium carbonates can readily release the calcium and magnesium, which buffers the lakes against the addition of acids. Lakes with high concentrations of such elements are called hard-water lakes. Lakes on sand or igneous rocks, such as granite, tend to lack sufficient buffering to neutralize acids and are more susceptible to acidification. In practice, a simple and fast index of a lake's hardness or buffering capacity is its electrical conductivity. Pure water is a poor conductor of electricity; water high in dissolved elements is a good conductor.[13]

Thousands of kilometers of rivers and thousands of lakes in the United States and Canada located in areas sensitive to acid rain are currently in various stages of acidification. In Nova Scotia, for example, at least a dozen rivers have acid contents sufficiently high that they no longer support healthy populations of Atlantic salmon. In the northeastern United States about 200 lakes in the Adirondacks are no longer able to support fish, and thousands more are slowly losing the battle with acid rain.

One solution to lake acidification is rehabilitation by the periodic addition of lime. This has been done in New York State as well as in Sweden and Ontario. This solution is not satisfactory over a long period, however, because it is expensive and re- quires a continuing effort. The only practical long- term solution to the acid rain problem is to ensure that the production of acid-forming components in the atmosphere is minimized.

Human Society

Acid rain damages not only forests and lakes but also many building materials, including steel, paint, plastics, cement, masonry, galvanized steel, and several types of rock, especially limestone, sandstone, and marble (Figure 22.16). Classical buildings on the Acropolis in Athens and in other cities show considerable decay (chemical weathering) that has accelerated in this century as a result of air pollution. The problem has grown to such an extent that buildings require restoration, and statues and other monuments need to have protective coatings replaced quite frequently, resulting in costs that reach billions of dollars a year. Particularly important statues in Greece and other areas have been removed and placed in protective glass containers, with replicas standing in their former outdoor locations for tourists to view.[12]

In the United States, cities along the eastern seaboard are more susceptible to acid rain today because emissions of sulfur dioxide and nitrogen oxide are more abundant there. However, as noted, the problem is moving westward; acid precipitation has been recorded in California. Even more alarming is the discovery that acid fog events in Los Angeles may have a pH as low as 3, which is over 10 times as acidic as the average acid rain in the eastern United States. In contrast to acid rain that may form

Figure 22.16 Damage to a statue in the city of Chicago resulting from an acid deposition (*left*) and the same statue following restoration (*right*).

relatively high in the atmosphere and travel long distances, acid fog forms when water vapor near the ground mixes with pollutants and turns into an acid. The acid evidently condenses around very fine particles of smog, and if the air is sufficiently humid, a fog may form. When the fog eventually dissipates, nearly pure drops of sulfuric acid may be left behind; tiny particles containing the acid may be inhaled deeply into people's lungs—a considerable health hazard.

Stone decay occurs about twice as rapidly in cities as it does in urban areas. This damage mainly comes from acid rain and humidity in the atmosphere, as well as corrosive groundwater.[14] This implies that measuring rates of stone decay will tell us something about changes in the acidity of rain and groundwater in different regions and ages. It is now possible, where the ages of stone buildings and other structures are known, to determine if the acid rain problem has changed through time.

22.5 CONTROL OF AIR POLLUTION

For both stationary and mobile sources of air pollutants, the most reasonable strategies for control have been to reduce, collect, capture, or retain the pollutants before they enter the atmosphere. From an environmental viewpoint, the reduction of emissions via energy efficiency and conservation measures (for example, burning less fuel) is *the* preferred strategy, with clear advantages over all other approaches (see Chapters 15–18).

Pollution problems vary in different regions of the world; reducing air pollution requires that strategies be tailored to specific sources and type of pollutants. There is great variance even within the United States. For example, in the Los Angeles basin, nitrogen oxides and hydrocarbons are particularly troublesome because they combine in the presence of sunlight to form photochemical smog. Furthermore, in Los Angeles most of the nitrogen oxides and hydrocarbons are emitted from automobiles, a collection of mobile sources.[15] In other urban areas, such as in Ohio and the Great Lakes region in general, air quality problems result from emissions of sulfur dioxide and particulates from industry and from coal-burning power plants, which are point sources. This is not to say that automobiles are not a problem in areas outside the Los Angeles area but rather to emphasize contrasting conditions. It is important also to keep in mind that automobiles produce many small particulates, whose pollution potential is only beginning to be understood. Thus, areas such as San Francisco, with a relatively cool climate, have a serious air pollution problem from automobile exhaust, even though the city usually lacks the necessary sunlight to produce photochemical smog. Of course, the small particulates are also abundant in the Los Angeles area, where the problem is more pronounced because of the photochemical smog.

In the United States, nitrogen dioxide is a problem primarily in southern California and is related to automobiles. Oxidants (mostly ozone), however, are a problem in the southwestern, southeastern, and northeastern United States, as are particulates.[16] Because air pollution problems vary so greatly from country to country and from region to region, it is often difficult to obtain both the money and the political consensus for effective control. Nevertheless, effective control strategies do exist for many air pollution problems.

Particulates

Particulates emitted from fugitive, point, or area stationary sources are much easier to control than are the very small particulates of primary or secondary origin released from mobile sources, such as automobiles. As we learn more about these very small particles, new methods will have to be devised to control them.

A variety of settling chambers or collectors are used to control emissions of coarse particulates from power plants and industrial sites (point or area sources) by providing a mechanism that causes particles in gases to settle out in a location where they may be collected for disposal in landfills. In recent decades there have been tremendous gains in control of particulates, such as ash, from power plants and industry. Cities in the eastern United States, where buildings used to turn black from soot and ash, are now much cleaner and no longer have the serious health risks from these pollutants that in recent years have plagued parts of Eastern Europe.

Particulates from fugitive sources (such as a waste pile) must be controlled on-site so that the wind does not blow them into the atmosphere. Methods include protecting open areas, dust control, and reducing the effect of wind. For example, waste piles may be covered by plastic or other material and soil piles may be vegetated to inhibit wind erosion; water or a combination of water and chemicals may be spread to hold dust down; and structures or vegetation may be positioned to lessen wind velocity near the ground, thus retarding wind erosion of particles.

Automobile Pollution

Control of pollutants such as carbon monoxide, nitrogen oxides, and hydrocarbons in urban areas is best achieved through pollution control measures for automobiles. Control of these materials will also regulate the ozone in the lower atmosphere, where it forms by reactions with nitrogen oxides and hydrocarbons in the presence of sunlight.

Control of nitrogen oxides from automobile exhausts is accomplished by recirculating exhaust gas, diluting the air-to-fuel mixture being burned in the engine. Dilution reduces the temperature of combustion and decreases oxygen concentration in the burning mixture (makes a richer fuel), thus producing fewer nitrogen oxides. Unfortunately, the same process increases hydrocarbon emissions, which are greater for rich fuels. Nevertheless, exhaust recirculation to reduce nitrogen oxide emissions has been in common practice in the United States for more than 20 years.[17]

The most common device that removes carbon monoxide and hydrocarbon emissions from automobiles is the exhaust system's catalytic converter.[17] In the converter, carbon monoxide is converted to carbon dioxide and hydrocarbons to carbon dioxide and water by passing exhaust gases over a catalyst (typically platinum, palladium). Outside air provides oxygen. Oxidation then removes carbon monoxide and hydrocarbons from the exhaust.

As government regulations are more restrictive on emissions control, it is difficult to meet new standards without the aid of a computer controlled engine system. Computer-controlled fuel injection began to replace carburetors in the 1980s, and have resulted in, among other things, lower fuel consumption and lower exhaust emissions.[17]

It has been argued that the automobile emission regulation plan in the United States has not been very effective in reducing pollutants. Pollutants may be reduced when a car is relatively new, but many people do not take care of their automobiles well enough to ensure that the emission control devices work. Also, some people disconnect smog control devices. Evidence suggests that these devices tend to become less efficient every year following purchase. Because of these adverse aspects of emission control, it has been suggested that effluent fees replace automobile controls as the primary method of regulating air pollution in the United States.[18] Under this scheme, vehicles would be tested each year for emission control, and fees would be assessed on the basis of test results. Fees would provide incentive for purchase of automobiles that pollute less, and annual inspections would ensure that pollution control devices are properly maintained. Although there is considerable controversy regarding enforced pollution inspections, inspections are common in a number of areas and expected to increase as air pollution abatement becomes essential.

Another approach to reducing urban air pollution produced by vehicles revolves around a number of options, most of which aim to reduce the number of cars on roads. Some of these methods, were mentioned earlier, and are being tried or discussed in Los Angeles and other areas. Other recent measures include development of cleaner automobile fuels by using fuel additives and reformulated fuel to help reduce pollutant emissions.

Acid Rain

As mentioned earlier, acid rain is a particularly troublesome problem because pollutants that cause it may be emitted long distances (sometimes across national boundaries) from where the actual acid rain falls. The cause of acid precipitation is known. We also know that the only long-term solution involves decreasing emissions of sulfur dioxide and nitrogen oxides. From an environmental point of view the best strategy is increasing energy efficiency and conservation measures that result in burning less coal and utilizing nonpolluting alternative energy sources. Another strategy is to utilize pollution abatement technology at power plants to lower emissions of air pollutants.[19] Such technology is expensive and an additional cost to producing energy.

Sulfur Dioxide

Sulfur dioxide emissions can be reduced by abatement measures performed before, during, or after combustion. Technology to clean up coal so it will burn cleanly is already available, although cost of removing sulfur makes fuel more expensive. If nothing is done, however, the consequences of burning sulfur-rich coal in the next decade will be very expensive.

Changing from high-sulfur coal to low-sulfur coal seems an obvious solution to reducing emissions of sulfur dioxide into the atmosphere. In some regions this change will work. Unfortunately, most low-sulfur coal in the United States is located in the western part of the country, whereas most coal is burned in the east. Thus transportation is an issue, and use of low-sulfur coal is a solution only in cases where it is economically feasible. Another possibility is cleaning up relatively high-sulfur coal by washing it to remove sulfur. In this process finely ground coal is washed with water, and iron sulfide (mineral pyrite) settles out because of its relatively high density. Although the washing process is effective in removing nonorganic sulfur from minerals such as pyrite (FeS_2), it is ineffective for removing organic sulfur bound up with carbonaceous material. Cleanup by washing is limited; it is also expensive.

Another option is **coal gasification,** which converts coal that is relatively high in sulfur to a gas in order to remove the sulfur. The gas obtained from coal is quite clean and can be transported relatively easily, augmenting supplies of natural gas. The synthetic gas produced from coal is now fairly expensive compared to gas from other sources, but its price may become more competitive in the future.[16]

Sulfur oxide emissions from stationary sources, such as power plants, can be reduced by removing the oxides from the gases in the stack before they reach the atmosphere. Perhaps the most highly developed technology for the cleaning of gases in tall stacks is flue gas desulfurization, or **scrubbing** (Figure 22.17). In this method the gases are treated with a slurry (a watery mixture) of lime (calcium oxide, CaO) or limestone (calcium carbonate, $CaCO_3$). The sulfur oxides react with the calcium to form insoluble calcium sulfides and sulfates, which are collected and then disposed of. However, there are environmentally preferable alternatives to sludge disposal. It can be used to make a product! Read on.

The technology to scrub sulfur dioxide and other pollutants out of emissions before release into the atmosphere was developed in the 1970s in the United States in response to passage of the Clean Air Act. However, the technology was not initially implemented in the United States because regulators chose to disperse pollutants using tall smokestacks rather than use this technology to remove them. This increased the regional acid rain problem. However, in 1980 a German company purchased the technology and improved on it as part of efforts to reduce air pollution and acid rain. The calcium sulfate sludge formed during the scrubbing process is used to produce building materials (sheet rock or wallboard) that are being sold worldwide.

Another innovative approach has been taken at a large coal-burning power plant near Mannheim, Germany, to remove sulfur before it enters the atmosphere. Smoke from combustion is cooled, then treated with liquid ammonia (NH_3), which reacts with the sulfur to produce ammonium sulfate. In this process, the sulfur-contaminated smoke from burning the coal is cooled in a heat-exchange process by outgoing clean smoke to a temperature that favors the reaction with ammonia. The cooled, newly cleaned, outgoing smoke is then heated by dirty smoke (from burning the coal, in the same sort of heat-exchange process) to force it out the vent. Waste heat from the cooling towers is used to heat nearby buildings, and the plant sells the ammonium sulfate to farmers as fertilizer in a solid granular form. In summary, Germany, in response to tough pollution control regulations, has succeeded in substantially reducing its sulfur dioxide emissions as well as many other pollutant emissions and in the process has boosted its economy.[20]

Clean Air Act Amendments of 1990

The **Clean Air Act Amendments of 1990** are comprehensive regulations (federal statute) that address acid rain, toxic emissions, ozone depletion, and automobile exhaust. In confronting acid deposition (acid rain), the amendments establish limits on the maximum permissible emissions of sulfur dioxide

FIGURE 22.17 Scrubber, used to remove sulfur oxides from the gases emitted by tall stacks.

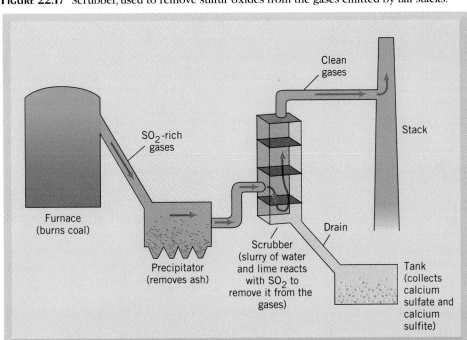

from utility companies burning coal. The legislation mandates that the emissions be reduced by about 50% to 10 million tons a year by the year 2000. In order to help reach this goal, utility companies will face choices such as burning coal with a low sulfur content, using a mixture of fuels, such as coal and natural gas, or adding scrubbers and other pollution abatement equipment to their power plants. An innovative aspect of the legislation is to provide incentives to utility companies to reduce emissions of sulfur dioxide early by providing allowances and credits (to pollute) that may then be sold to other companies with higher levels of emissions. Some environmentalists are also purchasing allowances to keep them from being bought by utility companies, forcing the utility companies to use more vigorous pollution abatement technology. The amendments also call for a reduction in emissions of nitrogen dioxides by approximately 2 million tons from the 1980 level. Greater reductions would be difficult because so much of the nitrogen oxide emissions are related to automobiles rather than to coal-burning power plants.[21]

Toxic emissions into the atmosphere are targeted to be reduced by as much as 90%. The toxins targeted are those thought to have the most potential for damaging human health, including causing cancer. Abatement will depend heavily on pollution control equipment that will be required for large manufacturers and small businesses alike. Although this requirement will undoubtedly result in an increase in the cost of many goods and services, there should be a compensating improvement in the health of people.

Regarding ozone depletion in the stratosphere (see chapter 24), the Clean Air Amendments have the goal of ending the production of all chlorofluorocarbons (CFCs) and other chlorine chemicals in steps from the year 2000 to the year 2030.[21]

Air pollution in urban areas is most commonly related to automobile exhaust. Strategies outlined in the legislation include more stringent emission controls on automobiles and requiring cleaner burning fuels. The aim is to reduce the occurrence of urban smog. Expected impacts of the legislation include increases in the cost of automobile fuels and in the price of new automobiles.

Air Quality Standards

Air quality standards are important because they are tied to emission standards that attempt to control air pollution. Many countries have developed air quality standards. France, Japan, Israel, Italy, Canada, Germany, Norway, and the United States, among others, have taken this first step. However, examination of the standards for different countries

shows a good deal of variability in acceptable levels of pollution. For example, the standard for sulfur dioxide varies from 50 in the former USSR to 200 in Norway, 380 in Italy, and 365 in the United States. The units are millionths of a gram per cubic meter of air (micrograms per cubic meter, or $\mu g/m^3$) over a 24-hour period, and the large range in the standards suggests a big difference in opinion concerning acceptable levels of sulfur dioxide in the atmosphere. One of the problems in establishing air quality standards is the lack of general agreement concerning what concentrations of pollutants cause environmental and health problems.[18]

In the United States, air quality in urban areas is often reported as good, moderate, unhealthy, very unhealthy, or hazardous (see Table 22.2). These levels, are derived from monitoring the concentration of five major pollutants: total suspended particulates, sulfur dioxide, carbon monoxide, ozone, and nitrogen dioxide. During a pollution episode in Los Angeles, hourly ozone levels are reported and a first-stage smog episode begins if the primary National Ambient Air Quality Standard (NAAQS) of 0.12 ppm (equivalent to 235 $\mu g/m^3$ in Table 22.2) is exceeded. This measure corresponds to unhealthy air with a PSI between 100 and 300. A second-stage smog episode is declared if the PSI exceeds 300, a point at which air quality is hazardous to all people. As air quality worsens, people are requested to remain indoors, minimize physical exertion, and avoid driving automobiles. Industry also may be requested to reduce emissions to a minimum during the episode. Recall the PSI was 800 during the Indonesian fires of 1997–98!

The national air quality standards are periodically reviewed, but they have not changed much, in spite of the fact that there is evidence that air pollutants at lower concentrations may cause health problems in certain segments of the population. For example, young children show adverse health effects when exposed to ozone at about one-half the concentration of the national air quality standard. In response, the U.S. Environmental Protection Agency (EPA), in December of 1996, proposed that the NAAQS for ozone be strengthened from 235 $\mu g/m^3$ (one-hour exposure) to 160 $\mu g/m^3$ (average 8-hour exposure). The EPA also proposed that particulates less than 2.5 μm in diameter (thought to be particularly damaging to human health) be regulated. It is hoped that the newer and more stringent regulations for ozone and particulates will save as many as 20,000 lives per year in the United States and greatly reduce the number of people hospitalized for respiratory illnesses such as pneumonia, bronchitis, and asthma. Nevertheless, the new regulations will be vigorously debated, and there will be political pressure and resistance from economic and industrial

TABLE 22.2 Pollutant Standards Index[a]

PSI Index Value	Air Quality Level	Cautionary Statements	Health Effect Label	Pollutant Level					General Health Effects
				TSP (24-hour), µg/m³	SO₂ (24-hour), µg/m³	CO (8-hour), µg/m³	O₃ (1-hour), µg/m³	NO₂ (1-hour), µg/m³	
500	Significant harm	All persons should remain indoors, keeping windows and doors closed; all persons should minimize physical exertion and avoid traffic.	—	1,000	2,620	57,500	1,200	3,750	Premature death of ill and elderly. Healthy people will experience adverse symptoms that affect their normal activity.
400	Emergency	Elderly and persons with diseases should stay indoors and avoid physical exertion; general population should avoid outdoor activity.	Hazardous (PSI > 300)	875	2,100	46,000	1,000	3,000	Premature onset of some diseases in addition to significant aggravation of symptoms and decreased exercise tolerance in healthy persons.
300	Warning	Elderly and persons with existing heart and lung disease should stay indoors and reduce physical activity.	Very unhealthful (PSI = 200–300)	625	1,600	34,000	800	2,260	Significant aggravation of symptoms and decreased exercise tolerance in persons with heart or lung disease with widespread symptoms in the healthy population.
200	Alert	Persons with existing heart or respiratory ailments should reduce physical exertion and outdoor activity.	Unhealthful (PSI = 100–200)	375	800	17,000	400	1,130	Mild aggravation of symptoms in susceptible persons, with irritation symptoms in the healthy population.
100	NAAQS[b]	—	Moderate (PSI = 50–100)	260	365	10,000	235	c	
50	50% of NAAQS[b]	—	Good (PSI = 0–50)	75[c]	80[d]	5,000	120	c	

SOURCE: Council on Environmental Quality.

[a]One measure of air quality is the Pollutant Standards Index. It is a highly summarized health-related index based on five of the criteria pollutants: carbon monoxide, ozone, sulfur dioxide, total suspended particulates, and nitrogen dioxide. The PSI for one day will rise above 100 in a Standard Metropolitan Statistical Area when one of the five pollutants at one station reaches a level judged to have adverse short-term effects on human health.

[b]NAAQS = National Ambient Air Quality Standard.

[c]There are no index values reported at concentrations below those specified by Alert criteria.

[d]Annual Primary NAAQS.

sectors of society to not strengthen air quality standards or to delay implementation.

Data from some major metropolitan areas in recent years suggest that the total number of unhealthful and very unhealthful days has declined. Although these data do not mean that air pollution has been eliminated, they do indicate that the nation's air quality is improving. However, most urban areas, such as New York and Los Angeles, still have unhealthful air much of the time.

Cost of Controls

The costs and benefits of air pollution control are controversial subjects. It has been argued that the present system of setting air quality standards is inefficient and unfair because regulations are tougher for new sources than for existing sources. Also, even if the benefits of pollution control exceed the total costs, there is tremendous variability in the cost of air pollution control from one industry to another. For example, consider the incremental control costs (cost to remove an additional unit of pollution beyond what is presently required) for utilities burning fossil fuels and for an aluminum plant. The utilities' cost for incremental control in a fossil fuel–burning utility is a few hundred dollars per additional ton of particulates removed. Compare this to an aluminum plant, where the cost may be as much as several thousand dollars to remove an additional ton of particulates.[22] Some economists would argue that it is wise to increase the standards for utilities and relax or at least not increase them for aluminum plants. This practice would lead to more cost-efficient pollution control while maintaining good air quality. However, the geographic distribution of various facilities will obviously determine the amount of prospective trade-offs possible.[22]

Another economic consideration is that, as the degree of control of a pollutant increases, a point is reached at which the cost of incremental control (reducing additional pollution) is very high in relation to the additional benefits of the increased control. Because of this and other economic factors, it has been argued that fees and taxes for emitting pollutants might be a better way to go than attempting to evaluate uncertain costs and benefits. The argument is that it makes more economic sense to enforce fees or taxes rather than standards. Another approach is to issue vouchers to emit a certain total amount of pollution in a region. These vouchers are bought and sold on the open market. All these economic alternatives are controversial and may be objectionable to people who believe that polluters should not be allowed to buy their way out of doing what is socially responsible (that is, not polluting our atmosphere).

Economic analysis of air pollution is not simple, and there are many variables, some of which are hard to quantify. We do know that

- with increasing air pollution controls, the capital cost for technology to control air pollution increases;
- as the controls for air pollution increase, the loss from pollution damages decreases; and
- the total cost of air pollution is the cost of pollution control plus the environmental damages of the pollution.

Although the cost of pollution abatement technology is fairly well known, it is difficult to adequately determine the loss from pollution damages, particularly when considering health problems and damage to vegetation, including food crops. For example, exposure to air pollution may cause or aggravate chronic respiratory diseases in human beings, with a very high cost. A recent study of the health benefits of cleaning up the air quality in the Los Angeles basin estimated that the annual cost associated with the air pollution in the basin is 1600 lives and about $10 billion.[23] Air pollution also leads to loss of revenue from people who choose not to visit areas such as Los Angeles and Mexico City because of known air pollution problems.

How do we determine the real and total benefits and costs of controlling or reducing air pollution? There are no easy answers to this question. In spite of the inability to determine all benefits and costs, it seems worthwhile to reduce the air pollution level below some particular standard. Thus, in the United States, the ambient air quality standards have been developed as a minimum acceptable air quality level. However, as discussed, it is also a good idea to consider alternatives, such as charging fees or taxes for emissions. If such charges are determined carefully and emissions are closely monitored, the charges would provide an incentive for the installation of control measures. The end result of either control measure would be better air quality.[24,25]

How Does Arctic Haze Affect the Environment?

A dark gray haze, full of industrial pollutants, hovers over the ground and extends to an altitude of 8 km (5 miles). It is not Los Angeles, but thousands of miles from heavy industry in the frozen Arctic. The polluted air mass includes all of the atmosphere above the Arctic Circle, as well as lobes extending into Eurasia and North America. The total area of polluted air mass, about as large as the African continent, persists during winter months and disappears in summer.

Scientists describe this as an aerosol, that is, microscopic particles dispersed in a gas, smoke, or fog. They have found dust from Mongolia and sea salt in the haze, but of greater concern are the pollutants, mainly sulfates, carbon soot, organic compounds, and toxic metals, including mercury, lead, and vanadium. The gaseous atmosphere itself contains elevated levels of carbon dioxide, methane, and carbon monoxide, as well as chemicals destructive of the ozone layer.

Like detectives searching for fingerprints, scientists use ratios of six elements (arsenic, antimony, zinc, indium, manganese, and vanadium) to a seventh, selenium, to track down sources of Arctic haze. Emissions from burning fossil fuels differ in amounts of these elements, depending on the type of fuel used. The ratios are so characteristic that scientists can tell whether pollution is from hard or soft coal.

Knowing the types of fuel used in various regions of the world, scientists use the six ratios to identify sources of pollution. For example, manganese was present in greater amounts and vanadium in lesser amounts in the haze than in emissions typical of North America. Combined with knowledge of air circulation, scientists identified Eastern Europe and Russia as major sources of Arctic air pollution, with a significant but lesser contribution from the United Kingdom and Western Europe. Two major routes (see map) are (1) from Eastern Europe and

Two airstreams out of the former Soviet Union (top arrows) carry most of the pollution that becomes Arctic haze. Europe is also an important source of smog (arrow at right). But the northeastern United States (bottom arrows) is not a big contributor to the haze because pollutants are often washed out by storms before they reach the Arctic. The dots represent air-sampling stations.

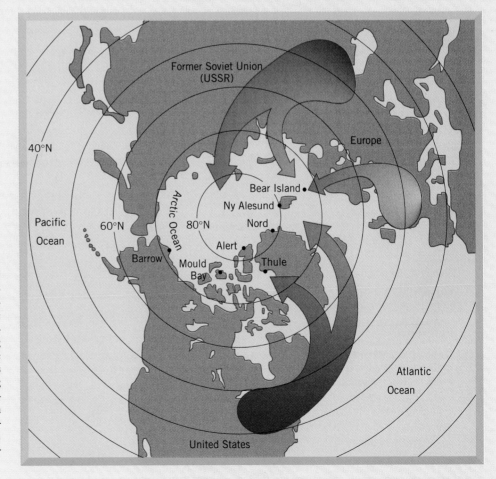

Russia across the Taimyr Peninsula, and North Pole, to the Alaskan Arctic; and (2) from the United Kingdom and Western Europe across Scandinavia to the Norwegian Arctic. However, a 1993 study argues that the pollution doesn't come from Western Europe or China, but from Eastern Europe and Russia, with some possibly originating within the Arctic Circle itself.

Major atmospheric forces driving pollution along these routes begin with large temperature differences between the equator and the poles in winter, creating strong air currents from 0° to 90° latitude. They are also propelled by seasonal lows in the North Atlantic and highs on the Eurasian continent. Once masses of pollutant-laden air reach the dry, stable air of Arctic winter, they form layers, which remain relatively intact. In spring, when northward flow of air diminishes, the haze disperses and is carried to higher levels in the atmosphere and back to the midlatitudes.

Now that scientists have a better idea of the sources of Arctic haze, they are beginning to speculate on the effects it has on the ecology of the Arctic and on global climate. Furthermore, they are calling for research into environmental impacts of the haze and for cooperation among countries in the Northern Hemisphere to reduce the amount of haze. In the meantime, humans are producing enough toxic emissions to carry as far as 10,000 km (6215 miles) to the North Pole and produce pollution levels as great as those found in medium-sized industrial cities.

Critical Thinking Questions

1. Although laden with pollutants, Arctic haze drops less of its aerosol on ground than would happen in other areas of the world. What is it in the Arctic environment that could account for this?

2. Elements such as mercury and lead are present in small amounts in Arctic haze, but often in levels comparable to those near industrial sites and 10–20 times those in Antarctica. If amounts are so low, why are scientists concerned about their effects? (You may wish to review Chapter 14.)

3. Sulfates in haze react with water to form acids. Scientists predict that the Arctic is particularly sensitive to additional acids. On what characteristic of Arctic tundra is this prediction based?

4. What would a diagram of possible effects of dark carbon particles on Arctic climate look like? Put the key term *dark carbon particles* in the center of a sheet of paper. Write as many direct consequences of dark carbon as you can around the key term and connect the consequences to the key term with arrows. What would be the consequences of the direct effects? Place them around the edge of the diagram, connecting them by arrows to the relevant direct effect. If you can think of third- and fourth-order effects, place these on the diagram in a similar manner. Your diagram should look something like this:

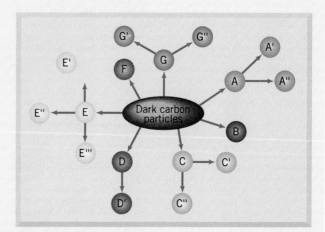

References

Khalil, M. and Rasmussen, R. A. 1993. "Arctic haze—patterns and relationships to regional signatures of trace gases," *Global Biogeochemical cycles*, vol. 7 no. 1, pp. 27–36.

Rahn, K. A. 1984. "Who's Polluting the Arctic?" *Natural History* 93(5): 31–38.

Shaw, G. E. 1995. "The Arctic haze phenomenon," *Bulletin of the American Meteorological Society*, vol. 76, no. 12, pp. 2403–2413.

Soroos, M. S. 1992. "The Odyssey of Arctic Haze," *Environment* 34(10): 6–27.

Young, O. R. 1990. "Global Commons: The Arctic in World Affairs," *Technology Review* 93: 52–61.

SUMMARY

- Every year approximately 250–300 million metric tons of primary pollutants enter the atmosphere above the United States from processes related to human activity. Considering the enormous volume of the atmosphere, this is a relatively small amount of material. If it were distributed uniformly, there would be little problem with air pollution. Unfortunately, the pollutants generally are not evenly distributed but are concentrated in urban areas or in other areas where the air naturally lingers.

- The two major types of pollution sources are stationary and mobile. Stationary sources have a relatively fixed position and include point sources, area sources, and fugitive sources.

- There are two main groups of air pollutants: primary and secondary. Primary pollutants are those emitted directly into the air: particulates, sulfur dioxide, carbon monoxide, nitrogen oxides, and hydrocarbons. Secondary pollutants are those produced through reactions between primary pollutants and other atmospheric compounds. A good example of a secondary pollutant is ozone, which forms over urban areas through photochemical reactions between primary pollutants and natural atmospheric gases.

- The effects of the major air pollutants are considerable. They include effects on visual resources, vegetation, animals, soil, water quality, natural and artificial structures, and human health.

- There are two major types of smog: photochemical and sulfurous. Each type of smog brings particular environmental problems that vary with geographic region, time of year, and local urban conditions.

- Meteorological conditions greatly affect whether polluted air is a problem in an urban area. In particular, restricted circulation in the lower atmosphere associated with temperature inversion layers may lead to pollution events.

- The combustion of large quantities of fossil fuels results in the emission of sulfur and nitrogen oxides into the atmosphere, creating a problem known as acid rain. Environmental degradations associated with acid rain include loss of fish and other life in lakes, damage to trees and other plants, leaching of nutrients from soils, and damage to stone statues and buildings in urban areas.

- From an environmental viewpoint the preferred method of reducing the emission of air pollutants is to practice energy efficiency and conservation so that smaller amounts of fossil fuels are burned. Another option is to increase the use of alternative energy sources, such as solar and wind power, that do not emit air pollutants.

- Methods to control air pollution are tailored to specific sources and types of pollutants. These methods vary from settling chambers for particulates to scrubbers or combustion processors that use lime to remove sulfur before it enters the atmosphere.

- Efforts to reduce air pollution in urban regions center on automobiles, buses, and other vehicles, because they account for the majority of pollutants that enter the urban atmosphere.

- Air quality in urban areas is usually reported in terms of whether the quality is good, moderate, unhealthy, very unhealthy, or hazardous. These levels or stages are defined in terms of the Pollution Standard Index (PSI) using the National Ambient Air Quality Standard (NAAQS).

- The relationships between emission control and environmental cost are complex. The minimum total cost is a compromise between capital costs to control pollutants and losses or damages resulting from pollution. If additional controls are necessary to lower the pollution to a more acceptable level, additional costs are incurred. Beyond a certain level of pollution abatement, these costs can increase rapidly.

REEXAMINING THEMES AND ISSUES

Human Population: Increase in human population is expected to continue to have a significant adverse effect on air pollution problems. This results because as human population increases, so does the total use of resources, many of which are related to emissions of air pollutants. This may be partially offset in the developed countries where the *per-capita* emissions of air pollutants have been reduced in recent years.

Sustainability: Ensuring that future generations inherit a quality environment with minimal air pollution is an important objective of sustainability. As a result, it remains an important objective to find and develop technology that minimizes air pollution.

Global Perspective: Atmospheric processes and pollution of the atmosphere occur by their very nature on a regional and global scale. Pollutants emitted into the atmosphere at a particular site may join the global circulation pattern and spread pollutants throughout the world. Air pollutants emitted from urban or agricultural areas may be dispersed to "pristine" areas far removed from human activities. Therefore, understanding of global atmospheric processes is critical to finding solutions to many of our globally present air pollution problems, including acid deposition.

Urban World: Cities and urban corridors are sites of intense human activity, many of which are associated with emission of air pollutants. Some of the most significant adverse effects of air pollution are found in our urban areas. Some large cities have air pollution problems to such an extent that the health and very lives of people are being affected.

Values and Knowledge: It is clear that people value a high-quality environment, and clean air is at the top of the list. The developed countries have an obligation to take a leadership role in finding ways to utilize resources while minimizing air pollution. Of particular importance is finding ways and technologies that will allow for reduction of air pollution while stimulating economies. What is considered waste in one part of the urban-industrial complex may be used as resources for another part. This is at the heart of what is sometimes called industrial ecology.

KEY TERMS

acid rain *479*
air quality standards *488*
area sources *469*
atmospheric inversion *474*
buffers *481*
Clean Air Act Amendments of 1990 *487*

coal gasification *486*
fugitive sources *469*
mobile sources *469*
photochemical smog *476*
point sources *469*
primary pollutants *470*

scrubbing *487*
secondary pollutants *470*
stationary sources *469*
sulfurous smog *476*

STUDY QUESTIONS

1. Compare and contrast the London 1952 fog event and smog problems in the Los Angeles basin.

2. Why do we have air pollution problems when the amount of pollution emitted into the air is a very small fraction of the total material in the atmosphere?

3. What is the difference between point and non-point sources of air pollution? Which type is easier to manage?

4. What are the differences between primary and secondary pollutants?

5. Examine carefully Figures 22.6, which shows a moving column of air through an urban area, and Figure 22.8, which shows relative concentrations of pollutants that develop on a typical warm day in Los Angeles. What are potential linkages between these two concepts that might be important in trying to identify and

learn more about potential air pollution in an area?

6. Why is acid deposition a major environmental problem and how can it be minimized?

7. Why will air pollution abatement strategies in developed countries probably be much different in terms of methods, process, and results from air pollution abatement strategies in developing countries?

8. Why is it so difficult to establish national air quality standards?

9. In a highly technological society, is it possible to have 100% clean air? Is it feasible or likely?

10. How good are the air quality standards being used by the United States? How might their usefulness be evaluated? Do you think the standards will change in the future? If so, what are the likely changes?

FURTHER READING

Boubel, R. W., Fox, D. L., Turner, D. B., and Stern, A. C. 1994. *Fundamentals of Air Pollution,* 3rd ed. New York: Academic. A thorough book covering the sources, mechanisms, effects, and control of air pollution.

Bryner, G. C. 1995. *Blue Skies, Green Politics: The Clean Air Act of 1990 and Its Implementation,* 2nd ed. Washington, DC: CQ Press. A good text of the Clean Air Act and air pollution in the United States.

Hesketh, H. E. 1991. *Air Pollution Control: Traditional and Hazardous Pollutants.* Lancaster, PA: Technomic. This is an in-depth text covering the methodology of controlling air pollution.

Hewitt, D. N., Sturges, W. T., and NOAA (eds.). 1995. *Global Atmospheric Chemical Change.* New York: Chapman & Hall. Aspects of global air pollution including chemical changes in the atmosphere, climate change, acidic deposition, and other anthropogenic pollutants.

Krupnick, A. J., and Portney, P. R. 1991. "Controlling Urban Air Pollution: A Benefit–Cost Assessment," *Science* 252: 522–528. The authors evaluate air pollution control proposals for Los Angeles and the United States as a whole and determine that the measures would cost less than the pollution damage they would alleviate.

Rose, J. (ed.). 1994. *Acid Rain: Current Situation and Remedies.* Philadelphia: Gordon and Breach Science. Essays covering issues and consequences of acid rain in Europe and the United States.

Shaw, R. W. 1987. "Air Pollution by Particles," *Scientific American* 257: 96–103. This article discusses the sources, effects, and mitigations of particulate pollution.

INTERNET RESOURCES

U.S. Environmental Protection Agency Office of Air and Radiation: *http://www.epa.gov/oar/oarhome.html*—This site offers an abundance of great current information and data on urban air pollution, acid rain, air toxics, and more. Browse their publications covering acid rain and emission trends for NO_x, SO_x CO_2, O_3 and others.

National Resources Defense Council Archives: *http://www.nrdc.org/nrdc/dire/inxarch.html*—This list of resources from the NRDC includes many related to air pollution. From here, view information on particulate air pollution, the Clean Air Act and any pending legislation related to it, air pollution and children's health, and city air pollution problems.

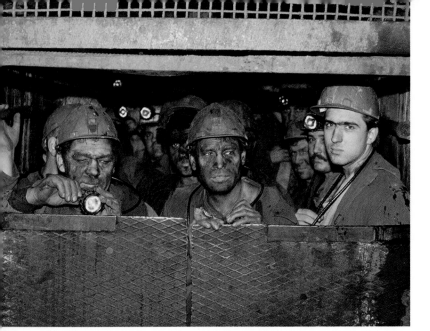

Coal miners in Eastern Europe covered with lung-damaging coal dust.

CHAPTER 23

INDOOR AIR POLLUTION

CASE STUDY

Massachusetts Registry of Motor Vehicles Building: Sick Building Syndrome

Employees of the Massachusetts Registry of Motor Vehicles moved into their newly constructed building on April 19, 1994. First hints of problems were reported in June of 1994, shortly after the building was fully occupied. Early reports were concerned with a variety of air quality problems, including unpleasant odors and a variety of symptoms that were experienced by a large number of the staff in the building. Some of the symptoms reported were respiratory problems, irritation of the eyes, nose, and throat, skin rashes, and central nervous system effects. The most often reported symptom in the spring of 1994 and February 1995 was fatigue, and this was followed closely by headaches and problems with the nasal and mucous membranes. Some type of respiratory problem was reported by 52% of the staff in the building compared to 17% having similar problems while working in their previous building.[1]

Studies by the Massachusetts Department of Public Health and private consultants suggested that the problem in the building was related to contaminated air due to a poorly constructed venti-lation system. The air circulation system in the building drew outside air into a space at the top of the structure (called the plenum). At that location the air was cooled and pumped through ventilation ducts throughout the building. This produced a problem. Because the outside air was cooled inside the building, the humidity (water) in the warmer outside air condensed onto ceiling tiles. A major component of the ceiling tiles is a starch that fermented, producing butyric acid, which smells like vomit. More serious than this was the discovery that the fireproofing material that was sprayed on all surfaces within the ventilation systems (plenum) was wet (due to the condensation of water) and falling apart. Mineral and wool fibers of the fireproofing material were then released and spread throughout the building in the ventilation system. This exposed people in the building to potentially dangerous particulate matter. After occupying the building only 15 months the staff of the Registry of Motor Vehicles moved out and the building was closed.[2]

The Massachusetts Registry of Motor Vehicles' new building is one recent example of an indoor air pollution problem that has been with us for many thousands of years, as long as people have constructed buildings and homes for protection from the elements and burned fuels to heat our human-made environments. A detailed autopsy of a fourth-century native american woman, frozen shortly after death, revealed that she suffered from **black lung disease** from breathing very polluted air over many years that included hazardous particles from lamps that burned seal and whale blubber.[3] This is the same disease that has long been recognized as a major health hazard for underground coal miners and has been called "coal miners' disease." As recently as the mid-1970s black lung disease was estimated to be responsible for about 4000 deaths per year in the United States.[4]

The employees of the Massachusetts Registry of Motor Vehicles and the coal miners breathed polluted air where they worked, whereas the native american woman breathed polluted air in her home. People today spend between 70% and 90% of their time indoors or in enclosed places (homes, workplace, automobiles, restaurants, and so forth). Only recently have we begun to fully study the indoor environment and how pollution of that environment affects our health.

LEARNING OBJECTIVES

Indoor air pollution from human fires for cooking and heating has existed and affected human health for thousands of years. Today lack of adequate ventilation in many energy-efficient homes and offices has increased the risk from pollutants. After reading this chapter, you should understand:

- What the major indoor air pollutants are and where they come from.
- Why concentrations of pollutants found in the indoor environment may be much greater than concentrations of the same pollutants generally found outdoors.

- What radon gas is and why it is perhaps one of our most serious environmental health problems.
- How radon gas enters homes and other buildings, and how its indoor concentration may be minimized.
- What the major strategies to control and minimize indoor air pollution are.
- Why indoor air pollutant is one of our most serious environmental health problems.

23.1 SOURCES OF INDOOR AIR POLLUTION

The potential sources of indoor air pollution are incredibly varied. They can arise from both human activity and natural processes. In recent years the public has been made aware of several of these sources, described in the following list:

- A bacteria, when inhaled, is responsible for outbreaks of Legionnaires' disease. Most commonly, this disease is spread by way of air-conditioning equipment, which harbors the disease-causing bacteria in air ducts and filters and transports it through a building as a bacte-

rial aerosol when heating or cooling units are in use. However, spread of the disease is not limited to this pathway. One epidemic occurred in a hospital as a result of contamination from an adjacent construction site.
- Some varieties of asbestos, used as an insulating material and fireproofing material in homes, schools, and offices, are known to cause a particular type of lung cancer (see Chapters 14 and 22).
- Formaldehyde is used in some foam insulation materials, as a binder in particleboard and wood paneling, and in many other materials found in homes and offices. These materials

can emit formaldehyde as a gas into buildings. Some mobile homes have been found to have high concentrations of formaldehyde because produts containing the chemical are used in their construction (e.g., wood paneling).

- Tobacco smoke (sometimes called second-hand smoke) is known to cause health problems, including lung cancer and heart disease.
- Radon gas seeps up naturally from soils and rocks below buildings and is thought to be the second most common cause of lung cancer (see A Closer Look 23.1, "Is Radon Gas Dangerous?").

Common indoor air pollutants and guidelines for allowable exposure are listed in Table 23.1. Many of the products and processes used in our homes and workplaces are sources of pollution. Common indoor air pollutants are often highly concentrated compared to outdoor levels. For example, carbon monoxide, particulates, nitrogen dioxide, radon, and carbon dioxide are generally found in much higher concentrations indoors than outdoors. This important concept is shown in more detail in Figure 23.1, which provides a comparison of indoor with outdoor pollution.

Why Are Concentrations of Indoor Air Pollutants Generally Greater Than Those Found Outdoors?

One obvious reason for the presence of air pollutants in our homes, offices, and other indoor areas is that there are so many potential indoor sources of pollutants. Another reason is somewhat ironic: The

---

TABLE 23.1 Sources, Concentrations, Occurrences, and Possible Health Effects of Indoor Air Pollutants

Pollutant	Source	Guidelines (Dose or Concentration)	Possible Health Effects
Asbestos	Fireproofing; insulation, vinyl floor, and cement products: vehicle brake linings	0.2 fibers/mL for fibers larger than 5 μm	Skin irritant, lung cancer
Biological aerosols/ microorganisms	Infectious agents; bacteria in heating, ventilation, and air-conditioning systems; allergens	None available	Diseases, weakened immunity
Carbon dioxide	Motor vehicles, gas appliances, smoking	1000 ppm	Dizziness, headaches, nausea
Carbon monoxide	Motor vehicles; kerosene and gas space heaters; gas and wood stoves; fireplaces; smoking	10,000 μg/m³ for 8 hours; 40,000 μg/m³ for 1 hour	Dizziness, headaches, nausea, death
Formaldehyde	Foam insulation; plywood; particleboard, ceiling tile; paneling and other construction materials	120 μg/m³	Skin irritant, carcinogen
Inhalable particulates	Smoking; fireplaces; dust; combustion sources (wildfire, burning trash, etc.)	55–110 μg/m³ annual; 350 μg/m³ for 1 hour	Respiratory and mucous irritant, carcinogen
Inorganic particulates Nitrates Sulfates	Outdoor air Outdoor air	None available 4 μg/m³ annual; 12 μg/m³ for 24 hours	
Metal particulates Arsenic Cadmium Lead Mercury	Smoking, pesticides, rodent poisons Smoking, fungicides Automobile exhaust Old fungicides; fossil fuel combustion	None available 2 μg/m³ for 24 hours 1.5 μg/m³ for 3 months 2μg/m³ for 24 hours	Toxic, carcinogen
Nitrogen dioxide	Gas and kerosene space heaters; gas stoves, vehicular exhaust	100 μg/m³ annual	Respiratory and mucous irritant
Ozone	Photocopying machines; electrostatic air cleaners; outdoor air	235 μg/m³ for 1 hour	Respiratory irritant, fatigue
Pesticides and other semivolatile organics	Sprays and strips; outdoor air	5 μg/m³ for chlordane	Possible carcinogens
Radon	Soil gas that enters buildings, construction materials, groundwater	4 pCi/L	Lung cancer
Sulfur dioxide	Coal and oil combustion, kerosene space heaters, outside air	80 μg/m³ annual; 365 μg/m³ for 24 hours	Respiratory and mucous irritant
Volatile organics	Smoking; cooking; solvents; paints; varnishes; cleaning sprays; carpets; furniture; draperies; clothing	None available	Possible carcinogens

Sources: N. L. Nagda, H. E. Rector, and M. D. Koontz, 1987; M. C. Baechler et al., 1991; E. J. Bardana Jr. and A. Montaro (eds.), 1997; M. Meeker, 1996; D. W. Moffatt, 1997.

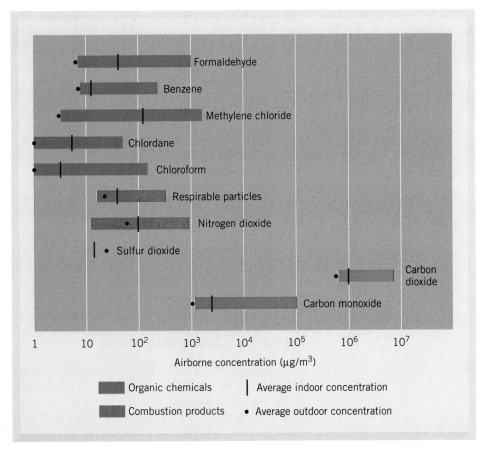

FIGURE 23.1 Concentrations of common indoor air pollutants compared to outdoor concentrations. [*Source:* A. V. Nero, Jr., 1988, "Controlling Indoor Air Pollution," *Scientific American,* 258(5), pp. 42–48.]

effectiveness of the steps we have taken to conserve energy in homes and other buildings has led to the trapping of pollutants inside. Two of the best ways to conserve energy in homes and other buildings are to increase the insulation and to decrease the infiltration of outside air. Prior to intensive energy conservation in our homes and offices, many more windows could be opened for ventilation.

An important function of ventilation is that it replaces the indoor air with outdoor air in which the concentrations of pollutants are generally much lower. Constructing our buildings with windows that do not open and applying extensive caulking and weather stripping does reduce energy consumption but also tends to affect the air quality of the building by reducing natural ventilation. As a result, we must depend more on the ventilation systems that are part of heating and air-conditioning systems. These require proper maintenance to ensure that they work effectively. When filters become plugged or contaminated with fungi, bacteria, or other potentially infectious agents, serious problems can result. Furthermore, the commonly used ventilation systems are not generally designed to reduce some types of indoor pollution, such as radon gas.[5,6]

23.2 HEATING, VENTILATION, AND AIR-CONDITIONING SYSTEMS

Heating, ventilation, and air-conditioning systems are designed to provide a comfortable indoor environment for people. Design of these systems depends on a number of variables, including the activity of people in the building, air temperature and humidity, and air quality. The interaction among these factors determines whether people are comfortable indoors. If the heating, ventilation, and air-conditioning system is designed correctly and functions properly, it will provide thermal comfort for people inhabiting the building. It will also provide the necessary ventilation (utilizing outdoor air) and remove common air pollutants via exhaust fans and filters.[5]

Personal comfort levels in terms of temperature and humidity vary depending on a person's age, physiology, and level of activity. However, different portions of buildings may have different temperatures and air quality because of their location in relation to heat sources, cold surfaces, and large windows. Humidity should be carefully controlled. High humidity may facilitate the growth of adverse mildews or molds, whereas low humidity may be a source of discomfort to some people.[5]

Regardless of the type of heating, ventilation, and air-conditioning system used in a home or other building, the effectiveness of that unit depends on the proper design of the equipment relative to the building, on installation, and on correct maintenance and operating procedures.[5] Indoor air pollution may result if any one of these factors concentrates pollutants from the many possible sources (see Table 23.1).

23.3 PATHWAYS, PROCESSES, AND DRIVING FORCES

Many air pollutants originate within buildings and may be concentrated there because of lack of proper ventilation with the outside atmosphere. Other air pollutants may enter a building by infiltration, either through cracks and other openings in the foundations and walls or by way of ventilation systems.

The driving forces that control or modify the flow of air in buildings result from a variety of processes related to both natural forces and human activity. Both natural and human processes in buildings create differential pressures that move air and contaminants from one area to another. As a very simple example, warm air tends to rise, and thus the temperature of the air in the ceiling portion of a room may be a degree or two warmer than that found below. Opening and closing doors also produces pressure differentials that induce air to move within buildings. Natural processes such as wind can affect the movement of air in a building, particularly if the structure is leaky.[5] A **chimney** (or **stack**) **effect** occurs whenever there is a temperature differential between the indoor and outdoor environments. For example, if the indoor air is warmer than that found outdoors, as the warmer air rises in the building to the upper levels, it is replaced in the lower portion of the building by outdoor air that is drawn in through a variety of openings, such as windows, doors, or cracks in the foundations and walls. Facilities such as elevator shafts and stairwells provide corridors through which air may move from one floor to another.[5]

Areas of high pressure may develop on the windward side of a building and be lower on the leeward, or protected, side. As a result, air is drawn into a building from the windward side. Because air is such a fluid medium, the possible interactions between the driving forces and the building are complex, and the distribution of potential air contaminants and pollutants is considerable. One outcome of this situation is that people in various parts of a building may complain about the air quality even if they are separated by considerable distance from each other and from potential sources of pollution.[5]

23.4 BUILDING OCCUPANTS

The building occupants are the people living or working in a particular indoor environment. Typically they do not all react, or do not react the same way, to pollutants. Several principles help to explain this variability:

- Some groups of people may be particularly susceptible to indoor air pollution problems.
- Symptoms reported by people relating discomfort from working or living in a building are variable.
- In some cases symptoms reported may be the result of factors other than air pollution.

Particularly Susceptible People

People have varying sensitivity to air pollutants. One person may be adversely affected by a particular pollutant whereas others in nearby areas may be seemingly unaffected. Some people, for example, are very sensitive to commonly used items, such as perfumes and deodorants. The problem may be a matter of concentration rather than sensitivity, and the person affected by a particular indoor air pollutant might be experiencing the greatest exposure. A person's susceptibility to a particular air pollutant will also depend on genetic factors, life-style, and age. Older people, those with impaired health, and children (because of their activity and developing lungs) will generally be more sensitive to air pollutants.[6] As a result of these individual differences, the response of one person bothered by a particular pollutant may differ from the response of another affected individual, making it difficult to evaluate indoor pollution problems. Nevertheless, certain people, especially those suffering from chronic lung or respiratory diseases such as chronic bronchitis, allergies, or asthma, are more likely to be affected adversely by poor indoor air quality. Another group more strongly affected comprises individuals who have suppressed immune systems owing to disease or medical treatment, such as chemotherapy or radiation therapy.[5]

Symptoms of Indoor Air Pollution

A great variety of symptoms can result from exposure to indoor air pollutants. Some chemical pollutants may cause irritation of the skin or eyes, nose, and throat. For example, chlorine tablets, which are often used in swimming pools and hot tubs, are a tremendous irritant if the dust from the tablets is inhaled;

shortness of breath and coughing result. Other pollutants cause dizziness or nausea. Exposure to carbon monoxide results in shortness of breath at low concentrations. At high concentrations, extreme toxicity and death can result. Tissues sensitive to carbon monoxide include the brain, heart, and muscles.[7]

The symptoms described here may have a quick onset after exposure. Other pollutants, including radon, asbestos, and chemicals such as benzene, may result in long-term chronic health effects and diseases such as cancer. Because of long lag times between exposure and disease, it may be difficult to establish relationships between a particular indoor air environment and disease in an individual.

Sick Building Syndrome

Sick building syndrome (an example of which opened this chapter) is a condition associated with a particular indoor environment that appears to be unhealthy. It describes a situation in which a number of people in a building report adverse health effects that they believe are related to the time they spend in the building. The range of complaints may vary from funny odors to more serious symptoms, such as headaches, dizziness, nausea, and so forth. Another complaint may be that a number of people in the building are sick a lot or a group of them may

FIGURE 23.2 Simplified diagram showing the radioactive decay chain for radon. Not all isotopes are shown. Half-lives and type of decay are shown for some.

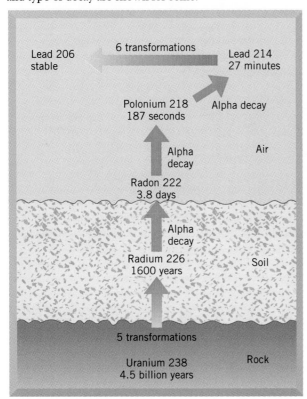

have contracted a disease, such as cancer. At one end of the spectrum there have been illnesses directly related to a building's environment. For example, outbreaks of Legionnaires' disease have been documented in convention centers and hospitals. In many cases, however, it is difficult to establish what may be causing a particular sick building syndrome. Sometimes the problem is as much related to poor management practices and worker morale as to exposure to toxins in the building. When the occupants of a building report adverse health effects and a study follows, often the cause is not detected. A number of things may be happening[5]:

- The complaints are the result of combined effects of a number of contaminants present in the building.
- Environmental stress other than air quality, such as noise, high or low humidity, poor lighting, or overheating, is responsible.
- Employment-related stress, such as poor relations between labor and management or overcrowding, may be leading to the symptoms reported.
- Other unknown factors may be responsible (for example, pollutants or toxins may be present but not identified).

Of course, sick building syndrome may also be the combined effect of various aspects of some or all of these factors. One common aspect of sick building syndrome is that often no one specific disease or cause is easily identified.[5]

Radon Gas

One of the interesting aspects about the radon gas hazard is that it is an environmental problem that comes from natural processes rather than human activities. Although radon gas is not related to public, industrial, or government activities, it has become apparent within the past decade or so that radon constitutes a significant environmental health problem in the United States.

Radon is a naturally occurring radioactive gas that is colorless, odorless, and tasteless. It is a part of the naturally occurring radioactive decay chain from radiogenic uranium to stable lead (Figure 23.2). Radon 222, which has a half-life of 3.8 days, is the product of radioactive decay of radium 226. Radon decays with emission of an alpha particle to polonium 218, which has a half-life of approximately 3 minutes. (The discussion of the phenomenon of radiation, radiation units, radiation doses, and health problems related to radiation in Chapter 18 should help in understanding the following materials.) Also see A Closer Look 23.1, "Is Radon Gas Dangerous?"

Closer Look 23.1

Is Radon Gas Dangerous?

People today are anxious and worried about radon gas in homes because studies have shown that exposure to elevated concentrations of radon is associated with increased risk of lung cancer. It is believed that the risk increases with the level of exposure to radon, the length of exposure, and a person's personal habits, such as smoking.[9] The Environmental Protection Agency (EPA) estimates that approximately 7,000 to 30,000 (best estimate is 14,000) lung cancer deaths per year in the United States are related to exposure to radon gas and its daughter products (products that result from its radioac-

tive decay), primarily polonium 218. By comparison, there are approximately 140,000 total lung cancer deaths in the United States each year. If these estimates are correct—and they are controversial—approximately 10% of the lung cancer deaths in the United States can be attributed to radon gas.

It is believed that when exposure to radon gas is combined with smoking, there is a synergistic effect that is particularly hazardous. One estimate is that the combination of exposure to radon gas and tobacco smoke is 10–20 times as hazardous as exposure to either pollutant by it-

self.[9] These inferences have been made even though there are few comprehensive studies directly linking radon gas exposure in houses to increased incidence of lung cancer. One study in Sweden did conclude that exposure to radon in homes is an important cause of lung cancer in the general population.[10] The linkage between radon and cancer is mostly based on studies of uranium miners, a group of people exposed to high concentrations of radon in mines.

It is believed that the health risk from radon gas is primarily related to its daughter products, such as

Figure 23.3 *(a)* Deposition of polonium 218 in the lungs. *(b)* Alpha radiation (particle) breaking one or both DNA strands at the cell level. (*Source:* Modified after D. J. Brenner, 1989, *Radon: Risk and Remedy,* W. H. Freeman & Co., New York. Reprinted with permission.)

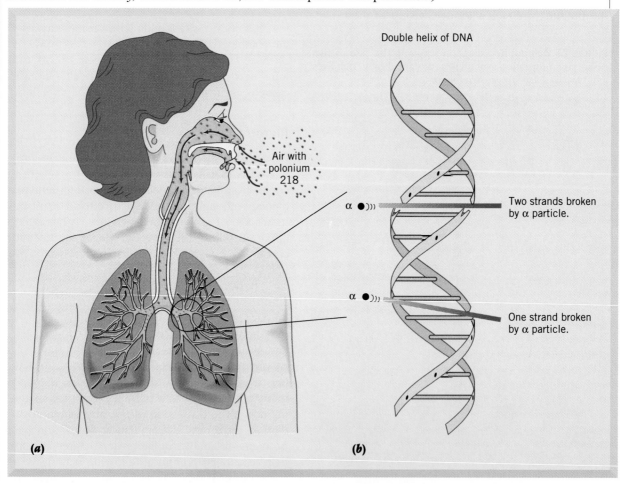

TABLE 23.2   Estimated Risk Associated with Radon

Radon Risk If You Smoke[a]

Radon Level (pCi/L)	If 1000 People Who Smoke Were Exposed to This Level Over a Lifetime	Risk of Cancer from Radon Exposure Compares to...	What to do
20	About 135 people could get lung cancer	←100 times the risk of drowning	Stop smoking and... Fix your home
10	About 71 people could get lung cancer	←100 times the risk of dying in a home fire	Fix your home
8	About 57 people could get lung cancer		Fix your home
4	About 29 people could get lung cancer	←100 times the risk of dying in an airplane crash	Fix your home
2	About 15 people could get lung cancer	←2 times the risk of dying in a car crash	Consider fixing between 2 and 4 pCi/L
1.3	About 9 people could get lung cancer	Average indoor radon level	
0.4	About 3 people could get lung cancer	Average outdoor radon level	Reducing radon levels below 2 pCi/L is difficult

Radon Risk If You Never Smoke[b]

Radon Level (pCi/L)	If 1000 People Who Smoke Were Exposed to This Level Over a Lifetime	Risk of Cancer from Radon Exposure Compares to...	What to do
20	About 8 people could get lung cancer	←The risk of being killed in a violent crime	Fix your home
10	About 4 people could get lung cancer		Fix your home
8	About 3 people could get lung cancer	←10 times the risk of dying in an airplane crash	Fix your home
4	About 2 people could get lung cancer	←The risk of drowning	Fix your home
2	About 1 person could get lung cancer	←The risk of dying in a home fire	Consider fixing between 2 and 4 pCi/L
1.3	Less than 1 person could get lung cancer	Average indoor radon level	
0.4	Less than 1 person could get lung cancer	Average outdoor radon level	Reducing radon levels below 2 pCi/L is difficult

[a]If you are a former smoker, your risk may be lower.
[b]If you are a former smoker, your risk may be higher.
Source: U.S. Environmental Protection Agency, 1992, *A Citizen's Guide to Radon,* (2nd ed.), ANR-464.

polonium 218, which is a particle and thus adheres to dust. It is hypothesized that the dust is then inhaled into lungs, where cell-damaging alpha radiation can occur when polonium 218 decays (it has a half-life of approximately 3 minutes) (Figure 23.3). The risk related to radon gas has been estimated by the EPA. Table 23.2 relates levels of exposure to radon gas to the estimated number of lung cancer deaths and gives comparable risks related to smoking or X rays.[9] Such comparisons have been criticized on the basis that there is insufficient evidence to support direct correlation of smoking habits and exposure to

radon gas with deaths from lung cancer. Exposure to radon gas has also been linked to other forms of cancer, such as melanoma (a deadly form of skin cancer) and leukemia, but such linkages are highly controversial.[11]

The average concentration of radon gas in the outdoor environment is approximately 0.4 pCi/L; the average indoor level is approximately 1 pCi/L. The EPA has set the action level for radon at 4 pCi/L. This level is the concentration below which exposure is thought (by the EPA) to be an acceptable risk. In its risk charts (see Table 23.2), the EPA equates 4 pCi/L (for a nonsmoker)

to about the risk of drowning. For a smoker this increases to about 100 times that of dying in an airplane crash. It should also be recognized that the risks shown in Table 23.2 are calculated in terms of long-term (lifetime) exposure. Furthermore, radon concentrations in one part of a home may be much different from those in another. If homes have basements, concentrations tend to be higher there than on upper level floors. As a result, the prediction of the number of deaths from lung cancer resulting from radon exposure is controversial. Nevertheless, all scientists agree that exposure to high levels of radon can cause cancer.

If the estimated risks from radon gas are anywhere close to the actual risk, then the hazard is a large one. The Surgeon General of the United States has stated that "indoor radon gas is a national health problem." It is comparable to the number of deaths from automobile accidents in the United States and is hundreds of times higher than risks resulting from outdoor pollutants present in air and water. Such pollutants are generally regulated to reduce the risk of premature death and disease to less than 0.001%. Risks from some indoor pollutants, such as organic chemicals, may be as high as 0.1%.[12] These still are very small compared to the risk for radon. For example, people who live in homes for about 20 years with an average concentration of radon of about 25 pCi/L are estimated to have a 1% to 2% chance of contracting lung cancer.[9,12]

Radon was discovered in the year 1900 by Ernest Dorn, a German chemist. The use and misuse of radon has an interesting history. In the early 1900s bathing in radon water became a health fad. During this period, when radon was thought to be beneficial to health, many products containing radium, and thus radon, hit the market. These included chocolate candies, bread, and toothpaste. As recently as 1953 a contraceptive jelly containing radium was marketed in the United States. In spite of the change in opinion about products like these, there are positive aspects of radiation; products with cancer-killing properties that result from radon decay are widely used as cancer therapy.[8]

Geology and Radon Gas

The concentration of radon gas that reaches the surface of Earth and thus may enter our dwellings is related to the concentration of radon in the rocks and soil as well as the efficiency of the transfer processes from the rocks or soil to the surface. Some regions in the United States contain bedrock with an above-average natural concentration of uranium. A large area that includes parts of Pennsylvania, New Jersey, and New York is now famous for elevated concentrations of radon gas. This area, known as the Reading Prong, contains a large number of homes with elevated concentrations of radon gas.[8] Areas with elevated concentrations of radon have also been identified in a number of other states, including Florida, Illinois, New Mexico, South Dakota, North Dakota, Washington, and California.

How Does Radon Gas Enter Homes?

Radon legend was born in Boyer Town, Pennsylvania, in 1984 when Stanley Watres, who had a job as a technical advisor to the Limerick Nuclear Power Station, set off radiation alarms on the way into the plant. The reactor in the power plant had not yet been turned on when the alarms went off, and extensive testing of Watres's clothing suggested that the contamination came not from where he worked but from where he lived. Investigators were astounded to find that the radiation level in his home was 3200 pCi/L, 800 times higher than the actual level of 4 pCi/L now set by the Environmental Protection Agency (EPA)! Until that time scientists did not believe that radon could occur naturally in concentrations high enough to be hazardous to anyone.[8,13–15] The Watres's home held the record until the latter part of the 1980s, when a home in Whispering Hills, New Jersey, was found to have a radiation level of 3500 pCi/L.[14]

It is very difficult to estimate the number of homes in the United States that may have elevated concentrations of radon gas. The U.S. EPA estimates that about 7% of homes in the United States have elevated radon levels, and recommends that all homes and schools be tested.[9] The test is simple and inexpensive.

Radon gas enters homes in three main ways (Figure 23.4):

1. It migrates up from soil and rock into basements and lower floors of houses.
2. Dissolved in groundwater, it is pumped into wells and then into homes.
3. Radon-contaminated materials, such as building blocks, are used in the construction of homes.

How Can Radon Be Reduced in Homes and Other Buildings?

The good news concerning the radon gas hazard is that, once identified, it is often easily fixed. That is, it is a hazard we do not have to live with. To put it in perspective, people living on floodplains have to contend with floods that are difficult to control, and people living in earthquake-prone areas also may have limited options to reduce the hazard. For people with radon in their homes, on the other hand, there are well-known ways to reduce the concentration of the gas. The simplest step is to locate the entry points of radon and seal them. This action, however, is often not sufficient, so additional ventilation to the home, using fans and other devices, may be necessary; increased ventilation is the primary remedy for radon problems. If these methods are not successful, a venting system may be constructed.[16,17]

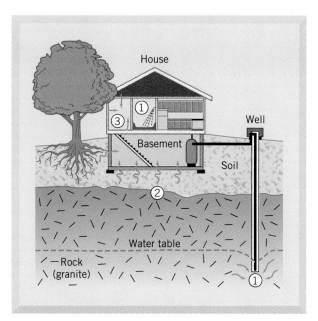

FIGURE 23.4 How radon may enter homes. (1) Radon in groundwater enters well and goes to house, where it is used for water supply, dish washing, showers, and other purposes. (2) Radon gas in rocks and soil migrates into basement through cracks in foundation and pores in construction. (3) Radon gas is emitted from construction materials used in building the house. (*Source:* Environmental Protection Agency.)

23.5 CONTROL OF INDOOR AIR POLLUTION

You might think that heating, ventilating, and air-conditioning systems, when operating properly and well maintained, will ensure good indoor air quality. Unfortunately these systems are not designed to maintain all aspects of air quality. Ventilation is one control strategy when faced with high concentrations of any indoor air pollutant, including radon. Other strategies, shown in Table 23.3, include

source removal, source modification, and air cleaning.[7] These three strategies do not constitute a complete list, and some combination of them may be the best approach.

One of the principal means for controlling the quality of indoor air is by dilution with fresh outdoor air via a ventilating air-conditioning system. Outside air is brought in and mixed with air in the return flow system from the building; the air is filtered, heated or cooled, and supplied to the building. A method that saves more of the heat in a building uses a countercurrent heat-exchange system, which warms incoming, clean, outdoor air by exchanging the heat with outgoing, warm, dirty air. Thus fresh air enters the building at a higher temperature, minimizing heat loss. In addition, various types of air-cleaning equipment for residential and nonresidential buildings are available. To handle a variety of potential pollutants, such as particles, vapors, and gases, a multistage system is used, consisting of mechanical filters, electronic air cleaners, and gas and vapor removal devices. These systems can be installed as part of the heating, ventilation, and air-conditioning system or as stand-alone appliances.[7]

Education also plays an important role in understanding and developing strategies for reducing indoor air pollution problems; it empowers people with knowledge necessary to make intelligent decisions. At one level this may involve deciding not to install unvented or poorly vented appliances. A surprising (and tragic) number of people are killed each year by carbon monoxide poisoning resulting from poor ventilation in homes, campers, and tents. At other levels, educated people are more aware of their legal rights with respect to product liability and safety. Furthermore, education provides people with the information necessary to make decisions concerning exposure to chemicals, such as paints and solvents, and strategies to avoid potentially hazardous conditions in the home and workplace.[7]

TABLE 23.3 Strategies to Control Indoor Air Pollution

Ventilation: general ventilation system; spot (zone or localized) ventilation (exhaust fans, etc.)

Source removal: material or product substitution; restrictions on source use (e.g., establishment of smoking areas; restrictions on sale of particular items and on activities that cause indoor air pollution

Source modification: change in combustion design (e.g., maximize efficiency of a gas stove; material substitution (use materials that don't cause air pollution); reduction in emission rates by intervention of barriers (e.g., apply coatings over lead paint or asbestos)

Air cleaning (pollutant removal): particle filtering; gas and vapor removal; passive scavenging or absorption

Education: consumer information on products and materials; public information on health, productivity, and nuisance effects; resolution of legal rights and liabilities of consumer, tenant, manufacturer, and so on, related to indoor air quality

Source: Modified after Committee on Indoor Pollutants, 1981, *Indoor Pollutants,* National Academy Press, Washington, DC, p. 489.

Are Airplanes Adequately Ventilated?

The environment at 10,667 m (35,000 ft) above Earth, a typical cruising altitude for jet airplanes, is not fit for humans. The temperature is −54°C (−65°F), the humidity is close to zero, and the air pressure is low. Outside air, which is compressed and heated by the jet engines, emerges at a higher pressure and temperature than would be comfortable. Before entering the cabin, therefore, the air is cooled by mixing it with outside air and the pressure is adjusted to that of an altitude of 1524 m (5000 ft). Humidity, between 5% and 20% at cruising altitudes, is supplied primarily by the breath and perspiration of passengers and crew.

Until the mid-1980s, commercial planes used 100% fresh air, which was recirculated every 3 minutes. But using fresh air taken in by the jet engines to provide air inside the plane reduces fuel efficiency. Providing fresh air in an airplane costs 22–37 times as much as supplying the same amount to the inside of a building in Washington, D.C., in January. By recirculating air already inside the cabin in combination with fresh air (usually in a 1:1 proportion), airlines found they could save money. This practice reduced the flow rate, so that there was a complete change of air only every 7 minutes and, frequently, even less often than that. One advantage of using recirculated air is a doubling of the humidity, from 5% to 10% to 10% to 20%. Higher humidity decreases the survival of viruses but increases the risk of bacterial contamination.

Air flow is measured in cubic feet per minute (cfm); ventilation rate is the air flow per person (cfm/number of people). Ventilation rates vary, from 150 cfm/person in the cockpit to 50 cfm/person in first class to 7 cfm/person in economy class. The latter is comparable to the rates found in trains and subways (5–7 cfm/person). Airline officials point out that the rate of air change in planes is much better than that found in office buildings (10–12 times an hour as opposed to 1–2 times).

Many flight attendants and passengers have complained about symptoms such as headaches, dry eyes, fatigue, nausea, and upper respiratory problems that could be associated with the air quality on planes. The carbon dioxide level, which is often used as an indicator of indoor air quality, averages 1500 ppm in planes but has been measured as high as 2000 ppm on 25% of flights studied. Even with ventilation rates as high as 35 cfm/person, carbon dioxide levels have been found at 1200 ppm or more. The American Society of Heating, Refrigerating, and Air-Conditioning Engineers (ASHRAE) recommends a maximum of 1000 ppm for indoor air,

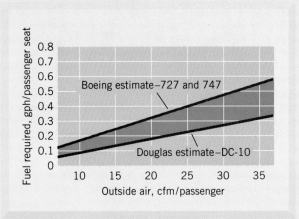

Fuel required for ventilation with outside air. (*Source:* Data from National Research Council, 1986, *The Airliner Cabin Environment,* National Academy Press, Washington, DC, p. 61.

and the Occupational Safety and Health Administration (OSHA) has a standard of 5000 ppm for industrial buildings. The Federal Aviation Administration (FAA) standard allows carbon dioxide levels up to 30,000 ppm.

Some experts are concerned that high carbon dioxide levels may be associated with other undesirable constituents, such as chemicals from cleaning fluids, pesticides, and fuels. Airborne contamination from sick passengers and animals below deck is another source of concern. Outbreaks of influenza have been tied to airplane air. Three passengers on one flight in 1992 became infected with tuberculosis, although it could not be clearly determined that infection occurred onboard. This occurrence and the fact that the incidence of tuberculosis has been on the rise worldwide since the 1980s sparked an inquiry by the U.S. Congress. Although medical experts concluded that long periods of close contact were required for transmission of tuberculosis, concern was great enough for the Centers for Disease Control (now called the Centers for Disease Control and Prevention) in Atlanta to undertake a study of the subject in 1993. The results of this study confirmed that although there is a possibility of tuberculosis transmission aboard airplanes, the possibility is so small that it is unlikely airplane cabin air circulation standards will be affected.

Critical Thinking Questions

1. The airlines contend that higher ventilation rates are necessary in the cockpit. What reasons can you think of to support their argument? Flight attendants claim that their requirements for air are different from those of passengers. What do you think is the basis for this claim?

2. If air exchange in airliners is as good or better than that in trains and subways, is there any reason to question the air quality? Develop a rationale for a position on this question.

3. What criteria would you accept for establishing a standard for the level of carbon dioxide in airplanes?

4. Three options exist for reducing carbon dioxide levels in airliner cabins: reducing emissions, increasing ventilation rates, or absorbing carbon dioxide from the ambient air. What are the advantages and disadvantages of each option?

5. Very little research has been done on air quality and the health of employees and passengers. Why, even if research is conducted, might it be hard to prove a connection?

References

Castleman, M. 1995. Clean Air Up There. *Sierra* 80(3): 16.

Douglass, W. C. 1992 Sept. "If You Fly, Don't Breathe, *Second Opinion,* pp. 1–5.

Kenyon, T. A. 1996. Transmission of multidrug-resistant *Myobacterium tuberculosis* during a long airplane flight. *The New England Journal of Medicine* 334(15): 933.

Manning, A. 1993 (June 22). "Airborne Ailments: Are Diseases Transmitted in Flight?" *USA Today,* pp. 1A, 2A.

Nagda, N. L. 1993 (July 29). Testimony before Committee on Science and Technology, U.S. House of Representatives Subcommittee on Technology, Environment and Aviation, Washington, DC.

Nagda, N. L., Koontz, M. D., and Konheim, A. G. 1991 (Aug.). "Carbon Dioxide Levels in Commercial Airliner Cabins," *ASHRAE Journal* 33: 35–38.

National Research Council. 1986. *The Airliner Cabin Environment: Air Quality and Safety.* Washington, DC: National Academy Press.

Tolchin, M. 1993 (June 21). "Exposures to Tuberculosis on Planes Are Investigated," *New York Times,* p. A7.

Tolchin, M. 1993 (June 25). "Inquiry Will Check Air Quality on Airplanes," *New York Times,* p. A16.

SUMMARY

- Indoor air pollution has been with us for thousands of years, since people first constructed structures and burned fuel indoors or worked underground in mines or other confined areas. It is one of our most serious environmental health problems.

- Sources of indoor air pollution are extremely varied. They may be associated with the materials with which we build our buildings, the furnishings we put in them, the types of equipment we use for heating and cooling, and natural processes that allow gases to seep into buildings.

- Concentrations of indoor air pollutants are generally greater than concentrations found outdoors for the same pollutants. Part of the reason for this is that in recent years we have attempted to conserve energy through better insulation of buildings. The common method for controlling indoor air pollution is ventilation. However, natural ventilation has been reduced through tighter construction of buildings, and commonly used ventilation systems are not generally designed to reduce certain types of indoor air pollutants. In addition, these systems require careful maintenance.

- There are a variety of pathways, processes, and driving forces that affect the air quality of a building. The most common natural processes involve the chimney, or stack, effect, which occurs when there is a temperature differential between indoor and outdoor environments, and the differential pressure produced by wind.

- Indoor air pollution has different effects on different people, and some groups of people are particularly susceptible to air pollution problems. Often the symptoms reported by people working in a building vary, and the symptoms may result from factors other than air pollution.

- Control of indoor air pollution involves several strategies, including source removal, source modification, and installation of air-cleaning equipment, as well as education.

- Radon gas that seeps into our homes may be one of the most serious environmental health hazards in the United States today. We are worried about radon gas because studies have suggested that exposure to elevated concentrations of radon is associated with an increased risk of lung cancer.

REEXAMINING THEMES AND ISSUES

Human Population: Indoor air pollution has been with us since we made the decision to build homes and move inside. However, as our population has grown and we have constructed a broader variety of homes, we have exposed ourselves to additional pollutants. As the number of people on Earth continues to increase, we will use more and different resources to build our homes and more people are likely to be living in smaller spaces. As a result, effects of indoor air pollution are expected to increase in the future and continue to be one of our most serious environmental health problems.

Sustainability: If we accept the premise that sustainability begins at home, then air quality in our homes and in our workplaces is an important part of building a sustainable future. Our health and those of future generations depend upon breathing pollution-free air where we spend most of our time, mainly indoors. There may be conflicts, however, because we also wish as part of sustainability to conserve resources and build energy-efficient homes. As we saw in this chapter, that may lead to restricted circulation of air and resulting indoor air pollution problems. This conflict may be solved through technological advances that minimize air pollution while maximizing energy efficiency.

Global Perspective: Indoor air pollution occurs everywhere in the world where people live inside or work in buildings or other human-constructed features such as mines. However, it is not strictly speaking a global problem. Indoor air pollution *is* a significant issue if we wish to address globally experienced human health concerns. Asphyxiation of people due to carbon monoxide poisoning and increase in the number of lung cancer cases due to exposure to high levels of radon gas are examples of toxicity and health problems found almost everywhere people live.

Urban World: Indoor air pollution problems may result anywhere people live inside. However, problems are more likely in urban areas such as cities that already have a significant outdoor air pollution problem. Therefore, in cities, if we wish to help eliminate indoor air pollution, we must also pay attention to our outdoor air quality.

Values and Knowledge: For too long we have not paid sufficient attention to our indoor environment and the consequences of poor air quality there. This issue is now moving to the forefront, however, and we believe that in the future technological advances will allow for better design of our homes, buildings, and other structures to maximize air quality for the people that spend their time there.

KEY TERMS

black lung disease *497*
chimney (stack) effect *500*
radon *501*
sick building syndrome *501*

STUDY QUESTIONS

1. What are some of the common sources of air pollutants where you live or in the school you attend?

2. Develop a research plan to complete an audit of the indoor air quality in your local library. How might that research plan differ if you were doing a similar audit for the science buildings on your campus?

3. What do you think about the concept of the sick building syndrome? If you were working for a large corporation and a number of your employees stated that they were becoming sick and listed a series of symptoms and problems, how would you react? What could you do? Play the role of the administrator and develop a plan to look at the potential problem.

4. Some people argue that the potential hazard from radon gas in homes is a lot less than suggested by the Environmental Protection Agency. Do you agree or disagree? How might potential differences of opinion be ultimately answered?

5. Develop a plan to study the potential radon hazard in your community. Where would you start? How would you gather data, and so on? If your community has undergone extensive testing already, review the results and decide if further testing is necessary.

FURTHER READING

Brenner, D. J. 1989. *Radon: Risk and Remedy*. New York: W. H. Freeman. This is a wonderful book concerning the hazard of radon gas. It covers everything from the history of the problem to what was happening in 1989, as well as solutions, and is highly recommended.

Brooks, B. O., and Davis, W. F. 1992. *Understanding Indoor Air Quality*. Ann Arbor: CRC Press. A comprehensive evaluation of indoor air pollution. It discusses most of the sources of indoor air pollutants as well as health effects and controls.

Kay, J. G., Keller, G. E., and Miller, J. F. 1991. *Indoor Air Pollution*. Chelsea, MI: Lewis. Essays on problems with biological and nonbiological air pollution in the indoor environment.

Marconi, M., Seifert, B., and Lindvall, T. 1995. *Indoor Air Quality: A Comprehensive Reference Book*. New York: Elsevier. A thorough text covering all aspects of indoor air pollution and air quality.

Nagda, N. L., Rector, H. E., and Koontz, M. D. 1987. *Guidelines for Monitoring Indoor Air Quality*. New York: Hemisphere. This book provides a good overview of indoor air pollution, including some of the factors affecting indoor air pollutants and how they might be measured.

U.S. Environmental Protection Agency. 1991. *Building Air Quality*. EPA/400/1-91/033, DHHS (NIOSH). Publication No. 91-114. This book is a guide for people building structures and managing them in terms of air quality issues. It has a good section on factors affecting indoor air quality and another on resolving problems.

U.S. Environmental Protection Agency. 1995. *The Inside Story: A guide to Indoor Air Quality*. A brief introduction to the concepts of indoor air pollution.

INTERNET RESOURCES

United Federation of Teachers Indoor Air Quality page: *http://www.uft.org/publicat/talk/air/index.html*—Good introduction to indoor air quality issues commonly encountered in a classroom setting.

U.S. Environmental Protection Agency Indoor Air home page: *http://www.epa.gov/iaq/*—Read files and publications on indoor air pollutants affecting humans including radon, biological contaminants, tobacco smoke, organic gases, asbestos, and others. Read other factual text on sick building syndrome, asthma caused by indoor pollutants, sources of indoor pollutants, and ways to reduce indoor air pollution.

U.S. National Institute for Occupational Safety and Health (NIOSH) Indoor Environmental Quality page: *http://www.cdc.gov/niosh/ieqfs.html*—Answers to frequently asked questions about indoor air pollution and indoor environmental quality, with links to other sites of interest.

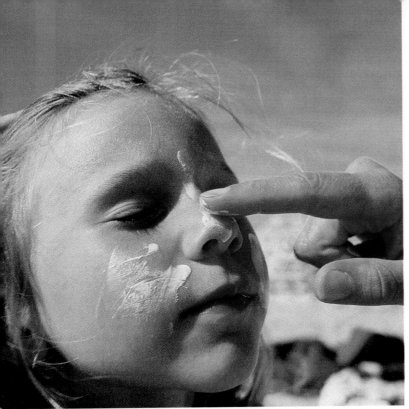

Sunscreen lotion being applied to a child's skin, reduces exposure to UVB which is known to cause skin cancer.

CHAPTER 24

OZONE DEPLETION*

CASE STUDY

Epidemic of Skin Cancers in the United States

There is little doubt that people in the United States are experiencing an epidemic of skin cancer. A worldwide increase in skin cancer has occured since early 1970, and increase in the United States has been about 90% over that period. It is popularly believed that increase in skin cancer rates is directly linked to ozone depletion.[1] Ozone depletion and, in particular, development of the "ozone hole" (discussed later in this chapter) is most dramatic since about 1975, and this correlates with increased incidence of skin cancers.

It is believed that as ozone depletion occurs, there is an increase in ultraviolet B (UVB) radiation that is damaging to living cells and a potential cause of skin cancer. It is believed that a 1% decrease in ozone causes an increase of UVB radiation of about 1 to 2%, and for each 1% increase in UVB radiation it is projected that skin cancers will increase 2%. With these numbers we can explore the skin cancer epidemic.

Since 1970 ozone depletion in midlatitudes that affect the United States has been about 10%.

Under a worst case scenerio this could cause an increase in skin cancer rates of 20 to 40%. However, the observed increase has been 90%! Thus, either we are incorrectly evaluating effects of ozone depletion and increased UVB radiation on skin cancers or there are other factors affecting the incidence of skin cancers. Because disease is seldom a one cause–one effect situation, the epidemic of skin cancers is probably due to multiple causes. For example, people are living longer and cancers are a disease of aging; second, cancer seems to be more a problem of urbanized and industrial societies. Today, people in the United States are more affluent than in the past and spend more time outdoors. There is little doubt that melanoma and other skin cancers are related to exposure to UVB radiation. Especially hazardous are sunburns that produce blistering, and severe sunburns in childhood are thought to increase risk of melanoma in later life.[1] In fact, skin cancers often take decades to develop. Therefore it has been argued that UVB exposure in the 1950s and 60s must have been responsible for skin cancers that developed in the 1970s. However, ozone depletion and the ozone hole occurred from about 1975 to present. Nevertheless, there remains a tenuous, causative link between ozone depletion and skin cancer, but not a relationship that has been absolutely proven.

If we assume that ozone depletion is contributing to more cases of skin cancer then it is prudent to better protect ourselves from exposure to UVB radiation. This includes wearing more clothing and hats as well as using sunblock ointment.

As we discuss later in the chapter, ozone depletion is likely to intensify in the next decade or two, and this will be a critical time to protect ourselves from UVB radiation. Additional work on understanding the causes of skin cancer is warranted because their increase is beyond what is thought to result from ozone depletion. If this is not done, the incidence of skin cancer will continue to create problems after ozone levels and UVB radiation have returned to normal.[1]

O zone depletion is a potentially serious global environmental problem. We understand the science of the depletion and have made positive policy to solve the problem.[2] Science and the policy of ozone depletion are the focus of this chapter.

LEARNING OBJECTIVES

Ozone depletion in the stratosphere is now recognized as a major environmental problem with potentially catastrophic effects. After reading this chapter, you should understand:

- What ozone is and how it is naturally formed and destroyed in the stratosphere, producing the so-called ozone shield.
- How chemical and physical processes and reactions link emissions of chlorofluorocarbons (CFCs) to stratospheric ozone depletion.

- What role polar stratospheric clouds play in ozone depletion.
- Why the problem of ozone depletion is not going to go away soon.
- What options are available to minimize ozone depletion.
- Why international cooperation, including significant economic aid from wealthy to less wealthy nations, will be necessary to encourage future reduction or elimination of emissions of ozone-depleting chemicals into the atmosphere.

24.1 OZONE

Air we breathe at sea level contains approximately 21% diatomic oxygen (O_2), which consists of two oxygen atoms bonded together. **Ozone** (O_3) is a triatomic form of oxygen, in which three atoms of oxygen are bonded in an uneasy union. Ozone is a strong oxidant and reacts with many materials in the atmosphere. In the lower atmosphere, ozone is a pollutant produced by photochemical reactions involving sunlight, nitrogen oxides, hydrocarbons, and diatomic oxygen. In the stratosphere, ozone provides an essential shield against damaging ultraviolet radiation. Figure 24.1 shows the structure of the atmosphere and concentrations of ozone. Approximately 90% of the ozone in the atmosphere is found

in the stratosphere, where peak concentrations are about 300 ppb. Altitude of peak concentration varies from about 25 km (15 miles) near the equator (Figure 24.1) to about 15 km (9 miles) in polar regions.[3]

The ozone layer in the stratosphere is often called the **ozone shield,** because it absorbs most ultraviolet radiation that is potentially damaging to life. Figure 24.2 shows part of the electromagnetic spectrum, discussed in Chapter 3. Ultraviolet radiation consists of wavelengths between 0.1 and 0.4 mm and is subdivided into **ultraviolet A (UVA), ultraviolet B (UVB),** and **ultraviolet C (UVC).** UltravioletC has the shortest wavelength and is the most energetic of the ultraviolet radiation. It has sufficient energy to break down diatomic oxygen (O_2) into two oxygen atoms. Each of these oxygen atoms

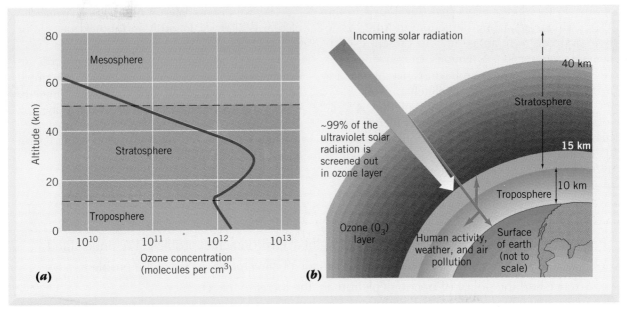

FIGURE 24.1 *(a)* Structure of the atmosphere and ozone concentration. *(b)* Reduction of ultraviolet radiation by ozone. (*Source:* Ozone concentrations from R. T. Watson, "Atmospheric Ozone," in J. G. Titus, ed., *Effects of Change in Stratospheric Ozone and Global Climate,* vol. 1, Overview, U.S. Environmental Protection Agency, p. 70.)

combines with an O_2 molecule to create ozone. UltravioletC is strongly absorbed in the stratosphere, and none reaches the surface of Earth.[3,4]

UltravioletA radiation has the longest wavelength of the ultraviolet radiation and can cause some damage to living cells. It is not affected by stratospheric ozone and is transmitted to the surface of Earth.[3] Most of the attention relative to the ozone problem is concerned with UVB radiation, which is fairly energetic and is strongly absorbed by stratos-

pheric ozone. Ozone is the only known gas that absorbs UVB. As a result, depletion of ozone in the stratosphere results in a significant increase in the UVB that reaches the surface of the Earth. This increase in UVB is the danger we are talking about when we discuss the ozone problem, because this radiation is known to be biologically damaging.[3,4]

As ozone absorbs ultraviolet radiation in the stratosphere, it breaks down again, to diatomic oxygen (O_2) and monatomic oxygen (O), with the re-

FIGURE 24.2 Part of the electromagnetic spectrum showing ultraviolet radiation with wavelengths between 0.1 and 0.4 μm.

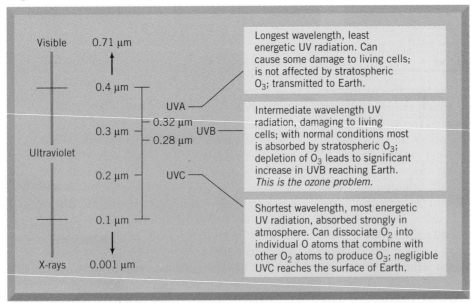

lease of heat. Natural conditions prevailing in the atmosphere result in a dynamic balance between the creation and destruction of ozone.

In summary, approximately 99% of all ultraviolet solar radiation (all UVC and most UVB) is absorbed or screened out in the ozone layer. The absorption of ultraviolet radiation by ozone is a natural service function of the ozone shield and protects us from the potentially harmful effects of ultraviolet radiation, particularly of UVB.

Measurement of Stratospheric Ozone

The concentration of atmospheric ozone was first measured in the 1920s, from the ground, using an instrument known as a Dobson ultraviolet spectrometer; the **Dobson unit** is still commonly used to measure the concentration of ozone. One Dobson unit is equivalent to a concentration of 1 ppb O_3. Today there is a nearly 30-year record of ozone con-

FIGURE 24.3 Average ozone during October at Halley Bay (1957–1984). Length of bar is estimated error in measurement and minimum values of ozone hole (1987–1995). (*Sources:* Data from J. C. Farman, B. G. Gardiner, and J. D. Shanklin, 1985, "Large Losses of Total Ozone in Antarctica Reveal Seasonal ClO_x/NO_x Interactions," *Nature,* 315, p. 207, reprinted with permission from *Nature.* Copyright 1985. Macmillan Magazines Limited: London, England, and NASA/GSFC, National Atmospheric Space Administration/ Goddard Space Flight Center.)

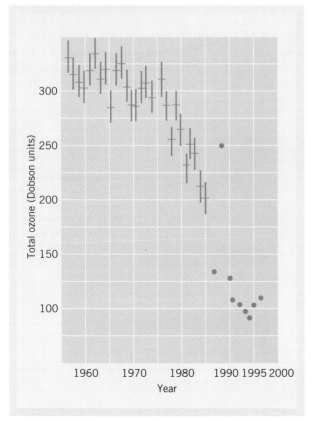

centrations from more than 30 locations around the world. Most of these stations are in the midlatitudes, but the accuracy of the data varies with different levels of quality control.[3] Satellite measurements of concentrations of atmospheric ozone began in 1970 and continue today.

Ironically, ground-based measurements first identified ozone depletion over the Antarctic. Members of the British Antarctic Survey began measurements of ozone in 1957 and in 1985 published the first data that suggested significant ozone depletion over Antarctica. Figure 24.3 shows the data from the British Antarctic Survey from the Halley Bay site. The data were taken during October of each year and show that from 1957 to about 1970 the concentration of ozone hovered around 300 Dobson units (DU) but then showed a sharp drop to a value of approximately 200 DU in 1984. Since 1984 the variability of the ozone concentration has been considerable, with a high of about 250 DU in 1988 and a low of about 88 DU in 1994 (Figure 24.3). Therefore, in spite of recent variations, the direction of change with minor exceptions, is clear; ozone concentrations in the stratosphere (during the Antarctic spring) have been decreasing since the mid-1970s.[5–8] Satellite measurements of ozone were also being recorded prior to 1984 and 1985, and although a significant reduction in ozone concentration was inferred, the values were so low that they were not believed. After the announcement of the decrease in ozone over Antarctica, the satellite measurements were reevaluated and found to confirm the observations reported by the British Antarctic Survey. This depletion in ozone was dubbed the *ozone hole.* However, there is not an actual hole in the ozone shield where all the ozone is depleted, but rather a relative depletion in the concentration of ozone that occurs during the Antarctic spring.

24.2 OZONE DEPLETION AND CFC

Early Hypothesis

The hypothesis that ozone in the stratosphere is being depleted by the presence of **chlorofluorocarbons (CFCs)** was first suggested in 1974 by Mario Molina and F. Sherwood Rowland.[9] This hypothesis, based for the most part on physical and chemical properties of CFCs and knowledge about atmospheric conditions, was immediately controversial. The idea received a tremendous amount of exposure both in newspapers and on television and was vigorously debated by scientists, companies producing CFCs, and other interested parties. The public became concerned because everyday products such as shaving cream, hair spray, deodorants, paints, and insecticides, were packaged in spray cans that carried CFCs as a propellant. The idea that these products could be responsible for threatening their health and well-being captured the imagination of the American people, many of whom responded by writing to their senators and representatives and making individual decisions to reduce the purchase of products containing CFCs.[10]

The major features of the Molina and Roland hypothesis are as follows[3]:

- The CFCs emitted in the lower atmosphere by human activity are extremely stable. They are unreactive in the lower atmosphere and therefore have a very long residence time (about 100 years). Another way of stating this is that there are no known significant tropospheric sinks for CFCs. A possible exception is soils, which evidently do remove an unknown amount of CFCs from the atmosphere at Earth's surface.[11]

- Because CFCs have a long residence in the lower atmosphere and because the lower atmosphere is very fluid with abundant mixing, the CFCs eventually (by the process of dispersion) wander upward and enter the stratosphere. Once they have reached altitudes above most of the stratospheric ozone, they may be destroyed by the highly energetic solar ultraviolet radiation. This process releases chlorine, a highly reactive atom.

- The reactive chlorine released may then enter into reactions that deplete ozone in the stratosphere.

- The result of the depletion of ozone is an in-crease in the amount of UVB radiation that reaches Earth's surface. UltravioletB is a cause of human skin cancers and is also thought to be harmful to the human immune system.

Uses and Emissions of Ozone-Depleting Chemicals

The major uses and emissions of the commonly used chemicals related to ozone depletion are shown in Table 24.1. From the late 1970s through the mid-1980s emissions of chemicals thought to destroy stratospheric ozone amounted to approximately 1.5 million metric tons per year, with CFCs accounting for approximately 60% of the total emissions. The chemicals and their uses varied. In addition to their use as aerosol propellants in spray cans, CFCs are used as a working gas in refrigeration and air-conditioning units and in the foam-blowing process for the production of Styrofoam. A variety of cleaning solvents, such as carbon tetrachloride and methyl chloroform, contain chlorine and thus destroy ozone, as does halon, which contains bromine (another chemical like chlorine) and is used in fire extinguishers.[3,10]

Table 24.1 shows the approximate atmospheric lifetime of these chemicals, which varies from about a decade to more than 100 years. The data in the table are for the year 1985. Given the long atmospheric lifetime of the chemicals, these data are still appropriate for this discussion. In 1985 the annual growth rate of the chemicals was appreciable, and for the CFCs (which probably are responsible for most of the stratospheric ozone depletion) the growth rate varied from 5% to 10%. Because these chemicals have such a long lifetime in the atmos-

TABLE 24.1 Use and Annual Emissions of Chemicals Associated with Stratospheric Ozone Depletion in 1989

Chemical	Emissions (thousand tons)	Atmospheric Lifetime[a] (years)	Applications	Annual Growth Rate (%)	Share of Contribution to Depletion (%)
CFC-12	454	139	Air conditioning, refrigeration, aerosols, foams	5	45
CFC-11	262	76	Foams, aerosols, refrigeration	5	26
CFC-113	152	92	Solvents	10	12
Carbon tetrachloride	73	67	Solvents	1	8
Methyl chloroform	522	8	Solvents	7	5
Halon 1301	3	101	Fire extinguishers	n.a.	4
Halon 1211	79	22	Refrigeration, foams	11	0

Source: C. P. Shea, January–February 1989, "Mending the Earth's Shield," World Watch, pp 27–34. Copyright 1989. Reprinted with permission of Worldwatch Institute, Washington, DC.

[a]Time it takes for 63% of the chemical to be washed out of the atmosphere.

phere, they will be with us for many years. One of the first restrictions on CFCs was its use as a propellant gas for spray cans. This practice was banned in the late 1970s in a number of countries, setting a trend that continued to 1990 with the result that CFCs as aerosol propellants are not likely to be a problem in the future.[3] On the other hand, the use of CFCs as a refrigerant has increased dramatically in recent years, especially in developing countries, such as China.

Simplified Stratospheric Chlorine Chemistry

The discussion here centers on CFCs because they are responsible for most of the ozone depletion. Earlier we noted that there are no tropospheric sinks for CFCs; that is, the processes that remove most chemicals in the lower atmosphere—destruction by sunlight, rain-out, and oxidation—do not break down CFCs because CFCs are transparent to sunlight, are essentially insoluble, and are nonreactive in the oxygen-rich lower atmosphere.[12] However, when they wander to the upper part of the stratosphere, reactions do occur. The highly energetic ultraviolet radiation splits up the CFC, releasing chlorine. When this happens, the following two reactions can take place[12]:

$$Cl + O_3 \rightarrow ClO + O_2$$
$$ClO + O \rightarrow Cl + O_2$$

These two equations define a chemical cycle that can deplete ozone. That is, the chlorine combines with ozone to produce chlorine monoxide, which in the second reaction combines with monatomic oxygen to produce chlorine again. Following this, the chlorine can enter another reaction with ozone and cause additional ozone depletion. This series of reactions is what is known as a *catalytic chain reaction*, because the chlorine is not removed but reappears as a product from the second reaction, so the process may be repeated over and over again. It has been estimated that each chlorine atom may destroy approximately 100,000 molecules of ozone over a period of 1 or 2 years before the chlorine is finally removed from the stratosphere through other chemical reactions and rain-out.[12] The significance of these reactions is apparent when we realize that the 0.9 million metric tons of CFCs emitted into the atmosphere annually (see Table 24.1) may account for ozone depletion 100 times larger than the original emissions. It should be noted that what actually happens in the stratosphere in terms of the chemical reactions is considerably more complex than the two equations shown here. The atmosphere is essentially a chemical soup that interacts with a variety of processes related to other aerosols and clouds, some of which are addressed in the discussion of the

ozone hole. Nevertheless, these equations are valuable in that they portray the concept of a chemical change reaction, the basic form of which occurs in the stratosphere to deplete ozone.

Diversion of Chlorine into Reservoirs in the Stratosphere

The catalytic chlorine chain reaction can be interrupted through storage of chlorine in other compounds in the stratosphere. Two possibilities are:

1. Ultraviolet light breaks down CFCs to release chlorine, which combines with ozone to form chlorine monoxide (ClO). This is the first reaction discussed. The chlorine monoxide may then react with nitrogen dioxide (NO_2) to form a chlorine nitrate ($ClONO_2$). If this reaction occurs, ozone depletion is minimal. The chlorine nitrate, however, is only a temporary reservoir for chlorine; the compound may be destroyed, and the chlorine released again.

2. Chlorine released from CFCs may combine with methane (CH_4) to form hydrochloric acid (HCl). The hydrochloric acid may then diffuse downward and, if it enters the troposphere, rain-out, removing the chlorine from the ozone-destroying chain reaction. This is the ultimate end for most chlorine atoms in the stratosphere. However, while the hydrochloric acid molecule is in the stratosphere, it may be destroyed by incoming solar radiation, releasing the chlorine for additional ozone depletion. It has been estimated that the chlorine chain reaction that destroys ozone may be interrupted by the processes described here as many as 200 times while a chlorine atom is in the stratosphere.[3,13]

The ozone depletion reactions are in part responsible for the decline in concentrations of ozone in both northern and southern temperate latitudes. Figure 24.4 shows the estimated decline in stratospheric ozone from 1969 to 1986 by latitude in the Northern Hemisphere. Similar declines are thought to be occurring in the Southern Hemisphere. These declines are much less than the massive destruction of ozone identified in the Antarctic ozone hole but nevertheless are a source of concern.[3]

24.3 THE ANTARCTIC OZONE HOLE

Since the Antarctic ozone hole was first reported in 1985 it has captured the imagination of many people around the world. Every year since then, Antarctic ozone depletion has occurred during the spring (October) in what is now known as the Antarctic ozone

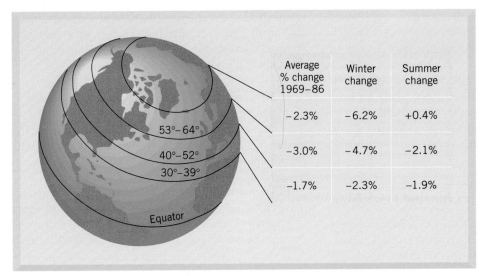

	Average % change 1969-86	Winter change	Summer change
53°-64°	-2.3%	-6.2%	+0.4%
40°-52°	-3.0%	-4.7%	-2.1%
30°-39°	-1.7%	-2.3%	-1.9%

FIGURE 24.4 Average change in concentrations of stratospheric ozone 1969-1986 by latitudinal belts. [*Source:* F. S. Rowland, 1990, "Stratospheric Ozone Depletion by Chlorofluorocarbons," *AMBIO*, 19(6-7), pp. 281-292.]

hole. The amount of depletion has varied from about 15 to 80% (Figure 24.5). In 1988 ozone depletion was 15%. The lesser amount of depletion is thought to be related to the fact that during 1988 there were fewer polar stratospheric clouds over the Antarctic. By contrast, for years when the ozone de-pletion was greater, the regions with the most ozone depletion were in the lower stratosphere at an altitude between 14 and 24 km (9-15 miles), where polar stratospheric clouds are present. The significance of these clouds is addressed in the next section. In summary, the thickness of the ozone layer

FIGURE 24.5 Ozone hole in October 1997. (*Source:* NOAA data archive accessed 8/19/98 at http://www.nic.fb4.noaa.gov:80/products/stratosphere/tovsto/archive/.)

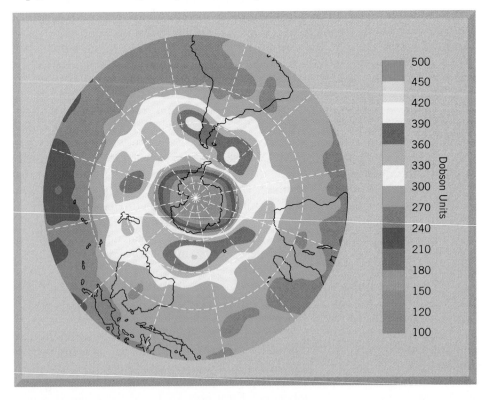

above the Antarctic during the springtime has been declining since the mid-1970s, and the geographic area covered by the ozone hole continues to increase.[14] The ozone hole was its largest ever in 1995, and it lasted longer than in any previous year. Depletion was greater than 80% for eight consecutive weeks.[15] The size of the area of depleted ozone is significant and has increased from a million or so square kilometers in the late 1970s and early 1980s to about 22 million square kilometers today! That exceeds the area of the United States (including Alaska) and Canada combined.

In discussing the global distribution of ozone it is important to remember that, under natural conditions, the highest concentration of ozone is found in the polar regions and the lowest near the equator. At first this may seem strange, because ozone is produced in the stratosphere by solar energy and more of such energy is found near the equator. Much of the world's ozone is produced near the equator, but the ozone in the stratosphere moves from the equator toward the poles with global air circulation patterns, which are not well understood.[7]

Polar Stratospheric Clouds

Polar stratospheric clouds have been observed for at least the past 100 years at altitudes of approximately 20 km (about 12 miles) above the polar regions. The clouds are approximately 10 to 100 km (6–60 miles) in length and several kilometers thick.[13] They have an eerie beauty, an iridescent glow with a color reminiscent of mother-of-pearl (Figure 24.6).[13]

Polar stratospheric clouds form during the polar winter (called the polar night because of the lack of sunlight owing to the tilt of Earth's axis). During the polar winter the Antarctic air mass is isolated from the rest of the atmosphere and circulates about the pole in what is known as the Antarctic **polar vortex.** The vortex, which rotates counterclockwise because of the rotation of Earth in the Southern Hemisphere, forms as the isolated air mass cools, condenses, and descends.[6] The cooling occurs because the isolated air mass continues to lose heat through radiation, and more heat is not supplied because of the lack of sunlight. The rotating air mass in the vortex cools and descends, and clouds are formed when the air mass reaches a temperature between 195 and 190 K (−78° to −83°C; −108° to −117°F). At these very low temperatures small sulfuric acid particles (approximately 0.1 μm) are frozen and serve as seed particles for nitric acid (HNO_3). These clouds are called Type I polar stratospheric clouds.

If temperatures drop below 190 K (−83°C; −117°F), water vapor condenses around some of the

FIGURE 24.6 Polar stratospheric clouds, February 12, 1989. Photographed from an aircraft cruising at an altitude of approximately 12 km in the polar region north of Stavanger, Norway. The red haze and thin orange or brown layers at lower altitudes are type I clouds. The red coloring is probably due to scattering from nitric acid particles. The higher white clouds are type II polar stratospheric clouds that consist mostly of water molecules frozen as ice.

earlier-formed Type I cloud particles, forming Type II polar stratospheric clouds, containing larger particles (10–100 μm). The first scenario, which produces Type I clouds, is for relatively slow cooling. If the cooling is rapid, the water vapor condenses and forms much smaller particles (approximately 2 μm). The mother-of-pearl Type II polar stratospheric clouds, formed by rapid cooling of the air mass and condensation of water vapor, are the clouds visible to people in polar areas. During the formation of both types of polar stratospheric clouds nearly all the nitrogen oxides in the air mass are tied up in the clouds as nitric acid. The nitric acid particles grow large enough to fall out by gravitational settling from the stratosphere. This phenomenon has the important result of leaving very little of the nitrogen oxides in the atmosphere in the vicinity of the clouds.[3,6,13] It is this process that in part facilitates the ozone-depleting reactions, which ultimately may reduce stratospheric ozone in the polar vortex by as much as 1% to 2% per day in the early spring when sunlight returns to the polar region.

An idealized diagram showing the polar vortex that forms over Antarctica is shown in Figure 24.7a. Ozone-depleting reactions that occur within the vortex are shown in Figure 24.7b. In the dark Antarctic winter almost all the available nitrogen oxides are tied up on the edges of particles in the polar stratospheric clouds or have settled out. The important reaction that partitions chlorine is $HCl + ClONO_2 \rightarrow Cl_2 + HNO_3$. That is, hydrochloric acid and chlorine

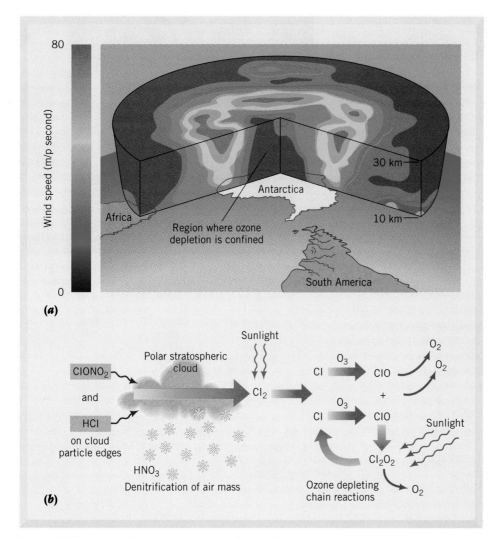

(a)

(b)

Figure 24.7 *(a)* Idealized diagram of the Antarctic polar vortex and *(b)* the role of polar stratographic clouds in the ozone depletion chain reaction. [*Source:* After O. B. Toon and R. P. Turco, 1991, "Polar Stratospheric Clouds and Ozone Depletion," *Scientific American,* 264(6), pp 68–74.]

nitrate (the two important sinks of chlorine) are partitioned on particles of polar stratospheric clouds to dimolecular chlorine and nitric acid.[16] The ozone-depleting reactions discussed earlier occur in the spring, when sunlight returns and breaks apart chlorine (Cl_2). Nitrogen oxides are absent from the Antarctic stratosphere in the spring; thus chlorine cannot be sequestered to form chlorine nitrate, one of its major sinks. Therefore chlorine is free to destroy ozone. In the early spring of the Antarctic (usually around October) these ozone-depleting reactions can be rapid, producing the 80% reduction in ozone observed in 1994. Ozone depletion in the Antarctic vortex ceases later in spring as the environment warms and the polar stratospheric clouds disappear, releasing nitrogen back into the atmosphere where it can combine with chlorine, thus removing it from ozone-depleting reactions. Stratospheric

ozone concentrations then increase as ozone-rich air masses again migrate to the polar region.

Arctic Ozone Hole?

A polar vortex also forms over the North Pole area, but it is generally weaker and does not last as long. Nevertheless, ozone depletion occurs over the North Pole as well, and as the vortex breaks up, it sends ozone-deficient air masses southward, where they may drift over populated areas of Europe and North America. In January 1992 satellite data indicated a large air mass containing high levels of chlorine monoxide (ClO) over northern Europe. The ClO is sometimes referred to as the "smoking gun" in the ozone problem because it is a major player in ozone depletion (recall the earlier discussion of ozone reactions). It is speculated that if the northern polar vortex persists for a month or more, ozone losses in the affected air mass may be on the order of 30% to 40%. A 1992 air mass with high concentrations of chlorine monoxide was detected over Europe stretching from Great Britain eastward over northern Europe.[17] One of the worries concerning the northern polar vortex and ozone depletion there is that it does tend to break up and may move ozone-deficient air masses southward, in contrast to the Antarctic polar vortex, which tends to remain more stationary. Nevertheless, in 1987 an air mass depleted in ozone that formed over Antarctica drifted northward over Melbourne, Australia. In December of that year the city reported record low concentrations of stratospheric ozone.[3]

In summary, although not as severe as the ozone depletion over the Antarctic, ozone depletion over the Arctic each winter is troublesome. Since the Arctic polar vortex is relatively weak, warmer air from midlatitudes is usually able to dissipate the vortex before ozone depletion becomes severe. However, in 1995, ozone levels were as much as 40% below normal. Scientists studying the developing ozone hole speculated that the unusually cold Arctic winter of 1995 triggered the record losses of ozone, leading to an ozone hole more like the one that forms over the Antarctic.[18]

24.4 TROPICAL AND MIDLATITUDE OZONE DEPLETION

It has been firmly established that ozone depletion in polar regions occurs as a result of reactions that take place on particles in the polar stratospheric clouds. Ice particles also occur in the stratosphere over the tropics, and at times there is an abundance of sulfuric acid aerosols resulting from the injection of sulfur into the stratosphere by volcanic eruptions. That these particles may cause ozone depletion is only a hypothesis; there is no substantial evidence. Nevertheless, following the 1982 eruption of the volcano El Chichon in Mexico a 10% reduction in ozone in the Northern Hemisphere was measured. The June 1991 eruption of Mount Pinatubo in the Philippines was the largest eruption of the century and injected approximately 20 million metric tons of sulfur dioxide into the atmosphere. This created a stratospheric aerosol cloud larger than all the polar stratospheric clouds combined.[19] It was hypothesized that the sulfur-based aerosol cloud contained particles that facilitated the occurrence of ozone-depleting reactions. Figure 24.8a shows the sulfur dioxide–rich cloud (orange band) that circled Earth in the tropics after the Mount Pinatubo eruption. Figure 24.8b shows a corresponding band of air relatively low in ozone (violet band). The correlation between the two suggests that they might be related. However, such correlations cannot be used to establish a cause-and-effect relationship without further supporting evidence.[19] In fact, additional data and analysis from the Pinatubo event suggest that anomalously low values of tropical ozone (Figure 24.8b) are probably due to effects of the eruptions on physical atmospheric circulation (the eruptions caused air at lower elevation and low ozone concentration to heat and rise to higher elevations where higher values of ozone are normally present) rather than chemical reactions that produce loss of ozone (J. W. Waters, 1994, personal written communication). Globally, stratospheric ozone is being lost at a rate of about 5% per decade, mostly at the poles.[20] However, recent evidence suggests an increase in ozone depletion at midlatitude over areas including the United States and Europe. During the unusually cold winter of 1996, ozone depletion over the coterminous United States was reported by NOAA scientists. In summary, although we know the most information about ozone depletion in polar regions (particularly Antarctica), depletion of ozone is a global concern, from the poles to the tropics.

24.5 THE FUTURE OF OZONE DEPLETION

A troubling aspect of ozone depletion is that if the manufacture, use, and emission of all ozone-depleting chemicals were to stop today, the problem would not go away, because millions of metric tons of those chemicals are now in the lower atmosphere, working their way up to the stratosphere. As illustrated in Table 24.1, several of the CFCs have atmospheric lifetimes of 75 to 140 years. Thus, about 35% of the CFC-12 molecules in the atmosphere are expected to be there still in the year 2100, and approximately 15% will be with us in the year 2200.[3] In addition, approximately 10% to 15% of the CFC molecules manufactured in recent years have not yet been admitted to the atmosphere, because they remain tied up in foam insulation, air-conditioning units, and refrigerators.[3]

Environmental Effects

Ozone depletion has several serious potential environmental effects, including damage to Earth's food chains on land and in the oceans and human health effects, including increases in all types of skin cancers and cataracts and suppression of immune systems.[21,22] As mentioned earlier, a 1% decrease in ozone can cause an increase in UVB radiation of 1 to 2% and increase incidence of skin cancer by 2%.[22]

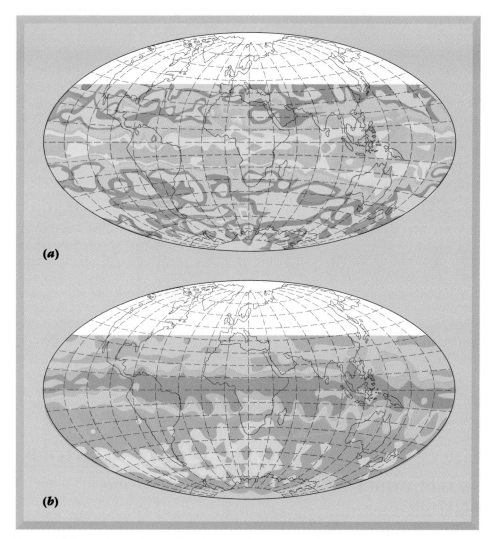

(a)

(b)

Figure 24.8 *(a)* Tropical belt of sulfur-dioxide-rich aerosol cloud (orange) and *(b)* corresponding belt of air with lower concentration of ozone (violet). [*Source:* J. B. Waters, Jet Propulsion Laboratory, in J. Horgan, 1992, "Volcanic Disruption," *Scientific American,* 266(3), pp. 28–29.]

There has been speculation for some time that ozone depletion might lead to a reduction of primary productivity in the world's oceans. Because ozone over the Antarctic area has been depleted by as much as 80% in recent years, more UVB radiation is reaching the surface of the ocean there. A loss of productivity of the phytoplankton (microscopic marine algae and photosynthetic bacteria that float near the surface of the ocean) would have a negative impact on a variety of other marine organisms, because they are at the base of the food chain. A recent study of the Antarctic waters beneath the mass of ozone-depleted air suggests that there is a minimum of 6% to 12% reduction of primary productivity associated with ozone depletion.[4] The disruption of the food chain may eventually affect people who consume marine resources, because fewer resources would be available for harvest. Also, because plankton are a sink for atmospheric carbon dioxide, their disruption

might increase the concentration of CO_2 in the atmosphere, thereby increasing global warming.

If ozone depletion becomes more widespread and affects major food crops (such as beans, wheat, rice, and corn), serious social disruption could occur. Even a small decrease in food production might have large social and political consequences around the world. A loss of 10% to 15% production could be catastrophic.

The range of human health effects of ozone depletion is being vigorously researched and debated. There is general agreement that the effects will be negative and will result in an increase in a variety of diseases, perhaps at an epidemic level compared to what would otherwise be expected. One of the most serious hazards anticipated is an increase in skin cancers of all types, including the often fatal melanoma. In 1987, record low ozone measurements over Antarctica in October were followed by record low readings over Australia and New Zealand in December.[23] As discussed earlier, this event is interpreted as a mass of ozone-depleted air that separated from the Antarctic vortex as it broke up in late spring. There is fear that such large ozone holes will increasingly threaten populated areas in mid-latitude regions. Around the world there has been an increase in the incidence of skin cancers. (See case study at beginning of this chapter.) For many years, a suntan was considered a healthy look, and people deliberately exposed their bodies to sunlight. Sun-block lotions and hats are now replacing the tanning oils of health-conscious people. Newspapers in 58 experimental regions of the United States are forecasting new EPA and National Weather Service Ultraviolet Index, which ranges from 0 (0–2, minimal exposure) to 15 (10+, very high.) Some news agencies also interpret the Index to recommend the level of sunblock. It is speculated that incidence of skin cancers, as a result of ozone depletion, will increase to about the year 2060 and then decline as the ozone shield recovers due to controls on CFC emissions.[20]

Ultraviolet radiation may damage eyes, causing cataracts, an eye disease characterized by the lens becoming opaque and impairing vision. People are now more often choosing glasses that block ultraviolet radiation. An increase in exposure to ultraviolet radiation may also damage or reduce efficiency of the human immune system.[21] For example, ozone depletion may weaken the immune systems of people exposed to AIDS, resulting in more cases of the deadly disease or shortening the time lag between infection and development of AIDS. It is further speculated that decrease in effectiveness of the human immune system may result in higher numbers of a variety of diseases. Also, a variety of environmental pollutants in air and water could have synergistic effects, increasing potential health risks (see synergistic effects in Chapter 10).

Management Issues

Prior to discussing management issues and strategies, it is important to know if observed ozone depletion is a natural or human-induced. If stratospheric ozone depletion were a natural process, then we would not see the continuous, dramatic reductions since the mid-1970s, when the real impact of production of CFCs began. World production of CFC-11 increased sevenfold from 1970 to 1994. During this period the most dramatic declines in ozone in Antarctica occurred. Thus, it appears chlorine from CFCs is the "smoking gun." Supporting this hypothesis, a recent investigation directly addressed this question. That study concluded that the concentration of stratospheric chlorine (which is responsi-

ble for ozone destruction) is more than five times what could be expected from natural emissions from oceans or other natural processes. The authors conclude that CFCs are beyond "a reasonable doubt" responsible for ozone depletion in the stratosphere.[24]

A diplomatic achievement of monumental proportion was completed with the signing of the Montreal Protocol in September 1987. The protocol, signed by 27 nations, outlined a plan for the eventual reduction of global emissions of CFCs to 50% of 1986 emissions. However, since the protocol was signed, it has become apparent that ozone depletion is a much more serious problem than originally thought. Even with the stipulations of the agreement met, it is likely that ozone depletion will result in more ultraviolet radiation reaching Earth in the next few decades, causing potential environmental disruption and human health problems.[21] Furthermore, not all the chemicals that cause ozone depletion are covered by the protocol, which originally called for elimination of the production of CFCs by 1999. Due to scientific evidence that stratospheric ozone depletion was occurring faster than predicted, the timetable for elimination of CFC production elimination was shortened. Most industrialized countries, including the United States, stopped production by the end of 1995, and the deadline for undeveloped countries is the end of 2005. An eventual phase-out of all CFC consumption is part of the Montreal Protocol; 119 more countries have signed the agreement since 1987.

It is apparent that the Montreal Protocol is working. Assessment of the protocol suggests that stratospheric concentrations of ozone-depleting substances (CFCs) are expected to return to pre–1980 levels by 2050. This assumes full compliance with the Montreal Protocol and its amendments. On the other hand, because of long residence times of CFCs in the stratosphere, maximum ozone depletion is expected to occur within the next two decades.[25,26] That's the bad news! The good news is that following maximum depletion there will be relief as ozone levels slowly increase again.

Unfortunately, all industrial nations are not responding to the urgency of this issue, partly because of the economic gap between wealthy and poorer nations. For example, China and India chose not to participate in the protocol. Their refusal was probably related to the substantial investments they are making in refrigeration and the fact that replacement chemicals currently considered for CFCs are approximately six times as expensive. It is evident that the poorer nations probably will not participate in the reduction of CFCs unless they are assisted by the wealthier countries.[27] The Montreal Protocol stipulates a transfer of both technology and funds from industrial to developing countries to speed up CFC phase-out. We return to this issue in the discussion of substitutes for CFCs.

A growing black market in CFC trade is countering the attempt to eliminate production and consumption of CFCs. Recently, illegal imports into Europe from Russia equaled 10% of the total legally allowed. Since the United States has placed such a large excise tax on imported CFCs, illegal imports have become a problem; an estimated 10,000 tons were illegally imported in 1994.[28]

Collection and Reuse

One way to lower emissions of CFCs into the atmosphere is to develop ways to collect and reuse CFCs. Every year approximately 50 million refrigerators are discarded worldwide, and each one contains approximately 1.2 kg (2.6 lb) of CFCs, mostly tied up in the foam plastic insulation that lines the refrigerators. Methods have been developed to liberate and collect these CFCs when refrigerators are recycled. The CFCs used as the coolant gas can also be collected. One company in Germany recycles approximately 6000 refrigerators a month.[29] The same techniques can be used to recover the CFCs in air conditioners used in automobiles and homes.

Substitutes for CFCs

Two substitutes for CFCs being experimented with today are the so-called **hydrofluorocarbons (HFCs)** and **hydrochlorofluorocarbons (HCFCs).** These chemicals are controversial but do have advantages. The advantage of HFCs is that they do not contain chlorine. They do contain fluorine, though, and when fluorine atoms are released into the stratosphere, they participate in reactions similar to those of chlorine and can cause ozone depletion. However, fluorine is approximately 1000 times less efficient in those reactions; thus ozone depletion is not thought to be a significant problem.[3,27]

The HCFCs contain an atom of hydrogen in place of a chlorine and may be broken down in the lower atmosphere and thus not inject chlorine into the stratosphere; they can cause ozone depletion if they do reach the stratosphere before being broken down. Although their atmospheric lifetime is considerably shorter than that of the CFCs, used in tremendous quantities, HCFCs would still cause ozone depletion.[3] Thus it may be necessary to control HCFCs also. The new HCFCs are expensive, about 3 to 5 times as costly as the CFCs now used in the refrigeration industry. Estimates (by one of the major firms that manufacture CFCs) of the worldwide cost to replace CFCs with environmentally safer HCFCs are:[30]

- $4 billion to the chemical industry over a 10-year period (this cost would be passed on to those using the chemicals) and

- $385 billion for necessary replacement of older refrigeration equipment that cannot be economically modified to use the new HCFCs.

Another option being considered is the use of helium as the working gas in refrigerators. At present, however, the world supply of helium is very limited. Furthermore, the use of helium is expected to be too expensive for home refrigeration because of the cost of making the equipment. Nevertheless, research in this area continues, and the price may come down, making its use more feasible.

Of primary importance in developing substitutes for CFCs is finding those that are both safe and effective. At least one major automobile manufacturer, Mercedes, offered a CFC-free air conditioner in 1992, and automobile manufacturers are now using substitute chemicals in car air conditioners. The hydrocarbon propane, a colorless gas easily separated from crude oil and natural gas, is a common fuel, and is easily liquefied and stored in containers ranging from large tanks to small cigarette lighters. It is also an inexpensive substitute for CFCs that has not received nearly as much attention as have HFCs and HCFCs. The cost of propane is about 10% that of CFCs and less than 2% of the cost of HFCs and HCFCs. A refrigerator in a London laboratory was inexpensively converted to use propane as the coolant, and oil refineries have used propane as a coolant in industrial processes.[30] Why, then, isn't propane being more seriously considered as a replacement for CFCs? There are two main reasons:[30]

- The chemical companies that manufacture CFCs and have patented HFCs and HCFCs are not in the refrigeration industry, and their focus is on their products rather than on finding an inexpensive substitute;
- Propane is thought to be dangerous in refrigerators and air conditioners because it is potentially explosive.

For most uses in household refrigerators and air conditioners the concern about the danger of propane is a bit of a smoke screen. The modern refrigerator would use about 100 g (3.5 oz) of propane in a sealed system, the equivalent of the amount of propane in a hand-held cigarette lighter. Because the cooling system is sealed, it is unlikely that a house fire could generate sufficient heat to rupture the system. If it did, the resulting risk should be weighed against the risk of CFCs in the same situation; combustion products associated with a fire in which CFCs (which will not burn by themselves) are present are also dangerous and include toxic nerve gas. In contrast, in commercial refrigeration systems that use large amounts of coolant, the use of propane would be a potential hazard and careful engineering would be required to ensure safety.

The use of propane as a refrigeration coolant is being tested in the United States and other places. The results will be important, especially to developing countries, where most CFC use is for refrigeration and air conditioning. Developing countries tend not to have large industrial uses for CFCs. For example, in India, about 75% of the CFCs are for refrigerators and air conditioners, a pattern characteristic of developing countries. To reduce or eliminate CFC use by replacement with HCFCs and HFCs, developing countries will need massive financial aid. They could do it on their own if an inexpensive, safe alternative were available. Research is being conducted to determine if propane or some other chemical could be that alternative.[30]

Reducing Antarctic Ozone Depletion through Injection of Chemicals

Work is just beginning on potential solutions to ozone depletion by injecting chemicals into the polar vortex where the ozone depletion occurs. The ozone depletion in the polar vortex occurs in a fairly short period of time. One idea is to inject a chemical, such as propane (C_3H_8), which will react with the chlorine to form hydrochloric acid, tying up the chlorine and not allowing it to enter into ozone-depleting reactions.[31] Early studies suggest that approximately 50,000 metric tons of propane might do

the job. The propane would be injected at an elevation of approximately 15 km (9 miles), utilizing several hundred large aircraft. The authors of this study point out that, before any real injection experiment could be initiated, many scientific and technical questions need to be answered. The possibility of unexpected or unintentional side effects must be considered.[31] The principle of environmental unity—*you can't do just one thing*—is important to remember here.

Short-Term Adaptation to Ozone Depletion

There is good news in the ozone depletion story. Concentrations of CFCs in the upper atmosphere where ozone depletion occurs are expected to peak in an estimated 3–5 years. Depletion of stratospheric ozone is likely to occur for the remainder of this decade; this will be followed by gradual recovery by the mid-twenty-first century.[20,28] Recovery will take place as a result of restrictions in the production of ozone-depleting chemicals such as CFCs.

Given the nature of the ozone depletion problem and the atmospheric lifetimes of the chemicals that produce the depletion, the major short-term adaptation by people will be learning to live with an increase in exposure to ultraviolet radiation. In the long term, achievement of "sustainability" with respect to stratospheric ozone will require management of human-produced ozone-depleting chemicals.

ENVIRONMENTAL ISSUE

Are Natural or Human-Made Chemicals Causing the Ozone Hole?

By 1993 scientists had accumulated enough evidence to support earlier predictions that stratospheric ozone was being depleted over the Antarctic. Most of them blamed the damage on organic chlorine compounds (those containing both carbon *and* chlorine) manufactured by humans, such as CFCs. But consensus among most of the scientists in the field did not prevent a continuing storm of controversy over these findings. Critics included meteorologists, scientists from other fields, amateur scientists, journalists, talk-show hosts, and authors of nontechnical books on the environment, who charged that natural sources of chlorine, not those

generated by humans, were responsible for ozone depletion and that the environmental and health threats of ozone depletion were greatly exaggerated.

Accusations leveled against scientists, NASA officials, and some industrialists included claims that the ozone scare was a sham, a scam, a hoax, or a conspiracy. Scientific uncertainties, such as the causes of ozone depletion in the Northern Hemisphere and whether depletion was allowing more ultraviolet radiation to reach Earth's surface, were emphasized. In his 1993 presidential address to the American Association for the Advancement of Science, F. Sherwood Rowland, principal architect of the hypothesis that CFCs damage the ozone layer, cited poor communication between scientists and non-scientists as the cause of the controversy. His colleague, Mario Molina, stated that the arguments put forth by critics about natural causes of ozone depletion had been tested; no results had been found to support these arguments in the 20 years since the hypothesis that CFCs were responsible for ozone depletion was proposed.

Others felt that the controversy fed on the lack of understanding of the nature of science, particularly the discomfort with uncertainty, on the part of most nonscientists. In making this point, science fiction writer Frederic Pohl quoted the late Nobel Prize physicist Richard Feynman: "Scientific knowledge is a body of statements of varying degrees of certainty—some most unsure, some nearly sure, but none absolutely certain."

The table that follows summarizes the major points of argument about whether ozone depletion is primarily a result of natural or of human-made chemicals.

Critical Thinking Questions

1. Evaluate the arguments in the table as best you can in light of what you have learned about ozone in this chapter. Which statements are (a) hypotheses; (b) inferences; (c) confirmed by observation or experiment (fact); (d) pseudoscience; (e) absolutely certain? If you feel the information is incomplete, what additional observations or experiments would you suggest? Based on your analysis, what is your position on the question of whether natural or human-made chemicals are destroying the ozone layer?

2. Without taking any preventative steps, ultraviolet radiation will increase by 10% over the next 20 years, according to scientists in the field. Some critics have pointed out that ultraviolet radiation increases by 50 times (5000%) in going from pole to equator. The expected increase in ultraviolet radiation, they say, is like moving from New York City to Philadelphia or from sea level to an elevation of 1500 ft. What is your reaction to this criticism?

3. In February 1992, NASA scientists reported unusually high levels of chlorine over the Northern Hemisphere and warned that this might lead to significant ozone loss over heavily populated areas. Congress reacted quickly to enact legislation to speed up discontinuing use of CFCs. When the ozone loss over the Northern Hemisphere was much less than the scientists had expected, they were attacked by some people as being unnecessarily alarmist and using scare tactics to obtain additional funding. Do you think these attacks were justified? Would it have been better if the scientists had presented a much more conservative scenario?

4. F. Sherwood Rowland identified two sources of public misunderstanding about scientific issues: poor communication about science and widespread scientific illiteracy. Do you agree or disagree with Rowland? Explain your position. What remedies do you suggest for dealing with the problems identified by Rowland and you?

Natural sources of chlorine are so large that CFCs are insignificant by comparison.	*Humans are producing chemicals that significantly deplete stratospheric ozone in the Antarctic and possibly in the Northern Hemisphere.*
1. CFCs are heavier than air and would not reach the stratosphere in significant amounts.	**1.** The atmosphere is in a state of constant churing of large masses of air.
2. No measurements have shown CFCs to be present in the stratosphere.	**2.** CFCs are found in stratospheric samples.
3. Volcanoes, which produce 20 times as much chlorine as CFCs, caused the Antarctic ozone hole. Mount Erebus in Antarctica has been erupting since 1973 and emits more than 1000 tons of chlorine a day.	**3.** Volcanic activity has been around for a long time, and recent eruptions have not been unusual. Mount Erebus does not erupt forcefully enough to propel chlorine into the stratosphere in significant amounts. Furthermore, it produces only 15,000 tons of chorine a year.
4. Evaporation of sodium chloride from ocean water is a source of stratospheric chlorine.	**4.** Sodium chloride, unlike CFCs, is water soluble and is washed out of the atmosphere when it rains. No sodium is found in the lower stratosphere.
5. Volcanoes release hydrogen chloride, which has increased in the stratosphere in the last 10 years.	**5.** Both hydrogen chloride and hydrogen fluoride, also water soluble, are increasing, which is what would be predicted if CFCs were reaching the stratosphere. There are almost no natural sources of hydrogen fluoride, and increases in it would not be consistent with a volcanic source for chlorine. Sulfuric acid aerosols from volcanoes that do eject into the stratosphere promote destruction of ozone.
6. Burning releases methyl chloride.	**6.** Twenty percent of the chlorine in the stratosphere occurs as methyl chloride, but only 25% of it comes from burning.

References

Brasseur, G., and Branier, C. 1992. "Mount Pinatubo. Aerosols, Chlorofluorocarbons, and Ozone Depletion," *Science* 257: 1239–1241.

Kerr, R. A. 1993. "Ozone Takes a Nose Dive after the Eruption of Mt. Pinatubo," *Science* 260: 490–491.

Pohl, F., and Hogan, J. P. 1993. "Ozone Politics: They Call This Science?" *Omni* 15(8): 34–42, 91.

Rowland, F. S. 1993. "President's Lecture: The Need for Scientific Communication with the Public," *Science* 260: 1571–1576.

Russel, J. M., Luo, M., Cicerone, R. J., and Deaver, L. E. 1996. Satellite confirmation of dominance of chlorofluorocarbons in the global stratospheric chlorine budget. *Nature* 379(6565): 526.

Tabazadeh, A., and Turco, R. P. 1993. "Stratospheric Chlorine Injection by Volcanic Eruptions: HCl Scavenging and Implications for Ozone," *Science* 260: 1082–1085.

Taubes, G. June 11, 1993. "The Ozone Backlash," *Science* 260: 1580–83.

SUMMARY

- Concentration of atmospheric ozone has been measured for more than 70 years. In the last decade measurements have been taken from instruments mounted on satellites. Evaluation of the available data suggests that the trend is quite clear: ozone concentrations in the stratosphere have been decreasing since the mid-1970s.

- In 1974, Mario Molina and F. Sherwood Rowland advanced the hypothesis that stratospheric ozone might be depleted as a result of emitting chlorofluorocarbons (CFCs) into the lower atmosphere. Major features of the hypothesis are: CFCs are extremely stable and have a long residence time in the atmosphere; eventually the CFCs reach the upper atmosphere or stratosphere, where they may be destroyed by highly energetic solar ultraviolet radiation, releasing chlorine; the chlorine may then enter into a catalytic chain reaction that depletes ozone in the stratosphere. An environmentally significant result of the depletion is that increased ultraviolet radiation reaches the lower atmosphere, where it can cause damage to living cells.

- The Antarctic ozone hole was first reported in 1985 and since has captured the imagination of people around the world. Of particular importance to understanding the ozone hole are the complex reactions that occur in the polar vortex and the development of polar stratospheric clouds. Reactions in the clouds tend to denitrify the air mass in the vortex, and in the polar spring chlorine is released to react in the catalytic ozone depletion cycle. The reactions can be very rapid, producing the observed 80% reduction in stratospheric ozone in only a few weeks.

- Tropical and midlatitude ozone depletion is also hypothesized. Polar stratospheric clouds may be present above the tropics; in addition, belts of sulfur dioxide–rich aerosol clouds may result from volcanic eruption. These clouds may be related to processes that denitrify the atmosphere and facilitate ozone depletion.

- Millions of tons of chemicals with the potential to deplete stratospheric ozone are now in the lower atmosphere and working their way to the stratosphere. As a result, if all production, use, and emissions of these chemicals were stopped today, the problem would continue and would worsen until sometime between the years 2000 and 2020.

- Potential environmental effects related to ozone depletion include damages to Earth's food chain, both on land and in the ocean, and human health effects, including increases in skin cancers, cataracts, and suppression of the immune system.

- Many nations around the world have agreed to the Montreal Protocol, which will reduce global emissions of CFCs to 50% of the 1986 levels. The agreement calls for elimination of production of the chemicals by 1996 for industralized nations and by 2006 for developing nations. This timetable for CFC production elimination was accelerated from an original schedule after ozone depletion was found to be more severe than originally believed. A serious hurdle to the compliance of all nations relates to the economic fact that most chemical replacements for CFCs are more expensive than CFCs. Therefore, it appears that financial aid will be required if the less wealthy nations are to eliminate CFC use.

- Potential management strategies for the ozone depletion problem include (1) collection and reuse of CFCs; (2) finding substitutes for CFCs; and (3) continued research with the objective of reducing ozone depletion, perhaps through injection of chemicals into the atmosphere. The idea that we might reduce the potential of chemicals to deplete ozone in the stratosphere through injection of other chemicals is only just being explored. There are very significant scientific and technical questions as well as legal and ethical issues related to the injection of any chemicals into the atmosphere.

- The adaptation that people most likely will face and accept is that ozone concentrations in the stratosphere are going to decrease in the next few decades. As a result, we will have to live with higher levels of exposure to ultraviolet radiation. Banning chemicals that can deplete stratospheric ozone is a step in the right direction and will ultimately result in the reduction of atmospheric ozone depletion.

REEXAMINING THEMES AND ISSUES

Human Population: Stratospheric ozone depletion has a strong link to human population because as the number of humans on Earth has increased, so has our use of chemicals that are currently causing ozone depletion. Of particular significance are developing countries wishing to increase their industrial output. Coupled with a rapid rise in human population, these countries could dramatically increase emissions of chemicals that are causing global change in the atmosphere. Thus, control of human population is an important objective in addressing ozone depletion in the future.

Sustainability: Controlling emissions of chemicals that cause ozone depletion is at the heart of sustainability, which has the objective of providing a quality environment for future generations. Potential environmental effects related to ozone depletion in the areas of biological productivity and disease are sufficiently significant that control and reduction of emissions of ozone-depleting chemicals are high priorities. Results of these reductions should become apparent during the first half of the twenty-first century when stratospheric ozone concentrations may recover to preindustrial levels.

Global Perspective: Stratospheric ozone depletion is a significant global environmental problem conclusively shown to result from human activity. Understanding the causes and potential solutions to the problem has required careful evaluation of Earth and its atmosphere at the global scale. Interactions between emissions of chemicals and chemical reactions occurring in the atmosphere as well as general circulation of the atmosphere have all been involved in understanding how our atmosphere works to produce processes that have ultimately resulted in ozone depletion in the stratosphere. Recommended management strategies to solve this problem have come directly through our understanding of the science and application of global systems science.

Urban World: Stratospheric ozone depletion has strong linkages to our urban world because a significant amount of CFCs have been used for refrigeration and air conditioning in urban environments. Urban areas are often places where the chemicals are produced, used, and emitted into the environment.

Values and Knowledge: The fact that human emission of chemicals is altering the stratosphere to the detriment of people and the environment in general comes as a tremendous shock to many people. Solutions to ozone depletion are costly and will require sacrifice on the part of most people on Earth. Most of the cost will fall on the wealthiest countries who, hopefully, will have the moral and ethical courage to take the leadership and pay the necessary costs for an improved global environment.

KEY TERMS

chlorofluorocarbons (CFCs) *513*

Dobson unit *513*

hydrofluorocarbons (HFCs) *522*

ozone *511*

ozone shield *511*

polar stratospheric clouds *517*

polar vortex *517*

UVA *511*

UVB *511*

UVC *511*

STUDY QUESTIONS

1. Do you believe that there is a perceptual gap between understanding the ozone problem and acting to find a solution to the ozone problem? Defend your position.

2. Given that primary productivity is reduced in the Arctic by ozone depletion, how could you define a problem to test the hypothesis that penguins might be adversely affected?

3. Is it possible to build a small chamber to test the hypothesis that polar stratospheric clouds facilitate ozone depletion? What might be some of the problems in attempting to do this? What are alternative approaches to studying the role of stratospheric clouds?

4. Suppose that next year all our understanding of ozone depletion is changed by the discovery that concentrations of stratospheric ozone have natural cycles and that our observations of the lower concentrations in the last few years have been a result of natural rather than human-induced processes. How would you put all the information of this chapter into perspective? Would you think that science had let you down?

5. What types of economic and political change will be necessary to encourage the poorer countries to support plans to eliminate chemicals responsible for ozone depletion? Do you believe that the richer countries have an obligation to help the poorer ones?

6. Do you agree or disagree that most of the people in the world will adjust to ozone depletion by bearing the loss related to increased exposure to ultraviolet radiation?

FURTHER READING

Environmental Protection Agency. 1995. *Stratospheric Ozone Depletion: A Focus on EPA's Research*. U.S. EPA Office of Research and Development, Washington, DC. A brief but good overview of ozone depletion, current status of ozone holes, and consequences of ozone depletion.

Hamill, P., and Toon, O. B. 1991. "Polar Stratospheric Clouds and the Ozone Hole," *Physics Today* 44(12): 34–42. This is a good review article of the ozone problem and important chemical and physical processes related to ozone depletion.

Jones, R. R., and Wigley, T. 1989. *Ozone Depletion: Health and Environmental Consequences*. New York: Wiley. Proceedings of the International Conference on the Health and Environmental Consequences of Stratospheric Ozone Depletion, held at the Royal Institute of British Architects, London, November 28–29, 1988.

Litfin, K. T. 1994. *Ozone Discourses: Science and Politics in Global Environmental Cooperation*. New York: Columbia University Press. Discussions on the importance of science and scientific discourse in shaping world politics and policy with regard to ozone depletion.

Makhijani, A. and Gurney, K. R. 1995. *Mending the Ozone Hole*. Cambridge, MA: MIT Press. A good overview of causes and consequences of stratospheric ozone depletion, sources and alternatives to ozone-depleting chemicals, and policy recommendations to reverse ozone damage.

Rowland, F. S. 1990. "Stratospheric Ozone Depletion by Chlorofluorocarbons," *AMBIO* 19(6–7): 281–292. This is an excellent summary article of stratospheric ozone depletion that discusses some of the major issues.

Shea, C. P. 1989. "Mending the Earth's Shield," *World Watch* 2(1): 28–34. This article focuses on solutions to the ozone problem. It is primarily concerned with control strategies to stop ozone-depleting chemicals from being emitted into the atmosphere.

Toon, O. B. and Turco, R. P. 1991. "Polar Stratospheric Clouds and Ozone Depletion," *Scientific American* 246(6): 68–74. This article provides valuable information concerning polar stratospheric clouds and their importance in ozone depletion; it offers a good explanation of the formation of polar stratospheric clouds and the chemistry that occurs there.

INTERNET RESOURCES

U.S. Environmental Protection Agency Stratespheric Ozone page: *http://www.epa.gov/docs/ozone*—Find current information on the science of ozone depletion, U.S. regulations and international agreements on the use and phasing out of ozone depleting chemicals, classification of these chemicals, CFC production, CFC atmospheric concentrations, and the latest measurements of ozone depletion.

NASA's Atmospheric Chemistry and Dynamics Branch: *http://geo.arc.nasa.gov/sgg.html*—See what type of research NASA is conducting on stratospheric ozone dynamics including research programs that examine the causes and chemical dynamics of ozone loss.

National Oceanic and Atmospheric Administration Climate Prediction Center, stratosphere: *http://nic.fb4.noaa.gov:80/products/stratosphere/*—The CPC provides up-to-date and maps of stratospheric ozone levels.

National Oceanic Atmospheric Administration Nitrous Oxide and Halocompounds Group: *http://www.cmdl.noaa.gov/noah—home/noah.html*—This program's site provides graphs of atmospheric CFC concentrations and trends, recent publications on dynamics of ozone depleting chemicals, data to view, and information on the many programs within the Nitrous Oxide and Halocompounds Group related to halocompounds in the atmosphere.

Stratospheric Ozone Law, Information and Science: *http://www. acd.ucar.edu/gpdf/ozone/*—This site has the latest version of the scientific assessment of ozone depletion report from the World Meterological Organization and the United Nations Environment Programme, as well as ozone depletion facts, summaries of recent Antarctic ozone surveys, and the latest changes to international and national ozone policy.

World Meteorological Organization Atmospheric Research and Environment Programme: *http://www.wmo.ch/web/arep/arep-home.html*—Check out some of the latest maps of northern hemisphere stratospheric ozone levels, and data trends in the Antarctic ozone hole.

PART VIII

Hunting and fishing. Etruscan wall painting, c. 500 B.C.

Environment and Society

Logging of Asian mahogany in the Philippines.

CHAPTER 25

ENVIRONMENTAL ECONOMICS

CASE STUDY

The Economics of Mahogany

The debate over how to harvest mahogany trees without serious damage to tropical forests illustrates the importance of economics in resolving environmental issues. Mahogany, which occurs naturally in Latin America, is valued for its beautiful grain and color, hardness, and durability. Since the Spanish landed in the Americas in the 16th century, mahogany has been used for shipbuilding, furniture, musical instruments, and coffins. As West Indian mahogany, whjich occurs in Florida and the West Indies, became scarce, commercial logging shifted to the big-leaf mahogany found from southern Mexico to Brazil.

Big-leaf mahogany (Figure 25.1), which can grow as tall as 45 meters (150 feet) with leaves up to 0.6 meter (2 feet) long, wholesales for approximately $1,500 a tree in the United States. Mahogany seedlings cannot grow in the shade of other trees and, under natural conditions, regenerate in clearings created by fires, hurricanes, or farmers. Traditional logging of mahogany is selective; that is, other species are left standing as long

FIGURE 25.1 Big-leaf mahogany tree.

FIGURE 25.2 Logging mahogany in Belize.

as they do not interfere with the mahogany operations. However, the size of the clearings left by selective logging is not large enough to promote regeneration, so the practice is not sustainable. Mahogany takes 120 years or more to reach marketable size, so maintaining a sustainable population in the face of extensive logging is difficult. Experts estimate that the rate at which mahogany is being logged—140,000 trees a year—is three times the rate that would be sustainable (Figure 25.2).

Because mahogany has a naturally low density (1–6 trees/ha or 2.5 acres), logging involves building an extensive system of roads through the forest. The greater access the roads provide to the forest interior makes it easier for people to clear the forest and convert the land to agriculture. The Rainforest Action Network and some other environmental groups argue that mahogany logging is, therefore, a threat to the entire rainforest ecosystem. They advocate a boycott of all mahogany products and restrictions on international trade in mahogany. Critics contend that, by lowering the economic value of the rainforest, a boycott will encourage conversion of the forest to agricultural land or other, more profitable, uses.

An alternative approach, advocated by some forest ecologists, is to maintain the economic value of the rainforest by continuing to log mahogany

and by developing commercial uses for other rainforest tree species. Some scientists are experimenting in the Rio Bravo Conservation and Management Area of Belize with harvest methods that remove more trees in order to provide clearings in which mahogany seedlings can grow. They are monitoring the ecological impact of each treatment by studying bird and butterfly populations that depend upon an intact forest ecosystem. At the same time, they are analyzing the economic costs and benefits of each method. Because all three methods involve removing more trees than would be done through traditional selective logging, this approach is highly controversial. Based on results of a study in the Bolivian Amazon, scientists at Conservation International estimate that the traditional method would damage the rainforest only one-third as much as management methods aimed at achieving sustainability.

A third approach, advocated by the International Institute of Tropical Forestry in Puerto Rico, is to grow mahogany on plantations and leave the old growth tropical forest intact. Indonesia, for example, is raising mahogany on 120,000 ha (300,000 acres). Institute scientists suggest that second growth forests and abandoned agricultural lands would be good sites for mahogany plantations, and a better solution than disturbing old growth. However, this approach might not help the people in Latin America whose livelihood depends upon the mahogany market.

REFERENCES

Brokaw, N. V. L., Wilson, R., Whitman, A. A., Hagan, J. M., Bird, N., Martins, P. J., Snook, L. K., Mallory, E. P., Novelo, D., White, D., & Losos, E. In press. *Toward sustainable forestry in Belize.*

Line, L. 1996. Advocates of sustainable mahogany harvests counter boycott. *The New York Times,* June 4, C-4.

Richards, E. M. 1991. The forest *ejidos* of south-east Mexico. *Commonwealth Forestry Review* 70: 290–311. Commonwealth Forestry Foundation.

Schneider, D. 1996. Good wood: can timber certification save the rain forest? *Scientific American,* June 36–37.

Snook, L. K. 1991. *Opportunities and constraints for sustainable tropical forestry: lessons from the Plan Piloto Forestal, Quintana Roo, Mexico.* Proceedings of the Humid Tropical Lowland Conference on Development Strategies and Natural Resource Management.

———.1993. From circumstantial to managed conservation: the mahogany forests of Quintana Roo, Mexico. Landis, S. ed. *Conservation by Design.* Rhode Island School of Design and Woodworkers Alliance for Forest Protection.

———. 1995. *Sustaining harvest of mahogany* Swientenia macrophylla *King) from Mexico's Yucatan forests: past, present and future.* Proceedings of the Conference on Conservation and Community Development in the Mayan Forest of Belize, Guatemala and Mexico, Chetumal, Quintana Roo, Mexico.

LEARNING OBJECTIVES

Other chapters in this text have explained the causes of environmental problems and discussed technical solutions. The scientific solutions, however, are only part of the answer. This chapter introduces some basic concepts of environmental economics and shows how these have been applied in the analysis of environmental issues. After reading this chapter, you should understand:

- What the "tragedy of the commons" is and how it leads to an overexploitation of resources.
- How the perceived future value of an environmental benefit affects our willingness to pay for it now.

- What externalities are, and why. Why it is important to evaluate them in determining the costs of actions that affect the environment.
- What factors may be involved in determining a level of acceptable risk.
- Why it is difficult, yet important to evaluate environmental intangibles, such as aesthetics.
- What issues are involved in determining who pays the direct and indirect costs of controlling pollution and minimizing environmental damage.
- What kinds of policy methods are available to control pollution or harvesting of resources.

25.1 THE IMPORTANCE OF ENVIRONMENTAL ECONOMICS

The United States spends about $115 billion a year to deal with pollution, about 2% of the nation's gross national product. The defense budget is only two and a half times larger. By the year 2000 the costs may rise to more than $170 billion.[1] Included in these figures are the amounts spent by consumers and by private companies as well as by the government. Clearly the total is much greater than the $6 billion budget of the Environmental Protection Agency (EPA), a budget that Congress attempted to cut substantially in 1996.[2] And this 2% of the nation's budget is related to pollution control only; it does not include expenses for other environmental programs, such as those for conservation of endangered species, wildlife management, or conservation of forests.

Populations subject to high levels of certain pollutants (people in inner cities, for example) have a lower average life expectancy or a higher incidence of certain diseases. Studies in the United States have indicated that particulate air pollution in cities alone is at least a significant contributor to 60,000 deaths annually.[3] Other studies in the United States have estimated that from 2 to 9% of total mortality in cities is associated with particulate air pollution.[4] Thus the economics of environment affects all of us, and a field has developed called environmental economics. Economic analysis involves two different kinds of environmental issues: the use of desirable resources (such as fish in the ocean, oil in the ground, or forests on the land) and the minimization of undesirable pollution. An understanding of basic concepts of environmental economics will provide insight into why environmental resources have been poorly conserved in the past. This insight may help us determine how best to modify our policies so that in the future conservation becomes a part of economic development.

Issues in Environmental Economics

Environmental decision making often involves analysis of tangible and intangible factors. A mudslide that results from altering the slope of land is an example of a tangible factor; the aesthetics of the altered slope is an example of an intangible factor. Of the two, the intangibles are obviously more difficult to deal with because they are hard to measure and to value. Nonetheless, evaluation of the intangibles is becoming more important in local, regional, and national land-use planning and environmental analysis. Therefore, one task of environmental economics is to develop a method of aesthetic evaluation that provides good guidelines, is easy to understand, and is quantitatively credible.

25.2 USE OF DESIRABLE RESOURCES

The Environment as a Commons

It often seems curious that those who use a natural resource or benefit in some way from it sometimes do not act in a way that protects that resource and its environment in a renewable state. When individuals benefit from the resource, it would seem to be in their best interests not to damage or destroy it. Economics shows that the profit motive, by itself, will not always lead a person to act in the best interest of the environment. Here we give two reasons why this may be so.

The first has to do with what the ecologist Garrett Hardin called "the tragedy of the commons.[5] When a resource is shared, an individual's personal share of profit from exploitation of the resource is usually greater than the individual's share of the resulting loss. The second has to do with the low growth rate, and therefore the low productivity, of a resource.

The Concept of the Commons

A **commons** is land owned publicly with public access for private uses. Historically a commons was a part of old English and New England towns where all the farmers could graze their cattle (Figure 25.3). The practice of sharing the grazing area works as long as the number of cattle is low enough to prevent overgrazing. As Hardin points out, in such a situation each herder is trying to maximize personal gain and must periodically consider whether to add more cattle to the herd. The addition of cattle has both a positive and a negative *utility* (in economic terms, value or worth). The positive utility is the benefit the herder receives by selling the extra cattle; the negative utility is the overgrazing caused by the

addition of too many animals. Hardin argues that because the benefit to the individual of selling a cow for personal profit is greater than that individual's share of the loss in the degradation of the commons, freedom of action in a commons inevitably brings ruin to all. According to Hardin, without some management or control, all natural resources treated like commons will inevitably be destroyed.

Some biological resources, such as much of the forest land in the United States, are on privately owned lands. Many other resources, including 38% of our nation's forests, are on publicly owned lands that are accessible to everyone. In addition, resources in international regions, such as ocean fisheries away from coastlines, are not controlled by any single nation. A society that controls resources such as public lands and waters has a number of social mechanisms to achieve its environmental goals. At one extreme, a society may simply rely on individual motivation, on the assumption that what people find best for themselves will also be best for society. However, a society can also enact laws that provide direct regulation of use, such as the U. S. Marine Mammal Protection Act, which allows no taking of marine mammals without a permit; it can set total harvest quotas while allowing open access to the resource, such as occurs with some fisheries; issue licenses, as is typical with sport fishing, so that control of harvest may be done by limiting the number

FIGURE 25.3 Classic village green or commons in Townshend, Vermont. The public land of the commons was typically surrounded by private homes and buildings used by the public, such as the church shown in this picture.

of licenses issued or it can limit the individual harvest while not directly limiting the total number harvested.

When biological resources occur in international or national areas with open access, they are like a commons. Thus they are threatened by the tragedy of the commons. There are many examples of commons, both past and present. An important one today is the deep oceanic seabed, where there may be valuable mineral deposits. At present, these areas are a true commons, since the high seas have always been considered areas open to all and not the property of any single nation. Another example is Antarctica. Although there are some national territorial claims on this continent, most of the continent is a commons. Negotiations have continued for a number of years concerning the conservation of Antarctica and possible use of its resources.

At a national level, the overcrowding of national parks and the pollution of the atmosphere are also examples of the problem of the commons. For example, for a number of years there has been a debate in Minnesota about the status of Voyageurs National Park. This park, located within the boreal forest biome (see Biome section) of North America, contains many lakes and islands and is an excellent place for fishing, hiking, canoeing, and viewing wildlife. Before the area became a national park, it was used for motor boating, ski-mobiling, and hunting; a number of people in the region made their living from tourism that was based on these kinds of recreation. Some environmental groups argue that the Voyageurs National Park is ecologically fragile and needs to be made one of the U.S. legally designated wilderness areas, to protect it from overuse and from adverse effects of motorized vehicles. Others argue that the nearby million area Boundary Waters Canoe Area (see Chapter 13) provides ample wilderness, that Voyageurs can withstand a moderate level of hunting and motorized transportation, and that these uses should be allowed. At the heart of the issue is the problem of the commons, which in this case can be phrased as: What is the appropriate public use of public lands? Should all public lands be open to all public uses? Should some public lands be protected from the people? At present, the United States has a policy of different uses for different lands. In general, national parks are open to the public and many kinds of recreation, whereas designated wildernesses have restricted visitorship and kinds of uses.

Low Growth Rate and Therefore Low Profit as a Factor in Exploitation

Another reason individuals tend to overexploit natural resources held in common, according to mathematician Colin Clark, is the low growth rate of the resource.[6] The example of whales presented in A Closer Look 25.1, "Whale Harvest and the Commons," looks at whale oil (a marketable product) as the capital investment of the industry. How can whalers get the best return on their capital? (Here we need to remember that whale populations, like any population, increase only if there are more births than deaths.) We will examine two: resource sustainability and maximum profit.

If whalers adopt a simple, one-factor resource sustainability policy, they will harvest only the net biological productivity each year and thus maintain the total abundance of whales at its current level (see Chapter 13); that is, they will stay in the whaling business indefinitely. If they adopt the simplest approach to maximize immediate profit policy, they will harvest all the whales now, sell the oil, get out of the whaling business, and invest the profits.

Suppose they adopt the first policy. What is the maximum gain they can expect? Whales, like other large, long-lived creatures, reproduce slowly; a calf born every 3 or 4 years per female is typical. The total net growth of a whale population is unlikely to be more than 5% per year, where the net growth is defined as the net increase in biomass (see the explanation of biological growth in Chapter 8). If all the oil in the whales in the ocean today represented a value of $10 million, then the most the whalers could expect to take in each year would be 5% of this amount, or $500,000. Meanwhile, they must pay the cost of upkeep on ships and other equipment, interest on loans, and salaries of employees, all of which decrease profit.

If they adopt the second policy and harvest all the whales, then they can invest the money from the oil. Although investment income varies, even a conservative investment of $10 million would very likely yield more than 5%, especially when this income could be received without the cost of paying a crew, maintaining the ships, buying fuel, marketing the oil, and so on.

It is quite reasonable and practical, *if one considers only direct profit,* to adopt the second policy: Harvest all the whales, invest the money, and relax. Whales simply are not a highly profitable long-term investment under the resource sustainability policy. It is no wonder that there are fewer and fewer whaling companies and that companies have left the whaling business when their ships became old and inefficient. (Recall the discussion in Chapter 12 about international treaties that affect the harvest of whales, designed to prevent the problems that arise from the simple profit motive described here.) Few nations support whaling; those that do have stayed with whaling for cultural reasons. For example, whaling is important to Eskimo culture and some harvest of bowheads takes place in Alaska; whale

CLOSER LOOK 25.1

WHALE HARVEST AND THE COMMONS

The harvesting of whales is an example of the economics of the commons. Many nations have argued for a reduction or elimination of commercial whaling in the interest of conserving the great whales. If whalers profit by harvesting whales, it seems logical to ask why they have not acted to protect the resource on which their livelihood depends. Ranchers, after all, do not intention-ally kill all their cattle just because they can make money by selling cattle. Whalers, however, have brought species of the great whales to the brink of extinction. One major difference, of course, is that ranchers know how many cattle they have when deciding how many to slaughter, whereas an accurate census of whales is not possible. The blue whale, for example, was reduced to an estimated few hundred individuals before harvesting was stopped in the 1960s (Figure 25.4) (see Chapter 13 for more information about whaling).

A fundamental issue in whaling and fishing is the lack of property rights. Ranchers refrain from killing off entire herds because in the future they themselves will reap the benefits of maintaining the herds (and the yet unborn). Fishermen or whalers, however, cannot be assured that they will reap future benefits. The offspring of the fish or whales they do not kill today are not necessarily theirs to harvest in the future.

Some people might argue in favor of private ownership of what have been commons. The rancher who owns a cattle herd and grazing land has a direct interest in assuring their sustainability, whereas the whaler, whose sustenance derives from the commons, cannot quantify the resource base and cannot determine a sustainable harvest. Others might argue that private ownership of public goods, such as fisheries and forests, is undemocratic and unfair. One thing is certain: As the human population grows and resources remain constant or decline, competition for their use will intensify and any decisions about political and economic control will be more hotly contested.

FIGURE 25.4 *(a)* Blue whale calf breaching. *(b)* Annual blue whale catch from 1925 to 1965. After 1965, the whales were protected. (*Source:* C. W. Clark, 1973, "The Economics of Overexploitation," *Science,* 181, 630-634.)

(a)

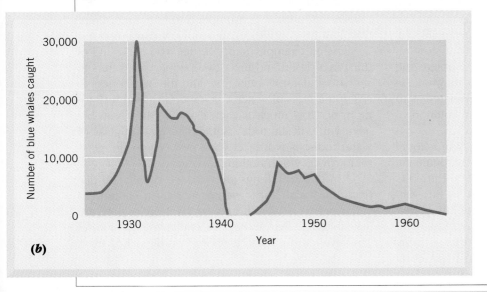

(b)

meat is a traditional Japanese food, and whale harvest is maintained for this reason.

From these examples we learn that if we want to conserve whales, we must think beyond the immediate, direct economic advantages of whaling. Policies that seem ethically good may not be the most profitable for an individual. Finally, in the example of whaling, the economic analysis clarifies how an environmental resource is used, and this brings us to the question of externalities.

Externalities

The gap in our thinking about the whales, an environmental economist would say, is that we must be concerned with externalities in whaling. An **externality,** also called **indirect costs,** is an effect not normally accounted for in the cost–revenue analysis of producers.[7] In this case, externalities include the loss of revenue to tourist boats used to view whales and the loss of an ecological role played by whales in marine ecosystems.

Air and water pollution provide other good examples of externalities. Consider production of nickel from ore at the Sudbury, Ontario, smelters, which has serious environmental effects, as discussed in Chapter 14. Traditionally, the economic costs associated with the production of commercially usable nickel from an ore are the **direct costs,** that is, those borne by the producer and passed directly on to the user or purchaser. In this case, direct costs include the costs of purchasing the ore, of energy to run the smelter, of building the plant, and of paying employees.

Traditionally costs associated with degradation of the environment from emissions from the plant have been considered externalities. For example, prior to implementation of pollution control, Sudbury smelter destroyed vegetation over a wide area, which led to an increase in erosion. Although air emissions from smelters have been substantially reduced and restoration efforts have begun a slow recovery of the area, pollution remains a problem and total recovery of the local ecosystem may take a century or more.[8] These are costs associated with the value of trees and soil, and with restoration of vegetation and land to a productive state.

An important societal question is: Who should bear the burden of these costs? Some environmental advocates suggest costs should be included in costs of production through taxation or fees. In this way expense would be borne by the corporation that benefits directly from sale of nickel or would be passed on, in increased sale prices, to users (purchasers) of nickel. Others suggest that these costs should be shared by the entire society and paid for by general taxation. How to deal with externalities and indirect costs remains a controversial and unresolved issue in environmental economics.

Other factors to consider in resource use are the relative scarcity of a necessary resource and its price. For example, if a whaler lived on a desert island and whales were the only food available, then the whaler would have to consider his interest in remaining alive as well as short-term profit. Therefore, he could not choose to sell off all whales to maximize profit. It would seem to make sense that he would maintain a harvest that would maintain the population of whales. Of course, even a whale-eating whaler might choose a policy somewhere between the two extremes. The whaler might decide that his own life expectancy was only 10 years and try to harvest the whales so that they would become extinct at the same time—a "you can't take it with you" attitude. However, this choice to cause the extinction of whales would not necessarily occur if ocean property rights existed. A whaler could then sell rights to future whalers, or mortgage against them, and thus reap the benefits of whales that could be caught after his death.

How is the Future Valued?*

The preceding example reminds us of the old saying, "A bird in the hand is worth two in the bush." That is, a profit now is worth much more than a profit in the future. This economic concept—the future value compared with the present value—is another important idea for environmental science. The value of some elements of the environment may increase, decrease, or remain the same (Figure 25.5).

Economists observe that it is an open question whether something promised in the future has less value than something today. The value is the result of the interaction of consumer's preferences for present and future consumption, the technology for transferring present consumption to future consumption, and how future consumers view consumption.

As an example, suppose that you are dying of thirst in a desert and meet two people; one offers to sell you a glass of water now and the other offers to sell you a glass of water if you can be at the well tomorrow. How much is each glass worth? If you believe you will die today without water, the glass of water today is worth all your money and the glass tomorrow is worth nothing.

In practice, things are rarely so simple and distinct. But we all know we are mortal, so we tend to value personal wealth and goods more if they are available now than if they are available in the future. This evaluation is made more complex, however,

*With assistance of Charles Kolstad.

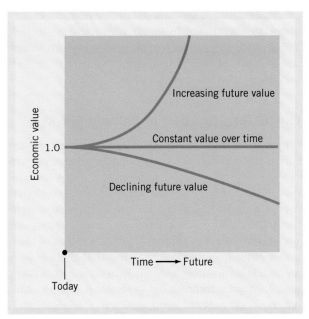

FIGURE 25.5 Economic value as a function of time; a way of comparing the value of having something now compared with having it in the future. A negative value means that there is more value attached to having something in the present than having it in the future. A positive value means that there is more value attached to having something in the future than having it today. We might attach a positive value for an endangered species (its survival in the future is worth more than its existence today); however, economists argue that a positive rate leads to an infinite value and cannot be used in practice.

because we are accustomed to thinking of the future, to planning a nest egg for retirement or for our children.

There are many people today who argue that we have a debt to future generations and must leave the environment in at least as good condition as we found it; these people would argue that the future environment is not to be valued less than the present.

Second, many people argue that humans place a higher value on a possession in hand today than on one promised tomorrow. In our example, you would rather have something today, while you are alive to enjoy it, than tomorrow when you might be dead. Our discussion of future value is obviously sensitive to time. The future time associated with some important global environmental topics such as stratospheric ozone depletion and global warming are closer to a century. This results because chlorofluorocarbon (CFC) has a residence time in the atmosphere of about 100 years. Similarly, the time to realize potential benefits from changing energy policy to offset potential global climate change is also on the order of 100 years. How to deal with future value in an analysis of environmental issues remains a complex, contentious problem.

Another tough philosophical issue is that it might be argued that conserving resources and other parts of the environment for the future is tantamount to taking from the poor at present and giving to the rich in the future.

One further complicating issue is that as we get wealthier, our value on many environmental assets (like wilderness areas) increases dramatically. Thus, if society continues to grow in wealth over the next century as it has over the past century, the environment will be worth far more to our great-grandchildren than our great-grandparents, at least in terms of willingness to pay to protect it.

If history is a guide, Americans in 2095 will be far better off than Americans in 1995. To what extent should we ask the average American today to sacrifice now for much richer great-great grandchildren? How can we know the future usefulness of today's sacrifices? Put another way, what would you have liked your ancestors in 1897 to have sacrificed for our benefit today? Should they have increased research and development on electric transportation? Should they have saved more tall-grass prairie or restricted whaling? The two issues are that (1) we are so much richer and better off than our ancestors so their sacrificing for us would have been inappropriate and (2) even if they wanted to sacrifice, how would they have known what sacrifices would be important to us? As a general rule, one answer to these thorny questions about future value is: Do not throw away or destroy something that cannot be replaced if you are not sure of its future value. For example, if we do not fully understand the value of the wild relatives of potatoes that grow in Peru but do know that their genetic diversity might be helpful in developing future strains of potatoes, then we ought to preserve those wild strains.

25.3 RISK–BENEFIT ANALYSIS

Death is the fate of all individuals, and every activity in life involves some risk of injury or death. What, then, does it mean to save a life by reducing the level of a pollutant? With some activities, the relative risk is clear. It is much more dangerous to stand in the middle of a busy highway than to stand on the sidewalk. Hang gliding has a much higher mortality rate than hiking. The effects of pollutants are often more subtle, so the risks are harder to pinpoint and quantify. Table 25.1 gives the risk associated with a variety of activities. The degree of risk is an important concept in our legal processes. For example, the U.S. Toxic Substances Control Act states that no one may manufacture a new chemical substance or process a chemical substance for a new use without

TABLE 25.1 Risk of Death for Various Activities

Activity	Risk of Death/Year
Sports	
Auto Racing	1.2 in 1,000
Biking	1 in 100,000
Football	4 in 100,000
Water sports	1.9 in 100,000
Travel	
Air travel (one transcontinental trip across United States)	3 in 1 million
Motor vehicles	2.2 in 10,000
Natural hazards	
Hurricanes	4 in 10 million
Lightning	4 in 10 million
Tornadoes	5 in 10 million
Pollution	
Air pollution	1.5 in 10,000

Source: R. Wilson, 1980, "Risk/Benefit Analysis for Toxic Chemicals," *Ecotoxicology and Environmental Safety,* 4, pp. 370–383. Copyright 1980. Reprinted by permission of Academic Press, Orlando, FL.

obtaining a clearance from the EPA. The act establishes procedures to estimate the hazard to the environment and to human health of any new chemical before it becomes widespread. The EPA examines the data provided and judges the degree of risk associated with all aspects of the production of the new chemical or the new process, including extraction of raw materials, manufacturing, distribution, processing, use, and disposal. The chemical can be banned or restricted in either manufacturing or use if the evidence suggests that it will pose an unreasonable risk of injury to human health or to the environment. But what is *unreasonable?* [9]

Calculating the Costs of Pollution Risks

When DDT was first used, no one understood the subtle environmental transport and ecological effects of the chemical; new scientific observations revealed these effects (see A Closer Look 25.2, "Risk–Benefit Analysis and DDT"). At that time, there was relatively little concern with environmental issues, and society was not yet willing to pay for many of these costs. Now people widely agree that the environment is a major concern, and there is less willingness to accept indirect environmental effects. What had been considered externalities to the use of DDT have become internal cost factors.

As seen in Chapter 22, the total cost of pollution is the sum of the costs to control pollution and the loss from pollution damages. In some cases, these two factors have opposite trends in terms of economic cost (as one goes down, the other goes up), and their intersection point is the minimum total cost, as shown by point *A* in Figure 25.6. In other cases, total costs may stabilize or even decline as companies utilizing pollution control become more efficient and external costs in the form of environmental damage are minimized. If the minimum total cost involves a pollution level that is too high a

 CLOSER LOOK 25.2

RISK-BENEFIT ANALYSIS AND DDT

A review of the history of the use of DDT illustrates the difficulty of completely eliminating a pollution risk. As noted in Chapter 11, DDT was first applied widely in the 1940s to control the spread of diseases such as malaria via insects and was subsequently used to control crop pests. At first, tests of the safety of DDT focused on human health effects, which were believed to be small. In the late 1950s, however, DDT was discovered in the livers of sharks. Because DDT had not been used in the ocean, at first it was believed that this observation was simply a measurement error or the result of an unknown dumping of DDT directly in the ocean.

Gradually, and to their surprise, scientists began to understand that DDT had spread from farmland through surface-water runoff and the air into the ocean and that the chemical had become a worldwide contaminant. DDT was found in the tissues of penguins in the Antarctic and seals in the Pribilof Islands of the Bering Sea. By 1970, it had become clear that DDT was everywhere and was a global environmental problem.[21] Although DDT is banned for use in the United States, U.S. corporations continue to be among the largest producers of this chemical, which is shipped to other nations. DDT is still used widely elsewhere, especially in developing nations of the tropics.

Complete elimination of DDT residues from all the environments of the world—and from all the sharks, penguins, seals, and birds—no longer seems feasible. Aside from the difficulty of eliminating pollutants, another lesson of this example is that what had seemed to be an economic externality of DDT (its indirect ecological effects on birds and ocean animals) became a major societal issue.

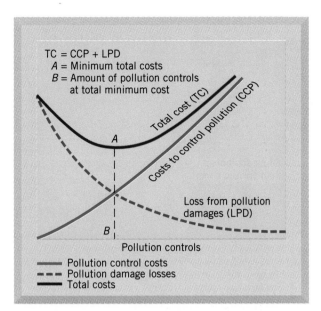

FIGURE 25.6 Total cost of pollution is the sum of the costs to control pollution and the loss from pollution damages. These two factors have opposite trends in terms of economic cost, and their intersection point forms a minimum total cost as shown by point A. If the minimum total cost involves a pollution level that is too high a risk, then additional control may add considerable expense.

risk, then additional control may add considerable expense. The level of acceptable pollution (and thus risk) is a social–economic–environmental trade-off. The level of acceptable risk changes over time in society, depending on changes in scientific knowledge, comparison with risks from other causes, the expense of decreasing the risk, and the social and psychological acceptability of the risk.

We must therefore ask several questions. What risk from a particular pollutant is acceptable? How much is a given reduction in risk from that pollutant worth to us? How much will each of us, as individuals or collectively as a society, be willing to pay for a given reduction in that risk? The answers depend not only on facts but on societal and personal values. What also must be factored into the equation is that costs associated with cleanup of pollutants and polluted areas and with restoration programs could be minimized or even eliminated if a recognized pollutant is controlled initially. The total cost of pollution control need not increase indefinitely.

Acceptability of Risks

Societies differ in socially, psychologically, and ethically acceptable levels of risk for any cause of death or injury. It is commonly believed that future discoveries will help to decrease the risk, perhaps eventually allowing us to approach a zero-risk environment. But complete elimination of risk is generally either technologically impossible or prohibitively expensive.

We can make some generalizations about the acceptability of various risks. Novel or new risks appear to be less acceptable than long-established or natural risks, and society tends to be willing to pay more to reduce novel risks. For example, France spends approximately $1 million to reduce the likelihood of one air-traffic death but only $30,000 for the same reduction in automobile deaths.[10] Some argue that the greater safety of commercial air travel compared with automobile travel is in part a function of the relatively novel fear of flying compared with the more ordinary fear of death from a road accident. That is, because the risk is newer to us and thus less acceptable, we are willing to spend more per life to reduce the risk from flying than to reduce the risk from driving.

Another factor in people's acceptance of risk is that willingness to pay for reducing a risk varies with how essential and desirable the activity is. For example, many people accept much higher than average risks for athletic or recreational activities than they would for risks associated with transportation or employment (see Table 25.1). The risks associated with playing a sport or utilizing transportation are assumed to be inherent in the activity, whereas the risks to human health from pollution are much more widespread, are linked to a much larger number of deaths, and in many cases are unavoidable and unseen.

In an ethical sense it is impossible to put a value on a human life. However, it is possible to determine how much people are willing to pay for a certain reduction in risk or a certain probability of an increase in longevity. For example, a study by the Rand Corporation considered measures that would save the lives of heart-attack victims, including increasing ambulance services and initiating pretreatment screening programs. According to the study, which identified the likely cost per life saved and the willingness of people to pay, people were willing to pay approximately $32,000 per life saved, or $1600 per year of longevity.[11]

Although information is incomplete, it is possible to estimate the cost of extending lives in terms of the dollars per person per year for various actions (Figure 25.7). For example, on the basis of direct effects on human health, it costs more to increase longevity by a reduction in air pollution than it would to directly reduce deaths by the addition of a coronary ambulance system. Such a comparison is useful as a basis for decision making. Clearly, when a society chooses to reduce air pollution, many factors beyond the direct measurable health benefits are considered. Pollution directly affects more than just our health, and ecological and aesthetic damage can also indirectly affect human health (see Section 25.4). We might want to choose a slightly higher risk of

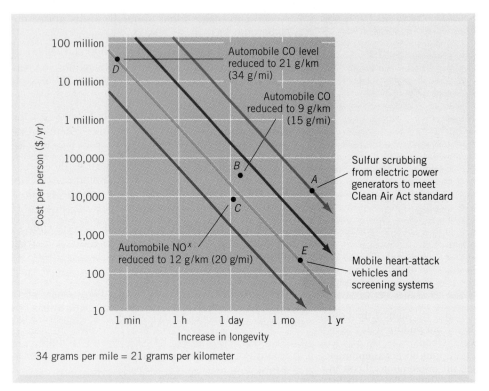

Figure 25.7 One way to rank the effectiveness of various efforts to reduce pollutants is to estimate the cost of extending a life in dollars per year. This graph shows that reducing sulfur emissions from power plants to the Clean Air Act level *(A)* would extend a human life 1 year at a cost of about $10,000. Similar restrictions applied to automobile emissions *(B, C)* would increase lifetimes by 1 day. More stringent automobile controls would be much more expensive. *(D)*, and mobile units and screening programs for heart problems would be much cheaper *(E)*. This graph represents only one step in an environmental analysis. (*Source:* After R. Wilson, 1980, "Risk/Benefit Analysis for Toxic Chemicals," *Ecotoxicology and Environmental Safety,* 4, 370–383.)

death in a more pleasant environment (spend money to clean up the air instead of on increased ambulance services) rather than increase the chances of living longer in a poor environment (spend the money on reducing heart attacks). Such comparisons may make you feel uncomfortable. But like it or not, we cannot avoid making choices of this kind. The issue boils down to whether we should improve the quality of life for the living or extend life expectancy regardless of the quality of life.[12]

Although pollution control may involve many dollars, the average cost per family in the United States is low, especially in comparison to other costs. It has been estimated that the per-family pollution control cost is between $30 and $60 per year for a family with a median income. In addition to the low cost per family, pollution control has many benefits whose quantitative value can be estimated. For example, federal air quality standards are estimated to reduce the risk of asthma by 3% and the risk to locally exposed adults of chronic bronchitis and emphysema by 10% to 15%. Air pollution contributes to inflation by reducing the number of productive workdays, reducing work efficiency, adding

to direct expenditures for health treatments, and necessitating restoration of nonhuman environmental damage. Estimates of the total cost of the direct and indirect effects on human health from stationary sources of air pollution are $250 per family per year. On this basis, air pollution control appears to be cost-effective; in fact, it has economic benefits.[13]

25.4 ENVIRONMENTAL INTANGIBLES

Environmental intangibles, such as the visual or other sensual pleasures of experiences, whether in the natural or urban environment, are extremely difficult to evaluate economically. This is the Achilles' heel of environmental economics. A nature photographer finding an elusive subject on a lonely mountaintop, a hunter in a blind on a crisp autumn morning, a beachcomber searching coastal tidal pools on a misty morning, picnickers relaxing in an urban park, a motorist out for a drive in the country, an office worker in a city daydreaming about a canoe trip in a wilderness area, all are enjoying different aspects of the environment. Their experiences and memories cannot be readily assigned an economic value; they may be priceless to each individual.

One of the perplexing problems associated with aesthetic evaluation is personal preference. One person may appreciate a high mountain meadow far removed from civilization, whereas a second person prefers visiting with others on a patio at a trail head lodge. A third person may prefer to visit a city park; another may prefer the austere beauty of a desert. If we are going to consider aesthetic factors in environmental analysis, we must develop a method of aesthetic evaluation that allows for individual differences.

The criteria necessary to judge aesthetic qualities include unity, vividness, and variety.[14] *Unity* refers to the quality or wholeness of the perceived landscape, not as an assemblage but as a single, harmonious unit. *Vividness* refers to that quality of landscape that reflects a visually striking scene and is nearly synonymous with intensity, novelty, or clarity. *Variety* refers to how different one landscape is from another. Variety includes ideas of diversity and uniqueness; however, it is not always the case that the greater the diversity, the greater the aesthetic value.

The aesthetic aspects of resources are intangible and thus difficult to evaluate quantitatively. In contrast, quantitative evaluation of tangible natural resources, such as air, water, forests, or minerals, prior to development or management of a particular area is standard procedure. Water resources for power or other uses may be evaluated by the amount of flow of the rivers and the quantity of water storage in rivers and lakes; forest resources may be evaluated by the number, type, and sizes of trees and their subsequent yield of lumber; and mineral resources may be evaluated by estimating the number of metric tons of economically valuable mineral material at a particular location. We can make a statement of the quality and quantity of each tangible resource compared to some known quality or quantity.

Ideally, we would like to make similar statements about the more intangible resources, such as scenery. That is, we would like to compare scenery to specific standards and compare one scene with another. Or we might want to compare the tangible economic value of a resource with the aesthetic value of the same resource. This is a very difficult task for which few standards are available.

Public Service Functions of Living Things

Another kind of environmental intangible, already discussed briefly, is the public service functions of the biota and of the environment. For example, it is estimated that bees pollinate $20 billion worth of crops in the United States (Figure 25.8). The cost of pollinating these crops by hand would be exorbi-

FIGURE 25.8 Public service functions of living things. Wild creatures and natural ecosystems perform public service functions for us—carrying out tasks important for our survival that would be extremely expensive for us to accomplish by ourselves. For example, bees pollinate millions of flowers important for food production, timber supply, and aesthetics.

tant, and a pollutant that eliminated bees would have large, indirect economic consequences. However, we rarely think of this benefit of bees. Recently an outbreak of parasites of bees in the United States has reduced the abundance of bees, bringing this once intangible factor to public attention as having more direct, tangible benefits than generally recognized. As another example, bacteria fix nitrogen in the oceans, lakes, rivers, and soils. The cost of replacing this function in terms of production and transport of artificially produced nitrogen fertilizers would be immense, but again we rarely think about this activity of the bacteria. Bacteria also clean water in the soil by decomposing toxic chemicals.

The atmosphere performs a public service by acting as a large disposal for toxic gases. For instance, carbon monoxide is eventually converted to nontoxic carbon dioxide, either by inorganic chemical reactions or by bacteria. Residents of cities such as Denver and Los Angeles, where the topography often slows down the mixing of the city's atmosphere with the rest of the atmosphere, become acutely aware of the public service functions of the environment. Only when our environment loses a public service function do we usually begin to recognize its economic benefits; then, what had been accepted as an economic externality suddenly becomes an economic internality.

Economics of Global Environmental Problems

Global environmental problems make us more aware of the public service functions of the environment of our planet and of life around us as well as raise new economic questions. An important case in point is the possibility of global warming. The problem is that our technological society is adding carbon dioxide and other greenhouse gases to the atmosphere, which have the potential to warm the climate (see the discussion of global warming in Chapter 21).[15,16] The direct solution is to decrease the release of these gases, but to do so there would have to be a worldwide decrease in the burning of fossil fuels. Although most of the production of greenhouse gases is from the industrial nations, in the future, developing nations, especially China and India, will contribute large quantities of these gases.

The economist Ralph D'Arge points out an economic problem arising from this global issue: The less developed countries did not share in the economic benefits of the burning of fossil fuels during the first two centuries of the industrial revolution, but they are sharing in the disadvantages of this activity.[17] Now the industrialized nations are suggesting that all nations, including the less developed ones, restrict their use of fossil fuels and therefore participate in future disadvantages without obtaining the benefits of cheap energy, although developed nations have had a considerable environmental cost with that cheap energy. Developing nations tend to think that industrial nations, which enjoyed the past benefits, should accept most of the future costs. At the same time, why shouldn't the developing nations proceed to develop and burn fossil fuels? This perspective, limited to what benefits individual nations, may be too restricted in light of the

global environmental effects of our technological civilization. It may be necessary to reduce the total production of greenhouse gases. If so, then the economic question is: Who pays, and how? One suggestion is that the developed nations pay for the reduction in greenhouse emissions of the less developed nations. At present, this is an unresolved issue in environmental economics. Another method of reducing total greenhouse gas emissions would be the sharing of technology that developed countries have already produced, thus helping developing countries to reduce both local and global pollution. These issues were a major concern at the recent Earth Summit in Rio de Janeiro.[18]

25.5 POLICY INSTRUMENTS

The Value of Public Goods

A *public good,* such as clean air, is something that cannot be sold by private sellers (Figure 25.9). Suppose you decided to go into the business of cleaning up the air. Because anyone could use the clean air, people would not voluntarily pay a fee for your cleanup. It is unlikely that you could make a successful business of this. As discussed earlier, to some extent natural ecosystems provide cleaning services that are public goods. Forests absorb particulates, salt marshes convert toxic compounds to nontoxic forms, and biological activity in streams removes sewage. Globally, we all profit from these public service functions of natural ecosystems, but we have no simple way to put a value on them or even to estimate the amount of pollution removal that takes place. How does a society achieve an environmental goal, such as preservation and use of a resource or reduction of a pollutant?

As mentioned earlier, any society has several methods to achieve these goals: persuasion, regulation (establishing laws, regulatory agencies, etc.), taxation and subsidies, and licenses (see Tables 25.2 and 25.3). Means to implement a society's policies are known among economists as **policy instruments.** These include moral suasion (which politicians call jawboning, i.e., persuading people by talk, publicity, and social pressure); direct controls, which include regulations; market processes, which affect the price of goods and processes and include taxation of various kinds; subsidies; licenses; and deposits. Society also has administrative mechanisms to ensure that the policy instruments chosen actually function.

Establishing Risk
The risk associated with a pollutant can be determined by the present levels of exposure and pre-

FIGURE 25.9 Housing development in Alexandria, Virginia, near Washington, D.C. The rural and wooded Virginia countryside is considered beautiful, and this beauty is lost when dense housing tracts are built. It is difficult to place an economic value on such scenic quality, as it is on other public goods. This raises several economic questions: How much is the quality of the environment worth? How much would it cost to prevent the loss of public goods? How much economic activity is lost by not developing an area? (See text for discussion.)

TABLE 25.2 Approaches to Environmental Policy

Policy Instruments

1. Moral suasion (publicity, social pressure, etc.)
2. Direct controls
   a. Regulations limiting the permissible levels of emissions
   b. Specification of mandatory processes or equipment
3. Market processes[a]
   a. Taxation of environmental damage
      i. Tax rates based on evaluation of social damage
      ii. Tax rates designed to achieve preset standards of environmental quality
   b. Subsidies
      i. Specified payments per unit of reduction of waste emissions
      ii. Subsidies to defray costs of damage control equipment
   c. Issue of limited quantities of pollution licenses
      i. Sale of licenses to the highest bidders
      ii. Equal distribution of licenses with legalized resale
   d. Refundable deposits against environmental damage
   e. Allocation of property rights to give individuals a proprietary interest in improved environmental quality
4. Government investment
   a. Damage prevention facilities (e.g., municipal treatment plants)
   b. Regenerative activities (e.g., reforestation, slum clearance)
   c. Dissemination of information (e.g., pollution control techniques, opportunities for profitable recycling)
   d. Research
   e. Education
      i. Of the general public
      ii. Of professional specialists (ecologists, urban planners, etc.)

Administrative Mechanisms

1. Administrative unit
   a. National agency
   b. Local agency
2. Financing
   a. Payment by those who cause the damage
   b. Payment by those who benefit from improvements
   c. General revenues
3. Enforcement mechanism
   a. Regulatory organization or police
   b. Citizen suits (with or without sharing of fines)

Source: From W. J. Baumol and W. E. Oates, 1979, *Economics, Environmental Policy and the Quality of Life,* Prentice-Hall, Englewood Cliffs, NJ.

[a]Subisdies and taxes can also be distinguished by using a property-rights framework. Per-unit subsidies implicitly confer ownership of the right to pollute on the polluter, and these rights are then purchased by the government via the subsidy. Taxes essentially say that there is public ownership of usage rights, which can be purchased from the public through its agent, the government, by private parties on payment of the tax (price).

dicted future trends. These trends depend on the production and origin of pollutant, pathways it follows through the environment, and changes it undergoes along these pathways. Dose–response curves establish risk to a population from a particular level of pollutant (see Chapter 14). Relative risks of different pollutants can be determined by comparing current levels and their dose–response curves.

In summary, it is possible to take scientific and technological steps to estimate risk and, once the level of risk is established, estimate cost of reducing risk and then compare the cost to the benefit. However, what constitutes an acceptable risk is more than a scientific or technical issue. Acceptability of a risk involves ethical and psychological attitudes of individuals and society. Risks that are voluntary appear more acceptable than those that are not voluntary. Risks that affect a small population (such as employees at nuclear power plants) are usually more acceptable than those that involve all members of a society (such as risk from radioactive fallout). Finally, familiar, long-established risks seem to be more acceptable than novel ones.

Control of Pollutants

In many cases, after initial expenditures for pollution controls, industry becomes more efficient in production, eliminates or greatly reduces the need for expensive disposal of pollutants, and minimizes external costs, leading to a reduction in total costs. Japan, whose economy is now the most energy efficient, has reduced its air pollution more than any other industrial nation. Japanese industry uses only 5 megajoules to produce $1 gross domestic product (GDP) while U.S. industry requires 12 megajoules. China is the least energy efficient country due primarily to the use of inefficient coal-burning boilers. However, since 1989 China's energy efficiency has improved significantly.[19] A large international chemical company, 3M, initiated a pollution prevention program and was able to stop the release of a billion pounds of toxic chemicals and save the company $500 million in the process.

In discussing the control of pollutants, the concept of the marginal cost is useful. **Marginal cost** is the cost to reduce one additional unit of pollutant. In some cases the marginal cost increases rapidly as the percentage of reduction increases. For example, the marginal cost of reducing the biological oxygen demand (see Chapter 20) in wastewater from petroleum refining increases exponentially. When 20% of the pollutants have been removed, the cost of removing an additional kilogram is 5 cents. When 80% of the pollutants have been removed, it costs 49 cents to remove an additional kilogram. Extrapolating from these results, it would cost an infinite amount to remove all the pollution.

TABLE 25.3 Performance of Various Policy Instruments

Policy Instrument	Reliability	Permanence	Adaptability to Growth	Resistance to Inflation	Incentive for Improved Effort	Economy	Feasibility without Metering	NonInterference in Private Decisions	Political Attraction Actual	Political Attraction Potential
Moral suasion	Good[a]	Poor	Good[b]	Good[b]	Fair	Poor[c]	Excellent	Excellent	Excellent	—
Direct controls										
By quota	Fair	Poor	Fair	Excellent	Poor	Poor	Poor	Poor	Excellent	—
By specification of technique	Fair	Poor	Good[b]	Good[b]	Poor	Poor	Excellent	Poor	Excellent	—
Fees	Excellent	Excellent	Fair	Fair	Excellent	Excellent	Poor	Excellent	Poor	Good
Sale of permits or licenses	Excellent	Excellent	Excellent	Excellent	Excellent	Excellent[d]	Poor	Excellent	Poor	Good
Subsidies										
Per-unit reduction	Fair[e]	Good	Fair	Fair	Excellent	Good	Poor	Excellent	Good	—
For equipment purchase	Fair	Good	Fair	Fair	Excellent	Good	Poor	Excellent	Good	—
Government investment	Good	?	?	?	—	?	Excellent	—	Good	—

Source: W. J. Baumol and W. E. Oates, 1979, *Economics, Environmental Policy and the Quality of Life,* Prentice-Hall, Englewood Cliffs, NJ.

[a]For short periods of time when urgency of appeal is made very clear.

[b]Baumol and Oates's judgment.

[c]Induces contributions from decision makers who are most cooperative, not necessarily from thos able to do the job most effectively (most expensively).

[d]Tends to allocate reduction quotas among firms in a cost-minimizing manner, but if the number of emissions permitted is too small, it will force the community to devote an excessive quantity of resources to environmental protection.

[e]Tends to allocate reduction quotas among firms in cost-minimizing manner but introduces inefficiency into the environmental protection process by attracting more polluting firms into the subsidized industry, so that aggregate response is questionable.

There are three common methods of direct control of pollution: (1) setting maximum levels of pollution emission; (2) requiring specific procedures and processes that reduce pollution; and (3) charging fees for pollution emission. In the first case, a political body could set a maximum level for the amount of sulfur emitted from the smokestack of an industry. In the second, it could restrict the kind of fuel the industry could use. Many areas have chosen the latter method by prohibiting the burning of high-sulfur coal. The problem with the first approach—controlling emissions—is that careful monitoring is required indefinitely to make certain the allowable levels are not exceeded. Such monitoring may be costly and may be difficult to carry out. The disadvantages of the second approach—requiring specific procedures—are that the required methodology may impose a severe financial burden on the producer of the pollutant, restrict the kinds of production methods open to an industry, and become technologically obsolete (see the discussion of air pollution and laws regulating it in Chapter 22). However, after an initial expenditure on procedures to reduce pollution, production efficiency may be increased and other costs such as monitoring and waste disposal can be reduced.

Although the United States has emphasized the use of direct regulation to control pollution, other countries have been successful in controlling pollution by charging effluent fees. For example, charges for effluents into the Ruhr River in Germany are assessed on the basis of both the concentration of pollutant and the total quantity of polluted water emitted into the river. In response, plants have introduced water recirculation and internal treatment to reduce emissions.[20]

Studies of the uses of different policy instruments for environmental matters have resulted in some ability to evaluate their relative success (see Table 25.3). For example, moral suasion is reliable but not very permanent in its effect. Sale of licenses or permits has been found to be among the more successful recourses.

In every environmental matter there is a desire on the one hand to maintain individual freedom of choice and on the other to achieve a specific social goal. In ocean fishing, for example, how does a society allow every individual to choose whether or not to fish and yet prevent everyone from fishing at the same time and bringing the fish species to extinction? This interplay between private good and public good is at the heart of environmental issues. Some argue that the market itself will provide the proper control. For example, it can be argued that people will stop fishing when there is no longer a profit to be made. We have already seen, however, that two factors interfere with this argument: (1) by the time the reduced fish population results in no economic gain for fishermen, it may be too late to avoid eventual extinction; and (2) even when it is not possible to make a sustained annual profit, there may be an advantage in harvesting the entire resource and getting out of the business (see A Closer Look 25.3, "Fishing Resources and Policy Instruments").

 CLOSER LOOK 25.3

FISHING RESOURCES AND POLICY INSTRUMENTS

The example of ocean fishing can be used to look at different policy instruments. The oceans outside of national territorial waters are commons, and thus the fish and mammals that live in them are common resources. What is a common resource may change over time, however. The move by many nations to define international waters as beginning 325 km (200 miles) from their coasts has turned some fisheries from a completely open common resource to a national resource open only to domestic fishermen.

In fisheries there are four main management options:[22]

1. Establish total catch quotas for the entire fishery and allow anybody to fish until the total is reached.

2. Issue a restricted number of licenses but allow each licensed fisherman to catch many fish.

3. Tax the catch (the fish brought in) or the effort (the cost of ships, fuel, and other essential items).

4. Allocate fishing rights; that is, assign each fisherman a transferable and salable quota.

With total-catch quotas, the fishery is closed when the quota is reached. Whales, Pacific halibut, tropical tuna, and anchovies have been regulated in this way. In Alaska, this practice resulted in all of the halibut being caught in a few days, with the result that restaurants no longer had halibut available for most of the year. This undesirable result has led to a change in policy; the total-catch approach has been replaced by the sale of licenses. Although regulating the total catch might be done in a way that helps the fish, it tends to increase the number of fishermen and the capacity of vessels, and the end result is a hardship on fishermen.

Recent economic analysis suggests that taxes that take into account the cost of externalities can work to the best advantage of fishermen and fish. Similar results are achieved by allocating a transferable and salable quota to each fisherman.

Determining which of these social methods achieves the best use of a desirable environmental resource is not simple. The answer varies with the specific attributes of both the re-

source and the users. The tools of economics can be used to determine the methods that will work best within a given social framework.

ENVIRONMENTAL ISSUE

How Can We Reconcile Environmental and Economic Interests?

Both overfishing and pollution have been blamed for the alarming decline in groundfish (cod, haddock, flounder, redfish, pollack, hakes) off the northeastern coast of the United States. Most scientists and fisheries managers have focused on overfishing, but attempts to regulate fishing have generated bitter disputes with fishermen, many of whom contend that restrictions on fishing make them scapegoats for pollution problems. The controversy has become a classic battle between short-term economic interests and long-term environmental concerns.

The Massachusetts Division of Marine Fisheries conducted a study in 1992 to settle the question of overfishing versus pollution. They concluded that key characteristics in declining flounder populations were consistent with overfishing, rather than pollution. Furthermore, if pollution were the cause, the skate and dogfish populations would be expected to decline as well. However, the authors of the study caution that the issue should not be seen as an either/or question and "overfishing may currently mask any pollution-induced population growth constraints." (Correia, p. 5)

In 1977, in response to concern about overfishing in U.S. waters by foreign factory ships, the U.S. government extended the nation's coastal waters from 12 to 200 miles (19–322 km). To encourage domestic fishermen, the National Marine Fisheries Service provided loan guarantees for replacing older vessels and equipment with newer boats with high-tech equipment for locating fish. During this same period, demand for fish increased as Americans became more concerned about cholesterol levels in red meat. Consequently, the number of fishing boats, the number of days at sea, and fishing efficiency increased sharply. As a result, 50–60% of the populations of some species were landed each year.

A decision over Canadian and American fishing rights in Georges Bank, the most prolific area in the North Atlantic, in favor of Canada in 1984 by the International Court of Justice in The Hague intensified competition for remaining fishing waters among U.S. fishermen. Overfishing knows no boundaries, however, and in 1992, Canada was forced to suspend all cod fishing to save the stock from complete annihilation.

In 1982, the New England Fisheries Management Council attempted to enforce harvest quotas but rescinded the order under pressure from the commercial fishing industry. In the wake of a bitter controversy that resulted from this decision, as well as further declines in fish populations, the council has since instituted a series of measures prohibiting fishing at certain times and in certain areas, mandating minimum net sizes, and enacting quotas on the catch.

In 1992, the council adopted a more ambitious plan intended to cut fishing effort in half by 1997, but by means other than quotas and removal of current fishermen. Recent stock assessments have indicated that such a time frame is inadequate to address critical stock declines. The council decided to issue a limited number of fishing permits, limit the number of days at sea, use high-tech monitoring equipment to ensure compliance, and establish trip limits on some fish. Even more recently, portions of the Georges Bank were closed indefinitely to fishing for some fish species including yellowtail, cod, and haddock. Landings of yellowtail in 1993 were 3800 metric tons, the lowest on record and a mere 6% of the historical maximum in 1969.[23]

The National Marine Fisheries Service advocates instead a system of Individual Transferable Quotas (ITQs) by which permits are issued to boat owners to allow them to harvest a fixed amount of fish each year. They can lease, sell, or bequeath the permits to others. Although ITQs have been successful in several U.S. fisheries and in some other countries, some small operators in New England fear that large corporations will buy up permits and dominate the industry.

Critical Thinking Questions

1. Using the example of the New England fishery, what are some arguments for and against the proposition that all natural resources treated like commons will inevitably be destroyed unless controls are instituted?

2. Which measures described attempt to convert the fishing industry from a commons system to private ownership? How might these measures help prevent overfishing? Is it right to institute private ownership of public resources?

3. What approach to future value (approximately) does each of the following people assume for fish?

   Fisherman: If you don't get it now, someone else will.

   Fisheries Manager: By sacrificing now, we can do something to protect fish stocks.

4. Develop a list of the environmental and economic advantages and disadvantages of Individual Transferable Quotas (ITQs). Would you support instituting ITQs in New England? Explain.

5. Do you think it possible to reconcile economic and environmental interests in the case of the New England fishing industry? If so, how? If not, why not?

References

Anthony, V. C. 1993. "The State of Groundfish Resources off the Northeastern United States," *Fisheries* 18(3): 12–17.

Correia, S. J. 1992 (3rd quarter). "Flounder Population Declines: Overfishing or Pollution?" *Division of Marine Fisheries News.* Boston: Massachusetts Division of Marine Fisheries.

"How to Fish," *The Economist* 309, no. 7580 (December 10, 1988): 93–96.

Keen, E. A. 1991. "Ownership and Productivity of Marine Fisheries Resources," *Fisheries* 16: 18–22.

Lawren, B. 1992. "Net Loss," *National Wildlife* 30(6): 47–52.

Leal, D. R. 1992 (July 30). "Using Property Rights to Regulate Fish Harvest," *Christian Science Monitor*, 84: 18.

Pierce, D. 1992 (2nd quarter). "New England Council to Cut Fishing Effort in Half over Next 5 Years," *Division of Marine Fisheries News.* Boston: Massachusetts Division of Marine Fisheries.

Satchell, M. 1992. "The Rape of the Oceans," *U.S. News and World Report* 112(24): 64–75.

SUMMARY

- An economic analysis can help us understand why environmental resources have been poorly conserved in the past and how we might achieve more effective means to achieve conservation in the future.

- Economic analysis is applied to two different kinds of environmental issues: the use of desirable resources (fish in the ocean, oil in the ground, forests on the land) and the minimization of undesirable pollution.

- Resources may be common property or privately controlled. The kind of ownership affects the methods available to achieve an environmental goal. There is a tendency to overexploit a common property resource and to harvest to extinction nonessential resources whose innate growth rate is low, as suggested in Hardin's "tragedy of the commons."

- Future worth compared to present worth can be an important determinant of the level of exploitation.

- The relation between risk and benefit affects our willingness to pay for an environmental good.

- Evaluation of environmental intangibles, such as landscape aesthetics and scenic resources, is becoming more common in environmental analysis. When quantitative, such evaluation balances the more traditional economic evaluation and helps separate facts from emotion in complex environmental problems.

- Societal methods to achieve an environmental goal include moral suasion, direct regulation, taxation and subsidies, licensing, and establishment of quotas. All five kinds of controls have been applied to the use of desirable resources.

REEXAMINING THEMES AND ISSUES

Human Population: The tragedy of the commons will increase as human population density increases because there will be more and more individuals to seek gain at the expense of community values. For example, there will be more and more individuals who will try to make a living from harvesting natural resources. How this can be done, while at the same time conserving those resources, requires an understanding of environmental economics.

Sustainability: From this chapter we learn why people sometimes are not interested, economically, in sustaining an environmental resource from which they are making a living. When the goal is simply to maximize profits, it is sometimes a rational decision to liquidate an environmental resource and put the money gained in a bank or another investment. To avoid such liquidation, we need to understand economic externalities and intangible values.

Global Perspective: Solutions to global environmental issues, such as global warming, require that we understand the different economic interests of developed and developing nations. These can lead to different economic policies and different valuation of global environmental issues.

Urban World: The tragedy of the commons began with grazing rights in small villages. As the world becomes increasingly urbanized, the pressures to use public lands for private economic gain is likely to increase. An understanding of environmental economics can help us find solutions to urban environmental problems.

Values and Knowledge: One of the central questions of environmental economics is how can we develop equivalent economic valuation for tangible and intangible factors. For example, how can we compare the value of timber that could be harvested with the beauty people attach to the scenery, trees intact? As another example, how can we compare the value of a dam that provides irrigation water and electrical power on the Columbia River with the scenery without the dam and the salmon that could inhabit that river?

KEY TERMS

commons *533*	marginal cost *543*
direct costs *536*	policy instruments *542*
externality *536*	risk–benefit analysis *537*
indirect costs *536*	

STUDY QUESTIONS

1. What is meant by the term "the tragedy of the commons"? Which of the following are the result of this tragedy: (a) the California condor, (b) the right whale, and (c) the high price of walnut wood used in furniture? (See Chapters 7 and 12 for additional information about the condor, the whale, and forest products.)

2. What is meant by risk–benefit analysis?

3. Cherry and walnut are valuable woods used to make fine furniture. Basing your decision on the information in the following table, would you invest in (a) a cherry plantation; (b) a walnut plantation; (c) a mixed stand of both species; or (d) an unmanaged woodland where you see some cherry and walnut growing? (Hint: Refer to the discussion in this chapter concerning whales.)

Species	Longevity	Maximum Size	Maximum Value
walnut	400 years	1 m	$15,000 per tree
cherry	100 years	1 m	$10,000 per tree

4. Flying over Los Angeles, you see smog below. Your neighbor in the next seat says, "That smog looks bad, but eliminating it would save only a few lives. Doing that isn't worth the cost. We should spend the money on other things, like new hospitals." Do you agree or disagree? Give your reasons.

5. Which of the following are intangible resources? Which are tangible?

   a. The view of Mount Wilson in California

   b. Owning property with a view of Mount Wilson

   c. Porpoises in the ocean

   d. Tuna fish in the ocean

   e. Clean air

   f. Owning property outside the smog area of Los Angeles

6. What kind of future value is implied by the statement "Extinction is forever"? Discuss how we might approach providing an economic

analysis for extinction (see Chapter 12 for additional information about extinction).

7. Which of the following can be thought of as "commons" in the sense discussed by Garrett Hardin in "The Tragedy of the Commons?" Explain.

a. Tuna fisheries in the open ocean
b. Catfish grown in artificial freshwater ponds
c. Grizzly bears in Yellowstone National Park
d. A view of Central Park in New York City
e. Air over Central Park in New York City

FURTHER READING

Block, W. (ed.) 1990. *Economics and the Environment: A Reconciliation.* Vancouver: Fraser Institute.

Bromley, D.W. 1995. *Handbook of Environmental Economics.* Cambridge, MA: Blackwell.

Clark, C. 1976. *Mathematical Bioeconomics: The Optimal Management of Renewable Resources.* New York: Wiley. This book employs analytic mathematical models to provide a theoretical basis for renewable resource management, with an emphasis on the commercial fishery and forestry industries.

Daly, H. E., and Townsend, K. N. (ed.). 1993. *Valuing the Earth. Economics, Ecology, Ethics.* Cambridge, MA: MIT Press. Essays that take a look at economic growth and come to the conclusion that sooner or later it is both physically and economically unsustainable as well as morally undesirable.

Hardin, G. 1968. "The Tragedy of the Commons," *Science* 162: 1243–1248. One of the most cited papers in both science and social science, this classic work outlines the differences between individual interest and the common good.

Hodge, I. 1995. *Environmental Economics: Individual Incentives and Public Choice.* New York: St. Martin's.

Jordan, C. F. 1995 *Conservation.* New York: Wiley.

Mercuro, N., Lopez, F. A., and Preston, K. P. 1994. *Ecology, Law and Economics. The Simple Analytics of Natural Resource and Environmental Economics.* Lanham, MD: University Press of America. Presents a conceptual model that describes the interrelations among the economy, the ecology of the natural world, and the state (government and law), including conventional remedies to environmental problems.

National Academy of Public Administration. 1994. *The Environment Goes to Market. The Implementation of Economic Incentives for Pollution Control.* Description and analysis of a number of implementation issues as evidenced in four case studies: air pollution trading; pollution charges; solid waste recycling; and a deposit-refund program. Washington, D.C.: National Academy Press.

Schnaiberg, A., and Gould, K. A. 1994. *Environment and Society. The Enduring Conflict.* New York: St. Martin's. An examination of several environmental myths related to economics and environmental problems.

Tietenberg, T. H. 1992. *Environmental and Natural Resource Economics.,* 3rd ed. New York: HarperCollins. This book provides a broad range of national and international examples of the application of economic principles as they relate to environmental issues and natural resource management.

Tisdell, C. A. 1991. *Economics of Environmental Conservation: Economics of Environmental and Ecological Management.* Amsterdam and New York: Elsevier Science. This book focuses on the ecological dimensions of environmental economics, concentrating on living or biological resources and their life support systems.

Tisdell, C. A. 1993. *Environmental Economics: Policies for Environmental Management and Sustainable Development.* Aldershot, Hants, England: Edward Elgar. Includes discussions of externalities, cost-benefit analysis, sustainable development and economic activity, ecological economics, and global resource conservation.

INTERNET RESOURCES

European Center for Nature Conservation, Sustainable Development and Human Impact on the Natural Environment: *http://www.ence.nl/doc/servers/sustaina.html*—Although not a U.S. organization, this center provides an abundance of links to internet resources on sustainable development and economics. Links to economics include the United Nations Environment Programme and its considerable efforts to promote sustainable resource use worldwide, as well as links to sites that describe efforts and offer views of publications covering environmental economics and of cost-benefit analyses.

An urban garden in New York City.

CHAPTER 26

URBAN ENVIRONMENTS

CASE STUDY

The Ecological Capital of Brazil

In 1950, the city of Curitiba in Brazil (see Figure 26.1) had 300,000 inhabitants, but by 1990 the population had grown to 2.3 million, making it the tenth largest city in Brazil. The growth of Curitiba came primarily as a result of migration by rural people displaced by mechanization of agriculture. The newcomers lived in squatter huts at the edge of the city in conditions of great poverty, poor sanitation, and frequent flooding caused by converting rivers and streams into artificial canals. By 1970, Curitiba was well on the way to becoming an example of environmental degradation and social decay. The story of how Curitiba turned itself from an urban disaster into a model of planning and sustainability by 1995 illustrates that cities can be designed in harmony with people and the environment.[1,2]

Much of the credit for the transformation of Curitiba goes to its three-time mayor, Jaime Lerner, who believed a workable transportation system was the key to making Curitiba an integrated city where people can live as well as work. Rather than a more expensive underground rail system, Lerner spearheaded development of a bus system with five major axes, each containing lanes dedicated to express buses (see Figure 26.2), with others carrying local traffic and high-speed automobile traffic. Forty-nine blocks of the historic center of Curitiba were reserved for pedestrians. Tubular bus stations were built in which passengers paid fares before boarding, an arrangement that avoids long delays caused by collecting fares after boarding. Circular routes and smaller feeder routes between the

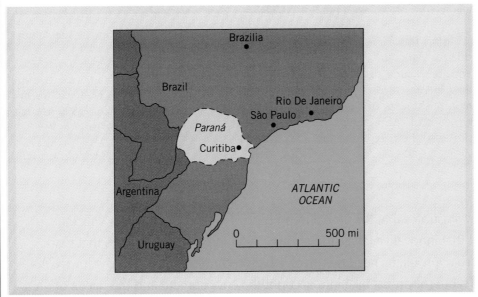

Figure 26.1 Located in southern Brazil, Curitiba provides an example of how a city can transform itself from an urban disaster into a model of planning and sustainability. (Courtesy of Karl Gude.)

major axes maintain vital connections between the central city and outlying areas. As a result, more than 1.3 million passengers ride busses each day. Although Curitiba has the second highest per capita number of cars in Brazil, it uses 30% less gas than eight comparable Brazilian cities and its air pollution is among the lowest in the country.[2,4]

To solve its serious garbage problem, Curitiba required each household to sort recyclables from garbage. As a result, two-thirds of the garbage,

more than 100 tons a day, is recycled, with 70% of the population participating. Where streets are too narrow to allow access by garbage trucks, residents are encouraged to bring garbage bags to the trucks, for which they are reimbursed with bus tokens, surplus food, or school notebooks. Through a low-cost housing program, 40,000 new homes were built, many placed so that residents have easy access to job sites. The city also embarked on a program to increase the amount of green space.

Figure 26.2 Express bus routes form the five spoke shaped axes of Curitiba's bus system; interdistrict and smaller feeder routes form connections between the main routes. (Courtesy of Karl Gude.)

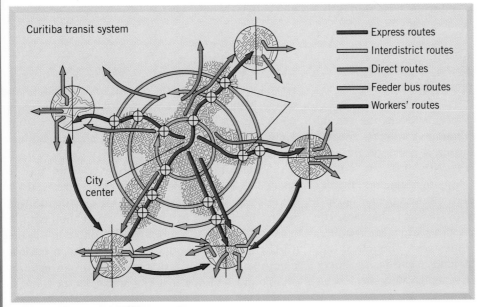

Artificial drainage channels were replaced with natural drainage, reducing the need for expensive flood control. Areas, including those around the river basins, were set aside for parks. In 1970, Curitiba had only 0.5 m² of green area per capita, while by 1990 the area had increased to 50 m² for each inhabitant. The accomplishments of Curitiba have led some to call it the "ecological capital of Brazil" and to hope that it is also the "city of the future."[2,3]

NOTES

1. Dobbs, F. 1995. "Curitiba: City of the Future?" (video). World Bank.
2. Rabinovitch, J. 1997. "Integrated Transportation and Land Use Planning Channel Curitiba's Growth," World Resources 1996–97, World Resources Institute.
3. Rabinovitch, J., and J. Leitman. 1996 (March). "Urban Planning in Curitiba," *Scientific American*, pp.46–53.
4. Hunt, J. 1994 (April). "Curitiba," *Metropolis Magazine*.

This case history illustrates that the global trend toward urbanization poses difficult challenges, but that innovative ideas and strong leadership can create cities that work. The solutions to urban problems need not involve high technology in order to improve the quality of urban life, but they must take into account the environment and the needs of the people.

LEARNING OBJECTIVES

Because the world is becoming increasingly urbanized, it is important to learn how to improve urban environments, to make cities more pleasant and healthier places in which to live and to reduce undesirable effects on the environment. After reading this chapter, you should understand:

- How to view a city from an ecosystem perspective.
- What features are important to a city's environmental site and geographic situation and the types of geographic situations that are desirable.
- How cities have changed with changes in technology and with ideas about city planning.

- How cities change their own environment and affect the environment of the surrounding areas and how we can plan cities to minimize some of these effects.
- Why trees and other vegetation in cities are important as pleasing elements and as habitats for animals and how we can alter the urban environment to encourage wildlife and to discourage pests.
- How cities can be designed to promote biological conservation and become pleasant environments for people.

26.1 CITY LIFE

In the past, the emphasis of environmental action has most often been on wilderness, wildlife, endangered species, and the impact of pollution on natural landscapes outside cities. Now it is time to turn more of our attention to city environments. In the development of the modern environmental movement in the 1960s and 1970s, it was fashionable to consider everything about cities bad and everything about wilderness good. Cities were thought of as polluted, lacking in wildlife and native plants, dirty, and artificial—and therefore bad. Wilderness was

considered as unpolluted, clean, full of wildlife and native plants, and natural—and therefore good. Although it was fashionable to disdain cities, the majority of people live in urban environments and have suffered directly from their decline.

Comparatively little public concern has focused on urban ecology; as a result, many urban people see environmental issues as outside their realm. However, city dwellers are at the center of some of the most important environmental issues. Today there is a rebirth of interest in urban environments and in the development of urban ecology. People are realizing that city and wilderness are in-

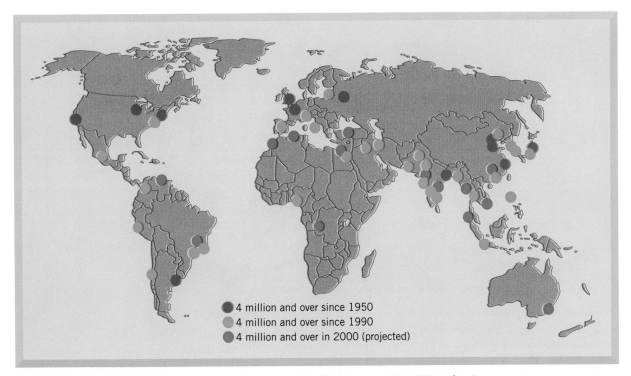

FIGURE 26.3 Location of cities with populations of more than 4 million in 1950, 1990, and estimated for the year 2000.

extricably connected. We cannot fiddle in the wilderness while our Romes burn from sulfur dioxide and nitrogen oxide pollution.

Worldwide, we are becoming an increasingly urbanized species. In the United States, about 75% of the population lives in urban areas and 25% lives in rural areas.[5] Today approximately 50% of the world's population, 3.3 billion people, live in cities and it is projected that 62% of the population, 6.5 billion people, will live in cities by the year 2025.[6] Economic development leads to urbanization; 73% of people in developed countries live in cities, but only 36% of the people in the poorest developing countries are city dwellers.

Not only is human population increasingly urbanized, but there is rapid growth of huge metropolitan areas with more than 10 million residents. In 1950 there were only two such areas in the world: New York City and its nearby New Jersey areas (12.2 million residents) and greater London (12.4 million). By 1975, Mexico City, Los Angeles, Tokyo, Shanghai, and São Paulo, Brazil, had joined this list. It is estimated that by the year 2000, 20 more cities and their surrounding areas will have grown to this size. Almost 400 million people will be concentrated in the largest 25 urban areas (Figure 26.3). It is clear that in the future most people will live in cities; in fact, in most nations, most *urban residents will live in the country's single largest city*. In the future, for most

people, living in an environment of good quality will mean living in a city that is managed carefully to maintain that environmental quality.

26.2 THE CITY AS A SYSTEM

One of the ways in which we can improve the management of a city environment is to analyze the city as an ecological system. Like any life-supporting system, *a city must maintain a flow of energy, provide necessary material resources, and have ways of removing wastes*. These ecosystem functions are maintained in a city by transportation and communication with outlying areas. A city is not a self-contained ecosystem; it depends on other cities and rural areas. A city takes in raw materials from the surrounding countryside: food, water, wood, energy, mineral ores, everything that a human society uses. In turn, the city produces and exports material goods and, if it is a truly great city, exports ideas, innovations, inventions, arts, and the spirit of civilization. A city cannot exist without a countryside to support it. As was said half a century ago, city and country, urban and rural, are one thing, one connected system of energy and material flows, not two things (Figure 26.4).[7]

As a consequence, if the environment of a city declines, almost certainly the environment of its surroundings will also decline. The reverse is also true: If

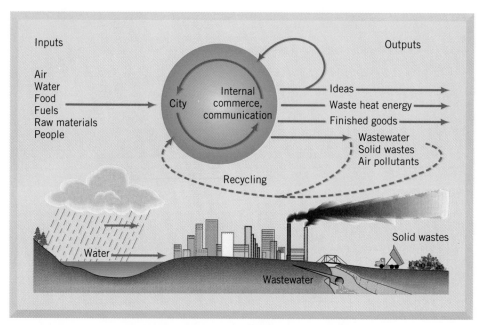

FIGURE 26.4 The city as a system: energy and materials flow. A city must function as part of a city–countryside ecosystem, with an input of energy and materials, internal cycling, and an output of waste heat energy and material wastes. As with any natural ecosystem, recycling of materials can reduce the need for input and the net output of wastes.

the environment around a city declines, the city itself will be threatened. Some people suggest, for example, that the ancient Native American settlement in Chaco Canyon, Arizona, declined after the environment surrounding that settlement was adversely affected.[8]

Cities also export waste products to the countryside, including polluted water, air, and solids. It has been estimated that the average city resident in an industrial nation annually uses (directly or indirectly) 208,000 kg (229 tons) of water, 660 kg (0.8 tons) of food, and 3146 kg (3.5 tons) of fossil fuels and produces 1,660,000 kg (1826 tons) of sewage, 660 kg (0.8 tons) of solid wastes, and 200 kg (440 lb) of air pollutants. If these are exported without care, they pollute the countryside, reducing the countryside's ability to provide necessary resources for the city and making life in the surroundings less healthy and less pleasant.

With such dependencies and interactions between city and surroundings, it is no wonder that relationships between people in cities and in the countryside have often been strained. Why, country dwellers want to know, should they have to deal with the wastes of those in the city? The answer is that many of our serious environmental problems occur at the interface between urban and rural areas. People who live outside but near a city have a vested interest in maintaining a good environment for that city and a good system for managing the city's resources.

A city can never be free of environmental constraints, even though its human constructions give us a false sense of security. Lewis Mumford, the historian of cities, wrote, "Cities give us the *illusion* of self-sufficiency and independence and of the possibility of physical continuity without conscious renewal."[9] But this security is only an illusion.

26.3 SITE AND LOCATION

As the case study of Venice illustrates, the location of a city is influenced by two factors: **site**, which is the summation of all the environmental features of that location; and **situation**, which is the placement of the city with respect to other areas. A good site includes a good geologic substrate, such as a firm rock base and well-drained soils; dry land for buildings; good nearby supplies of water; good local agricultural land; abundant timber and other natural resources; and a benign local climate.

The environmental situation strongly affects the development and importance of a city, particularly with regard to transportation and defense. Waterways are important for transportation. Especially in early times, before railroads, automobiles, and airplanes, cities depended on water for transportation. Most early cities were located on or near waterways. In the ancient Roman Empire, *all* important cities were located near waterways. Waterways have continued to influence the locations of cities; most major cities of the eastern United States are situated either at major ocean harbors or at the fall line on major rivers. The siting of cities at the fall line in the past was no accident. The discussion in A Closer Look 26.1, "Cities

A CLOSER LOOK 26.1

CITIES AND THE FALL LINE

The **fall line** on a river occurs where there is an abrupt drop in elevation of the land and there are numerous waterfalls (Figure 26.5). The fall line typically is located where streams pass from harder, more erosion-resistant rocks to softer rocks. Cities have frequently been established at fall lines for a number of reasons. Fall lines provide good sites for water power, which was an important source of energy in the eighteenth and nineteenth centuries when the major eastern cities of the United States were established or rose to importance. At this time it was the farthest inland that larger ships could navigate, and just above the fall line was the farthest downstream that the river could be easily bridged. Not until the development of steel bridges in the late nineteenth century did it become practical to span the wider regions of the rivers below the fall line.[33] The proximity of a city to a river has another advantage, because river valleys have rich, water-deposited soils that are good for agriculture. In early times rivers also provided an important means of waste disposal, which today has become a serious problem.

FIGURE 26.5 Most major cities of the eastern and southern United States lie either at the sites of harbors or along the fall line, marking locations of waterfalls or rapids on most major rivers in the area (shown on the colored line). The location of cities is thus strongly influenced by the characteristics of the environment. (*Source:* C. B. Hunt, 1974, *Natural Regions of the United States and Canada,* Freeman. Copyright 1974 by W. H. Freeman & Co.)

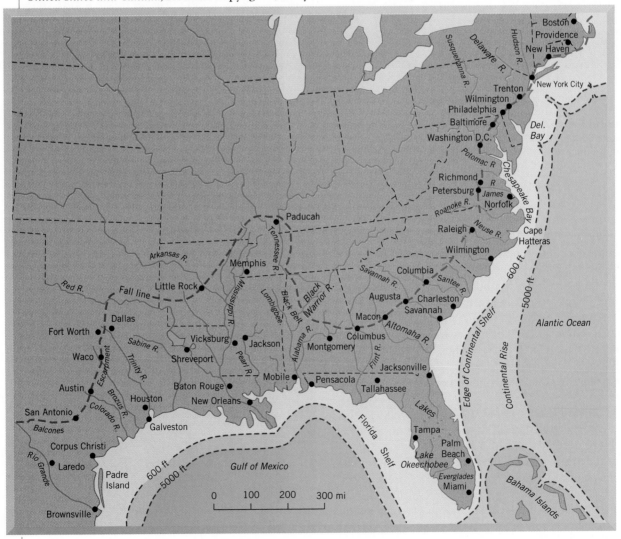

(a)

(b)

FIGURE 26.6 (*a*) Landsat satellite image of St. Louis, Missouri, showing the confluence of the Mississippi and Missouri Rivers and the city. (*b*) The famous castle of Edinburgh, Scotland, built on top of bedrock, overlooking the city, and in a readily defended location. (*c*) Bruges, Belgium, where canals were developed providing transportation to the sea.

and the Fall Line" illustrates how much environment affects the location and success of cities.

Cities are often founded at other crucial transportation points, growing up around a market, a river crossing, or a fort. Newcastle, England, and Budapest, Hungary, are located at the lowest bridging points on their rivers; other cities, such as Geneva, are located where a river enters or leaves a major lake. Some well-known cities are located at the confluence of major rivers: Saint Louis lies at the confluence of the Missouri and Mississippi rivers (Figure 26.6*a*); Manaus, Brazil, Pittsburgh, Pennsylvania, Koblenz, Germany, and Khartoum, Sudan, are located at the confluence of several rivers. Many famous cities are located at crucial defensive locations, such as on or adjacent to easily defended rock outcrops. Examples include Edinburgh (Figure 26.6*b*), Athens, and Salzburg, Austria. Other cities are situated on peninsulas, for example, Monaco and Istanbul.

(c)

Cities are often founded close to a mineral resource, such as salt (Salzburg, Austria); metals (Kalgoorlie, Australia); and medicated waters and thermal springs (Spa, Belgium; Bath, Great Britain; Vichy, France; and Saratoga Springs, New York).

A successful city can grow and spread over surrounding terrain so that its original purpose may be obscured to a resident; its original market or fort may have evolved into a minor square or a historical curiosity. In most cases, though, cities originated where the situation led to a natural meeting point for people.

An ideal location for a city has both a good site and a good situation, but such a place is difficult to find. Paris is perhaps one of the best examples of a perfect location for a city, one with both a good site and a good situation.

Site Modification

Site begins as something provided by the environment, but technology and environmental change can alter the characteristics of a site for better or worse. People can improve the site of a city and have done so when the situation of a city made it important and when its citizens could afford large projects. An excellent situation can sometimes compensate for a poor site, as we learned in the case study of Venice that opened this chapter. However, improvements are almost always required to the site so that the city can persist. New Orleans has a good situation but a poor site. An important transportation center at the mouth of the Mississippi River, it lies on the low mud flats of the delta, which flood frequently and provide a poor substrate for construction. Backwaters and swamps offer little as a local resource for agriculture but provide breeding habitats for mosquitoes. Fishing in the Gulf of Mexico is a plus, however. Modern construction methods, including levees to prevent flooding, have improved the site.

Decline in Site–Decline in a City

The qualities of a site can change over time, with adverse effects on a city. For example, Bruges, Belgium, developed as an important center for commerce in the thirteenth century because of its harbor on the English Channel, which permitted trade with England and other European nations. By the fifteenth century, however, the harbor had seriously silted, and the limited technology of the time did not make dredging possible on the scale necessary to save the harbor. This problem, combined with political events, led to a decline in the importance of Bruges, a decline from which it has never recovered. Today Bruges still lives, a beautiful city with many fine examples of medieval architecture. Ironically,

the fact that these buildings were never replaced with modern ones is testimony to the decline of commerce there (Figure 26.6c). Ghent, Belgium, and Ravenna, Italy, are examples of other cities whose harbors silted; Ghent responded by building a canal to artificially improve its situation.[10]

As human effects on the environment extend to global change, there may be rapid, serious changes in the sites of many cities. For example, if global warming occurs and sea levels rise, many coastal cities will be subject to flooding.

26.4 CITY PLANNING AND THE ENVIRONMENT

City planning has a long history. At various times city planners have taken environmental factors carefully into consideration (see A Closer Look 26.2, "An Environmental History of Cities").

A danger in city planning is that a city contains the seeds of its own destruction. There is a tendency to transform a city center from natural to artificial features and to replace grass and soil with pavement, gravel, houses, and temples, creating an impression that civilization has dominated the environment. Ironically, the very artificial aspects of the city that make it seem so independent of the rest of the world actually make it more dependent on its rural surroundings for all resources. Although such a city appears to its inhabitants to grow stronger and more independent, it actually becomes more fragile.[11]

In this way, a city grows at the expense of the surrounding countryside, destroying the surrounding landscape on which it ultimately depends. As the nearby areas are ruined for agriculture and as the transportation network extends, the use, misuse, and destruction of the environment increase. Many of the classic cities of history were built, prospered, and declined in this way.

The City as a Beautiful Environment

Ideas of the **fortress city** and the **park city** influenced the planning of cities in North America (see A Closer Look 26.3, "A Brief History of City Planning"). The importance of aesthetic considerations is illustrated in the plan of Washington, DC, designed by Pierre-Charles L'Enfant, who mixed a traditional rectangular grid pattern of streets (which can be traced back to the Romans) with broad avenues set at angles. The intention was to design a city of beauty, with many parks, including small ones at the intersections of avenues and streets. This design has made Washington one of the most pleasant cities in the United States.

CLOSER LOOK 26.2

AN ENVIRONMENTAL HISTORY OF CITIES

The Rise of Towns

The first cities emerged on the landscape thousands of years ago during the New Stone Age with the development of agriculture, which provided the excess of food resources that is necessary for the maintenance of a city.[34] In this first stage, the density of people per square kilometer was much higher than in the surrounding countryside, but the density was still too low to cause rapid, serious disturbance to the land. In fact, the waste provided by the city dwellers and their animals was an important fertilizer for the surrounding farmlands. In this stage, the city's size was restricted by the primitive transportation methods that brought food and necessary resources into the city and removed waste. Because of such limitations, no European medieval town served only by land transportation had a population greater than 15,000.[35]

The Urban Center

In the second stage, more efficient transportation made possible the development of much larger urban centers, with a totally urban social core. Boats, barges, canals, and wharves, as well as roads, horses, carriages, and carts, made it possible for cities to be located farther from agricultural areas. Rome, originally dependent on local produce, became a city fed by granaries in Africa and the Near East. The population of a city is limited by how far a person can travel in one day to and from work and by the density (number of people per unit area) with which people can be packed into an area. In the second stage, the internal size of a city was limited by pedestrian travel. The city could be no larger in area than the distance a worker could walk to work, do a day's work, and walk home. The density of people per square kilometer was limited by architectural techniques and primitive waste disposal. These

cities never exceeded a population of 1 million, and there were few cities near this size, most notably Rome and some cities in China.

The Industrial Metropolis

The modern industrial revolution allowed greater modification of the environment than had been possible before. Two technological advances that had significant effects on the city environment are improved sanitation methods, which have led to the control of many diseases, and modern transportation methods (Figure 26.7). Improved transportation makes a larger city possible: Workers can live farther from their place of work and commerce and communication can extend over larger areas. Air travel has freed cities even more from the traditional limitation of situation. We now have thriving urban areas where previously transportation was poor: in the far north (Fairbanks, Alaska) or on islands (Honolulu). These changes increase the urban dwellers' sense of separateness from their natural environment.

Rapid transportation, such as subways and commuter trains, have also allowed the development of suburbs.

In some cities, the negative effects of urban sprawl are leading many people back to the urban centers or to the development of smaller, satellite cities surrounding the central city. The drawbacks of suburban commuting and the destruction of the landscape in suburbs have brought new appeal to the city center.

The Center of Civilization

We are at the beginning of a new stage in the development of cities (see A Closer Look 26.2, "An Environmental History of Cities," and 26.3 "A Brief History of City Planning") . With modern telecommunications, people can work at home or

FIGURE 26.7 The New York City aqueduct, built in the 1860s, brought clean, clear water from the Croton Reservoir, miles away, into the heart of the city. It was one of the most important factors in creating a livable environment for that city.

long distances apart. Perhaps, as telecommunication frees us from the necessities of certain kinds of commercial travel and related activities, the city can become a cleaner, more pleasing center of civilization.

An optimistic future for cities requires a continued abundance of energy and material resources, which are certainly not guaranteed, and wise use of these resources. If energy resources are rapidly depleted, modern mass transit may fail. Fewer people will be able to live in suburbs and the cities will become more crowded. Reliance on coal and wood will increase air pollution. The continued destruction of the land within and near cities could compound transportation problems, making local production of food impossible. The future of our cities depends on our ability to plan and to conserve and use our resources wisely.

 CLOSER LOOK 26.3

A BRIEF HISTORY OF CITY PLANNING

Although many cities in history have grown without any conscious plan, formal plans for new cities in modern times can be traced back as far as the fifteenth century. Sometimes cities have been designed for specific social purposes, with little consideration of the environment; in other cases, the environment and its effect on city residents have been major planning considerations.

Two dominant themes in formal city planning have been planning for defense and planning for beauty. Roman cities were typically designed along simple geometric patterns that had both practical and aesthetic benefits. During the height of Islamic culture, in the first millennium A.D., Islamic cities typically contained beautiful gardens, often within the grounds of royalty. One of the most famous urban gardens in the world is the Alhambra of the Islamic Alhambra Palace in Granada, Spain. The garden was created when this city was a Moorish capital and was maintained after the fall of Islamic control of Granada in 1492. Today, as a tourist attraction that receives 2 million visitors a year, the Alhambra garden demonstrates the economic benefits of aesthetic considerations in city planning.

After the fall of the Roman Empire, the earliest planned towns and cities in Europe were walled fortress cities designed for defense. But even in these, city planners considered the aesthetics of the town. In the fifteenth century, one such planner, Leon Battista Alberti, argued that large and important towns should have broad and straight streets; smaller, less fortified towns should have winding streets to increase their beauty. He also advocated the inclusion of town squares and recreational areas, which continue to be important considerations in city planning.[36] Today, walled cities have become major tourist attractions, again illustrating the economic benefits of good aesthetic planning in urban development.

The usefulness of walled cities essentially ended with the invention of gunpowder, and the Renaissance sparked an interest in the ideal city, which in turn led to the development of the park city. A preference for gardens and parks, emphasizing recreation, developed in Western civilization in the seventeenth and eighteenth centuries, culminating in the plan of Versailles, France, with its famous parks of many sizes and tree-lined walks.

The Alhambra gardens of Granada, Spain, illustrate how vegetation can be used to create beauty within a city.

The City Park

Over time, parks have become more and more important in cities (see A Closer Look 26.3, "A Brief History of City Planning"). An important advance for cities in the United States was the nineteenth-century planning and construction of Central Park in New York City by Frederick Law Olmsted. Olmsted was one of the most important modern experts on city planning. He took site and situation into account and attempted to blend improvements to a site with the aesthetic qualities of the city.

Central Park was the first large public park in the United States. In his plan, Olmsted carefully considered the opportunities and limitations of the topography, geology, hydrology, and vegetation. He placed recreational areas in the southern part of the park, where there were flat meadows. He created depressed roadways that allowed traffic to cross the park without detracting from the vistas seen by park visitors. Central Park is an example of design with nature, a term coined much later. The design of Central Park influenced other U.S. city parks, and Olmsted remained a major figure in American city planning throughout the nineteenth century. The firm he founded continued to be important in city planning into the twentieth century.[12,13]

Olmsted's skill in creating designs that combined the physical and aesthetic needs of a city is illustrated by his work in Boston. The original site of Boston had certain advantages: a narrow peninsula with several hills that could be easily defended; a good harbor; and a good water supply. As Boston grew, however, there was increasing demand for more land for buildings, a larger area for docking ships, and a better water supply as well as a need to control ocean floods and to dispose of solid and liquid wastes. Much of the original tidal flats, which were too wet to build on and too shallow to navigate, had been filled in (Figure 26.8). Hills were leveled and the marshes filled with soil. The largest project was the filling of Back Bay, which began in 1858 and continued for decades. Once filled, however, the area suffered from flooding and water pollution.[14] The solution to these problems was Olmsted's water control project called the fens.

Olmsted's goal was to "abate existing nuisances" by keeping sewage out of the streams and ponds and providing artificial banks for the streams to prevent flooding, and to do this in a natural-looking way. His solution: Create artificial watercourses by digging shallow depressions in the tidal flats, following a meandering pattern like a natural stream; set aside other artificial depressions as holding ponds for tidal flooding; restore a natural salt marsh planted with vegetation tolerant of the brackish water; and plant the entire area to serve as a recreation park

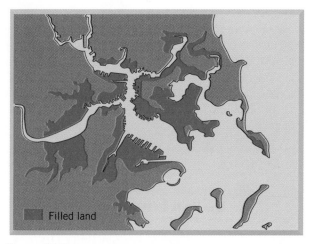

Figure 26.8 Boston is an example of a city whose site has been modified over time to improve the environment and provide more building locations. This map of Boston shows land filled in to provide new building sites as of 1982. Although such landfill allows for expansion of the city, it can also create environmental problems, which then must be solved. (*Source:* From A. W. Spirn, 1984, *The Granite Garden: Urban Nature and Human Design*, Basic Books, New York.)

when not in flood. The result of his vision was that control of water became an aesthetic addition to the city. The blending of several goals made the development of the fens a landmark in city planning. Although it appears to the casual stroller to be simply a park for recreation, the area serves an important environmental function in flood control.

An important extension of the park town idea was the **garden city,** a term coined in 1902 by Ebenezer Howard. Howard's idea was that city and countryside should be planned together. A garden city was one that was surrounded by a **greenbelt.** The idea was to locate garden cities in a set connected by these greenbelts, forming a system of countryside and urban landscapes. The garden city idea caught on, and garden cities were planned and developed in Great Britain and the United States. Greenbelt, Maryland, just outside Washington, D.C., is one of these cities, as is Lecheworth, England.

Olmsted's use of the natural landscape in designing city parks and Howard's garden city concept continue to influence city planning today. They are two important approaches to making modern cities livable.

26.5 THE CITY AS AN ENVIRONMENT AND CITY DESIGN

Because a city changes the landscape, it also changes the relationship between biological and physical aspects of the environment. Many of these changes were discussed in earlier chapters as aspects of pollution, water management, or climate. They are mentioned again as appropriate in the following sections, generally with a focus on how effective city planning can reduce the problems.

Pollution in the City

Everything is concentrated in a city, including pollutants. City dwellers are exposed to more kinds of toxic chemicals in higher concentrations and to more human-produced noise, heat, and particulates than are their rural neighbors. This environment makes life riskier. Lives are shortened by an average of one to two years in the most polluted cities in the United States. The city with the greatest number of early deaths is Los Angeles, with an estimated 5,973 early deaths per year followed by New York with 4,024, Chicago with 3,479, Philadelphia with 2,590, and Detroit with 2,123.[15] The primary sources of particulate air pollution, which consist of smoke and soot and tiny particles formed from emissions of sulfer dioxide and voltile organic compounds, are older, coal-burning power plants, industrial boilers, and gas and diesel powered vehicles.[16]

The Urban Atmosphere and Climate

Cities affect the local climate; as the city changes, so does its climate (see Chapter 21). Cities are generally less windy than nonurban areas, because buildings and other structures obstruct the flow of air. But city buildings also channel the wind, sometimes creating local wind tunnels with high wind speeds. The actual flow of wind around one building is influenced by nearby buildings. The total wind flow through a city is the result of the relationships among all the buildings. In planning a new building, its shape as well as its location among other buildings must be taken into account. In some cases when this has not been done, dangerous winds around tall buildings have resulted in blown-out windows.

Recall that a city can receive less sunlight than the countryside because particulates in the atmosphere over cities are often 10 or more times greater than in surrounding areas.[17] In spite of the reduced sunlight, cities are warmer than surrounding areas (a city is a **heat island**) because of increased heat production (the burning of fossil fuels and other industrial and residential activities) and also because there is a decreased rate of heat loss, partly owing to the abundance of building and paving materials that act as solar collectors (Figure 26.9).[18]

Solar Energy in Cities

Until modern times, it was common to make use of solar energy to heat city houses. Our century is one major exception, because cheap and easily accessible fossil fuels have led us to forget certain fundamental lessons. Cities in ancient Greece, Rome, and China were designed so that houses and patios faced south and solar energy was accessible to each household.[19] Today, we are beginning to appreciate the impor-

FIGURE 26.9 Sketch of a typical urban heat island profile: This graph shows temperature changes correlated to the density of development and trees. (*Source:* Andrasko and Huang, 1990. In: Akbari, H. et al. 1992, *Cooling Our Communities: A Guidebook on Tree Planting and Light-Colored Surfacing.*)

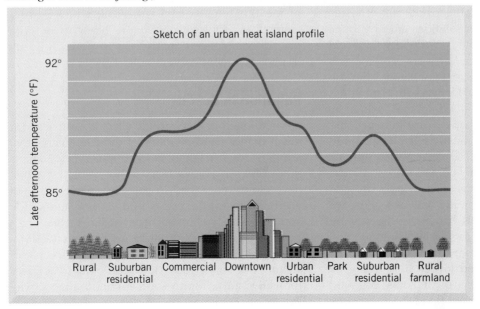

tance of solar energy once again; some cities have enacted solar energy ordinances that make it illegal to shade another property owner's building in such a way that it loses solar heating capability (see Chapter 17 for a discussion of solar energy).

Water in the Urban Environment

The construction of modern cities affects the water cycle greatly (see Chapter 19), in turn affecting soils and, consequently, plants and animals in the city. Paved city streets and city buildings prevent water infiltration. As a result, most rain runs off directly and is channeled into storm sewer systems. Hard city surfaces prevent water in the soil from evaporating to the atmosphere. In natural ecosystems, evaporation is an important way of cooling the surface. City pavement increases the chances of local flooding within the city, and the increased runoff from the city to the countryside can increase the chances of flooding downstream. Midlatitude cities generally record a lower relative humidity (2% lower in winter to 8% lower in summer) than the surrounding countryside.

Cities can have higher local rainfall than their surroundings, because dust above a city provides particles for condensation of raindrops. Some urban areas have 5% to 10% more precipitation and considerably more cloud cover and fog than do surrounding areas. Fog is particularly troublesome in the winter and may impede ground and air traffic.

Most cities have a single underground sewage system. During times of no rain or light rain, this system handles sewage alone. But during periods of heavy rain, the runoff is mixed with the sewage and can exceed the capacity of sewage treatment plants. During heavy rains, sewage is emitted downstream without sufficient treatment. It is too expensive to build a completely new and separate runoff system in an existing city, so other solutions must be found. One such example is found in the city of Woodlands, Texas. It was designed by Ian McHarg, who originated the phrase, "design with nature," the subject of A Closer Look 26.4, "Design with Nature."[20]

The problem with flooding and overtaxing storm-sewage systems is made worse in many cities built on floodplains. As mentioned earlier, floodplains are often chosen as sites for cities because the land is flat and easy to build on and river transportation is available nearby. This practice leads to a conflict, because cities are typically built as if there will never be flooding. Floods damage buildings and other property, cause the loss of lives, and are considered a natural catastrophe. More often than not, the river is channeled and levees built along the shores. But as the citizens of St Louis, Missouri, learned in the heavy rains and floods of the summer of 1993, these levees have their limits. They concentrate the waters in a narrower channel, which increases the speed of water flow and the destructive force of the waters when they do rise above the levee. The levee is an example of the idea expressed earlier that the artificial structures of cities give us a sense of independence from the environment, while in fact these structures make us more dependent on the environment. An alternative to artificial structures to control flooding is to allow for more areas that have multiple uses and can flood without severe damage during times of heavy rains, along the lines of the Olmsted design for Back Bay in Boston, as discussed earlier.

In urban environments long stretches of major rivers are channelized. Recall from Chapter 19 that channelization has two negative effects. First, when the river is maintained in an artificial channel, its sediment load is not deposited on the land, and the land's fertility is not renewed (Figure 26.11). Second, this sediment load passes downriver and is lost at the mouth, where it causes siltation and may fill in important harbors and do damage to cities at the ocean side. It is important to note that these negative effects are an indirect result of building cities on floodplains without proper planning and then trying to prevent the inevitable floods.

Soils in the City

A modern city has a great impact on soils. Most soil is covered by cement, asphalt, or stone; the soil no longer has its natural cover of vegetation and the natural exchange of gases between the soil and air is greatly reduced. Such soils lose organic matter, because they are no longer replenished by vegetation growth. Soil organisms die from lack of food and oxygen. The process of construction and the weight of the buildings compact the soil, which restricts water flow. City soils are more likely to be compacted, waterlogged, impervious to water flow, and lacking in organic matter.

Pollution in the City

As mentioned earlier, city dwellers are subjected to much higher concentrations of most pollutants than are their rural neighbors (see Chapter 14). Some of this pollution comes from motor vehicles, which have contributed lead in gasoline (where it is still used), nitrogen oxides, ozone, carbon monoxide, and other air pollutants from exhaust. Stationary power sources also produce air pollutants. Home heating is a third source, contributing particulates, sulfur oxides, nitrogen oxides, and other toxic gases. Industries are a fourth source, contributing a wide variety of chemicals.

A CLOSER LOOK 26.4

DESIGN WITH NATURE

When a new city is developed, our knowledge of the problems of flooding and water runoff can be used to plan a better flow of water. The new town of Woodlands, a suburb of Houston, Texas, is illustrative of such planning (Figure 26.10). Woodlands was designed so that most houses and roads were placed on ridges, and the lowlands were left as natural open space. The lowlands provide areas for temporary storage of floodwater and, because the land is unpaved, allow the rain to penetrate the soil and recharge the aquifer for Houston. Preserving the natural lowlands has other environmental benefits. In this region of Texas, low-lying wetlands are the habitat for native wildlife, such as deer. Large, pleasant trees, such as magnolias, grow there, providing food and a habitat for birds. The innovative city plan has economic as well as aesthetic and conservational benefits. It is estimated that a conventional drainage system would have cost $14 million more than the amount spent to maintain and develop the wetlands.[37]

FIGURE 26.10 Woodlands, Texas, a planned community. Lowlands are preserved as open space, while houses are kept to the uplands.

Although it is impossible to eliminate exposure to pollutants in a city, it is possible to reduce the exposure by careful design, planning, and development. For example, when lead was used in gasoline, exposure to lead was greater near a road than away from it. Exposure to lead can be reduced by placing

FIGURE 26.11 Once a meandering river that deposited its sediment over a broad flood plain, the Mississippi is channelized over much of its length, as shown here. Sediments no longer can fertilize the surrounding land; instead, deposited at the river mouth, they create new environmental problems.

houses and recreational areas away from roadways and by developing a buffer zone that makes use of trees resistant to the pollutant. Trees absorb pollutants and slow the rate of spread. In addition, such tree buffer zones can reduce the amount of noise in residential areas of a city.

26.6 BRINGING NATURE TO THE CITY

A practical problem for planners and managers of cities is how to bring nature to the city, that is, how to make plants and animals a part of a city landscape. This activity has evolved into several specialized professions, including **urban forestry** (whose professionals are often called tree wardens), landscape architecture, city planning, and city engineering. Most cities have an urban forester on the payroll who determines the best sites for planting trees and the best species of trees to suit the environment. These professionals take into account climate, soils, and the general influences of the urban setting, such as the shading imposed by tall buildings and the pollution from motor vehicles.

Vegetation in Cities

Planting of trees, shrubs, and flowers improves the beauty of a city.[21] Plants provide for different needs in different locations.[22]Trees provide shade that reduces the need for air conditioning and makes travel much more pleasant in hot weather.

In parks, vegetation provides places for quiet contemplation, trees and shrubs can block some of the city sounds, and the complex shapes and structures create a sense of solitude. Plants also provide habitats for wildlife such as birds and squirrels, which many urban residents consider pleasant additions to a city.

The use of trees in cities has expanded since the time of the European Renaissance. In earlier times, trees and shrubs were set apart in gardens, where they were viewed as scenery but not experienced as part of ordinary activities. Street trees were first used in Europe in the eighteenth century; among the first tree-lined streets were the rue de Rivoli in Paris and Bloomsbury Square in London (Figure 26.12). In many cities trees are now considered an essential element of the urban visual scene, and major cities have large tree-planting programs. For example, in New York City 11,000 trees are planted each year; in Vancouver, Canada, 4000 are planted per year.[23] Today there is growing use of trees to soften the effects of climate near houses. In colder climates, rows of conifers planted to the north of a house can protect the house from winter winds. Deciduous trees to the south can provide shade in the summer, reducing requirements for air conditioning, and yet allow sunlight to warm the house in the winter (Figure 26.13).[24] The Global Releaf project of American Forests seeks to replant trees in the urban environment; this project has resulted in tree planting in many cities, from Moscow to Los Angeles.

FIGURE 26.12 Paris was one of the first modern cities to use trees along streets to provide beauty, shade, and other aesthetic qualities, as shown in this picture of the famous Champs-Elysées.

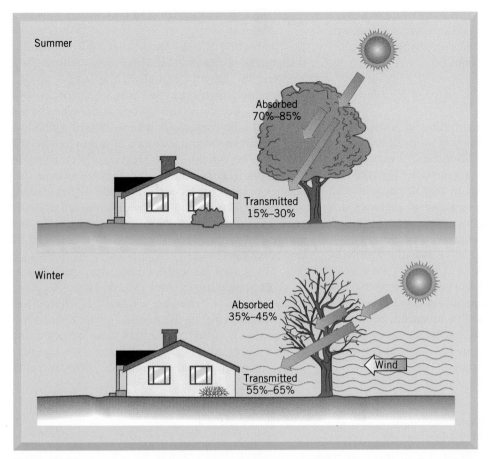

FIGURE 26.13 Trees can improve the microclimate near a house, protecting the house from winter winds and providing shade in the summer. (*Source:* J. Huang and S. Winnett, 1992, *Cooling Our Communities: A Guidebook on Tree Planting and Light Colored Surfacing,* U.S. EPA Office of Policy Analysis, U.S. Superintendent of Documents, Washington, DC.)

Urban Stress on Vegetation

Vegetation in cities is under special kinds of stress. Trees along city streets are often surrounded by cement, which prevents normal access to water and air. The root systems are more likely to experience extremes of dryness and soil saturation (immediately following or during a rainstorm). Because city soils tend to be compacted and do not drain well, trees planted in a city sidewalk are often overwatered and the roots may die from lack of oxygen. A solution, suggested by landscape architect Anne Spirn, is to plant trees in strips so the soil is connected and so a larger volume of soil is available to each of them and water can drain between them.[25] Other solutions involve careful use of soil, artificial structures, and special containers.[26]

Many species of trees and plants are very sensitive to air pollution. For example, the Eastern white pine of North America is extremely sensitive to ozone pollution and does not do well in cities with heavy motor vehicle traffic or along highways. Dust can interfere with the exchange of oxygen and carbon dioxide necessary for photosynthesis and respiration of the trees. City trees also suffer direct damage from physical impact of bicycles, cars, and trucks and from vandalism. Trees subject to such stresses are more susceptible to attacks by fungus diseases and insects. The lifetime of trees in a city is generally shorter than in their natural woodland habitats unless they are given considerable care.

Some species of trees are more useful and successful in cities than are others. An ideal urban tree would be resistant to all forms of urban stress; have a beautiful form and foliage; and produce no messy fruit, flowers, or leaf litter that requires cleaning. In most cities, only a few species of trees are used for street planting. However, the reliance on one or a few species results in an ecologically fragile urban planting, as we learned when Dutch elm disease spread throughout the eastern United States, destroying urban elms. It is prudent to use a greater diversity of trees to avoid outbreaks of insect pests and tree diseases.[27]

Wild plants that do particularly well in cities are those characteristic of disturbed areas and of early stages in ecological succession (see Chapter 9).

City roadsides in Europe and North America have wild mustards, asters, and other early successional plants. Disturbances in cities promote the occurrence of certain kinds of plants.

Wildlife in Cities

With the exception of some birds and small, docile mammals such as squirrels, most forms of wildlife in cities are considered pests. But there is much more wildlife in cities, much of it unnoticed. In addition, there is a growing recognition that urban areas can be modified to provide additional habitats for wildlife that people can enjoy; this can be an important method of biological conservation.[28,29]

We can divide city wildlife into the following categories: (1) those species that cannot persist in an urban environment and disappear; (2) those that tolerate an urban environment but do better elsewhere; (3) those that have adapted to urban environments, are abundant there, and are either neutral or beneficial to human beings; (Figure 26.14) and (4) those that are so successful that they become pests.

The City as Wildlife and Endangered Species Habitat

Peregrine falcons once hunted pigeons above the streets of Manhattan. Unknown to most New Yorkers, the falcons nested on the ledges of skyscrapers and dived on their prey in an impressive display of predation. The falcons disappeared when DDT and other organic pollutants caused a thinning of their eggshells and a failure in reproduction, but they have been reintroduced recently into the city. In New York City's Central Park approximately 260 species of birds have been observed—100 in a single day. Foxes live in London, feeding on garbage and road kill (animals run over by motor vehicles; see Figure 26.14b); shy and nocturnal, they are seen by few Londoners.[30]

We do not associate wildlife with cities, but as these examples show, cities provide homes to many forms of wildlife. Cities are a habitat, albeit artificial. They can provide all the needs—physical structures and necessary resources such as food, minerals, and water—for many plants and animals. We can also identify ecological food chains in cities (Figure 26.14).

FIGURE 26.14 (*a*) An urban food chain based on plants of disturbed places and insect herbivores. (*b*) An urban food chain based on road kill.

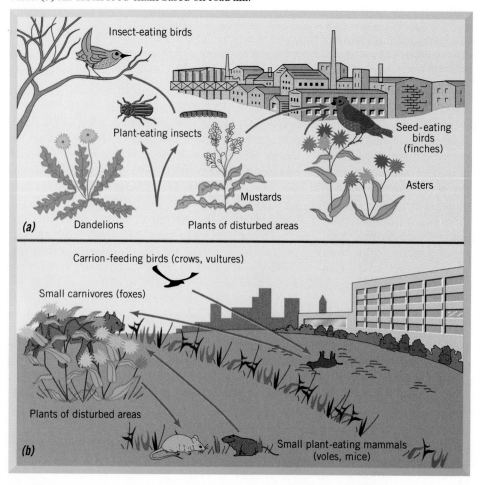

For some species, citys' artificial structures are sufficiently like their original habitat to be home. For example, chimney swifts once lived in hollow trees but are now common in chimneys and other vertical shafts. Their nests are glued to walls with saliva. A city can easily have more chimneys per square kilometer than a forest has hollow trees. Cities include natural habitats in parks and preserves. Modern parks provide some of the world's best wildlife habitats, and importance of parks will increase as truly wild areas shrink (Figure 26.15). Cities that are harbors often have many species of marine wildlife at their doorsteps. New York City's waters include sharks, bluefish, mackerel, tuna, striped bass, and nearly 250 other species of fish.[31]

Urban drainage structures can be designed as wildlife habitats. A typical urban runoff design depends on concrete-lined ditches that speed flow of water from city streets to lakes, rivers, or the ocean. However, these can be planned in a manner similar to Boston's Back Bay design discussed earlier, so stream and marsh habitats are maintained or created, with meandering waterways and storage areas that do not interfere with city processes. These can become habitats for fish and mammals (Figure 26.16).

Modified to promote wildlife, cities provide urban corridors that allow wildlife to migrate, following natural routes of movement in spite of the imposition of the city.[32] Urban corridors also serve to prevent some of the effects of ecological islands and are increasingly important to biological conservation.

Cities can also provide habitat for endangered plants. For example, in Lakeland, Florida, endangered plants are used in local landscaping with considerable success.

Animal Pests and Their Control

Pests are familiar to urban dwellers. City pests include cockroaches, fleas, termites, rats, and pigeons, but there are many more, especially species of insects. In gardens and parks, pests include insects, birds, and mammals that feed on fruit and vegetables and destroy foliage of shade trees and plants. Pests compete with people for food and spread diseases. Before modern sanitation and medicine, such diseases played a major role in limiting human population density in cities. Plague is spread by fleas found on rodents. Mice and rats in cities promoted the spread of the Black Death (see Chapter 1). Although cities are still less healthy than rural areas in terms of pollutants, control of disease has been an important improvement. An animal is a pest when it is in the wrong place at the wrong time doing the wrong thing. For example, a termite in a woodland helps in the natural regeneration of wood by hastening decay and speeding return of chemical

Figure 26.15 Cities can be planned so wildlife and their habitats are part of the landscape, as in Jamaica Bay, New York, shown here. A park in New York City, within view of the skyscrapers of Manhattan, Jamaica Bay had become polluted by sewage and was turning into a wasteland. Restoration, which began in the 1960s and extends over 7000 ha, has enabled the Bay to function once again as habitat for many species of birds, mammals, and fish. During spring and fall, many people come to see migrating birds, such as avocets, dowitchers, and sandpipers. Jamaica Bay is unusual among city parks in that it was planned as a naturalistic park, emphasizing native vegetation and habitats.

elements to the soil so they are available to living plants. But termites in a house are pests because they threaten the physical structure of the house.

Animals that survive best in cities have certain common characteristics. Animals that are urban pests usually are generalists in their food choice, so they can share the diet of people, have a high reproductive rate, and have a short average lifetime.

Controlling Pests

We can best control pests by recognizing how they fit their natural ecosystem and identifying their natural controlling factors. It is often assumed the only way to control animal pests is poisons, but there are limitations to this approach. Early poisons used in pest control were generally toxic to people. Another problem with reliance on one toxic compound is that, through evolution, a species can develop resistance, which can lead to rebound. If a pesticide is used once and spread widely, it will greatly reduce the population of the pest. However, when the pesticide's effectiveness is lost, the population can increase rapidly as long as habitat is suitable and food plentiful. This situation occurred when an attempt was made to control the Norway rat in Baltimore: There was plenty of waste food for rats to eat

FIGURE 26.16 How water drainage systems in a city can be modified to provide wildlife habitat. In the community on the right, concrete-lined ditches result in rapid runoff and have little value to fish and wildlife. In the community on the left, the natural stream and marsh were preserved; water is retained between rains and an excellent habitat is provided. (*Source:* D. L. Leedly and L. W. Adams, 1984, *A Guide to Urban Wildlife Management,* National Institute for Urban Wildlife, Columbia, MD, pp. 20–21.)

and an abundance of areas in which to breed.

The best way to control rats is to make the city a bad habitat for them by reducing the amount of open garbage, and eliminating areas to hide and nest. Common access areas used by rats are the space between walls and openings between buildings where pipes and cables enter. Houses can be constructed to restrict access by rats. In older buildings, areas of access can be sealed (see Chapter 11).

SUMMARY

- As an urban society, we must recognize the city's relation to the environment. A city influences and is influenced by its environment and is an environment itself. Like any life-supporting system, a city must provide for a flow of energy and cycling of chemical elements necessary for life. Because it is not self-sufficient, a city must have a source of energy and material resources and a sink for waste disposal. These require a transportation network.

- Because cities depend on outside resources, they developed only when human ingenuity resulted in modern agriculture and thus excess food production. The history of cities divides into four stages: (1) rise of towns; (2) era of classic urban centers; (3) industrial metropo-

lises; and (4) age of mass telecommunication, computers, and new forms of travel.

- Locations of cities are strongly influenced by environment. It is clear that cities are not located at random but in places of particular importance and environmental advantage. A city's site and situation are both important.

- A city creates an environment that is different from surrounding areas. Cities change local climate; they are commonly cloudier, warmer, rainier, and less humid. In general, life in a city is riskier because of higher concentrations of pollutants and pollutant-related diseases.

- Cities favor certain animals and plants. Natural habitats in city parks and preserves will become more important as wilderness decreases.

How Can Urban Sprawl Be Controlled?

As the world becomes increasingly urbanized, individual cities are growing (see p. 553) in area as well as population. Residential areas and shopping centers move into undeveloped land near cities, impinging on natural areas and creating a chaotic, unplanned human environment. More than 7.6 million ha of rural land were developed in the United States between 1970 and 1990, a process that is continuing at the rate of 160,000 ha per year. Urban sprawl has become a serious concern in communities all across the United States. In 1998, voters said "yes" to over 200 state and local ballot initiatives designed to control urban sprawl.

The city of Boulder, Colorado, has been in the forefront of this effort since 1959, when it created the "blue line"—at an elevation of 1761 m (the city itself is at 1606 m)—above which it would not extend city water or sewer services. Boulder's citizens felt, however, that the blue line was insufficient to control development and maintain its scenic beauty in the face of rapid population growth. (In the decade before 1959, Boulder had grown from a population of 29,000 to 66,000 and would reach 96,000 by 1998). To prevent uncontrolled development in the area between the city and the blue line, Boulder began in 1967 using a portion of the city sales tax to purchase land, creating a 10,800 ha greenbelt around the city proper.

In 1976, Boulder went one step further and set a limit of a 2% increase a year on new residences. Two years later, recognizing that planned development requires a regional approach, the city and surrounding Boulder County adopted a coordinated development plan. By the early 1990s, it became apparent that further growth control would need to come in the area of nonresidential building. The plan finally adopted by the city reduced the allowable density of many commercial/industrial properties; in effect, limiting jobs rather than building space.

The benefits of Boulder's controlled growth initiatives have been a defined urban/rural edge; rational, planned development; protection of sensitive environmental areas and scenic vistas; and large areas of open space within and around the city for recreation. And, in spite of its growth control measures, its economy has remained strong. But with restraints on residential growth, many people who found jobs in Boulder were forced to find affordable housing in adjoining communities. As a result, for example, the population of Superior, Colorado, grew from 225 in 1990 to 6,500 in 1998.

As commuting workers—40,000 a day—tried to get to and from their jobs in Boulder, traffic congestion and air pollution increased. In addition, developers had not built stores in the outlying areas, so shoppers flocked into Boulder's downtown mall. But when plans for a competing mall in the suburbs were announced, Boulder officials worried about the loss of revenue to the city if the new mall drew shoppers away from the city. At the same time, sprawl from Denver (only 48 km from Boulder), as well as its infamous "brown cloud" of polluted air, began to spill out along the highway connecting the two communities.

Critical Thinking Questions

1. Is a city an open or closed system (Chapter 3)? Use examples from the case of Boulder to support your answer.

2. As Boulder takes steps to limit growth, it becomes an even more desirable place to live, which subjects it to even greater growth pressures. What ways can you suggest to avoid such a positive feedback loop?

3. Some people in Boulder think that the next step is for the city to increase residential density within the city. How do you think people living there will accept this plan? What advantages and disadvantages are there to increasing density?

4. To some, Boulder is the story of a heroic battle against commercial interests that would destroy environmental resources and a unique quality of life. To others, it is the story of an elite group building an island of prosperity and the good life for themselves, while shifting the more unpleasant aspects of modern life elsewhere. How do you view Boulder's story, and why?

References

Egan, T. 1996 (December 30). Drawing a Hard Line Against Urban Sprawl. *The New York Times.*

Egan, T. 1996 (December 29). Urban Sprawl Strains Western States. *The New York Times.*

Pollock, P. 1998 (January). Controlling Sprawl in Boulder: Benefits and Pitfalls. *Land Lines,* Lincoln Institute of Land Policy, vol. 10, no. 1.

Pryne, E. 1998 (August 23). How Curbing Development Shaped a City. *The Seattle Times.*

- Trees are an important part of urban environments. Cities, however, create stresses on trees, and attention must be paid to condition of urban soils and supply of water for trees.

REEXAMINING THEMES AND ISSUES

Human Population: As the world human population increases, we are becoming an increasingly urbanized species. Present trends indicate that in the future most citizens of most nations will live in their country's single largest city. Thus a concern with urban environments will become increasingly more important.

Sustainability: Cities contain the seeds of their own destruction: The very artificiality of a city gives its inhabitants the sense that they are independent of their environment. But the opposite is the case: The more artificial a city, the more it depends on its surrounding environments for resources, and the more susceptible it becomes to major disasters, unless this susceptibility is recognized and planned for. The keys to sustainable cities are an ecosystem approach to urban planning and a return to a concern with the aesthetics of urban environments. Cities affect the sustainability of many endangered species, especially as cities spread and affect wetlands, rivers, and coastal areas. Good city planning can benefit the sustainability of endangered species.

Global Perspective: The great urban centers of the world are having global effects. As an example, because people are concentrated in cities and because many cities are located at the mouths of rivers, most major river estuaries of the world are severely polluted.

Urban World: The primary message of this chapter is that Earth is becoming an urban world, and environmental science must deal more and more with urban issues.

Values and Knowledge: A tendency in the twentieth century has been to focus environmental conservation on wilderness, large parks, and preserves outside of cities. Meanwhile, city environments have been allowed to decay. As the world becomes increasingly urbanized, a change in values is necessary. The conservation of biological diversity requires that an increased value be placed on urban environments. Indirectly, more pleasant urban environments can benefit natural areas outside of cities: The more pleasant city environments are, the more recreation people can find in them, and the less pressure there will be on the countryside.

KEY TERMS

city planning 557
fall line 555
fortress city 557
garden city 560

greenbelt 560
heat island 561
park city 557

site 554
situation 554
urban forestry 564

STUDY QUESTIONS

1. How does the environment influence the location of cities?

2. What types of cities are most likely to become ghost towns in the next 30 years? In answering this question, make use of your knowledge of changes in resources, transportation, and communication.

3. Some futurists picture a world that is one giant biospheric city. Is this possible? Under what conditions?

4. Among ancient Greeks it was said that a city should have no more people than the number that can hear the sound of a single voice. Would you apply this rule today? If not, how would you decide how to plan the size of a city?

5. Standing on top of the Sears Tower in Chicago, you overhear someone say, "Planning never works. The most interesting cities just grow. Planned cities are always dull and sterile." The speaker points to large, low-income housing developments far in the distance. How would you respond?

6. You are the manager of Central Park in New York City and receive the following requests. Which would you approve? Explain your reasons.

   a. A gift of $1 million to plant trees from all the eastern states.

   b. A gift of $1 million to set aside half the park to be forever untouched, thus producing an urban wilderness.

c. A gift of the construction of an asphalt jogging track and a gym for physical fitness. The donor says that lack of physical fitness is a major urban health problem.

d. A request to install an ice skating rink with artificially made ice. Facilities include an elegant restaurant with many views of the park.

7. Your state asks you to locate and plan a new town. The purpose of the town is to house people who will work at a wind farm, a large area of many windmills, all linked to produce electricity. You must first locate the site for the wind farm, and then plan the town. How would you proceed? What factors would you take into account?

8. Visit your town center. What changes, if any, would make better use of the environmental location? How could the area be made more livable?

9. In what ways does air travel alter the location of cities? The value of land within a city?

10. You are put in charge of ridding your city's parks of slugs, which eat up the vegetable gardens rented to residents. How would you approach controlling this pest?

FURTHER READING

Akbari, H., Davis, S., Dorsano, S., Huang, J., and Winnett, S. 1992. *Cooling Our Cities: A Guidebook on Tree Planting and Light-Colored Surfacing.* Washington DC: U.S. EPA Office of Policy Analysis, U. S. Superintendent of Documents. One of the most complete reviews of research on the uses of trees in cities.

Bornkamm, R., Lee, J. A., and Seaward, M. R. D., (eds.). 1982. *Urban Ecology.* Oxford: Blackwell Scientific Publications. A collection of papers presented at the Second European Ecological Symposium held in Berlin in 1980, covers a broad range of subjects, from the constitution of urban ecosystems to the impact of human activity in urban areas to the application of ecological knowledge in urban environments.

Burton, J. A. 1977. *Worlds Apart: Nature in the City.* Garden City, NY: Doubleday. An intriguing descriptions of wildlife including foxes, that inhabit temperate zone cities, little known to the people in the cities.

Butti, K., and Perlin, J. 1980. *The Golden Thread: 2500 Years of Solar Architecture and Technology.* New York: Cheshire. The authors present a history and outline of the use of solar energy from ancient Greece to modern times.

Cronon, W. 1991. *Nature's Metropolis: Chicago and the Great West.* New York: W. W. Norton. Discusses the interdependency of a city and its countryside, as exemplified by Chicago.

Hengeveld, H., and De Vocht, C., (eds.). 1982. *Role of Water in Urban Ecology. Developments in Landscape Management and Urban Planning,* vol. 5. Amsterdam, Oxford, New York: Elsevier Scientific. The proceedings of the Second International Environmental Symposium, held in Amsterdam in 1979, this publication addresses the role of water in urban ecology.

Leedly, D. L., and Adams, L. W. 1984. *A Guide to Urban Wildlife Management.* Columbia, MD: National Institute for Urban Wildlife. The proceedings of the National Symposium on Urban Wildlife held in Cedar Rapids, Iowa, in 1990, this book presents papers by many authors, covering wildlife conservation, ecology, and management.

Moll, G., and Ebernreck, S. 1992. *Shading Our Cities: A Resource Guide for Urban and Community Forests.* Washington, DC: Island Press. An overview of urban forests, this book includes discussions of the history and value of urban trees, followed by examples of programs designed to encourage urban tree planting and types of trees for various requirements.

Spirn, A. W. 1984. *The Granite Garden: Urban Nature and Human Design.* New York: Basic Books. Comprehensive strategies for designing cities in concert with natural processes are presented in this book, which focuses on the look and shape of the city rather than on economics or public policy.

INTERNET RESOURCES

American Cancer Society: *http://www.cancer.org*—National nonprofit organization dedicated to education, research in the fight against cancer. Web site has research publications, new reports, educational information about cancer and the environment.

Natural Resources Defense Council: *http://www.nrdc.org*—NRDC Web site provides information about wildlife and the environment and access to publications, briefings, frequently asked questions and answers, and a search engine.

National Association of Physicians for the Environment (NAPE): *http://napenet.org*—NAPEnet provides news about the National Association of Physicians for the Environment (NAPE) and its activities and makes available scientific information about health and the environment.

U.S. Department of Agriculture Forest Service, Urban National Forest home page: *http://www.fs.fed.us/outernet/urban—nf/welcome.htm.*—Information about national forests near major urban centers and the problems they face.

U.S. Department of Housing and Urban Development: *http://www.huduser.org/habitat.html*—Facts on the state of U.S. cities and urban sustainable development.

United Nations Development Programme Habitat II page: *http://www.undp.org/un/habitat/*—Results of the 1996 City Summit and information on urban heat islands, sustainable cities, and links to internet resources on urban sustainability.

United Nations Division for Sustainable Development: *http://www.un.org/dpcsd/dsd/*—Details of UN sustainable development programs and UN sponsored meetings covering sustainability and environmental concerns.

United Nations Population Information Network: *http://www.undp.org/popin/popin.htm*—Up-to-date information on urban population trends and the environment.

School children in Chicago learning about recycling of aluminum cans.

CHAPTER 27

WASTE MANAGEMENT

◢ CASE STUDY

Fresh Kills Landfill, New York City

Located on a mixture of wetlands (salt marsh), woodlands and grasslands on the western shore of Staten Island, the Fresh Kills landfill (Figure 27.1), which opened in 1948, is the only landfill operating in New York City. Owned and operated by the New York City Department of Sanitation, the landfill is comprised of approximately 7500 ha (3000 acres). At its peak in 1986, Fresh Kills received more than 21,000 tons of waste per day. During the 1990s the flow of waste began to slow down, however, as the city eliminated commercial deliveries, and New Yorkers began to recycle returnable bottles, plastic containers, and newspapers. Currently Fresh Kills receives between 12,000 and 14,000 tons of solid waste at a cost of approximately $44 per ton.[1]

Although this landfill is one of the largest in the world, space for waste generated by the city of New York is running out. Today, less than 2000 ha of Fresh Kills lands are actually used for land filling. The city is negotiating deals to begin trucking garbage out of New York at a cost of $50 to $70 per ton as the landfill enters its final closure stages. Closing the landfill will cost the city of New York more than $1 billion, including the 30 years of monitoring after final closure of the landfill in December 2001. Transforming one of the world's largest landfills into an environmentally sound and aesthetically pleasing park or other public place is a huge task. The elevation of part of the landfill when closed will be about 80m—A sizable hill for coastal New York! A slurry wall (underground concrete barrier) containment system to prevent the migration of untreated leachate (noxious, polluted liquid produced when water infiltrates through organic and other waste material) outside of the landfill has already been constructed. There is a leachate collection trench averaging 5m in depth placed one meter inside the containment wall, and 32 wells collect leachate for treatment. A leachate treatment center will neutralize over 1 million gallons (3800m³) of leachate per day. Furthermore, storm water runoff is diverted into retention ponds around the perimeter of each landfill section. Currently, about 150 wells collect 10 million cu ft (283,000 m³) of methane daily from the two active sections of the landfill. The gas is purified and sold to Union Gas in Brooklyn.

The goal is to ensure that the Fresh Kills site following closure of the landfill will be environmentally safe; however, another (perhaps bigger) challenge is transforming the site into an area that is aesthetically pleasing. Maintaining a "lawn" of grass the size of the landfill site was estimated to be too costly at more than $20 million over the 30 year monitoring period, so the city Sanitation Department developed a series of test plots and experiments. To re-establish native woodland communities, three sites were planted with about 3,000 shrubs and 523 trees, many rescued from sites on part of Staten Island scheduled to be developed. A 40 ha meadow was also planted next to the woodland. The trees require minimal care, and the plan is to let nature take over. The trees have grown moderately well, and the shrubs are doing great. Another benefit of this replanting experiment is that many birds came to perch in the trees, further dispersing seeds and thereby adding new species of plants to the site. Lastly, it was feared that by replanting a landfill site the roots might reach down and break the clay cap used to contain water borne pollutants; however, so far this has not been a problem at the Fresh Kills

site. The replanting project at Fresh Kills indicates that there may actually be an upside to having a landfill in your city. Landfills may offer a way for crowded urban areas to gain open space guaranteed to remain undeveloped.[2] The land reclamation of the site to a mixture of marsh, woodland and grassland as it was before the landfill is a positive action that is being attempted at other closed landfills in the U.S. today. However, preservation of the original land 150 years ago as a nature preserve would obviously have been environmentally preferred. Today a large preserve would be a treasure as is Central Park in New York City. This emphasizes the failure of our past waste management policy, based on a throwaway mentality.

The message from Fresh Kills, aside from the positive aspects of land reclamation, which are significant but represents a tremendous financial burden to society, is that we have failed in the past 50 years to move from a throwaway waste-oriented society to a society that sustains natural resources through improved materials management. We are finally now moving in that direction—a wasteless society as a real and necessary goal. With this in mind we introduce in this chapter concepts of waste management applied to urban waste, hazardous chemical waste, and waste in the marine environment.

FIGURE 27.1 Fresh Kills landfill, perhaps the largest waste disposal site in the world, accepts approximately 15,000 tons per day of municipal and commercial waste from the city of New York. Yet this is only about one-half of the city's total waste.

The message from Fresh Kills, aside from the positive aspects of land reclamation, is that we have failed in the past years to move from a throwaway waste-oriented society to sustaining natural resources. This chapter introduces concepts of waste management and their application to urban waste, hazardous waste, and waste in the marine environment.

LEARNING OBJECTIVES

The old "dilute and disperse" concept of waste management no longer works. The newer concept of "concentrate and contain" is giving way to concepts of waste management focusing on managing materials and eliminating waste. After reading this chapter, you should understand:

- The advantages and disadvantages of each of the major methods that constitute integrated waste management.
- The ways in which the physical and hydrologic conditions at a site affect its suitability for a landfill.

- The concept of multiple barriers for landfills and how landfill sites can be monitored.
- That management of hazardous chemical waste is one of our most serious environmental concerns.
- The various methods of managing hazardous chemical waste.
- The major pathways by which hazardous-waste pollutants from a disposal site may enter the environment.
- Problems related to ocean dumping and why these problems are likely to persist for some time.

27.1 EARLY CONCEPTS OF WASTE DISPOSAL

During the first century of the industrial revolution, the volume of waste produced in the United States was relatively small and could be handled by a concept of dilute and disperse. Factories were located near rivers because the water provided a number of benefits, including easy transport of materials by boat, sufficient water for processing and cooling, and easy disposal of waste into the river. With few factories and a sparse population, dilute and disperse seemed to remove the waste from the environment.[3]

As industrial and urban areas expanded, the concept of dilute and disperse became inadequate and a new concept, known as concentrate and contain, became popular. It has become apparent, however, that containment was and is not always achieved. Containers, whether landfills or drums, natural or artificial, may leak or break and allow waste to escape. Perceived hazards related to waste disposal have led to the present situation, where many people have little confidence in government or industry to preserve and protect public health as it relates to waste disposal.[4]

Waste disposal sites are necessary if society is to function smoothly. However, no one wants to live near a waste disposal site, be it a sanitary landfill for municipal waste, an incinerator that burns urban waste, or a hazardous-waste disposal operation for chemical materials.

The problem of waste disposal in New York, presented in the opening case story of this chapter, is not unique. In the United States, as well as in many other parts of the world, people are facing a serious solid-waste disposal problem. The problem results because we are producing too much waste, and there is too little acceptable space for permanent disposal. It has been estimated that within the next few years approximately half the cities in the United States may run out of landfill space. Philadelphia, for example, is essentially out of landfill space and is bargaining with other states on a month-by-month or yearly basis to dispose of its trash; the Los Angeles area has landfill space for only about 7–10 years. To say we are actually running out of space for landfills isn't really true. Land used for landfills is minute compared to the land area of the United States. Rather, existing sites are being filled and it is difficult to site new landfills. Another major limiting factor is the cost of disposal. Not much more than 10 years ago the cost of disposal of 1 metric ton of urban refuse was approximately $5–$10. Today the average cost is about $32, and some cities, such as Philadelphia, pay as much as $75/metric ton for waste disposal.[4,5] And these costs are only a small part of the total waste management picture; disposal or treatment of liquid and solid waste costs about $20 billion every year and is one of our most costly environmental expenditures.[6]

27.2 MODERN TRENDS

The environmentally preferable concept with respect to waste management is to consider wastes as resources out of place. Although we may not soon be able to reuse and recycle all waste, it seems apparent that the increasing cost of raw materials, energy, transportation, and land will make it financially feasible to reuse and recycle more resources. Moving toward this objective is moving toward an environmental view that there really is no such thing as waste, only resources. Under this concept, waste would not exist, because it would not be produced or, if produced, would be a resource to be used again. This concept is referred to as the "zero waste" movement. Zero waste is the ideal essence of what is known as **industrial ecology** in which our industrial society would function more like an ecological system where waste from one part of the system would be a resource for another part.

Of particular importance is the growing awareness that many of our waste management programs involve simply moving waste from one site to another and not really managing it. For example, waste from urban areas may be placed in landfills, but eventually these may cause new problems if they produce methane gas or noxious liquids that leak from the site and contaminate the surrounding areas. Managed properly, however, methane produced from landfills is a resource that may be burned as a fuel (an example of industrial ecology).

The dominant concept today in managing our waste is known as **integrated waste management (IWM),** which is best defined as a set of management alternatives including **reuse**, **source reduction**, **recycling**, **composting**, **landfill**, and **incineration**.[4]

Reduce, Reuse, Recycle

The *three R's of IWM* are **reduce, reuse,** and **recycle.** The ultimate objective of the three R's is to reduce the amount of urban and other waste that must be disposed of in landfills, incinerators, or other waste management facilities. Study of the waste stream in areas that utilize IWM technology suggests that by early in the 21st century the weight of urban refuse disposed of in landfills or incinerated could be reduced by at least 50% and perhaps as much as 70%.[4] A 50% reduction by weight of urban waste could be facilitated by:[4]

- better design of packaging to reduce waste, an element of source reduction (10% reduction);
- establishment of recycling programs (30% reduction); and
- large-scale composting programs (10% reduction).

This list suggests that recycling is a major player in the reduction of the urban waste stream. Can recycling in fact reduce the waste stream by the 30% suggested? Recent work suggests that the 30% goal is reasonable and will in fact be reached before the year 2000 in some parts of the United States. Interestingly, the potential upper limit for recycling is considerably higher. It is estimated that as much as 80–90% of the U.S. waste stream might be recovered through what is known as intensive recycling.[7] A pilot study involving 100 families in East Hampton, New York, achieved a level of 84%. More realistic for many communities is partial recycling, which targets a specified number of materials, such as glass, aluminum cans, plastic, organic material, and newsprint. Partial recycling can provide the 30% reduction. Nationwide, the recycling rate was 27% (by weight) in 1996 as recycling programs and services continued to expand. On a regional basis, New England states in 1995 led the nation with a rate of 28% while Rocky Mountain states report recycling at 10%.[8] Most states have set ambitious goals for recycling and waste reduction ranging to 70%, and a few have reached their goals. New Jersey reported the nation's highest recycling rate in 1995 at 52%, and several others including Wisconsin, Minnesota, Washington, and Florida were not far behind. However, many states have had to push back their deadline after failing to meet original goals. Overall, recycling rates have begun to slow nationwide, and some states still have not established recycling and waste reduction goals.[9]

Public Support

An encouraging sign associated with public support for the environment is an increase in the willingness of industry and business to support recycling on a variety of scales. For example, some fast-food restaurants are using less packaging for their products and providing on-site bins for recycling used paper and plastic. Grocery stores are encouraging the recycling of plastic and paper bags by providing bins for their collection and recycling. Some food stores offer inexpensive canvas shopping bags to people who prefer them over disposable plastic and paper bags. Companies are redesigning products so that they can be easily disassembled after use and the various parts recycled. As this concept catches on, small appliances, such as electric frying pans and toasters, may be recycled rather than ending up in landfills. The automobile industry is also responding by designing automobiles with coded parts so that they may be more easily disassembled (by professional recyclers) and recycled, rather than left to become rusting eyesores in junkyards.

On still another front, consumers are now more likely to purchase products that may be recycled or that come in containers that are more easily recycled or composted. People also may purchase small home appliances that crush bottles and aluminum cans, reducing their volume and facilitating recycling. The entire industry is rapidly changing, and innovations and opportunities in this field will undoubtedly continue.

Markets for Recycled Products

As with many other environmental solutions, implementing the IWM concept successfully can be a complex undertaking. In some communities where recycling has been successfully initiated, it has resulted in glutted markets for the recycled products, which has sometimes required temporary stockpiling or suspension of recycling of some items. It is apparent that if recycling is to be successful, markets and processing facilities will also have to be developed to ensure that recycling is a sound financial venture as well as an important part of IWM.

The recycling option of IWM has been seriously attempted for nearly two decades, and has been responsible for generating entire systems of waste management that have produced tens of thousands of jobs and reduced the amount of urban waste from homes (in the United States) that is sent to landfills from 90% in the 1980s to about 65% today. Many firms have combined waste reduction with recycling to reduce by 50% to 90% the waste they deliver to landfills. However, in spite of this success IWM is being criticized for not effectively advancing policy to prevent waste production and/or overemphasizing recycling. Forward waste management planning has the goal of "zero production of waste." What is now thought of as waste will become a resource! This visionary goal will require more sustainable use of materials combined with resource conservation in what is being termed, **materials management**. It is believed that this can be established by:[10]

- Elimination of subsidies for extraction of virgin materials such as minerals, oil, and timber;
- Establish "green building" incentives that use recycled-content materials and products in new construction;
- Establish financial penalties for production of products with negative materials management practices;

- Provide financial incentives for industrial practices and products that benefit the environment by enhancing sustainability (for example, reducing waste production and using recycled materials); and
- Increase the production of new jobs in the technology and practice of reuse and recycling of resources. This is the essence of materials management and sustainable resource utilization.

27.3 SOLID-WASTE MANAGEMENT

Waste management is a problem in urban and rural areas of the United States as well as other countries of the world. Many areas, particularly in developing countries, still have inadequate waste management; poorly controlled open dumps and illegal roadside dumping remain a problem. Such dumping spoils scenic resources, pollutes soil and water resources, and is a potential health hazard to plants, animals, and people. This situation is probably a social problem as much as a physical one; many people apparently are simply disposing of their waste as inexpensively and as quickly as possible. Many, in fact, may not see dumping their garbage as an environmental problem. If nothing else, this is a tremendous waste of resources; much of what is dumped could be recycled or reused. In areas where illegal dumping has been reduced, the keys are awareness, education, and alternatives; environmental problems of unsafe, unsanitary dumping are made known and funds are provided for cleanup and inexpensive collection and recycling of trash at sites of origin.

Figure 27.2 illustrates the generalized composition of solid waste likely to end up at a disposal site in the United States. It is no surprise that paper is by far the most abundant of these solid wastes; however, this is only an average content, and considerable variation can be expected because factors such as land use, economic base, industrial activity, climate, and season of the year vary. In some areas, infectious wastes from hospitals and clinics can create problems if they are not properly sterilized before disposal. Some hospitals have facilities to incinerate such wastes, and incineration is probably the surest way to manage infectious medical waste. In urban areas a large amount of toxic materials may also end up at disposal sites; many older urban landfills are now being considered hazardous waste sites and will require costly cleanup.

People have many misconceptions about our waste stream.[11] There has been much publicity concerning urban waste associated with fast-food packaging, polystyrene foam, and disposable diapers. Therefore, many people assume that these products

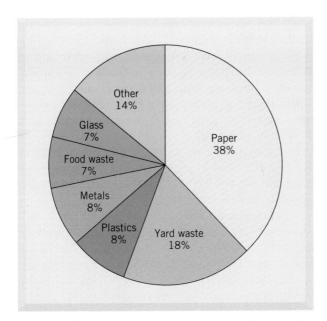

FIGURE 27.2 Composition of U.S. urban solid waste (by weight), 1998. (*Source:* U.S. Environmental Protection Agency, Office of Solid Waste. Accessed 10/9/98 at http://www.epa.gov/epaoswer/non-hw/recycle/index.htm.)

make up a large percentage of the total waste stream and are responsible for the rapid filling of landfills. However, excavations into modern landfills using archeological tools have cleared up some of the misconceptions concerning these items. We now know that fast-food packaging accounts for only about 0.25% of the average landfill; disposable diapers account for approximately 0.8%; and polystyrene products add another 0.9%.[12] People using disposable diapers (and most parents still do) may feel guilty that their children's disposable diapers end up in landfills. The billions of dirty diapers per year sent to landfills are a potential sanitation problem; however, they constitute only a small percentage of the volume in a landfill. On the other hand, paper, as shown in Figure 27.2, is the major constituent in landfills (perhaps as much as 50% by volume and 40% by weight). The largest single item is newspaper, which accounts for as much as 18% by volume.[12] Newspaper is one of the major items targeted for recycling because big environmental dividends can be expected. However (and this is a value judgment), the need to deal with the major contributors does not mean that we need not reduce disposable diapers, polystyrene, and other paper products. These are also made from resources that might be better managed.

On-Site Disposal

A common on-site disposal method in urban areas is the mechanical grinding of kitchen food waste.

Garbage disposal devices are installed in the wastewater pipe system at the kitchen sink, and the garbage is ground and flushed into the sewer system. This effectively reduces the amount of handling and quickly removes food waste. Final disposal is transferred to sewage treatment plants, where solids remaining as sewage sludge still must be disposed of.[13]

Composting

Composting is a biochemical process in which organic materials such as lawn clippings and kitchen scraps decompose to a rich, soil-like material. It is a process of rapid, partial decomposition of moist, solid, organic waste by aerobic organisms. Although simple backyard compost piles may come to mind, as a waste management option large-scale composting is generally carried out in the controlled environment of mechanical digesters.[13] This is a popular technique in Europe and Asia, where intense farming creates a demand for the compost.[13] A major drawback of composting is the necessity to separate organic material from other waste. Therefore, it is probably economically advantageous only when organic material is collected separately from other waste. Nevertheless, composting is an important component of IWM and its contribution will undoubtedly grow in the future.

Incineration

In **incineration** combustible waste is burned at temperatures high enough (900–1000°C, or 1650–1830°F) to consume all combustible material, leaving only ash and noncombustibles to dispose of in a landfill. Under ideal conditions, incineration may reduce the volume of waste by 75% to 95%.[13] In practice, however, the actual decrease in volume is closer to 50%, because of maintenance problems as well as waste supply problems. This is approximately the same savings that could probably be realized from waste reduction and recycling.[11] Besides reducing a large volume of combustible waste to a much smaller volume of ash, incineration has another advantage in that the process of incineration can be used to supplement other fuels and generate electrical power.

Incineration of urban waste is not necessarily a clean process. Incineration may produce air pollution and toxic ash. For example, incineration in the United States apparently is a significant source of environmental dioxin, a carcinogenic toxin (see Chapter 14) and a controversy over incineration has resulted.[14] Smokestacks from incinerators may emit oxides of nitrogen and sulfur that lead to acid rain; heavy metals such as lead, cadmium, and mercury; and carbon dioxide, which is hypothetically related to global warming. In modern incineration facilities, smokestacks are fitted with special devices to trap pollutants, but the process of pollutant abatement is expensive. Furthermore, the plants themselves are expensive and government subsidization may be needed to aid in their establishment. Recent evaluation of the urban waste stream suggests that with an investment of $8 billion a sufficient number of incinerators could be constructed in the United States today to burn approximately 25% of the solid waste that is generated. However, a similar investment in source reduction, recycling, and composting could result in diversion from landfills of as much as 75% of the nation's urban waste stream.[7] Currently, about 10% of municipal solid waste is disposed of using incinerators, amounting to approximately 30 million tons per year.[8] Finally, the economic viability of incinerators depends on revenue from the sale of the energy produced by burning the waste. As recycling and composting are increased, they will compete with incineration for their portion of the waste stream, and sufficient waste (fuel) to generate a profit from incineration may not be available. The main conclusion that may be drawn from considering IWM principles is that a combination of reusing, recycling, and composting may reduce the volume of waste requiring disposal at a landfill by at least as much as incineration.[7]

Open Dumps

In the past, solid waste was usually accumulated in open dumps, where the refuse was piled up without being covered or otherwise protected. Although thousands of open dumps have been closed in recent years and new open dumps are banned in the United

Figure 27.3 Urban garbage dump in Rio de Janeiro, Brazil. At this site people are going through the waste and recycling materials that can be reused or resold.

States and many other countries, many are still being used worldwide (Figure 27.3). Dumps have been located wherever land is available, without regard to safety, health hazards, and aesthetic degradation. Common sites are abandoned mines and quarries, where gravel and stone have been removed (sometimes by ancient civilizations); natural low areas, such as swamps or floodplains; and hillside areas above or below towns. The waste is often piled as high as equipment allows. In some instances, the refuse is ignited and allowed to burn. In others, the refuse is periodically leveled and compacted.[13]

As a general rule, open dumps create a nuisance by being unsightly, providing breeding grounds for pests, creating a health hazard, polluting the air, and sometimes polluting groundwater and surface water. Fortunately, open dumps are giving way to the better planned and managed sanitary landfills.

Sanitary Landfills

A **sanitary landfill** is designed to concentrate and contain refuse without creating a nuisance or hazard to public health or safety. The idea is to confine the waste to the smallest practical area, reduce it to the smallest practical volume, and cover it with a layer of compacted soil at the end of each day of operation or more frequently if necessary. Covering the waste is what makes the landfill sanitary. The compacted layer restricts (but does not eliminate) continued access to the waste by insects, rodents, and other animals, such as seagulls. It also isolates the refuse, minimizing the amount of surface water entering into and gas escaping from the waste.[15]

Leachate

The most significant hazard from a sanitary landfill is pollution of groundwater or surface water. If waste buried in a landfill comes into contact with water percolating down from the surface or with groundwater moving laterally through the refuse, **leachate**—noxious, mineralized liquid capable of transporting bacterial pollutants—is produced.[16] For example, two landfills dating from the 1930s and 1940s in Long Island, New York, have produced subsurface leachate trails (plumes) several hundred meters wide that have migrated kilometers from the disposal site.

The nature and strength of leachate produced at a disposal site depends on composition of the waste, amount of water that infiltrates or moves through the waste, and length of time that infiltrated water is in contact with the refuse.[13]

Site Selection

The siting of a sanitary landfill is very important. A number of factors must be taken into consideration, including topography, location of the groundwater table, amount of precipitation, type of soil and rock, and location of the disposal zone in the surface-water and groundwater flow system. A favorable combination of climatic, hydrologic, and geologic conditions helps to ensure reasonable safety in containing the waste and its leachate.[17]

The best sites are in arid regions, where disposal conditions are relatively safe because little leachate is produced in a dry environment. In a humid environment some leachate is always produced; therefore, an acceptable level of leachate production must be established to determine the most favorable sites. What is acceptable varies with local water use, local regulations, and the ability of the natural hydrologic system to disperse, dilute, and otherwise degrade the leachate to harmless levels.

Elements of the most desirable site in a humid climate are shown in Figure 27.4. The waste is buried above the water table in relatively impermeable clay and silt soils, material through which water cannot easily move. Any leachate produced remains in the vicinity of the site and degrades by natural filtering action and chemical reactions between clay and leachate. This also holds true for high water table conditions often found in humid areas, provided that impermeable material is present.[18]

There are also important social considerations concerning siting of waste disposal facilities. Often planners choose sites where they expect local resistance to be minimal or where they perceive land to have little value. Waste disposal facilities are frequently located in areas where residents tend to have low socioeconomic status or belong to a particular race or ethnic group. Study of social issues of siting waste facilities, chemical plants, and other facilities to which many people object based on potential environmental problems is an emerging field in social sciences known as **environmental justice.**[19]

Monitoring Pollution

Once a site is chosen for a sanitary landfill and before filling starts, **monitoring** the movement of groundwater should begin. The monitoring is accomplished by periodically taking samples of water and gas from specially designed monitoring wells. Monitoring the movement of leachate and gases should be continued as long as there is any possibility of pollution. This procedure is particularly important after the site is completely filled and a final, permanent cover material is in place. Continued monitoring is necessary because a certain amount of settlement always occurs after a landfill is completed, and if small depressions form, surface water may collect, infiltrate, and produce leachate. Monitoring and proper maintenance of an abandoned landfill reduce its pollution potential.[15]

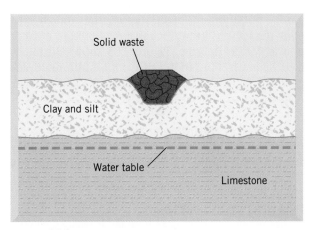

FIGURE 27.4 Most desirable landfill site in a humid environment. Waste is buried above the water table in a relatively impermeable environment. (*Source:* W. J. Schneider, 1970, *Hydraulic Implications of Solid-Waste Disposal,* U.S. Geological Survey Circular 601F.)

How Pollutants Enter the Environment

Hazardous-waste pollutants from a solid-waste disposal site may enter the environment by as many as six paths:[20]

1. Methane, ammonia, hydrogen sulfide, and nitrogen gases may be produced from compounds in the soil and the waste and may enter the atmosphere.

2. Heavy metals, such as lead, chromium, and iron, may be retained in the soil.

3. Soluble materials, such as chloride, nitrate, and sulfate, may readily pass through the waste and soil to the groundwater system.

4. Overland runoff may pick up leachate and transport it into streams and rivers.

5. Some plants (including crops) growing in the disposal area may selectively take up heavy metals and other toxic materials to be passed up the food chain as people and animals eat them.

6. If plant residue from crops left in fields contains toxic substances, these return to the soil.

Modern sanitary landfills are engineered to include multiple barriers (double-lined), such as clay and plastic liners to limit movement of leachate; surface and subsurface drainage to collect leachate; systems to collect methane gas produced as waste decomposes; and groundwater-monitoring to detect leaks of leachate below and adjacent to the landfill. A thorough monitoring program considers all six possible paths by which pollutants enter the environment. In practice, monitoring all six pathways is not often done. It is particularly important to monitor water in vadose zone (above water table, where soil and rock pores are unsaturated) to identify potential pollution problems before they reach and contaminate groundwater resources, where correction is very expensive. Figure 27.5 shows an idealized diagram of a landfill that utilizes the multiple-barrier approach and a landfill site under construction.

Federal Legislation

New landfills opened in the United States after October 1993 must comply with stricter requirements under the Resource Conservation and Recovery Act of 1980. States may choose from two options.

1. Seek EPA approval of solid-waste management plans, which allows greater flexibility; or

2. Compliance with federal standards.

Legislation is intended to strengthen and standardize design, operation, and monitoring of sanitary landfills. Landfills that cannot comply with regulations face closure. Provisions of regulations include:

FIGURE 27.5 (*a*) Idealized diagram of solid-waste facility (sanitary landfill) illustrating multiple-barrier design, monitoring system, and leachate collection system. (*b*) Rock Creek landfill, Calaveras County, California, under construction. This municipal solid-waste landfill is underlain by a compacted clay liner (exposed light brown slope in the center left portion of the photograph). The black slopes, covered with gravel piles, overlie the compacted clay layer. These are a vapor barrier designed to keep moisture in the clay and help avoid cracking of the clay line. The sinuous gray trench is lined with plastic and is part of the leachate collection system for the landfill. The excavated squared pond (upper part of photograph) is a leachate evaporation pond under construction. The landfill is also equipped with a system to monitor the vadose zone below the leachate collection system. Photograph courtesy of John Kramer.

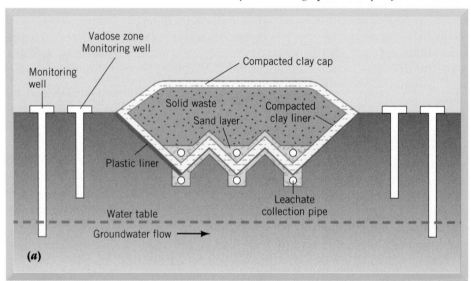

- Landfills may not be sited on floodplains, wetlands, earthquake zones, unstable land or airports (birds at sites are a hazard to aircraft);
- Landfills must have liners;
- Landfills must have a leachate collection system;
- Landfill operators must monitor groundwater for many specified toxic chemicals; and
- Landfill operators must meet financial assurance criteria that monitoring continues for 30 years after closure of the landfill.

As mentioned, states with EPA approval of their landfill program are provided more flexibility:

- Groundwater monitoring may be suspended if the landfill operator can demonstrate that hazardous constituents are not migrating from the landfill;
- Alternative types of daily cover over the waste may be used;
- Alternative groundwater protection standards are allowed;
- Alternative schedules for documentation of groundwater monitoring are allowed;
- Under certain circumstances, landfills in wetlands and fault zones are allowed; and
- Alternative financial assurance mechanisms are allowed.

Given the added flexibility, it appears advantageous for states to develop EPA-approved waste management plans.

27.4 HAZARDOUS CHEMICAL WASTE MANAGEMENT

Creation of new chemical compounds has proliferated in recent years. In the United States, approximately 1000 new chemicals are marketed each year and about 70,000 chemicals are currently on the market. Although many of the chemicals have been beneficial to people, approximately 35,000 chemicals used in the United States are classified as definitely or potentially hazardous to the health of people (see Table 27.1). The United States currently generates about 250 million metric tons of hazardous chemical waste per year, referred to more commonly as **hazardous waste.**[21] In the recent past, as much as half the total volume of wastes was indiscriminately dumped.[22] (See A Closer Look 27.1.) Another source of hazardous chemicals is buildings destroyed by events such as hurricanes. Unless collected chemicals such as paints, solvents, pesticides stored in destroyed buildings may be released into the environment when debris is burned or buried. One home may not be a significant problem, but thousands of buildings are.

Uncontrolled Sites

In the United States, there are 32,000–50,000 uncontrolled waste disposal sites, of these probably 1200–2000 contain sufficient hazardous waste to be a threat to public health and the environment. For this reason, many scientists believe management of hazardous chemical materials may be the most serious environmental problem in the United States.

Uncontrolled dumping of chemical waste has polluted soil and groundwater in several ways:

- Chemical waste stored in barrels, either stacked on ground or buried, eventually, corrode and leak, polluting surface water, soil, and groundwater;
- Liquid chemical waste dumped in an unlined lagoon, from which contaminated water percolates through soil and rock to the groundwater table; and
- Liquid chemical waste may be illegally dumped in deserted fields or even along roads.

Table 27.1 Products and the Potentially Hazardous Waste They Generate	
Products We Use	*Potentially Hazardous Waste*
Leather	Heavy metals, organic solvents
Medicines	Organic solvents and residues, heavy metals (e.g., mercury and zinc)
Metals	Heavy metals, fluorides, cyanides, acid and alkaline cleaners, solvents, pigments
Oil, gasoline, and other petroleum products	Oil, phenols and other organic compounds, heavy metals, ammonia salts, acids
Paints	Heavy metals, pigments, solvents, organic residues
Pesticides	Organic chlorine compounds, organic phosphate compounds
Plastics	Organic chlorine compounds
Textiles	Heavy metals, dyes, organic chlorine compounds, solvents

Source: U.S. Environmental Protection Agency, SW-826, 1980.

A **CLOSER LOOK 27.1**

LOVE CANAL

In 1976, in a residential area near Niagara Falls, New York, trees and gardens began to die. Rubber on tennis shoes and bicycle tires disintegrated. Puddles of toxic substances began to ooze through the soil; a swimming pool popped from its foundation and floated in a bath of chemicals.

The area had been excavated in 1892 to make way for Love Canal, designed as a transportation route but never completed. The ditch (canal) remained unused for decades and became a dump for wastes. From 1920 to 1950, 20,000 tons of more than 80 chemicals were dumped in the ditch. In 1953, the company that owned the land and had dumped the chemicals donated the land to the city of Niagara Falls for $1. Eventually several hundred homes and an elementary school were built on and near the site. However, in 1976–77 heavy rains and snows triggered events that made Love Canal a household word.

A study of the site identified many substances suspected of being carcinogens, including benzene, dioxin, dichlorethylene, and chloroform. Although officials admitted that

little was known about the impact of these chemicals, there was grave concern for people living in the area. Eventually concern centered on alleged high rates of miscarriages, blood and liver abnormalities, birth defects, and chromosome damage. However, a study by New York health authorities suggests that no chemically caused health effects have been absolutely established.[23–25]

Cleanup of Love Canal is an important case that demonstrates the technology for hazardous waste treatment. The objective has been to contain waste, stop migration of wastes through the groundwater flow system, and remove and treat contaminated soil and sediment in streambeds and sewers.[26]

To minimize further contamination, the area has been covered with 1 m (3.3 ft) of compacted clay and a polyethylene plastic cover to reduce infiltration of surface water. Water is inhibited from entering the site by specially designed barriers. These procedures greatly reduce subsurface seepage of water, and water that does seep out is collected and treated.[23–26]

By 1990 $275 million had been spent for cleanup and relocation projects. Homes in an adjacent area were purchased, and about 200 homes and a school had to be destroyed by the government. The U.S. Environmental Protection Agency (EPA) declared the area clean, and about 280 remaining homes were marketed for resale. They were expected to sell in spite of all the adverse publicity. By 1994, 193 homes had sold, many with only a 15% price reduction.

A court-mediated settlement in 1994 resulted in the company responsible for the chemical waste paying the State of New York $98 million and taking responsibility for $25 million for treatment operations.[27] In 1995, the company agreed to pay an additional $129 million to the federal government as reimbursement for its costs.[27,28]

What went wrong in Love Canal? How can we avoid such disasters in the future? The real tragedy of Love Canal is that it is probably not an isolated incident; there are many hidden Love Canals across the country, each a potential time bomb waiting to explode.[23,24]

Responsible Management

In the United States, management of hazardous waste began in 1976 with passage of the Resource Conservation and Recovery Act. At the heart of the act is identification of hazardous wastes and their life cycles. Regulations require stringent record keeping and reporting be maintained to verify wastes do not present a public nuisance or a health problem. The act also classifies hazardous wastes in several categories: materials highly toxic to people and other living things; wastes that may explode or ignite when exposed to air; extremely corrosive wastes; and unstable wastes.

Recognizing that many waste disposal sites are hazards, Congress passed the Comprehensive Environmental Response Compensation and Liability Act (CERCLA) in 1980, defining policy and procedures

for release of hazardous substances into the environment (i.e., landfill regulations). CERCLA mandated development of a priorities list of sites where hazardous substances are likely to or do produce the most serious environmental problems and established a revolving fund (*Superfund*) to cleanup the worst abandoned hazardous-waste sites. In 1984 and 1986 CERCLA was strengthened by amendments that:

- improved and tightened standards for disposal and cleanup of hazardous waste (i.e., requiring double liners, monitoring landfills);
- banned land disposal of certain hazardous chemicals, including dioxins, polychlorinated biphenyls (PCBs), and most solvents;
- initiating a timetable for phasing out disposal

of all untreated liquid hazardous waste in land-fills or surface impoundments; and

- increasing the size of the fund. Superfund was allocated at about $8 billion in 1993. Congress approved a $7.36 billion bill to the EPA for fis-cal year 1998, which provides for almost dou-bling the Superfund budget. However, stipula-tions included that Superfund must be reformed.[29]

Although Superfund has experienced signifi-cant management problems and cleanup efforts are way behind schedule, a small number of sites have been treated. Unfortunately, funds available are not sufficient to pay for decontamination of all targeted sites, which would cost many times more, as much as $100 billion. Furthermore, there is concern that present technology is not sufficient to treat all aban-doned waste disposal sites; it may be necessary to simply try to confine waste to those sites until better disposal methods are developed. It seems apparent that abandoned disposal sites are likely to persist as problems for some time to come.

Federal legislation has changed the ways in which real estate does business. For example, there are provisions by which property owners may be li-able for costly cleanup of hazardous waste present on their property even if they did not directly cause the problem. As a result, banks and other lending institutions might be liable for release of hazardous materials by their tenants. The Superfund Amend-ment and Reauthorized Act (SARA) of 1986 provides possible defense for those who purchase real estate against such liability, provided they have completed an **environmental audit** prior to the purchase of property. Such an audit involves the study of past land use at the site, usually determined from analyz-ing old maps, aerial photographs, and reports. It may also involve drilling and sampling of ground-water and soil to determine if hazardous materials are present. Environmental audits are now standard operating procedure completed on a routine basis prior to purchase of property for development.

SARA legislation also required that certain in-dustries report all releases of hazardous materials, and a list of companies releasing hazardous sub-stances became public. This list was known as the "Toxic 500." No property owner or industry wants his or her company to be on such a list, and the list is thought to have provided some pressure to de-velop safer handling of hazardous materials by in-dustries formerly identified as polluters.[30]

In 1990, the U.S. Congress reauthorized haz-ardous-waste control legislation. Priorities include:

- establishing who is responsible (liable) for ex-isting hazardous-waste problems;

- when necessary, assisting in or providing fund-ing for cleanup at sites identified as having a hazardous-waste problem;
- providing measures whereby people who suf-fer damages from the release of hazardous ma-terials are compensated; and
- improving the required standards for disposal and cleanup of hazardous waste.

Management of hazardous chemical waste in-volves several options, including recycling, on-site processing to recover by-products with commercial value, microbial breakdown, chemical stabilization, high-temperature decomposition, incineration, and disposal by secure landfill or deep-well injection. A number of technological advances have been made in toxic-waste management, and as land disposal be-comes more expensive, the recent trend toward on-site treatment is likely to continue. However, on-site treatment will not eliminate all hazardous chemical waste; disposal of some waste will remain neces-sary. Table 27.2 compares hazardous-waste reduc-tion technologies for treatment and disposal. Notice that all available technologies cause some environ-mental disruption. There is no simple solution for all waste management issues.

Secure Landfill

A **secure landfill** for hazardous waste is designed to confine the waste to a particular location, control the leachate that drains from the waste, collect and treat the leachate, and detect possible leaks. This type of landfill is similar to the modern sanitary landfill; it is an extension of the sanitary landfill for urban waste. Because in recent years it has become apparent that urban waste contains a lot of haz-ardous materials, the design of sanitary landfills and that of secure landfills for hazardous waste have converged to some extent.

Design of a secure landfill is shown in Figure 27.6. A dike and liner (made of clay or other imper-vious material such as plastic) confines waste, and a system of internal drains concentrates leachate in a collection basin from which it is pumped and trans-ported to a wastewater treatment plant. Designs of modern facilities include multiple barriers consisting of several impermeable layers and filters. The func-tion of impervious liners is to ensure that leachate does not contaminate soil and groundwater re-sources. However, this type of waste disposal proce-dure, like the sanitary landfill from which it evolved, must have several monitoring wells to alert person-nel if and when leachates leak out of the system and threaten water resources.

It has recently been argued that there is no such thing as a really secure landfill, implying that

Table 27.2 Comparison of Hazard Reduction Technologies

	Disposal			Treatment	
Parameter Compared	**Landfills and Impoundments**	**Injection Wells**	**Incineration and Other Thermal Destruction**	**Emerging High-temperature Decomposition**[a]	**Chemical Stabilization**
Effectiveness: how well it contains or destroys hazardous characteristics	Low for volatiles, questionable for liquids; based on lab and field tests	High, based on on theory; limited field data available	High, based on field tests; little data on specific constituents	Very high, based on commercial scale tests	High for many metals, based on lab tests
Reliability issues	Siting, construction, and operation Uncertainties: long-term integrity of cells and cover, linear life less than life of toxic waste	Site history and geology, well depth, construction, and operation	Monitoring uncertainties with respect to high degree of DRE: surrogate measures, PICs, incinerability[b]	Limited experience Mobile units; on-site treatment avoids hauling risks Operational simplicity	Some inorganics still soluble Uncertain leachate test, surrogate for weathering
Environment media most affected	Surface water and groundwater	Surface water and groundwater	Air	Air	Groundwater
Least compatible wastes[c]	Linear reactive, highly toxic, mobile, persistent and bioaccumulative	Reactive; corrosive; highly toxic, mobile, and persistent	Highly toxic and refractory organics, high heavy-metal concentration	Some inorganics	Organics
Relative costs	Low to moderate	Low	Moderate to high	Moderate to high	Moderate
Resource recovery potential	None	None	Energy and some acids	Energy and some metals	Possible building material

Source: Council on Environmental Quality, 1983.

[a]Molten salt, high-temperature fluid well, and plasma arc treatments.

[b]DRE = destruction and removal efficiency; PIC = product of incomplete combustion.

[c]Wastes for which this method may be less effective for reducing exposure, relative to other technologies. Wastes listed do not necessarily denote common usage.

they all leak to some extent. This is true; impervious plastic liners, filters, and clay layers can fail, even with several backups, and drains can become clogged and cause overflow. Animals, such as gophers, ground squirrels, woodchucks, and muskrats, can chew through plastic liners and some may burrow through clay liners, thus promoting or accelerating leaks. Yet careful siting and engineering can minimize problems. As with sanitary landfills, preferable sites are those with good natural barriers to migration of leachate: thick clay deposits, an arid climate, or a deep water table. Nevertheless, land disposal should be used only for specific chemicals compatible with and suitable for the method.

Land Application

Intentional application of waste materials to the surface soil is referred to as **land application,** land spreading, or land farming. Land application of waste may be a desirable method of treatment for certain biodegradable industrial waste, such as oily petroleum waste and some organic chemical-plant wastes. A good indicator of the usefulness of land application of a particular waste is the *biopersistence* (the measure of how long a material remains in the biosphere). The greater or longer the biopersistence, the less suitable the wastes are for land application procedures. Land application is not an effective treatment or disposal method for inorganic substances such as salts and heavy metals.[31]

Land application of biodegradable waste works because, when such materials are added to the soil, they are attacked by microflora (bacteria, molds, yeasts, and other organisms) that decompose the waste material. The soil thus may be thought of as a microbial farm that constantly recycles organic and inorganic matter by breaking it down into more fundamental forms useful to other living things in the soil. Because the upper soil zone contains the largest microbial populations, land application is restricted to the uppermost 15–20 cm (6–8 in.) of the soil.[31]

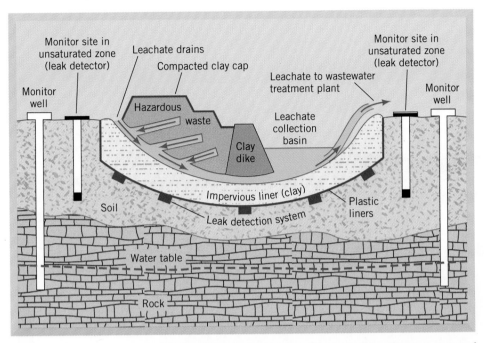

FIGURE 27.6 A secure landfill for hazardous chemical waste. The impervious liners, systems of drains, and leak detectors are an integral part of the system to ensure that leachate does not escape from the disposal site. Monitoring in the unsaturated zone is important and involves periodic collection of soil water.

Surface Impoundment

Both natural topographic depressions and human-made excavations have been used to hold hazardous liquid waste. These are primarily formed of soil or other surface materials but may be lined with manufactured materials such as plastic. The **surface impoundment** is designed to hold the waste; examples include aeration pits and lagoons at hazardous-waste facilities. Surface impoundments have been criticized because they are especially prone to seepage, resulting in pollution of soil and groundwater. Evaporation from surface impoundments can also produce an air pollution problem. This type of storage or disposal system for hazardous waste is controversial, and many sites have been closed.

Deep-Well Disposal

Deep-well disposal, another controversial method of waste disposal, involves injection of waste into deep wells. A deep well must penetrate to rock (not soil) that is below and completely isolated from all freshwater aquifers, thereby assuring that injection of waste will not contaminate or pollute existing or potential water supplies. Typically, the waste is injected into a permeable rock layer several thousand meters below the surface, in geologic basins topped by relatively impervious, fracture-resistant rock such as shale or salt deposits.[3]

Deep-well injection of oil-field brine (salt water) has been important in the control of water pollution in oil fields for many years, and huge quantities of liquid waste (brine) pumped up with oil have been injected back into the rock.[32]

Deep-well disposal of industrial wastes should not be viewed as a quick and easy solution to industrial waste problems.[33] Even where geologic conditions are favorable for deep-well disposal, there are a limited number of suitable sites and within these sites there is limited space for disposal of waste. Finally, disposal wells must be carefully monitored by additional wells, known as monitoring wells, that are required to determine if the waste is remaining in the disposal site.

Summary of Land Disposal Methods

Direct land disposal of hazardous waste is often not the best initial alternative. There is consensus that even with extensive safeguards and state-of-the-art designs, land disposal alternatives cannot guarantee that the waste is contained and will not cause environmental disruption in the future. This concern holds true for all land disposal facilities, including landfills, surface impoundments, land application, and injection wells. Pollution of air, land, surface water, and groundwater may result from failure of a land disposal site to contain hazardous waste. Pollution of groundwater is perhaps the most significant

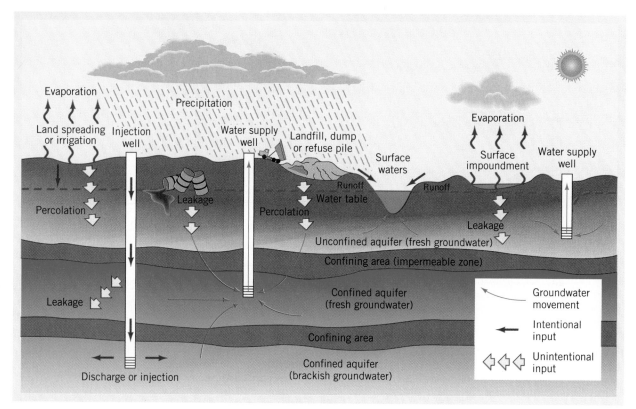

FIGURE 27.7 Examples of how land disposal and treatment methods of managing hazardous waste may contaminate the environment. (*Source:* Modified after C. B. Cox, 1985, *The Buried Threat,* California Senate Office of Research, No. 115-5.)

risk, because groundwater provides a convenient route for pollutants to reach humans and other living things. Figure 27.7 shows some of the paths that pollutants may take from land disposal sites to contaminate the environment. These paths include leakage and runoff to surface water or groundwater from improperly designed or maintained landfills; seepage, runoff, or air emissions from unlined lagoons; percolation and seepage from failure of surface land application of waste to soils; leaks in pipes or other equipment associated with deep-well injection; and leaks from buried drums, tanks, or other containers.

Alternatives to Land Disposal of Hazardous Waste

The philosophy of handling hazardous chemical waste should be multifaceted. In addition to the disposal methods just discussed, chemical waste management should include such processes as source reduction, recycling and resource recovery, treatment, and incineration.[34] Recently it has been argued that these alternatives to land disposal are not being utilized to their full potential; that is, the volume of waste could be reduced and the remaining waste could be recycled or treated in some form

prior to land disposal of the residues of the treatment processes.[34] Advantages to source reduction, recycling, treatment, and incineration include:

- The actual waste that must eventually be disposed of is reduced to a much smaller volume;
- Useful chemicals may be reclaimed and reused;
- Treatment of wastes may make them less toxic and therefore less likely to cause problems in landfills; and
- Because a reduced volume of hazardous waste is finally disposed of, there is less stress on the dwindling capacity of waste disposal sites.

Although some of these techniques have been discussed as part of IWM, the techniques have special implications and complications.

Source Reduction

The object of **source reduction** in hazardous waste is to reduce the amount of hazardous waste generated by manufacturing or other processes. For example, changes in the chemical processes involved, equipment used, raw materials used, or maintenance measures may successfully reduce the amount or toxicity of the hazardous waste produced.[34]

Recycling and Resource Recovery

Hazardous chemical waste may contain materials that can be recovered for future use. For example, acids and solvents collect contaminants when they are used in manufacturing processes. These acids and solvents can be processed to remove the contaminants and can then be reused in the same or in different manufacturing processes.[34]

Treatment

Hazardous chemical waste may be treated by a variety of processes to change the physical or chemical composition of the waste to reduce its toxic or hazardous characteristics. For example, acids can be neutralized, heavy metals can be separated from liquid waste, and hazardous chemical compounds can be broken up through oxidation.[34]

Incineration

Hazardous chemical waste can be successfully destroyed by high-temperature incineration. Incineration is considered a waste treatment rather than a disposal method because the process produces an ash residue that must be disposed of in a landfill operation. Hazardous waste has also been incinerated offshore on ships, creating potential air pollution and ash disposal problems for the marine environment. The technology used in incineration and other high-temperature decomposition or destruction is changing rapidly. Figure 27.8 shows a generalized diagram of one type of high-temperature incineration system that may be used to burn toxic waste. Waste (as liquid, solid, or sludge) enters the rotating combustion chamber, where it is rolled and burned. Ash from this burning process is collected in a water tank while the remaining gaseous materials move into a secondary combustion chamber, where the process is repeated. Finally, the remaining gas and particulates move through a scrubber system that eliminates surviving particulates and acid-forming components. Carbon dioxide, water, and air are then emitted from the stack. As shown in Figure 27.8, ash particulates and wastewater are produced at various parts of the incineration process and these must be treated or disposed of in a landfill.

More advanced techniques for the incineration and thermal decomposition of waste are being developed. One of these utilizes a molten salt bed that should be useful in destroying certain organic materials. Finally, other incineration techniques include liquid-injection incineration on land or sea and multiple-hearth furnaces. Which incineration method is used for a particular waste depends on the nature and composition of the waste and the temperature necessary to destroy the hazardous components. For example, the generalized incineration system shown in Figure 27.8 could be used to destroy PCBs.

FIGURE 27.8 Generalized diagram of a high-temperature incineration system.

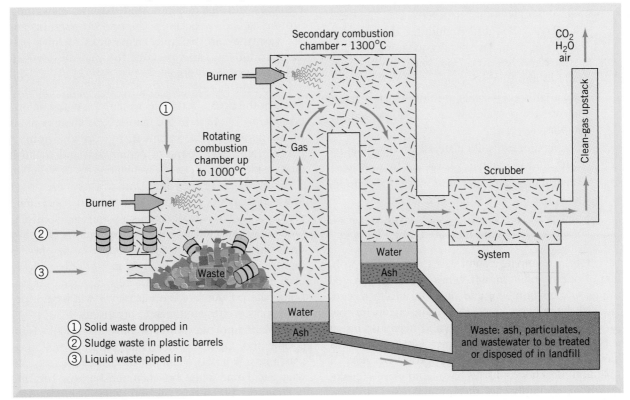

27.5 OCEAN DUMPING

Oceans cover more than 70% of Earth. They play a part in maintaining our global environment and are of major importance in the cycling of carbon dioxide that helps regulate global climate. Oceans are also important in cycling many chemical elements important to life, such as nitrogen and phosphorus, and are a valuable resource to people because they provide necessities such as foods and minerals.

It seems reasonable that such an important resource would receive preferential treatment, yet oceans continue to be dumping grounds for waste. In 1972, the Ocean Dumping Act was passed in the United States to provide for meeting U.S. commitments under the 1972 Convention on the Prevention of Marine Pollution by Dumping of Wastes and Other Matters, an international treaty signed by 80 countries. The law bans ocean dumping of radiological, chemical, and biological warfare agents and high-level radioactive waste. Amendments in 1988 extended this ban to sewage sludge, industrial waste, and medical wastes. In 1992, provisions were added to establish a national coastal water quality monitoring program. Although federal legislation enacted in 1972 by the U.S. EPA has reduced the number of ocean dumping sites, ocean dumping continues.[6] Furthermore, if population growth in coastal regions continues as expected, increased amounts of waste will end up in the oceans.

The types of wastes dumped in the oceans include the following:[35]

- Dredge spoils—solid materials, such as sand, silt, clay, rocks, and pollutants, deposited in the ocean from industrial and municipal discharges and removed from water bodies, generally to improve navigation;
- Industrial waste—acids, refinery wastes, paper mill wastes, pesticide wastes, and assorted liquid wastes;
- Sewage sludge—solid material (sludge) that remains after municipal wastewater treatment;
- Construction and demolition debris—cinder block, plaster, dirt, stone, and tile; and
- Solid waste—refuse, garbage, explosives, radioactive waste, and untreated urban sewage.

Ocean dumping contributes to the larger problem of ocean pollution, which has seriously damaged the marine environment and caused a health hazard. Shellfish have been found to contain organisms that produce diseases, such as polio and hepatitis, and at least 20% of the nation's commercial shellfish beds have been closed (mostly temporarily) because of pollution. Beaches and bays have been closed (again, mostly temporarily) to recreational uses. Lifeless zones in the marine environment have been created. Heavy kills of fish and other organisms have occurred and profound changes in marine ecosystems have taken place (see Chapters 9, 13, and 20.[35,36]

Marine pollution has a variety of effects on oceanic life, including the following:

- death or retarded growth, vitality, and reproductivity of marine organisms;
- reduction in the dissolved oxygen content necessary for marine life because of increased biochemical oxygen demand;
- eutrophication caused by nutrient-rich waste in shallow waters of estuaries, bays, and parts of the continental shelf, resulting in depletion of oxygen and subsequent killing of algae that may wash up and pollute coastal areas; and
- habitat change caused by waste disposal practices that subtly or drastically change entire marine ecosystems.[35]

Marine waters of Europe are in particular trouble, in part the result of urban and agricultural pollutants that raised concentrations of nutrients in seawater. Blooms (heavy, sudden growth) of toxic algae are becoming more common; in 1988 in the waterway connecting the North Sea to the Baltic Sea, a bloom was responsible for killing nearly all marine life to a depth of about 15 m (50 ft). It is believed that urban waste and agricultural runoff contributed to the toxic bloom. There is concern that some ecosystems in the oceans, such as coral reefs, estuaries and salt marshes, and mangrove swamps, are threatened by ocean pollution. Although oceans are vast, they are basically giant sinks for materials from continents, and parts of the marine environment are extremely fragile.[36]

Recent works suggest that toxic materials threaten the ocean bottom as well as the entire marine ecosystem. The base of the marine food chain consists of planktonic life abundant in the upper 3 mm of ocean water. The young of certain fish and shellfish also reside in the upper few millimeters of water in the early stages of their life. Unfortunately, the upper few millimeters of the ocean also tend to concentrate pollutants, such as toxic chemicals and heavy metals. One study reported that the concentrations of heavy metals including zinc, lead, and copper in the upper 3 mm (or *microlayer*) are *from 10 to 1000 times higher* than in the deeper waters. There is fear that disproportionate pollution of the microlayer will have especially serious effects on marine organisms.[36]

Marine pollution can also have major impacts on people and society. Contaminated marine organisms may transmit toxic elements or diseases to people who eat them. When beaches and harbors become polluted by solid waste, oil, and other materials, there

Should We Dispose of Waste in the Ocean?

The New York Bight, the coastal water off the Long Island and New Jersey shores, includes Newark Bay, an important commercial shipping port (see map). To keep this $20 billion industry open to large vessels, accumulated sediments must be dredged periodically to maintain a minimum depth of 12.8 m (42 ft). Since 1976, 5.3 million m^3 (7 million yd^3) a year of dredged material has been dumped at the Mud Dump site, an area of about 5.2 k^2 (2 mi^2) about 9.6 km (6.0 miles) from shore, with an average depth of 22 m (72 ft). The site is in the midst of rich fishing grounds; only 1.6 km (1 mile) away is the famous 17 Fathoms, a prime fishing spot for bluefish, bonitos, false albacore, blackfish, and flukes and the focus of a $100 million a year industry.

Much of the sediment originates as runoff from the three rivers feeding the estuary in the bight: the Hudson, Passaic, and Raritan. But with the sediment come many metallic and organic toxic wastes from industries along the rivers, among them dioxin, one of the most toxic substances known. Primarily a by-product in the manufacture of Agent Orange (used as a defoliant in the Vietnam War), dioxin continues to be produced in the manufacture of paper and herbicides and in many other industrial processes (see Chapter 14 for a discussion of dioxin use). Because dioxin has been associated with cancer and immune deficiencies in animals and is suspected to have many other adverse health effects, dredging the dioxin-contaminated sediments became highly controversial in the early 1990s.

A request by the Port Authority of New York to dredge and dump 383,000 m^3 (500,000 yd^3) of contaminated sediment at the Mud Dump site was held up while questions regarding the impact on the marine environment and human health were debated. Disposal of moderately contaminated sediments in the ocean requires that they be capped, or covered with noncontaminated materials, to attempt to isolate the toxic materials from the marine environment (see drawing). Highly contaminated materials, which pose a significant health and environmental risk, are not allowed to be dumped in the ocean. Differing opinions as to the lower and upper limits of moderate dioxin contamination complicated the decision on disposal of the dredged materials. Concentrations greater than 4 parts per trillion (pptr) would have required capping, and ocean disposal was prohibited for concentrations above 25 pptr. Some experts felt these limits were too liberal, but others pointed out that Europe and Canada accepted even higher levels.

Major federally maintained navigation channels in the port of New York and New Jersey. (*Source:* U.S. Army Corps of Engineers, New York District, December 1989, *Managing Dredged Materials,* p. 22.)

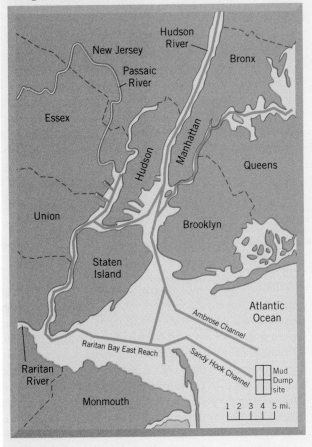

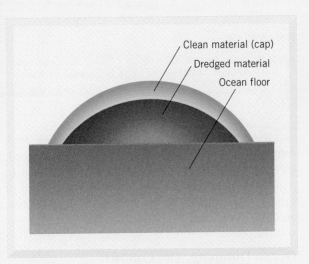

Source: U.S. Army Corps of Engineers, New York District, December 1989, *Managing Dredged Materials,* p. 74.

By 1991 scientists had reached a consensus on the mechanism by which dioxin acts on living cells and agreed on the need to question existing guidelines for safe exposure. While some scientists worked to translate the new understanding of how dioxin works into revised guidelines, others investigated its effects on marine organisms.

The average dioxin concentration varies for different parts of Newark Bay, from 39.4 to 110.6 pptr. Sediments at the Mud Dump site average 24.1 pptr, and the average level in sandworms there is 27 pptr of dry weight. Dioxin is highly fat soluble and accumulates in fatty tissue (see Chapter 14). Predators of the sandworms and other marine organisms can be expected to accumulate concentrations as much as 1000 times greater than those in their prey. As dioxin is passed up a food chain, concentrations in fatty tissues of fish can be as much as 10,000 times those in the surrounding water.

Alternative	Estimated Cost per Cubic Yard	Estimated Capacity (MCY)[a]	Ecological Effects
Ocean disposal: Mud Dump site	$5.00	large volumes	
Land disposal	$5.60–$13.20	2.6–8.5	
Creation of special 500-acre containment island	$10.90–$36.10	6.6–28.2	

Source: U.S. Army Corps of Engineers, New York District, December 1989, *Managing Dredged Material*, p. 95.

[a]MCY = million cubic yards.

Critical Thinking Questions

1. In an attempt to study the effectiveness of capping, 900,000 m³ (1.2 x10⁶ yd³) of clean sediment was used to cap 383,000 m³ (500,000 yd³) of contaminated material in an area of the Mud Dump site in 1980. Six years later the cap was estimated to contain only 612,000 m³ (800,000 yd³), and 99,000 m³ (130,000 yd³) of material was found outside the site. Assuming that all the measurements were accurate, how can you explain the discrepancy? What ecological effects might occur as a result of your explanation?

2. Complete the following table by filling in the Ecological Effects column. Evaluate each of the alternatives. Which alternative do you favor? Why?

3. How does the idea of the tragedy of the commons apply to the problem of ocean disposal of contaminated materials? (See Chapter 25 for a discussion of the tragedy of the commons.)

References

Bruno, K. (Feb. 23)1993. "E.P.A. Plays a Dioxin Numbers Game in Newark Bay Dredging," *New York Times*, letter, p. A20.

Cernadas, A. 1993 (Feb. 10). "Why Dredging in Newark Bay Must Proceed," *New York Times*, letter, p. A22.

Clark, S. L. 1993 (March 30). Testimony of the Environmental Defense Fund before the Subcommittee on Oceanography, Gulf of Mexico, and the Outer Continental Shelf of the Merchant Marine and Fisheries Commission, House of Representatives.

Muszynski, W. J. and J. A. Capo. 1993 (Feb.16). "An Awful Problem Gets Dredged Up," *Wall Street Journal,* letter, p. A15.

Schmidt, K. F. 1992. "Dioxin's Other Face," *Science News* 141(2): 24–27.

Strum, C. 1993 (Jan. 27). "U.S. Suspends Dredging Permit for Newark Bay," *New York Times*, pp. B1, B4.

Strum, C. 1993 (Jan. 30). "From Port to 17 Fathoms, Views on Dredging Differ," *New York Times*, pp. 23, 27.

Strum, C. 1993 (Feb. 13). "Limited Dredging Allowed in Newark Bay," *New York Times*, p. 24.

U.S Army Corps of Engineers, New York District. 1989 (Dec.). *Managing Dredged Material*. New York: Author.

U.S. Environmental Protection Agency, Environmental Research Laboratory, Office of Research and Development. 1993 (March). *Interim Report on Data and Methods for Assessment of 2,3,7,8-Tetrachlorodibenzo-p-dioxin Risks to Aquatic Life and Associated Wildlife*. Duluth, MN: U.S. Environmental Protection Agency.

is a loss of visual appeal and other amenities. Economic loss is considerable: Not only does loss of shellfish from pollution in the United States amount to many millions of dollars per year but a great deal of money is spent cleaning up solid waste, liquid waste, and other pollutants in coastal areas.[31]

Ocean Dumping: The Conflict

The ultimate solution to the problem of ocean dumping is the development of economically feasible and environmentally safe alternatives. Difficulties in doing this arise because ocean areas near the shore are both most desirable as fisheries and most subject to ocean dumping. The city of Los Angeles provides a classic example of the pollution of near-shore waters.

For more than 30 years, sewage and sewage sludge have been dumped several miles offshore into Santa Monica Bay. The rate of flow in recent years has been about 1.5 billion liters (400 million gallons) per day, only about 25% of which receives secondary treatment. Los Angeles successfully fought state and federal regulations for 14 years to avoid providing secondary treatment for all sewage.

It is now recognized that the bay is seriously polluted by the sewage and by earlier waste disposal dating to the 1940s, when oil refinery wastes, cyanide, and PCBs were introduced into the marine environment. Concerns over potential health-related issues have convinced local government officials that all the sewage should receive secondary treatment. The Los Angeles City Council's decision to spend an additional $172 million for secondary treatment ended a long environmental battle.

Dredge Spoils

Disposal of dredge spoils (sediment and any material incorporated within the sediment) represents the vast majority of all ocean dumping. Dredging is done primarily to improve navigation, and the spoils are usually disposed of (dumped) only a few kilometers away. Approximately one-third is seriously polluted with heavy metals, such as cadmium, chromium, lead, and nickel, as well as with other industrial, municipal, and agricultural wastes. As a result, disposal in the marine environment can be a significant source of pollution.[31]

Although the long-range alternative to disposal of dredge spoils is to phase out ocean disposal, it is not currently possible to do so because of the great volume of sediment involved. Until land-based disposal can handle the necessary volume, interim techniques should be developed. For instance, polluted spoils can be identified before disposal and then hauled farther from the dredging site to a safe disposal site on land. The main disadvantages are the increased cost of longer hauls and the potential pollution of soil and groundwater resources if the landfill should fail.[31]

SUMMARY

- The history of waste disposal practices since the industrial revolution has progressed from a practice of dilution and dispersion to a new concept of integrated waste management (IWM).

- The most common method for disposal of urban waste is the sanitary landfill. However, around many large cities, space for landfills is hard to find and few people wish to live near any waste disposal operation.

- We are headed toward a disposal crisis if new methods are not developed soon. One trend is toward more recycling or incineration of waste, processes that reduce the volume of material sent to landfills.

- Physical and hydrologic conditions of a site greatly affect its suitability as a landfill. These include landform, topography, rock and soil type, depth to groundwater, and amount of precipitation.

- Hazardous chemical waste management is one of the most serious environmental problems in the United States. Hundreds or even thousands of uncontrolled disposal sites could be time bombs that will eventually cause serious public health problems.

- We know that we will continue to produce some hazardous chemical wastes. Therefore, it is imperative to develop and use safe disposal methods.

- Management of hazardous chemical wastes includes several options, such as on-site processing to recover by-products with commercial value; microbial breakdown; chemical stabilization; incineration and disposal by secure landfill; and deep-well injection.

- Ocean dumping can be a significant source of marine pollution, and efforts to control indiscriminate dumping are in effect. Alternatives to ocean dumping of materials such as polluted dredge spoils are being developed, but in many cases such alternatives are not yet practical or economically feasible.

REEXAMINING THEMES AND ISSUES

Human Population: Waste management strategies are inextricably linked to human population. As population increases, so does the waste generated because the total amount of waste is the product of *per-capita* waste generation times the population. Furthermore, in developing countries where population increase is the most dramatic, increase in industrial output, when linked to poor environmental control, produces or aggravates waste management problems.

Sustainability: The objective of providing for a quality environment for future generations is closely linked to waste management. Of particular importance here are the concepts of integrated waste management and industial ecology so that waste generated in one sector of the economy is considered a resource for another sector. Carried to its natural conclusion, the concept of waste would no longer be present but becomes an issue of resource management.

Global Perspective: Management of waste is becoming a global problem; inappropriate management of waste contributes to air and water pollution, with the potential of causing environmental disruption at a regional or global scale. For example, waste generated from large inland cities and disposed of in river systems may eventually enter the oceans, where it may become dispersed by the global circulation patterns of ocean currents. Similarly, soils polluted by hazardous materials may erode and the particles enter the atmosphere or water systems, to be dispersed widely.

Urban World: Because so much of our waste is generated in the urban environment, it is a focus of special attention for waste management. Where population densities are high, it is easier for the principles behind reduce, reuse, and recycle to be implemented. There are greater financial incentives for management of waste where they are more concentrated.

Values and Knowledge: People today value a quality, pollution-free environment. Past activities related to waste management have contributed to health and other environmental problems; an understanding of these problems has resulted in a considerable amount of work and research aimed at reducing or eliminating the impact of how we manage our waste. We are also more conscious of environmental justice issues related to how we deal with waste management.

KEY TERMS

composting 577	industrial ecology 575	reduce 575
deep-well disposal 585	integrated waste management (IWM) 575	reuse 575
environmental audit 583	land application 584	sanitary landfill 578
environmental justice 579	leachate 579	secure landfill 583
hazardous waste 582	monitoring 579	source reduction 586
incineration 577	recycling 575	surface impoundment 585

STUDY QUESTIONS

1. Have you ever contributed to the hazardous-waste problem through disposal methods practiced in your home, school laboratory, or other location? If so, how big a problem do you think such events are? For example, how bad is it to dump paint thinner down a drain?

2. Why is it so difficult to ensure safe land disposal of hazardous waste?

3. Would you approve the siting of a waste disposal facility in your part of town? If not, why not, and where do you think such facilities should be sited?

4. Why does there seem to be a trend toward on-site disposal rather than land disposal of hazardous waste? Consider physical, biological, social, legal, and economic aspects of the question.

5. Is government doing enough to clean up abandoned hazardous waste dumps? Do private citizens have a role in choosing where cleanup funds should be allocated?

6. Considering how much waste has been dumped in the near-shore marine environment, how safe is it to swim in bays and estuaries near large cities?

7. Do you think we should collect household waste and burn it in special incinerators to make electrical energy? What problems and what advantages do you see compared with other disposal options?

8. Lots of jobs will be available in the next few years in the field of hazardous-waste monitoring and disposal. Would you take such a job? If not, why not? If yes, do you feel secure that your health will not be jeopardized?

9. Should companies that dumped hazardous waste years ago when the problem was not understood or recognized be held liable today for health-related problems their dumping may have contributed to?

10. Suppose you found that the home you had been living in for 15 years is located over a buried waste disposal site. What would you do? What kinds of studies could be done to evaluate the potential problem?

FURTHER READING

Anderson, D. D., and Burnham, L. 1992. "Toward Sustainable Waste Management," *Issues in Science and Technology* 9(1): 65–72. The authors see technological advances, free-market economics, and social consciousness as solutions to current solid-waste disposal problems.

Blumberg, L., and Gottlieb, R. 1989. *War on Waste: Can America Win Its Battle with Garbage?* Washington, DC: Island Press. A broad discussion of the social, economic, and political aspects of the system of throwaway consumption.

Kovacs, W. L. 1993. "Solid Waste Management: Historical and Future Perspectives," *Resources, Conservation, and Recycling* 8: 113–130. According to the author, if this nation cannot preserve and increase its solid-waste disposal capacity, it will gradually cease to be a competitive industrial nation, which will mean it will be less able to protect itself and its environment.

Kreith, F. (ed.). 1994. *Handbook of Solid Waste Management*. New York: McGraw-Hill. Thorough coverage of municipal waste management including waste characteristics, federal and state legislation, source reduction, recycling, and landfilling.

Rhyner, C. R., Schwartz, L. J., Wenger, R. B., and Kohrell, M. G. 1995. *Waste Management and Resource Recovery*. Boca Raton, FL: CRC, Lewis. Up-to-date discussions on the archeology of waste, waste generation, source reduction and recycling, wastewater treatment, incineration and energy recovery, hazardous waste, and costs of waste systems and facilities.

Zirm, K. L., and Mayer, J. 1990. *The Management of Hazardous Substances in the Environment*. New York: Elsevier Applied Science.

INTERNET RESOURCES

U.S. Environmental Protection Agency Office of Emergency and Remedial Response, Superfund page: *http://www.epa.gov/superfnd/*—Find out the current status of Superfund, details of its many programs and goals, a current list of Superfund sites, the Superfund budget, and more.

U.S. Environmental Protection Agency Office of Solid Waste and Emergency Response: *http://www.epa.gov/epaoswer/*—Check out what programs the EPA has for solid waste prevention and reduction both for corporate and private America. Read facts and figures on the volume of solid waste generated and recycled in the U.S., and on hazardous waste generation; also read examples of how corporations have saved money and resources through waste generation reduction.

Open pit iron mine at Para, Brazil.

CHAPTER 28

MINERALS AND THE ENVIRONMENT

CASE STUDY

Palo Alto Golden Sludge

It was discovered in the 1970s that ash from the incineration of sewage sludge in Palo Alto, California, contained large concentrations of gold (30 ppm), silver (660 ppm), and copper (8000 ppm). Each metric ton of the ash contained approximately 1 troy ounce of gold and 20 oz of silver. The concentration of gold was 7500 times greater than naturally occurring concentrations; thus the concentration of gold in the sewage sludge was well above that found in the average grade of rock mined for gold. Silver was present in the ash in a concentration similar to that of rich ore deposits in Idaho. Copper was concentrated in the ash by a factor of 145, similar to that of a common ore

grade. The ash in the Palo Alto dump represented a silver and gold deposit with a 1980 value of about $10 million! Once this resource was recognized, gold and silver worth approximately $2 million were concentrated and delivered each year for several years.[1] By 1991 pretreatment standards had improved and the amount of gold and silver was reduced somewhat. Each metric ton of ash then contained approximately 0.4 oz of gold and 15 oz of silver. The market value of the metals (which are extracted by a private company under contract with the city) was approximately $400,000 per year based on the production of about 4 metric tons of ash per day.

The most likely sources of the metals in the Palo Alto sewage are the large electronics and photographic industries located in the area. Gold in significant amounts has been found in the sewage of only one other city, and silver is usually present in much smaller concentrations than at Palo Alto. By the mid 1990s, industry in Palo Alto was treating their wastewater and recovering nearly all the gold and silver. As a result, the city shut down their recovery program.[1]

The urban ore of Palo Alto is an unusual mineral deposit, but so are many naturally occurring mineral deposits. This chapter discusses the origin of mineral deposits as well as environmental consequences of mineral development.

Learning Objectives

Modern society depends on the availability of mineral resources, which can be considered a nonrenewable heritage from the geologic past. After reading this chapter, you should understand:

- Why the standard of living in modern society is related in part to the availability of natural resources.
- Why minerals are not uniformly distributed throughout Earth's crust.

- What processes are responsible for the distribution of mineral deposits.
- What the differences are between mineral resources and reserves.
- What factors control the environmental impact of mineral exploitation.
- How wastes generated from the use of mineral resources affect the environment.
- What the social impacts of mineral exploitation are.

28.1 THE IMPORTANCE OF MINERALS TO SOCIETY

Modern society depends on the availability of mineral resources.[2] Many mineral products are found in a typical American home (see Table 28.1). Consider your breakfast this morning. You probably drank from a glass made primarily of sand, ate food from dishes made from clay, flavored your food with salt mined from Earth, ate fruit grown with the aid of fertilizers such as potassium carbonate (potash) and phosphorus, and used utensils made from stainless steel, which comes from processing iron ore and other minerals.

Minerals are so important to people that the standard of living increases with the availability of

---

TABLE 28.1 Mineral Products in a Typical U.S. Home

Building materials: sand, gravel, stone, brick (clay), cement, steel, aluminum, asphalt, glass

Plumbing and wiring materials: iron and steel, copper, brass, lead, cement, asbestos, glass, tile, plastic

Insultaing materials: rock wool, fiberglass, gypsum (plaster and wallboard)

Paint and wallpaper: mineral pigments (such as iron, zinc, and titanium) and fillers (such as talc and asbestos)

Plastic floor tiles, other plastics: mineral fillers and pigments, petroleum products

Appliances: iron, copper, and many rare metals

Furniture: synthetic fibers made from minerals (principally coal and petroleum products); steel springs; wood finished with rottenstone polish and mineral varnish

Clothing: natural fibers grown with mineral fertilizers; synthetic fibers made from minerals (principally coal and petroleum products)

Food: grown with mineral fertilizers; processed and packaged by machines made of metals

Drugs and cosmetics: mineral chemicals

Other items: windows, screens, light bulbs, porcelain fixtures, china utensils, jewelry all made from mineral products

Source: U.S. Geological Survey, 1975, U.S. Geological Survey Professional Paper 940.

minerals in useful forms. The availability of mineral resources is one measure of the wealth of a society. Those who have been successful in locating and extracting or importing and using minerals have grown and prospered. Without mineral resources to grow food, construct buildings and roads, and manufacture everything from the tape recorder and computer I am writing this with to televisions and automobiles, modern technological civilization as we know it would not be possible. In maintaining our standard of living in the United States, every person requires about 10 tons of nonfuel minerals per year.[3]

A Limited Supply

Minerals can be considered our nonrenewable heritage from the geologic past. Although new deposits are still forming from present Earth processes, these processes are producing new mineral deposits too slowly to be of use to us today. Because mineral deposits are generally located in small, hidden areas, they must be discovered. Unfortunately, most of the easy-to-find deposits have been exploited; if modern civilization were to vanish, our descendants would have a harder time discovering rich mineral deposits than we did. It is interesting to speculate that they might mine landfills for metals thrown away by our civilization. Unlike biological resources, minerals cannot be easily managed to produce a sustained yield; the supply is finite. Recycling and conservation will help, but eventually the supply will be exhausted.

28.2 HOW MINERAL DEPOSITS ARE FORMED

Metals in mineral form are generally extracted from naturally occurring, anomalously high concentrations of Earth materials. When metals are concentrated in anomalously high amounts by geologic processes (or by urban processes, such as in Palo Alto), **ore deposits** are formed. The discovery of natural ore deposits allowed early peoples to exploit copper, tin, gold, silver, and other metals while slowly developing skills in working with metals.

The origin and distribution of mineral resources is intimately related to the history of the biosphere and to the entire geologic cycle (see Chapter 4). Nearly all aspects and processes of the geologic cycle are involved to some extent in producing local concentrations of useful materials.

Earth's outer layer, or crust, is silica rich, made up mostly of rock-forming minerals containing silica, oxygen, and a few other elements. The elements are not evenly distributed in the crust: Eight elements account for over 99% by weight (oxygen, 46.4%; sili-

con, 28.2%; aluminum, 8.2%; iron, 5.6%; calcium, 4.2%; sodium, 2.4%; potassium, 2.1%; and titanium, 0.6%). Remaining elements are found (on the average) in trace concentrations. The geologic cycle occasionally concentrates elements in a local environment to a greater degree than average.

The ocean, covering nearly 71% of Earth, is another reservoir for many materials. Most elements in the ocean have been weathered from crustal rocks on the land and transported to the oceans by rivers. Other elements are transported to the ocean by wind or glaciers. Ocean water contains about 3.5% dissolved solids, most of which is chlorine (55.1% by weight). Each cubic kilometer of ocean water contains about 2.0 metric tons of zinc, 2.0 metric tons of copper, 0.8 metric ton of tin, 0.3 metric ton of silver, and 0.01 metric ton of gold. These concentrations are low compared with those in the crust, where corresponding values (in metric tons/km³) are: zinc, 170,000; copper, 86,000; tin, 5700; silver, 160; and gold, 5. If the rich crustal ore deposits are depleted, we will be more likely to extract metals from lower grade ore deposits or even from common rock than from ocean water. On the other hand, if mineral extraction technology becomes more efficient, this prognosis could change.

Why are there local concentrations of minerals? Planetary scientists now believe that Earth, like the other planets in the solar system, formed by condensation of matter surrounding the sun. Gravitational attraction brought together the matter dispersed around the forming sun. As the mass of the proto-Earth increased, the material condensed and was heated by the process. The heat was sufficient to produce a molten liquid core, consisting primarily of iron and other heavy metals, which sank toward the center. The crust formed of generally lighter elements and is a mixture of many different elements. It does not have a uniform distribution of elements because geologic processes and some biological processes selectively dissolve, transport, and deposit elements and minerals.

Plate Boundaries

Plate tectonics is responsible for the formation of some mineral deposits. According to the theory of plate tectonics (see Chapter 4), the continents (which are crustal rocks and part of the lithosphere) are composed mostly of relatively light rocks. As the tectonic plates of the lithosphere slowly move across Earth's surface, so do the continents. Metallic ores are thought to be deposited in the crust both where the tectonic plates separate, or diverge, and where they come together, or converge. At divergent plate boundaries cold ocean water comes in contact with hot molten rock. The heated water is lighter and more active chemically. It rises through fractured rocks and leaches metals from them. The metals are carried in solution and deposited as metal sulfides when the water cools.

At convergent plate boundaries, rocks saturated with seawater are forced together, heated, and subjected to intense pressure, which causes partial melting. The combination of heat, pressure, and partial melting mobilizes metals in the molten rocks, or magma. Most major mercury deposits, for example, are associated with the volcanic regions that occur close to convergent plate boundaries. Geologists believe that the mercury is distilled out of the tectonic plate as the plate moves downward; as the plate cools, the mercury migrates upward and is deposited at shallower depths, where the temperature is lower.

Igneous Processes

Ore deposits may form when magma cools. As the molten rock cools, heavier minerals that crystallize (solidify) early may slowly sink or settle toward the bottom of the magma, whereas lighter minerals that crystallize later are left at the top. Deposits of an ore of chromium, called chromite, are thought to be formed in this way. When magma containing small amounts of carbon is deeply buried and subjected to very high pressure during slow cooling (crystallization), diamonds (which are pure carbon) may be produced (Figure 28.1).[4]

Hot waters moving within the crust are perhaps the source of most ore deposits. It is speculated that circulating groundwater is heated and enriched with minerals on contact with deeply buried rocks. This water then moves up or laterally to other, cooler rocks, where the cooled water deposits the dissolved minerals.[5]

Sedimentary Processes

Sedimentary processes often concentrate materials in amounts sufficient for extraction. As sediments are transported, running water and wind help segregate the sediments by size, shape, and density. Thus, the best sand or sand and gravel deposits for construction purposes are those in which the finer materials have been removed by water or wind. Sand dunes, beach deposits, and deposits in stream channels are good examples. The sand and gravel industry amounts to about $3 billion annually, and in terms of the total volume of materials mined it is one of the largest nonfuel mineral industries in the United States.[3]

Stream processes transport and sort all types of materials according to size and density. Therefore, if the bedrock in a river basin contains heavy metals, such as gold, streams draining the basin may concentrate the metals in areas where there is less water turbulence or velocity. These concentrations, called placer deposits, are often found in open crevices or fractures at the bottoms of pools, on the inside

FIGURE 28.1 Diamond mine near Kimberley, South Africa. This is the largest hand-dug excavation in the world.

curves of bends, or on riffles, where shallow water flows over rocks. Placer mining of gold (which was known as a poor man's method because a miner needed only a shovel, a pan, and a strong back to work the streamside claim) played an important role in the settling of California, Alaska, and other areas of the United States. The gold in California attracted miners who acquired the expertise necessary to locate and develop other resources in the western states and Alaska.

Rivers and streams that empty into the oceans and lakes carry tremendous quantities of dissolved material derived from the weathering of rocks. Over geologic time, a shallow marine basin may be isolated by tectonic activity that uplifts its boundaries. In other cases, climatic variations, such as the ice ages, produce large inland lakes with no outlets. Both types eventually dry up. As evaporation progresses, the dissolved materials precipitate (drop out of solution), forming a wide variety of compounds, minerals, and rocks that have important commercial value. Most of these *evaporites* (deposits originating by evaporation) can be grouped into one of three types[6]:

marine evaporites (solids)—potassium and sodium salts, gypsum, and anhydrite.

nonmarine evaporites (solids)—sodium and calcium carbonate, sulfate, borate, nitrate, and limited iodine and strontium compounds.

brines (liquids derived from wells, thermal springs, inland salt lakes, and seawaters)—bromine, iodine, calcium chloride, and magnesium.

Heavy metals (such as copper, lead, and zinc) associated with brines and sediments in the Red Sea, Salton Sea, and other areas are important resources that may be exploited in the future.

Evaporite minerals are widely used in industrial and agricultural activities, and their annual value is about $1 billion.[6] Evaporite and brine resources in the United States are substantial, ensuring an ample supply for many years.

Biological Processes

Some mineral deposits are formed by biological processes; many are formed under conditions of the biosphere that have been greatly altered by life. Examples include phosphates (discussed in Chapter 4) and iron ore deposits.

The major iron ore deposits exist in sedimentary rocks that were formed more than 2 billion years ago.[7] There are several types of iron deposits. Gray beds, an important type, contain unoxidized

iron. Red beds contain oxidized iron (the red color is the color of iron oxide). Gray beds formed when there was relatively little oxygen in the atmosphere, and red beds formed when there was relatively more oxygen. Although the processes are not completely understood, it appears that major deposits of iron stopped forming when the atmospheric concentration of oxygen reached its present level. This phenomenon suggests that early life was important in beginning and ending the ore-forming processes for iron.[8]

Organisms are able to form many kinds of minerals, such as the calcium minerals in shells and bones. Some of these minerals cannot be formed inorganically in the biosphere. Thirty-one different biologically produced minerals have been identified. Minerals of biological origin contribute significantly to sedimentary deposits.[9]

Weathering Processes

Weathering (chemical and mechanical decomposition of rock) concentrates some minerals in the soil. When insoluble ore deposits, such as native gold, are weathered from rocks, they may accumulate in the soil unless removed by erosion. Accumulation occurs most readily when the parent rock is relatively soluble, as is limestone. Intensive weathering of certain soils derived from aluminum-rich igneous rocks may concentrate oxides of aluminum and iron. (The more soluble elements, such as silica, calcium, and sodium, are selectively removed by soil and biological processes.) If sufficiently concentrated, residual aluminum oxide forms an ore of aluminum known as bauxite. Important nickel and cobalt deposits are also found in such soils developed from iron- and magnesium-rich igneous rocks.

Weathering produces sulfide ore deposits from low-grade primary ore through **secondary enrichment** processes. Near the surface, primary ore containing minerals such as iron, copper, and silver sulfides is in contact with slightly acidic soil water in an oxygen-rich environment. As the sulfides are oxidized, they dissolve, forming solutions that are rich in sulfuric acid and silver and copper sulfate. These solutions migrate downward, producing a leached zone devoid of ore minerals (Figure 28.2). Below the leached zone and above the groundwater table, oxidation continues, and sulfate solutions continue their downward migration. Below the water table, if oxygen is no longer available, the solutions are deposited as sulfides, enriching the metal content of the primary ore by as much as 10 times. In this way, low-grade primary ore is rendered more valuable, and high-grade primary ore is made even more attractive.[10,11]

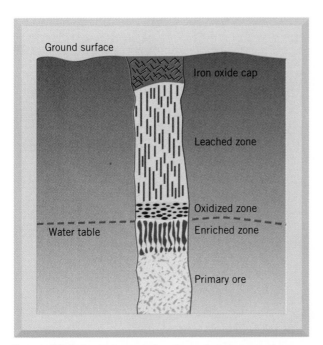

FIGURE 28.2 Typical zones that form during secondary enrichment processes. Sulfide ore minerals in the primary ore vein are oxidized and altered and then are leached from the oxidized zone by descending groundwater and redeposited in the enriched zone. The iron oxide cap is generally a reddish color and may be helpful in locating ore deposits that have been enriched. (*Source:* R. J. Foster, 1983, *General Geology,* 4th ed., Charles E. Merrill, Columbus, OH.)

Several disseminated copper deposits are concentrated to economic levels because of secondary enrichment, which concentrates dispersed metals. For example, secondary enrichment of a disseminated copper deposit at Miami, Arizona, increased the grade of the ore from less than 1% copper in the primary ore to as much as 5% in some areas.[10]

Minerals from the Sea

Mineral resources in seawater or on the bottom of the ocean are vast and, in some cases, such as magnesium, nearly unlimited.

Magnesium

In the United States, magnesium was first extracted from seawater in 1940. By 1972, one company in Texas, utilizing seawater as its raw material source, produced 80% of domestic magnesium. Now companies in Alabama, California, Florida, Mississippi, and New Jersey are also extracting magnesium from seawater.

The deep-ocean floor may be the site of the next big mineral rush. Two types of deposits have been identified: massive sulfide deposits associated with hydrothermal vents and manganese oxide nodules.

Sulfide Deposits

The massive sulfide deposits, containing zinc, copper, iron, and trace amounts of silver, are produced at divergent plate boundaries by the processes of plate tectonics. Cold seawater enters rock fractures at midocean ridges, where it is heated by upwelling magma and emerges as hot springs. The circulating water leaches the rocks, removing metals that are deposited where the hot, mineral-rich water is ejected at temperatures up to 350°C (660°F) into the cold sea. Sulfide minerals precipitate, or drop out of solution, near vents known as *black smokers* because of the color of the ejected mineral-rich water and form massive towerlike formations rich in metals (Figure 28.3). The hot vents are of particular biological significance because they support a unique assemblage of animals (discussed in Chapter 8), including giant clams, tube worms, and white crabs. Communities of these animals are based on sulfide compounds extruded from black smokers and exist through a process called chemosynthesis, as opposed to photosynthesis, which supports all other known ecosystems on Earth.

The extent of sulfide mineral deposits along oceanic ridges is poorly known, and although mining is a possibility and leases to some possible deposits are being considered, it seems unlikely that such deposits will be extracted at a profit in the near future. The potential environmental effects, including serious physical and biological impacts associated with water quality and sediment pollution, will have to be carefully evaluated prior to any mining activity.

Manganese Oxide Nodules

These nodules contain manganese (24%) and iron (14%) with secondary copper (1%), nickel (1%), and cobalt (0.25%) and cover vast areas of the deep-ocean floor. Some nodules are found in the Atlantic Ocean off Florida, but the richest and most extensive accumulations occur in large areas of the northeastern, central, and southern Pacific, where the nodules cover 20% to 50% of the ocean floor.[12]

The average size of the manganese nodules varies from a few millimeters to a few tens of centimeters in diameter. Composed primarily of concentric layers of manganese in iron oxides mixed with a variety of other materials, each nodule originally formed around a nucleus of broken nodules, fragments of volcanic rock, and sometimes fossils. The estimated rate of growth is 1–5 mm (about 0.04–0.2 in.) per million years. The nodules are most abundant in those parts of the ocean where sediment accumulation is at a minimum, generally at depths of 2500–6000 m (8200–19,700 ft).[13]

The origin of the nodules is not well understood. Most likely, material weathers from the conti-

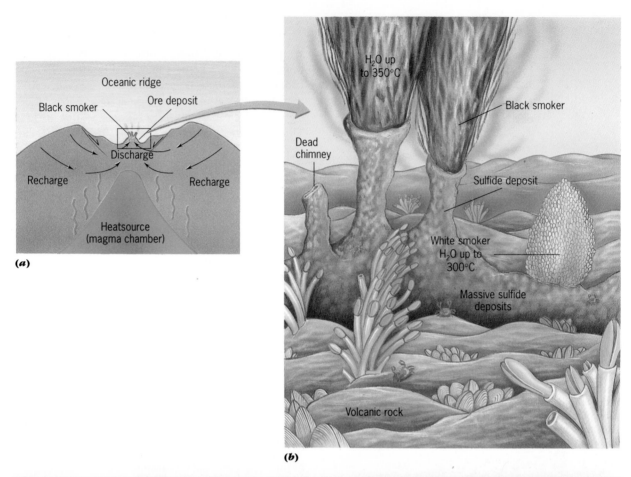

FIGURE **28.3** (*a*) Oceanic ridge hydrothermal environment. (*b*) Detail of black smokers, where massive sulfide deposits form. (*c*) Black smoker. Photographed from minisubmarine *Alvin* during 1991 cruise to the East Pacific Rise.

nents and is transported in solution by rivers to the oceans, where it drops out of solution and forms nodules. Both inorganic precipitation and bacteria-induced precipitation are probably important in the formation of nodules.

Expenditures for mining and metallurgical research to recover the nodules have surpassed several hundred million dollars. At least 20 corporations in several countries have examined metallurgical systems to process the nodules. Some would produce cobalt, copper, nickel, and manganese, and others would produce combinations of only copper and nickel.[14] However, commercial exploitation of the nodules is not expected to be economically and technically viable until well into the 21st century.[15]

Law of the Seabed

The mining of the seabed generates a lot of controversy. On the one hand, there are environmental concerns about disrupting and degrading the bottom and near-bottom environments. On the other hand, the potential exploitation of seabed resources raises questions about who has the right to do the mining. Developing countries argue that the mineral wealth on the seabed is a common heritage to all people and should not just benefit the wealthy nations that can exploit the resource. International consortia of companies who have spent the hundreds of millions of dollars want a return on their investments. The United Nations has sponsored Law of the Seabed conferences aimed at passing international laws (treaties) concerning ocean mining. The International Law of the Sea became a reality in 1994 and

states that seabed resources outside national limits are a common heritage of mankind. Seabed resources may be exploited (with International Seabed Authority) either privately by mining companies or publicly by an international enterprise.[15]

Industrial nations have threatened to mine the seabed without an international agreement. The U.S. Congress has passed legislation authorizing U.S. companies to mine the ocean floor. Some developing countries have responded by threatening to withhold other resources, such as oil and metals, from those countries that mine the seabed without approval of the United Nations. Threats aside, mining of the seabed has raised important ethical questions that must be addressed by all countries interested in the oceans' resources.

28.3 RESOURCES AND RESERVES

Mineral resources are broadly defined as elements, chemical compounds, minerals, or rocks concentrated in a form that can be extracted to obtain a usable commodity. It is assumed that a resource can be extracted economically or at least has the potential for economic extraction. A **reserve** is that portion of a resource that is identified and from which usable materials can be legally and economically extracted *at the time of evaluation* (Figure 28.4). Whether a mineral deposit is classified as part of the resource base or as a reserve may be a question of economics. For example, if an important metal becomes scarce, the price may rise, which would en-

FIGURE 28.4 Classification of mineral resources used by the U.S. Geological Survey and the U.S. Bureau of Mines. (*Source: Principles of a Resource Preserve Classification for Minerals,* U.S. Geological Survey Circular 831, 1980.)

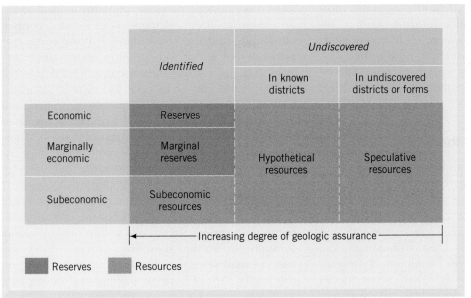

CLOSER LOOK 28.1

SILVER

Silver provides an example that illustrates some important points about resources and reserves. Earth's crust (to a depth of 1 km, or 0.6 miles), based on geochemical estimates of the concentration of silver in rocks, contains almost 2×10^{12} metric tons of silver, an amount much larger than annual world use, which is approximately 10,000 metric tons. If this silver existed as pure metal concentrated in one large mine, it would represent a supply sufficient for several hundred million years at current levels of use. Most of the silver, however, exists in extremely low concentrations, too low to be extracted economically with current technology. The known reserves of silver, reflecting the amount we could obtain immediately with known techniques, is about 200,000 metric tons, or a 20-year supply at current-use levels.

The problem with silver, as with all mineral resources, is not with its total abundance but with its concentration and relative ease of extraction. When an atom of silver is used, it is not destroyed but remains an atom of silver. It is simply dispersed and may become unavailable. In theory, given enough energy, all mineral resources could be recycled, but this is not possible in practice. Consider lead, which is mined from minerals in which it is concentrated. The lead that was used in gasoline for many years is now scattered along highways across the world and deposited in low concentrations in forests, fields, and salt marshes close to these highways. Recovery of this lead is, for all practical purposes, impossible.

courage exploration and extraction (mining). As a result of the price increase, previously uneconomic deposits (part of the resource base before the scarcity and price rise) may become profitable, and those deposits would be reclassified as reserves.

The main point about resources and reserves is that *resources are not reserves.* An analogy from a student's personal finances may help clarify this point. A student's reserves are the liquid assets, such as money in the pocket or bank, whereas the student's resources include the total income the student can expect to earn during his or her lifetime. This distinction is often critical to the student in school because resources that may become available in the future cannot be used to pay this month's bills.[16]

Regardless of potential problems, it is important for planning to estimate future resources, requiring a continual reassessment of all components of a total resource by considering new technology, the probability of geologic discovery, and shifts in economic and political conditions.[2] (See A Closer Look 28.1, "Silver".)

Availability of Mineral Resources

Earth's mineral resources can be divided into several broad categories, depending on our use of them: elements for metal production and technology; building materials; minerals for the chemical industry; and minerals for agriculture. Metallic minerals can be classified according to their abundance. The abundant metals include iron, aluminum, chromium, manganese, titanium, and magnesium. Scarce metals include copper, lead, zinc, tin, gold, silver, platinum, uranium, mercury, and molybdenum.

Some mineral resources, such as salt (sodium chloride), are necessary for life. Primitive peoples traveled long distances to obtain salt when it was not locally available. Other mineral resources are desired or considered necessary to maintain a certain level of technology.

When we think about mineral resources, we usually think of the metals, but, with the exception of iron, the predominant mineral resources are not metallic. Consider the annual world consumption of a few selected elements. Sodium and iron are used at a rate of approximately 100–1000 million metric tons per year. Nitrogen, sulfur, potassium, and calcium are used at a rate of approximately 10 to 100 million metric tons per year, primarily as soil conditioners or fertilizers. Elements such as zinc, copper, aluminum, and lead have annual world consumption rates of about 3–10 million metric tons, and gold and silver have annual consumption rates of 10,000 metric tons or less. Of the metallic minerals, iron makes up 95% of all the metals consumed, and nickel, chromium, cobalt, and manganese are used mainly in alloys of iron (as in stainless steel). Thus, with the exception of iron, the nonmetallic minerals are consumed at much greater rates than are elements used for their metallic properties.

The basic issue with mineral resources is not actual exhaustion or extinction but the cost of maintaining an adequate stock within an economy through mining and recycling. At some point, the costs of mining exceed the worth of material. When the availability of a particular mineral becomes a limitation, there are four possible solutions:

1. find more sources;

2. recycle what has already been obtained;

3. find a substitute; or

4. do without.

Which choice or combination of choices is made depends on social, economic, and environmental factors.

The availability of a mineral resource in a certain form, in a certain concentration, and in a certain amount is a geologic issue determined by Earth's history. What is considered a resource and at what point a resource becomes limited are ultimately social questions. Before metals were discovered, they could not be considered resources. Before smelting was invented, the only metal ores were those in which the metals appeared in their pure form. For example, originally gold was obtained as a pure, or native, metal. Now gold mines are deep beneath the surface, and the recovery process involves reducing tons of rock to ounces of gold.

Mineral resources are limited, which raises important questions. How long will a particular resource last? How much short-term or long-term environmental deterioration are we willing to accept to ensure that resources are developed in a particular area? How can we make the best use of available resources? These questions have no easy answers. We are now struggling with ways to better estimate the quality and quantity of resources.

Mineral Resource Use

We can use a particular mineral resource in several ways: rapid consumption, consumption with conservation, or consumption and conservation with recycling. Which option is selected depends in part on economic, political, and social criteria. Figure 28.5 shows the hypothetical depletion curves corresponding to these three options. Historically, with the exception of precious metals, rapid consumption

FIGURE 28.5 Several hypothetical depletion curves.

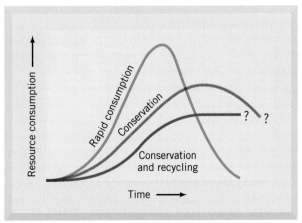

has dominated most resource utilization. However, as the supply of more resources becomes short, increased conservation and recycling are expected. Certainly the trend toward recycling is well established for metals such as copper, lead, and aluminum.

As the world population and the desire for a higher standard of living increase, the demand for mineral resources expands at a faster and faster rate. From a global viewpoint, our limited mineral resources and reserves threaten our affluence. This results, in part, because the more developed countries, with about 16% of Earth's population, consume a disproportionate amount of the mineral resources extracted. For example, the United States, western Europe, and Japan collectively use about 70% of all aluminum, copper, and nickel that is extracted from the earth.[5] Predicted increases in the world use of iron, copper, and lead, when linked with expected population increases, suggest that the rate of production of these metals would have to increase by several times if the world per-capita consumption rate were to rise to the level of developed countries today. Such an increase is very unlikely; affluent countries will have to find substitutes for some minerals or use a smaller proportion of the world's annual production. This situation parallels the Malthusian predictions discussed in Chapter 5, that it is impossible in the long run to support an ever-increasing population on a finite resource base.

U.S. Supply

Domestic supplies of many mineral resources in the United States and many other affluent nations are insufficient for current use and must be supplemented by imports from other nations. Many of the minerals we need for our complex military and industrial system, called strategic minerals (examples include bauxite, manganese, graphite, cobalt, strontium, and asbestos), are mostly imported. Of particular concern to industrial countries is the possibility that the supply of a much desired or needed mineral may be interrupted by political, economic, or military instability in the supplying nation. For example, in recent years a civil war in Zaire caused such concern that the price in cobalt temporarily increased by 800%.[17] Cobalt is a metal used to strengthen steel that is in great demand in many industrial countries.

Today the United States, along with many other countries, depends on a steady supply of imports to meet the mineral demand of industries. Of course the fact that a mineral is imported into a country does not mean that it does not exist within the country in quantities that could be mined. Rather, it suggests that there are economic, political, or environmental reasons that make it easier, more practical, or more desirable to import the material.

This situation has resulted in political alliances that otherwise would be unlikely; industrial countries often need minerals from countries with whose policies they do not necessarily agree and as a result make political concessions on human rights and other issues that they would not otherwise make.

28.4 ENVIRONMENTAL IMPACT OF MINERAL DEVELOPMENT

The impact of mineral exploitation on the environment depends on such factors as ore quality, mining procedures, local hydrologic conditions, climate, rock types, size of operation, topography, and many more interrelated factors. The impact varies with the stage of development of the resource. For example, the exploration and testing stages involve considerably less impact than do the mining and processing stages. In addition, our use of mineral resources has a significant social impact.

Exploration activities for mineral deposits vary from collecting and analyzing remote-sensing data gathered from airplanes or satellites to fieldwork involving surface mapping and drilling. Generally, exploration has a minimal impact on the environment, provided that care is taken in sensitive areas, such as arid lands, marshes, and areas underlain by permafrost. Some arid lands are covered by a thin layer of pebbles over fine silt several centimeters thick. The layer of pebbles, called desert pavement, protects the finer material from wind erosion. When the desert pavement is disturbed by road building or other activities, the fine silts may be eroded, impairing physical, chemical, and biological properties of

FIGURE 28.6 Aerial photograph of Bingham Canyon Copper Pit, Utah. This is one of the largest artificial excavations in the world.

the soil and possibly scarring the land for many years. Marshes and other wetlands, such as the northern tundra, are very sensitive to even seemingly small disturbances such as vehicular traffic.

The mining and processing of mineral resources generally have a considerable impact on land, water, air, and biological resources; they have a social impact as well, because of the increased demand for housing and services in mining areas. As it becomes necessary to use lower and lower grade ores, we are more often faced with the problem of how to minimize mining's negative effects on the environment. For example, there is concern about asbestos fibers in the drinking water (from Lake Superior) of Duluth, Minnesota, as a result of the disposal of iron mining waste (tailings) from mining low-grade iron ore.

A major practical issue is whether surface or subsurface mines should be developed in an area. Surface mining is cheaper but has more direct environmental effects. The trend in recent years has been away from subsurface mining and toward large, open-pit (surface) mines, such as the Bingham Canyon copper mine in Utah (Figure 28.6). The Bingham Canyon mine is one of the world's largest human-made excavations, covering nearly 8 km^2 (3 mi^2) to a maximum depth of nearly 800 m (2600 ft).

Surface mines and quarries today cover less than one-half of 1% of the total area of the United States. Even though the impact of these operations is a local phenomenon, numerous local occurrences will eventually constitute a larger problem. Environmental degradation tends to extend beyond the excavation and surface plant areas of both surface and subsurface mines. Large mining operations disturb the land by directly removing material in some areas and dumping waste in others, thus changing topography. At the very least, these actions produce severe aesthetic degradation. In addition, dust at mines may affect air resources, even though care is often taken to reduce dust production by sprinkling water on roads and on other sites that generate dust.

A potential problem associated with mineral resource development is the possible release of harmful trace elements to the environment. Water resources are particularly vulnerable to such degradation, even if drainage is controlled and sediment pollution is reduced. Surface drainage is often altered at mine sites, and runoff from precipitation (rain or snow) may infiltrate waste material, leaching out trace elements and minerals. Trace elements (cadmium, cobalt, copper, lead, molybdenum, and others), when leached from mining wastes and concentrated in water, soil, or plants, may be toxic or may cause diseases in people and other animals who drink the water, eat the plants, or use the soil. Specially constructed ponds to collect such runoff

FIGURE 28.7 Tailings from a lead, zinc, and silver mine in Colorado. White streaks on the slope are mineral deposits apparently leached from the tailings.

help but cannot be expected to eliminate all problems. The white streaks in Figure 28.7 are mineral deposits apparently leached from tailings from a zinc mine in Colorado. Similar-looking deposits may cover rocks in rivers for many kilometers downstream from some mining areas.

Groundwater may also be polluted by mining operations when waste comes into contact with slow-moving subsurface waters. Surface water infiltration or groundwater movement causes leaching of sulfide minerals that may pollute groundwater and eventually seep into streams to pollute surface water. Groundwater problems are particularly troublesome because reclamation of polluted groundwater is very difficult and expensive (see Chapter 20 for a discussion of acid mine drainage).

Physical changes in the land, soil, water, and air associated with mining directly and indirectly affect the biological environment. Kills caused by mining activity or contact with toxic soil or water are examples of direct impacts. Indirect impacts include changes in nutrient cycling, total biomass, species diversity, and ecosystem stability owing to alterations in groundwater or surface water availability or

quality. Periodic or accidental discharge of low-grade pollutants through failure of barriers, ponds, or water diversions or through breach of barriers during floods, earthquakes, or volcanic eruptions also may damage local ecological systems to some extent.

Social Impacts

Social impacts associated with large-scale mining result from a rapid influx of workers into areas unprepared for growth. Stress is placed on local services: water supplies, sewage and solid-waste disposal systems, schools, and housing. Land use shifts from open range, forest, and agriculture to urban patterns. The additional people also increase the stress on nearby recreation and wilderness areas, some of which may be in a fragile ecological balance. Construction activity and urbanization affect local streams through sediment pollution, reduced water quality, and increased runoff. Air quality is reduced as a result of more vehicles, dust from construction, and generation of power.

Adverse social impacts also occur when mines are closed; towns surrounding large mines come to depend on the income of employed miners. Closures of mines produced the well-known ghost towns in the old American West, and today the price of coal and other minerals directly affects the livelihood of many small towns. This relationship is especially evident in the Appalachian Mountain region of the United States, where closures of coal mines are taking their toll. These mine closings are partly the result of lower prices for coal and partly the result of rising mining costs. One of the reasons mining costs are rising is the increased level of environmental regulation of the mining industry. Of course, regulations have also helped make mining safer and have facilitated land reclamation. Some miners, however, believe the regulations are not flexible enough, and there is some truth to their arguments. For example, some mined areas might be reclaimed for use as farmland now that the original hills have been leveled. Regulations, however, may require the restoration of the land to its original hilly state, even though hills make inferior farmland.

28.5 MINIMIZING ENVIRONMENTAL IMPACT OF MINERAL DEVELOPMENT

Minimization of environmental impacts of mineral development requires consideration of the entire cycle of mineral resources as idealized in Figure 28.8. Inspection of this diagram reveals that many components of the cycle are related to the genera-

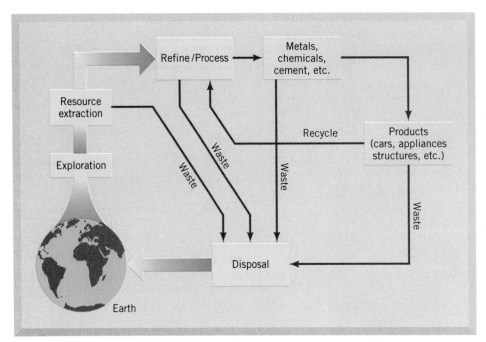

FIGURE 28.8 Simplified flowchart of the resource cycle.

tion of waste material. In fact, the major environmental impacts of mineral resource utilization are related to waste products. Waste produces pollution that may be toxic to humans; may harm natural ecosystems and the biosphere; and may be aesthetically undesirable. Waste may attack and degrade resources such as air, water, soil, and living things. Waste also depletes nonrenewable mineral resources and when simply disposed of provides no offsetting benefits for human society.

Minimization of environmental effects associated with mineral development may take several paths:

- *Environmental regulations at the federal, state, and local levels that address topics such as sediment, air, and water pollution resulting from all aspects of the mineral cycle.* Other regulations may address reclamation of land used for mining. Today in the United States approximately 50% of the land utilized by the mining industry has been reclaimed.
- *On-site and off-site treatment of waste.* Minimizing on-site and off-site problems by controlling sediment, water, and air pollution through good engineering and conservation practices is an important goal. Of particular interest is the development of biotechnological processes such as biooxidation, bioleaching, and biosorption as well as genetic engineering of microbes. These practices have enormous potential for both extracting metals and minimizing environmental degradation. For exam-

ple, engineered (constructed) wetlands are being used at several sites where acid-tolerant plants in the wetlands remove metals from mine wastewaters and neutralize acids by biological activity.[18] At the Homestake Gold Mine in South Dakota biooxidation is being used to convert contaminated water from the mining operation to substances that are environmentally safe; bacteria, with a natural capacity to oxidize cyanide to harmless nitrates, are used.[19]

- *Practicing the three R's of waste management.* That is, reduce the amount of waste produced; reuse materials in the waste stream as much as feasible; and maximize recycling opportunities.

Waste from some parts of the mineral cycle may themselves be referred to as an ore, because they contain materials that might be recycled and used again to provide energy or useful products.[20,21] That these can be valuable is clear from the case study of Palo Alto's urban ore at the beginning of this chapter.

The notion of reusing waste materials is not new, and such metals as iron, aluminum, copper, and lead have been recycled for many years. For example, in 1990 the value of recycled metals was $37 billion. Of this, recycling and reuse of iron and steel amounted to approximately 90% by volume and 68% of the total value of recycled metals. The reasons iron and steel are recycled in such large volumes are threefold[22]: First, the market for iron and steel is huge, and as a result there is a large scrap collection and processing industry; second, an enor-

Will Mining with Microbes Help the Environment?

Mining is an ancient technology practiced for at least 6500 years. Modern mining methods are more technologically sophisticated but use the same basic processes (digging and smelting) to isolate valuable metals. To be economic, these methods require high-grade ore and cheap sources of energy as well as tolerance toward damage to the environment. Although these conditions have prevailed for most of human history, they are changing. Earlier exploitation of mineral resources is pushing the mining industry toward lower grade ores; nonrenewable energy sources are expensive and disappearing; and concern over degradation of the environment and health threats to humans and other species is growing. Demand for minerals, however, is increasing, because of both population growth and technological development.

As an example, the average grade of copper ore has dropped from 6 to 0.6% over the last century, making copper mining more energy intensive and more wasteful. Five metric tons of coal is required to produce 1 metric ton of copper, and every kilogram of copper produced represents 89 kg (198 lb) of waste. Open-pit mining of copper causes acids and heavy metals, such as arsenic, to contaminate surface water and groundwater. Smelting produces sulfur dioxide and other gaseous compounds as well as particles, which contribute to air pollution.

Microscopic organisms produced by biotechnology offer an entirely new approach to mining. By 1989, more than 30% of copper mined in the United States depended on a biochemical process that begins with a microbe, *Thiobacillus ferrooxidans*. Biological processes have also been used in mining uranium and gold. Research is underway to use microbes to remove sulfur from coal and cyanide from mining waste. The union of biological processes and mining is called **biohydrometallurgy.**

In the future it may be possible to use microbes on ores without removing them from Earth. Metallurgists envision drilling wells into the ore and fracturing it, then injecting bacteria into the wells and fractures. The ore could be removed by flooding the wells with water, removing the ore, and recycling the water. Biotechnologists envision using genetic engineering to produce bacteria to mine specific metals when naturally occurring bacteria do not exist.

The disadvantages of biohydrometallurgy are that it is slow, requiring decades rather than years, and methods for breaking ores into small enough particles for efficient extraction are not yet available. Already, biological methods are economically feasible for low-grade ores that elude conventional methods, but further technological innovations may make them competitive in more situations.

Critical Thinking Questions

1. What are the environmental advantages of biohydrometallurgy over conventional methods? What are the possible disadvantages?

2. How do you assess the possible dangers from genetic engineering of organisms for mining compared to those engineered for use in agriculture and medicine?

3. Some experts believe that without economic pressure (e.g., a decline in high-grade ore; an increase in energy requirements for extracting metal; governmental regulations on clean air and water; economic recessions) the mining industry would not continue to explore biochemical methods. How could the government encourage the industry to devote more research and development efforts to expanding biochemical mining? Develop a proposal for a government policy.

Reference

Debus, K. H. 1990 (Aug./Sept.). "Mining with Microbes," *Technology Review*, 93: 50.

mous economic burden would result if recycling was not done; and third, significant environmental impacts related to disposal of over 55 million tons of iron and steel would result if we did not recycle.

Other metals that are recycled in large quantities include lead (70%), aluminum (30%), and copper (25%).[3] Recycling aluminum reduces our need to import raw aluminum ore and saves approximately 95% of the energy required to produce a new alu-

minum from bauxite.[23] It is estimated that each ton of recycled steel saves 1136 kg (2500 pounds) of iron ore, 455 kg (1000 pounds) of coal, and 18 kg (40 pounds) of limestone. In addition, only one-third as much energy is required to produce steel from recycled scrap as from native ore.[24] Finally, of the millions of automobiles that are discarded annually in the United States, the metal from 94% of them was recycled in 1993.[24]

SUMMARY

- Mineral resources are usually extracted from naturally occurring, anomalously high concentrations of earth materials. These natural deposits allowed early peoples to exploit minerals while slowly developing technological skills.

- The origin and distribution of mineral resources is intimately related to the history of the biosphere and to the entire geologic cycle. Nearly all aspects and processes of the geologic cycle are involved to some extent in producing local concentrations of useful materials.

- An important concept in analyzing resources and reserves is that resources are not reserves. Unless discovered and developed, resources cannot be used to solve present shortages.

- The availability of mineral resources is one measure of the wealth of a society. Modern technological civilization as we know it would not be possible without the exploitation of mineral resources. However, it is important to recognize that mineral deposits are not infinite and that we cannot maintain exponential population growth on a finite resource base.

- The United States and many other affluent nations have insufficient domestic supplies of many mineral resources for current use and must supplement their supplies with imports from other nations.

- As other nations industrialize and develop, such imports may be more difficult to obtain, and affluent countries may have to find substitutes for some minerals or use a smaller portion of the world's annual production.

- The environmental impact of mineral exploitation depends on many factors, including mining procedures, local hydrologic conditions, climate, rock types, size of operation, topography, and many more interrelated factors.

- The mining and processing of mineral resources greatly affect the land, water, air, and biological resources and create social impacts as a result of the increased demand for housing and services in mining areas.

- Because the demand for mineral resources will increase in the future, we must strive to minimize both on-site and off-site problems by controlling sediment, water, and air pollution through good engineering and conservation practices.

- Reducing consumption, reusing, and recycling of mineral resources are environmentally preferable ways to delay or partly alleviate a possible crisis caused by the convergence of a rapidly rising population and a limited resource base.

REEXAMINING THEMES AND ISSUES

Human Population: Increasing human population is a major issue in the availability and use of mineral resources. We would be in a major minerals crisis if the consumption rate for all the people of the world was anywhere close to that of developed countries today. Therefore, as human population increases and developing countries increase their own demands, we must find ways to reduce our *per-capita* mineral consumption. Satisfying the mineral needs of an ever-expanding population on a finite resource base is impossible in the long term.

Sustainability: There are many sustainability issues related to our use of mineral resources. This results because so many parts of the mineral cycle are related to waste management problems involving pollution of soil, air, and water. Therefore, if we are to provide a quality environment for future generations, we must find ways to minimize adverse environmental effects of mineral exploitation, development,

and use. Of particular importance will be conservation measures related to reducing our use of minerals, reusing materials whenever possible, and recycling as much as is feasible.

Global Perspective: Our mineral resources are scattered in various locations on Earth where they were concentrated by geologic processes. We then centralize these and use them for a variety of processes, mostly related to agriculture and industry. The origin of many mineral deposits is directly related to global tectonics, and so understanding how our world works gives us a better perspective on our mineral resources.

Urban World: Ultimately the use of mineral resources are concentrated. Therefore, linkages between our urban environment and mineral use are direct. It is in urban areas where recycling efforts are likely to be most economically viable and where reduction in total use of minerals will have the greatest impact.

Values and Knowledge: Because we value a quality environment in terms of air, water, and land and because mining activities and other activities related to the mineral cycle can cause environmental disruption, we have a great interest in developing the knowledge necessary to minimize adverse effects of mineral utilization.

Reduction of environmental problems related to mineral utilization has resulted in a variety of new technologies in recent years, including biotechnology that utilizes the biological environment to assist in solving environmental problems related to the mineral cycle, particularly as it relates to waste management.

KEY TERMS

biohydrometallurgy *607*
brines *598*
marine evaporites *598*

mineral resources *601*
nonmarine evaporites *598*
ore deposits *596*

reserve *601*
secondary enrichment *598*

STUDY QUESTIONS

1. What is the difference between a resource and a reserve?

2. Under what circumstances might sewage sludge be considered a mineral resource?

3. If surface mines and quarries cover less than 0.5% of the land surface of the United States, why is there so much environmental concern about them?

4. When is recycling of a mineral a viable option?

5. Which biological processes can influence mineral deposits?

6. A deep-sea diver claims that the oceans can provide all our mineral resources with no negative environmental effects. Do you agree or disagree?

7. What factors determine the availability of a mineral resource?

8. Utilizing a mineral resource involves four phases: (a) exploration, (b) recovery, (c) consumption, and (d) disposal of waste. Which phase do you think has the greatest environmental effect?

FURTHER READING

Brookins, D. G. 1990. *Mineral and Energy Resources.* Columbus, OH: Charles E. Merrill. A good summary of mineral resources.

Earney, F. C. F. 1990. *Marine Mineral Resources.* Columbus, OH: Charles E. Merrill. A good summary of mineral resources.

Park, C. F. Jr., and MacDiarmid, R. A. 1975. *Ore Deposits,* 3rd ed. San Francisco: W. H. Freeman. A good, basic book on geologic aspects of ore deposits.

U.S. Bureau of Mines. 1995. *FY Annual Research Report.* U.S. Department of the Interior. Describes the Bureau's efforts to meet our mineral needs and to mitigate the associated economic, human, and environmental costs of mining and mineral processing.

INTERNET RESOURCES

U.S. Geological Survey Mine Drainage Information: *http://water. wr.usgs.gov/mine/home.html*—Read informative articles and facts about acid mine drainage and other environmental problems associated with mining.

U.S. Geological Survey Minerals Information: *http://minerals. er.usgs.gov/minerals/*—Check out the most recent information on the worldwide supply, demand, and flow of minerals, the

importance of minerals to the U.S. economy and security, and statistics on what minerals the U.S. must import.

U.S. Department of the Interior Office of Surface Mining Reclamation and Enforcement: *http://www.osmre.gov/*—Read about legislation and programs that aim to regulate surface mining practices and to clean up old mines and polluted areas as a result of mining operations in the U.S.

A home in Sedona, Arizona, using solar energy.

CHAPTER 29

ENVIRONMENTAL IMPACT AND PLANNING*

CASE STUDY

FLORIDA HOUSE

Investigations of ancient architecture suggest that earlier societies paid careful attention to the environment. For example, approximately a thousand years ago in Mesa Verde, Arizona, Native Americans built their houses with solar energy in mind. Homes were set in shallow recesses along the steep sides of the mesas so that they were shaded from summer sun by the rocks above and warmed directly by winter sun. The Native Americans built their homes using the concept that today we call "design with nature" (recall the discussion of design with nature in Chapter 26).[1]

In contrast, many homes built in the United States during the twentieth century, when central heating and air conditioning were common and fossil fuels were cheap, were designed with little concern for their location relative to sunlight and other environmental factors. Fortunately, more environmentally sound development is now being fostered in a number of locations around the world. One good example of this is Florida House, completed and opened to the public in May of 1994.

Southwestern Florida is a rapidly growing area. Sarasota, Florida, is no exception to this; the city has experienced growth pains related in particular to water problems. Pollution from urban runoff and drought in the late 1980s was thought

to be responsible for degradation of Sarasota Bay. In 1991 a two-year building moratorium appeared on the ballot, and athough the initiative failed, it brought attention to the environmental problems of the area. Environmentalists, developers, and others joined together in an attempt to develop a new type of home that would be a model of resource efficiency (Figure 29.1). The objective was not to develop a home that would necessarily be the

FIGURE 29.1 Florida House, Sarasota, Florida. This normal looking home incorporates many special design features generally available today to conserve energy and water resources. For example the banana trees (center) use gray water from laundry and showers.

*With assistance from Marc J. McGinnes.

most energy efficient and use the least amount of water resources, but rather to develop a home using resource efficiency technology that would be widely accessible to the community.

Florida House incorporates a number of interesting environmental features. In addition to both passive and active solar systems (see Chapter 17) and the use of many recycled materials for construction, conservation of water resources is an important objective. For example:

- Irrigation systems use "gray water" rich in nutrients from home laundries, sinks, and baths to supply water to vegetation, including banana trees.
- Rain water from the roof is collected and stored in cisterns to augment the irrigation system.
- Use of porous concrete and other materials around the home allow for more infiltration of water into the soil, reducing surface runoff.

- Surface water is collected and routed to a small on-site artificial wetland that serves as a sediment trap and further facilitates infiltration of water into the soil.

In summary, Florida House is a successful venture promoted by both environmentalists and developers to provide readily accessible technology for resource conservation at the community level. The home also plays an important role in environmental education by providing information to other communities interested in what are sometimes termed "green choices" with respect to development. Thus Florida House has become a learning center, and an additional home is being planned in the Florida Keys, where water resources and water pollution problems are perhaps even more significant than in Sarasota. Demonstration projects such as Florida House are helping to develop ideas and concepts necessary in establishing sustainable development for future generations.

Oliver Wendell Holmes wrote, "Every year, if not every day, we have to wager our solution upon some prophecy based upon imperfect knowledge." This statement certainly applies to environmental planning, as we often do not have all the data necessary to make the best possible choices. Even so, we do our best. Today, evaluating the landscape for environmental impact and land-use planning is common practice. The evaluation is usually completed by a team of people in disciplines such as planning, geography, geology, biology, civil engineering, architecture, and law.

LEARNING OBJECTIVES

Evaluating the landscape for environmental impact and land-use planning is common practice. After reading this chapter, you should understand:

- The major components of an environmental impact statement (EIS).
- The processes of scoping and mitigation in environmental impact assessment.
- The important elements of a Geographic Information System (GIS).
- Mediation as a tool in environmental law.

- The problems and controversy concerning the development of international environmental agreements.
- Steps in land-use planning.
- Major and environmentally preferable adjustments to natural hazards.
- The relationship between population increase and the effects of natural hazards.
- The controversy, problems, and opportunities of global forecasting.

29.1 ENVIRONMENTAL IMPACT ANALYSIS

Environmental impact analysis is the process of determining the probable effects of human use of (and interest in) the land. Such analysis became common

in the United States following the passage of the National Environmental Policy Act (NEPA) in 1969 (see the later discussion of environmental law). The legislation required that all major federal projects that might affect the quality of the human environment be preceded by an evaluation of the project and its potential impact on the environment. Guidelines to assist in the preparation of **environmental impact statements (EISs)** were prepared by the Council on Environmental Quality. The major components of the EIS as revised in 1979 are[2]

- a summary of the EIS;
- a statement concerning the purpose of and need for the project;
- a rigorous comparison of the reasonable alternatives;
- a succinct description of the environment affected by the proposed project; and
- a discussion of the environmental consequences of the proposed project and of the possible alternatives.

The discussion must include both direct and indirect effects (applying the principle of environmental unity, see Chapter 3); energy requirements; conservation potential; possible resource depletion; impact on cultural and/or historical resources and urban quality; possible and potential conflicts with local or state land-use plans, policies, and controls; and mitigation measures that might reduce or eliminate potential environmental degradation.

During the first 10 years under NEPA the process of environmental impact analysis was criticized because it produced a tremendous volume of paperwork that tended to obscure the important issues. In response to that criticism, the revised regulations of 1979 introduced two important changes: scoping and the record of decision.

Scoping is the process of early identification of important environmental issues that require detailed evaluation. Citizen groups as well as federal, state, and local agencies are asked to participate in the scoping process by identifying issues and alternatives that they believe should be addressed as part of the environmental analysis.

The **record of decision** is a concise statement prepared by the agency planning the proposed project that outlines the alternatives considered and discusses which alternatives are environmentally preferable. The record of decision also addresses the rationale for a particular decision and how environmental degradation might be minimized or avoided for the alternative chosen. Finally, the agency has the responsibility of monitoring the project to ensure that the decisions are in fact carried out as the project is initiated and completed; this responsibility puts teeth into the environmental impact analysis process. For example, if an EIS for a proposed pipeline commits an agency to a particular right-of-way and a specific design to minimize environmental degradation, it is the agency's responsibility to

see that the firms doing the project incorporate those designs and plans. For this reason, special consultants are often hired to monitor the work to ensure that the procedures outlined in the record of decision are executed. *during planning*

Mitigation

Mitigation is the process that identifies actions that will avoid, lessen, or compensate for anticipated adverse environmental impacts resulting from a particular project. For example, if a project in the coastal zone will damage wetlands, a possible mitigation might be the creation or enhancement of wetlands at another site. The process of mitigation is fast becoming a common requirement of many environmental impact reports and statements. However, there is a danger that the process of mitigation may be overused, and sometimes no mitigation is possible for a particular environmental problem. Requiring mitigation to reduce or compensate for environmental degradation is a useful endeavor in many instances, but it must not be considered a standard operating procedure to circumvent the adverse environmental impacts associated with a particular project.

Methodology for Assessing Environmental Impact

There is no one methodology for assessing the environmental impact that results from a particular action or project. No single method of impact assessment is appropriate for the broad spectrum of projects, which may range from construction of large reservoirs, highway construction, or urban renewal to the hunting of migratory birds. The important principle is that those responsible for preparing the EIS produce a report that minimizes personal bias and maximizes objectivity. The analysis must be prepared according to highly objective standards and must be technically and legally defensible.[3]

Geographic Information Systems

A technology that has proved useful in environmental impact analysis and other environmental work is the **Geographic Information System (GIS).** The GIS is a relatively new technology capable of storing, retrieving, transferring, and displaying environmental data.[4,5] The data used in a GIS may be points, lines, areas, or numbers. Intersections of roads and locations of particular facilities, such as hospitals or fire stations, are points; roads, rivers, and landscape features, such as the shoreline, are lines; and land use, rock type, or soil type are areas.

The GIS is much more than a method to categorize and store data; it has the capability to both manipulate data and create new products. The creation of new products is important in environmental impact analysis. For example, the product may be a land-use map or an analysis of important environmental variables, such as relationships between topography, geology, hydrology, and population density.

Figure 29.2 outlines the process used in a GIS, using an example in which the product is a land capability map. The process involves an inventory of data sources, such as topography, geology, and hydrology, and analysis of issues such as slope stability and flooding, to produce the computer-generated capability map. Depending on how data are analyzed, a variety of maps are produced as part of the environmental analysis process. This basic idea can be extended to various types of projects in which a number of environmental data sources and factors must be analyzed and interpreted.

State Environmental Policy Acts

After the passage of NEPA in 1969 the need for state and local governments to have more active control over activities not covered by federal legislation became apparent. Today, approximately one-half of the states have enacted legislation requiring environmental review prior to initiation of projects that may adversely affect the environment. Nineteen states and the District of Columbia have enacted State Environmental Policy Acts (SEPAs) patterned after the federal legislation.[6,7]

The California Environmental Quality Act (CEQA), passed in 1970, requires the completion of an **environmental impact report (EIR)** for a wide spectrum of projects that may affect the environment. Following a court decision, additional legislation now requires that all private as well as public projects in California undergo environmental review. (Examples of private projects include a hotel convention center or a shopping center; examples of public projects are construction of a dam or reservoir or implementation of regulations concerning timber harvesting.)

The California law potentially provides more protection for the environment than does the federal legislation. It specifically addresses environmental quality, whereas the federal legislation is concerned with environmental policy. The EIRs required by CEQA contain information similar to that required by the federal legislation but in addition require evaluation of growth-inducing impacts and mitigation measures proposed to minimize significant adverse environmental effects.[8]

The CEQA is similar to NEPA in that it allows for the filing of a **negative declaration** or a **mitigated negative declaration.** A negative declaration may be filed if an agency has determined that a particular project will not have significant adverse impacts on the environment. The declaration may be filed only following the completion of an environmental inventory and assessment. The negative declaration includes a description of the project, along with specific information supporting the argument that the project will not have significant environmental effects. Negative declarations need not consider alternatives to the project but must provide a comprehensive statement concerning potential environmental problems.[8] A mitigated negative declaration may be filed if initial study of the project suggests that, although significant environmental problems may occur, the project may be modified in such a way as to reduce the impacts to near insignificance.[8]

Both the negative declaration and the mitigated negative declaration are important additions to the environmental review process. They provide a process whereby projects that will not cause environmental disruption may avoid the preparation of a lengthy environmental impact report. However, they are not intended to avoid the important process of public review and comment prior to approval of a project.

In general, the environmental review process at both the state and federal level has accomplished the following:

- focused attention on potential environmental problems related to proposed projects;
- provided a framework useful for evaluation of the environmental consequences of a project; and
- significantly increased environmental protection efforts in the United States.

29.2 ENVIRONMENTAL LAW*

The legal system of the United States has historic origins in the British system. When our legal system was formed, it preserved and strengthened the tendency of British law to protect the individual from society rather than the reverse. Individual freedom and nearly unlimited discretion to use property as the owner pleased were given high priority, and the powers of the federal government were strictly limited. (At that time, natural resources seemed limitless and little need to control their development or exploitation was apparent.) However, when behavior was so egregious that it infringed on the property or the well-being of others, the **common law** (i.e., law derived from custom, judgment, and decrees of the

---

*Written in part by Harold Ward.

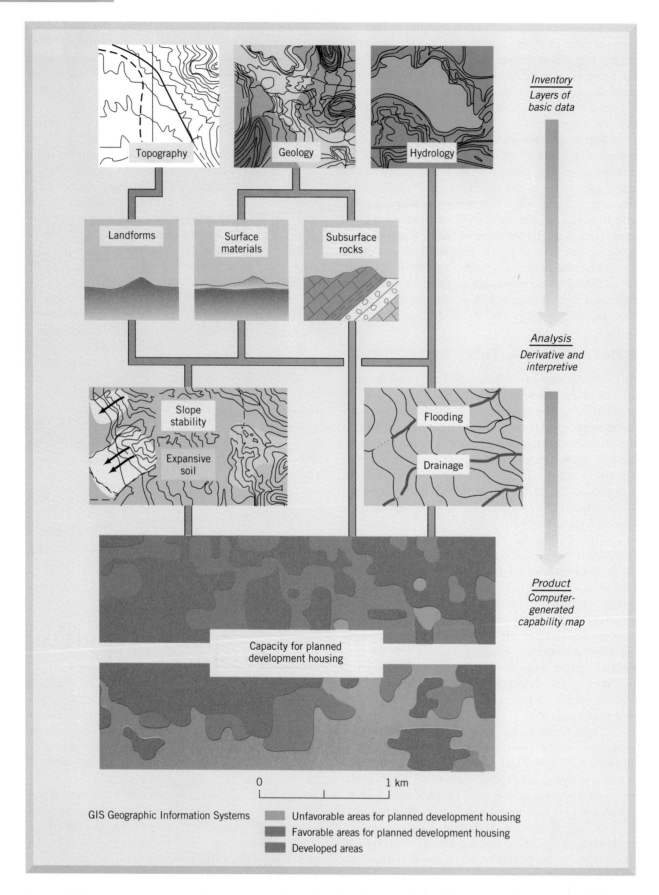

Figure 29.2 Idealized diagram showing the use of Earth science information to develop a Geographical Information System (GIS) for land use planning.

courts, as distinguished from legislation) provided protection by doctrines prohibiting trespass and nuisance. An individual who suffered damage, for example, from erosion or flooding caused by improper management of adjacent land had recourse under common law. If the harm was more widely spread to the community, then only the government had the authority to take action, for instance to limit certain air and water pollution to abate a public nuisance.

The common law provides another doctrine, that of **public trust,** which both grants and limits the authority of government over certain natural areas of special character. Beginning with Roman law, navigable and tidal waters were entrusted to the government to hold in trust for public use. More generally, "the Public Trust Doctrine makes the government the public guardian of those valuable natural resources which are not capable of self-regeneration and for which substitutes cannot be made by man."[9] For such resources, the government has the strong responsibility of a trustee to provide protection and is not permitted to transfer such properties into private ownership. This doctrine was considerably weakened by the exaltation of private property rights and strong development pressures in this country but in more recent times has shown increased vitality, especially concerning the preservation of coastal areas.

Thus, although the common law provides a few examples of environmental protection possibilities, the vast majority of the activities we regulate today in the name of the environment would proceed unrestricted if left to common law. The need for regulation of these activities has been filled by legislation enacted at all levels of government. In the late 1960s, public perception in the United States that our environment was deteriorating reached a high level of awareness and concern. Congress responded with NEPA in 1969, and in the next 20 years major environmental legislation was passed at the national level (Figure 29.3).

International Environmental Law and Diplomacy*

International law is different in basic concept from domestic law because there is no world government with enforcement authority over nations. As a result, international law must depend on the agreement of the parties concerned to bind themselves to behavior that many residents of a particular nation may oppose. Certain issues of multinational concern are addressed by a collection of policies, agreements, and treaties that are loosely called international envi-

*Written in part by Marc McGinnes.

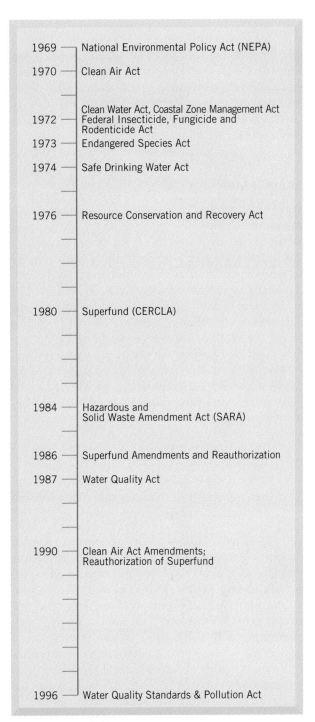

1969	National Environmental Policy Act (NEPA)
1970	Clean Air Act
1972	Clean Water Act, Coastal Zone Management Act Federal Insecticide, Fungicide and Rodenticide Act
1973	Endangered Species Act
1974	Safe Drinking Water Act
1976	Resource Conservation and Recovery Act
1980	Superfund (CERCLA)
1984	Hazardous and Solid Waste Amendment Act (SARA)
1986	Superfund Amendments and Reauthorization
1987	Water Quality Act
1990	Clean Air Act Amendments; Reauthorization of Superfund
1996	Water Quality Standards & Pollution Act

FIGURE 29.3 Major federal environmental legislation and the year enacted. Most important environmental legislation was adopted from 1969 to 1996. Some laws were enacted earlier in a much less comprehensive form (e.g., the Clean Air Act in 1963) and most have been amended subsequently.

ronmental law. There have been encouraging developments in this area, such as agreements to reduce air pollutants that destroy stratospheric ozone (the Montreal Protocol of 1987). The pace of diplomatic efforts to reach legal agreements addressing a host of other global environmental issues has increased substantially in recent years, reflecting the realization that global environmental degradation poses a dire threat to the security of all nations and peoples. Agreements to control human population, reduce emissions of greenhouse gases, preserve biodiversity, prevent destruction of the world's rain forests, control pollution of the oceans, and protect endangered species are all needed.

Summit Meetings

The United Nations sponsored an international environmental conference in 1992 in Rio de Janeiro, Brazil, popularly called the Earth Summit. Many of the world's leaders attended, along with tens of thousands of other people, including politicians, diplomats, scientists, ecologists, and journalists. The conferences did not produce sweeping international agreements on the major environmental problems facing the world. However, it did bring to the surface conflicts between environmental concerns and economic issues. One issue discussed is the need for the rich, developed countries to provide financial help to the developing countries to combat environmental problems. As a result of the Earth Summit:

* Attempts will be made to reduce greenhouse gas emissions by the year 2000 to the 1990 levels.
* Developing countries will be given more and easier access to technology that minimizes environmental damage.

The United Nations in 1997 sponsored a summit on global warming in Kyoto, Japan. The purpose was to negotiate a treaty to reduce greenhouse gases, especially carbon dioxide, to 5% below 1990 emissions by the year 2010. The primary, long-term objective was to prevent emissions from human sources from interferring with the natural climate system. Some now consider the Kyoto conference to be a failure. However, the idea of forced control of carbon dioxide emissions now has become a political issue (see Chapter 21). Therefore, the summit meetings were a step in the right direction. If we are going to move toward a sustainable society on a global basis, we need to start by developing equitable policies to ensure that all people on Earth have opportunities to work and live in a quality environment.

Environmental Litigation

Environmental groups working through the courts have been a powerful force in shaping the direction of environmental quality control since the early 1970s. Their influence arose in part because the courts, appearing to respond to the national sense of environmental crisis of that time, took a more activist stance and were less willing to defer to the judgment of agencies. At the same time, citizens were granted an unprecedented access to the courts and through them to environmental policy.

Citizen Actions

Even without specific legislative authorization for citizens' suits, courts have allowed citizen actions in environmental cases as part of a trend to liberalize standing requirements. One of the earliest examples, and the most significant, was the willingness of the courts to entertain lawsuits to enforce NEPA. The statute itself and its legislative history do not mention either judicial review or standing, and the courts were left to decide whether a judicial remedy should be applied. Without such a remedy, and in the face of very substantial reluctance on the part of administrative agencies to change their habitual ways of doing business, it is unlikely that the EIS requirement would ever have had a significant influence on agency actions. With the courts' involvement, administrators were no longer able to take an action without explaining the reasoning that led them to conclude that the action was desirable. This development occurred in a general climate of suspicion of government, arising from protests against U.S. military involvement in Vietnam and from the political scandals of Watergate. An example of the effectiveness of citizen action is found in A Closer Look 29.1, "The Florissant Fossil Beds."

In the 1980s a new type of environmentalism (that some people would label radical) arose, based in part on the premise that when it comes to the defense of wilderness there can be no compromise. Methods used by these new environmentalists have included sit-ins to block roads that provide access to forest areas where mining or timber-harvesting activities are scheduled to take place; climbing trees and sitting in them to block timber harvesting; implanting large steel spikes in trees to discourage timber harvesting; and sabotaging equipment, such as bulldozers (a practice known as *ecotage*).

Civil disobedience and ecotage have undoubtedly been responsible for millions of dollars of damage to a variety of industrial activities related to the use of natural resources in wilderness areas. One result of civil disobedience by some environmental groups is that other environmental groups, such as the Sierra Club, are now considered moderate in their approach to protecting the environment. There is no doubt that civil disobedience has been successful in defending the environment in some circumstances. For example, in 1983 members of Earth First successfully halted the construction of a road being built to allow access for timber harvesting in an area of southwestern Oregon. Tactics used in-

A **CLOSER LOOK 29.1**

THE FLORISSANT FOSSIL BEDS

A case reported by Yannacone, Cohen, and Davison in Colorado emphasizes the power of citizen groups to use the law.[10] The conflict surrounded the use of 7.3 km^2 (about 2.8 mi^2) of land near Colorado Springs, a part of the Florissant Fossil Beds, where insect bodies, seeds, leaves, and plants were deposited in an ancient lake bed about 30 million years ago. Today the deposits are remarkably preserved in thin layers of volcanic shale. The fossils are delicate and, unless protected, tend to disintegrate when exposed. Many people consider the fossils unique and irreplaceable. At the time of the controversy, a bill had been introduced in Congress to establish a Florissant Fossil Beds National Monument. The bill had passed the Senate, but the House of Representatives had not yet acted on it.

While the House of Representatives was deliberating the bill, a land development company that had con-

tracted to purchase and develop recreational homesites on 7.3 km^2 of the ancient lake bed announced that it was going to bulldoze a road through a portion of the proposed national monument site to gain access to the property it wished to develop. A citizens' group formed to fight the development until the House acted on the bill. The group tried to obtain a temporary restraining order, which was first denied because there was no law preventing the company that owned the property from using that land in any way it wished, provided that existing laws were upheld.

The conservationists then went before an appeals court and argued that, even though there was no law protecting the fossils, they were subject to protection under the Ninth Amendment to the U.S. Constitution and that, furthermore, because the property had tremendous public interest, it was also protected by the

Trust Doctrine. An analogy used by the plaintiffs was that if a property owner were to find the Constitution of the United States buried on the land and wanted to use it to mop the floor, certainly that person would be restrained. After several more hearings on the case, the court issued a restraining order to halt development; shortly thereafter, the bill to establish a national monument was passed by Congress and signed by the president.[10]

The court order prohibiting destruction of the fossil beds may have deprived the landowner of making the most profitable use of the property, but it did not prohibit uses consistent with protecting the fossils. For instance, the property owners are free to develop the land for tourism or scientific research. Although such use might not result in the largest possible return on the property owner's investment, it probably would return a reasonable profit.[10]

cluded blockading the road by people sitting down or standing in front of the bulldozer blades. Owing to the persistence of the protesters in the face of tremendous bodily harm (some of the protesters were injured), the pace of road building was slowed considerably. In conjunction with this incident a lawsuit was filed against the U.S. Forest Service. Similar activities occurred in 1996.

Environmentalists in the 1990s have relied more on the law when arguing for ecosystem protection. The Endangered Species Act has been used as a tool in attempts to halt activities such as timber harvesting and development. Although rarely is the presence of an endangered species responsible for stopping a proposed development, those species are increasingly being used as a weapon in attempts to save remaining portions of relatively undisturbed ecosystems.

Mediation

The expense and delay of litigation have led to a search for other methods to resolve disputes. An alternative that has recently received considerable

attention in environmental conflicts is environmental mediation, a negotiation process between the adversaries guided by a neutral facilitator. The task of the mediator is to clarify the issues in contention, help each party understand the position and the needs of other parties, and attempt to arrive at a conclusion where each party gains enough from a compromise to prefer a settlement to the risks and costs of litigation. Often citizens' suits or the possibility that a suit might be filed gives an environmental group a place at the table in such a mediation. Litigation, which may delay a project for years, becomes something that can be bargained away in order to gain concessions of decreased environmental impact (mitigation) from a developer. There have been some successes with mediation where bargaining positions are approximately equal and where participants who truly represent the conflicting interests can be identified and persuaded to devote the considerable effort that the process demands. In some states, mediation is required by legislation as an alternative or a precedent to litigation in the highly contentious siting of waste treatment facilities. For example, in

Rhode Island a developer who wishes to construct a hazardous-waste treatment facility must negotiate with representatives of the host community and submit to arbitration of any issues not resolved by negotiation. Costs of the negotiation process are borne by the developer.

The Storm King Mountain case is a classic example of a situation in which mediation could have saved millions of dollars in legal expenses and years of litigation. The case involved a conflict between a utility company and conservationists. In 1962 the Consolidated Edison Company of New York announced plans for a new hydroelectric project in the Hudson River highlands, an area considered to have many unique aesthetic values as well as thriving fisheries. The utility company argued that it needed the new facility, and the environmentalists fought to preserve the beautiful landscape and fisheries resource. The first lawsuit was filed in 1965, and following 16 years of intense courtroom battles, the litigation ended in 1981. Incredibly, the paper trail exceeded 20,000 pages, and after millions of dollars

were spent, the various parties finally managed to get together and settle the case with the assistance of an outside mediator. Could the outcome have been decided earlier? If the parties had been able to sit down and talk about the issues at an early stage, mediation might have settled the issue much earlier and at much less cost to the individual parties and to society.[11] The Storm King Mountain case is often cited as a major victory for environmentalists, but the cost was great to both sides.

Environmental Review

The environmental review process was developed in response to the recognized need to consider the environmental consequences of a particular action prior to its implementation. The overriding objective of environmental review is to recognize potential impacts, conflicts, and problems as early as possible so that they may be evaluated and environmental degradation minimized. The cases of the Cape Hatteras National Seashore in North Carolina and the San Joaquin Valley in California are considered next.

Case History: Cape Hatteras National Seashore

The Outer Banks of North Carolina consists of long, thin barrier islands that appear as ribbons of sand on satellite images (Figure 29.4) and have for generations been inhabited by people living and working in a marine-dominated environment. Until recently their way of life depended on raising livestock and on fishing, hunting, and boat building and other marine pursuits.[12]

Characteristically, the landscape of the Outer Banks can change in a very short time in response to major storms, such as hurricanes (Figure 29.5) and northeasters, that strike the islands periodically. Of the two types of storms, the more frequent northeasters probably cause the most erosion. On the other hand, infrequent hurricanes can cause major changes, including extensive dune overwash and formation of new inlets.[13]

Historically, the people of the Outer Banks have lived with and adjusted to a changing landscape. In recent times, however, their philosophy has changed because of economic pressure to develop coastal real estate. Because a stable landscape is necessary to encourage commercial development (people are not likely to buy property that could be eroded by the action of ocean waves), attempts have been made to stabilize the coastal environment through erosion controls.

An act establishing the Cape Hatteras National Seashore as the first national seashore was approved by Congress in 1937 and amended in 1940. The park consists of 115 km² (about 45 mi²) along a 120-km

FIGURE 29.4 The Outer Banks of North Carolina appear in this image from the *Apollo-9* mission as a thin white ribbon of sand. The large cape in the center of the image is Cape Hatteras. The lower hooked cape is Cape Lookout. The body of water to the left (west) of Cape Hatteras is Pamlico Sound. The distance from Cape Hatteras to Cape Lookout is on the order of 100 km. The light color in the water of the sound and seaward of the barrier islands is sediment suspended in the water column.

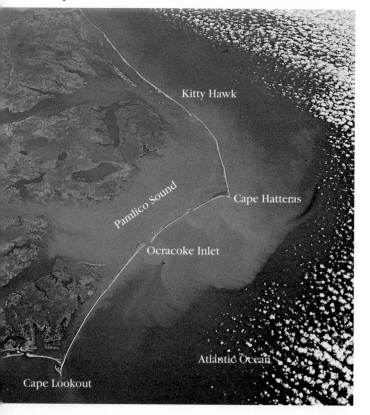

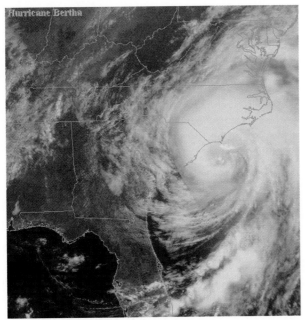

FIGURE 29.5 (*a*) Hurricane Bertha, July 12, 1996, striking the coast of North Carolina. (*b*) Damage to homes from Hurricane Fran, which hit the coast of North Carolina in September 1996.

(75-mile) portion of the Outer Banks; it includes portions of three islands of the more than 240 km (150 miles) of the barrier island system. The barrier islands bound and protect the largely undeveloped, coastal plains lowland of North Carolina.[12]

Eight unincorporated villages are bounded by the Cape Hatteras National Seashore and are spaced along nearly its entire length. The legislation authorizing the park provided for the continued existence of these villages, including the beach in front of each village facing the Atlantic Ocean. This legislation has been interpreted by many as creating an obligation of the federal government to maintain and stabilize the beach access and frontage of these villages. Unfortunately, stabilization has been difficult. In some villages, less than 60 m (200 ft) of beach remains be-

cause of recent coastal erosion. The philosophy of protection brought changes in building patterns. The early development of the islands occurred mainly on the sound (inland) side of the islands, away from the ocean. However, with the construction of the first dune systems in the 1930s and the opening of a road link in the 1950s, people began to count on protection from the elements. The basic configuration of the villages started to change as communities began to spread toward the ocean.[12]

It is assumed that at one time much of the Outer Banks was heavily forested and that logging and overgrazing destroyed these forests, although evidence for this is still inconclusive. The wooded areas were located in an area that is protected by a natural dune line, so the logical conclusion was to construct an artificial barrier dune system to help the area return to a natural state by inhibiting beach erosion. Of course, the dune line would also protect the roads and communities. An artificial dune system was constructed along the entire Cape Hatteras National Seashore, with an average dune height of 5 m (about 16 ft) at an average distance of 100 m (about 330 ft) from the ocean.[12] This artificial erosion control measure, along with the programs to keep sand on the beach artificially, could cost several million dollars per year. The expense of this measure, along with the questions as to whether the artificial dune lines will jeopardize the future of the barrier islands, led to a controversy on how best to manage the park.

The controversy about how dynamic Earth processes create and maintain barrier islands is at the philosophical heart of the U.S. Park Service's dilemma in developing a long-range program to maintain the natural environment for the use and enjoyment of present and future generations. Although there are several ideas and known ways in which coastal processes can develop barrier islands, there is heated debate on the processes needed to maintain them.

Debate on the nature and extent of the geologic processes that maintain the naturally changing barrier islands centers on whether the periodic overwash of the frontal dune system by storm waves is essential to maintaining the islands. If the overwash is essential, then the building of a dune line is contrary to natural processes, and the final result over a period of years could be the deterioration of the islands as a natural system. This would be contrary to the Park Service's objectives to preserve the islands in a natural state.

Regardless of the historic role of overwash, it is apparent that overwash is a natural process and as such is probably significant in the geologic history of the barrier islands.[13] Nevertheless, it is not necessarily always bad to attempt to selectively control rapid coastal erosion by maintaining a protective frontal dune system. Proper placement of dunes and sound conservation practices for maintaining them are a feasible alternative to some short-range coastal erosion problems, particularly near settlements and critical communication lines.

Faced with the problem of selecting a management policy for Cape Hatteras National Seashore, in 1974 the Park Service presented five alternatives, ranging from essentially no control of natural processes to attempting complete protection. Each alternative was analyzed to determine the spectrum of possible impacts. Few people wanted either no control or complete control of the seashore.[12] The position of the Park Service was to attempt to reach a compromise in which the barrier islands could be preserved in a nearly natural state while recognizing the need to maintain a transportation link with the mainland. People in these communities would have to live with and adjust to the dynamic high-risk environment in which they chose to reside, much as people of the Outer Banks historically had done. This approach is contrary to prevailing trends in coastal development, which assume a more stable environment, and acknowledge that natural processes play a significant role in preserving natural environment.[12]

A general management plan for Cape Hatteras National Seashore was completed after several years of careful environmental impact work and public review. The plan proposes several actions[14]:

- allowing natural seashore processes and dynamics to occur except in instances when life, health, or significant cultural resources on the major transportation link are jeopardized;
- controlling the use of off-road vehicles;
- expanding and allowing easier access to recreation sites on the beach and sound;
- controlling spread of exotic vegetation;

- ensuring that significant natural and cultural resources are preserved and maintained; and
- encouraging state and local governments to participate in mutually beneficial planning.

For planning purposes, the national seashore was divided into four environmental resource units (ERUs): ocean–beach, vegetated sand flats, interior dunes–maritime forest, and marsh–sound. Table 29.1 summarizes the planning objectives for each ERU, and Figure 29.6 illustrates the principle of management zoning for Hatteras Island.

Implications of the management plan are threefold. First, Cape Hatteras National Seashore will be preserved in a natural state. Second, residents of villages will have to live with and adjust to effects of natural events to a greater extent. Third, villages may change; although the road will be maintained, it will be subject to more periodic damage. For example, the Park Service does not perform routine dune maintenance or beach nourishment (artificially adding sand to the beach). Only in rare cases are dunes rebuilt after overwash.

The management plan is controversial, and coastal erosion has continued to be a significant problem in several areas of the national seashore in the 1990s. The major goal of shoreline stabilization that has continued is to maintain the highway that extends along the entire seashore. A five-year multi-agency investigation aimed at investigating and evaluating alternatives to long-term maintenance of the highway began in 1995. More controversial have been conflicts about protection of private property and public facilities and allowing natural coastal processes to operate. A focal point has been the Cape Hatteras Lighthouse, the tallest brick lighthouse in the United States (Figure 29.7), which was originally constructed in 1870 approximately 0.5 km from the shoreline. Today it is nearly in the surf zone, and there is fear that it might soon be de-

TABLE 29.1 Planning Objectives for Cape Hatteras National Seashore Environmental Resource Units (ERU)

ERU	Characteristics	Planning Objectives
Ocean beach	Shifting sands, frequent overwash, limited vegetation on dunes	Allow natural processes to continue unhampered; allow for wide range of unstructured recreational activities by visitors; no construction allowed
Vegetated sand flats	Located between dune line and edge of saltwater marsh	Continue use as transportation corridor; allow development necessary to support visitor activities and resource protection
Interior dunes/ maritime forest	Found in relatively few locations; variable topography, remoteness, dense vegetation	Maintain in natural state; allow passive recreation; design any construction to minimize impact on natural processes and systems
Marsh/sound	Includes the sound, sound shore, and associated marshes	Maintain in natural state; provide limited access to the sound; allow limited development of passive recreational activities

Source: National Park Service.

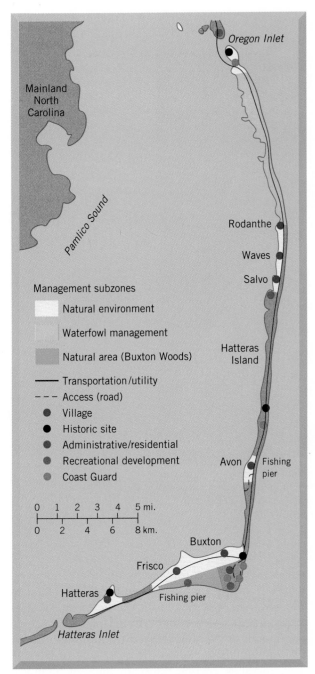

FIGURE 29.7 The famous Cape Hatteras lighthouse, originally constructed (in the late 19th century) about 0.5 km from the sea, is now threatened by coastal erosion. The beach groins (black structures in the surf) are designed to trap sand and retard erosion by widening the beach. Note also the light yellow seawall constructed to retard erosion in front of the lighthouse. If not moved (or greatly defended) the lighthouse will eventually be destroyed by the sea.

FIGURE 29.6 Proposed management subzones for part of the Cape Hatteras National Seashore. (*Source:* U.S. Department of the Interior, *Cape Hatteras Shoreline Erosion Policy Statement.*)

- Move the lighthouse out of harm's way.[15]

The decision has now been made to move the lighthouse inland about 500 m from the eastern shore of Hatteras Island at a cost of about $12 million. Given the present rate of sea level rise, the lighthouse should be safe in its new location from coastal erosion until the middle or end of the 21st century. This decision is consistent with the philosophy of flexible coastal planning and avoiding hazardous zones rather than trying to control natural processes. The decision is also consistent with policy to preserve historical objects (such as the lighthouse) for the enjoyment of future generations.

Case History: Selenium Toxicity in the San Joaquin Valley, California

The problem of selenium toxicity in Kesterson Wildlife Refuge discussed in Chapter 11 is revisited here to illustrate the environmental review process.

San Joaquin Valley in California is one of the richest farm valleys in the world. On the west side of the valley (Figure 29.8), with an annual precipitation of less than 25 cm (10 in.), extensive irrigation is necessary for agriculture. When irrigation waters are applied, however, the crop-root zone may become saturated and drown plants.[16] As a result, subsurface drainage systems are installed to remove excess water (Figure 29.9). Tile drains, pipes with holes to let water in, are placed beneath fields and connected to larger drains (ditches or canals) that

stroyed by coastal erosion. Only luck in the mid-to-late 1990s prevented this, as several big storms and hurricanes came dangerously close. Several solutions to the lighthouse dilemma were considered:

- Do nothing and let nature take its course, coastal erosion would destroy the lighthouse;
- Reverse policy of yielding to erosion and artificially control coastal erosion by constructing a seawall around the lighthouse;

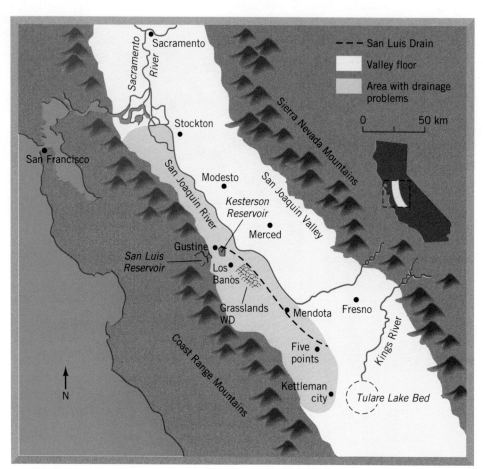

FIGURE 29.8 The San Joaquin valley and San Luis drain that terminates at Kesterson Reservoir. (*Source:* Modified from R. Tanji, A. Lauchli, and J. Meyer, 1986, "Selenium in the San Joaquin Valley," *Environment,* 28(6): pp. 6–11. Copyright 1986 by Heldref Publications. Reprinted with permission of the Helen Dwight Reid Educational Foundation.)

are tied into a master drainage system, usually a canal.

The San Luis drain is a 135-km-long (about 85-miles) concrete-lined canal constructed between 1968 and 1975. Its initial purpose was to carry agricultural drainage water north to San Francisco Bay. However, the drain to the bay was not completed because of limitations in funding and uncertainty over potential environmental degradation associated with discharging water into the bay. As a result, the terminal point for drainage water became Kesterson Reservoir, then part of the much larger Kesterson Wildlife Refuge consisting of 12 ponds with an average depth of approximately 1.3 m (about 4.3 ft) and total surface area of about 486 ha (1200 acres).

When the Kesterson Reservoir was first completed, it was filled primarily with fresh water purchased from water agencies to provide water for the Kesterson Wildlife Refuge. In 1978 the reservoir began receiving agricultural drainage, and by 1981 this was its main source of water. Initially the canal and reservoir were viewed as a beneficial system

that provided a sink for wastewater from farming activities. Unfortunately, however, the water infiltrated the soil, where it picked up the heavy metal selenium as well as other salts. The selenium is a natural constituent of the soils derived from the weathering of sedimentary rocks on the west side of the San Joaquin Valley. In 1982 unusually high selenium levels in fish from Kesterson Reservoir were identified. From 1983 to 1985, dead and deformed wildfowl chicks (Figure 29.10) were reported, and analysis of the water revealed that the selenium in the reservoir ponds ranged from 60 to 390 ppb. Further study revealed that selenium concentrations in the agricultural drain waters were as high as 4000 ppb, a concentration of selenium many times that of the drinking water standard of 10 ppb set by the EPA. Subsequent work suggested that selenium toxicity was the probable cause of the death and deformities of the wildfowl. Finally, in 1985, the Kesterson Reservoir and the water in the San Luis drain were classified as hazardous and a threat to public health. The California State Water Resources Board then or-

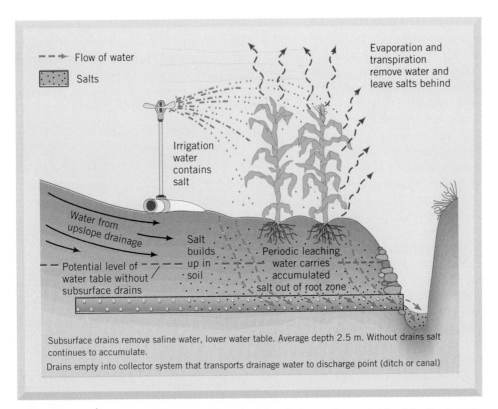

FIGURE 29.9 Diagram indicating the need for subsurface drains on the west side of the San Joaquin valley. (*Source:* Modified after U.S. Department of the Interior, Bureau of Reclamation, 1984, *Drainage and Salt Disposal,* Information Bulletin 1, San Luis Unit, Central Valley Project, CA.)

dered the U.S. Bureau of Reclamation, which was managing the wildlife refuge and reservoir, to alleviate the hazardous conditions.[17] The bureau was

FIGURE 29.10 Three-day-old black-necked stilt born without eyes. The location is the Tulare Lake Basin, San Joaquin Valley, California. Deformities such as these are presumably related to selenium toxicity.

faced with developing a cleanup plan and implementing it.

The environmental planning and review process for the Kesterson Reservoir provides insight into the environmental impact process. The scoping process identified the following issues:

- the disposition of Kesterson Reservoir and cleanup of the San Luis drain;
- the environmental, social, legal, and economic impact of not allowing Kesterson to receive agricultural drain water;
- the potential impacts of selenium and other contaminants in the drain water on public health;
- the development of alternative methods to clean up Kesterson and the San Luis drain; and
- the potential for contaminated groundwater to migrate away from the Kesterson area.

The Bureau of Reclamation of the U.S. Department of Interior issued its final EIS and record of decision in October 1986. The document addressed the potential impacts of the alternative methods to clean up the Kesterson Reservoir as well as the San Luis ditch. The no-action alternative was also considered but it was not accepted, and then several alternative plans were developed. These varied from flooding reservoir ponds with fresh water to constructing an on-site waste disposal facility. In March 1987 the Water Resources Control Board ordered Bureau of Reclamation to dispose of contaminated soil and vegetation in an on-site, monitored, lined, and capped landfill. That is, the site was treated as a toxic waste disposal facility. The board also required Bureau of Reclamation to mitigate loss of Kesterson Wildlife Refuge by providing alternative wetlands for habitats.

Environmental review and research at Kesterson illustrates the process of environmental analysis at several levels of government. Scoping was an important process, as was evaluation of alternatives and mitigation to minimize environmental degradation.

Capping ponds at Kesterson has been successful in reducing impacts on wildfowl. Monitoring of the site has been ongoing since the ponds were filled. Wetlands have provided valuable wildlife habitat in an area where wetlands had been lost. Selinium-rich agricultural drainwater from 17,000 ha of irrigated agriculture land has been eliminated. However, the selenium problem in the broader valley hasn't disappeared, as selenium-contaminated waters from agriculture drains in other areas continue to cause concern.

29.3 LAND-USE PLANNING

Land use in the United States is dominated by agriculture and forestry; only a small portion of land (about 3%) is used for urban purposes. However, rural lands are being converted to nonagricultural uses at about 9000 km² (about 3500 mi²) per year. About half the conversion is for wilderness areas, parks, recreational areas, and wildlife refuges. The other half is for conversion to urban development, transportation networks, and other facilities. On a national scale conversion of rural lands to urban uses is relatively small. In rapidly growing urban areas, increasing urbanization may be viewed as destroying agricultural land and exaggerating urban environmental problems. Urbanization that takes place in remote areas with high scenic and recreational values may be viewed as potentially damaging to important ecosystems.

Land-use planning is an important and controversial environmental issue. Although people are concerned about environmental problems, rights of property owners are also an issue, and herein lies the controversy. People are beginning to realize that good land-use planning is essential for sound economic development. People at community, regional, and state levels are recognizing the role of good land-use planning in avoiding conflicts and maintaining a high quality of life. An analogy that has been cited is that when a firm manages its resources efficiently, we call it good business; when a city, county, or state efficiently manages its land resources, we call it good planning.[18] The philosophy of good land-use planning is to conserve natural resources, avoid natural hazards, and protect the environment through use of proven ecological principles.

The process of land-use planning is illustrated in Figure 29.11 and includes the following steps[19]:

- identification and definition of objectives, goals, issues, and problems;
- collection, analysis, and interpretation of data;
- development and testing of alternatives;
- formulation of land-use plans;
- review and adoption of plans;
- implementation of plans; and
- revision and amendment of plans.

Comprehensive Planning

A **comprehensive plan** is an official plan adopted by local government. The plan formally states general and long-range policies concerning future development.[18] The planning process includes development of an environmental inventory, forming the basis for planning and zoning by local governments.

Only a few states in the United States have developed statewide planning programs that require development and approval of a comprehensive plan. Oregon, a leader in the development of rigorous statewide planning, requires cities and counties to develop comprehensive plans that reflect and implement statewide goals. Comprehensive plans in Oregon address 19 statewide goals, including:

- citizen involvement in the planning;
- air and water quality;
- land resources; and
- protection from natural hazards.

Oregon state law requires that all land is planned and zoned on the basis of environmental inventories of resources and hazards. However, it is the responsibility of cities and counties to prepare plans according to state goals and guidelines and to issue permits for development.[18] Comprehensive planning in Oregon is innovative, but controversial as it relates to individual property rights. Nevertheless, it appears to be working, and other states, such as Florida, have developed similar programs.

California has also developed a system of comprehensive planning. Plans in California must address several elements identified by state law, including land use, housing, conservation, open space, and safety (natural hazards).[19] Other elements may be added that are relevant to specific needs.

Because California has the State Environmental Quality Act, environmental review is an important part of the planning process. Local governments

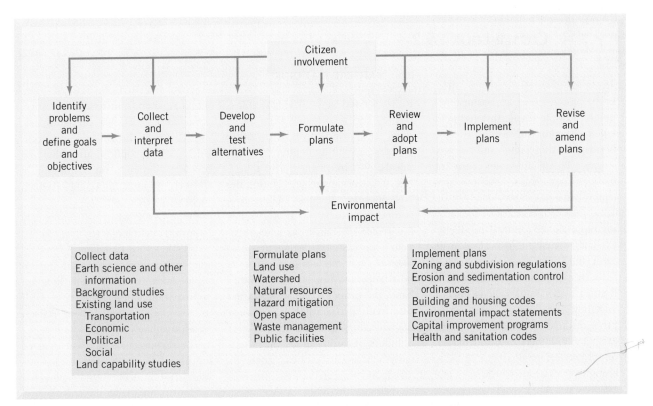

FIGURE 29.11 The land-use planning process. (*Source:* Modified after William Spangle and Associates; F. Beach Leighton and Associates; and Baxter McDonald and Company, 1976, *Earth-Science Information in Land Use Planning—Guidelines for Park Scientists and Planners,* U.S. Geological Survey Circular 721, 1976.)

often combine the comprehensive plan itself and the environmental impact report on the plan to save both time and money.[19]

The land-use element is at the heart of planning in California. It is a statement of land-use policy that

- serves as a guide for public investment in land;

- protects natural and environmental resources;

- alerts planners and the public to environmental hazards;

- provides a framework for planning, budgeting, and constructing facilities, such as roads, sewers, water systems, and schools; and

- assists in the coordination of regulatory policies and decisions.

The land-use element of a comprehensive plan contains maps that designate intensity and uses of the land for a variety of purposes, including housing, industry, open space, recreation, public facilities, and other uses.[18] Ideally, the land-use maps reflect the goals and objectives of the comprehensive plan and are the most visible part of the plan; they are also the most often used part.[19]

Conservation and safety elements are also important in comprehensive planning. The conserva-

tion element has the objective of setting goals and policies concerning the development and use of natural resources, such as air, water, wildlife, forests, and mineral deposits. The safety element is designed to protect people from natural hazards, such as earthquakes, landslides, floods, and wildfires (see A Closer Look 29.2, "Natural Hazards").[19] Included with this element are maps that identify hazardous areas, such as active faults and floodplains. The safety element may also cover such issues as disposal of hazardous wastes, transport of hazardous materials, and failure of utility services.[19]

Regional Planning

Land-use planning occurs at a variety of levels, from local to county to state and less frequently at the regional (such as multistate) level. Probably one of the earliest examples of regional planning in the United States goes back to 1933, when Franklin D. Roosevelt introduced his New Deal, which included a number of programs, such as the Tennessee Valley Authority (TVA). The regional plan for the Tennessee River established a semi-independent authority whose responsibility was to promote economic growth and social well-being for the people of the region. The area included parts of seven states and

CLOSER LOOK 29.2

NATURAL HAZARDS

In the past 20 years, approximately 3 million people on Earth have been killed by natural disasters, such as floods, earthquakes, and violent storms. Financial losses resulting from damage to property and agriculture exceeded $300 billion in the same period. If social losses such as unemployment, reduced productivity, and mental anguish are considered, the losses are much larger.[20] However, we must understand that Earth is a dynamic evolving system, and there have always been natural processes that are hazardous to people. On both personal and institutional levels, our goal should be to recognize hazardous processes that produce disasters, avoid them where possible, and minimize their threat to human life and property.

A look at mean annual losses in the United States from several natural hazards provides insight into the magnitude of the problem. Table 29.2 summarizes information about selected natural hazards or processes for the United States. The largest loss of life every year results from tornadoes and windstorms, but other processes, such as lightning strikes, floods, and hurricanes, also take a heavy toll in human life. During the past 50 years, the loss of life from natural hazards in the United States has declined sharply as a result of better warning, forecasting, and prediction of some hazards, such as hurricanes and floods. On the other hand, the property damage resulting from natural hazards has increased dramatically as a result of the larger numbers of people living in potentially hazardous zones, such as floodplains, steep slopes prone to landslides, low-lying coastal areas vulnerable to hurricanes, and lands likely to experience periodic wildfire.

Adjustments to Hazards

Major adjustments to natural hazards and processes include land-use planning, construction of structures to control natural processes, insurance, evacuation, disaster preparedness, and bearing the loss (doing nothing). Which option an individual chooses depends on a number of factors, the most important of which is hazard perception.

One of the most environmentally sound adjustments to hazards involves land-use planning. For example, people can avoid building on floodplains, in areas where there are active landslides or active fault traces, and in areas where coastal erosion is likely to occur.

In recent years a good deal of effort has been directed toward understanding how people perceive various natural hazards. This is obviously an important endeavor because the success of hazard reduction programs depends on the attitudes of the people likely to be affected by the hazard. For example, it may be difficult to develop an effective earthquake hazard reduction program where strong earthquakes occur only once every few generations. Similarly, it is difficult to tell an individual who has lived many years in a particular home on a floodplain that the area is very dangerous because the floodplain is inundated on average every 100 years. Because flooding at a particular site may occur infrequently, the individual may not perceive flooding as a serious hazard. Although there may be an adequate perception of hazards at the institutional level, this perception may not filter down to the general population. This situation is particularly true for hazards that occur infrequently.

Population Increase and Natural Hazards

We have said that population increase is one of our most serious environmental problems. As the population increases and puts greater

TABLE 29.2	Natural Hazards in the United States		
Hazard	**Deaths per Year**	**Occurrence Influenced by Human Use**	**Catastrophe Potential[a]**
Flood	86	Yes	H
Earthquake[b]	50+?	Yes	H
Landslide	25	Yes	M
Volcano[b]	<1	No	H
Coastal erosion	0	Yes	L
Expansive soils	0	No	L
Hurricane	55	Perhaps	H
Tornado and windstorm	218	Perhaps	H
Lightning	120	Perhaps	L
Drought	0	Perhaps	M
Frost and freeze	0	Yes	L

Source: Modified after G. F. White and J. E. Haas, 1975, *Assessment of Research on Natural Hazards,* MIT Press, Cambridge, MA.

[a]Catastrophe potential: H = high; M = medium; L = low.

[b]Estimate based on recent or predicted loss over 150-year period. Actual loss of life and/or property could be much greater.

demands on our land and resources, the need for planning to minimize losses from natural hazards and processes also increases. This need is dramatically illustrated by the recent loss of thousands of lives in Colombia and more than 100,000 lives in Bangladesh.

The Colombian volcano Nevado del Ruiz erupted in February 1845, producing a mudflow that roared down the east slope of the mountain. The mudflow killed approximately 1000 people in the town of Ambalema, located on the banks of the Lagunilla River, 80 km (50 miles) from the volcano's summit. Deposits from the mudflow produced rich soils at a site 32 km (20 miles) up the river valley, and an agricultural center developed there. The town supported by the area was known as Armero, and by 1985 it had a population of about 22,500.

On November 13, 1985, a mudflow buried the newer town, leaving about 20,000 people dead or missing (Figure 29.12). The death toll from the mudflow was 20 times higher than the toll 140 years earlier because of the population increase over that period. Ironically, the same force that stimulated development and population growth in the area later decimated it.[21]

A cyclone struck the low-lying Bangladesh coastal area in 1970, killing approximately half a million people (as discussed in Chapter 5). (A cyclone is the same as a hurricane; it is called a cyclone when it occurs in the Southern Hemisphere.)

The hazard in Bangladesh is very much associated with how people perceive the hazard and their economic and cultural backgrounds. The poorer people living in the most hazardous areas know that the hazard is real, but because of economic factors, they are unable to live in safer areas. Even when shelter is available from a storm, people are often reluctant to leave their homes. For example, during the 1970 cyclone, 22 people in a single family died only 400 m (about 0.25 mile) from a community center that offered safety. The greater problem, however, is that about 10 million people are living in low-lying areas of the country and evacuation is very difficult. This situation was dramatized again in 1991 when another very high magnitude cyclone struck Bangladesh. Although thousands of people were able to make their way to shelters, more than 100,000 people lost their lives and property damage was extensive. The problem in Bangladesh is at least as much related to the numbers of people as to the hazard itself. If there were fewer people, they could live in less hazardous areas and still farm the flood-prone lowlands. Since the 1991 storm over 1000 brick and concrete shelters have been constructed to provide space for people and livestock during storms. As a result, a cyclone in 1994 claimed fewer than 200 lives because many of the people took shelter.[22]

FIGURE 29.12 The town of Armero, Colombia, following a mudflow produced by a volcanic eruption. Much of the town was buried in mud that claimed as many as 20,000 lives.

was economically depressed at the time the authority was established. There had been rampant exploitation of timber and petroleum resources, and the people living in the region were some of the poorest in the entire country.[23] Today the TVA is reported to be one of the best examples of regional planning in the world. It is characterized by multidimensional and multilevel planning to manage land and water resources. The authority is involved in the production and regulation of electrical power, flood control, navigation, and outdoor recreation.

Green Plans: National Planning

People around the world are becoming concerned with global environmental problems, such as ozone depletion, potential global warming, and acid rain. On local and regional scales, attention is focused on air and water pollution, management of toxic materials, and management of resources. Governments around the world are spending billions of dollars in attempts to solve pressing environmental problems.

Green plans are long-term strategies for identifying and solving global and regional environmental problems. At the philosophical heart of green plans is sustainability, the notion that we can live on this planet and meet our present needs and at the same time ensure that future generations will inherit a quality environment.

The development of national policies for environmental protection and restoration will involve cooperative relationships between various groups, such as government agencies, industry, and labor. Those in favor of developing green plans point out that for the plans to be successful nations will have to place a high priority on environmental issues. Green plans are being implemented today in several countries, including Canada and the Netherlands, and planning is underway in a number of other nations around the world. Nevertheless, environmental plans at the national level are often fragmental rather than comprehensive. The United States, for example, still has no comprehensive land-use planning legislation but has implemented planning for certain land uses, particularly those that affect very sensitive areas, such as wilderness areas and coastal zones.

Coastal Zone Planning

Although coastal areas vary in topography, climate, and vegetation, they are all generally dynamic environments. As we saw in the Cape Hatteras example, continental and oceanic processes converge along coasts to produce a landscape that is characteristically capable of rapid change. The impact of hazardous coastal processes is considerable, because many populated areas are located near the coast. In the United States it is expected that most of the population will eventually be concentrated along the nation's 150,000 km (93,000 miles) of shoreline, including the Great Lakes. Today, the nation's largest cities lie in the coastal zone and approximately 75% of the population lives in coastal states.[24] Because of existing and potential conflicts in the coastal zone, the U.S. Congress in 1972 passed the Coastal Zone Management Act, which was amended in 1976.

The purpose of the act was to establish a federal program to assist states in developing land-use plans for coastal areas. In particular, the act mandated that individual states define the boundaries of the coastal zone, specify permissible land uses within the zone, and address issues of particular public concern (such as public access to the coastal zone and protection of fragile environments or endangered species). All coastal states are now involved to a lesser or greater degree in coastal zone management.

The people of North Carolina, recognizing the importance of planning for their fragile coastal environments along barrier islands, passed the Coastal Area Management Act of 1974. This act recognizes the people's need to enjoy the aesthetic, cultural, and recreational qualities of the natural coastline. One major objective of the act is to provide a management system that allows preservation of estuaries, barrier islands, sand dunes, and beaches so that their natural productivity and biological, economic, and aesthetic values are safeguarded. The second objective is to ensure that development in the coastal areas does not exceed the capabilities of the land and water resources.

The North Carolina act establishes a cooperative program of coastal management between local and state governments in which local government has the initiative and does the planning. The role of state government is to set standards and to review and support local government in the planning program. There are three main steps in planning: development of state planning guidelines for coastal areas; development and adoption of a land-use plan for each county in the coastal area; and use of plans with criteria for issuing or denying permits to develop land or water resources within the coastal areas.

An interesting case in coastal management occurs in Alaska, where some of the boundaries for the coastal zone are set many kilometers inland, at boundaries of watersheds. The rationale is that the streams that flow to the ocean are natural migratory routes for fish, such as salmon, and thus there is a natural tie between the rivers and the coastal zone. Development, such as urbanization or timber harvesting, within river basins must be reviewed by the coastal zone management system as well. Although the boundaries in Alaska seem rather remote from the ocean in some instances, the program appears to be effective in protecting marine and coastal resources (in this case, the fish).

Recreation and Planning

Planning for recreational activities on government lands (including national forests and national parks) is becoming a controversial issue. Some regulation seems necessary if environmental degradation resulting from the recreational activities of people is to be minimized. In some popular areas such as Yosemite National Park and the Grand Canyon, protecting the environment may simply require limiting the number of permits to trails or to rivers. For example, the number of people allowed to raft in the Grand Canyon is limited to protect the river environment from overzealous recreational land use. Limiting people in national parks is fairly straightforward, and most visitors understand the necessity to protect parklands. In fact, it is a congressional mandate that park environments be maintained for future generations. On national forest lands, however, there is more controversy concerning controlling land use for recreational purposes. Managers for national forests tend to accommodate various interest groups through a concept known as *multiple use* (see Chapter 13). This practice has had variable success because some uses are basically incompatible, as illustrated in A Closer Look 29.3, "Off-Road Vehicles."

Planning for National Parks

Plans for many of the national forest and national park lands in the United States have been or are being developed. Developers of these plans tend to consider a spectrum of recreational activities and attempt to balance the desires of several user groups. At the extremes, certain areas have been set aside for intensive off-road vehicle use and other areas have been closed entirely. For example, severe 1996–1997 winter floods in Yosemite National Park damaged roads, bridges, campgrounds, and other human structures. Closing the park for a period has caused some rethinking of goals and objectives of park management. There is discussion of returning some land claimed by the floods to natural ecosystems, further limiting or eliminating private vehicles in parts of the park, and further limiting the number of visitors to the park. In wilderness areas limits are placed on the number of people admitted, and coastal areas may have regulations limiting such activities as jet skiing and surfing in swimming areas. In regions where endangered species exist, there may be more stringent regulations to govern the activities of people. For example, in Yellowstone National Park in Wyoming and Montana, special consideration is given to the grizzly bear habitat by controlling where humans may venture.

Other recreational activities that are or may become subject to regulation include hiking, camping, fishing, boating, skiing, snowmobiling, and such recently popularized activities as treasure hunting, which includes panning for gold.

Activities that occur on government lands may be more easily regulated than those occurring elsewhere, but the preservation of a quality environment is a concern common to all areas, regardless of land ownership. Finally, park management may be difficult if goals are not clear and natural processes understood. See A Closer Look 29.4, "Chankanab Lagoon National Park."

29.4 GLOBAL FORECASTING

Global effects of our modern technological civilization have been discussed in earlier chapters. The burning of fossil fuels is adding a significant amount of carbon dioxide to the atmosphere; lead used in gasoline fuels has spread to the glaciers of Greenland; and DDT and other pesticides and artificial organic compounds are found in marine organisms inhabiting the Antarctic. We can imagine other effects, such as those of a nuclear war or of a large-scale release of a toxic compound. Few have yet tried to plan on a global level. However, in the last decade procedures for **global forecasting** and evaluation have been developed.

An international group of business executives, intellectuals, and government officials founded the Club of Rome, an organization whose objectives are to promote a better understanding of humanity's predicament as it relates to these global concerns; to disseminate information about this predicament; and to stimulate the development of new attitudes, policies, and institutions to redress the present situation.

From the concerns of this organization came the development of computer simulation models: models of global phenomena (such as human population growth), the use of the world's resources, and human impacts on the biosphere. The results of the use of these models were reviewed in *The Limits to Growth,* a controversial book.[27] The book's forecasts showed some dire consequences of our current activities. Supporters of *The Limits to Growth* hailed it as a new approach that could help save us from ourselves by forcing us to recognize the true limits to our uses of the biosphere. Opponents condemned it as another example of the GIGO rule of computer simulation—garbage in, garbage out (garbage here means incorrect data or assumptions). They claimed that it said nothing more than Malthus said hundreds of years before (see Chapter 5).

A CLOSER LOOK 29.3

OFF-ROAD VEHICLES

The widespread use of off-road vehicles (ORVs) has had a significant impact on the environment. There are now more than 12 million ORVs, many of which are invading the deserts, coastal dunes, and forested lands of the United States. Intensive use of ORVs causes soil erosion, changes in hydrology, and damage to plants and animals. The problem is severe. A single motorcycle need travel only 7.9 km (4.9 miles) to affect 1000 m² (about 11,000 ft²); a 4-wheel-drive vehicle affects the same area by traveling only 2.4 km (1.5 miles). In some desert areas the tracks produce scars that remain part of the landscape for hundreds of years.[25,26]

The ORVs cause mechanical erosion and facilitate wind and water erosion of materials loosened by their passing (see Chapter 3). Runoff from ORV sites is as much as eight times greater than that for adjacent unused areas, and sediment yields are comparable to those found on construction sites in urbanizing areas.[26] Figure 29.13 shows an ORV site near Frazier Mountain, California.

Hydrologic changes from ORV activity are primarily the result of soil compaction, which reduces the soil's ability to absorb water, makes the water less available to plants and animals, and changes the variability of soil temperature. The result of intensive ORV activity is a combination of soil erosion, compaction, temperature change, and moisture content change.[26]

Environmental planning that encompasses the use of ORVs is a difficult task. There is little doubt that some land must be set aside for ORV use. The problems are how much land should be involved and how to minimize environmental damage. The possible effects of airborne soil removal (by wind) must also be evaluated carefully, as must the sacrifice of nonrenewable cultural, biological, and geologic resources.[25] Intensive ORV use is incompatible with nearly all other land uses, and it is very difficult to restrict ORV damages to a specific site. Material removed by mechanical, water, and wind erosion will always have an impact on other areas and activities.[25,26]

In recent years there has been a trend in some areas toward the use of all-terrain bicycles (ATBs, or mountain bikes) that may be ridden on trails. Controversy is growing over how to manage ATBs, particularly in areas frequently used by pedestrians and people on horseback. Riders of ATBs argue that they cause little environmental disruption and damage trails less than horses do. As long as the number of mountain bikes remains small, this may be true. However, because bicycles are not nearly as expensive as horses and they do not have to be fed, we can expect many more ATBs than horses on trails. There is little doubt that if the mountain bike sport continues to grow, intensive use in some areas will be associated with environmental degradation.

The management of public lands for recreational activities requires planning at a variety of levels, with considerable public input. For example, when a national forest is developing management plans, there is often a series of public meetings in which people are advised of the process by which the plan is developed and asked for ideas and suggestions. Maximizing public input provides better communication between those managing resources and those using them for recreational purposes. However, government officials and scientists involved in developing use plans are faced with complex land-use problems at a variety of levels. Often the complexity of a problem is such that no easy answers can be found. Nonetheless, because action or inaction today can have serious consequences tomorrow, it seems best to have conservative plans to protect and preserve a quality environment for future generations.

FIGURE 29.13 Trails and gullies on this site are caused by extensive use of off-road vehicles (mostly motorcycles). Location is Frazier Mountain in the western Transverse Ranges, California.

A CLOSER LOOK 29.4

CHANKANAB LAGOON NATIONAL PARK

The Yucatan Peninsula is a low-lying limestone plateau at the southern tip of Mexico, bordered on the west by the Gulf of Mexico and on the east by the Caribbean Sea (Figures 29.14*a* and *b*). Most of the rain that falls on the limestone quickly infiltrates into the rock, so there are few surface streams or rivers. Movement of groundwater through the limestone dissolves the rock along preferred directions, producing systems of caverns and caves. These occasionally collapse, forming sinkholes, which the Maya people used as natural wells. Because limestone is also a relatively soft stone, it is easily cut and carved. It provided an abundant building material for the large pyramids and buildings constructed by the Mayan between the years A.D. 500 and 1000. The island of Cozumel is a small island approximately 50 km long by 15 km wide (30 by 9 miles) located approximately 25 km (about 15 miles) off the coast of the peninsula.

Cozumel Island, like the nearby mainland, is composed of limestone and has a very recent geologic origin, probably emerging from the sea during the last million years or so. The island also has a very interesting record of human history. The Mayan who inhabited the island prior to the arrival of the Spanish in 1518 believed the land was a sacred shrine and referred to it as the island of the swallows. It is believed that many Mayan women journeyed to the island by boat to worship the Goddess of Fertility, and numerous archaeological sites on the island contained small dolls sacrificed during fertility ceremonies. The island was decimated in 1519, when many of the Mayan temples were destroyed by the invading Spanish. Outbreaks of smallpox and other diseases reduced the Mayan population from about 40,000 in 1519 to only about 300 in 1570. The people completely abandoned the island. It was used in the seventeenth century by pirates and was finally resettled again in about 1848.

From an environmental viewpoint the island faces many challenges

FIGURE 29.14 The location of (*a*) Cozumel Island and (*b*) Chankanab Lagoon in the Caribbean Sea, near Cancun, Mexico. (*c*) The coast at Chankanab Lagoon.

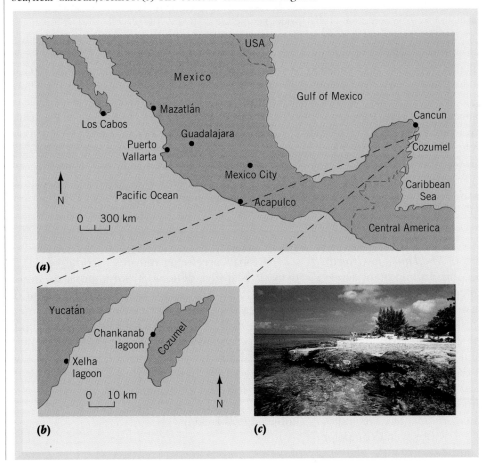

today. The water supply is groundwater from underground limestone rocks. That supply is extremely vulnerable to pollution, and waste disposal practices are not up to high standards. Nevertheless the people have a strong sense of environmental consciousness and are thinking seriously about the future welfare of the land. In particular there have been national parks set aside, including Chankanab Lagoon, located approximately 10 km (6 miles) south of the only city on the island, San Miguel. The government manages the site as a nature reserve, and no fishing or collecting is allowed. The clarity of the water is incredible at Cozumel, and the waters abound with multicolored fish. The people evidently have decided that it is more advantageous to conserve resources in parks than to exploit them. There is more concern given to long-term planning for the conservation of resources than to short-term exploitation. Although Cozumel does not have the extensive high-rise development that is seen on the Mexican mainland at Cancun, there is some evidence that high-rise development is likely to occur south of the city of San Miguel. This situation may intensify in the future, and there is some concern that the island may go down the path of nearby Cancun—intensive development and alteration of the natural environment to the extent that it scarcely resembles the natural state. However, the people of Cozumel are environmentally aware and are attempting to develop long-range comprehensive land-use planning that may better protect this beautiful island.

Chankanab Lagoon is an interesting national park because the government is trying to restore the lagoon and, at the same time, provide a tourist attraction. Several decades ago the lagoon was in very bad shape; trash and other materials had accumulated in it. Originally a series of caves connected the lagoon with the ocean and allowed circulation of the ocean water into the lagoon. When a road was built, the caves collapsed, causing the circulation to cease. As a result, the lagoon died. Approximately 15 years ago, a restoration project began that opened the cave system again, allowing circulation of seawater. Much of the trash was removed. At that time, a beachfront was developed, consisting of restaurants and other facilities for tourists (Figure 29.14c). Because the natural coastline at the site is rocky and contains very little sand, the beach was built by constructing small seawalls and then filling the land side with sand to produce a beach approximately 30 m (100 ft) wide. People are allowed to swim in the ocean side of the lagoon, and the fish and marine life are completely protected. However, the creation of the sandy beach turned out to be a major problem. In 1989, Hurricane Gilbert struck the island. The high winds and waters moved the beach sand into the shallow lagoon, nearly filling it and burying the coral, once again killing the lagoon. Restoration work began in 1991 to dredge out part of the lagoon, and seawater again circulates from the ocean to the lagoon and back with the tides. Unfortunately, the long-term restoration plan is likely to fail because hurricanes can be expected again, and they will again wash sand into the lagoon.

The Chankanab Lagoon story emphasizes problems that may arise when conflicts of management occur. It may be difficult to both satisfy the desires of the tourist industry and protect the natural environment. Nevertheless a management plan could be developed to work harmoniously with the natural system and preserve the lagoon. A better design might have been to build wooden walkways with seats from which people could observe the reef and ocean, rather than to import the sand. If true restoration of the lagoon is to be successful, the management plan will have to change.

The situation at Xelha Lagoon on the Yucatan's east coast is quite different. That lagoon is much larger, and there has been very little importation of sand. Thus the lagoon has been preserved and remains a healthy ecosystem. At that site there are also restaurants and other facilities, but they do not intrude directly on much of the total lagoon environment. It is evident that the people of Mexico and the Yucatan Peninsula have a strong growing consciousness for environmental concerns. However, as in many other places, additional expertise concerning natural processes is needed when developing management plans.

Since the publication of *The Limits to Growth,* global forecasting has become an important aspect of modern civilization's attempt to deal with its own global effects. The Council on Environmental Quality was directed by the president of the United States to conduct a study of the probable changes in the world's population, natural resources, and environment through the end of the century. This study resulted in *The Global 2000 Report to the President,* which used several methods to assess the status of population resources and environment by the end of the century, including a review of several computer models for global forecasting.[28]

Global forecasting remains a controversial activity. It is always difficult, if not impossible, to predict the future. The more we understand about the processes that govern change in the environment, the more likely it is that our forecasts will be helpful. Such forecasts are perhaps most valuable in showing us general effects, or the consequences of what we assume governs these processes. In this way the forecasts force us to recognize the meaning of what we think we know and suggest to us where our knowledge is lacking.

What Is the Impact of Introduced Species?

Lake Victoria in East Africa, with a surface area of almost 70,000 km² (27,000 mi²), a length of more than 338 km (210 miles), and a mean depth of 40 m (131 ft), is the largest body of fresh water in the tropics. Until recently, it was renowned for its variety of fish species, 350 or more, most of which existed nowhere else. One group, the *haplochromine cichlids,* underwent such extensive diversification in only 750,000 years that Lake Victoria has been a natural laboratory for the study of evolution. Different species have evolved to feed on different types of food (algae, plants, insects, mollusks, arthropods, other fish) and at various depths and locations in the lake.

More than 30 million people in the surrounding nations of Tanzania, Uganda, and Kenya have depended on the fish as their main source of protein. Since the 1980s, however, the cichlid populations have declined disastrously, threatening the food supply. Scientists attribute the decline of the cichlid populations to the Nile perch, first introduced to Lake Victoria in the 1950s, and to overfishing.

Although the Nile perch is found in other African waters, it was new to Lake Victoria. An aggressive and voracious predator that produces numerous eggs, it has preyed on the cichlids, which typically produce relatively small numbers of eggs. In place of hundreds of species, only three, the Nile perch, a native minnow, and the Nile *tilapia* (another introduced species), now dominate the ecosystem. From the 1950s until 1978, cichlids were numerous and perch represented less than 2% of the lake's biomass. In the early 1980s, however, the perch underwent a population explosion, so that by 1986 they constituted 80% of the biomass and the native cichlids constituted less than 1% (see graphs); tilapias and minnows made up the other 20%. Having destroyed most of the cichlids, the perch have turned to feeding on a tiny native shrimp and the young of their own species.

Flourishing populations of the perch, which can grow to be 2 m (6.5 ft) long and weigh 91 kg (200 lb), have made Lake Victoria the world's most productive lake, with a yield of 200,000–300,000 metric tons per year. This productivity is now the basis for an export fishing industry. Technologically more advanced fishing equipment has replaced the fine-mesh nets used for the smaller, native fish. Refrigeration and modern processing have replaced traditional processing methods: sun drying, smoking, and frying. The indigenous people, who depended on the native fish for protein, now find the perch too expensive and the cichlids too few. They are forced to live on a meager catch and discarded heads and tails of perch. The economic boom from exporting perch has put the livelihood of small-scale fishermen and fish vendors at risk.

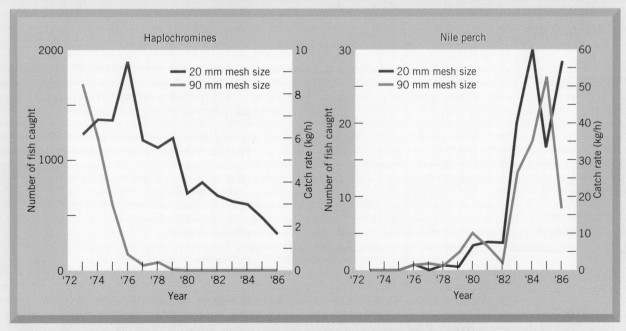

Source: F. Witte, T. Goldschmidt, J. Wanink, M. van Oijen, K. Goudswaard, E. Witte-Maas, and N. Bouton, 1992, *Environmental Biology of Fishes,* 19(1), p. 10.

Even before the introduction of the perch, the lake had experienced dense algal blooms, but oxygen levels declined and eutrophication increased in the 1980s. Because perch require high oxygen levels, many of them died off (recall the discussion of biochemical oxygen demand and eutrophication in Chapter 20). Masses of dead perch, changes in their diet, and the smaller size of those caught are evidence of overfishing of the perch. Consequently, scientists have questioned whether the perch population itself may be headed for a crash. Other signs of trouble for Lake Victoria are increased turbidity, decline of papyrus swamps, increase in snails, larger and more prevalent lake flies, decline of other fish, and an invasion of water hyacinth, a floating plant that chokes out native vegetation.

Critical Thinking Questions

1. There is some controversy about the relative contributions of the perch and overfishing to the decline of the cichlids. Describe a controlled study that could be used to settle the question.

2. Based on the information given, draw a food web for each of the following times in the history of Lake Victoria: (a) prior to the introduction of the perch; (b) shortly after the introduction of the perch; and (c) after the population of cichlids crashed. What would be the effect of the perch on the food web?

3. Draw two food chains, each based on the first two food webs. How does the number of trophic levels in the second compare to the first? It has been estimated that there is an 80% energy loss in the second food chain compared to the first. Why would this be? (Hint: Review Chapter 15.)

4. What are the possible causes of eutrophication before and after introduction of the perch? In the case of indirect causes, what are the pathway(s) by which they could lead to nutrient enrichment of Lake Victoria?

5. Nile perch were introduced into Lake Victoria in spite of protests from some scientists. Develop an outline for an EIS that should have been conducted before introducing the perch.

References

Baskin, Y. 1992. "Africa's Troubled Waters," *Bioscience* 42(7): 476–481.

Bruton, M. N. 1990. "The Conservation of the Fishes of Lake Victoria, Africa: An Ecological Perspective," *Environmental Biology of Fishes* 27: 161–175.

Kaufman, L. 1992. "Catastrophic Change in Speciesrich Freshwater Ecosystems," *Bioscience* 42(11): 846–858.

Ribbink, A. J. 1987. "African Lakes and Their Fishes: Conservation Scenarios and Suggestions," *Environmental Biology of Fishes* 19(1): 3–26.

Witte, F., Goldschmidt, T., Wanink, J., van Oijen, M., Goudswaard, K., Witte-Maas, E., and Bouton, N. 1992. "The Destruction of an Endemic Species Flock: Quantitative Data on the Decline of the Haplochromine Cichlids of Lake Victoria," *Environmental Biology of Fishes* 34: 1–28.

SUMMARY

- The major components of an environmental impact statement (EIS) are (1) summary; (2) purpose and need of project; (3) rigorous comparison of alternatives; (4) succinct description of the environment affected; and (5) discussion of the environmental consequences of the proposed project and possible alternatives.

- Scoping is the process of early identification of important environmental issues requiring detailed evaluation. Mitigation is the process that identifies actions that will avoid, lessen, or compensate for anticipated adverse environmental impacts.

- Elements and capabilities of Geographic Information Systems (GIS) are data acquisition; data manipulation and analysis; and generation of products, such as statistical analysis and maps.

- Mediation is a method of conflict or dispute resolution that seeks to avoid expensive and lengthy litigation by arriving at a conclusion whereby each party gains enough from a compromise to prefer an agreed settlement to litigation.

- International environmental law is difficult because there is no world government with enforcement authority; it depends on agreements and treaties. Often it is difficult to gain acceptance for the contents of such agreements and to fund and enforce them.

- Land-use planning is a controversial topic with important environmental implications. The planning process includes defining goals and objectives; collection and analysis of data; development of alternatives; formulation of plans; and review, implementation, and revision of plans.

- The most preferable adjustment to natural hazards is land-use planning to avoid the hazard.

Other adjustments are construction of structures to control hazards; insurance; evacuation; disaster preparedness; and bearing the loss (doing nothing).

- As the human population increases, more people are forced to live in areas likely to be frequented by hazardous processes, such as flooding and storms. Therefore, greater loss of life and property damage can be expected, especially in developing countries where population increase is the greatest.

- Global forecasting is controversial because the models that predict the future make controversial assumptions and are premature at best. Nevertheless, global forecasting provides an opportunity to better understand general effects or the consequences of our actions and to suggest where more knowledge is needed.

REEXAMINING THEMES AND ISSUES

Human Population: As human population continues to increase on a regional and global level, careful planning will become even more important to ensure that we make wise decisions concerning where we live, how we conserve and preserve ecosystems, and how we construct our urban environments.

Sustainability: Environmental planning is at the heart of sustainable development. Planning decisions we make today and in the future will greatly impact the quality of the environment we leave for future generations. Concepts of environmental impact analysis, mitigation of adverse environmental impact, and development of environmentally preferable adjustments to natural hazards are integral land-use planning methods for obtaining sustainability.

Global Perspective: We are becoming a global community, and our planning needs must include a significant global perspective. Development of international environmental agreements is becoming increasingly important as are continuing efforts in global forecasting. We are only beginning to plan on a global level.

Urban World: Because human migration is generally toward cities and urban areas, these areas should be the focus of much of our attention when it comes to environmental issues. One of our most pressing environmental needs is to ensure that our urban environments are carefully planned to maximize human well-being and minimize effects of natural hazards and pollution of our land, air, and water.

Values and Knowledge: Because we value our environment—from the local levels to global systems—we must continue to develop the knowledge, technology, and tools to assist in the environmental planning process. Of particular importance is that we recognize where knowledge is lacking in the solving of particular environmental problems so that we might focus on those areas. We must also recognize that people's values with respect to resource use and land use are variable, resulting in periodic conflicts between those who wish to use the environment and those who wish to preserve it. Environmental mediation can help resolve conflicts when values collide.

KEY TERMS

common law *613*
comprehensive plan *624*
environmental impact report (EIR) *613*
environmental impact statements
 (EISs) *612*

Geographic Information System
 (GIS) *612*
global forecasting *629*
green plans *628*
land-use planning *624*
mitigated negative declaration *613*

mitigation *612*
negative declaration *613*
public trust *615*
record of decision *612*
scoping *612*

STUDY QUESTIONS

1. Make a map of your neighborhood. Does it seem planned with the environment in mind?

2. What is meant by "mitigation of an environmental impact"?

3. An expert on impact analysis tells you, "Mitigation is the impact. Once you have determined the mitigating factors, you have determined the impact." Do you agree?

4. What major steps must be taken in the assessment of any environmental impact?

5. Discuss the elements of a land-use plan. Which element is likely to be most controversial? Why?

6. Why is planning immediately following disasters loaded with potential problems?

7. Discuss the major components of an environmental impact statement. Which are the most necessary? Why?

8. What are the major advantages of the scoping process in environmental impact analysis?

9. What are some potential conflicts in the management of recreational activities?

10. What are the advantages and disadvantages of regional national planning and what are some potential problems?

11. Discuss the role of mediation in environmental law. What is necessary for it to work?

12. How has population increase resulted in natural hazards taking a greater toll in human life than would be expected by the simple increase in population?

FURTHER READING

Arnold, F. S. 1995. *Economic Analysis of Environmental Policy and Regulation.* New York: Wiley. Presents a wide variety of practical applications of economics to environmental regulatory and policy analysis.

Bass, R. E., and Herson, A. I. 1993. *Mastering NEPA: A Step-by-Step Approach.* Point Arena, CA: Solano. For those with the interest, this book includes the background and implementation of NEPA, written by environmental consultants with a thorough understanding of the act.

Bregman, J. I., and Mackenthum, K. M. 1992. *Environmental Impact Statements.* Boca Raton, FL.: Lewis. This book provides an up-to-date examination of the many scientific and technical aspects of EIS preparation.

Jain, R. K., Urban, L. V., Stacey, G. S., and Balbach, H. E. 1993. *Environmental Assessment.* New York: McGraw-Hill. A good book covering the importance of and steps involved with investigating environmental considerations for a proposed action.

Kubasek, N. K. and Silverman, G. S. 1994. *Environmental Law.* Englewood Cliffs, NJ: Prentice-Hall. A look at how our legal system is used to assist in environmental policy and preservation.

McHarg, I. L. 1971. *Design with Nature.* Garden City, NY: Doubleday. This is a classic book on environmental planning.

Orlof, N. 1979. *The Environmental Impact Statement Process: A Guide to Citizen Action.* Washington, DC: Information Resources. This is a simple-to-understand explanation of the entire EIS process, including both political and scientific aspects. It emphasizes the role that local citizens can play in the process.

Rohse, M. 1987. *Land-Use Planning in Oregon.* Corvallis, OR: Oregon State University Press. The author provides a good discussion of basic land-use planning.

Teubner, G., Farmer, L., and Murphy, D. (eds.). 1994. *Environmental Law and Ecological Responsibility: The Concept and Practice of Ecological Self-organization.* New York: Wiley. Argues that accepting that the old-style command-and-control regulation of business has reached its limits; a new, externally induced, internal self-organizing process may render business more sensitive to environmental concerns.

INTERNET RESOURCES

Cornell University Law School: *http://www.law.cornell.edu/topics/environmental.html*—A great resource for information on federal environmental regulations, NEPA, state environmental regulations, and international environmental law.

Council on Environmental Quality NEPA Net: *http://ceq.eh.doe.gov/nepa/nepanet.htm*—From the political body created to administer NEPA, read the full text of NEPA, regulations for implementing NEPA, and frequently asked questions about NEPA. Scan environmental impact statements of federal projects.

Today's swimmers enjoy Walden Pond, where Henry David Thoreau once sought nature's solitude.

CHAPTER 30

INTEGRATING VALUES AND KNOWLEDGE

CASE STUDY

Sea Lions and Steelhead Trout: A Conflict of Values

The conflict between the conservation of sea lions and the maintenance of fisheries in the Pacific Northwest illustrates the importance of values in resolving environmental issues. One of the truly amazing fish in the Pacific Northwest is the steelhead trout (Figure 30.1). A variety of the rainbow

FIGURE 30.1 Protected by the Marine Mammal Protection Act, some California sea lions feed on endangered stocks of salmon, thus creating a new kind of environmental conflict.

trout, the steelhead is born in rivers, but unlike its strictly freshwater relative, it spends its adult life in the ocean, returning to the river to spawn.

Steelhead trout are highly valued by the people of the Pacific Northwest. They are economically valuable (fishing for steelhead is a recreational activity that also provides economic return) and they are food for other wildlife. In addition, their abundance is used by some people as a measure of the environmental health of the area.

In recent decades there has been a pronounced decrease in the productivity of steelhead trout. The causes of the decline have been a matter of controversy. Fishermen claim that the decline is due to the destruction of the freshwater spawning and rearing habitats as a result of logging, agriculture, and urbanization. Loggers claim that the decline is the result of overfishing, both by people and by wild predators, including marine mammals. Some oceanographers point out that the abundance of the types of fish that live in both fresh and salt water varies greatly with changes in ocean upwellings and that the decline may be affected by factors over which we have no control.[1]

What happens when the population of steelhead is threatened by another species we care about? People also value sea lions, but for different reasons. They are valued for their continued existence as part of the biological diversity on Earth and because people enjoy seeing them. They are not, however, a food for people or the source of a commercial product. Sea lions, like all marine mammals, are protected from such commercial uses in the United States by the 1972 Marine Mammal Protection Act. This law requires that an optimum sustainable population be maintained and that no sea lions (or other marine mammals) may be killed without a permit from the federal government.

The protection of sea lions has led to a rapid increase in their numbers. In the San Francisco Bay area they have become a nuisance to tourists and boaters; farther north, in the state of Washington, they have begun to feed on steelhead trout, whose numbers have been carefully nurtured as an important resource.

Interestingly, it is the management of steelhead trout that has in part created the problem. With the construction of dams in the Northwest, the corridors for migration of the trout have been altered. Often, the trout have to go through a confined area that leads to a fish ladder (a structure that allows the trout to swim upstream around a dam). Sea lions congregate near these structures, where they easily catch fish.

Efforts to remove the sea lions from the area have proved unsuccessful. In one attempt, several sea lions were captured and moved over a thousand kilometers (622 miles) south to the Channel Islands off southern California. Within a relatively short time the sea lions were again seen basking on rocks in the Washington area and presumably feeding on steelhead trout.

Some argue that both sea lions and steelhead, along with all other forms of life, have an intrinsic right to live, and it is our moral obligation to promote their continued existence (see Chapters 7 and 12). However, the attempts to promote the abundance of both the trout and its predator, the sea lion, create a moral conflict. How do we promote the abundance of both when one feeds on the other? How can we achieve a sustainable population of both? Behind the belief that we can maximize the abundance of both is the old idea of a balance of nature. As previously discussed, this is the idea that, left alone, all living things will come into a balance with one another and that the abundance of all life will continue at a maximum.[2] Apparently, sea lions lying in wait by fish ladders are not aware of these old ideas.

Others argue that our first obligation is to provide food and jobs for people and that the continued, uncontrolled growth of the sea lion is a luxury we cannot afford. They contend that we must give first priority to an economic value (see Chapter 25), that food and tourism income provided by the trout take precedence over the conservation of what has become a pest species. They point to studies that show that people move to, retire to, and take vacations in fishing villages, and therefore the presence of a fishing industry, commercial or sport, has many economic benefits.[3]

From our study of population dynamics (see Chapter 5), we know that a continued, uncontrolled increase in the abundance of both trout and sea lions cannot be achieved simultaneously, nor can either of these species increase indefinitely without detriment to the environment. How we resolve these conflicts, which have both a philosophical (moral, ethical, and aesthetic) aspect and a practical (utilitarian) aspect, depends in part on our values and in part on our knowledge and the methods available to us.

LEARNING OBJECTIVES

Now that we have explored the great diversity of subjects that constitute environmental sciences, it is time to review common themes and connections in order to synthesize and integrate what has been learned. The purpose of this chapter is to identify some of the linkages, especially the connection of the different subjects to the larger issue of how we value our environment. After reading this chapter, you should understand:

- Linkages among some of the diverse topics discussed in this book.
- Some of the issues of environmental ethics.
- The various ways in which people have perceived the idea of wilderness throughout history.
- The value that people have placed on wilderness in times past and in the present.
- Ways in which an individual can improve the environment of the future.

30.1 AN INTEGRATED APPROACH

As a first step in attempting to develop an integrated perspective on environmental science, we return to the threads of thought introduced in the first chapter: the human population problem, sustainability, a global perspective, an urban world, and values and knowledge.

Especially relevant to the discussion in this chapter is the importance of values and knowledge. The concept was stated thus: *Although environmental issues are often portrayed as simply a question of facts, central conflicts about the environment have to do with values and knowledge.* As we attempt to deal with the problems of an exploding population, of sustaining our resources for future generations, of preserving our environment on a global level, and of meeting the needs of an urban world, we need to consider values and knowledge. We must know how we are changing the environment and how we can fix it, and we must determine what factors are most important to us.

The solution to the sea lion–steelhead conflict requires a broad base of knowledge beyond the scope of any single traditional academic discipline. Ethics, biology, hydrology, and environmental engineering are among the fields that can contribute to a resolution of this and similar conflicts. In pursuit of the resolution of such environmental conflicts, we begin with the fundamental question, "How can one place a value on aspects of the environment?"

The discussion of critical thinking can help us understand how to analyze an issue and how to present arguments in a logical, and therefore persuasive, manner (see Chapter 2). It should be clear from the opening case study of the present chapter that we must adopt a perspective of the whole environment as a unified system. We see more and more examples of the interconnection between environmental and social and economic issues. For example, it is becoming increasingly clear that scarcity of resources is often at the core of civil instability and strife. Examples range from loss of agricultural land to indigenous people in Mauritania to conflicts between Israelis and Palestinians over access to water resources in the Mideast.[4]

If we place a value on each aspect of the environment separately, without seeing the whole picture, conflicts invariably result. Environmental economics can clarify how to place a monetary or societal value on environmental factors and arrive at a just division of the benefits among peoples and between people and the rest of the environment (see Chapter 25). The scientific fundamentals explained in Parts II and III provide a basis from which we can consider the sustainability of renewable and nonrenewable resources, looking at both their benefits and the problems associated with their uses.

30.2 PLACING A VALUE ON THE ENVIRONMENT

How valuable is the environment? The simplest answer is that a proper environment is essential to sustain all renewable resources—food, fibers, wood, air, water, and even ourselves and our own health. But, as the case study of the sea lion and steelhead trout demonstrates, people have placed many other kinds of values on the environment. The debate over its ethical value is as old as civilization.

The Balance of Nature as an Ancient Idea

Although environmentalism may seem a relatively recent interest, its roots lie deep within human history, society, and psychology. All cultures have had to deal with their environment, which not only provided the resources necessary for life but created challenges and threats as well. Throughout Western civilization, three central questions have been asked about people and nature:

1. What is the condition of nature when undisturbed by human influence?
2. What is the influence of nature on people?
3. What is the effect of humanity on nature, and what is humanity's role within nature?

These are social and personal issues of moral, ethical, religious, and metaphysical importance, and they must be interpreted and reexamined in every age. At many times in history these three questions have aroused controversy, just as they do in our time.

Every human society has had a set of beliefs about nature, the effects of nature on human beings, and the effects of human beings on their natural surroundings. These beliefs have reflected attempts to find order and harmony in nature, to discern a design and purpose for this natural order, and to define the role of humanity in nature.

Technology and the Environment

Although our modern civilization has had many negative effects on the environment, we recognize that modern civilization has also made the environment more livable in many ways. With medical and technological advances, we have better health care, an increased standard of living that allows significant public health and sanitation measures, better control of parasites, such as mosquitoes (and of the diseases they spread), and better ability to protect ourselves from natural hazards, such as earthquakes, hurricanes, and tornadoes. We are able to feed more people, and feed them better, than ever before.

In the past few years, we have begun to learn how we can use our technologies to live in closer harmony with our environment than we have in the recent past. For example, we are attempting to control some pests using more benign methods than were used earlier in the twentieth century, substituting natural, biological control for toxic chemicals. We are experimenting with new ways of growing crops that minimize soil erosion and require fewer chemicals. We are finding new ways to use renewable, nonpolluting sources of energy. We are learning to choose carefully among alternative technologies. At the heart of these new developments is a change in how we perceive our natural environment and how we define our roles and responsibilities.

Whether the benefits of technology will outweigh negative effects in the long run is an open question. The choices we make now will lead us in one of two directions. We can move forward to a future in which our population ceases to grow and we live in harmony with our environment, maintaining our renewable resources and conserving and reusing our nonrenewable ones. Or we can act in ways that will lead to a continued increase in the human population, an impoverished and highly polluted landscape depleted of resources, with soils, forests, and fisheries exhausted and many important species extinct. Our choices will depend in part on our knowledge of the environment, in part on our values.

Ethics and the Environment

In the 1970s, philosophers began to formulate a new field called *environmental ethics,* a study concerned with the value of the physical and biological environment. The focus of this field of study contrasts with traditional ethical studies, which had to do with the relationships among people.[5] (Those familiar with the subject of environmental ethics will recognize that it is much broader than discussed here. Only a few aspects of the academic subject are introduced in this book. In recent years, the study of environmental ethics has grown into a large and complex field.)

As discussed in Chapter 12, there are both practical and moral reasons for placing a value on the environment. Recall that there are four categories of justification: utilitarian, ecological, aesthetic, and moral.

Why do we need a new set of ethics for the environment? The answer includes three factors:

1. *New effects on nature.* Because our modern technological civilization affects nature greatly, we must examine the ethical consequences of these new actions.
2. *New knowledge about nature.* Modern science demonstrates how we have changed and are changing our environment in ways not previously understood, thus raising new ethical issues. For example, until the past decade, few people believed that human activities could be changing Earth's global environment. Now, however, scientists believe that burning fossil fuels and clearing forests have changed the amount of carbon dioxide in the atmosphere and that this may change our climate. Hence we have emphasized a global perspective (see Chapters 3, 4, 15, and 21).[6] This new perspective raises new moral issues.
3. *Expanding moral concerns.* Some people argue that animals, trees, and even rocks have moral and legal rights and that it is a natural extension of civilization to begin including the environment in ethics. These expanded concerns lead to a need for a new ethic.[7]

The Land Ethic

We can see from the preceding discussion that a concern with environmental ethics leads to a discussion of the rights of animals and plants, of nonliving structures, and of large systems that are important to our life support.

The *land ethic* put forward by Aldo Leopold in 1949 affirms the right of all resources, including plants, animals, and earth materials, to continued existence and, at least in certain locations, to continued existence in a natural state.[8–10] This ethic effectively alters the human role—from conqueror of the land to citizen to protector of the environment. This new role requires that we revere and love the land and not see it solely as an economic commodity to be used up and thrown away.

Leopold emphasized our changing sense of ethics through the story of Odysseus, who, on returning from the Trojan War, hanged a dozen slave women for suspected misbehavior during his absence. His right to do this was unquestioned; the women were property, and the disposal of property was entirely up to the owner. Although ethical values have since been extended so that humans are no longer considered the personal property of others, only within this century have moral considerations been extended to include our physical environment.

Ecological ethics limit social as well as individual freedom of action in the struggle for existence.[11] A land ethic assumes that we are ethically responsible not only to other individuals and society but also to the larger community that includes all life, soils, atmosphere, and water, that is, to the environment.

There is a potential source of confusion in distinguishing between an ideal and a realistic land ethic. Giving rights to plants and animals and landscapes might be interpreted as granting to individual plants and animals the fundamental right to live, as in the Indian religion Jainism. However, we must eat to live. Not being able to make our own food from sunlight and minerals, as do trees and crops, we must consume other organisms. Therefore, although the land ethic assigns rights for animals to survive as species, it does not necessarily assign rights to an individual deer, cow, or chicken for that survival. The same argument may be given to justify using stream gravel for construction material or mining other resources necessary for our well-being. However, unique landscapes with high aesthetic value or ecosystems that sustain endangered species need to be protected within this ethical framework.

The land ethic places us in the role of stewards of nature, with a moral responsibility to sustain nature for ourselves and for future generations. According to this view, wilderness has both an intrinsic value and should be maintained for itself and an extrinsic value, necessary to our own survival.

Because our effects on the environment today have consequences for the future, discussions of environmental ethics also involve the rights of future generations. The arguments for and against various principles in environmental ethics are made more complex because of conflicting values. The resolution of the resulting conflict requires that we recognize differing values and that we have a basic scientific knowledge about the environment as well as the ability to clearly formulate a logical argument.

Does Nature Have Rights?

Does nature have rights? This question is one of the growing concerns of those interested in environmental ethics. The question arose as a legal issue in the 1970s, in a case involving the proposed development of Mineral King Valley (Figure 30.2), a wilderness area in the Sierra Nevada Mountains of California. Christopher D. Stone, an attorney, discussed this idea in his article "Should Trees Have Standing? Toward Legal Rights for Natural Objects."[12]

Disney Enterprises proposed to develop Mineral King Valley into a ski resort with a multimillion-dollar complex of recreational facilities. The Sierra

FIGURE 30.2 Within Mineral King Valley, a wilderness area in the Sierra Nevada, California. In the 1970s, a proposal to develop this valley led to a new environmental legal question: Does nature have rights, separate from the rights of human beings? Shown here is the east fork of the Kaweah River in the Mineral King Valley.

Club, arguing that this development would adversely affect both the aesthetics and the ecological balance of the wilderness area, brought a suit against the government. But the case raised a curious question: If a wrong was being done, who was wronged? The California courts ruled that the Sierra Club itself could not claim direct harm from the development. Moreover, because the land was government owned and the government represented the people, it was difficult to argue that the people in general were wronged. Stone suggested that the Sierra Club might base its case on common law, or precedent, citing particular cases, such as one involving ships, in which inanimate objects have been treated as having legal standing. Stone suggested that trees also should have legal standing. The Sierra Club was not able to claim direct damage to itself but instead argued on behalf of the nonhuman wilderness.

The case was taken to the U.S. Supreme Court, which concluded that the Sierra Club itself did not have sufficient "personal stake in the outcome of the controversy" to bring the case to court. But in a famous dissenting statement, Justice William O. Douglas addressed the question of legal standing. Douglas proposed the establishment of a new federal rule that would allow "environmental issues to be litigated before federal agencies or federal courts in the name of the inanimate object about to be despoiled, defaced, or invaded by roads and bulldozers and where injury is the subject of public outrage" (*Sierra Club v. Morton,* 1972). If such a rule were enacted, trees would have legal standing. Although trees did not achieve legal standing in that decision, it was a landmark case in that ethical values and legal rights for wilderness and natural systems were explicitly discussed.

This subject in ethics is still a lively, controversial one. Should our ethical values be extended to nonhuman, biological communities and even to the life-support system of Earth? What position you take will depend in part on your values and in part on your understanding of the characteristics of wilderness, natural systems, and other environmental factors and features.

Deep Ecology: Putting Nature First

The idea that nonhuman organisms might have legal rights has been extended even further in an ideological–political–ethical movement that has become known as "deep ecology." This movement starts with the scientific observation, as discussed in Chapter 6, that the persistence of life is a characteristic of an ecological system, and that the biosphere—the Earth's global life containing and life supporting system—is ultimately necessary to sustain life on the Earth, as discussed in Chapter 3. It raises the status of the biosphere to what has been called a "quasi-divine entity." Because it is necessary for life to persist, and people are not, the biosphere ranks highest in the moral order. Next in the moral order are the nonrational, nonthinking organisms that are simply carrying out their life functions. Unable to make choices, they are considered innocent of environmental destruction. And at the very bottom of the moral pyramid are people who, because of our ability to reason, invent new technologies, and make choices, have been able to mess up the environment. This complete inversion of the traditional moral order that has persisted since the beginning of the scientific-industrial revolution has been called the greatest opposing idea to the belief in rationality, individualism, humanism, and democracy since the time of Descartes.[13] The deep ecology movement shows how important environment and environmentalism has become.

Obligations to the Future

Another major development in environmental ethics has been a concern with the question of what we owe future generations. Although most of us think at one time or another about the future and our descendants, this question is being given increasing precedence because we know that our modern technology is affecting the environment in ways that will last hundreds and thousands of years and that we are producing chemicals that can remain active even longer. Of particular concern are:

- radioactive wastes from nuclear power plants;
- the environmental effects of thermonuclear war;
- long-term climatic change resulting from land-use changes and technological activities;
- the worldwide spread of nonradioactive toxic chemicals;
- extinctions of large numbers of species as a result of human activities;
- the direct effects of rapid increases in human population;
- the destruction of forests and fertile agricultural soils, impoverishing them for very long periods; and
- the long-term impacts of apparently short-term technological benefits, such as the impact on natural systems caused by rapid advances in genetic engineering.

As with other aspects of environmental ethics, what we know about these effects influences the value judgments we make. As Ernest Partridge, philosopher of environmental ethics, has written, "Our moral responsibility grows with foresight."[14] As an example, in 1983 a meeting of scientists in the United States determined that a major thermonuclear war could lead to a nuclear winter. Because of extensive fires and the transport of dust into the high atmosphere, the sky might be darkened for months or years, possibly leading to freezing temperatures over much of Earth, putting a halt to the growth of green plants. This new information (based on mathematical and computer analysis and projections) placed thermonuclear war in a new ethical light, suggesting that the chances that such a war would end much of life on Earth were even greater than previously believed. It should be noted that some scientists disputed the mathematical and computer analysis and projections on which this information was based. However, for many this new information deepens the ethical quandary that faces all of us. One hundred years ago no one could imagine that a device made by humans could end most of life. But today we realize that our power over the environment could make us or our children the last generation of *Homo sapiens* on Earth.

Philosophers in the 1970s began to grapple with the complex issues that arise from concern about the rights of future generations and the idea of stewardship of Earth. At the heart of this ethical viewpoint is the belief that we do not really possess Earth during our lifetime, that we are merely another group in the line of human beings who are the shepherds or stewards of Earth while we are here.

The overwhelming impact of these issues has led some people to reject all science, technology, and progress and to seek a return to a simpler age. But this head-in-the-sand approach can only be a short-term answer. The long-term solution, if there is to be one, will come from the best use of science and technology within the framework of an evolving ethic. To quote Partridge, "It is a fundamental paradox of our age that scientific knowledge and discipline, supplemented by critical moral sense and passionate moral purpose, will be needed to save the future."[14]

30.3 WILDERNESS AS A CONCEPT AND A REALITY

To put a value on what we are doing to our environment, we must first understand what it is we are doing to that environment. To do so, we must know what nature is like without human influence. One of the first tasks in environmental sciences is to establish what nature undisturbed is like, to understand what wilderness really is. Wilderness as a concept is important to all aspects of environmental sciences. In addition, wilderness has taken on significance as a reality, having intrinsic value. In the United States

there are legally designated wilderness areas, and people travel to these areas for recreation.

Today it is popular to think of wilderness in a positive way, as something to be valued and preserved. But this was not always so. In the past, wilderness has been perceived in a variety of ways:

- as dangerous and threatening and perhaps a place to test one's bravery and heroism;
- as a place to discover one's true self and one's purpose in life;
- as capricious and harmful to people;
- as a chaotic place to be put into order;
- as a resource to be used; and
- as empty land to be converted to more productive uses.

Nature has also been perceived as simply beautiful, not necessarily useful, but to be viewed much as sculptures and paintings are viewed.

In some preagricultural and preindustrial societies, there was no concept of wilderness as separate from a person's immediate surroundings: Everything was habitat. As one Native American, Chief Luther Standing Bear, said, "We did not think of the great open plain, the beautiful rolling hills and the winding streams with their tangled growth as 'wild.'" Only to the European settlers was nature a wilderness and "the land infested with wild animals and savage people."[15]

Nature as Dangerous

We have many myths about nature and wilderness. The idea of wilderness as separate from people and their habitats can be found rather early in European civilization. The word itself is derived from the Anglo-Saxon word *wild*(d)oer ("wild beast") and means literally the place of wild creatures. In the great Anglo-Saxon epic poem *Beowulf,* wilderness was viewed as a place of danger and discomfort. It was the home of strange and terrible creatures, like the monster Grendel, a place where a brave person could prove to be a hero, as did Beowulf when he ventured from the warm hearth of the king's castle to slay the evil Grendel and the monster's mother. From this point of view, the environment threatened people. This attitude is found mainly among cultures that have low population densities and little technology. In many primitive societies, when civilization had yet to exert much influence over the environment, this is how uninhabited nature was seen, as a wild and dangerous place where people could test themselves against the challenges of nature. People who are starving and forced to look for food in a wild area might tend to think of wilderness as dangerous and as a threat to their survival.

Nature as Chaotic

A related view of wilderness is common in Western civilization and appears even in eighteenth-century Europe. In a classical work of that century, *Natural History, General and Particular,* Count de Buffon described nature untouched by human beings as "melancholy deserts [that are] overrun with briars, thorns and trees which are deformed, broken and corrupted [or wetlands] occupied with putrid and stagnating water [covered with] stinking aquatic plants [that] serve only to nourish venomous insects, and to harbour impure animals." A person in a wilderness has to "watch perpetually lest he should fall victim to wild animals' rage, terrified by the occasional roarings and even struck with the awful silence of those profound solitudes."[16]

To those who see wilderness in this sense—as chaotic and uncontrolled—environmental problems are the result of disorder and can be corrected when people add the order that is missing. It is a human being who "cuts down the thistle and the bramble, and he multiplies the vine and the rose."[16] From this point of view, nature is disordered, and the role and purpose of people is to tame and manage it.

Nature as Ordered

The idea that people should control and order nature contrasts with another view that runs throughout Western civilization and is particularly common in our own time. This is the view that nature undisturbed is perfectly ordered, balanced, and harmonious. According to this belief, environmental problems occur simply because human beings upset the natural order; we are the great destroyers of nature's balance. The solution to all environmental problems is simply hands off: Leave nature alone, and perfect harmony will be restored. This viewpoint is evident in the writings of many classical Greek and Roman philosophers.

For these philosophers, perfect order seemed to imply that there was a purpose behind the order, that the perfect order could not have occurred by chance, and that this order must have had a creator.

Aristotle, who some call the great-grandfather of the study of ecology, perceived this order in many aspects of biology as he observed it. Cicero summarized many of the classical beliefs in *The Nature of the Gods.*[17] He saw order in the food habits of animals and wrote of the amazing adaptations of living creatures for their needs. Lacking a theory of biological evolution, Cicero and other classical writers believed these adaptations were part of a divinely wise and purposeful plan. For instance, the elephant, Cicero observed, "has a trunk, as otherwise the size of his body would make it difficult for him to reach his food." Cicero marveled at these adaptations and asked, "What power is it which preserves them all according to their kind?" He added, "Who cannot wonder at this harmony of things, at this symphony of nature, which seems to will the well-being of the world?"

Assuming a purpose behind the order in nature, the next question was, for whom was the world so well-ordered? The answer, naturally enough, had to be human beings, who alone among all living things were intelligent enough to appreciate it. This interpretation of nature, then, is that nature is ordered, balanced, and harmonious; that human beings, like all living things, have a place and purpose in the order; and that the divine purpose of nature's order is for human benefit.

Nature as Capricious

Some people have a sense of nature as simply capricious, unpredictable and not well suited to people. For example, the Roman poet Lucretius wrote that nature was not designed for human benefit; nature gave human beings only a hard life, and a person had to struggle to survive and obtain the necessities of life against the natural workings of things. He saw nature as capricious. "How many a time the produce of great agonies of toil burgeons and flourishes," wrote Lucretius in *De Rerum Natura,* "and then the sun is much too hot and burns it to a crisp; or sudden cloudbursts, zero frosts, or winds of hurricane force are, all of them, destroyers."[18] Although nature provides for every need of other creatures, human beings alone must struggle for their existence. This view characterizes nature as unpredictable and uncaring. Environmental problems occur because nature does not care about us. In other words, human problems are due to the capriciousness of nature, and human beings must struggle for survival against this capriciousness.

Nature as Separate

At least some people in hunting and gathering societies do not make a distinction between human beings, their habitat, and a separate wilderness. With the development of agriculture and herding, wilderness was seen as separate from home and human habitat. Once that separation began to develop, and the less control people had over nature as they attempted to grow crops and herd animals, the more likely that they would view nature undisturbed by human influence as dangerous, capricious, and an obstacle to survival. It is not surprising to find such views among societies with less technological control over their environment, and therefore less assurance about the production of the necessities of life, than our own.

Nature as Machine

During the European Renaissance the ideas for and against order in nature were reexamined and the controversy over the balance and harmony of nature was renewed. Exploration and the discovery of strange creatures, the development of science and an increased understanding of the processes involved in nature, and growth in technology and civilization all promoted such a reexamination. The old issues were restated, and the same arguments presented, with new pieces of evidence.

With the development of the new physics of Newton, Galileo, and others, many people optimistically believed that the workings of the world could be understood from physical laws, that the world and the creatures who lived in it could be understood like a mechanical device. They believed that nature was a system following inexorable laws of the universe and that they could learn to understand and control it. Their view was that nature is like a machine and that human beings can learn to be nature's engineers, captains of the great ship Earth.[19]

Nature as Beautiful

In the nineteenth century, another great change in ideas took place, best known to us through the writings of the English romantic poets. Part of this view is that wildness is to be appreciated and that the power and unpredictable grandeur of nature is beautiful, sublime, and a demonstration of the power and glory of God. The romantic view was that nature's wild power was grand and that the experience of wild nature was of great significance, a source of inspiration, peace, and beauty. In this view, experiencing wild nature was a way for an individual to discover the self and the meaning of existence. Nature's primary value was aesthetic. A person who likes to watch sea lions even as they in turn hunt steelhead, as discussed at the beginning of this chapter, is an example of someone who values nature for aesthetic reasons.

Nature as Commodity

With the discovery of the Americas, the largest wilderness ever known to Western civilization was opened to exploration. Early explorers and settlers saw wilderness as a commodity or a resource to mold into something usable to improve their economic well-being. Progress, as seen in a nineteenth-century American painting, was a beautiful lady subduing the wilderness, driving away the wild animals and primitive buffalo, and bringing with her the wondrous new inventions of technology.

In early New England, for example, the wilderness was forest to be cleared, to be burned up or

FIGURE 30.3 The summit of New Hampshire's Mt. Monadnock, once forested to the top, was burned in nineteenth-century fires lit to clear lands in the valleys. Trees and soils were removed by the fires, leaving a barren, rocky summit that has not revegetated, except in drainages and lower areas where soil can accumulate. This clearing of forested lands has always been part of the effect of civilization on the environment.

cut down and transformed into farms (Figure 30.3). In the Big Woods of Michigan, wilderness was 8 million ha (20 million acres) of white pine, a commodity to be cut to build houses across the eastern and midwestern United States. Today a single preserve of less than 25 ha (62 acres) of that original, uncut white pine exists in a small park in central Michigan. The American wilderness was Buffalo Bill killing Indians; it was Daniel Boone; it was the Big Sky. It was "The People, Yes" country of Carl Sandburg's poem, a place for optimism, of big things to conquer.[20] The American western wilderness was the land of the strange and the big and the land of the tall tale, where the wind blew so strong you had to make a kite of an iron shutter and a chain; where fog was so thick you could put shingles on it; where plants grew so fast that "the boy who climbed a cornstalk would have starved to death if they hadn't shot biscuits up to him"; and where a herd of cattle got lost in a hollow redwood tree.[21]

A person who wants to remove sea lions so that steelhead can be maintained as a productive resource and who also argues that sea lions themselves should be viewed as a resource for fur and should be harvested, especially when there are so many of them that they are a pest, is arguing from the perspective of nature as a commodity.

Nature as Necessary for Survival

Following not far behind the belief in progress and subduing the wilderness for personal benefit was the recognition that technology might kill soil that grew golden corn and ruin streams full of trout and salmon. At first the wilderness seemed too big to destroy. Early loggers in Michigan are said to have believed they would never run out of wood; by the time the last of the virgin forests were cut, the first ones would have grown back and could be cut again.

The view that wilderness was a commodity to be used up conflicted with a new belief that wilderness and human beings were intimately tied together and that survival of one required survival of the other. In 1864 the first important American statement of these concerns appeared in *Man and Nature* by George Perkins Marsh, a native of Vermont and U.S. Ambassador to Italy and Egypt.[22] Struck by differences between soils, forests, and general appearance of landscape in Europe and North Africa, which had been used by civilized human beings for thousands of years, and those of the barely touched wilderness of Vermont and New Hampshire, Marsh proposed that the rise and fall of civilizations were linked to use and misuse of nature.

He suggested misuse of farmland contributed to the fall of Rome; as Romans exhausted soil, they expanded their empire to obtain new sources of food. Repetition of this process led to an increasing network of transportation and government. The empire finally collapsed when distances required to transport food exceeded technological capabilities. Although today's historians argue that such a view is oversimplified, Marsh was the first modern writer to state the possibility that sustaining human life depends on nature's balance. He became the mid-nineteenth century's classic proponent of the idea that nature undisturbed achieves a permanency of form and substance, a harmony only human beings destroy. The view that wilderness provides a mechanism for survival of human beings began in the United States with writers such as George Perkins Marsh and Henry David Thoreau.

The usefulness of uninhabited, wild areas has been widely recognized in recent years. For example, it has become popular to suggest that forests could help reduce buildup of greenhouse gases in our atmosphere and we should try to maintain forests wherever possible and plant new ones.

Nature as Scenic Wonder

Opening of the American West led to discoveries of scenic grandeur in Yosemite, Yellowstone, and the Grand Canyon, which were eventually set aside as national parks. As historian Alfred Runte has observed, monumental scenery was America's answer to architectural and sculptural wonders of civilizations in Europe.[23] National parks were seen not as biological or ecological units to be preserved but as places where the average citizen could view the peculiar (geysers, rock formations) and gain a sense of peace, beauty, or even religious experience. In the late nineteenth-century, the idea developed of wilderness as a place of the strange and monumental. In this view, wilderness is beyond practical concerns; it provides pleasure and entertainment, like a circus, or spiritual inspiration, like a church.

The first major conservation movement in the United States began in the late-nineteenth and early-twentieth centuries. It was stimulated by nineteenth-century writings of Henry David Thoreau, Ralph Waldo Emerson, and Marsh. It was also influenced by ideas of the founders of national parks and forests, such as John Wesley Powell, who in 1869 was first to travel down Grand Canyon's Colorado River. Finally, it was influenced by the deep reverence for nature found in the writings of John Muir.

30.4 PERSONAL INVOLVEMENT

Individuals can become involved in improving the environment. Like any social and political movement, environmentalism encompasses a wide range of approaches. There is the conservative style of the Nature Conservancy, whose function is to help purchase lands that are important for conservation and ensure these lands are maintained as nature preserves. At the opposite extreme is the radical activism of organizations such as Greenpeace, whose activities include maneuvering small boats between whaling ships and whales in an attempt to prevent and draw attention to the practice of whaling.

At a deeper level, environmentalism encompasses a broad range of political and philosophical approaches. The emphasis of this book, and of environmental sciences, is on a rational approach involving the application of scientific and technical information. By using the best available scientific instruments and methods, we can best understand, conserve, and manage our environment and its resources.

As we have emphasized throughout this text, environmental problems are in part the result of the number of human beings on Earth. This means individual actions, summed over large numbers of people, can have great influence on the environment.

Given the wide range of environmental issues, a person interested in contributing to improving the environment faces a distressing dilemma: In which of the many environmental issues should one participate? Our advice to individuals who ask this question is to choose one of the problems that has the most personal meaning and then participate directly in solutions. Today, there are local organizations, or

How Can We Set Priorities
for Dealing with Environmental Problems?

Two main points have emerged from our study of environmental science: (1) everything in the environment is connected and (2) there are limits. Because there are limits in all spheres, not only the environmental, we cannot solve all these problems immediately. Also, any attempts we make to solve some problems will affect others, for better or worse. Perhaps we can take advantage of the interconnectedness of the environment; by solving some problems, we may be able to ameliorate others. If we are careful, we may be able to avoid making some conditions worse as we try to improve others.

The obvious conclusion is that we need to identify key areas, namely, those that have the greatest potential for improving the environment. Selecting key areas is a difficult task that can be controversial. After studying environmental science, you are in a better position than you were before to try to select the key areas and to think about priorities.

Critical Thinking Questions

1. Think about the values discussed in this chapter. Which are most reflective of your own values? Select three to five that are most important to you.

2. Review the list of chapter headings and make a list of environmental problems. Which 10 would you consider to be most basic or important? Explain your choices.

3. Combine your values, selected in question 1, with the list of environmental problems in question 2. How would you rank the 10 problems, from the most urgent (1) to the least urgent (10)? Explain your ranking.

4. Which problem discussed in this chapter would you consider most urgent? Explain your reason for selecting it. What needs to be done to solve it? How does it relate to other environmental problems? Write a short paper (two to three pages).

local chapters of national organizations, that are concerned with a variety of environmental problems. We have listed many of these organizations in the Internet Resources sections of this textbook. People looking for a way to help can seek the organization that takes the philosophical approach most consistent with their own and that focuses on the issues that have the most meaning to them.

Science is an open-ended process of finding out about the natural world. The essence of science is questioning and critical examination of accepted truths. Rather than viewing science as a collection of facts, it is the collection of these facts into coherent pictures (models, theories) of the world that is important. At the beginning of this text, we encouraged you to learn environmental science in an active mode, to be critical of what you heard, saw, and read. We hope you will carry on in this mode and that you will continue to see, not isolated facts, but connections among the facts that reflect the interdependence and unity in the environment. Armed with this knowledge, you will be better able to participate in the efforts to address the environmental issues challenging us today.

SUMMARY

- Deciding how to place a value on the environment is a fundamental issue that underlies all environmental topics.

- Environmental sciences require a synthesis of ideas, drawing from a broad range of major traditional fields of study.

- How we value the environment depends on long-held cultural values, one of which is the idea of the balance of nature.

- How we perceive nature undisturbed by human influence—wilderness—influences how we place a value on the environment.

- If we place a value on each aspect of the environment individually, without seeing the whole picture, conflicts are the invariable result.

REEXAMINING THEMES AND ISSUES

This chapter is a discussion of how the major themes and issues pervaid attempts to solve environmental problems. Perhaps most importantly, this chapter reaffirms the basic point made throughout this text: There cannot be long-term solutions to the world's environmental problems, nor true **sustainability** of our resources, if the **human population** continues to grow. Environmental issues are **global**, especially because of the worldwide effects of life on Earth and the recent global effects of human beings and our technology. As we learned from Chapter 26 on **urban environments**, the world is becoming increasing urbanized, and this brings with it some new effects and new environmental pressures. Finally, the debate over its ethical value is as old as civilization. **How valuable is the environment?** The simplest answer is that a proper environment is essential to sustain all renewable resources: food, fibers, wood, air, water, and even ourselves and our own health. But as the case study of the sea lion and steelhead trout demonstrates, people have placed many other kinds of values on the environment, and environment is important to our religions, our cultures, and our perceptions of ourselves.

STUDY QUESTIONS

1. Consider the case study of sea lions and steelhead trout. What are some arguments for the conservation of both? Attempt to work out a solution that will allow both to persist.

2. How could Leopold's idea of the land ethic be applied to the introductory case study?

3. Today there are legally designated wilderness areas. What could be meant by a "managed wilderness" for these areas? Is a managed wilderness a self-contradictory term? Refer to the different ideas of wilderness discussed in this chapter. Make a list and decide which ones are consistent with active management, which would allow some management, and which would require no human intervention.

4. What is the connection between sustainability and the rights of future generations?

5. Which is more likely to be consistent with sustainability of renewable resources, an increasingly urban world or a world in which cities become less and less important and most people live in suburbs and towns or scattered throughout the rural countryside?

6. Many popular television programs deal with nature. Consider one of them. Which ideas of nature, as described in this chapter, are:
   a. Implied or discussed?
   b. Accepted as true, either through a direct statement or implied by the subject?

7. In what way could the underlying problem of human population growth be said to be a factor in the case of Mineral King Valley, discussed in this chapter?

FURTHER READING

Botkin, D. B. 1995. *Our Natural History: The Lessons of Lewis and Clark.* New York: Putnam. What nature was really like before it was changed by European settlement and modern technology.

Botkin, D. B. 1990. *Discordant Harmonies: A New Ecology for the 21st Century.* New York: Oxford University Press. This book presents an analysis of the myths that lie behind attempts to solve environmental issues.

Elliot, R. (ed.). 1995. *Environmental Ethics.* New York: Oxford University Press. A significant contribution to human-centered environmental ethics with a range of views and issues on the subject.

Ferry, Luc. 1996. *The New Ecological Order.* Discusses the legal and philosophical implications of deep ecology and the radical environmentalism.

Hayward, T. 1994. *Ecological Thought. An Introduction.* Cambridge. As increasingly destructive practices are negatively altering our environment and ourselves, this book calls for radical change in the view of nature, our place within it, environmental economics, and political ecology that is required to shape future perspectives and politics on the environment.

Leopold, A. 1949. *A Sand County Almanac.* New York: Oxford University Press. An environmental classic, this book of essays chronicles Leopold's thoughts and observations on his land ethic, the seasons of the year on his sand farm in Wisconsin, and episodes from his life as a wildlife manager.

Marietta, D. E., Jr. 1995. *For People and The Planet: Holism and Humanism in Environmental Ethics.* Philadelphia, PA: Temple University Press. A look at reconciliation of environmental concerns with traditional humanistic ethical concerns.

Nash, R. F. 1989. *The Rights of Nature: A History of Environmental Ethics.* Madison: University of Wisconsin Press. The intellectual history of environmental consciousness is presented in this book, in particular the idea that morality should also include the relationships between people and nature.

Westra, L. 1994. *An Environmental Proposal for Ethics: The Principle of Integrity.* Lanham, MD: Rowman & Littlefield. A discussion on the integrity of the natural world and our ethical concerns in maintaining or restoring that integrity.

INTERNET RESOURCES

Center for Applied Ethics at the University of British Columbia: *http://www.ethics.ubc.ca/resources/environmental*—The Center for Applied Ethics maintains excellent links to a variety of sources in its page on Environmental Ethics.

Center For Environmental Philosophy: *http://www.cep.unt.edu*— Affiliated with the University of North Texas, this nonprofit organization provides access to Internet resources throughout the world which pertain to or focus on environmental ethics and environmental philosophy. It contains links to numerous other resources.

Center for the Study of the Environment: *http://www.naturestudy.org*—A nonprofit scientific research in education organization that provides critical information, analysis, and scientifically based solutions to environmental problems.

Ethics Updates, Environmental Ethics: *http://pwa.acusd.edu/index.html*—Maintained by Lawrence Hinman, professor of philosophy, this page is an excellent resource for information about environmental ethics. Find articles to read, links to other internet sources, lists of recommended books, and contemporary environmental ethics topics.

APPENDIX A

Special Feature: EMR Laws

Properties of Waves

- Direction of wave propagation

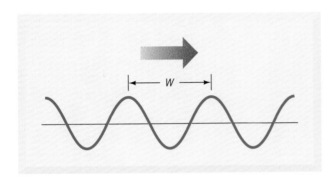

- W = wavelength (distance from one wave crest to the next)
- A wave travels at the speed of light in a vacuum—300,000 km/s.

Absolute Temperature Scale (kelvin, K)

- Zero is really zero; there are no negative values of K
- Temperature in K = temperature in °C + 273
  $$K = °C + 273$$
- Example: water freezes at °C = 0 = 273 K
  water boils at °C = 100 = 373 K

Stefan–Boltzmann Law

- All bodies with a temperature greater than absolute zero radiate EMR. These bodies are called thermal radiators. The amount of energy per second radiated from thermal radiators is called *intensity* and is given by the Stefan–Boltzmann law

$$E = aT^4$$

where E is the energy per second (intensity); T is the absolute temperature; and a is a constant (the nature of this constant involves physical ideas beyond the scope of this text).

- The Stefan–Boltzmann law states that the intensity of EMR coming from a thermal radiator is directly proportional to the fourth power of its absolute temperature.

Wien's Law

$$W_P = a/T$$

- where W_P is the wavelength of the peak intensity of a thermal radiator; T is temperature in K; and a is a constant. For example, Figure 3.8 shows that W_P for the earth is about 10 μm. Wien's law states in a general way that the hotter a substance is, the shorter the wavelength of the emitted predominant electromagnetic radiation. That is, wavelength is inversely proportional to temperature.

APPENDIX B

Prefixes and Multiplication Factors

Number	10^x, Power of 10	Prefix	Symbol
1,000,000,000,000,000	10^{18}	exa	E
1,000,000,000,000,000	10^{15}	peta	P
1,000,000,000,000	10^{12}	tera	T
1,000,000,000	10^{9}	giga	G
1,000,000	10^{6}	mega	M
10,000	10^{4}	myria	
1,000	10^{3}	kilo	k
100	10^{2}	hecto	h
10	10^{1}	deca	da
0.1	10^{-1}	deci	d
0.01	10^{-2}	centi	c
0.001	10^{-3}	milli	m
0.000 001	10^{-6}	micro	μ
0.000 000 001	10^{-9}	nano	n
0.000 000 000 001	10^{-12}	pico	p
0.000 000 000 000 001	10^{-15}	femto	f
0.000 000 000 000 000 001	10^{-18}	atto	a

APPENDIX C

Common Conversion Factors

Length

1 yard = 3 ft, 1 fathom = 6 ft					
	in.	*ft*	*mi*	*cm*	*m*	*km*
1 inch (in.) =	1	0.083	1.58×10^{-5}	2.54	0.0254	2.54×10^{-5}
1 foot (ft) =	12	1	1.89×10^{-4}	30.48	0.3048	—
1 mile (mi) =	63,360	5,280	1	160,934	1,609	1.609
1 centimeter (cm) =	0.394	0.0328	6.2×10^{-6}	1	0.01	1.0×10^{-5}
1 meter (m) =	39.37	3.281	6.2×10^{-4}	100	1	0.001
1 kilometer (km) =	39,370	3,281	0.6214	100,000	1,000	1

Area

1 square mi = 640 acres, 1 acre = 43,560 ft² = 4046.86 m² = 0.4047 ha 1 ha = 10,000 m² = 2.471 acres					
	in.²	*ft²*	*mi²*	*cm²*	*m²*	*km²*
1 in.² =	1	—	—	6.4516	—	—
1 ft² =	144	1	—	929	0.0929	—
1 mi² =	—	27,878,400	1	—	—	2.590
1 cm² =	0.155	—	—	1	—	—
1 m² =	1,550	10.764	—	10,000	1	—
1 km² =	—	—	0.3861	—	1,000,000	1

Volume

	in.³	*ft³*	*yd³*	*m³*	*qt*	*liter*	*barrel*	*gal (U.S.)*
1 in.³ =	1	—	—	—	—	0.02	—	—
1 ft³ =	1,728	1	—	0.0283	—	28.3	—	7.480
1 yd³ =	—	27	1	0.76	—	—	—	—
1 m³ =	61,020	35.315	1.307	1	—	1,000	—	—
1 quart (qt) =	—	—	—	—	1	0.95	—	0.25
1 liter (l) =	61.02	—	—	—	1.06	1	—	0.2642
1 barrel (oil) =	—	—	—	—	168	159.6	1	42
1 gallon (U.S.) =	231	0.13	—	—	4	3.785	0.02	1

Mass and Weight

1 pound = 453.6 grams = 0.4536 kilogram = 16 ounces

1 gram = 0.0353 ounce = 0.0022 pound

1 short ton = 2000 pounds = 907.2 kilograms

1 long ton = 2240 pounds = 1008 kilograms

1 metric ton = 2205 pounds = 1000 kilograms

1 kilogram = 2.205 pounds

Energy and Power[a]

1 kilowatt-hour = 3413 Btus = 860,421 calories

2 Btu = 0.000293 kilowatt-hour = 252 calories = 1055 joules

1 watt = 3.413 Btu/hr = 14.34 calorie/min

1 calorie = the amount of heat necessary to raise the temperature of 1 gram (1 cm^3) of water 1 degree Celsius

1 quadrillion Btu = (approximately) 1 exajoule

1 horsepower = 7.457×10^2 watts

1 joule = 9.481×10^{-4} Btu = 0.239 cal = 2.778×10^{-7} kilowatt-hour

[a]Values from Lange, N. A., 1967, *Handbook of Chemistry*, McGraw Hill: New York.

Temperature

$$F = \tfrac{9}{5}C + 32$$

F is degrees Fahrenheit.
C is degrees Celsius (centigrade).

Fahrenheit		Celsius
32	Freezing of H$_2$0 (Atmospheric Pressure)	0
50	———————	10
68	———————	20
86	———————	30
104	———————	40
122	———————	50
140	———————	60
158	———————	70
176	———————	80
194	———————	90
212	Boiling of H$_2$0 (Atmospheric Pressure)	100

Other Conversion Factors

1 ft^3/sec = 0.0283 m^3/sec = 7.48 gal/sec = 28.32 liter/sec

1 acre-foot = 43,560 ft^3 = 1233 m^3 = 325,829 gal

1 m^3/sec = 35.32 ft^3/sec

1 ft^3/sec for one day = 1.98 acre-feet

1 m/sec = 3.6 km/hr = 2.24 mi/hr

1 ft/sec = 0.682 mi/hr = 1.097 km/hr

1 atmosphere = 14.7 lb(in.$^{-2}$) = 2116 lb(ft^{-2}) = 1.013×10^5 N(m^{-2})

APPENDIX D

Geologic Time Scale and Biologic Evolution

TABLE D.1 Geologic Time Scale and Biologic Evolution

Era	Approximate Age in Millions of Years Before Present	Period	Epoch	Life Form
	Less than 0.01		Recent (Holocene)	
	0.01–2	Quaternary	Pleistocene	Humans
	2			
Cenozoic	2–5		Pliocene	
	5–24		Miocene	
	24–38	Tertiary	Oligocene	
	38–55		Eocene	Mammals
	55–63		Paleocene	
	63			
Mesozoic	63–138	Cretaceous		
	138–205	Jurassic		Flying reptiles, birds
	205–240	Triassic		Dinosaurs
	240			
Paleozoic	240–290	Permian		Reptiles
	290–360	Carboniferous		Insects
	360–410	Devonian		Amphibians
	410–435	Silurian		Land plants
	435–500	Ordovician		Fish
	500–570	Cambrian		
	570			
	700			Multicelled organisms
	3,400			One-celled organisms
	4,000	Approximate age of oldest rocks discovered on Earth		
Precambrian	4,500	Approximate age of the Earth and meteorites		

Notes

Chapter 1

1. Everett, G. D. 1961. One man's family. *Population Bulletin* 17:153–169.
2. Population Reference Bureau, 1998. *World Population Data Sheet*. Washington, D.C. Population Reference Bureau.
3. Ehrlich, P. R., Ehrlich, A. H., and Holdren, P. H. 1977. Ecoscience: population, resources, environment. 3d ed. San Francisco: W. H. Freeman.
4. Deevey, E. S. 1960. The human population. *Scientific American* 203:194–204.
5. Keyfitz, N. 1989. The growing human population. *Scientific American* 261:118–126.
6. Gottfried, R. S. 1983. The black death: natural and human disaster in medieval Europe. New York: Free Press.
7. Glantz, M. H., ed. 1987. Drought and hunger in Africa: denying famine a future. Cambridge: Cambridge University Press.
8. Seavoy, R. E. 1989. Famine in East Africa: food production and food politics. New York: Greenwood Press.
9. Field, J.O. (Ed.), 1993. *The Challenge of Famine: Recent Experience, Lessons Learned*. Hartford, Conn. Kumarian Press, Inc.
10. Gower, B. S. 1992. What do we owe future generations? In D. E. Cooper, and J. A. Palmer, eds., The environment in question: ethics and global issues, pp. 1–12. London and New York: Routledge.
11. Brown, L. R. and Jacobson, J. L. 1987. The future of urbanization: facing the ecological and economic constraints: In K. Davis, M. S. Bernstam, and H. M. Sellers, eds., *Population and resources in a changing world*. Stanford, Calif.: The Morrison Institute for Population and Resource Studies, Stanford University.
12. Population Reference Bureau, Inc. 1998. *World and Regional Population*. Washington, D.C. Population Reference Bureau, Inc.
13. World Resources Institute 1998. *Teacher's Guide to World Resources: Exploring Sustainable Communities*. Washington, D.C. World Resources Institute.
14. Brookfield, H. 1992 (Jan.). The numbers crunch. Is it possible to measure the population "carrying capacity" of our planet? *UNESCO Courier,* pp. 25–29.
15. Van Hyning, J. M. 1973. Factors affecting the abundance of fall chinook salmon in the Columbia River. *Research Reports of the Fish Commission of Oregon* 4:1–87.
16. Margulis, L., and J. E. Lovelock. 1989. Gaia and geognosy. In M. B. Rambler, L. Margulis, and R. Fester, eds., Global ecology: towards a science of the biosphere. pp. 1–30. Boston: Academic Press.
17. Botkin, D. B., Caswell, M., Estes, J. E., and Orio, A., eds. 1989. Changing the global environment: perspectives on human involvement. New York: Academic Press.
18. Thompson, J. 1991. East Europe's dark dawn. The iron curtain rises to reveal a land tarnished by pollution. *National Geographic Magazine* 179:36–69.
19. Beveridge, C. E. 1990. Introduction to the landscape design reports: the California origins of Olmsted's landscape design principles for the semiarid American west. The Papers of Frederick Law Olmsted, vol. 5. Baltimore: Johns Hopkins Univ. Press.
20. Rolston, H., III. 1992. Challenges in environmental ethics. In D. E. Cooper and J. A. Palmer, eds., *The environment in question: ethics and global issues,* pp. 135–146. London, New York: Routledge.
21. Nash, R. F. 1988. *The rights of nature: a history of environmental ethics*. Madison: University of Wisconsin Press.

Chapter 2

1. Schmidt, W. E. 1991 "Jovial con men" take credit (?) for crop circles. *New York Times*. Sept. 10, p. B1; Tuohy, W. 1991 (Sept. 10). "Crop circles" their prank, 2 Britons say. *Los Angeles Times,* Sept. 10, p. A14.
2. Taylor, F. S. 1949. *Science and scientific thought*. New York: W. W. Norton.
3. Reif, F., and Larkin, J. R. 1991. Cognition in scientific and everyday domains: comparison and learning implications. *Journal of Research in Science Teaching* 28:733–760.
4. Gibbs, A., and Lawson, A. E. 1992. The nature of scientific thinking as reflected by the work of biologists and by biology textbooks. *The American Biology Teacher* 54:137–152.
5. Pease, C. M., and Bull, J. J. 1992. Is science logical? *Bioscience* 42:293–298.
6. Lerner, L. S., and Bennetta, W. J. 1988 (April). The treatment of theory in textbooks. *The Science Teacher,* pp. 37–41.
7. Kuhn, T. S. 1970. *The structure of scientific revolutions*. Chicago: University of Chicago Press.
8. Trefil, J. S. 1978. A consumer's guide to pseudoscience. *Saturday Review* 4:16–21.
9. Hastings and Hastings, 1992. *Telepathy*. American Institute of Public Opinion poll from June 1990.
10. Tyser, R. W., and Cerbin, W. J. 1991. Critical thinking exercises for introductory biology courses. *Bioscience* 41:41–46.

Chapter 3

1. Western, D., and Van Prat, C. 1973. Cyclical changes in habitat and climate of an East African ecosystem. *Nature* 241(549):104–106.
2. Dunne, T., and Leopold, L. B. 1978. *Water in environmental planning*. San Francisco: W. H. Freeman.
3. Van Koevering, T. E., and Sell, N. J. 1986. *Energy: a conceptual approach* Englewood Cliffs, N.J.: Prentice-Hall, p. 271.
4. Bartlett, A. A. 1980. Forgotten fundamentals of the energy crisis. *Journal of Geological Education* 28:4–35.
5. Meadows, D. H., Meadows, D. L., and Randers, J. 1992. *Beyond the limits*. Post Mills, Vermont: Chelsea Green Publishers.
6. Wootton, J. T., Parker, M. S. and Power, M. E. 1996. Effects of disturbances on river food webs. *Science* 273:1558–1561.
7. Leach, M. K., and Givnich, T. J. 1996. Ecological determinants of species loss in remnant prairies. *Science* 273:1555–1558.
8. Lovelock, J. 1995. *The Ages of Gaia: a Biography of Our Living Earth,* revised and expanded edition. New York. W. W. Norton.

Chapter 4

1. Lehman, J. T. 1986. Control of eutrophication in Lake Washington. In *Ecological knowledge and environmental problem solving*, ed. G. H. Orians, pp. 302–316. Washington, D.C.: National Academy of Science.

2. Henderson, 1913. *The Fitness of the Environment.* New York: Macmillan, and Boston: Beacon Press, 1966.

3. Le Pichon, X. 1968. Seafloor spreading and continental drift. *Journal of Geophysical Research* 73:3661–3697.

4. Isacks, B., Oliver, J., and Sykes, L. 1968. Seismology and the new global tectonics. *Journal of Geophysical Research* 73:5855–5899.

5. Dewey, J. F. 1972. Plate tectonics. *Scientific American* 22:56–68.

6. Botkin, D. B. 1990. *Discordant harmonies: a new ecology for the 21st century.* New York: Oxford University Press.

7. Ehrlich, P. R., Ehrlich, A. H., and Holdren, J. P. 1970. *Ecoscience: population, resources, environment.* San Francisco: W. H. Freeman, p. 1051.

8. Post, W. M., Peng, T., Emanuel, W. R., King, A. W., Dale, V. H., and De Angelis, D. L. 1990. The global carbon cycle. *American Scientist* 78:310–326.

9. Keeling C. D., Whorf, T. P., Wahlen, M., and van der Plicht, J. 1995. Interannual extremes in the rate of rise of atmospheric carbon dioxide since 1980. *Nature* 375:666–670.

10. Hudson, R. J. M., Gherini, S. A., and Goldstein, R. A. 1994. Modeling the global carbon cycle: nitrogen fertilization of the terrestrial biosphere and the "missing" CO_2 sink. *Global Biogeochemical Cycles* 8:307–333.

11. Chameides, W. L. and Perdue, E. M. 1997. *Biogeochemical Cycles.* New York. Oxford University Press.

12. Agren, G. I. and Bosatta, E. 1996. *Theoretical Ecosystem Ecology.* Cambridge University Press. New York.

13. Kasting, J. F., Toon, O. B., and Pollack, J. B. 1988. How climate evolved on the terrestrial planets. *Scientific American* 258:90–97.

14. Smil, V. 1997. Global population and the nitrogen cycle. *Scientific American,* 277 (1): 76-81.

15. Carter, L. J. 1980. Phosphate: debate over an essential resource. *Science* 209:4454.

16. Longwell, C. L., Flint, R. F., and Sanders, J. E. 1969. *Physical geology.* New York: John Wiley & Sons.

Chapter 5

1. Population Reference Bureau, 1998, *World Population Data Sheet.* Washington, D.C.: Population Reference Bureau.

2. Fazi, A. 1991 Bangladesh bears the brunt. *World Health,* Jan.–Feb. pp. 26–27.

3. Kent, M. M., and Crews, K. A. 1998. *World Population: fundamentals of growth.* Washington, D.C.: Population Reference Bureau.

4. World Resources Institute. 1990. *World Resources 1990–91.* New York: Oxford University Press, table 16.2, p. 257, and table 16.4, p. 261.

5. Hardin, G. 1971. Nobody ever dies of overpopulation. *Science* 171:527.

6. United Nations. 1991. *World Population Prospects 1990.* Report no. 120. New York.

7. Malthus, T. R. 1803. *An essay on the principle of population.* Selected and introduced by Donald Winch. Cambridge: Cambridge University Press. 1992.

8. Ehrlich, P. R. 1971. *The population bomb.* Rev. ed. New York: Ballantine Books.

9. Keyfitz, N. 1989. The growing human population. *Scientific American* 261:118–126.

10. Gottfried, R. S. 1983. *The black death: natural and human disaster in medieval Europe.* New York: Free Press.

11. Graunt, J. 1662. *Natural and political observations made upon the bill of mortality.* London: Roycraft. (Reprint. Germany: Gregg International Publishers. 1973.)

12. World Bank. 1984. *World Development Report 1984.* New York: Oxford University Press.

13. Dumond, D. E. 1975. The limitation of human population: a natural history. *Science* 187:713–721.

14. World Bank. 1992. *World Development Report.* New York: Oxford University Press.

15. Keyfitz, N. 1992. Completing the worldwide demographic transition: the relevance of past experience. *Ambio* 21:26–30.

16. U.S. Bureau of the Census. 1990. *Statistical Abstract of the United States 1990.* The National Data Book. U.S. Dept. of Commerce, 4 pp.

17. Haupt, A., and Kane, T. T. 1978. *The Population Reference Bureau's population handbook.* Washington, D.C.: Population Reference Bureau.

18. Population Division, 1998. *World Population Growing Despite AIDS Spread.* United Nations Department of Economic and Social Affairs.

19. World Bank. 1985. *World Development Report.* New York: Oxford University Press.

20. Wu, L. S., and Botkin, D. B. 1980. Of elephants and men. *American Naturalist* 116:831–849.

21. Guz, D., and Hobcraft, J. 1991. Breastfeeding and fertility: a comparative analysis. *Population Studies* 45:91–108.

22. Fathalla, M. F. 1992. Family planning: future needs. *Ambio* 21:84–87.

23. Centers for Disease Control and Prevention. *CDC Surveillance Summaries,* May 3,1996. MMWR 1996;45(no. SS-3).

24. Frejka, T. 1973. The prospects for a stationary world population. *Scientific American* 228:15–23.

25. World Bank. 1998. *Improving Reproductive Health: The Role of the World Bank.* Washington, D.C.: The World Bank.

26. Planned Parenthood Federation of America Public Policy Division. *International Family Planning: The Need for Services,* June 1997. Planned Parenthood Federation of America, Inc.

Chapter 6

1. Line, L. 1996 (April 16). "Ticks and Moths, Not Just Oaks, Linked to Acorns," *The New York Times.*

2. Ostfield, R. S., Jones, C. G., and Wolff, J. O. 1996 (May). "Of Mice and Mast: Ecological Connections in Eastern Deciduous Forests" *BioScience* 46(5):323–330.

3. Morowitz, H. J. 1979. *Energy flow in biology.* Woodbridge, Conn.: Oxbow Press.

4. Brock, T. D. 1967. Life at high temperatures. *Science* 158:1012–1019; Brock, T. D. 1971. Life in the geysers basins. Washington, D.C.: National Park Service.

5. Ibid.

6. Lavigne, D. M., Barchard, W., Innes, S., and Oritsland, N. A. 1976. *Pinniped bioenergetics.* ACMRR/MM/SC/112. Rome: U.N. Food and Agriculture Organization.

7. Estes, J. A., and Palmisano, J. F. 1974. Sea otters: their role in structuring nearshore communities. *Science* 185:1058–1060.

8. Kenyon, K. W. 1969. The sea otter in the eastern Pacific Ocean. *North American Fauna,* no. 68. Washington, D.C.: Bureau of Sport Fisheries and Wildlife, U.S. Department of the Interior.

9. Duggins, D. O. 1980. Kelp beds and sea otters: an experimental approach. *Ecology* 61:447–453.

10. Kvitek, R. G., Oliver, J. S., DeGange, A. R., and Anderson, B. S. 1992. Changes in Alaskan soft-bottom prey communities along a gradient in sea otter predation. *Ecology* 73:413–428.

11. Paine, R. T. 1969. A note on trophic complexity and community stability. *American Naturalist* 100:65–75.

12. Clements, F. E. 1936. Nature and structure of the climax. *J. Ecol.* 24:252–284.

13. Roughgarden, J., and Diamond, J. 1986. Overview: the role of species interactions in community ecology. In J. Diamond and T. J. Case, eds., *Community ecology*, pp. 333–343. New York: Harper & Row.

14. Gleason, H. A. 1926. The individualistic concept of the plant association. *Bulletin of the Torrey Botanic Club* 44:7–26.

15. Miles, J. 1987. Vegetation succession: past and present perceptions. In A. J. Gray, M. J. Crawley, and P. J. Edwards, eds., *Colonization, succession and stability*. British Ecol. Soc. 26th Symp. Oxford: Blackwell Scientific Publications; Strong, D. R., Simberloff, D., Abele, L. G., and Thistle, A. B., eds. 1984. Ecological communities: conceptual issues and the evidence. Princeton, N.J.: Princeton University Press.

16. Bormann, F. H., and Likens, G. E. 1979. *Pattern and process in a forested ecosystem*. New York: Springer-Verlag.

Chapter 7

1. Beck, K. 1994. Natural control for the purple loosestrife. *North Coast Newsletter,* Ohio Lake Erie Commission, 1–2.

2. Carroll, D. 1994. Subduing purple loosestrife. *The Conservationist* 49:6–9.

3. Malecki, R. A., Blossey, B., Hight, S. D., Schroeder, D., Kok, L. T., and Coulson, J. R. 1993. Biological control of purple loosestrife. *BioScience* 43:680–686.

4. Hutchinson, G. E. 1965. *The ecological theater and the evolutionary play*. New Haven: Yale University Press.

5. May, R. M. 1988. *How many species are there on earth?* Science 241:1441–1449.

6. Olson, E. O. 1985. The biological diversity crisis. *BioScience* 35:700–706.

7. Hardin, G. 1960. The competitive exclusion principle. *Science* 131:1292–1297.

8. Rogers, C. 1996. *Red Squirrel: Sciurus vulgaris*. The Wild Screen Trust.

9. Schoener, T. W. 1983. Field experiments in interspecific competition. *American Naturalist* 1222:240–285.

10. Elton, C. S. 1927. *Animal ecology*. New York: Macmillan.

11. Hutchinson, G. E. 1958. Concluding remarks. *Cold Spring Harbor Symposium in Quantitative Biology* 22:415–427.

12. Miller, R. S. 1967. Pattern and process in competition. *Advances in Ecol. Research* 4:1–74.

13. Hubbell, S. P., and Foster, R. B. 1987. The spatial contest of generation in a neotropical forest. In A. J. Gray, M. J. Crawley, and P. J. Edwards, eds, *Colonization, succession and stability*, pp. 395–412. British Ecol. Soc. 26th Symp. Oxford: Blackwell Scientific Publications.

14. Botkin, D. B. 1985. The need for a science of the biosphere. *Interdisciplinary Science Reviews*. 10:267–278.

15. Forman, Richard T. T. 1995. *Land mosaics: the ecology of landscapes and regions*. New York: Cambridge University Press.

16. Valentine, J. W. 1973. Plates and provinces, a theoretical history of environmental discontinuity. In N. F. Hughes, ed., *Organisms and continents through time*, pp. 79–92. Special Papers in Paleontology 12.

17. Wallace, A. R. 1896. *The geographical distribution of animals* Vol. 1. Reprint. New York: Hafner. 1962.

18. Pielou, E. C. 1979. *Biogeography*. New York: John Wiley & Sons.

19. Good, R. 1974. *The geography of the flowering plants*. 4th ed. London: Longman Group Ltd.

20. Takhtadzhian, A. L. 1986. *Floristic regions of the world*. Berkeley: University of California Press.

21. Udvardy, M. 1975. A classification of the biogeographical provinces of the world. *IUCN Occasional Paper* 18. Morges, Switzerland: IUCN.

22. Lentine, J. W. 1973. Plates and provinces, a theoretical history of environmental discontinuity. In N. F. Hughes, ed., *Organisms and continents through time*, pp. 79–92. Special Papers in Paleontology 12.

23. Hallam, A. 1975. Alfred Wegener and the hypothesis of continental drift. *Scientific American,* 232:88–97.

24. Hurley, P. M. 1968. The confirmation of continental drift. *Scientific American* 218:52–64.

25. Mather, J. R., and Yoshioka, G. A. 1968. The role of climate in the distribution of vegetation. *Ann. Ass. American Geography* 58:29–41.

26. Prentice, I. C., Cramer, W., Harrison, S. P., Leemans, R., Monserud, R. A., and Solomon, A. M. 1992. A global biome model based on plant physiology and dominance, soil properties and climate. *Journal of Biogeography* 19:117–134.

27. Darwin, C. R. 1859. *The origin of species by means of natural selection or the preservation of favored races in the struggle for life*. London: Murray.

28. Grant, P. R. 1986. *Ecology and evolution of Darwin's finches*. Princeton, N.J.: Princeton University Press.

29. Cox, C. B., Healey, I. N., and Moore, P. D. 1973. *Biogeography*. New York: Halsted.

30. MacArthur, R. H., and Wilson, E. O. 1967. *The theory of island biogeography*. Princeton, N.J.: Princeton University Press.

31. Botkin, D. B. 1977. The vegetation of the west. In H. R. Lamar, ed., *The reader's encyclopedia of the American west*, pp. 1216–1224. New York: Thomas Y. Crowell.

32. Woese, C. R., Kandler, O. and Wheelis, M. L. 1990, Towards a natural system of organisms: Proposal for the domains Archaea, Bacteria, and Eucarya, Proc. Natl. Acad. Sci., USA 87:4576–4579.

33. Pace, N. R., 1996. New Perspective on the Natural Microbial World: Molecular Microbial Ecology, 1996, *American Society For Microbiology News* 62:463–470.

34. Botkin, D. B. 1977. The vegetation of the west. In H. R. Lamar, ed., *The reader's encyclopedia of the American west*, pp. 1216–1234. New York: Thomas Y. Crowell.

35. Ibid.

36. Tallis, J. H. 1991. *Plant community history*. London: Chapman and Hall.

37. Kettlewell, H. B. D. 1959. Darwin's missing evidence. *Scientific American* 200:48–53.

Chapter 8

1. Rackham, O., 1986. *The history of the countryside*. London: Dent & Sons, p. 63.

2. Perlin, J., 1989. *A forest journey: the role of wood in the development of civilization*. New York: W. W. Norton.

3. Schrödinger, E., 1942. *What is life?* Cambridge: Cambridge University Press.

4. Schrödinger. *Opus cited.*

5. Slobodkin, L. B. 1960. Ecological energy relations at the population level. *American Naturalist* 95:213–236.

6. Peterson, R.O. 1995. *The Wolves of Isle Royale: A Broken Balance.* Minocqua, WI: Willow Creek Press.

7. Jordan, J. D., Botkin, D. B., and Wolf, M. I. 1971. Biomass dynamics in a moose population. *Ecology* 52:147–152.

8. Kozlovsky, D. G. 1968. A critical evaluation of the trophic level concept: I. Ecological efficiencies. *Ecology* 49:147–160.

9. Schaefer, M., 1991. Secondary production and decomposition. In E. Rohrig and B. Ulrich, eds., *Temperate deciduous forests.* Ecosystems of the world, vol. 7. Amsterdam: Elsevier.

10. Golley, F. B. 1989. Energy dynamics of a food chain of an old-field community. *Ecol. Monographs* 30:187–291.

11. Bagley, P. B. 1989. Aquatic environments in the Amazon basin, with an analysis of carbon sources, fish production and yield. In D. P. Dodge, ed., *Proc. Int. Large Rivers Symp. Can. Spec. Publ. Fish. Aquat. Sci.* 106:385–398.

12. Bagley. *Opus cited.*

13. Fasham, M. J. R., ed. 1984. *Flows of energy and materials in marine ecosystems: theory and practice.* New York and London: Plenum Press.

14. Gaill, F., Shillito, B., Menard, F., Goffinet, g., Childress, J., 1997. Rate and process of tube production by the deep sea hydrothermal vent tubeworm *Riftia pachyptila. Marine Ecology Progress Series* 148: 135-143.

15. Wills, J., 1996. Upwelling. *FMF Glossary.* First Millennial Foundation.

Chapter 9

1. Walthern, P. 1986. Restoring derelict lands in Great Britain. In G. Orians, ed., *Ecological knowledge and environmental problem-solving: concepts and case studies,* pp. 248–274. Washington, D.C.: National Academy Press.

2. Walthern, *ibid.*

3. Perlin, J. 1989. *A forest journey: the role of wood in the development of civilization.* New York: W. W. Norton.

4. Botkin, D. B., Caswell, M., Estes, J. E., and Orio, A., eds. 1989. *Changing the global environment: perspectives on human involvement.* New York: Academic Press.

5. Forest Management Institute 1995. *Report on Forestry of the Czech Republic, 1995.* Brandys nad Labem, Czech Republic. Ministry of Agriculture.

6. Russell, J. 1990. *Environmental Issues in Eastern Europe: setting an environmental agenda.* London. Royal Institute of International Affairs.

7. Environmental Action Program for Central and Eastern Europe 1993. *Transboundary Issues: Regional and Global Concerns.* Lucerne, Switzerland. Environmental Action Programme for Central and Eastern Europe Ministerial Conference.

8. Hall, F. G., Botkin, D. B., Strebel, D. E., Woods, K. D., and Goetz, S. J. 1991. Large-scale patterns in forest succession as determined by remote sensing. *Ecology* 72:628–640.

9. Botkin, D. B. 1990. *Discordant harmonies: a new ecology for the 21st century.* New York: Oxford University Press.

10. Forman, R. T. T., 1995. *Landscape mosaics.* Cambridge: Harvard University Press.

11. Bazzaz, F. A. 1979. The physiological ecology of plant succession. *Ann. Rev. Ecol. Syst.* 10:351–371.

12. Gorham, E., Vitousek, P. M., and Reiners, W. A. 1979. The regulation of chemical budgets over the course of terrestrial ecosystem succession. *Ann. Rev. Ecol. Syst.* 10:53–84.

13. Connell, J. H., and Slatyer, R. O. 1977. Mechanisms of succession in natural communities and their role in community stability and organization. *American Naturalist* 111:1119–1144.

14. Pickett, S. T. A., Collins, S. L., and Armesto, J. J. 1987. Models, mechanisms and pathways of succession. *Botanical Review* 53:335–371.

15. Gomez-Pompa, A., and Vazquez-Yanes, C. 1981. Successional studies of a rain forest in Mexico. In D. C. West, H. H. Shugart, and D. B. Botkin, eds., *Forest succession: concepts and application,* pp. 246–266. New York: Springer-Verlag.

16. Gorham, Vitousek, and Reiners, *ibid.*

17. Gomez-Pompa, A., and Vazquez-Yanes, C. 1981. Successional studies of a rain forest in Mexico. In D. C. West, H. H. Shugart, and D. B. Botkin, eds., *Forest succession: concepts and application,* pp. 246–266. New York: Springer-Verlag.

18. MacMahon, J. A. 1981. Successional processes: comparison among biomes with special reference to probable roles of and influences on animals. In D. C. West, H. H. Shugart, and D. B. Botkin, eds., *Forest succession: concepts and application,* pp. 277–304. New York: Springer-Verlag.

19. Luken, J. O. 1990. *Directing ecological succession.* London and New York: Chapman and Hall.

20. Vitousek, P. M., and Walker, L. R. 1987. Colonization, succession and resource availability: ecosystem-level interactions. In A. J. Gray, M. J. Crawley, and P. J. Edwards, eds., *Colonization, succession and stability,* pp. 207–223. British Ecol. Soc. 26th Symp. Oxford: Blackwell Scientific Publications.

21. Odum, E. P. 1969. The strategy of ecosystem development. *Science* 164:262–270.

22. Gorham, Vitousek, and Reiners, *ibid.*

23. Vitousek, P. M., and White, P. S. 1981. Process studies in forest succession. In D. C. West, H. H. Shugart, and D. B. Botkin, eds., *Forest succession: concepts and application,* pp. 267–276. New York: Springer-Verlag.

24. Vitousek, P. M., and Reiners, W. A. 1975. Ecosystem succession and nutrient retention: a hypothesis. *BioScience* 25:376–381.

25. Walker, J., Thompson, J. H., Fergus, I. F., and Tunstall, B. R. 1981. Plant succession and soil development in coastal sand dunes of subtropical eastern Australia. In D. C. West, H. H. Shugart, and D. B. Botkin, eds., *Forest succession: concepts and application,* pp. 107–131. New York: Springer-Verlag.

26. Franklin, J. F., MacMahon, J. A., Swanson, F., and Sedell, J. R. 1985. Ecosystem responses to the eruption of Mount St. Helens. *National Geographic Research* 1:217–235.

27. Merrill, E. H., Raedeke, K. J., Knoutson, K. L., and Taber, R. D. 1986. In S. A. C. Keller, ed., *Mount St. Helens: five years later,* pp. 359–368. Cheney, Wash.: Eastern Washington University Press.

28. Lucas, R. E. 1986. Recovery of game fish populations impacted by the May 18, 1980, eruption of Mount St. Helens: winter-run steelhead in the Toutle River watershed. In S. A. C. Keller, ed., *Mount St. Helens: five years later,* pp. 276–292. Cheney, Wash.: Eastern Washington University Press.

29. Frenzen, P., 1998. *Volcano Review.* USDA, U.S. Forest Service, Summer/Fall 1998.

30. Whittaker, R. H. 1975. *Communities and ecosystems.* 2d ed. New York: Macmillan.

31. Cowles, H. C. 1911. The causes of vegetative cycles. *Botanical Gazette* 51:161–183.

Chapter 10

1. Brown, Lester R. 1995. *Who will feed China? Wake-up call for a small planet.* New York: W. W. Norton.

2. Hawthorne, P. 1998. *Rebirth.* Time 100/Africa April 13, 1998 Vol. 151 No. 15 Time 100/Leaders & Revolutionaries.

3. Johnson, G. L., and Wittwer, S. H. 1984. *Agricultural technology until 2030: prospects, priorities, and policies*. Michigan State University Agricultural Experiment Station, Special Report 12.

4. Raven, P.H., R.F. Evert and S.E. Eichhorn. 1999. Biology of Plants. W.H. Freeman and Company/Worth Publishers, New York.

5. U.N. Food and Agriculture Organization. 1998. *Global Information and Early Warning System on Food and Agriculture*. Global Watch, Food Outlook, No. 4, September 1998. Rome, Italy. Food and Agriculture Organization of the United Nations.

6. U.N. Food and Agriculture Organization. 1998. *FAOSTAT Database*. Food and Agriculture Organization of the United Nations.

7. Bardarch, J. E. 1968. Aquaculture. *Science* 161:1098–1106.

8. Bardarch, *ibid.*

9. Bardarch, *ibid.*

10. Smil, V. 1984. *The bad Earth: environmental degradation in China*. Armonk, N. Y.: M. E. Sharpe.

11. Bardarch, *ibid.*

12. California Department of Food and Agriculture. 1997. *Celebrating More Than 50 Years as the Nation's #1 Agriculture State*. California Department of Food and Agriculture.

13. U.N. Food and Agriculture Organization. 1998. *FAOSTAT Database*. Food and Agriculture Organization of the United Nations.

14. Biological Resources Division USGS. 1998. Historical interrelationships between population settlement and farmland in the conterminous united states, 1790 to 1992, *Land Use History of North America*. U.S. Geological Survey.

15. National Wildlife Federation. 1997. *Wetlands Status and Trends, 1985-1995: Annual rate of wetlands destruction down 60%*. National Wildlife Federation.

16. World Watch Institute. 1995. Rising food prices threaten political stability, *World Watch Magazine*. Washington, DC. World Watch Institute.

17. Bishaw, Z. and Turner, M. 1998. *A Regional Perspective on Seed Security*. Aleppo, Syria. Seed Unit, ICARDA.

18. Murdock, W. M. 1980. *The poverty of nations: the political economy of hunger and population*. Baltimore: Johns Hopkins University Press.

19. Smith, J. B., and Tirpak, D. 1989. *The potential effects of global climate change on the United States*. Report to Congress. U.S. Environmental Protection Agency, EPA-230-05-89-050.

20. Botkin, D. B., Nisbet, R. A., and Reynales, T. E. 1989. Effects of climate change on forests of the Great Lake states. In J. B. Smith and D. A. Tirpak, eds., *The potential effects of global climate change on the United States*, pp. 2.1–2.31. Washington, D.C.: U.S. Environmental Protection Agency, EPA-203-05-89-0.

21. Field, J.O. (ed.), 1993. *The Challenge of Famine: Recent Experience, Lessons Learned*. Hartford, Conn. Kumarian Press, Inc.

22. World Food Programme, 1998. *Tackling Hunger in a World Full of Food: Tasks Ahead for Food Aid*. Prepared for the World Food Summit in 1996, revised March 19, 1998. Food and Agriculture Organization of the United Nations.

23. Flannery, K. V. 1965. The ecology of early food production in Mesopotamia. *Science* 147:1247–1256.

24. Borland, N. E. 1983. Contributions of conventional plant breeding to food production. *Science* 147:689–693.

25. U.N. Food and Agriculture Organization. *ibid.*

26. Hinman, W. 1984. New crops for arid lands. *Science* 225:1445–1448.

27. Riley, B. 1987. Personal communication.

Chapter 11

1. Kansas Rural Center, 1996. *More Clean Water Farm Demonstrations Selected*. March, 1996, p.5.

2. World Resources Institute. *Annual report 1992–93*. 1992. Washington, D.C.

3. Vitousek, P. M. 1987. Personal communication.

4. Pimentel, D., Terhune, E. C., Dyson-Hudson, R., Rochereau, S., Samis, R., Smith, E. A., Denman, D., Reifschneider, D., and Shepard, M. 1976. Land degradation: effects on food and energy resources. *Science* 194:149–155.

5. Committee for the National Institute of the Environment. 1998. *Rangeland Health*. Accessed 11/14/98 http://www.cnie.org

6. Buschbacher, R. J. 1986. Tropical deforestation and pasture development. *BioScience* 36:22–28.

7. Grainger, A. 1982. *Desertification: how people make deserts, how people can stop and why they don't*. Earthscan Books, 2d ed. London: Russell Press, Ltd.

8. United Nations. 1978. *United Nations conference on desertification: roundup, plan of action and resolutions*. New York: United Nations.

9. U.N. Food and Agricultural Organization, 1998. *The United Nations Convention to Combat Desertification: An Explanatory Leaflet*. Food and Agriculture Organization of the United Nations.

10. U.N. Food and Agricultural Organization, 1998. *What Is Desertification*. Food and Agricultural Organization of the United Nations.

11. Grainger, *ibid.*

12. Sheridan, D. 1981. *Desertification of the United States*. Washington, D.C.: Council on Environmental Quality.

13. Lashof, J. C., ed. 1979. *Pest management strategies in crop protection*. Vol. 1. Washington, D.C.: Office of Technology Assessment, U.S. Congress.

14. Baldwin, F. L., and Santelmann, P. W. 1980. Weed science in integrated pest management. *BioScience* 30:675–678.

15. May, R. M. 1985. Evolution of pesticide resistance. *Nature* 315:12–13.

16. Barfield, C. S., and Stimac, J. L. 1980. Pest management: an entomological perspective. *BioScience* 30:683–688.

17. Barfield and Stimac, *ibid.*

18. Lashof, *ibid.*

Chapter 12

1. Cadieux, C. L. 1991. *Wildlife extinction*. Washington, D.C.: Stone Wall Press.

2. Cadieux, C. L., *ibid.*

3. Miller, R. S., and Botkin, D. B. 1974. Endangered species: models and predictions. *American Scientist* 62: 172–181. Binkley, C. S., and Miller, R. S. 1988. Recovery of the whooping crane, Grus americana. *Biol. Conserv.* 45:11–20.

4. Wilbur, S. R. 1978. The California condor, 1966–1976: a look at its past and future. *U.S. Fish and Wildlife Service North American Fauna* No. 72. Washington, D.C.: U.S. Department of the Interior.

5. Los Angeles Zoo Education Department. Nov. 17, 1992. Personal communication.

6. *Bulletin of the Santa Barbara Museum of Natural History*, No. 157 (Jan. 1992), p. 1.

7. Reed, M. 1992. California condor dies after one year in the wild. *Los Angeles Times* (Oct. 10), p. A22.

8. Halpern, S. 1992. Losing ground. Whooping cranes have made a comeback. It's their refuge that's eroding. *Audubon* (July–August 1992), pp. 71–79.

9. Wilcox, B., ed. 1988. *1988 IUCN red list of threatened animals.* Gland, Switzerland: IUCN.

10. International Union for the Conservation of Nature 1997. *1997 IUCN Red List of Threatened Plants.* Cambridge, England.

11. International Union for the Conservation of Nature 1994. *The World List of Threatened Trees.* Cambridge, England.

12. Defenders of Wildlife 1995. The case for saving species. *Defenders* 70(3).

13. Reid, W. V., and Miller, K. R. 1989. *Keeping options alive: the scientific basis for conserving biodiversity.* Washington, D.C.: World Resources Institute, p. 27.

14. Principe, P. P. 1989. The economic significance of plants and their constituents as drugs. In H. Wagner, H. H. Hikino, and N. R. Farnsworth, eds., *Economic and medicinal plant research.* Vol. 3, pp. 1–17. New York: Academic Press.

15. World Resources Institute 1993. A Short List of Plant-Based Medicinal Drugs. Washington, DC. World Resources Institute.

16. Myers, N. 1983. *A wealth of wild species.* Boulder, Colo: Westview Press.

17. Botkin, D. B., and Talbot, L. M. 1992. Biological diversity and forests. In N. Sharma, ed., *Contemporary issues in forest management: policy implications.* Washington, D.C.: World Bank.

18. Peters, C. M., Gentry, A. H., and Mendelsohn, R. O. 1989. Commentary: valuation of an Amazonian rain forest. *Nature* 339:655–656.

19. Lettau, H. L., and Molion, L. C. B. 1979. Amazonia's hydrologic cycle and the role of atmospheric recycling in assessing deforestation effects. *Monthly Weather Review* 107:227–238.

20. Price, Mark. 1996. *Taming the Tourists.* People and the Planet, Vol 5:1.

21. Wright, Pamela A. 1996. *North American Ecotourists: Market Profile and Trip Characteristics.* Journal of Travel Research. Spring, 24(4): 2-10.

22. U.S. Congress Office of Technology Assessment. 1987. *Technologies to maintain biological diversity.* Washington, D.C.: U.S. Government Printing Office, OTA-330, p. 45.

23. Leopold, A. 1949. *A Sand County almanac and sketches here and there.* New York: Oxford University Press.

24. Beck, W.T., Cass, C., and Houghton, P. 1997. Taxol. In J.F. Holland, R.C. Bast, D.C. Morton, E. Frei, D.W. Kufe, A. Weichel, R.R. Baum, eds., *Cancer Medicine.* Vol. 1, pp.1014-1015. Baltimore, Maryland: William & Wilkins.

25. Sidle, R. C., Pearce, A. J., and O'Loughlin, C. L. 1985. Hillslope stability and land use. *American Geophysical Union Water Resources Monograph* 11. Washington, D.C.

26. Villa Nova, N. A., Salati, E., and Matsui, E. 1976. Estimative de evaportranspiracao na Bacia Amazonica. *Acta Amazonica* 6:215–228.

27. Salati, E., Marques, J., and Molion, L. 1978. Origem e distribucao das chuvas na Amazonia. *Interciencia* 3:200–205.

28. Shukla, Nombre and Sellars, *ibid.*

29. Salati, E., Vose, P. B., and Lovejoy, T. E. 1986. Amazon rainfall, potential effects of deforestation and plans for future research. In G. T. Prance, ed., *Tropical rainforests and the world atmosphere,* pp. 61–64. AAAS Selected Symposium 101. Boulder, Colo.: Westview Press.

30. Botkin and Talbot, *ibid.*

31. Raup, D. M. 1988. Diversity crises in the geological past. In E. O. Wilson, ed., *Biodiversity,* pp. 51–57. Washington, D.C.: National Academy Press.

32. Wallace, A. R. 1876. *The geographical distribution of animals.* New York: Hafner.

33. Martin, P. S. 1963. *The last 10,000 years.* Tucson: University of Arizona Press.

34. Dennis, B., Munholland, P. L., and Scott, J. M. 1991. Estimation of growth and extinction parameters for endangered species. *Ecol. Monogr.* 61:115–143.

35. Ehrlich, P., and Ehrlich, A. 1981. *Extinction: the causes and consequences of the disappearance of species.* New York: Random House.

36. Cobhentz, B. E. 1990. Exotic organisms: a dilemma for conservation biology. *Conservation Biology* 4:261–265.

37. Ehrenfeld, D. W. 1972. *Conserving life on Earth.* New York: Oxford University Press.

38. Rauber, P. 1996. An end to evolution. *Sierra* 81:28.

39. Moulton, G. E., and Dunlay, T. W. 1986. *The journals of Lewis and Clark: Vol. 2, August 30, 1803–August 24, 1804.* Lincoln: University of Nebraska Press.

40. Leppäkoski, E. J. 1991. Introduced species—resource or threat in brackish-water seas? Examples from the Baltic and the Black Sea. *Marine Pollution Bulletin* 23:219–223.

41. Vitousek, P. M. 1988. Diversity and biological invasions of oceanic islands. In E. O. Wilson, ed., *Biodiversity,* pp. 181–190. Washington, D.C.: National Academy Press.

42. Van Vuren, D., and Coblentz, B. E. 1989. Population characteristics of feral sheep on Santa Cruz Island, California, USA. *J. Wildlife Management* 53:306–313.

43. Fisher, J., Simon, H., and Vincent, V. 1969. *Wildlife in danger.* New York: Viking Press.

44. World Wildlife Fund. 1984. *Annual report,* p. 32. Washington, D.C.

45. Tillett, S., 1998. FWS system arms inspectors with fresh data. *Federal Computer Week.* FCW Government Technology Group.

46. Mares, M. A., and Ojeda, R. A. 1984. Faunal commercialization and conservation in South America. *BioScience* 34:580–584.

47. Leopold, *ibid.*

48. Talbot, L. M. 1978 (May). Wildlife quotas sometimes ignored the real world. *Smithsonian,* pp. 116–124.

49. World Wildlife Fund. 1995. Africa's elephants again face poaching threat. *Focus* 17:1–2.

50. World Resources Institute. 1992. *World Resources 1992–93.* New York: Oxford University Press, p. 133.

51. Robinson, S. K., F. R. Thompson III, T. M. Donovan, D. R. Whitehead, and J. Faaborg. 1995. Regional forest fragmentation and the nesting success of migratory birds. *Science* 267:1987–1990.

52. Faaborg, J., Brittingham, M., Donovan, T., and Blake, J. 1995. Habitat fragmentation in the temperate zone. In T. E. Martin and D. M. Finch eds., *Ecology and management of neotropical migratory birds,* pp. 357–379. Oxford University Press.

53. Petit, D. R., Lynch, J. F., Hutto, R. L., Blake, J. G., and Waide, R. B. 1995. Habitat use and conservation in the neotropics, pp. 145–196. In T. E. Martin and D. M. Finch, eds., *Ecology and management of neotropical migratory birds.* Oxford University Press.

54. Watson, J. 1996. Charisma. *International Wildlife* 16(1).

55. World Wildlife Fund. 1995. Black rhinos recovering well in Kenya. *Focus* 17:1.

56. National Wildlife Federation. 1996. Vigilant protection and public support bring the southern white rhino back from the brink of extinction. *International Wildlife* 26(1).

57. Ehrenfeld, *ibid.*

58. Myers, N. 1983. *A wealth of wild species.* Boulder, Colo.: Westview Press.

59. Diamond, J. 1986. The design of a nature reserve for Indonesian New Guinea. In M. E. Soule, ed., *Conservation biology: the science of scarcity and diversity,* pp. 485–503. Sunderland, Mass.: Sinauer Associates.

60. Myers, *ibid.*

61. Botkin, D. B. 1977. Strategies for the reintroduction of species into damaged ecosystems. In J. Cairns, ed., *Recovery and restoration of damaged ecosystems,* pp. 241–260. Charlottesville: University of Virginia Press.

62. World Wildlife Fund. 1984, *ibid.*

63. Mares and Ojeda, *ibid.*

64. Sierra Club. 1995. Personal communication.

65. Defenders of Wildlife. 1995. Species bill would gut current law. *Defenders* 70:6.

66. Eisner, T. J., Lubchenco, E. O., Wilson, D. S., Wilcove, and M. J., Bean. 1995. Building a scientifically sound policy for protecting endangered species. *Science* 268:1231–1232.

67. Defenders of Wildlife. 1995. *Saving America's wildlife: renewing the Endangered Species Act.* Defenders of Wildlife, Washington, D.C.

68. Mann, C., and Plummer, M. 1995. Is the Endangered Species Act in danger? *Science* 267:1256–1258.

69. Martin, *ibid.*

70. Braithwaite, W. R., Dudzinski, M. L., Ridpath, M. G., and Parker, B. S. 1984. The impact of water buffalo on the monsoon forest ecosystem in Kakadu National Park. *Australian J. Ecology* 9:309–322.

71. Freeland, W. J. 1990. Large herbivorous mammals: exotic species in northern Australia. *J. Biogeography* 17:445–449.

Chapter 13

1. A crucial test founders in Clayoquot Sound. 1996 (Aug./Sept.) *Truck Logger,* p. 24.

2. Food and Agriculture Organization of the United Nations, 1995. *Forest Resources Assessment 1990,* Global Synthesis, FAO Forestry Paper #124, p. ix. Rome, Italy. Food and Agriculture Organization of the United Nations.

3. World Resources Institute, 1996-97. State of the World's Forests. *World Resources: A Guide to the Global Environment.* World Resources Institute.

4. Dombeck, M., 1997. *Toward Sustainable Forest Management.* Speech in April 1997. USDA Forest Service

5. Busby, F.E., et.al., 1994. *Rangeland Health: New Methods to Classify Inventory and Monitor Rangelands.* Washington, D.C. National Academy Press.

6. Kimmins, H. 1995. Proceedings of the conference on certification of sustainable forestry practices, Malaysia (in press).

7. Council on Environmental Quality and the U.S. Department of State. 1981. *The global 2000 report to the President: entering the twenty-first century.* Washington, D.C.: Council on Environmental Quality.

8. Botkin, D. B., and Simpson, L. 1990. The first statistically valid estimate of biomass for a large region. *Biogeochemistry* 9:161–174.

9. Perlin, J. 1989. *A forest journey: the role of wood in the development of civilization.* New York: W. W. Norton.

10. World Resources Institute, 1998-99. *Deforestation: The Global Assault Continues.* Global Trends, Resources at Risk, World Resources 1998-99. Washington, D.C. World Resources Institute.

11. Bryant, D., Nielsen, D., Tangley, L. 1997. *The Last Frontier Forests: Ecosystems and Economies on the Edge.* Washington D.C. World Resources Institute.

12. World Resources Institute, 1992-93. *World Resources* 1992-93. New York, NY. Oxford University Press.

13. Manandhar, A., 1997. *Solar Cookers As A Means For Reducing Deforestation In Nepal.* Nepal. Center for Rural Technology.

14. Office of Technological Assessment. 1983. *Technologies to sustain tropical forest resources.* Washington, D. C.: Congress of the United States.

15. Botkin, D. B., Nisbet, R. A., and Reynales, T. E. 1989. Effects of climate change on forests of the Great Lake states. In J. B. Smith and D. A. Tirpak, eds., *The potential effects of global climate change on the United States,* pp. 2.1–2.31. Washington, D.C.: U.S. Environmental Protection Agency, EPA-203-05-89-0.

16. Zabinski, C., and Davis, M. B. 1989. Hard times ahead for Great Lakes forests: a climate threshold model predicts responses to CO_2-induced climate change. In J. B. Smith and D. A. Tirpak, eds., *The potential effects of global climate change on the United States: Appendix D—Forests,* pp. 5.1–5.19. Washington, D.C.: U.S. Environmental Protection Agency.

17. Botkin, D. B., Woodby, D. A., and Nisbet, R. A. 1991. Kirtland's warbler habitats: a possible early indicator of climatic warming. *Biological Conservation* 56:63–78.

18. World Resources Institute. 1986. *Tropical forests: a call to action.* Washington, D.C.: World Resources Institute.

19. Likens, G. E., Borman, F. H., Pierce, R. S., Eaton, J. S., and Johnson, N. M. 1977. *The biogeochemistry of a forested ecosystem.* New York: Springer-Verlag.

20. Swanson, F. J., and Dyrness, C. T. 1975. Impact of clearcutting and road construction on soil erosion by landslides in the western Cascade Range, Oregon. *Geology* 3:393–396.

21. Fredriksen, R. L. 1971. Comparative chemical water quality—natural and disturbed streams following logging and slash burning. In *Forestland uses and stream environment,* pp. 125–137. Corvallis: Oregon State University.

22. Richter, D. D., Ralston, C. W., and Harmes, W. R. 1982. Prescribed fire: effects on water quality and forest nutrient cycling. *Science* 215:661–663.

23. Sedjo, Roger A. 1983. The comparative economics of plantation forestry: a global assessment, resources for the future, inc. Research Paper. Baltimore: Johns Hopkins University Press.

24. Runte, A. 1979. *National parks, the American experience.* Lincoln: University of Nebraska Press.

25. Runte, *ibid.*

26. Parks: how big is big enough? 1984. *Science* 225: 611–612.

27. Davis, M. 1990. *The green guide to France.* London: The Merlin Press.

28. Botkin, D. B. 1992. Global warming and forests of the Great Lakes states. In J. Schmandt, ed., *The regions and global warming: impacts and response strategies.* New York: Oxford University Press.

29. Nash, R. 1978. International concepts of wilderness preservation. In J. C. Hendee, G. H. Stankey, and R. C. Lucas, eds., *Wilderness management,* pp. 43–59. U.S. Forest Service Misc. Pub. No. 1365.

30. Nash, *ibid.*

31. Hendee, J. C., Stankey, G. H., and Lucas, R. C. 1978. *Wilderness management.* U.S. Forest Service Misc. Pub. No. 1365.

32. Hendee, Stankey, and Lucas, *ibid.*

33. Jackson, J. B. C. 1991. Adaptation and diversity of reef corals. *BioScience* 41:475–482.

34. Bell, F. W. 1978. *Food from the sea: the economics and politics of ocean fisheries.* Boulder, Colo.: Westview Press.

35. World Resources Institute, 1996-97. Water and Fisheries. *World Resources: A Guide to the Global Environment.* World Resources Institute.

36. United Nations Food and Agricultural Organization, Web Site data.

37. Bell, *ibid.*

38. Cushing, D. 1975. *Fisheries resources of the sea and their management.* London: Oxford University Press.

39. Rosenthal, D. 1996. *Environmental case studies: Northeastern regions.* New York: John Wiley & Sons, Inc., p. 32.

40. *The whaling question: the inquiry by Sir Sidney Frost of Australia.* San Francisco: Friends of the Earth, 1979.

41. Bockstoce, J. R., and Botkin, D. B. 1980. *The historical status and reduction of the western Arctic bowhead whale* (Balaena mysticetus) *population by the pelagic whaling industry, 1848–1914.* New Bedford, Conn.: Old Dartmouth Historical Society.

42. Bockstoce and Botkin, *ibid.*

43. U.N. Food and Agriculture Organization. 1978. *Mammals in the seas.* Report of the FAO Advisory Committee on Marine Resources Research, Working Party on Marine Mammals. FAO Fisheries Series 5, vol. 1. Rome: U.N. Food and Agriculture Organization.

44. The whaling question, *ibid.*

45. Perry. M., 1996. *Climate Change Biggest Risk to Whales, Says IWC.* Sydney, Australia. Reuter.

46. U.N. Food and Agriculture Organization, *ibid.*

47. U.N. Food and Agriculture Organization, *ibid.*

Chapter 14

1. Needleman, H. L., Riess, J. A., Tobin, M. J., Biesecker, G. E., and Greenhouse, J. B. 1996. Bone lead levels and delinquent behavior. *Journal of American Medical Association* 275:363–369.

2. Center for Disease Control. 1991. *Preventing lead poisoning in young children.* U.S. Department of Health and Human Services. Atlanta, GA: Public Health Service, Centers for Disease Control.

3. Goyer, R. A. 1991. Toxic effects of metals. In M. O. Amdur, J. Doull, and C. D. Klaassen, eds. *Toxicology,* pp. 623–680, New York: Pergamon Press.

4. Bylinsky, G. 1972. Metallic nemesis. In B. Hafen, ed. *Man, Health and Environment,* pp. 174–185. Minneapolis: Burgess.

5. Hong, S., Candelone, J., Patterson, C. C., and Boutron, C. F. 1994. Greenland ice evidence of hemispheric lead pollution two millennia ago by Greek and Roman civilizations. *Science* 265:1841–1843.

6. Hopps, H. C. 1971. Geographic pathology and the medical implications of environmental geochemistry. In H. L. Cannon and H. C. Hopps, eds., *Environmental geochemistry in health,* pp. 1–11. Geological Society of America Memoir 123. Boulder, Colo.: Geological Society of America.

7. Warren, H. V., and DeLavault, R. E. 1967. A geologist looks at pollution: mineral variety. *Western Mines* 40:23–32.

8. Evans, W. 1996. Lake Nyos. Knowledge of the fount and the cause of disaster. *Science* 379:21.

9. Gunn, J. (ed.). 1995. *Restoration and recovery of an industrial region: progress in restoring the smelter-damaged landscape near Sudbury, Canada.* New York: Springer-Verlag.

10. Ehrlich, P. R., Ehrlich, A. H., and Holdren, J. P. 1970. *Ecoscience: population, resources, environment.* San Francisco: W. H. Freeman.

11. Waldbott, G. L. 1978. *Health effects of environmental pollutants.* 2d ed. Saint Louis: C. V. Mosby.

12. Main, J. 1983. Dow vs. the dioxin monster. *Fortune,* May 30, pp. 83–90.

13. Carlson, E. A. 1983. International symposium on herbicides in the Vietnam War: an appraisal. *BioScience* 33:507–512.

14. Grady, D. 1983. The dioxin dilemma. *Discover,* May, pp. 78–83.

15. Roberts, L. 1991. Dioxin risks revisited. *Science* 251:624–626.

16. Johnson, J. 1995. SAB advisory panel rejects dioxin risk characterization. *Environmental Science & Technology* 29:302A.

17. Thomas, V. M., and Spiro, T. G. 1996. The U.S. dioxin inventory: are there missing sources? *Environmental Science & Technology* 30:82A–85A.

18. U.S. Environmental Protection Agency. 1994. *Health assessment document for 2,3,7,8-tetrachlorodibenzo-p-dioxin (TCDD) and related compounds* (external review draft). Vols. I–III. Washington, D.C.

19. Johnson, J. 1995. Dioxin risk assessment stalls: EPA to create new review panel. *Environmental Science & Technology* 29:492A.

20. Chanlett, E. T. 1979. *Environmental protection.* 2d ed. New York: McGraw-Hill.

21. Ross, M. 1990. *Hazards associated with asbestos minerals.* In B. R. Doe, ed., Proceedings of a U.S. Geological Survey Workshop on Environmental Geochemistry, pp. 175–176. U.S. Geological Survey Circular 1033.

22. Pool, R. 1990. Is there an EMF-cancer connection? *Science* 249:1096–1098.

23. Francis, B.M., 1994. *Toxic substances in the environment.* New York, John Wiley & Sons, Inc. A Wiley Interscience Publication.

24. Poisons and Poisoning. 1997. *Encyclopedia Britannica.* Volume 25, p. 913. Chicago, Encyclopedia Britannica, Inc.

25. Air Risk Information Support Center (Air RISC), U.S. Environmental Protection Agency. 1989. *Glossary of terms related to health exposure and risk assessment.* EPA/450/3-88/016. Research Triangle Park, N.C.

26. Blumenthal, D.S. and Ruttenber. 1995. Introduction to Environmental Health. 2nd edition. New York. Springer Publishing Co. 372 pp.

Chapter 15

1. Butti, K., and Perlin, J. 1980. *A golden thread: 2500 years of solar architecture and technology.* Palo Alto, Calif.: Cheshire Books.

2. Darmstadter, J., Landsberg, H. H., Morton, H. C., and Coda, M. J. 1983. *Energy today and tomorrow: living with uncertainty.* Englewood Cliffs, N.J.: Prentice-Hall.

3. Morowitz, H. J. 1979. *Energy flow in biology.* New Haven, Conn.: Oxbow Press.

4. Ehrlich, P. R., Ehrlich, A. H., and Holdren, J. P. 1970. *Ecoscience: population, resources, environment.* San Francisco: W. H. Freeman.

5. Feynman, R. P., Leighton, R. B., and Sands, M. 1964. *The Feynman lectures on physics.* Reading, Mass: Addison-Wesley.

6. Cuff, D. J., and Young, W. J. 1986. *The United States energy atlas.* 2d ed. New York: Macmillan.

7. Steinhart, J. S., Hanson, M. E., Gates, R. W., Dewinkel, C. C., Briody, K., Thornsjo, M., and Kambala, S. 1978. A low energy scenario for the United States: 1975–2000. In L. C. Ruedisili and M. W. Firebaugh, eds., *Perspectives on energy.* 2d ed., pp. 553–588. New York: Oxford University Press.

8. Olkowski, H., Olkowski, B., and Javits, T. (Farallones Institute). 1979. *The integral urban house: self reliant living in the city.* San Francisco: Sierra Club Books.

9. Flavin, C. 1984. *Electricity's future: the shift to efficiency and small-scale power.* Worldwatch Paper 61. Washington, D.C.: Worldwatch Institute.

10 Consumers' Research. 1995. Fuel economy rating: 1996 mileage estimates. *Consumers' Research* 78:22–26.

11. Lovins, A. B. 1979. *Soft energy paths: towards a durable peace*. New York: Harper & Row.

12. Energy Information Administration. 1992. *Annual energy review 1991*. Washington, D.C.: Energy Information Administration.

13. Brown, L. R., Flavin, C., and Postel, S. 1991. *Saving the planet: how to shape an* environmentally sustainable global economy. Worldwatch Institute. New York: W. W. Norton.

14. California Energy Commission. 1991. *California's energy plan. Biennial report*. Sacramento, Calif.

Chapter 16

1. Van Koevering, T. E., and Sell, N. J. 1986. *Energy: a conceptual approach*. Englewood Cliffs, N.J.: Prentice-Hall.

2. McCulloh, T. H. 1973. In D. A. Brobst and W. P. Pratt, eds., *Oil and gas in United States mineral resources*, pp. 477–496. U.S.: Geological Survey Professional Paper 820.

3. British Petroleum Company. 1998. *B. P. statistical review of world energy*. London: British Petroleum Company.

4. Knapp, D. H. 1995. Non-OPEC oil supply continues to grow. *Oil & Gas Journal* 93:35–45. Special from the International Energy Agency, Paris.

5. Darmstadter, J., Landsberg, H. H., Morton, H. C., and Coda, M. J. 1983. *Energy today and tomorrow: living with uncertainty*. Englewood Cliffs, N.J.: Prentice-Hall.

6. Duncan, D. C., and Swanson, V. E. 1965. Organic-rich shale of the United States and world land areas. U.S. Geological Survey Circular 523.

7. Committee on Environmental and Public Planning. 1974. Development of oil shale in the Green River formation. *The Geologist*. Supplement to vol. 9, no. 4.

8. U.S. Congress. 1980. *An assessment of oil shale technologies*. OTA-M-118. Washington, D.C.: Office of Technology Assessment.

9. Rahn, P. H. 1996. *Engineering geology: an environmental approach*, 2nd edition. New York: Elsevier.

10. U.S. Environmental Protection Agency. 1973. *Processes, procedures and methods to control pollution from mining activities*. EPA-430/9-73-001. Washington, D.C.

11. Council on Environmental Quality. 1978. *Progress in environmental quality*. Washington, D.C.

12. Miller, E. W. 1993. *Energy and American society, a reference handbook*. Santa Barbara, Calif: ABC-CLIO.

13. Corcoran, E. 1991. Cleaning up coal. *Scientific American* 264:106–116.

14. Energy Information Administration, February 1995. *Coal data: a reference*. Washington, D.C.: U.S. Department of Energy.

15. Committee on Environmental and Public Planning. 1974. Environmental impact on conversion from gas or oil to coal for fuel. *The Geologist*. Supplement to vol. 9, no. 4.

Chapter 17

1. Becker, N. D. 1992. The demise of Luz: A case study. *Solar Today*, Jan./Feb., pp. 24–26.

2. Shinnar, R. 1993. The rise and fall of Luz. *Chemtech* 23:50–53.

3. Duffield, W. A., Sass, J. H., and Sorey, M. L. 1994. *Tapping the Earth's natural heat*. U.S. Geological Survey Circular 1125.

4. Miller, E. W. 1993. *Energy and american society. A reference handbook. Santa Barbara, Calif.: ABC-CLIO*.

5. Eaton, W. W. 1978. Solar energy. In L. C. Ruedisili and M. W. Firebaugh, eds., *Perspectives on energy*. 2d ed., pp. 418–436. New York: Oxford University Press.

6. Energy Information Administration. 1990. *Annual energy review 1989*. Washington, D.C.

7. Johnson, J. T. 1990. The hot path to solar electricity. *Popular Science*, May, pp. 82–85.

8. Demeo, E. M., and Steitz, P. 1990. The U.S. electric utility industry's activities in solar and wind energy. In Böer, K. W., ed., *Advances in solar energy*. Vol. 6, pp. 1–218. New York: American Solar Energy Society, Inc.

9. Schatz Solar Hydrogen Project. (n.d.) Pamphlet. Arcata, Calif.: Humboldt State University.

10. Kartha, S. and Grimes, P. 1994. Fuel cells: energy conversion for the next century. *Physics Today* 47:54–61.

11. Haggin, J. 1995. Fuel-cell development reaches demonstration stage. *Chemical & Engineering News* 73:28–30.

12. Alward, R., Eisenbart, S., and Volkman, J. 1979. *Micro-hydro power: reviewing an old concept*. Butte, Mont.: National Center for Appropriate Technology, U.S. Department of Energy.

13. Nova Scotia Department of Mines and Energy. 1981. *Wind power*.

14. U.S. Congress, Office of Technology Assessment. 1993. *Potential environmental impacts of bioenergy crop production—background paper*. Washington, D.C.: U.S. Government Printing Office.

15. Council on Environmental Quality. 1979. *Environmental quality*.

16. Sterzinger, G. 1995. Making biomass energy a contender. *Technology Review* 98:34–40.

17. Wihersaari, M. 1996. Energy consumption and greenhouse gas emissions from biomass production chains. *Energy Conversion and Management* 37:1217.

Chapter 18

1. Rosa, E. A. and Dunlap, R. E. 1994. Nuclear power: three decades of public opinion. *Public Opinion Quarterly* 58(2):295-324.

2. Churchill, A. A. 1993. Review of WEC Commission: energy for tomorrow's world. *World Energy Council Journal*, July 1993 pp. 19-22.

3. Ehrlich, P. R., Ehrlich, A. H., and Holdren, J. P. 1970. *Ecoscience: Population, resources, environment*. San Francisco: W. H. Freeman.

4. Brenner, D. J. 1989. *Radon: risk and remedy*. New York: W. H. Freeman.

5. Duderstadt, J. J. 1978. Nuclear power generation. In L. C. Ruedisili and M. W. Firebaugh, eds., *Perspectives on energy*. 2d ed., pp. 249–273. New York: Oxford University Press.

6. Cohen, B. L. 1990. *The nuclear energy option: an alternative for the 90s*. New York: Plenum.

7. U.S. Department of Energy. 1980. *Magnetic fusion energy*. DOE/ER-0059. Washington, D.C.

8. U.S. Department of Energy. 1979. *Environmental development plan, magnetic fusion*. DOE/EDP-0052. Washington, D.C.

9. Cordey, J. G., Goldston, R. J., and Parker, R. R. 1992. Progress toward a Tokamak fusion reactor. *Physics Today* 45:22–30.

10. Greenberg, P. A. 1993. Dreams die hard. *Sierra* 78:78.

11. Van Koevering, T. E., and Sell, N. J. 1986. *Energy: a conceptual approach*. Englewood Cliffs, N.J.: Prentice-Hall.

12. Waldbott, G. L. 1978. *Health effects of environmental pollutants*. 2d ed. Saint Louis: C. V. Mosby.

13. Hanson, W. G. 1967. Cesium-137 in Alaskan lichens, caribou, and Eskimos. *Health Physics* 13:383–389.

14. University of Maine and Maine Department of Human Ser-

vices. 1983. *Radon in water and air.* Resource Highlights. February.

15. MacLeod, G. K. 1981. Some public health lessons from Three Mile Island: a case study in chaos. *Ambio* 10:18–23.

16. Anspaugh, L. R., Catlin, R. J., and Goldman, M. 1988. The global impact of the Chernobyl reactor accident. *Science* 242:1513–1518.

17. Nuclear News. 1996. Chernobyl: WHO conference confirms excess of thyroid cancers. *Nuclear News* 39:39.

18. Balter, M. 1995. Chernobyl's thyroid cancer toll. *Science* 270:1758.

19. Skuterud, L., Goltsova, N. I., Naeumann, R., Sikkeland, T., and Lindmo, T. 1994. Histological changes in *Pinus sylvestris* L. in the proximal-zone around the Chernobyl power plant. *The Science of the Total Environment* 157:387–397.

20. Williams, N. 1995. Chernobyl: life abounds without people. *Science* 269:304.

21. Office of Industry Relations. 1974. *Development, growth and state of the nuclear industry.* Washington, D.C.: U.S. Congress, Joint Committee on Atomic Energy.

22. Weisman, J. 1996. Study inflames Ward Valley controversy. *Science* 271:1488

23. Roush, W. 1995. Can nuclear waste keep Yucca Mountain dry—and safe? *Science* 270:1761.

24. Bredehoeft, J. D., England, A. W., Stewart, D. B., Trask, J. J., and Winograd, I. J. 1978. *Geologic disposal of high-level radioactive wastes—Earth science perspectives.* U.S. Geological Survey Circular 779. Arlington, Va.: U.S. Department of the Interior.

25. Flavin, C. 1991. The case against reviving nuclear power. In L. R. Brown, ed., *The Worldwatch Reader,* pp. 205–220. New York: W. W. Norton.

26. British Petroleum. 1998. *BP 1997.* England. 41 pp.

Chapter 19

1. Graf, W. L. 1985. *The Colorado River: instability and basin management.* Resource Publications in Geography. Washington, D.C.: Association of American Geographers.

2. Nash, R. 1986. Wilderness values and the Colorado River. In G. D. Weatherford and F. L. Brown, eds., *New courses for the Colorado River,* pp. 201–214. Albuquerque: University of New Mexico Press.

3. Hundley, N., Jr. 1986. The West against itself: the Colorado River—an institutional history. In G. D. Weatherford and F. L. Brown, eds., *New courses for the Colorado River,* pp. 9–49. Albuquerque: University of New Mexico Press.

4. Dolan, R., Howard, A., and Gallenson, A. 1974. Man's impact on the Colorado River and the Grand Canyon. *American Scientist,* 62:392–401.

5. Hecht, J. 1996. Grand Canyon flood a roaring success. *New Scientist,* vol. 151, no. 2045, p. 8.

6. Ballard, S. C., Michael, D. D., Chartook, M. A., Clines, M. R., Dunn, C. E., Hock, C. M., Miller, G. D., Parker, L. B., Penn, D. A., and Tauxe, G. W. 1982. *Water and western energy: impacts, issues, and choices.* Boulder, Colo.: Westview Press.

7. Henderson, L. J. 1913. *The fitness of the environment: an inquiry into the biological significance of the properties of matter.* New York: Macmillan.

8. Council on Environmental Quality and the U.S. Department of State. 1980. *The global 2000 report to the president: entering the twenty-first century.* Vol. 2. Washington, D.C.

9. Water Resources Council. 1978. *The nation's water resources.* 1975–2000. Vol. 1. Washington, D.C.

10. Solley, W. B., Pierce, R. R., and Perlman, H. A. 1993. *Esti-mated use of water in the United States in 1990.* U.S. Geological Survey Circular 1081.

11. Alexander, G. 1984. Making do with less. Special report. *National Wildlife,* Feb./March, pp. 11–13.

12. Gleick, P. H., Loh, P., Gomez, S. V., and Morrison, J. 1995. *California water 2020, a sustainable vision.* Oakland, Calif.: Pacific Institute for Studies in Development, Environment and Security.

13. Leopold, L. B. 1977. A reverence for rivers. *Geology* 5:429–430.

14. Holloway, M. 1991. High and dry. *Scientific American* 265:16–20.

15. Levinson, M. 1984. Nurseries of life. Special report. *National Wildlife,* Feb./March, pp. 18–21.

16. Nichols, F. H., Cloern, J. E., Luoma, S. N., and Peterson, D. H. 1986. The modification of an estuary. *Science* 231:567–573.

17. Hileman, B. 1995. Rewrite of Clean Water Act draws praise, fire. *Chemical & Engineering News* 73:8.

18. Gurardo, D., Fink, M. L., Fontaine, T. D., Newman, S., Chinmey, M., Bearzotti, R., and Goforth, G. 1995. Large-scale constructed wetlands for nutrient removal from stormwater runoff: an Everglades restoration project. *Environmental Management* 19:879–889.

19. U.S. Congress House Committee Report No. 93-530. 1973. *Stream channelization: what federally financed draglines and bulldozers do to our nation's streams.* Washington, D.C.: U.S. Government Printing Office.

Chapter 20

1. Smith, R. A. 1994. Water quality and health. *Geotimes* 39:19–21.

2. MacKenzie, W. R., and 11 others. 1994. A massive outbreak in Milwaukee of Cryptosporidium infection transmitted through the public water supply. *The New England Journal of Medicine* 331:161–167.

3. Centers for Disease Control and Environmental Protection Agency. 1995. Excerpts from: Assessing the public health threat associated with waterborne Cryptosporidiosis: report of a workshop. *Journal of Environmental Health* 58:31.

4. Lewis, S. A. 1995. Trouble on tap. *Sierra* 80:54–58.

5. Groundwater: issues and answers 1985. Arvada, Colo.: American Institute of Professional Geologists.

6. Gleick, P. H. 1993. An introduction to global fresh water issues. In P. H. Gleick(ed) Water in Crisis. New York. Oxford Univ. Press. p. 3–12.

7. Hileman, B. 1995. Pollution tracked in surface- and groundwater. *Chemical & Engineering News* 73:5.

8. Kluger, J. 1998. Anatomy of an outbreak, *Time,* vol. 152, no. 5, pp. 56-62.

9. Puckett, L. J. 1994. *Nonpoint and point sources of nitrogen in major watersheds of the United States.* National Water Quality Assessment, U.S. Geological Survey.

10. Maugh, T. H. 1979. Restoring damaged lakes. *Science* 203:425–427.

11. Hinga, K. R. 1989. Alteration of phosphorus dynamics during experimental eutrophication of enclosed marine ecosystems. *Marine Pollution Bulletin* 20:624–628.

12. Richmond, R. H. 1993. Coral reefs: present problems and future concerns resulting from anthropogenic disturbance. *American Zoologist* 33:524–536.

13. Bell, P. R. 1991. Status of eutrophication in the Great Barrier Reef Lagoon. *Marine Pollution Bulletin* 23: 89–93.

14. Hunter, C. L., and Evans, C. W. 1995. Coral reefs in Kaneohe Bay, Hawaii: two centuries of western influence and two decades of data. *Bulletin of Marine Science* 57:499.

15. Special Issue. 1989. *Alaska Fish and Game.* Vol. 21, no. 4.

16. Hollway, M. 1991. Soiled shores. *Scientific American* 265:102–106.

17. Robinson, A. R. 1973. Sediment, our greatest pollutant? In R. W. Tank, ed., *Focus on environmental geology*, pp. 186–192. New York: Oxford University Press.

18. Yorke, T. H. 1975. Effects of sediment control on sediment transport in the northwest branch, Anacostia River basin, Montgomery County, Maryland. *Journal of Research* 3:487–494.

19. Poole, W. 1996. Rivers run through them. *Land and People* 8:16–21.

20. Carey, J. 1984. Is it safe to drink? Special Report. *National Wildlife*, Feb./March pp. 19–21.

21. Pye, U. I., and Patrick, R. 1983. Groundwater contamination in the United States. *Science* 221:713–718.

22. Foxworthy, G. L. 1978. Nassau County, Long Island, New York—water problems in humid country. In G. D. Robinson and A. M. Spieker, eds., *Nature to be commanded*, pp. 55–68. U.S. Geological Survey Professional Paper 950. Washington, D.C.: U.S. Government Printing Office.

23. Van der Leeden, F., Troise, F. L., and Todd, D. K. 1990. *The water encyclopedia*, 2d ed. Chelsea, Mich.: Lewis Publishers.

24. Environmental Protection Agency, Drinking Water Committee of the Science Advisory Board. 1995. *An SAB report: safe drinking water. Future trends and challenges*. Washington, D.C.: U.S. EPA.

25. Jobling, S., Nolan, M., Tyler, C. R., Brighty, G., and Sumpter, J. P. 1998. "Widespread sexual disturbance in wild fish," *Environmental Science and Technology*, vol. 32, no. 17, p. 2498-2506.

26. Jewell, W. J. 1994. Resource-recovery wastewater treatment. *American Scientist* 82:366–375.

27. Task Force on Water Reuse. 1989. *Water reuse: manual of practice SM-3*. Alexandria, Va.: Water Pollution Control Federation.

28. Sutherland, J. C. 1982. Michigan wetland wastewater tertiary treatment systems. In E. J. Middlebrooks, ed., *Water reuse*. Ann Arbor, Mich.: Ann Arbor Science.

29. Kasperson, R. E. 1977. Water re-use: need prospect. In R. E. Kasperson and J. X. Kasperson, eds., *Water re-use and the cities*, pp. 3–25. Hanover, N.H.: University Press of New England.

30. Hileman, B. 1995. Rewrite of Clean Water Act draws praise, fire. *Chemical & Engineering News* 73:8.

Chapter 21

1. NOAA. 1998. *What is an El Niño?* Accessed October 2, 1998. http://www.elnino.noaa.gov.

2. Anthes, R. A. 1992. *Meteorology*. 6th ed. New York: Macmillan.

3. *Marsh, W. M., and Dozier, J. 1981*. Landscape. New York: John Wiley & Sons.

4. Detwyler, T. R., and Marcus, M. G., eds. 1972. *Urbanization and the environment: the physical geography of the city*. Belmont, Calif.: Duxbury Press.

5. Steig, E. J., Brook, E. J., White, J. W. C., Sucher, C. M., Bender, M. L., Lehman, S. J., Morse, D. L., Waddington, E. D., and Clow, G. D. 1998. "Synchronous climate changes in Antarctica and the North Atlantic," *Science*, vol. 282, p. 92-94.

6. Union of Concerned Scientists. 1989. *The greenhouse effect*. Cambridge, Mass.

7. Kerr, R. A. 1996. 1995 the Warmest Year? Yes and No. *Science* 271:137–138.

8. Kerr, R. A. 1995. It's Official: First Glimmer of Greenhouse Warming Seen. *Science* 270:1565–1566.

9. Moss, M. E., and Lins, H. F. 1989. *Water resources in the twenty-first century: a study of the implications of climate uncertainty*. U.S. Geological Survey Circular 1030. Washington, D.C.: U.S. Department of the Interior.

10. Campbell, I. D., Campbell, C., Apps, N. J., Rutter, N. W. and Bush, A. B. G. 1998. "Late Holocene approximately 1500 year climatic periodicities and their implications," *Geology*, vol. 26, no. 5, p. 471-473.

11. Broecker, W. 1997. "Will our ride into the greenhouse future be a smooth one?" *GSA Today*, vol. 7, no. 5, pp. 2-6.

12. Post, W. M., Peng, T., Emanuel, W. R., King, A. W., Dale, V. H., and De Angelis, D. L. 1990. The global carbon cycle. *American Scientist* 78:310–326.

13. Titus, J. G., Leatherman, S. P., Everts, C. H., Moffatt and Nichol Engineers, Kriebel, D. L., and Dean, R. G. 1985. *Potential impacts of sea level rise on the beach at Ocean City, Maryland*. Washington, D.C.: U.S. Environmental Protection Agency, Office of Policy Planning and Evaluation.

14. Rodhe, H. 1990. A comparison of the contribution of various gases to the greenhouse effect. *Science* 248:1217–1219.

15. Kerr, R. A. 1994. Methane increases put on pause. *Science* 263:751.

16. Dlugokencky, E. J., Steele, L. P., Lang, P. M., and Masarie. K. A. 1994. The growth rate and distribution of atmospheric methane. *Journal of Geophysical Research* 99(D8):17021–17043.

17. Council on Environmental Quality. 1990. *Environmental trends 1989*. Washington, D.C.

18. Hansen, J., Lacis, A., and Prather, M. 1989. Greenhouse effect of chlorofluorocarbons and other trace gases. *Journal of Geophysical Research* 94(D13):16,417–16,421.

19. Luoma, J. R., and Hiser, D. 1991. Gazing into our greenhouse future. *Audubon* 93(2).

20. Earth System Sciences Committee. 1988. *Earth system science: a preview*. Boulder, Colo.: University Corporation for Atmospheric Research.

21. Friis-Christensen, E., and Lassen, K. 1991. Length of solar cycle: an indicator of solar activity closely associated with climate. *Science* 254:698–700.

22. Stuiver, M., Grootes, P.M., and Braziunas, T. F. 1995. *Quarternary Research*, November 1995, vol. 44, no. 3, pp. 341-354.

23. Reid, G. C. 1993. "Solar forcing of global climate change since the mid-17th century," *Climatic Change*, vol. 37, pp. 391-405.

24. Charlson, R. J., Schwartz, S. E., Hales, J. M., Cess, R. D., Coakley, J. A. J., Hansen, J. E., and Hofmann, D. J. 1992. Climate forcing by anthropogenic aerosols. *Science* 255:423–430.

25. Kerr, R. A. 1995. Study unveils climate cooling caused by pollutant haze. *Science* 268:802.

26. Karl, T. R. 1991. Global warming: evidence for asymmetric diurnal temperature change. *Geophysical Research Letters* 18:2253–2256.

27. Mohnen, V. A., Goldstein, W., and Wang, W. 1991. The conflict over global warming. *Global Environmental Change* 1:109–123.

28. Kerr, R. A. 1995. U.S. climate tilts toward the Greenhouse. *Science* 268:363–364.

29. Kerr, R. A. 1995. Greenhouse report foresees growing global stress. *Science* 270:731.

30. Titus, J. G., and Narayanan, V. K. 1995. *The probability of sea level rise*. U.S. Environmental Protection Agency.

Chapter 22

1. Simons, L. M. 1998. "Plague of fire," *National Geographic*, vol. 194, no. 2, pp. 100-119.

2. National Park Service. 1984. *Air resources management manual.*

3. American Lung Association. 1998. *American Lung Association Outdoor Fact Sheet.* Accessed September 18, 1998. http://www.lungusa.org

4. Seitz, F., and Plepys, C. 1995. Monitoring air quality in healthy people 2000. *Healthy people 2000: statistical notes No. 9.* Atlanta: Centers for Disease Control and Prevention, National Center for Health Statistics.

5. Moore, C. 1995. Poisons in the air. *International Wildlife* 25:38–45.

6. Pope, C. A. III, Bates, D. V., and Raizenne, M. E. 1995. Health effects of particulate air pollution: Time for reassessment? *Environmental Health Perspectives* 103:472–480.

7. Pittock, A. B., Frakes, L. A., Jenssen, D., Peterson, J. A., and Zillman, J. W., eds. 1978. *Climatic change and variability: a southern perspective.* New York: Cambridge University Press.

8. Lents, J. M., and Kelly, W. J. 1993. Clearing the air in Los Angeles. *Scientific American* 269:32–39.

9. Blake, D. R., and Rowland, F. S. 1995. Urban leakage of liquefied petroleum gas and its impact on Mexico City air quality. *Science* 269:953.

10. Tyson, P. 1990. Hazing the Arctic. *Earthwatch* 10:23–29.

11. Office of Technology Assessment. 1984. Balancing the risks. *Weatherwise* 37:241–249.

12. Canadian Department of the Environment. 1984. *The acid rain story.* Ottawa: Minister of Supply and Services.

13. Lippmann, M., and Schlesinger, R. B. 1979. *Chemical contamination in the human environment.* New York: Oxford University Press.

14. Winkler, E. M. 1998. "The complexity of urban stone decay," *Geotimes,* September 1998, pp. 25-29.

15. Gates, D. M. 1972. *Man and his environment: climate.* New York: Harper & Row.

16. Anthes, R. A., Cahir, J. J., Fraser, A. B., and Panofsky, H. A. 1981. *The atmosphere,* 3d ed. Columbus, Ohio: Charles E. Merrill.

17. Pountain, D. 1993. "Complexity on wheels," *Byte,* May 1993, pp. 213-220.

18. Stern, A. C., Boubel, R. T., Turner, D. B., and Fox, D. L. 1984. *Fundamentals of air pollution,* 2d ed. Orlando, Fla.: Academic Press.

19. How many more lakes have to die? 1981. *Canada Today* 12, No. 2.

20. Moore, C. 1995. Green revolution in the making. *Sierra* 80:50.

21. Molnia, B. F. 1991. Washington Report. *GSA Today* 1:33.

22. Crandall, R. W. 1983. *Controlling industrial pollution: the economics and politics of clean air.* Washington, D.C.: The Brookings Institution.

23. Hall, J. V., Winer, A. M., Kleinman, M. T., Lurmann, F. W., Brajer, V., and Colome, S. D. 1992. Valuing the health benefits of clean air. *Science* 255:812–816.

24. Krupnick, A. J., and Portney, P. R. 1991. Controlling urban air pollution: a benefits–cost assessment. *Science* 252:522–528.

25. Lipfert, F. W., Morris, S. C., Friedman, R. M., and Lents, J. M. 1991. Air pollution benefit–cost assessment. *Science* 253:606.

Chapter 23

1. Massachusetts Department of Public Health, Bureau of Environmental Health Assessments. 1995. *Symptom prevalence survey related to indoor air concerns at the Registry of Motor Vehicles building, Ruggles Station.*

2. Horton, W. G. B. 1995. *NOVA: can buildings make you sick?* Video Production, WGBH, Boston.

3. Zimmerman, M. R. 1985. Pathology in Alaskan mummies. *American Scientist* 73:20–25.

4. Ehrlich, P. R., Ehrlich, A. H., and Holdren, J. P. 1970. *Ecoscience: population, resources, environment.* San Francisco: W. H. Freeman.

5. U.S. Environmental Protection Agency. 1991. *Building air quality: a guide for building owners and facility managers.* EPA/400/1-91/033, DHHS (NIOSH) Pub. No. 91-114. Washington, D.C.

6. Zummo, S. M. and Karol, M. H. 1996. Indoor air pollution: Acute adverse health effects and host susceptibility. *Environmental Health* 58:25–29.

7. Committee on Indoor Air Pollution. 1981. *Indoor pollutants.* Washington, D.C.: National Academy Press.

8. Brenner, D. J. 1989. *Radon: risk and remedy.* New York: W. H. Freeman.

9. U.S. Environmental Protection Agency. 1992. *A citizen's guide to radon: the guide to protecting yourself and your family from radon,* 2nd ed. ANR-464. Washington, D.C.

10. Pershagen, G., Akerblom, G., Axelson, O., Clavensjo, B., Damber, L., Desai, G., Enflo, A., Lagarde, F., Mellander, H., Svartengren, M., and Swedjemark, G. A. 1994. Residential radon exposure and lung cancer in Sweden. *New England Journal of Medicine* 330:159–164.

11. Henshaw, D. L., Eatough, J. P., and Richardson, R. B. 1990. Radon as a causative factor in induction of myeloid leukaemia and other cancers. *The Lancet* 335:1008–1012.

12. Nero, A. V., Jr. 1988. Controlling indoor air pollution. *Scientific American* 258:42–48.

13. Hurlburt, S. 1989. Radon: a real killer or just an unsolved mystery? *Water Well Journal,* June, pp. 34–41.

14. Egginton, J. 1989. Menace of Whispering Hills. *Audubon* 91:28–35.

15. University of Maine and Maine Department of Human Services. 1983. Radon in water and air. *Resource Highlights* February.

16. U.S. Environmental Protection Agency. 1986. *Radon reduction techniques for detached houses: technical guidance.* EPA 625/5-86-019. Research Triangle Park, N.C.: Air and Energy Engineering Research Laboratory, Office of Research and Development, U.S. Environmental Protection Agency.

17. Osborne, M. C. 1988. *Radon-resistant residential new construction.* EPA 600/8-88/087. Research Triangle Park, N.C.: Air and Energy Engineering Research Laboratory, Office of Research and Development, U.S. Environmental Protection Agency.

Chapter 24

1. Kane, R. P. 1998. "Ozone depletion, UV-B changes, and increased cancer incidence," *International Journal.*

2. Brown, L. R., Flavin, C., and Postel, S. 1991. *Saving the planet: how to shape an environmentally sustainable global economy.* New York, W. W. Norton.

3. Rowland, F. S. 1990. Stratospheric ozone depletion of chlorofluorocarbons. *AMBIO* 19:281–292.

4. Smith, R. C., Prezelin, B. B., Baker, K. S., Bidigare, R. R., Boucher, N. P., Coley, T., Karentz, D., Macintyre, S., Matlick, H. A., Menzies, D., Ondrusek, M., Wan, Z., and Waters, K. J. 1992. Ozone depletion: ultraviolet saturation and phytoplankton biology in Antarctic waters. *Science* 255:952–959.

5. Environmental Protection Agency. 1995. *Protection of the ozone layer.* U.S. EPA Office of Policy, Planning and Evalua-

tion and Office of Air and Radiation. EPA 230-N-95-00. Washington, D.C.

6. Anonymous, 1996. Ozone hole begins to shrink. Global Environmental Change Report V. VIII, No. 21, P.S. Cutter Information Co. Dunster, B.C. Canada.

7. Hamill, P., and Toon, O. B. 1991. Polar stratospheric clouds and the ozone hole. *Physics Today* 44:34–42.

8. Stolarski, R. S. 1988. The Antarctic ozone hole. *Scientific American* 258:30–36.

9. Molina, M. J., and Rowland, F. S. 1974. Stratospheric sink for chlorofluoro-methanes: chlorine-atom catalyzed distribution of ozone. *Nature* 249:810–812.

10. Brouder, P. 1986. Annals of chemistry in the face of doubt. *New Yorker*, June, pp. 20–87.

11. Khalil, M. A. K., and Rasmussen, R. A. 1989. The potential of soils as a sink of chlorofluorocarbons and other man-made chlorocarbons. *Geophysical Research Letters* 16:679–682.

12. Rowland, F. S. 1989. Chlorofluorocarbons and the depletion of stratospheric ozone. *American Scientist* 77:36–45.

13. Toon, O. B., and Turco, R. P. 1991. Polar stratospheric clouds and ozone depletion. *Scientific American* 264:68–74.

14. Kerr, R. A. 1994. Antarctic ozone hole fails to recover. *Science* 266:217.

15. Home page on the Worldwide Web of the World Meteorological Organization's Atmospheric Research and Environment Programme (http://www.wmo.ch/web/arep/arephome.html).

16. Webster, C. R., May, R. D., Toohey, D. W., Avallone, L. M., Anderson, J. G., Newman, P., Lait, L., Schoeberl, M. R., Elkins, J. W., and Chay, K. R. 1993. Chlorine chemistry on polar stratospheric cloud particles in the Arctic winter. *Science* 261:1130–1134.

17. Kerr, R. A. 1992. New assaults seen on Earth's ozone shield. *Science* 255:797–798.

18. Zurer, P. 1995. Record low ozone levels observed over Arctic. *Chemical & Engineering News* 73:8.

19. Horgan, J. 1992. Volcanic disruption. *Scientific American* 266:28–29.

20. Anonymous, 1996. Reports discuss present and future state of ozone layer. Global Environmental Change Report V. VIII, 21No. 22, pp 1–3. Cutter Information Co. Dunster, B.C. Canada.

21. Shea, C. P. 1989. Mending the Earth's shield. *World Watch* 2:28–34.

22. Kerr, J. B., and McElroy, C. T. 1993. Evidence for large upward trends of ultraviolet-B radiation linked to ozone depletion. *Science* 262:1032–1034.

23. Atkinson, R. J., Matthews, W. A., Newman, P. A., and Plumb, R. A. 1989. Evidence of the mid-latitude impact of Antarctic ozone depletion. *Nature* 340:290–294.

24. Russell, J. M., Luo, M., Cicerone, R. J., and Deaver, L. E. 1996. Satellite confirmation of dominance of chlorofluorocarbons in the global stratospheric chlorine budget. *Nature* 379:526.

25. Showstack, R. 1998. "Ozone layer is on slow road to recovery, new science assessment indicates," *Eos*, 79(27):317-318.

26. Spurgeon, D. 1998. "Surprising success of the Montreal protocol," *Nature*, 389(6648):219.

27. Makhijani, A., Bickel, A., and Makhijani, A. 1990. Still working on the ozone hole. *Technology Review* 93:52–59.

28. Brown, L. R., Lenssen, N., and Kane, H. 1995. CFC production plummeting. In: *Vital Signs 1995*. Worldwatch Institute. New York: W. W. Norton.

29. Shea, C. P. 1991. Disarming refrigerators. *World Watch* 4:36.

30. MacKenzie, D. 1990. Cheaper alternatives for CFCs. *New Scientist* 126:39–40.

31. Cicerone, R. J., Elliot, S., and Turco, R. P. 1991. Reduced Antarctic ozone depletions in a model with hydrocarbon injections. *Science* 254:1191–1194.

Chapter 25

1. Roberts, L. 1991. Costs of a clean environment. *Science* 251:1182.

2. Fairley, P. 1995. Compromise limits EPA budget cut, removes House riders. *Chemical Week* 157:17.

3. Moore, C. A. 1995. Poisons in the Air. *International Wildlife* 25:38–45.

4. Pope, C. A III, Bates, D. V., and Raizenne, M. E. 1995. Health effects of particulate air pollution: time for reassessment? *Environmental Health Perspectives* 103:472–480.

5. Hardin, G. 1968. The tragedy of the commons. *Science* 162:1243–1248.

6. Clark, C. W. 1973. The economics of overexploitation. *Science* 181:630–634.

7. Clark, *ibid.*

8. Gunn, J. M., ed. 1995. *Restoration and recovery of an industrial region. Progress in restoring the smelter-damaged landscape near Sudbury, Canada.* New York Springer-Verlag.

9. Cairns, J. Jr. 1980. Estimating hazard. *BioScience* 20:101–107.

10. Schwing, R. C. 1979. Longevity and benefits and costs of reducing various risks. *Technological Forecasting and Social Change* 13:333–345.

11. Schwing, *ibid.*

12. Gori, G. B. 1980. The regulation of carcinogenic hazards. *Science* 208:256–261.

13. Ostro, B. D. 1980. Air pollution, public health, and inflation. *Environmental Health Perspectives* 345:185–189.

14. Litton, R. B. 1972. Aesthetic dimensions of the landscape. In J. V. Krutilla, ed., *Natural environments.* Baltimore: Johns Hopkins University Press.

15. Office of Technology Assessment. 1991. *Changing by degrees: steps to reduce greenhouse gases.* Washington, D.C.: U.S. Superintendent of Documents.

16. Schneider, S. H. 1989. *Global warming: are we entering the greenhouse century?* San Francisco: Sierra Club Books.

17. D'Arge, R. 1989. Ethical and economic systems for managing the global commons. In D. B. Botkin, M. Caswell, J. E. Estes, and A. Orio, eds., *Changing the global environment: perspectives on human involvement*, pp. 327–337. New York: Academic Press.

18. Rogers, A. 1993. *The Earth summit: a planetary reckoning.* Global View Press.

19. World Resources Institute, 1993. *World Resources 1992-93.* World Resources Institute in collaboration with the United Nations Environment Programme and the United Nations Development Programme. New York, NY: Oxford University Press.

20. Baumol, W. J., and Oates, W. E. 1979. *Economics, environmental policy, and the quality of life.* Englewood Cliffs, N.J.: Prentice-Hall.

21. Cairns, *ibid.*

22. Clark, C. W. 1981. Economics of fishery management. In T. L. Vincent, and J. M. Skowronski, eds., *Renewable resource management: lecture notes in biomathematics,* pp. 95–111. New York: Springer-Verlag.

23. National Marine Fisheries Service. 1995. *Status of the fishery resources off the northeastern United States for 1994.* Woods Hole, Mass.: U.S. Department of Commerce, NOAA, NMFS Northeast Fisheries Science Center.

Chapter 26

1. Dobbs, F. 1995. *Curitiba: City of the Future?* (Video). World Bank.

2. Rabinovitch, J. 1997. *Integrated Transportation and Land Use Planning Channel Curitiba's Growth,* World Resources 1996-97, World Resources Institute.

3. Rabinovitch, J. and J. Leitman. 1996 (March). "Urban planning in Curitiba," *Scientific American,* pp. 46-53.

4. Hunt, J. 1994 (April). "Curitiba," *Metropolis Magazine.*

5. Roberts, Sam 1995. *Who We Are: A Portrait of America Based on the Latest U.S. Census.* New York. Times Books.

6. Population Reference Bureau, Inc. 1998. *World and Regional Population.* Washington, D.C. Population Reference Bureau, Inc.

7. Mumford, L. 1972. The natural history of urbanization. In R. L. Smith, ed., *The ecology of man: an ecosystem approach,* pp. 140–152. New York: Harper & Row.

8. Cronon, William. 1991. *Nature's metropolis: Chicago and the great west.* New York: W. W. Norton.

9. Mumford, *ibid.*

10. Leibbrand, K. 1970. *Transportation and town planning.* Translated by N. Seymer. Cambridge, Mass.: The MIT Press.

11. Mumford, *ibid.*

12. McLaughlin, C. C., ed. 1977. The formative years 1822–1852. *The papers of Frederick Law Olmsted, vol. 1.* Baltimore: Johns Hopkins University Press.

13. Miller, L. B. 1987. Miracle on 104th St. *American Horticulturalist* 66:14–17.

14. Spirn, A. W. 1984. *The granite garden: urban nature and human design.* New York: Basic Books.

15. Spirn, *ibid.*

16. Ford. A.B. and Bialik, O. 1980. Air pollution and urban factors in relation to cancer mortality. *Archives of Environmental Health* 35: 350-359.

17. Detwyler, T. R., and Marcus, M. G., eds. 1972. *Urbanization and the environment: the physical geography of the city.* North Scituate, Mass: Duxbury Press.

18. Butti, K., and Perlin, J. 1980. *A golden thread: 2500 years of solar architecture and technology.* New York: Cheshire.

19. Butti and Perlin, *ibid.*

20. McHarg, I. L., 1971. *Design with nature.* Garden City, N.Y.: Doubleday.

21. Akbari, H., Davis, S., Dorsano, S., Huang, J., and Winnett, S. 1992. *Cooling our communities: a guidebook on tree planting and light-colored surfacing.* Washington, D.C.: U.S. EPA Office of Policy Analysis, U.S. Superintendent of Documents.

22. Moll, G. 1989. Designing the ecological city. *American Forests* 85:61–64.

23. Nadel, I. B., Oberlander, C. H., and Bohm, L. R. 1977. *Trees in the city.* New York: Pergamon Press.

24. Moll, G., Rodbell, P., Skiera, B., Urban, J., Mann, G., and Harris, R. 1991. Planting new life in the city. *Urban Forests,* Apr./May, pp. 10–20. Washington, D.C.: American Forestry Association.

25. Spirn, *ibid.*

26. Moll et al., *ibid.*

27. Dreistadt, S. H., Dahlsten, D. L., and Frankie, G. W. 1990. Urban forests and insect ecology. *BioScience* 40:192–198.

28. Leedly, D.L. and Adams, L.W. 1984. *A guide to urban wildlife management.* Columbia, Md.; National Institute for Urban Wildlife.

29. Tylka, D. 1987. Critters in the city. *American Forests* 93:61–64.

30. Burton, J.A. 1977. *Worlds apart: nature in the city.* Garden City, N.Y.: Doubleday.

31. Burton, *ibid.*

32. Adams, L. W., and Dove, L. E. 1989. *Wildlife reserves and corridors in the urban environment.* Columbia, Md.: National Institute for Urban Wildlife.

33. Hunt, C. B. 1974. *Natural regions of the United States and Canada.* San Francisco: W. H. Freeman.

34. Mumford, *ibid.*

35. Leibbrand, *ibid.*

36. Reps, J. W. 1965. *The making of urban America: a history of city planning in the United States,* 2d ed. Princeton, N.J.: Princeton University Press.

37. Spirn, *ibid.*

Chapter 27

1. Wright, A.G. 1998. "Big apple is dumping its last active landfill. *ENR,* vol. 240, no. 7, pp. 28-31.

2. Young, W. 1995. "A dump no more," *American Forests,* Autumn, pp. 59-62.

3. Galley, J. E. 1968. Economic and industrial potential of geologic basins and reservoir strata. In J. E. Galley, ed., *Subsurface disposal in geologic basins: a study of reservoir strata,* pp. 1–19. American Association of Petroleum Geologists Memoir 10. Tulsa, Okla.: American Association of Petroleum Geologists.

4. Relis, P., and Dominski, A. 1987. *Beyond the crisis: integrated waste management.* Santa Barbara, Calif.: Community Environmental Council.

5. Repa, D. W., and Blakey, A. 1996. Municipal solid waste disposal trends: 1996 update. *Waste Age* 27:42–54.

6. Council on Environmental Quality. 1973. *Environmental quality—1973.* Washington, D.C.: U.S. Government Printing Office.

7. Relis, P. and H. Levenson, H. *Discarding Solid Waste as We Know It: Managing Materials in the 21st Century.* Policy Paper, Community Environmental Council, Santa Barbara, CA. 11 p.

8. Young, J. E. 1991. Reducing waste saving materials. In L. R. Brown, ed., *State of the world, 1991,* pp. 39–55. World Watch Institute. New York: W. W. Norton.

9. Steuteville, R. 1995. The state of garbage in America: Part I. *BioCycle* 36:54.

10. McGreery, P. 1995. Going for the goals: Will states hit the wall? *Waste Age* 26:68–76.

11. Rathje, W. L., and Murphy, C. 1992. Five major myths about garbage, and why they're wrong. *Smithsonian* 23:113–122.

12. Rathje, W. L. 1991. Once and future landfills. *National Geographic* 179(5):116–134.

13. Schneider, W. J. 1970. *Hydrologic implications of solid-waste disposal* 135(22). U.S. Geological Survey Circular 601F. Washington, D.C.: U.S. Geological Survey.

14. Thomas, V. M., and Spiro, T. G. 1996. The U.S. dioxin inventory: are there missing sources? *Environmental Science & Technology* 30:82A–85A.

15. Turk, L. J. 1970. Disposal of solid wastes—acceptable practice or geological nightmare? In *Environmental Geology,* pp. 1–42. Washington, D.C.: American Geological Institute Short Course, American Geological Institute.

16. Hughes, G. M. 1972. Hydrologic considerations in the citing and design of landfills. *Environmental Geology Notes,* no. 51. Urbana: Illinois State Geological Survey.

17. Bergstrom, R. E. 1968. Disposal of wastes: scientific and ad-

ministrative considerations. *Environmental Geology Notes,* no. 20. Urbana: Illinois State Geological Survey.

18. Cartwright, K., and Sherman, F. B. 1969. Evaluating sanitary landfill sites in Illinois. *Environmental Geology Notes,* no. 27. Urbana: Illinois State Geological Survey.

19. Bullard, R. D. 1990. *Dumping in Dixie: race, class and environmental quality.* Boulder, Colo.: Westview Press.

20. Walker, W. H. 1974. Monitoring toxic chemical pollution from land disposal sites in humid regions. *Ground Water* 12:213–218.

21. Liptak, B. G. 1991. *Municipal waste disposal in the 1990s.* Chilton Book Company.

22. Wilkes, A. S. 1980. *Everybody's problem: hazardous waste.* SW-826. Washington, D.C.: U.S. Environmental Protection Agency, Office of Water and Waste Management.

23. Elliot, J. 1980. Lessons from Love Canal. *J. American Medical Association* 240:2033–2034, 2040.

24. Kufs, C., and Twedwell, C. 1980. Cleaning up hazardous landfills. *Geotimes* 25:18–19.

25. Albeson, P. H. 1983. Waste management. *Science* 220:1003.

26. New York State Department of Environmental Conservation. 1994. *Remedial chronology: the Love Canal hazardous waste site.* New York State.

27. Kirschner, E. 1994. Love Canal settlement: OxyChem to pay New York state $98 million. *Chemical & Engineering News* 72:4–5.

28. Westervelt, R. 1996. Love Canal: OxyChem settles federal claims. *Chemical Week* 158:9.

29. Cooney, C. M. 1997. "New EPA budget boost particulate matter research, superfund cleanups," *Environmental Science and Technology,* vol. 31, no. 12, p. A555.

30. Bedient, P. B., Rifai, H. S., and Newell, C. J. 1994. *Ground water contamination.* Englewood Cliffs, NJ: Prentice-Hall.

31. Huddleston, R. L. 1979. Solid-waste disposal: land farming. *Chemical Engineering* 86:119–124.

32. McKenzie, G. D., and Pettyjohn, W. A. 1975. Subsurface waste management. In G. D. McKenzie and R. O. Utgard, eds., *Man and his physical environment: readings in environmental geology,* 2nd ed., pp. 150–156. Minneapolis: Burgess Publishing.

33. National Research Council, Committee on Geological Sciences. 1972. *The earth and human affairs.* San Francisco: Canfield Press.

34. Cox, C. 1985. *The buried threat: getting away from land disposal of hazardous waste.* California Senate Office of Research No. 115-5.

35. Council on Environmental Quality. 1970. *Ocean dumping: a national policy; a report to the President.* Washington, D.C.: U.S. Government Printing Office.

36. Lenssen, N. 1989. The ocean blues. *World Watch,* July/Aug., pp. 26–35.

Chapter 28

1. Gulbrandsen, R. A., Rait, N., Dries, D. J., Baedecker, P. A., and Childress, A. 1978. *Gold, silver, and other resources in the ash of incinerated sewage sludge at Palo Alto, California—a preliminary report.* Washington, D.C.: U.S. Geological Survey Circular 784. U.S. Department of the Interior.

2. McKelvey, V. E. 1973. Mineral resource estimates and public policy. In D. A. Brobst and W. P. Pratt, eds., *United States mineral resources,* pp. 9–19. U.S. Geological Survey Professional Paper 820.

3. U.S. Department of Interior, Bureau of Mines. 1993. *Mineral commodity summaries, 1993.* I 28.149:993. Washington, D.C.

4. Meyer, H. O. A. 1985. Genesis of diamond: A mantle saga. *American Mineralogist* 70:344–355.

5. Kesler, S. F. 1994. *Mineral resources, economics and the environment.* New York: Macmillan.

6. Smith, G. I., Jones, C. L., Culbertson, W. C., Erickson, G. E., and Dyni, J. R. 1973. Evaporites and brines. In D. A. Brobst and W. P. Pratt, eds., *United States mineral resources,* pp. 197–216. U.S. Geological Survey Professional Paper 820.

7. Awramik, S. A. 1981. The pre-Phanerozoic biosphere—three billion years of crises and opportunities. In M. H. Nitecki, ed., *Biotic crises in ecological and evolutionary time,* pp. 83–102. Spring Systematics Symposium. New York: Academic Press.

8. Margulis, L., and Lovelock, J. E. 1974. Biological modulation of the Earth's atmosphere. *Icarus* 21:471–489.

9. Lowenstam, H. A. 1981. Minerals formed by organisms. *Science* 211:1126–1130.

10. Bateman, A. M. 1950. *Economic mineral deposits,* 2d ed. New York: John Wiley & Sons.

11. Park, C. F., Jr., and MacDiarmid, R. A. 1970. *Ore deposits,* 2d ed. San Francisco: W. H. Freeman.

12. Cornwall, H. R. 1973. Nickel. In D. A. Brobst and W. P. Pratt, eds., *United States mineral resources,* pp. 437–442. U.S. Geological Survey Professional Paper 820.

13. Van, N., Dorr, J., Crittenden, M. D., and Worl, R. G. 1973. Manganese. In D. A. Brobst and W. P. Pratt, eds., *United States mineral resources,* p. 385–399. U.S. Geological Survey Professional Paper 820.

14. U.S. Secretary of the Interior. 1975. *Mining and mineral policy, 1975.* Washington, D.C.

15. George, D. 1994. Framework for deep ocean mining fixed despite lack of technology. *Offshore* 54:52.

16. Brobst, D. A., Pratt, W. P., and McKelvey, V. E. 1973. *Summary of United States mineral resources.* U.S. Geological Survey Circular 682.

17. U.S. Geological Survey. 1984. *Yearbook, fiscal year 1983.* Washington, D.C.

18. Jeffers, T. H. 1991. Using microorganisms to recover metals. *Minerals Today.* Washington, D.C.: U.S. Department of Interior, Bureau of Mines, June, pp. 14–18.

19. Haynes, B. W. 1990. Environmental technology research. *Minerals Today.* Washington, D.C.: U.S. Bureau of Mines, May, pp. 13–17.

20. Sullivan, P. M., Stanczyk, M. H., and Spendbue, M. J. 1973. *Resource recovery from raw urban refuse.* U.S. Bureau of Mines Report of Investigations 7760. Washington, D.C.

21. Davis, F. F. 1972. Urban ore. *California Geology,* May, pp. 99–112.

22. Staff, Division of Mineral Commodities. 1994. *Recycled metals in the United States.* Washington, D.C.: U.S. Department of the Interior, Bureau of Mines, Special Publication I 28.151:M56.

23. U.S. Bureau of Mines. 1992. *Minerals in 1992.* Washington, D.C.: U.S. Department of the Interior, Office of Public Information.

24. Brown, L., Lenssen, N., and Kane, H. 1995. Steel recycling rising. In *Vital Signs 1995.* Worldwatch Institute.

Chapter 29

1. McHarg, I. L. 1971. *Design with nature.* Garden City, N.Y.: Doubleday.

2. Council on Environmental Quality. 1979. *Environmental quality.* Washington, D.C.

3. Brew, D. A. 1974. *Environmental impact analysis: the example of the proposed trans-Alaska pipeline*. U.S. Geological Survey Circular 695. Reston, Va.: U.S. Department of the Interior.

4. Parker, H. D. 1987. What is a geographic information system? In *GIS '87*, pp. 72–79. Second Annual International Conference, Exhibits and Workshops on Geographic Information Systems. American Society for Photogrammetry and Remote Sensing.

5. Star, J., and Estes, J. 1990. *Geographic information systems: an introduction*. Englewood Cliffs, N.J.: Prentice-Hall.

6. Callies, D. L. 1984. *Regulating paradise: land use controls in Hawaii*. Honolulu: University of Hawaii Press.

7. Jain, R. K., Urban, L. V., Stacey, G. S., and Balbach, H. E. 1993. *Environmental assessment*. New York: McGraw-Hill.

8. Remy, M. H., Thomas, T. A., and Moose, J. G. 1991. *Guide to the California Environmental Quality Act*, 5th ed. Point Arena, Calif.: Solano Press Books.

9. Cohen, B. S. 1970. *The constitution, the public trust doctrine and the environment*. Utah Law Review 388.

10. Yannacone, V. J., Jr., Cohen, B. S., and Davison, S. G. 1972. *Environmental rights and remedies*. Rochester, N.Y.: Lawyers Co-operative Pub., pp. 39–46.

11. Bacow, L. S., and Wheeler, M. 1984. *Environmental dispute resolution*. New York: Plenum Press.

12. National Park Service. 1974. *Cape Hatteras shoreline erosion policy statement*. Washington, D.C.: Department of the Interior.

13. Leatherman, S. P. 1983. Barrier dynamics and landward migration with Holocene sea-level rise. *Nature* 301:415–418.

14. National Park Service. 1984. *General management plan, development concept plan, and amended environmental assessment*, Cape Hatteras National Seashore.

15. McDonald, K. A. 1993. A geology professor's fervent battle with coastal developers and residents. *The Chronicle of Higher Education* 40:A8.

16. Tanji, K., Lauchli, A., and Meyer, J. 1986. Selenium in the San Joaquin Valley. *Environment* 28:6–11, 34–39.

17. U.S. Department of the Interior, Bureau of Reclamation. 1987. *Kesterson program*. Fact Sheet No. 4. Sacramento, Calif.

18. Rohse, M. 1987. *Land-use planning in Oregon: a no nonsense handbook in plain English*. Corvallis: Oregon State University Press.

19. Curtin, D. J., Jr. 1991. *California land-use and planning law*, 11th ed. Point Arena, Calif.: Solano Press Books.

20. Advisory Committee on the International Decade for Natural Hazard Reduction. 1989. *Reducing disaster's toll*. Washington, D.C.: National Academy Press.

21. Russell, G. 1985. Colombia's mortal agony. *Time* 126:46–52.

22. Weintraub, B. 1996. Havens in Bangladesh from killer storms. *National Geographic* 189(4).

23. Steiner, F. 1983. Regional planning: historic and contemporary examples. *Landscape Planning* 10:297–315.

24. Coats, D. R., ed. 1973. *Coastal geomorphology*. Binghamton, N.Y.: Publications in Geomorphology, State University of New York.

25. Wilshire, H. G., and Nakata, J. K. 1976. Off-road vehicle effects on California's Mojave Desert. *California Geology* 29:123–132.

26. Wilshire, H. G., Bodman, G. B., Broberg, D., Kockelman, W. J., Major, J., Malde, H. E., Snyder, C. T., and Stebbs, R. C. 1977. *Impacts and management of off-road vehicles*. Report of the Committee on Environment and Public Policy. Boulder, Colo.: Geological Society of America.

27. Meadows, D. H., Meadows, D. L., Randers, J., and Behrens, W. W. III. 1972. *The limits to growth: a report for the Club of Rome's Project on the Predicament of Mankind*. New York: Universe Books (Potomac Associates).

28. Council on Environmental Quality and the U.S. Department of State. 1980. *The global 2000 report to the president: entering the twenty-first century*. Washington, D.C.

Chapter 30

1. Botkin, D. B., Cummins, K., Dunne, T., Regier, H., Sobel, M. J., and Talbot, L. M. 1995. *Status and Future of Anadromous Fish of Western Oregon and Northern California, Findings and Options*. Santa Barbara, Calif.: Center for the Study of the Environment.

2. Botkin, D. B. 1990. *Discordant harmonies: a new ecology for the 21st century*. New York: Oxford University Press.

3. Radtke, H. 1992. *Economic contribution of salmon to Oregon's coastal communities*. Paper presented at the Governor's Salmonid Initiative Conference, Nov. 1992.

4. Homer-Dixon, T. F., Boutwell, J. H., and Rathjens, G. W. 1993. *Environmental change and violent conflict*. Scientific American 268:38–45.

5. Strong, D. H., and Rosenfeld, E. S. 1976. Ethics or expediency: an environmental question. *Environmental Affairs* 5:255–270.

6. Schneider, S. H. 1989. *Global warming: are we entering the greenhouse century?* San Francisco: Sierra Club Books.

7. Nash, R. F. 1989. *The rights of nature: a history of environmental ethics*. Madison: University of Wisconsin Press.

8. Leopold, A. 1949. *A Sand County almanac*. New York: Oxford University Press.

9. Callicott, B. 1987. *Companion to "A Sand County Almanac."* Madison: University of Wisconsin Press.

10. Callicott, J. B., ed. 1989. *In defense of the land ethic: essays in environmental philosophy*. Albany: State University of New York Press.

11. Nash, R. 1979. Wilderness is all in your mind. *Backpacker* 31:39.

12. Stone, C. D. 1972. Should trees have standing? Toward legal rights for natural objects. *California Legal Review* 45:450.

13. Ferry, L. 1996. *The New Ecological Order*.

14. Partridge, E. 1981. Responsibilities to future generations: environmental ethics. Buffalo, N.Y.: Prometheus Books.

15. Nash, 1979, *ibid*.

16. Leclerc, G. L. 1812. *Natural history, general and particular*, vol. 3. Translated by W. Smellie. London: C. Wood.

17. Cicero. *The Nature of the Gods*. 1972. Translated by H.C.P. McGregor, Aylesbury: Penguin Books.

18. 1968, trans. R. Humphries, Bloomington, Indiana University Press.

19. Leclerc, G. L. 1812. *Natural history, general and particular*, vol. 3. Translated by W. Smellie. London: C. Wood.

20. Botkin, B. A. 1944. *A treasure of American folklore*. New York: Crown Publishers.

21. Botkin, B. A., *ibid*.

22. Marsh, G. P. 1967. *Man and nature*. Cambridge, Mass.: Belknap Press.

23. Runte, A. 1979. *National parks: the American experience*. Lincoln: University of Nebraska Press.

24. Ferry, L. 1996. *The New Ecological Order*.

25. Cicero. *The Nature of the Gods*. 1972. Translated by H. C. P. McGregor. Aylesbury: Penguin Books.

Abortion rate The estimated number of abortions per 1000 women aged 15 to 44 in a given year. (Ages 15 to 44 are taken to be the limits of ages during which women can have babies. This of course is an approximation, made for convenience.)

Abortion ratio The estimated number of abortions per 1000 live births in a given year.

Acid mine drainage Does not refer to an acid mine but to acidic water that drains from mining areas (mostly coal but also metal mines). The acidic water may enter surface water resources, causing environmental damage.

Acid rain Rain made artificially acid by pollutants, particularly oxides of sulfur and nitrogen. (Natural rainwater is slightly acid owing to the effect of carbon dioxide dissolved in the water.)

Active solar energy systems Direct use of solar energy that requires mechanical power; usually consists of pumps and other machinery to circulate air, water, or other fluids from solar collectors to a heat sink where the heat may be stored.

Adaptive radiation The process that occurs when a species enters a new habitat that has unoccupied niches and evolves into a group of new species, each adapted to one of these niches.

Advanced wastewater treatment Treatment of wastewater beyond primary and secondary procedures. May include sand filters, carbon filters, or application of chemicals to assist in removing potential pollutants such as nutrients from the wastewater stream.

Aerobic Characterized by the presence of free oxygen.

Aesthetic arguments Reasons for preserving the beauty of nature and natural wilderness areas.

Aesthetic justification for the conservation of nature An argument for the conservation of nature on the grounds that nature is beautiful and that beauty is important and valuable to people.

Age dependency ratio The ratio of dependent-age people (those unable to work) to working-age people. It is customary to define working-age people as those aged 15 to 65.

Age structure (of a population) Structure of a population divided into groups by age. Sometimes the groups represent the actual number of each age in the population; sometimes the group represents the percentage or proportion of the population of each age.

Agroecosystem An ecosystem created by agriculture. It has low genetic, species, and habitat diversity.

Air Quality Standards Levels of air pollutants that delineate acceptable levels of pollution over a particular time period. Valuable because they are often tied to emission standards that attempt to control air pollution.

Allowance trading Approach to managing coal resources and reducing pollution through buying, selling, and trading of allowances to emit pollutants from burning coal. The idea is to control pollution by controlling the number of allowances issued.

Alpha particles One of the major types of nuclear radiation, consisting of two protons and two neutrons (a helium nucleus).

Alternative energy Renewable and nonrenewable energy resources that are alternatives to the fossil fuels.

Anaerobic Characterized by the absence of free oxygen.

Aquaculture Production of food from aquatic habitats.

Aquifer An underground zone or body of earth material from which groundwater can be obtained from a well at a useful rate.

Area sources Sometimes also called nonpoint sources. These are diffused sources of pollution such as urban runoff or automobile exhaust. These sources include emissions that may be over a broad area or even over an entire region. They are often difficult to isolate and correct because of the widely dispersed nature of the emissions.

Asbestos A term for several minerals that have the form of small elongated particles. Some types of particles are believed to be carcinogenic or to carry with them carcinogenic materials.

Atmosphere Layer of gases surrounding Earth.

Atmospheric inversion A condition in which warmer air is found above cooler air, restricting air circulation; often associated in urban areas with a pollution event.

Autotroph An organism that produces its own food from inorganic compounds and a source of energy. There are photoautotrophs (photosynthetic plants) and chemical autotrophs.

Average residence time A measure of the time it takes for a given part of the total pool or reservoir of a particular material in a system to be cycled through the system. When the size of the pool and rate of throughput are constant, average residence time is the ratio of the total size of the pool or reservoir to the average rate of transfer through the pool.

Barrier island An island separated from the mainland by a salt marsh. It generally consists of a multiple system of beach ridges and is separated from other barrier islands by inlets that allow the exchange of seawater with lagoon water.

Becquerel The unit commonly used for radioactive decay in the International System (SI) of measurement.

Beta particles One of the three major kinds of nuclear radiation; electrons that are emitted when one of the protons or neutrons in the nucleus of an isotope spontaneously changes.

Biochemical oxygen demand (BOD) A measure of the amount of oxygen necessary to decompose organic material in a unit volume of water. As the amount of organic waste in water increases, more oxygen is used, resulting in a higher BOD.

Biogeochemical cycle The cycling of a chemical element through the biosphere; its pathways, storage locations, and chemical forms in the atmosphere, oceans, sediments, and lithosphere.

Biogeography The large-scale geographic pattern in the distribution of species, and the causes and history of this distribution.

Biohydromettalurgy Combining biological and mining processes, usually involving microbes to help extract valuable metals such as gold from the ground. May also be used to remove pollutants from mining waste.

Biological control A set of methods to control pest organisms by using natural ecological interactions, including predation, parasitism, and competition. Part of integrated pest management.

Biological diversity Used loosely to mean the variety of life on the Earth, but technically this concept consists of three components: (1) genetic diversity—the total number of genetic characteristics; (2) species diversity; and (3) habitat or ecosystem diversity—the number of kinds of habitats or ecosystems in a given unit area. Species diversity in turn includes three concepts: *species richness, evenness,* and *dominance.*

Biological evolution The change in inherited characteristics of a population from generation to generation, which can result in new species.

Biomagnification Also called *biological concentration.* The tendency for some substances to concentrate with each trophic level. Organisms preferentially store certain chemicals and excrete others. When this occurs consistently among organisms, the stored chemicals increase as a percentage of the body weight as the material is transferred along a food chain or trophic level. For example, the concentration of DDT is greater in herbivores than in plants and greater in plants than in the nonliving environment.

Biomass The amount of living material, or the amount of organic material contained in living organisms, both as live and dead material, as in the leaves (live) and stem wood (dead) of trees.

Biomass fuel A new name for the oldest fuel used by humans. Organic matter, such as plant material and animal waste, that can be used as a fuel.

Biome A kind of ecosystem. The rain forest is an example of a biome; rain forests occur in many parts of the world but are not all connected with each other.

Bioremediation A method of treating groundwater pollution problems that utilizes microorganisms in the ground to consume or break down pollutants.

Biosphere That part of a planet where life exists. On Earth it extends from the depths of the oceans to the summit of mountains, but most life exists within a few meters of the surface.

Biota A general term for all the organisms of all species living in an area or region up to and including the biosphere, as in "the biota of the Mojave Desert" or "the biota in that aquarium."

Biotic province A region inhabited by a characteristic set of taxa (species, families, orders), bound by barriers that prevent the spread of the distinctive kinds of life to other regions and the immigration of foreign species.

Birth rate The rate at which births occur in a population, measured either as the number of individuals born per unit of time or as the percentage of births per unit of time compared with the total population.

Black lung disease Often called coal miner disease because it is caused by years of inhaling coal dust, resulting in damage to lungs.

Body burden The amount of concentration of a toxic chemical, especially radionuclides, in an individual.

Breeder reactor A type of nuclear reactor that utilizes between 40% and 70% of its nuclear fuel and converts fertile nuclei to fissile nuclei faster than the rate of fission. Thus breeder reactors actually produce nuclear fuels.

Brines With respect to mineral resources, refers to waters with a high salinity that contain useful materials such as bromine, iodine, calcium chloride, and magnesium.

Buffers Materials (chemicals) that have the ability to neutralize acids. Examples include the calcium carbonate that is present in many soils and rocks. These materials may lessen potential adverse effects of acid rain.

Burner reactors A type of nuclear reactor that consumes more fissionable material than it produces.

Capillary action The rise of water along narrow passages, facilitated and caused by surface tension.

Carbon cycle Combined biochemical cycles of carbon, oxygen, and hydrogen. Carbon combines with and is chemically and biologically linked with the cycles of oxygen and hydrogen that form the major compounds of life.

Carbon dioxide (CO_2) Molecule of carbon and oxygen present in the atmosphere at approximately 350 ppm. Emissions of carbon dioxide resulting from burning of fossil fuels are thought to be contributing to potential global warming through an enhanced greenhouse effect.

Carbon monoxide (CO) Colorless, odorless gas that at very low concentrations is extremely toxic to humans and animals.

Carbonate-silicate cycle A complex biogeochemical cycle over time scales as long as one-half billion years. Included in this cycle are major geologic processes, such as weathering, transport by ground and surface waters, erosion, and deposition of crustal rocks. The carbonate-silicate cycle is believed to provide important negative feedback mechanisms that control the temperature of the atmosphere.

Carcinogen Any material that is known to produce cancer in humans or other animals.

Carnivores Organisms that feed on other live organisms; usually applied to animals that eat other animals.

Carrying capacity The maximum abundance of a population or species that can be maintained by a habitat or ecosystem without degrading the ability of that habitat or ecosystem to maintain that abundance in the future.

Cash crops Crops grown to be traded in a market.

Channelization An engineering technique that consists of straightening, deepening, widening, clearing, or lining existing stream channels. The purpose is to control floods, improve drainage, control erosion, or improve navigation. It is a very controversial practice that may have significant environmental impacts.

Chaparral A dense scrubland found in areas with Mediterranean climate (a long warm, dry season and a cooler rainy season).

Chemoautotrophs Autotrophic bacteria that can derive energy from chemical reactions of simple inorganic compounds. The most common use for inorganic sulfur compounds.

Chemosynthesis Synthesis of organic compounds by energy derived from chemical reactions.

Chimney (or stack) effect Process whereby warmer air rises in buildings to upper levels and is replaced in the lower portion of the building by outdoor air drawn through a variety of openings, such as windows, doors, or cracks in the foundations and walls.

Chlorofluorocarbons (CFCs) Highly stable compounds that have been or are being used in spray cans as aerosol propellants and in refrigeration units (the gas that is compressed and expanded in a cooling unit). Emissions of chlorofluorocarbons have been associated with potential global warming and stratospheric ozone depletion.

Chronic disease A disease that is usually present in a population, typically occurring in a relatively small but constant percentage of the population.

Chronic hunger Availability of sufficient food to stay alive, but not enough to lead a satisfactory and productive life.

Chronic patchiness A situation where ecological succession does not occur. Characteristic of harsh environments such as deserts. In this case, one species may replace another, or an individual of the first species may replace it, and no overall general temporal pattern is established.

City planning Conscious design of the growth and development of an urban area.

Clay May refer to a mineral family or to a very fine-grained sediment. It is associated with many environmental problems, such as shrinking and swelling of soils and sediment pollution.

Clean Air Act Amendments of 1990 Comprehensive regulations (federal statute) that address acid rain, toxic emissions, ozone depletion, and automobile exhaust.

Clear-cutting In timber harvesting, the practice of cutting all trees in a stand at the same time.

Climate The representative or characteristic conditions of the atmosphere at particular places on Earth. *Climate* refers to the average or expected conditions over long periods; *weather* refers to the particular conditions at one time in one place.

Climatic change Change in mean annual temperature and other aspects of climate over periods of time ranging from decades to hundreds of years to several million years.

Climax stage (or ecological succession) The final stage of ecological succession and therefore an ecological community that continues to reproduce itself over time, or a stage in ecological succession during which an ecological community achieves the greatest biomass or diversity. (The first definition is the classical definition.)

Climax state A hypothetical steady-state stage at the end of ecological succession. Traditionally, it was believed to be self-sustaining and also to have maximum organic matter, maximum storage of chemical elements, and maximum biological diversity. These ideas are now largely rejected among scientists doing basic ecological research, but they still have many adherents among practitioners.

Closed system A type of system in which there are definite boundaries to factors such as mass and energy such that exchange of these factors with other systems does not occur.

Closed-canopy forest Forests in which the leaves of adjacent trees overlap or touch, so that the trees form essentially continuous cover.

Coal Solid, brittle carbonaceous rock that is by far the world's most abundant fossil fuel. It is classified according to energy content as well as carbon and sulfur content.

Coal gasification Process that converts coal that is relatively high in sulfur to a gas in order to remove the sulfur.

Cogeneration The capture and use of waste heat; for example, using waste heat from a power plant to heat adjacent factories and other buildings.

Cohort All the individuals in a population born during the same time period. Thus all the people born during the year 1980 represent the world human cohort for that year.

Commensalism A relationship between two kinds of organisms in which one benefits from the relationship and the other is neither helped nor hurt.

Commercial-grade forest Forestland believed or known to be an economically profitable source of timber, often defined as capable of producing at least 1.4 m³/ha (200 ft³/acre) per year of wood.

Common law Law derived from custom, judgment, or decrees of courts.

Commons Land that belongs to the public, not to individuals. Historically a part of old English and New England towns where all the farmers could graze their cattle.

Community, ecological A group of populations of different species living in the same local area and interacting with one another. A community is the living portion of an ecosystem.

Community effect (community-level effect) When the interaction between two species leads to changes in the presence or absence of other species or in a large change in abundance of other species, then a community effect is said to have occurred. Here, a two-species interaction affects the entire community.

Competition The situation that exists when different individuals, populations, or species compete for the same resource(s) and the presence of one has a detrimental effect on the other. Sheep and cows eating grass in the same field are competitors.

Competitive exclusion principle The idea that two populations of different species with exactly the same requirements cannot persist indefinitely in the same habitat—one will always win out and the other will become extinct. Which one wins depends on the exact environmental conditions. Referred to as a principle, the idea has some basis in observation and experimentation.

Composting Biochemical process in which organic materials, such as lawn clippings and kitchen scraps, are decomposed to a rich, soillike material.

Comprehensive plan Official plan adopted by local government formally stating general and long-range policies concerning future development.

Cone of depression A cone-shaped depression in the water table caused by withdrawal of water at rates greater than the rates at which the water can be replenished by natural groundwater flow.

Conservation With respect to resources such as energy refers to changing our patterns of use or simply getting by with less demand. In a pragmatic sense the term means adjusting our needs to minimize the use of a particular resource, such as energy.

Consumptive use A type of off-stream water use. This water is consumed by plants and animals or in industrial processes and enters human tissue or products or evaporates during use. It is not returned to its source.

Contamination Presence of undesirable material that makes something unfit for a particular use.

Continental drift The movement of continents in response to seafloor spreading. The most recent episode of continental drift supposedly started about 200 million years ago with the breakup of the supercontinent Pangaea.

Continental shelf Relatively shallow ocean area between the shoreline and the continental slope that extends to approximately a 600-foot (~200 m) water depth surrounding a continent.

Contour plowing Plowing land along topographic contours, perpendicular to the slope—as much in the horizontal plane as possible, thereby decreasing the erosion rate.

Controlled burning Using prescribed fire to reduce the risk from wildfires, control tree diseases, increase food and habitat for wildlife, and manage forests for greater production of desirable tree species.

Controlled experiment A controlled experiment is designed to test the effects of independent variables on a dependent variable by changing only one independent variable at a time. For each variable tested, there are two set-ups (an experiment and a control) that are identical except for the independent variable being tested. Any difference in the outcome (dependent variable) between the experiment and the control can then be attributed to the effects of the independent variable tested.

Convection The transfer of heat involving the movement of particles; for example, the boiling water in which hot water rises to the surface and displaces cooler water, which moves toward the bottom.

Convergent evolution The process by which species evolve in different places or different times and, although they have different genetic heritages, develop similar external forms and structures as a result of adaptation to similar environments. The similarity in the shapes of sharks and porpoises is an example of convergent evolution.

Convergent plate boundary Boundary between two lithosphere plates in which one plate descends below the other (subduction).

Cosmopolitan species A species with a broad distribution, occurring wherever in the world the environment is appropriate.

Costs, direct In economics, costs borne by the producer and passed directly on to the user or purchaser.

Costs, indirect In environmental economics, costs associated with the degradation of the environment.

Crop rotation A series of different crops planted successively in the same field, with the field occasionally left fallow, or grown with a cover crop.

Crude oil Naturally occurring petroleum, normally pumped from wells in oil fields. Refinement of crude oil produces most of the petroleum products we use today.

Curie Commonly used unit to measure radioactive decay; the amount of radioactivity from 1 gram of radium 226 that undergoes about 37 billion nuclear transformations per second.

Death rate The rate at which deaths occur in a population, measured either as the number of individuals dying per unit time or as the percentage of a population dying per unit time.

Decomposer An organism that feeds on dead organic matter and in this way reduces complex organic compounds to simpler organic compounds or smaller particles or to inorganic compounds.

Decomposer An organism that obtains its energy and nutritional requirements by feeding on dead organisms; or, a feeder on dead organisms.

Deductive reasoning Drawing a conclusion from initial definitions and assumptions by means of logical reasoning.

Deep-well disposal Method of disposal of hazardous liquid waste that involves pumping the waste deep into the ground below and completely isolated from all freshwater aquifers. A controversial method of waste disposal that is being carefully evaluated.

Deforestation Harvesting trees for commercial and other uses and burning forest to convert lands to agricultural purposes.

Demand for food The amount of food that would be bought at a given price if it were available.

Demand-based agriculture Agriculture with production determined by economic demand and limited by that demand rather than by resources.

Demographic transition The pattern of change in birth and death rates as a country is transformed from undeveloped to developed. There are three stages: (1) in an undeveloped country birth and death rates are high, and the growth rate low; (2) the death rate decreases, but the birthrate remains high and the growth rate is high; (3) the birthrate drops toward the death rate and the growth rate therefore also decreases.

Demography The study of populations, especially their patterns in space and time.

Dependent variable See **Variable, dependent**

Denitrification The conversion of nitrate to molecular nitrogen by the action of bacteria—an important step in the nitrogen cycle.

Density-dependent population effects Factors whose effects on a population change with population density. The term is usually restricted to population growth, reproduction, and mortality. For example, during a famine the mortality rate increases. In this case the food supply can be said to have a density-dependent population effect.

Density-independent population effects Changes in the size of a population due to factors that are independent of the population size. For example, certain climatic factors that are not affected by the size of a specific population can affect the entire population. A storm that knocks down all trees in a forest, no matter how many there are, is a density-independent population effect.

Desalination The removal of salts from seawater or brackish water so that the water can be used for purposes such as agriculture, industrial processes, or human consumption.

Desertification The process of creating a desert where there was not one before. Farming in marginal grasslands, which destroys the soil and prevents the future recovery of natural vegetation, is an example of desertification.

Dioxin An organic compound composed of oxygen, hydrogen, carbon, and chlorine. About 75 types are known. Dioxin is not normally manufactured intentionally but is a by-product resulting from chemical reactions in the production of other materials, such as herbicides. Known to be extremely toxic to mammals, its effects on the human body are being intensively studied and evaluated.

Direct costs See **Costs, direct**

Divergent evaluation Organisms with the same ancestral genetic heritage migrate to different habitats and evolve into species with different external forms and structures, but typically continue to use the same kind of habitats. The ostrich and the emu are believed to be examples of divergent evolution.

Divergent plate boundary Boundary between lithospheric plates characterized by the production of new lithosphere; found along oceanic ridges.

Diversity, genetic The total number of genetic characteristics, sometimes of a specific species, subspecies, or group of species.

Diversity, habitat The number of kinds of habitats in a given unit area.

Diversity, species Used loosely to mean the variety of species in an area or on the Earth. Technically, it is composed of three components: species richness—the total number of species; species evenness—the relative abundance of species; and species dominance—the most abundant species.

Dobson unit Commonly used to measure the concentration of ozone. One Dobson unit is equivalent to a concentration of 1 ppb ozone.

Dominant species Generally, the species that are most abundant in an area, ecological community, or ecosystem.

Dominants Species that are most abundant or otherwise most important within the community.

Dose dependency Dependence on the dose or concentration of a substance for its effects on a particular organism.

Dose response The principle that the effect of a certain chemical on an individual depends on the dose or concentration of that chemical.

Doubling time The time necessary for a quantity of whatever is being measured to double.

Drainage basin The area that contributes surface water to a particular stream network.

Drip irrigation Irrigation by the application of water to the soil from tubes that drip water slowly, greatly reducing the loss of water from direct evaporation and increasing yield.

Early successional species Species that occur only or primarily during early stages of succession. With vegetation, these are typically rapidly growing and short-lived with high reproductive rates.

Earth system science Science of studying the Earth as a system. Includes understanding of processes and linkages between the lithosphere, hydrosphere, biosphere, and atmosphere.

Ecological community This term has two meanings. (1) A conceptual or functional meaning: a set of interacting species that occur in the same place (sometimes extended to mean a set that interacts in a way to sustain life). (2) An operational meaning: a set of species found in an area, whether or not they are interacting.

Ecological gradient A change in the relative abundance of a species or group of species along a line or over an area.

Ecological island An area that is biologically isolated so that a species occurring within the area cannot mix (or only rarely mixes) with any other population of the same species.

Ecological justification for the conservation of nature An argument for the conservation of nature on the grounds that a species, an ecological community, an ecosystem, or the Earth's biosphere provides specific functions necessary to the persistence of our life or of benefit to life. The ability of trees in forests to remove carbon dioxide produced in burning fossil fuels is such a public benefit and an argument for maintaining large areas of forests.

Ecological niche The general concept is that the niche is a species' "profession"—what it does to make a living. The term is also used to refer to a set of environmental conditions within which a species is able to persist.

Ecological succession The process of the development of an ecological community or ecosystem, usually viewed as a series of stages—early, middle, late, mature (or climax), and sometimes postclimax. Primary succession is an original establishment; secondary succession is a reestablishment.

Ecology The science of the study of the relationships between living things and their environment.

Ecosystem An ecological community and its local, nonbiological community. An ecosystem is the minimum system that includes and sustains life. It must include at least an autotroph, a decomposer, a liquid medium, a source and sink of energy, and all the chemical elements required by the autotroph and the decomposer.

Ecosystem effect Effects that result from interactions among different species, effects of species on chemical elements in their environment, and conditions of the environment.

Ecosystem energy flow The flow of energy through an ecosystem—from the external environment through a series of organisms and back to the external environment.

Ecotourism Tourism based on an interest in observation of nature.

ED-50 The effective dose, or dose that causes an effect in 50% of the population on exposure to a particular toxicant. It is related to the onset of specific symptoms, such as loss of hearing, nausea, or slurred speech.

Edge effect An effect that occurs following the forming of a forest island; in the early phases the species diversity along the edge is greater than in the interior. Species escape from the cut area and seek refuge in the border of the forest, where some may last only a short time.

Efficiency The ratio of output to input. With machines, usually the ratio of work or power produced to the energy or power used to operate or fuel them. With living things, efficiency may be defined as either the useful work done or the energy stored in a useful form compared with the energy taken in.

Efficiency improvements With respect to energy, refers to designing equipment that will yield more energy output from a given amount of energy input.

Effluent Any material that flows outward from something. Examples include wastewater from hydroelectric plants and water discharged into streams from waste-disposal sites.

Effluent stream Type of stream where flow is maintained during the dry season by groundwater seepage into the channel.

El Niño Natural perturbation of the physical earth system that effects global climate. Characterized by development of warm oceanic waters in the eastern part of the tropical Pacific Ocean, a weakening or reversal of the trade winds, and a weakening or even reversal of the equatorial ocean currents. Reoccurs periodically and effects the atmosphere and global temperature by pumping heat into the atmosphere.

Electromagnetic fields (EMF) Magnetic and electrical fields produced naturally by our planet and also by appliances such as toasters, electric blankets, and computers. There currently is a great deal of controversy concerning potential adverse health effects related to exposure to EMF in the workplace and home from such artificial sources as power lines and appliances.

Electromagnetic spectrum All the possible wavelengths of electromagnetic energy, considered as a continuous range. The spectrum includes long wavelength (used in radio transmission), infrared, visible, ultraviolet, X rays, and gamma rays.

Endangered species A species that faces threats that might lead to its extinction in a short time.

Endemic Referring to all factors confined to a given region, such as an island or a country; the whooping crane is endemic to North America.

Endemic species A species that is native to a particular area.

Energy An abstract concept referring to the ability or capacity to do work.

Energy flow The movement of energy through an ecosystem from the external environment through a series of organisms and back to the external environment. It is one of the fundamental processes common to all ecosystems.

Entropy A measure in a system of the amount of energy that is unavailable for useful work. As the disorder of a system increases, the entropy in a system also increases.

Environment All factors (living and nonliving) that actually affect an individual organism or population at any point in the life cycle. Environment is also sometimes used to denote a certain set of circumstances surrounding a particular occurrence (environments of deposition, for example).

Environmental audit Process of determining the past history of a particular site, with special reference to the existence of toxic materials or waste.

Environmental ethics A school, or theory, in philosophy that deals with the ethical value of the environment, including especially the rights of nonhuman objects and systems in the environment, for example, trees and ecosystems.

Environmental geology The application of geologic information to environmental problems.

Environmental impact The effects of some action on the environment, particularly action by human beings.

Environmental impact report (EIR) Similar to the environmental impact statement (EIS), a report describing potential environmental impacts resulting from a particular project, often at the state level.

Environmental impact statement A written statement that assesses and explores possible impacts associated with a particular project that may affect the human environment. The statement is required in the United States by the National Environmental Policy Act of 1969.

Environmental justice The principle of dealing with environmental problems in such a way as to not discriminate against people based upon socioeconomic status, race, or ethnic group.

Environmental law A field of law concerning the conservation and use of natural resources and the control of pollution.

Environmental risk Used in discussions of endangered species to mean variation in the physical or biological environment, including variations in predator, prey, symbiotic, or competitor species that can threaten a species with extinction.

Environmental unity A principle of environmental sciences that states that everything affects everything else, meaning that a particular course of action leads to an entire potential string of events. Another way of stating this idea is that you can't only do one thing.

Environmentalism A social, political, and ethical movement concerned with protecting the environment and using its resources wisely.

Epidemic disease A disease that appears occasionally in the population, affects a large percentage of it, and declines or almost disappears for a while only to reappear later.

Equilibrium A point of rest. A system that does not tend to undergo any change of its own accord but remains in a single, fixed condition is said to be in equilibrium. Compare with **steady state.**

Eukaryote Organism whose cells have nuclei and certain other characteristics that separate it from the prokaryotes. The eukaryotes include flowering plants, animals, and many single-cell organisms.

Eutrophic Referring to bodies of water having an abundance of the chemical elements required for life.

Eutrophication Increase in the concentration of chemical elements required for living things (for example, phosphorus). Increased nutrient loading may lead to a population explosion of photosynthetic algae and blue-green bacteria that become so thick that light cannot penetrate the water. Bacteria deprived of light beneath the surface die; as they decompose, dissolved oxygen in the lake is lowered and eventually a fish kill may result. Eutrophication of lakes caused by human-induced processes, such as nutrient-rich sewage water entering a body of water, is called cultural eutrophication.

Even-aged stands Forest area where all live trees began growth from seeds and roots planted in about the same year.

Evolution biological The change in inherited characteristics of a population from generation to generation, sometimes resulting in a new species or populations.

Evolution, nonbiological Outside the realm of biology, the term *evolution* is used broadly to mean the history and development of something.

Exotic species Species introduced into a new area by human action.

Experimental errors There are two kinds of experimental errors, random and systematic. Random errors are those due to chance events, such as air currents pushing on a scale and altering a measurement of weight. In contrast, a miscalibration of an instrument would lead to a systematic error. Human errors can be either random or systematic.

Exponential growth Growth in which the rate of increase is a constant percentage of the current size; that is, the growth occurs at a constant rate per time period.

Externality In economics, an effect not normally accounted for in the cost-revenue analysis of producers.

Extinction Disappearance of a life-form from existence; usually applied to a species.

Facilitation During succession, one species prepares the way for the next (and may even be necessary for the occurrence of the next).

Fact Something that is known based on actual experience and observation.

Fall line The point on a river where there is an abrupt drop in elevation of the land and where numerous waterfalls occur. The line is typically located where streams pass from harder to softer rocks.

Fallow A field allowed to grow with a cover crop without harvesting for at least one season.

Fecal coliform bacteria Standard measure of microbial pollution and an indicator of disease potential for a water source.

Feedback A kind of system response that occurs when output of the system also serves as input leading to changes in the system.

First law of thermodynamics The principle that energy may not be created or destroyed but is always conserved.

First-law efficiency The ratio of the actual amount of energy delivered where it is needed to the amount of energy supplied in order to meet that need; expressed as a percentage.

Fission The splitting of an atom into smaller fragments with the release of energy.

Flooding The natural process whereby waters emerge from their stream channel to cover part of the floodplain. Natural flooding is not a problem until people choose to build homes and other structures on floodplains.

Floodplain Flat topography adjacent to a stream in a river valley that has been produced by the combination of overbank flow and lateral migration of meander bends.

Fluidized-bed combustion A process used during the combustion of coal to eliminate sulfur oxides. Involves mixing finely ground limestone with coal and burning it in suspension.

Food chain The linkage of who feeds on whom.

Food-chain concentration See **Biomagnification.**

Food web A network of who feeds on whom or a diagram showing who feeds on whom.

Force A push or pull that affects motion.

Fortress city Earliest planned cities in Europe designed for defense; influenced the development of cities in North America.

Fossil fuels forms of stored solar energy created from incomplete biological decomposition of dead organic matter. Include coal, crude oil, and natural gas.

Fuel cell A device that produces electricity directly from a chemical reaction in a specially designed cell. In the simplest case the cell uses hydrogen as a fuel, to which an oxidant is supplied. The hydrogen is combined with oxygen as if the hydrogen were burned, but the reactants are separated by an electrolyte solution that facilitates the migration of ions and the release of electrons (which may be tapped as an energy source).

Fugitive sources Type of stationary air pollution sources that generate pollutants from open areas exposed to wind processes.

Fusion, nuclear Combining of light elements to form heavier elements with the release of energy.

Gaia The Gaia hypothesis states that the surface environment of the Earth, with respect to such factors as the atmospheric composition of reactive gases (for example, oxygen, carbon dioxide, and methane), the acidity-alkalinity of waters, and the surface temperature are actively regulated by the sensing, growth, metabolism and other activities of the biota. Interaction between the physical and biological system on the Earth's surface has led to a planetwide physiology which began more than 3 billion years ago and the evolution of which can be detected in the fossil record.

Game ranching Practice of maintaining wild herbivores in their native habitat to be harvested for meat, leather, and other products.

Gamma rays One of the three major kinds of nuclear radiation. A type of electromagnetic radiation emitted from the isotope similar to X rays but more energetic and penetrating.

Garden city Term for planning a city and countryside together, coined by Ebenezer Howard (1902).

Genetic drift Changes in the frequency of a gene in a population as a result of chance rather than of mutation, selection, or migration.

Genetic risk Used in discussions of endangered species to mean detrimental change in genetic characteristics not caused by external environmental changes. Genetic changes can occur in small populations from such causes as reduced genetic variation, genetic drift, and mutation.

Geochemical cycles The pathways of chemical elements in geologic processes, including the chemistry of the lithosphere, atmosphere, and hydrosphere.

Geographic Information System (GIS) Technology capable of storing, retrieving, transferring, and displaying environmental data.

Geologic cycle The formation and destruction of earth materials and the processes responsible for these events. The geologic cycle includes the following subcycles: hydrologic, tectonic, rock, and geochemical.

Geometric growth See **Exponential growth.**

Geopressurized systems Geothermal system that exists when the normal heat flow from the Earth is trapped by impermeable clay layers that act as an effective insulator.

Geothermal energy The useful conversion of natural heat from the interior of the Earth.

Global circulation models (GCM) A type of mathematical model used to evaluate global change, particularly related to climatic change. GCMs are very complex and require supercomputers for their operation.

Global extinction Disappearance or extinction of a species everywhere.

Global forecasting Process of predicting or forecasting future change in environmental areas such as world population, natural resource utilization, and environmental degradation.

Global warming Natural or human-induced increase in the average global temperature of the atmosphere near the Earth's surface.

Gravel Unconsolidated, generally rounded fragment of rocks and minerals greater than 2 mm in diameter.

Green plans Long-term strategies for identifying and solving global and regional environmental problems. The philosophical heart of green plants is sustainability.

Green revolution Name attached to post–World War II agricultural programs that have led to the development of new strains of crops with higher yield, better resistance to disease, or better ability to grow under poor conditions.

Greenbelt The idea of locating garden cities in a set connected by undeveloped areas, forming a system of countryside and urban landscapes.

Greenhouse effect Process of trapping heat in the atmosphere. Water vapor and several other gases warm the Earth's atmosphere because they absorb and remit radiation, that is, they trap some of the heat radiating from the Earth's atmospheric system.

Gross production (biology) Production before respiration losses are subtracted.

Groundwater Water found beneath the Earth's surface within the zone of saturation.

Growth efficiency Gross production efficiency (P/C), or ratio of the material produced (P = net production) by an organism or population to the material ingested or consumed (C).

Growth rate The net increase in some factor per unit time. In ecology, the growth rate of a population is sometimes measured as the increase in numbers of individuals or biomass per unit time and sometimes as a percentage increase in numbers or biomass per unit time.

Habitat Where an individual, population, or species exists or can exist. For example, the habitat of the Joshua tree is the Mojave Desert of North America.

Half-life The time required for half a substance to disappear; the average time required for one-half of a radioisotope to be transformed to some other isotope; the time required for one-half of a toxic chemical to be converted to some other form.

Hard path Energy policy based on the emphasis of energy quantity generally produced from large, centralized power plants.

Hazardous waste Waste that is classified as definitely or potentially hazardous to the health of people. Examples include toxic or flammable liquids and a variety of heavy metals, pesticides, and solvents.

Heat energy Energy of the random motion of atoms and molecules.

Heat island Warmer air of a city than surrounding areas as a result of increased heat production and decreased rate of heat loss caused by the abundance of building and paving materials, which act as solar collectors.

Heat island effect Urban areas are several degrees warmer than their surrounding areas. During relatively calm periods there is an upward flow of air over heavily developed areas accompanied by a downward flow over nearby greenbelts. This produces an air-temperature profile that delineates the heat island.

Heat pumps Devices that transfer heat from one material to another, such as from groundwater to the air in a building.

Heavy metal Refers to a number of metals, including lead, mercury, arsenic, and silver (among others) that have a relatively high atomic number (the number of protons in the nucleus of an atom). They are often toxic at relatively low concentrations, causing a variety of environmental problems.

Herbivore An organism that feeds on an autotroph.

Heterotroph An organism that feeds on other organisms.

Heterotrophs Organisms that cannot make their own food from inorganic chemicals and a source of energy and therefore live by feeding on other organisms.

High-level nuclear waste Extremely toxic nuclear waste, such as spent fuel elements from commercial reactors. A sense of urgency surrounds determining how we may eventually dispose of this waste material.

Homeostasis The ability of a cell or organism to maintain a constant environment. Results from negative feedback, resulting in a state of dynamic equilibrium.

Hot igneous systems A geothermal system that involves hot, dry rocks with or without the presence of near-surface molten rock.

Human demography The study of human population characteristics, such as age structure, demographic transition, total fertility, human population and environment relationships, death-rate factors, and standard of living.

Hutchinsonian niche The idea of a measured niche, a set of environmental conditions within which a species is able to persist.

Hydrocarbons Compounds containing only carbon and hydrogen are a large group of organic compounds, including petroleum products, such as crude oil and natural gas.

Hydrologic cycle Circulation of water from the oceans to the atmosphere and back to the oceans by way of evaporation, runoff from streams and rivers, and groundwater flow.

Hydrology The study of surface and subsurface water.

Hydroponics The practice of growing plants in a fertilized water solution on a completely artificial substrate in an artificial environment such as a greenhouse.

Hydrothermal convection systems A type of geothermal energy characterized by circulation of steam and/or hot water that transfers to the surface.

Hypothesis In science, an explanation set forth in a manner that can be tested and is capable of being disproved. A tested hypothesis is accepted until and unless it has been disproved.

Igneous rocks Rocks formed from the solidification of magma. They are extrusive if they crystallize on the surface of the Earth and intrusive if they crystallize beneath the surface.

Incineration Combustion of waste at high temperature, consuming materials and leaving only ash and noncombustibles to dispose of in a landfill.

Independent variable See **Variable, independent**

Indirect costs See **Costs, indirect**

Indirect deforestation Death of trees from pollution whose source is outside the forest.

Inductive reasoning Drawing a general conclusion from a limited set of specific observations.

Industrial ecology Process of designing industrial systems to behave more like ecosystems where waste from one part of the system is a resource for another part.

Industrial melanism A form of natural selection where the color of a living thing helps it to blend in with an urban, industrial environment.

Inference (1) A conclusion derived by logical reasoning from premises and/or evidence (observations or facts), or (2) a conclusion, based on evidence, arrived at by insight or analogy, rather than derived solely by logical processes.

Influent stream Type of stream that is everywhere above the groundwater table and flows in direct response to precipitation. Water from the channel moves down to the water table, forming a recharge mound.

In-stream use A type of water use that includes navigation, generation of hydroelectric power, fish and wildlife habitat, and recreation.

Integrated energy management Use of a range of energy options that vary from region to region, including a mix of technology and sources of energy.

Integrated pest management A term applied to a variety of practices whose overall goal is to contain biological pests through minimum use of artificial chemicals and minimum disruption of natural ecological processes. Biological control typically augments or replaces artificial pesticides.

Integrated waste management (IWM) Set of management alternatives including reuse, source reduction, recycling, composting, landfill, and incineration.

Interference During succession, one species prevents the entrance of later successional species into an ecosystem. For example, some grasses produce such dense and thick mats that seeds of trees cannot reach the soil to germinate. As long as these grasses persist, the trees that characterize later stages of succession cannot enter the ecosystem.

Island arc A curved group of volcanic islands associated with a deep-oceanic trench and subduction zone (convergent plate boundary).

Isotope Atoms of an element that have the same atomic number (the number of protons in the nucleus of the atom) but vary in atomic mass number (the number of protons plus neutrons in the nucleus of an atom).

Keystone species Loosely speaking, a species, such as the sea otter, that has a large effect on its community or ecosystem so that its removal or addition to the community leads to major changes in the abundances of many or all other species.

Kinetic energy The energy of motion. For example, the energy in a moving car that results from the mass of the car traveling at a particular velocity.

Kwashiorkor Lack of sufficient protein in the diet, which leads to a failure of neural development in infants and therefore to learning disabilities.

Land application Method of disposal of hazardous waste that involves intentional application of waste material to surface soil. Useful for certain biodegradable industrial waste, such as oil and petroleum waste, and some organic chemical waste.

Land ethic A set of ethical principles that affirm the right of all resources, including plants, animals, and earth materials, to continued existence and, at least in some locations, to continued existence in a natural state.

Land-use planning Complex process involving development of a land-use plan to include a statement of land-use issues, goals, and objectives; a summary of data collection and analysis; a land-classification map; and a report that describes and indicates appropriate development in areas of special environmental concern.

Late successional species Species that occur only or primarily in, or are dominant in late stages in succession. With plants, these are typically slower growing and longer-lived species.

Law of the minimum (Liebig's law of the minimum) The concept that the growth or survival of a population is directly related to the life requirement that is in least supply and not to a combination of factors.

LD-50 A crude approximation of a chemical toxicity defined as the dose at which 50% of the population dies on exposure.

Leachate Noxious, mineralized liquid capable of transporting bacterial pollutants. Produced when water infiltrates through waste material and becomes contaminated and polluted.

Leaching Water infiltration from the surface, dissolving soil materials as part of chemical weathering processes and transporting the dissolved materials laterally or downward.

Lead A heavy metal that is an important constituent of automobile batteries and other industrial products. A toxic metal capable of causing environmental disruption and producing a health problem to people and other living organisms.

Liebig's law of the minimum See **Law of the minimum.**

Life expectancy The estimated average number of years (or other time period used as a measure) that an individual of a specific age can expect to live.

Limiting factor The single requirement for growth available in the least supply in comparison to the need of an organism. Originally applied to crops but now often applied to any species.

Lithosphere Outer layer of Earth, approximately 100 km thick, of which the plates that contain the ocean basins and the continents are composed.

Littoral drift Movement caused by wave motion in nearshore and beach environment.

Local extinction The disappearance of a species from part of its range, but continued persistence elsewhere.

Logistic carrying capacity In terms of the logistic curve, the population size at which births equals deaths and there is no net change in the population.

Logistic equation The equation that results in a logistic growth curve, that is, the growth rate $dN/dt = rN[(K-N)/N]$, where r is the intrinsic rate of increase, K is the carrying capacity, and N is the population size.

Logistic growth curve The S-shaped growth curve that is generated by the logistic growth equation. In the logistic, a small population grows rapidly, but the growth rate slows down, and the population eventually reaches a constant size.

Low-level nuclear waste Waste materials that contain sufficiently low concentrations or quantities of radioactivity so as not to present a significant environment hazard if properly handled.

Luz solar electric generating system Solar energy farms comprising a power plant surrounded by hundreds of solar collectors (curved mirrors) that heat a synthetic oil, which flows through heat exchangers to drive steam turbine generators.

Macronutrients Elements required in large amounts by living things. These include the big six—carbon, hydrogen, oxygen, nitrogen, phosphorus, and sulfur.

Magma A naturally occurring silica melt, a good deal of which is a liquid state.

Malnourishment The lack of specific components of food, such as proteins, vitamins, or essential chemical elements.

Manipulated variable See **Variable, independent**

Marasmus Progressive emaciation caused by a lack of protein and calories.

Marginal cost In environmental economics, the cost to reduce one additional unit of a type of degradation, for example, pollution.

Marginal land An area of the Earth with minimal rainfall or otherwise limited severely by some necessary factor, so that it is a poor place for agriculture and easily degraded by agriculture. Typically, these lands are easily converted to deserts when used for light grazing and crop production.

Mariculture Production of food from marine habitats.

Marine evaporites With respect to mineral resources, refers to materials such as potassium and sodium salts resulting from the evaporation of marine waters.

Matter Anything that occupies space and has mass. It is the substance of which physical objects are composed.

Maximum lifetime Genetically determined maximum possible age to which an individual of a species can live.

Maximum sustainable yield (MSY) The maximum usable production of a biological resource that can be obtained in a specified time period. The MSY level is the population size that results on maximum sustainable yield.

Mediation Negotiation process between adversaries, guided by neutral facilitator.

Methane (CH₄) Molecule of carbon and hydrogen, which is a naturally occurring gas in the atmosphere. One of the so-called greenhouse gases.

Microclimate The climate of a very small local area. For example, the climate under a tree, near the ground within a forest, or near the surface of streets in a city.

Micronutrients Chemical elements required in very small amounts by at least some forms of life. Boron, copper, and molybdenum are examples of micronutrients.

Migration The movement of an individual, population, or species from one habitat to another or more simply from one geographic area to another.

Migration corridor Designated passageways among parks or preserves for allowing in-migration among several of these areas.

Mineral Naturally occurring inorganic material with a definite internal structure and physical and chemical properties that vary within prescribed limits.

Mineral resources Elements, chemical compounds, minerals, or rocks concentrated in a form that can be extracted to obtain a usable commodity.

Minimum viable population The minimum number of individuals that have a reasonable chance of persisting for a specified time period.

Mitigated negative declaration Special type of negative declaration that suggests that the adverse environmental aspects of a particular action may be mitigated through modification of the project in such a way as to reduce the impacts to near insignificance.

Mitigation Process that identifies actions to avoid, lessen, or compensate for anticipated adverse environmental impacts.

Mobile sources Sources of air pollutants that move from place to place, for example, automobiles, trucks, buses, and trains.

Monitoring Process of collecting data on a regular basis at specific sites to provide a database from which to evaluate change. For example, collection of water samples from beneath a landfill to provide early warning should a pollution problem arise.

Monoculture The planting of large areas with a single species or even a single strain or subspecies in farming.

Moral justification for the conservation of nature An argument for the conservation of nature on the grounds that aspects of the environment have a right to exist, independent of human desires, and that it is our moral obligation to allow them to continue or to help them persist.

Multiple use Literally, using the land for more than one purpose at the same time. For example, forestland can be used to produce commercial timber but at the same time serve as wildlife habitat and land for recreation. Usually multiple use requires compromises and trade-offs, such as striking a balance between cutting timber

for the most efficient production of trees at a level that facilitates other uses.

Mutation Stated most simply, a chemical change in a DNA molecule. It means that the DNA carries a different message than it did before, and this change can affect the expressed characteristics when cells or individual organisms reproduce.

Mutualism See **Symbiosis.**

Natural catastrophe Sudden change in the environment, not the result of human actions.

Natural gas Naturally occurring gaseous hydrocarbon (predominantly methane) generally produced in association with crude oil or from gas wells; an important efficient and clean-burning fuel commonly used in homes and industry.

Natural selection A process by which organisms whose biological characteristics better fit them to the environment are better represented by descendants in future generations than those whose characteristics are less fit for the environment.

Nature preserve An area set aside with the primary purpose of conserving some biological resource.

Negative declaration Document that may be filed if an agency has determined that a particular project will not have significant adverse effect on the environment.

Negative feedback A type of feedback that occurs when the system's response is in the opposite direction of the output. Thus negative feedback is self-regulating.

Net growth efficiency Net production efficiency (*P/A*), or ratio of the material produced (*P*) to the material assimilated (*A*) by an organism. The material assimilated is less than the material consumed, because some food taken is egested as waste (discharged) and never used by an organism.

Net production (biology) The production that remains after utilization. In a population, net production is sometimes measured as the net change in the numbers of individuals. It is also measured as the net change in biomass or in stored energy. In terms of energy, it is equal to the gross production minus the energy used in respiration.

New forestry The name for a new variety of timber harvesting practices to increase the likelihood of sustainability, including recognition of the dynamic characteristics of forests and of the need for management within an ecosystem context.

Niche (1) The "profession," or role, of an organism or species; or (2) all the environmental conditions under which the individual or species can persist. The *fundamental niche* is all the conditions under which a species can persist in the absence of competition; the *realized niche* is the set of conditions as they occur in the real world with competitors.

Nitrogen cycle A complex biogeochemical cycle responsible for moving important nitrogen components through the biosphere and other Earth systems. This is an extremely important cycle because nitrogen is required by all living things.

Nitrogen fixation The process by which atmospheric nitrogen is converted to ammonia, nitrate ion, or amino acids. Microorganisms perform most of the conversion but a small amount is also converted by lightning.

Nitrogen oxides Occur in several forms (NO, NO_2, and NO_3). Most important as an air pollutant is nitrogen dioxide, which is a visible yellow brown to reddish brown gas. It is a precursor of acid rain and produced through the burning of fossil fuels.

Noise pollution A type of pollution characterized by unwanted or potentially damaging sound.

Nonmarine evaporites with respect to mineral resources, refers to useful deposits of materials such as sodium and calcium bicarbonate, sulfate, borate, or nitrate produced by evaporation or surficial waters on the land, as differentiated from marine waters in the oceans.

Nonpoint sources Sources of pollutants that are diffused and intermittent and are influenced by factors such as land use, climate, hydrology, topography, native vegetation, and geology.

Nonrenewable energy Alternative energy sources, including nuclear and geothermal, that are dependent on fuels or a resource that may be used up much faster than it is replenished by natural processes.

Nonrenewable resource A resource that is cycled so slowly by natural Earth processes that once used, it is essentially not going to be made available within any useful time framework.

No-till agriculture Combination of farming practices that include *not* plowing the land, using herbicides to keep down the weeds, and allowing some weeds to grow.

Nuclear cycle The series of processes that begins with the mining of uranium to be processed and used in nuclear reactors and ends with the disposal of radioactive waste.

Nuclear energy The energy of the atomic nucleus that, when released, may be used to do work. Controlled nuclear fission reactions take place within commercial nuclear reactors to produce energy.

Nutrients Chemicals such as phosphorus and nitrogen that, when released into water sources, may cause pollution events such as eutrophication.

Observations Information obtained through one or more of the five senses or through instruments that extend the senses. For example, some remote sensing instruments measure infrared intensity, which we do not see, and convert the measurement into colors, which we do see.

Ocean thermal conversion Direct utilization of solar energy using part of a natural oceanic environment as a gigantic solar collector.

Off-site effect An environmental effect occurring away from the location of the causal factors.

Off-stream use Type of water use where water is removed from its source for a particular use.

Oil shale A fine-grained sedimentary rock containing organic material known as kerogen. On distillation yields significant amounts of hydrocarbons including oil.

Old growth A nontechnical term often used to mean a virgin forest (one never cut) but also used to mean a forest that has been undisturbed for a long, but usually unspecified, time.

Oligotrophic Referring to bodies of waters having a low concentration of the chemical elements required for life.

Omnivores Organisms that eat both plants and animals.

On-site effect An environmental effect occurring at the location of the causal factors.

Open dumps Area where solid waste is disposed of by simply dumping it. Often causes severe environmental problems, such as water pollution, and creates a health hazard. Illegal in the United States and in many other countries around the world.

Open system A type of system in which exchanges of mass or energy occur with other systems.

Open woodlands Areas in which trees are a dominant vegetation form but the leaves of adjacent trees generally do not touch or overlap, so that there are gaps in the canopy. Typically, grasses or shrubs grow in these gaps among the trees.

Operational definitions Definitions that tell you what you need to look for or do in order to carry out an operation, such as measuring, constructing, or manipulating.

Optimal carrying capacity This term has several meanings, but the major idea is the maximum abundance of a population or species that can persist in an ecosystem without degrading the ability of the ecosystem to maintain: (1) that population or species; (2) all necessary ecosystem processes; and (3) the other species found in that ecosystem.

Optimum sustainable population (OSP) The population level that results in an optimum sustainable yield; or the population level that is in some way best for the population, its ecological community, its ecosystem, or the biosphere.

Optimum sustainable yield (OSY) The largest yield of a renewable resource achievable over a long time period without decreasing the ability of the population or its environment to support the continuation of this level of yield.

Ore deposits Earth materials in which metals are concentrated in high concentrations, sufficient to be mined.

Organic compound A compound of carbon; originally used to refer to the compounds found in and formed by living things.

Outbreak The occurrence of a disease that may quickly and surprisingly occur.

Overdraft Groundwater withdrawal when the amount pumped from wells exceeds the natural rate of replenishment.

Overgrazing When the carrying capacity of land for an herbivore, such as cattle or deer, is exceeded.

Ozone (O_3) Form of oxygen in which three atoms of oxygen occur together. Is chemically active and has a short average lifetime in the atmosphere. Forms a natural layer high in the atmosphere (stratosphere) that protects us from harmful ultraviolet radiation from the sun. Is an air pollutant when present in the lower atmosphere above the National Air Quality Standards.

Ozone shield Stratospheric ozone layer that absorbs ultraviolet radiation.

Park city Earliest planned cities in Europe designed for beauty and recreation; influenced the development of cities in North America.

Particulate matter Small particles of solid or liquid substances that are released into the atmosphere by many activities, including farming, volcanic eruption, and burning fossil fuels. Particulates affect human health, ecosystems, and the biosphere.

Passive solar energy system Direct use of solar energy through architectural design to enhance or take advantage of natural changes in solar energy that occur throughout the year without requiring mechanical power.

Pasture Land plowed and planted to provide forage for domestic herbivorous animals.

Pebble A rock fragment between 4 and 64 mm in diameter.

Pedology The study of soils.

Pelagic whaling Practice of whalers taking to the open seas and searching for whales from ships that remained at sea for long periods.

Per-capita availability The amount of a resource available per person.

Per-capita demand The economic demand per person.

Per-capita food production The amount of food produced per person.

Permafrost Permanently frozen ground.

Pesticides, broad spectrum Pesticides that kill a wide variety of organisms. Arsenic, one of the first elements used as a pesticide, is toxic to many life-forms, including people.

Phosphorus cycle Major biogeochemical cycle involving the movement of phosphorus throughout the biosphere and lithosphere. This cycle is important because phosphorus is an essential element for life and often is a limiting nutrient for plant growth.

Photochemical oxidants Result from atmospheric interactions of nitrogen dioxide and sunlight. Most common is ozone (O_3).

Photochemical smog Sometimes called L.A.-type smog or brown air. Directly related to automobile use and solar radiation. Reactions that occur in the development of the smog are complex and involve both nitrogen oxides and hydrocarbons in the presence of sunlight.

Photosynthesis Synthesis of sugars from carbon dioxide and water by living organisms using light as energy. Oxygen is given off as a by-product.

Photovoltaics Technology that converts sunlight directly into electricity using a solid semiconductor material.

Physiographic province A region characterized by a particular assemblage of landforms, climate, and geomorphic history.

Pioneer species Species found in early stages of succession.

Placer deposit A type of ore deposit found in material transported and deposited by agents such as running water, ice, or wind; for example, gold and diamonds found in stream deposits.

Plate tectonics A model of global tectonics that suggests that the outer layer of Earth, known as the *lithosphere,* is composed of several large plates that move relative to one another. Continents and oceans basins are passive riders on these plates.

Point sources Sources of pollution such as smokestacks, pipes, or accidental spills that are readily identified and stationary. They are often thought to be easier to recognize and control than are area sources. This is true only in a general sense, as some very large point sources emit tremendous amounts of pollutants into the environment.

Polar stratospheric clouds Clouds that form in the stratosphere during the polar winter.

Polar vortex Arctic air masses that in the winter become isolated from the rest of the atmosphere and circulate about the pole. The vortex rotates counterclockwise because of the rotation of the Earth in the Southern Hemisphere.

Policy instruments The means to implement a society's policies. Include moral suasion (jawboning—persuading people by talk, publicity, and social pressure); direct controls, including regulations; and market processes affecting the price of goods and processes, such as subsidies, licenses, and deposits.

Pollutant In general terms, any factor that has a harmful effect on living things or their environment.

Pollution The process by which something becomes impure, defiled, dirty, or otherwise unclean.

Pool (geology) Common bed form produced by scour in meandering and straight channels.

Population A group of individuals of the same species living in the same area or interbreeding and sharing genetic information.

Population age structure The number of individuals or proportion of the population in each age class.

Population dynamics The study of changes in population sizes and the causes of these changes.

Population momentum or lag effect The continued growth of a population that occurs after achievement of replacement level fertility is reached.

Population regulation See **Density-dependent population effects** and **Density-independent population effects.**

Population risk Used in discussions of endangered species to mean random variation in population rates—birthrates and death rates—possibly causing species in low abundance to become extinct.

Positive feedback A type of feedback that occurs when an increase in output leads to a further increase in output. This is sometimes known as a vicious cycle, since the more you have the more you get.

Potential energy Energy that is stored. Examples include the gravitational energy of water behind a dam; chemical energy in coal, fuel oil, and gasoline; and nuclear energy (in the forces that hold atoms together).

Power The time rate of doing work.

Predation-parasitism Interaction between individuals of two species in which the outcome benefits one and is detrimental to the other.

Predator An organism that feeds on other, live organisms, usually of other species. The term is usually applied to animals that feed on other animals, but sometimes it is used to mean herbivore.

Premises In science, initial definitions and assumptions.

Primary pollutants Air pollutants emitted directly into the atmosphere. Included are particulates, sulfur oxides, carbon monoxide, nitrogen oxides, and hydrocarbons.

Primary production See **Production, primary**

Primary succession The initial establishment and development of an ecosystem.

Primary treatment (of wastewater) Removal of large particles and organic materials from wastewater through screening.

Probability The relative probability that an event will occur.

Production, ecological The amount of increase in organic matter, usually measured per unit area of land surface or unit volume of water, as in grams per square meter (g/m^2). Production is divided into *primary* (that of autotrophs) and *secondary* (that of heterotrophs). It is also divided into *net* (that which remains stored after use) and *gross* (that added before any use).

Production, primary The production by autotrophs.

Production, secondary The production by heterotrophs.

Productivity, ecological The *rate* of production; that is, the amount of increase in organic matter per unit time (for example, grams per meter squared *per year*).

Prokaryote A kind of organism that lacks a true cell nucleus and has other cellular characteristics that distinguish it from the *eukaryotes*. Bacteria and blue-green algae are prokaryotes.

Protocooperation A symbiotic relationship that is beneficial, but not obligatory, to both species.

Pseudoscientific Ideas which are claimed to have scientific validity but are inherently untestable and/or lack empirical support and/or were arrived at through faulty reasoning or poor scientific methodology.

Public trust Grants and limits the authority of government over certain natural areas of special character.

Qualitative data Data that are distinguished by qualities or attributes that cannot or are not expressed as quantities. For example, blue and red are qualitative data about the electromagnetic spectrum.

Quantitative data Data that are expressed as numbers or numerical measurements. For example, the wavelengths of specific colors of blue and red light (460 and 650 nanometers, respectively) are quantitative data about the electromagnetic spectrum.

Radiation absorbed dose Energy retained by living tissue that has been exposed to radiation.

Radioactive decay A process of decay of radioisotopes that change from one isotope to another and emit one or more forms of radiation.

Radioactive waste Type of waste produced in the nuclear fuel cycle; generally classified as high level or low level.

Radioisotope A form of a chemical element that spontaneously undergoes radioactive decay.

Radon Naturally occurring radioactive gas. Radon is colorless, odorless, and tasteless and must be identified through proper testing.

Range Land used for grazing.

Rare species Species with a small total population size, or restricted to a small area, but not necessarily declining or in danger of extinction.

Record of decision Concise statement prepared by the agency planning a proposed project that outlines the alternatives considered and discusses which alternatives are environmentally preferable.

Recycle Integral part of waste management that attempts to identify resources in the waste stream that may be collected and reused.

Reduce, reuse, and recycle The three Rs of integrated waste management.

Renewable energy Alternative energy sources, such as solar, water, wind, and biomass, that are more or less continuously made available in a time framework useful to people.

Renewable resource A resource, such as timber, water, or air, that is naturally recycled or recycled by artificial processes within a time framwork useful for people.

Replacement level fertility Fertility rate required for the population to remain a constant size.

Representative natural areas Parks or preserves set aside to represent presettlement conditions of a specific ecosystem type.

Reserves Known and identified deposits of earth materials from which useful materials can be extracted profitably with existing technology and under present economic and legal conditions.

Resource-based agriculture Agriculture with production limited by the availability of resources.

Resources Reserves plus other deposits of useful earth materials that may eventually become available.

Respiration The complex series of chemical reactions in organisms that make energy available for use. Water, carbon dioxide, and energy are the products of respiration.

Responding variable See **Variable, dependent**

Reuse With respect to waste mangement, refers to finding ways to reuse production and materials so they need not be disposed of.

Riffle A section of stream channel characterized at low flow by fast, shallow flow. Generally contains relatively coarse bed-load particles.

Risk assessment The process of determining potential adverse environmental health effects to people following exposure to pollutants and other toxic materials. Generally includes the four steps of identification of the hazard, dose-response assessment, exposure assessment, and risk characterization.

Risk–benefit analysis In environmental economics, the riskiness of the future that influences the value we place on things in the present.

Rock (engineering) Any earth material that has to be blasted in order to be removed.

Rock (geologic) An aggregate of a mineral or minerals.

Rock cycle A group of processes that produce igneous, metamorphic, and sedimentary rocks.

Rotation time Time between cuts of a stand or area of forest.

Rule of climatic similarity Similar environments lead to the evolution of organisms similar in form and function (but not necessarily in genetic heritage or internal makeup) and to similar ecosystems.

Ruminants Animals having a four-chambered stomach within which bacteria convert the woody tissue of plants to proteins and fats that, in turn, are digested by the animal. Cows, camels, and giraffes are ruminants; horses, pigs, and elephants are not.

Sand Grains of sediment between 1/16 and 2 mm in diameter; often sediment composed of quartz particles of this size.

Sand dune A ridge or hill of sand formed by wind action.

Sanitary landfill A method of disposal of solid waste without creating a nuisance or hazard to public health or safety. Sanitary landfills are highly engineered structures with multiple barriers and collection systems to minimize environmental problems.

Savanna An area with trees scattered widely among dense grasses.

Scientific method The systematic methods by which scientists investigate natural phenomena, including gathering data, formulating and testing hypotheses, and developing scientific theories and laws.

Scoping The process of early identification of important environmental issues that require detailed evaluation.

Scrubbing A process of removing sulfur from gases emitted from power plants burning coal. The gases are treated with a slurry of lime and limestone, and the sulfur oxides react with the calcium to form insoluble calcium sulfides and sulfates that are collected and disposed of.

Second growth Forest that has been clear-cut and regrown.

Second law of thermodynamics A fundamental principle of energy that states that energy always tends to go from a more usable (higher quality) form to a less usable (lower quality) form. When we say that energy is converted to a less useful form we mean that entropy (a measure of the energy unavailable to do useful work) of the system has increased.

Secondary enrichment A weathering process of sulfide ore deposits that may concentrate the desired minerals.

Secondary pollutants Air pollutants produced through reactions between primary pollutants and normal atmospheric compounds. An example is ozone that forms over urban areas through reactions of primary pollutants, sunlight, and natural atmospheric gases.

Secondary production See **Production, secondary**

Secondary succession The reestablishment of an ecosystem where there are remnants of a previous biological community.

Secondary treatment (of wastewater) Use of biological processes to degrade wastewater in a treatment facility.

Second-law efficiency The ratio of the minimum available work needed to perform a particular task to the actual work used to perform that task. Reported as a percentage.

Secure landfill A type of landfill designed specifically for hazardous waste. Similar to a modern sanitary landfill in that it includes multiple barriers and collection systems to ensure that leachate does not contaminate soil and other resources.

Sediment pollution By volume and mass, sediment is our greatest water pollutant. It may choke streams, fill reservoirs, bury vegetation, and generally create a nuisance that is difficult to remove.

Seed tree cutting A logging method in which mature trees with good genetic characteristics and high seed production are preserved to promote regeneration of the forest. It is an alternative to clear-cutting.

Seismic Referring to vibrations in the Earth produced by earthquakes.

Selective cutting In timber harvesting, the practice of cutting some, but not all, trees, leaving some on the site. There are many kinds of selective cutting. Sometimes the biggest trees with the largest market value are cut, and smaller trees are left to be cut later. Sometimes the best trees are left to provide seed for future generations. Sometimes trees are left for wildlife habitat and recreation.

Shelterwood cutting A logging method in which dead and less desirable trees are cut first; mature trees are cut later. This ensures that young, vigorous trees will always be left in the forest. It is an alternative to clear-cutting.

Sick building syndrome Condition associated with a particular indoor environment that appears to be unhealthy to the human occupants.

Silicate minerals The most important group of rock-forming minerals.

Silt Sediment between 1/16 and 1/256 mm in diameter.

Silviculture The practice of growing trees and managing forests, traditionally with an emphasis on the production of timber for commercial sale.

Sinkhole A surface depression formed by the solution of limestone or the collapse over a subterranean void such as a cave.

Site A factor considering the summation of all environmental features of a location that influences the placement of a city. For example, New Orleans is built on low-lying muds, which form a poor site, while New York City's Manhattan is built on an island of strong bedrock, an excellent site.

Site quality Used by foresters to mean an estimator of the maximum timber crop the land can produce in a given time.

Situation The relative geographic location of a site that makes it a good location for a city. For example, New Orleans has a good situation because it is located at the mouth of the Mississippi River and is therefore a natural transportation junction.

Soft path Energy policy that relies heavily on renewable energy resources as well as other sources that are diverse, flexible, and matched to the end-use needs.

Soil The top layer of a land surface where the rocks have been weathered to small particles. Soils are made up of inorganic particles of many sizes, from small clay particles to large sand grains. Many soils also include dead organic material.

Soil (in engineering) Earth material that can be removed without blasting.

Soil (in soil science) Earth material modified by biological, chemical, and physical processes such that the material will support rooted plants.

Soil fertility The capacity of a soil to supply the nutrients and physical properties necessary for plant growth.

Soil horizon A layer in soil (A,B,C) that differs from another layer in chemical, physical, and biological properties.

Solar cell (photovoltaic) Device that directly converts light into electricity.

Solar collector Device for collecting and storing solar energy. For example, home water heating is done by flat panels consisting of a glass cover plate over a black background on which water is circulated through tubes. Short-wave solar radiation enters the glass and is absorbed by the black background. As long-wave radiation is emitted from the black material, it cannot escape through the glass, so the water in the circulating tubes is heated, typically to temperatures of 38° to 93°C.

Solar energy Collecting and using energy from the sun directly.

Solar pond Shallow pond filled with water and used to generate relatively low-temperature water.

Solar power tower A system of collecting solar energy that delivers the energy to a central location where the energy is used to produce electric power.

Source reduction Process of waste management, the object of which is to reduce the amounts of materials that must be handled in the waste stream.

Species A group of individuals capable of interbreeding.

Stable equilibrium A condition in which a system will remain if undisturbed and to which it will return when displaced.

Stationary sources Air pollution sources that have a relatively fixed location, including point sources, fugitive sources, and area sources.

Steady state When input equals output in a system, there is no net change and the system is said to be in a steady state. A bathtub with water flowing in and out at the same rate maintains the same water level and is in a steady state. Compare with *equilibrium*.

Stress Force per unit area. May be compression, tension, or shear.

Strip cutting In timber harvesting, the practice of cutting narrow rows of forest, leaving wooded corridors.

Strip mining Surface mining in which the overlying layer of rock and soil is stripped off to reach the resource. Large strip mines are some of the largest excavations caused by people in the world.

Subduction A process in which one lithospheric plate descends beneath another.

Subsidence A thinking, settling, or otherwise lowering of parts of the crust of the Earth.

Subsistence crops Crops used directly for food by a farmer or sold locally where the food is used directly.

Sulfur dioxide (SD$_2$) Colorless and odorless gas normally present at the Earth's surface in low concentrations. An important precursor to acid rain. Major anthropogenic source is burning fossil fuels.

Sulfurous smog Produced primarily by burning coal or oil at large power plants. Sulfur oxides and particulates combine under certain meteorological conditions to produce a concentrated form of this smog.

Surface impoundment Method of disposal of some liquid hazardous waste. This method is controversial and many sites have been closed.

Sustainability A concept that is emerging in the environmental sciences. With respect to resources it involves management that has the objective of ensuring that future generations will have the opportunity to use their fair share of resources and will inherit a quality environment. In an economic sense the concept means development that will not cause irreparable damage to the environment while ensuring that future generations will inherit their fair share of all Earth's resources.

Sustainable ecosystem An ecosystem that is subject to some human use, but at a level that leads to no loss of species or of necessary ecosystem functions.

Sustainable energy A type of energy management that provides for reliable sources of energy while not causing environmental degradation and ensuring that future generations will have a fair share of the Earth's resources.

Sustainable forest ecosystem An ecosystem in which all properties of the forest are maintained.

Sustainable harvest An amount of a resource that can be harvested at regular intervals indefinitely. There are two concepts here. The first is the maximum abundance of a population or species that can be harvested without degrading the ability of that population or species to withstand the same level of harvest in the future. The second concept extends sustainability to the ecosystem. It is maximum abundance of a population or species that can be harvested without degrading (1) the ability of that population or species and (2) its ecosystem to withstand the same level of harvest in the future.

Sustainable timber harvest Amount of timber that can be removed periodically from a forest without decreasing the capacity of the forest ecosystem to sustain that level of harvest in the future.

Sustainable water use Use of water resources that does not harm the environment and provides for the existence of high-quality water for future generations.

Symbiont Each partner in symbiosis.

Symbiosis An interaction between individuals of two different species that benefits both. For example, lichens contain an alga and a fungus that require each other to persist. Sometimes this term is used broadly, so that domestic corn and people could be said to have a symbiotic relationship—domestic corn cannot reproduce without the aid of people, and some peoples survive because they have corn to eat.

Symbiotic Relationships that exist between different organisms that are mutually beneficial.

Synergism Cooperative action of different substances such that the combined effect is greater than the sum of the effects taken separately.

Synergistic effect When the change in availability of one resource affects the response of an organism to some other resource.

Synfuels Synthetic fuels, which may be liquid or gaseous, derived from solid fuels, such as oil from kerogen in oil shale or oil and gas from coal.

System Any part of the universe that may be isolated in thought or in deed for the purpose of study.

Tar sands Sedimentary rocks or sands impregnated with tar oil, asphalt, or bitumen.

Taxon A grouping of organisms according to evolutionary relationships.

Tectonic cycle The processes that change Earth's crust, producing external forms such as ocean basins, continents, and mountains.

Tertiary treatment (of wastewater) Advanced form of wastewater treatment involving chemical treatment or advanced filtration. An example is chlorination of water.

Theories Scientific models that offer broad, fundamental explanations of related phenomena and are supported by consistent and extensive evidence.

Thermal (heat energy) The energy of the random motion of atoms and molecules.

Thermal pollution A type of pollution that occurs when heat is released into water or air and produces undesirable effects on the environment.

Thermodynamics, first law of See **First Law of Thermodynamics.**

Thermodynamics, second law of See **Second Law of Thermodynamics.**

Thinning The timber harvesting practice of selectively removing only smaller or poorly formed trees.

Threatened species Species experiencing a decline in the number of individuals to the degree that a concern is raised about the possibility of extinction of that species.

Threshold A point in the operation of a system at which a change occurs. With respect to toxicology it is a level below which effects are not observable and above which effects become apparent.

Tidal power Form of water utilizing ocean tides in places where favorable topography allows for construction of a power plant.

Tolerance The ability to withstand stress resulting from exposure to a pollutant or harmful condition.

Total fertility Average number of children expected to be born to a woman during her lifetime. (Usually defined as the number born to a woman between the ages 15 to 44, taken conventionally as the lower and upper limit of reproductive ages for women.)

Toxic Harmful, deadly, or poisonous.

Toxicology The science concerned with study of poisons (or toxins) and their effects on living organisms. The subject also includes the clinical, industrial, economic, and legal problems associated with toxic materials.

Trophic level In an ecological community, all the organisms that are the same number of food-chain steps from the primary source of energy. For example, in a grassland the green grasses are on the first trophic level, grasshoppers are on the second, birds that feed on grasshoppers are on the third, and so forth.

Trophic level efficiency The ratio of the biological production of one trophic level to the biological production of the next lower trophic level.

Tundra The treeless land area in alpine and arctic areas characterized by plants of low stature and including bare areas without any plants and covered areas with lichens, mosses, grasses, sedges, and small flowering plants, including low shrubs.

Ubiquitous species Species that are found almost anywhere on the Earth.

Undernourishment The lack of sufficient calories in available food, so that one has little or no ability to move or work.

Uneven-aged stands Forest area with at least three distinct age classes.

Unified soil classification system A classification of soils, widely used in engineering practice, based on the amount of coarse particles, fine particles, or organic material.

Uniformitarianism The principle that processes that operate today operated in the past. Therefore observations of processes today can explain events that occurred in the past and leave evidence, for example, in the fossil record or in geologic formations.

Urban dust dome Polluted urban air produced by the combination of lingering air and abundance of particulates and other pollutants in the urban air mass.

Urban forestry The practice and profession of planting and maintaining trees in cities, including trees in parks and other public areas. Involves determining the best species and sites for urban tree planting, taking into account climate, soil, shading from tall buildings, and motor vehicle pollution.

Utilitarian justification for the conservation of nature An argument for the conservation of nature on the grounds that the environment, an ecosystem, habitat, or species, provides individuals with direct economic benefit or is directly necessary to their survival.

Utility Value or worth in economic terms.

UVA Least energetic form of ultraviolet radiation. It is capable of causing some damage to living cells, is not affected by stratospheric ozone, and is therefore transmitted to the Earth.

UVB Intermediate wavelength ultraviolet radiation, damaging to living cells. Most is absorbed by stratospheric ozone, and therefore depletion of ozone leads to significant increase of this radiation. This is the ozone problem.

UVC Shortest wavelength and most energetic of the ultraviolet radiation. It is strongly absorbed in the atmosphere and negligible amounts reach the surface of the Earth.

Variable, dependent A variable that changes in response to changes in an independent variable; a variable taken as the outcome of one or more other variables.

Variable, independent In an experiment, the variable that is manipulated by the investigator. In an observational study, the variable that is believed by the investigator to affect an outcome, or dependent, variable.

Variable, manipulated See **Variable Independent.**

Variable, responding See **Variable Dependent.**

Virgin forest Forest that has never been cut.

Vulnerable species Another term for *threatened species*—species experiencing a decline in the number of individuals.

Waldsterben German phenomenon of forest death as the result of acid rain, ozone, and other air pollutants.

Wallace's realms Six biotic provinces, or biogeographic regions, divided on the basis of fundamental inherited features of the animals found in those areas, suggested by A. R. Wallace (1876). His realms are Nearctic (North America), Neotropical (Central and South America), Palearctic (Europe, nothern Asia, and northern Africa), Ethiopian (central and southern Africa), Oriental (the Indian subcontinent and Malaysia), and Australian.

Wastewater renovation and conservation cycle Practice of applying wastewater to the land. In some systems treated wastewater is applied to agricultural crops, and as the water infiltrates through the soil layer it is naturally purified. Reuse of the water is by pumping it out of the ground for municipal or agricultural uses.

Wastewater treatment Process of treating wastewater (primarily sewage) in specially designed plants that accept municipal wastewater. Generally divided into three categories: primary treatment, secondary treatment, and advanced wastewater treatment.

Water budget Inputs and outputs of water for a particular system (a drainage basin, region, continent, or the entire Earth).

Water conservation Practices designed to reduce the amount of water we use.

Water power Alternative energy source derived from flowing water. One of the world's oldest and most common energy sources. Sources vary in size from microhydropower systems to large reservoirs and dams.

Water reuse The use of wastewater following some sort of treatment. Water reuse may be inadvertent, indirect, or direct.

Water table The surface that divides the zone of aeration from the zone of saturation, the surface below which all the pore space in rocks is saturated with water.

Watershed An area of land that forms the drainage of a stream or river. If a drop of rain falls anywhere within a watershed, it can flow out only through the same stream.

Weathering Changes that take place in rocks and minerals at or near the surface of Earth in response to physical, chemical, and biological changes; the physical, chemical, and biological breakdown of rocks and minerals.

Wetlands Comprehensive term for landforms such as salt marshes, swamps, bogs, prairie potholes, and vernal pools. Their common feature is that they are wet at least part of the year and as a result have a particular type of vegetation and soil. Wetlands form important habitats for many species of plants and animals, while serving a variety of natural service functions for other ecosystems and people.

Whole-tree harvesting Timber harvesting practice of removing all aboveground parts of a tree, most of which are chipped into small fragments for making paper.

Wilderness An area unaffected now or in the past by human activities and without a noticeable presence of human beings.

Wind power Alternative energy source that has been used by people for centuries. More recently thousands of windmills have been installed to produce electric energy.

Work (physics) Force times the distance through which it acts.

Zero population growth A population in which the number of births equals the number of deaths so that there is no net change in the size of the population.

Zone of aceration The zone or layer above the water table in which some water may be suspended or moving in a downward migration toward the water table or laterally toward a discharge point.

Zone of saturation Zone or layer below the water table in which all the pore space of rock or soil is saturated.

Zooplankton Small aquatic invertebrates that live in the sunlit waters of steams, lakes, and oceans and feed on algae and other invertebrate animals.

Chapter 1: *Page 2:* Jeff Rotman/Tony Stone Images/New York, Inc. *Page 3 (left):* Frans Lanting/Minden Pictures, Inc. *Page 3 (right):* Tom Bean. *Page 7:* Corbis-Bettmann. *Page 8 (left):* Peter Turnley/Black Star. *Page 8 (right):* Viviane Moos/SABA. *Page 10:* National Snow and Ice Data Center/Photo Researchers. *Page 12:* Lee Malis/Gamma Liaison.

Chapter 2: *Page 16 (top):* Larry Goldstein/Tony Stone Images/New York, Inc. *Page 16 (bottom left):* Gideon Mendel/Magnum Photos, Inc. *Page 16 (bottom right):* D. Hudson/Sygma. *Page 19:* Alan Root/Okapia/Photo Researchers. *Page 20 (top left):* Jim Brandenburg/Minden Pictures, Inc. *Page 20 (top right):* Norman Owen Tomalin/Bruce Coleman, Inc. *Page 20 (center):* Alvin E. Staffan/Photo Researchers.

Chapter 3: *Page 32:* Guido A. Rossi/The Image Bank. *Page 33 (top):* Guido Alberto Rossi/The Image Bank. *Page 33 (bottom):* M. Renaudeau/Hoa-Qui. *Page 34:* Chris Johns. *Page 35:* Mark Segal/Tony Stone Images/New York, Inc. *Page 39:* Ed Keller. *Page 42:* Charles A. Mauzy/Tony Stone Images/New York, Inc.

Chapter 4: *Page 49:* Andy Caulfield/The Image Bank. *Page 50 (top):* Terra Nova International/Photo Researchers. *Page 50 (center and left):* Joel Rogers/Tony Stone Images, Seattle/ PNI. *Page 50 (center and right):* Ed Keller. *Page 58:* Kim Heacox/Peter Arnold, Inc. *Page 59:* Manfred Gottschalk/Tom Stack & Associates. *Page 70:* William Felger/Grant Heilman Photography.

Chapter 5: *Page 74:* Alain Evrard/Photo Researchers. *Page 75:* Chip Hires/Gamma Liaison.

Chapter 6: *Page 97:* Stephen J. Krasemann/Photo Researchers. *Page 98 (top left):* Alvin E. Staffan/Photo Researchers. *Page 98 (top and center):* Bill Ivy/Tony Stone Images/New York, Inc. *Page 98 (top right):* O. Spielman/CNRI/Phototake. *Page 98 (bottom left):* Jeanne Drake/Tony Stone Images/ New York, Inc. *Page 98 (bottom and center):* Runk/Schoenberger/Grant Heilman Photography. *Page 98 (bottom right):* Scott Nielsen/Bruce Coleman, Inc. *Page 100:* Farrell Grehan/Photo Researchers. *Page 107 (left):* Fridmar Damm/Leo de Wys, Inc. *Page 107 (right):* Jeff Gnass. *Page 108 (top):* Richard Hartmier/First Light, Toronto. *Page 108 (bottom):* Courtesy USDA Forest Service.

Chapter 7: *Page 112:* Ferry/Gamma Liaison. *Page 113 (left and bottom right):* Courtesy Stephan C. White, Kansas Department of Agriculture. *Page 113 (top right):* Stephen G. Maka/DRK Photo. *Page 123 (left):* Alvin E. Staffan/Photo Researchers. *Page 123 (right):* H. Reinhard/Bruce Coleman, Inc. *Page 130 (top left):* Hans & Judy Deste/Animals Animals. *Page 130 (right):* Carl Purcell/Photo Researchers. *Page 130 (bottom left):* Stephen J. Krasemann/Tony Stone Images/New York, Inc. *Page 135 (top):* Stephen J. Krasemann/DRK Photo. *Page 135 (bottom left):* Kim Heacox/DRK Photo. *Page 135 (bottom right):* Tim Laman/The Wildlife Collection. *Page 136 (top left):* Ferrero/Labat/Auscape International Pty. Ltd. *Page 136 (top right):* Toni Angermayer/Photo Researchers. *Page 136 (bottom):* Martin Harvey/The Wildlife Collection.

Chapter 8: *Page 143:* James Balog/Tony Stone Images/New York, Inc. *Page 144 (left):* John & Ann Mahan. *Page 144 (right):* Heather Angel. *Page 148:* Daniel Botkin.

Chapter 9: *Page 157:* George Tucker/Photo Researchers. *Page 158:* David Tomlinsin/Windrush Photos. *Page 159 (top):* Kim Heacox/Peter Arnold, Inc. *Page 159 (bottom):* Robert Ragaini. *Page 160 (top):* Breck Kent/Earth Scenes. *Page 160 (center):* Michael P. Gadomski/Earth Scenes. *Page 160 (bottom):* Michael P. Gadomski/Earth Scenes. *Page 161:* Courtesy NASA Goddard Space Flight Center and the authors of Hall, F.G., D.B. Botkin, D.E. Strbel, K.D. Woods and S.J. Goetz, 1991; "Large Scale Patterns in Forest Succession As Determined by Remote Sensing," *Ecology,* 72: 628-640. *Page 162:* Don Johnston/Photo/Nats. *Page 165:* Daniel Botkin. *Page 166:* Terry Donnelly/Tom Stack & Associates.

Earth's Biomes: *Page 177 (left):* John Shaw/Bruce Coleman, Inc. *Page 177 (bottom):* Charles Glatzer/The Image Bank. *Page 178 (top left):* Tom & Pat Leeson. *Page 178 (bottom):* Courtesy NASA. *Page 179:* David Muench Photography. *Page 180 (top):* Frans Lanting/Minden Pictures, Inc. *Page 180 (bottom):* Steve Kaufman/Peter Arnold, Inc. *Page 181 (top):* David Muench Photography. *Page 181 (bottom):* Andre Jenny/Stock South. *Page 182:* William Johnson/Stock, Boston.

Chapter 10: *Page 184:* Herb Schmitz/Tony Stone Images/New York, Inc. *Page 188 (top left):* Kevin Morris/Tony Stone Images, New York, Inc. *Page 188 (top right):* JC Carton/Bruce Coleman, Inc. *Page 188 (bottom left):* Thomas Hovland/Grant Heilman Photography. *Page 188 (bottom and center):* William E. Ferguson. *Page 188 (bottom right):* Hans Reinhard/Okapia/Photo Researchers. *Page 192:* J. Victolero/International Rice Research Institute. *Page 195:* Doug Plummer/Photo Researchers. *Page 199:* Courtesy Food and Agriculutre Organization.

Chapter 11: *Page 207:* Ovak Arslanian/Gamma Liaison. *Page 208 (left):* Grant Heilman Photography. *Page 208 (right):* Joy Sprurr/Bruce Coleman, Inc. *Page 211 (top):* Science Vu/Visuals Unlimited. *Page 211 (bottom):* Jeri Gleiter/Peter Arnold, Inc. *Page 212 (top):* Bill Bachman/Photo Researchers. *Page 212 (center):* Grant Heilman Photography. *Page 212 (bottom):* Corbis-Bettmann. *Page 213 (left):* R. de la Harp/Biological Photo Service. *Page 213 (right):* Alex Von Koschembahr/Photo Researchers. *Page 216:* Scott Anger. *Page 220 (center):* Larry A. Hull. *Page 220 (bottom):* Matthew McVay/Tony Stone Images/New York, Inc.

Chapter 12: *Page 225:* Nicholas Parfitt/Tony Stone Images/New York, Inc. *Page 226 (left):* David Clendenen/USFWS. *Page 226 (right):* Larry Brock/Tom Stack & Associates. *Page 227:* David Muench Photography. *Page 228:* P. Przwalski/Photo Researchers. *Page 231:* Kevin Schafer/Peter Arnold, Inc. *Page 232:* Tom & Pat Leeson/Photo Researchers. *Page 233:* Tom & Pat Leeson. *Page 236:* Dr. Alfred O. Gross/Photo Researchers. *Page 237 (left):* Stephen Krasemann/Tony Stone Images/New York, Inc. *Page 237 (top right):* N. N. Birks/Auscape International Pty. Ltd. *Page 237 (bottom right):* Michael George/Bruce Coleman, Inc. *Page 247 (center left):* Richard Elliott/Tony Stone Images/New York, Inc. *Page 247 (center right):* Tom & Pat Leeson/DRK Photo. *Page 247 (bottom left):* Dianne Blell/Peter Arnold, Inc. *Page 247 (bottom right):* R.O. Bierregaard

Chapter 13: *Page 256:* Chip Vinai/Gamma Liaison. *Page 257:* David Muench/David Muench Photography. *Page 259 (left):* Lee Rentz/Bruce Coleman, Inc. *Page 259 (top right):* Martin

Wendler/Peter Arnold, Inc. *Page 259 (bottom right):* James Martin/Tony Stone Images/New York, Inc. *Page 263:* NASA/Science Source/Photo Researchers. *Page 264 (left):* Steve McCurry/Magnum Photos, Inc. *Page 264 (right):* Robert Frerck/Odyssey Productions. *Page 271:* James Randklev/Tony Stone Images/New York, Inc. Page 274 (left): David Muench Photography. *Page 274 (top right):* © Swiss National Tourist Office. *Page 274 (bottom right):* David Weintraub/Stock, Boston. *Page 279:* Natalie Fobes. *Page 280 (left):* Stephen J. Krasemann/Tony Stone Images/New York, Inc. *Page 280 (right):* Animals Animals.

Chapter 14: *Page 286:* AP/Wide World Photos. *Page 288:* T. Orban/Sygma. *Page 289:* Bill Brooks/Masterfile. *Page 290:* Ed Keller. *Page 293:* O. Franken/Sygma. *Page 294:* Michael Yamashita. *Page 295:* Martin Bond/Science Photo Library/Photo Researchers.

Chapter 15: *Page 308:* George Grigoriou/Tony Stone Images/New York, Inc. *Page 309:* FPG International.

Chapter 16: *Page 325:* Russell D. Curtis/Photo Researchers. *Page 330 (left):* George Hunter/Tony Stone Images/New York, Inc. *Page 330 (right):* Ken Graham/Bruce Coleman, Inc. *Page 334:* Grant Heilman/Grant Heilman Photography. *Page 335:* Ed Keller. *Page 336:* William P. Hines/The Scranton-Times-Tribune Library.

Chapter 17: *Page 342:* Charles Krebs/Tony Stone Images/New York, Inc. *Page 343:* Courtesy Luz International. *Page 345:* Courtesy Pacific Gas & Electric Company. *Page 348:* Tom Bean. *Page 350 (top):* H. Gruyaeart/Magnum Photos, Inc. *Page 350 (bottom):* T. J. Florian/Rainbow. *Page 356:* C. Delis/Explorer. *Page 357:* Glen Allison/Tony Stone Images/New York, Inc.

Chapter 18: *Page 363:* Hans Wolf/The Image Bank. *Page 369 (left):* Graham Finlayson/Tony Stone Images/New York, Inc. *Page 369 (right):* J. Schlegel/Leo de Wys, Inc. *Page 370:* Roger RessmeyerRoger Ressmeyer/©CORBIS. *Page 372:* Courtesy Princeton Plasma Physics Laboratory. *Page 374:* Corbis-Bettmann. *Page 379:* Igor Kostin/Sygma.

Chapter 19: *Page 388:* Peter Cole/Bruce Coleman, Inc. *Page 391:* Rich Buzzelli/Tom Stack & Associates. *Page 397:* Jim Tuten/Black Star. *Page 404 (top left):* Stephen Krasemann/Tony Stone Images/New York, Inc. *Page 404 (top right):* Mike Price/Bruce Coleman, Inc. *Page 404 (bottom left):* Gregory G. Dimijian, M.D./Photo Researchers. *Page 404 (bottom right):* Jim Brandenburg/Minden Pictures, Inc. *Page 409 (top left):* Sterling Dimmitt. *Page 409 (top right):* Courtesy Jacksonville Corp. of Engineers. *Page 409 (bottom left):* Stewart Halperin/Earth Scenes. *Page 409 (bottom right):* Cameron Davidson/Comstock, Inc.

Chapter 20: *Page 414:* Ben Osborne/Tony Stone Images/New York, Inc. *Page 420:* McAllister/Gamma Liaison. *Page 421:* William E. Ferguson. *Page 423:* Michelle Barnes/Gamma Liaison. *Page 424:* Charles Mason/Black Star. *Page 425 (top):* Jim Strawser/Grant Heilman Photography. *Page 425 (bottom):* John Cancalosi/DRK Photo.

Chapter 21: *Page 440:* European Space Agency/Science Photo Library/Photo Researchers. *Page 458:* Shawn Henry/SABA.

Chapter 22: *Page 466:* Michael Yamashita. *Page 473:* Bartholomew/Gamma Liaison. *Page 477:* Jim Mendenhall. *Page 484:* Don & Pat Valenti/Tony Stone Images/New York, Inc.

Chapter 23: *Page 496:* Blanche/Gamma Liaison.

Chapter 24: *Page 510:* Peter Cade/Tony Stone Images/New York, Inc. *Page 517:* Courtesy NASA.

Chapter 25: *Page 530:* Ted Spiegel/PNI. *Page 531 (left):* Robert A. Lubeck/Animals Animals/Earth Scenes. *Page 531 (right):* Ken Wagner/Phototake. *Page 533:* Mae Scanlan. *Page 535:* François Gohier/Auscape International Pty, Ltd. *Page 541:* C. Bradley Simmons/Bruce Coleman, Inc. *Page 542:* Photri.

Chapter 26: *Page 550:* Rotolo/Gamma Liaison. *Page 551:* Luigi Tazzari/Gamma Liaison. *Page 556 (top left):* Frank Rossotto/Tom Stack & Associates. *Page 556 (top right):* Steve Vidler/Leo de Wys, Inc. *Page 556 (bottom):* Rohan/Tony Stone Images/New York, Inc. *Page 558:* Collection of ESL Information Services, Engineering Societies Library, New York City. *Page 559:* J. Messerschmidt/Leo de Wys, Inc. *Page 563 (top):* Courtesy of the Woodlands Corporation. *Page 563 (bottom):* Nathan Benn/Woodfin Camp & Associates. *Page 564:* Suzanne & Nick Geary/Tony Stone Images/New York, Inc. *Page 567:* Arthur Morris/Visuals Unlimited.

Chapter 27: *Page 572:* Marc Pokempner/Tony Stone Images/New York, Inc. *Page 574:* SABA. *Page 578:* Alex Quesada/Matrix. *Page 580:* Courtesy John H. Kramer.

Chapter 28: *Page 594:* Jacques Jangoux/Tony Stone Images/New York, Inc. *Page 597:* Helen Thompson/Earth Scenes. *Page 600:* Courtesy Rachel Haymon. *Page 604:* Royce Bair/TSS/ProFiles West, Inc. *Page 605:* Ed Keller.

Chapter 29: Page *610 (top):* Bob Wallace/Stock, Boston. *Page 610 (bottom):* Courtesy Florida House Learning Center, Sarasota County, Florida. *Page 618:* Courtesy NASA. *Page 619 (top):* Courtesy NOAA. *Page 619 (bottom):* A. Tannenbaum/Sygma. *Page 621:* Bruce H. Morrison/Phototake. *Page 623:* Scott Anger. *Page 627:* ©Sygma. *Page 630:* Ed Keller. *Page 631:* Dave G. Houser.

Chapter 30: *Page 637 (top):* Rob Crandall/Stock, Boston. *Page 637 (bottom):* Tim Davis/Photo Researchers. *Page 641:* David Muench Photography. *Page 645:* Clark Linehan/New England Stock Photo.

Part Opening Credits: *Part I:* Button blanket, Village Island. Courtesy Royal British Columbia Museum/Catalog Number 13685. *Part II:* Haitian painting, "Giraffes" by Montas Antoine/SUPERSTOCK. *Part III:* Egyptian Pavement Fragment. Malqata. Dynasty 18, reign of Amenhotep III. ca. 1390-1353 B.C. The Metropolitan Museum of Art, Rogers Fund, 1920 (20.2.2). *Part IV:* Kalila wa Dimna (book of the fables of Bidpai), The Ascetic and the Jar of Honey. India (Gujarat); mid-16th century. The Metropolitan Museum of Art, The Nasli Heeramaneck Collection, Gift of Alice Heeramaneck, 1981. *Part V:* Page 307: Untitled, Uta Uta Tjangala, 1984. Courtesy Anthony Wallis, Aboriginal Artists' Agency, North Sydney. *Part VI:* Minoan vase, from Palaikastro, c. 1500 B.C. Archaeological Museum, Heraklion, Crete, Greece/Art Resource. *Part VII:* Kingmeata/Art Gallery of Ontario, Toronto, Gift of the Klamer Family, 1978. *Part VIII:* Page 529: Hunting and Fishing. Etruscan wall painting, C. 500 B.C. Scala/Art Resource, NY.